SYNAPTIC PLASTICITY AND TRANSSYNAPTIC SIGNALING

Patric K. Stanton
Clive Bramham
Helen E. Scharfman
(Eds.)

Synaptic Plasticity and Transsynaptic Signaling

With 119 Figures

Patric K. Stanton
Department of Neurology
New York Medical College
Valhalla, NY 10595
USA

Clive Bramham
Department of Physiology
University of Bergen
Bergen 5009
Norway

Helen E. Scharfman
CNRRR Department
Helen Hayes Hospital
West Haverstraw, NY 10993
USA

Library of Congress Control Number: 2005923310

ISBN-10: 0-387-24008-X
ISBN-13: 978-0387-24008-4
eISBN: 0-387-25443-9
Printed on acid-free paper.

Printed in the United States of America. (SBA)

9 8 7 6 5 4 3 2 1

springeronline.com

ERRATA

Due to a printing error the figures on pages 35, 36, 40, 42, 44, and 292 were not reproduced properly. The corrected figures are included in this errata booklet.

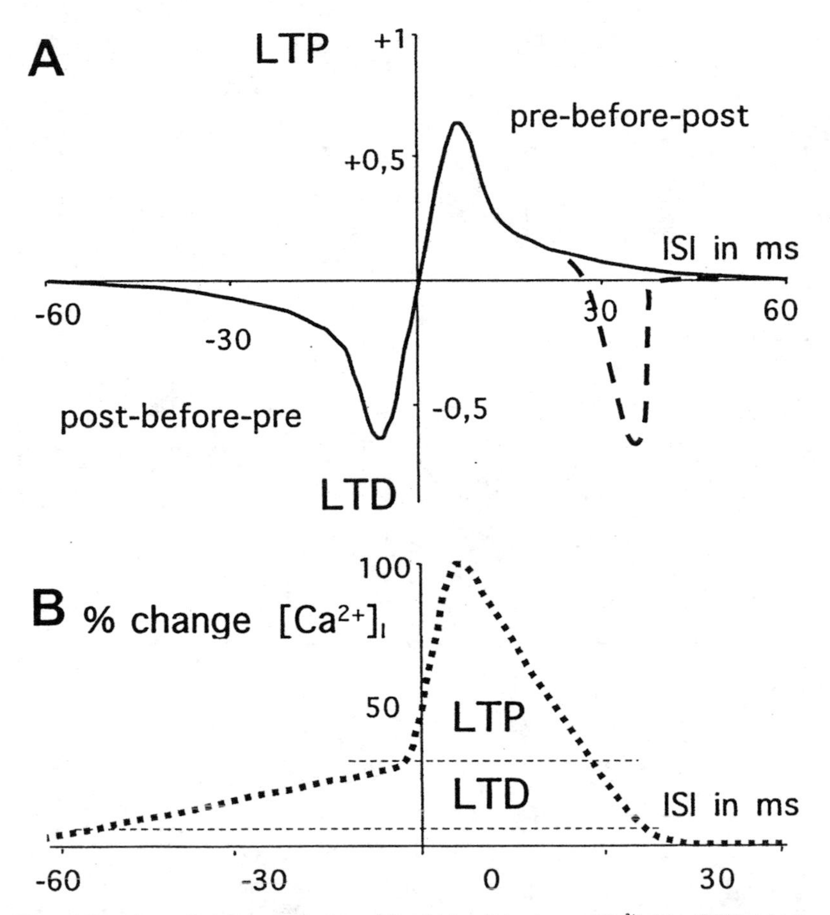

Figure 1. Dependence of relative synaptic change (A) and intracellular change of $[Ca^{2+}]_i$ (B) in STDP on inter-spike interval (ISI = time of postsynaptic spike - time of presynaptic spike). The dashed line in A shows a second LTD window at long positive ISIs due to weak Ca^{2+}-increases (see B and text).

STDP crucially depends on the back-propagation of the postsynaptic action potential to the dendritic site of synaptic contact (see below) to unblock NMDA receptors by depolarization-dependent unbinding of Mg^{2+}. The STDP relation "LTD/LTP versus ISI" can be understood in the context of the Bienenstock-Cooper-Munro (BCM) model of synaptic plasticity derived from theoretical issues of network stability, and experimental properties of low frequency stimulus-induced LTD and high frequency stimulation inducing LTP (Bienenstock, et al., 1982). In this model, LTD increases with the relevant Ca^{2+}-concentration in the dendritic spine above a certain Ca^{2+}-concentration threshold, followed by a second threshold (LTP threshold 1 in Fig. 2), where LTD apparently decreases and/or LTP increases with a further increase of $[Ca^{2+}]$. LTP appears to reach a

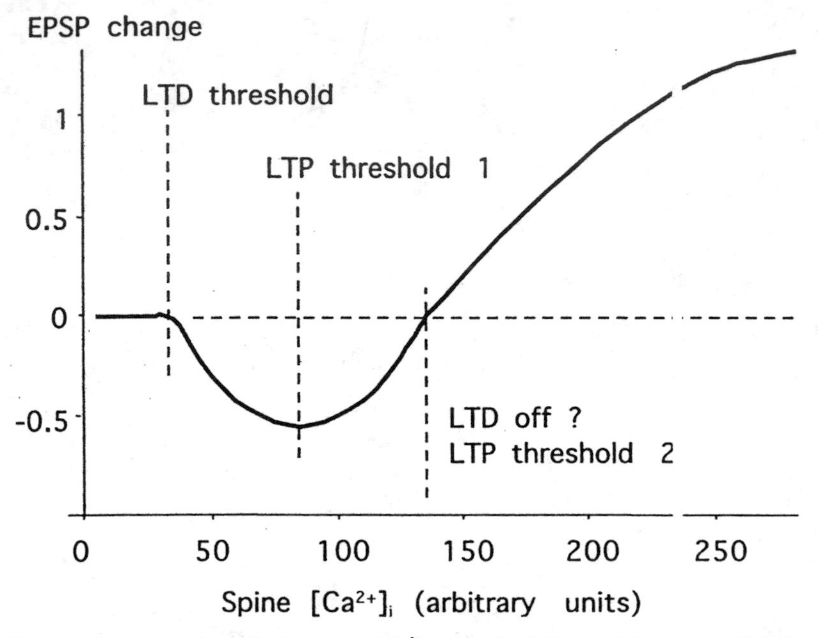

Figure 2. Dependence of synaptic change on peak Ca^{2+}-concentration (arbitrary units) at the sensor(s) in the dendritic spine. Peak Ca^{2+}-concentration depends on the precise timing of paired pre- and postsynaptic spikes. Several thresholds for synaptic modification processes are hypothesized.

saturating level for very high Ca^{2+}-concentrations (Fig. 2; (Bienenstock et al., 1982, Abarbanel, et al., 2002, Karmarkar, et al., 2002, Lisman, 1989). Classically, either a high or low frequency train stimulation that produces a long lasting sustained increase of free Ca^{2+} concentration, a weak one lasting several minutes for LTD and a strong one lasting 500 ms or longer for LTP, has been considered essential for the induction of long-term plasticity. In contrast, STDP induction protocols are associated with very brief rise times of free Ca^{2+} concentration in spines of $\leq$4.5 ms for back-propagating action potentials and 9-20 ms for EPSPs (20 to 80 % rise time, Koester and Sakmann, 1998).

Whether LTD can still be induced at higher Ca^{2+} concentrations above LTP threshold 2 is unclear, as it would be masked by simultaneous strong LTP. The timing of action potentials determines the time course and amplitude of Ca^{2+}-influx into the spine. This amplitude time course is determined approximately by the time course of Mg^{2+} unblock of the NMDA receptor by the postsynaptic dendritic spike, multiplied by the time course of glutamate activated NMDA receptors. The latter time course has been recorded as the

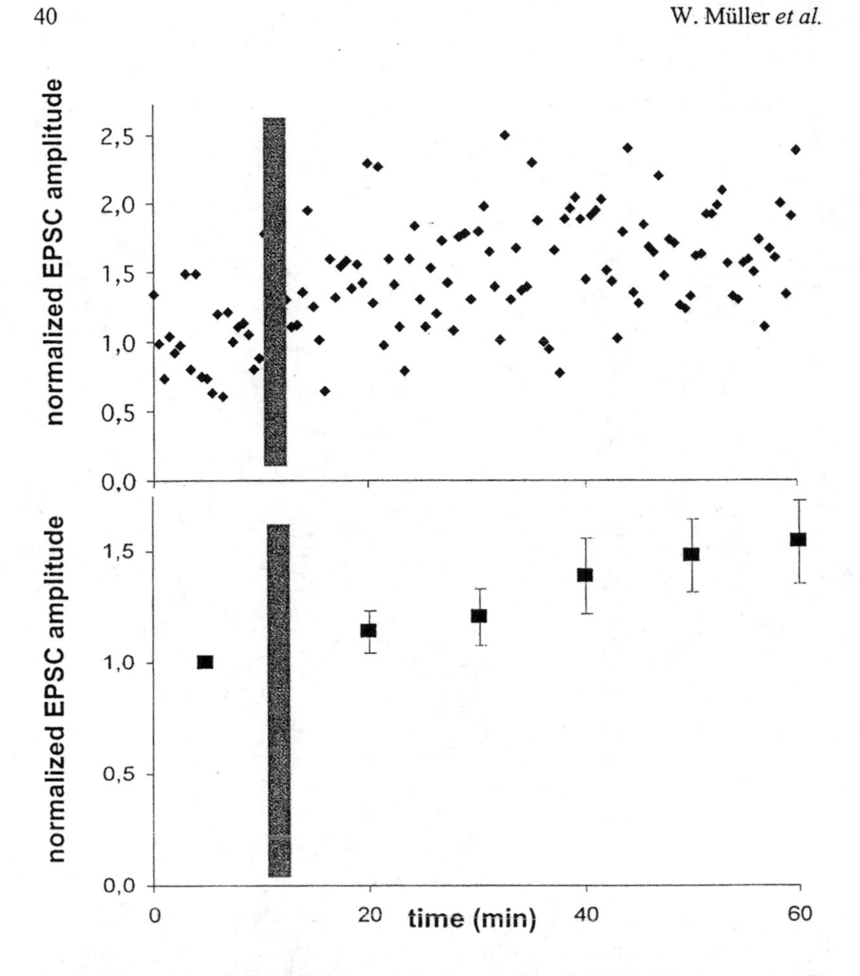

Figure 3. Time courses of EPSC amplitudes in an hippocampal CA1 pyramidal neuron and averages ± SEM from 8 cells (bottom) before and after spike-pairing induction of long-term potentiation (vertical bar, Schaffer collateral stimulation), in control conditions, demonstrate slow but robust expression of LTP (75 pairings at a frequency of 0.33 Hz using fiber stimulation with an extracellular electrode in combination with voltage/current clamp recording from a CA1 pyramidal neuron; adapted from (Adams, et al., 2004).

activated K^+-channels or reduced NMDA receptor activity (but see Markram and Segal (1992).In this way, the results could still be in agreement with the BCM model. In line with this argumentation, elimination of IP_3R1 mediated signaling enhances LTP in response to the spike timing LTP induction and produces, in addition, heterosynaptic LTP, i.e. specificity of LTP is lost (Nishiyama, et al., 2000).

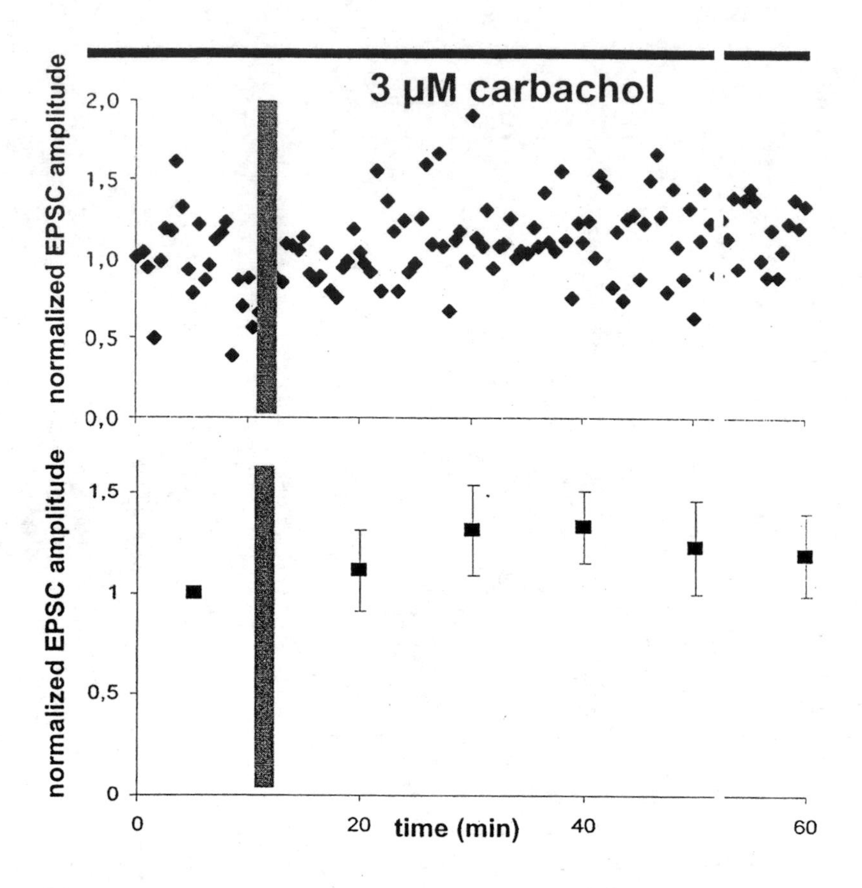

Figure 4. Time courses of EPSC amplitudes in an example neuron and averages ± SEM from five cells (bottom) before and after spike-pairing induction of long-term potentiation (vertical bar) in th‹ presence of the stable cholinergic agonist carbachol (3 μM) demonstrate only transient potentiation of EPSCs 75 pairings at a frequency of 0.33 Hz; adapted from (Adams, et al., 2004).

All these conductance changes may contribute to cholinergic-muscarinic ıctivation of theta rhythmic oscillations. These oscillations favor synchronized firing ac ;vity at theta frequencies across cell populations. In addition they presumably depolaı ze dendrites rhythmically by local synaptic input, thereby enhancing dendritic excitability rhythmically.

During cholinergic induction of a theta rhythm oscillatory state, reqı irements for synaptic plasticity are dramatically altered; a single burst given at the ı ɜak of theta induces homosynaptic LTP, but induces homosynaptic *LTD* of previousl · potentiated

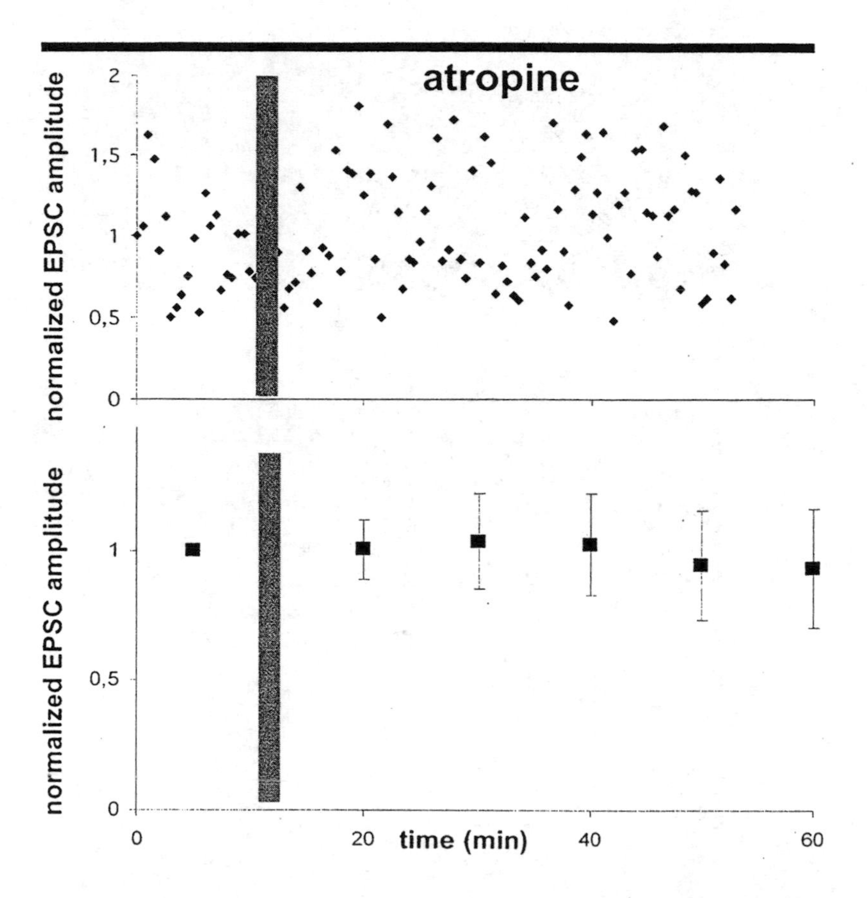

Figure 5. Time courses of EPSC amplitudes in an example neuron and of averages ± SEM from six cells (bottom) before and after spike-pairing induction of long-term potentiation (vertical bar) in the presence of the muscarinic receptor antagonist atropine (10 µM) show a clear failure to express LTP (75 pairings at a frequency of 0.33 Hz; adapted from (Adams, et al., 2004).

receptors type 1 (IP_3R1) has an important role for homo- and heterosynaptic spike timing-dependent LTD, respectively (Nishiyama, et al., 2000).

With long-term plasticity being induced by single spike pairings, stability of synaptic weight changes will be disturbed by ongoing spontaneous activity, processing of novel information by the same neuronal networks and even by activity being employed for recalling previously stored information. Artificial neuronal network simulations usually

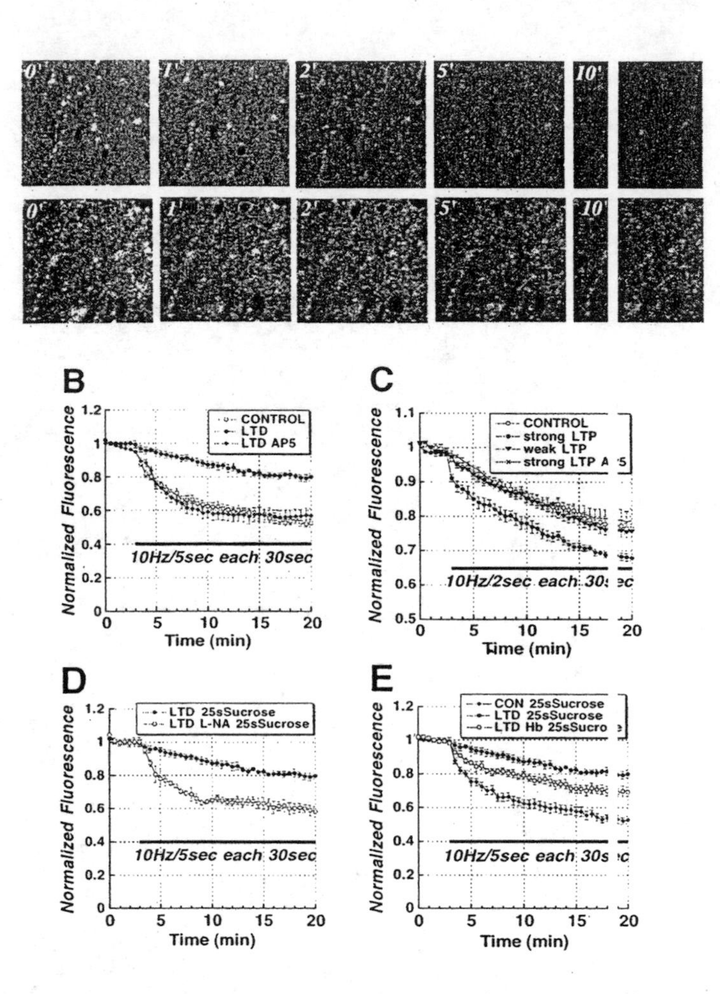

Figure 2. Presynaptic LTD and LTP of release from the readily-releasable vesicle pool of Schafffer collateral terminals. **A:** Two-photon excitation fluorescent images of readily-releasable vesicle poo puncta in stratum radiatum of field CA1 in a control slice (Upper Row), versus a slice where LTD was ind ced (Lower Row). Numbers represent time in minutes following start of unloading stimulation. **B:** Time ourses of Schaffer collateral stimulus-evoked (solid bar; 10 Hz/5 s bursts each 30 s) FM1-43 destaining from the RRP in control slices (○; n=7) , LTD slices (●; n=6), and slices where LTD was blocked with D-AP5 (1 µM). **C:** **D:** Time course of RRP release (solid bar; 10 Hz/5 s bursts each 30s) in slices treated with th NOS inhibitor L-nitroarginine (L-NA;10 µM;○; n=4) compared to controls (●; n=5). **E:** The NO scaveng r hemoglobin (Hb; 10 µM;○; n=4) partially blocked the reduction in RRP release seen in LTD slices comparec to control LTD (●; n=5) and unstimulated slices (◆; n=6). (modified from Stanton et al. (2003))

PREFACE

This volume serves two purposes; addressing what we believe is the beginning of a paradigm shift in our understanding of plasticity of synaptic function, and honoring the scientific contributions of John Michael Sarvey. These purposes are inextricably intertwined, because John Sarvey made a number of seminal contributions to our understanding of activity-dependent synaptic plasticity.

John began his training with Dr. Edson Albuquerque, studying cholinergic receptors mediating skeletal muscle contraction. His postdoctoral work in collaboration with Manfred Klee and Ulrich Misgeld led to some of the earliest demonstrations of the utility of brain slices in the study of synaptic function. One of their experiences illustrates the excitement, and controversy, of that era, and how their persistence moved our field forward. Sir John Eccles visited the laboratory and, after witnessing some intracellular recordings from pyramidal neurons first hand, he pronounced the slice useless because there was no evidence of intact synaptic inhibition. When Sir John returned a year later, Sarvey and Misgeld gleefully showed him recordings of robust i.p.s.p.s and, from then on, Eccles was one of the most active proponents of the new *in vitro* slice.

Almost as soon as he established his own laboratory, John Sarvey began making important novel contributions to the field of synaptic plasticity. He was among the first to demonstrate the importance of postsynaptic action potentials, neuromodulatory receptors that raise intracellular [cyclic AMP], and ongoing protein synthesis, for the induction of long-term potentiation (LTP) of synaptic strength. This was the beginning of a career of methodical, thorough and reliable investigations into the mechanisms underlying LTP in the hippocampus. Among these were demonstrations of heterosynaptic LTP in field CA3, and norepinephrine-induced long-lasting potentiation and depression of perforant path synaptic transmission. John, and the many people who have worked in his laboratory, have played key roles in identifying neuromodulators influencing LTP, and the mechanisms of their actions, including roles for opiate, noradrenergic, muscarinic, GABAergic and metabotropic glutamate receptors.

During his career, John Sarvey also contributed significantly to our knowledge of hippocampal and neocortical synaptic physiology relevant to cognition, learning, and epilepsy. Early studies of the properties of GABAergic responses of visual cortical neurons improved the basis for our understanding of processing in visual cortical circuits. His investigations of how epileptiform activity is produced by direct actions of cholinesterase inhibitors, and by lowering extracellular Ca^{2+} and Mg^{2+} concentrations, have supplied important information. Perhaps most impressive is the understood reliability and integrity of his work, transcending the number of papers published. Moreover, the help, advice and support he always gave freely to colleagues, both junior and senior, stands as one of his greatest contributions, of which he would be the proudest.

The most recent work from Dr. Sarvey's lab is a perfect illustration of the creativity, rigor, tenacity and integrity he has exhibited throughout his career. This work addressed an area of great interest for decades; the functional roles of synaptic release and transsynaptic diffusion of Zn^{2+} ions. Past interpretations of the roles of Zn^{2+} have included stabilization of vesicular opioid peptides, synaptogenesis, and heterosynaptic regulation of NMDA receptors. In this context, John entered the field, joining with a superb team of collaborators and, as usual, brought his creativity, technical and intellectual talents, and careful rigor to the question. This resulted in compelling evidence that Zn^{2+} is rapidly translocated from presynaptic terminals into postsynaptic pyramidal neurons, and that this translocation can be a necessary event in the induction of LTP at mossy fiber-CA3 synapses. Furthermore, work of John and colleagues has also shown that periods of ischemia are associated with Zn^{2+} release that may be a key factor in the extent of long-term brain damage following stroke and other types of traumatic injury. The thorough nature of their work has clarified developmental and other factors that may help explain divergent findings, and will continue to fuel interest in understanding the role of Zinc in transsynaptic signaling and synaptic plasticity.

We hope that two things will be evident as you peruse the chapters in this volume. The first is that neuroscience is in a time of numerous exciting advances in our understanding of the complexity and subtlety of electrochemical signaling across the synapse. The second is that the interests and work of John Sarvey have significantly influenced the research of most of these contributors, some of the finest scientists in our field. The range of John's influence is one of his unique contributions; this volume includes chapters that consider the importance of neuromodulatory transmitters such as norepinephrine, histamine and acetylcholine in regulating long-term activity-dependent synaptic plasticity, and how these modulators regulate memory formation. Classical LTP, and much newer models of synaptic plasticity such as long-term depression (LTD), spike-timing plasticity and metaplasticity of the threshold for induction of LTP, are all considered. There are chapters that address the functional roles of new/unusual messengers such as hydrogen peroxide, nitric oxide, brain-derived neurotrophic factor, zinc and endocannabinoids in inducing or regulating various forms of activity-dependent plasticity. This volume also contains chapters addressing the basic properties of GABAergic synaptic transmission and their impact on vesicular release, plasticity and epileptiform discharge.

While the neurochemical steps in the induction of long-term plasticity are one continuing research focus, the types of changes that express long-term plasticity in its different phases are of equal importance, and are addressed as well. The many mechanisms by which various types of glutamate and neuromodulatory receptors regulate NMDA receptor activity, leading to calcium influx and release that activates Ca^{2+}-calmodulin dependent protein kinase and Ca^{2+}-activated adenylyl cyclases, and the roles of CREB and rapid nuclear responses in LTP are all addressed. No volume addressing recent advances in our understanding of memory-related plasticity would be complete without considering how rapid, activity and kinase-dependent changes are consolidated into persistent physical alterations in neural circuits, so we have chapters on topics such as synapse preservation and elimination, the transsynaptic bridging molecules integrins, and alterations in dendritic spine morphology. A final chapter addresses the physiological functions of beta amyloid as they might relate to Alzheimer's disease.

It is, of course, impossible for any such volume to be anywhere near exhaustive. The overarching theme of this volume is the vast richness of the bidirectional transsynaptic communication that occurs over time scales both rapid and slow. In this sense, we mean bidirectional both in the sense of messengers moving both pre to postsynaptic and post to presynaptic, and the strength of synapses moving both up and down. We are just beginning to uncover these mechanisms, and it is this knowledge that will crucially inform and constrain our understanding of learning and memory, and higher cognitive function.

John Sarvey had strong personal and scientific connections to many of the authors in this volume. One of his greatest gifts was his ability to stimulate and encourage a love for science in the people who worked with him and with whom he communicated. Many contributors supplied reminiscences of John that emphasized both his impressive intellectual talents and his gentle ability to help, teach and encourage. Some remembered him as the calmest, most composed person in the room, a levelheaded thinker, suggesting he must have had a superabundance of some natural benzodiazepine receptor agonist. Others thought his unfailing enthusiasm must be a sign of high serotonin levels. In discussing the often great frustration associated with allowing graduate students to write drafts of their first papers, Johns' response was that, if we wrote them ourselves, we would have only papers to show for our life's work, while this way, we produce scientists *and* papers.

John Sarvey always relished a challenge much more than the credit associated with overcoming it. One friend remembered his arriving at the Grand Canyon after the Phoenix Society for Neuroscience meeting, sans overnight camping pass, intent upon hiking to the Colorado River. His friends, all outfitted with passes, foolishly attempted to dissuade him from doing the hike. Of course, he left before dawn, flew down the Bright Angel trail to reach the river, turning back up the Kaibab trail for the arduous return trip before lunch. By the time those friends returned from their more leisurely hike the evening of the next day, John had completed the 48km round-trip, slept on a colleagues hotel room floor, and slipped quietly away. Another reminisced about scuba diving together at the Winter Conference on Neural Plasticity in St. Lucia. Having been warned about the strong currents flowing just outside the bay, they nevertheless found themselves perilously close to being swept into them. After swimming as hard as they could for 10-15 minutes, the nervous friend asked John if they were, in fact, getting closer to their entry point. He drolly replied "We had better, or it's next stop - the Grenadines!"

The stories go on and on, but they all come back to John. He was a quiet, caring, very good man who happened to also be a superb scientist. He lived a life to be proud of, touched us all deeply in so many ways, and we are the richer for having known him.

Patric K. Stanton, Ph.D.
Valhalla, NY

Helen E. Scharfman, Ph.D.
West Haverstraw, NY

Clive Bramham, M.D., Ph.D.
Bergen, Norway

CONTENTS

THE THREE FACES OF NOREPINEPHRINE: PLASTICITY AT THE PERFORANT PATH-DENTATE GYRUS SYNAPSE

Carolyn W. Harley, Susan G. Walling, and Robert A.M. Brown*

1. INTRODUCTION

John Sarvey and his collaborators made major contributions to our understanding of the noradrenergic contribution to plasticity in the dentate gyrus of the hippocampus. The present chapter considers, first, the conclusions reached using *in vitro* models, and is based primarily on the work of John Sarvey and his collaborators, and, then, describes *in vivo* data and the interrelationships between the two approaches. These considerations lead us to the proposal that norepinephrine (NE) mediates three distinct forms of plasticity: immediate increases in cell excitability (i.e., likelihood of cell firing to a given synaptic input), immediate increases in synaptic strength and delayed increases in synaptic strength. All of these changes could contribute to the enduring functional alteration of information flow in the dentate gyrus and, thus, could act as memory mechanisms.

A decade after the first report of tetanic stimulation-induced homosynaptic (i.e., presumed glutamate-mediated modulation at glutamate synapses) long-term potentiation (LTP) of the perforant path evoked potential in the dentate gyrus *in vivo* (Bliss and Gardner-Medwin, 1973; Bliss and Lømo, 1973), we observed that iontophoretic application of NE in the dentate gyrus could induce a potentiation of the perforant path evoked population spike that greatly outlasted the 1- 5 min iontophoresis period (Neuman and Harley, 1983). In ~40% of the experiments population spike potentiation lasted more than 30 min, an accepted criterion for LTP at that time. Potentiation in one rat was monitored for 11 hours. These observations were christened NE-induced long-lasting potentiation to distinguish the phenomenon from the earlier homosynaptic LTP. The heterosynaptic (i.e., NE-activation modulating a glutamate-mediated response) modulation of glutamate synapses by NE in the perforant path of the dentate gyrus did not require high frequency electrical input, and suggested that the co-activation of

---

* Carolyn W. Harley, Memorial University, St. John's, Newfoundland, A1B 3X9. Susan G. Walling, Mount Allison University, Sackville, New Brunswick, E1G 1H7. Robert A.M. Brown, University of California, Los Angeles, Los Angeles, California 90095. E-mail: charley@play.psych.mun.ca

glutamatergic and noradrenergic signals could alter brain information processing for extended periods. The effect appeared to be primarily an alteration of EPSP to spike coupling (E-S potentiation), since the EPSP was not consistently increased. This implied an increase in cell excitability.

2. NEP AND NELLP IN THE FIRST SET OF SARVEY *IN VITRO* STUDIES

In 1985, Stanton and Sarvey carried out a pioneering series of *in vitro* studies to better characterize NE effects in dentate gyrus and to examine their relationship to LTP. In their first report they applied NE to the slice for 30 min and produced two forms of NE-induced potentiation of population spike amplitude (Stanton and Sarvey, 1985a). These two forms were distinguished by their dependence on protein synthesis. The first had the acronym NEP (NE-induced potentiation) and was seen even in the presence of a protein synthesis inhibitor (emetine) that completely blocked LTP in area CA1. NEP occurred when NE was present in the bath and disappeared during the 30 min of NE wash out. The second, NELLP, for NE-induced long-lasting potentiation, occurred only in the absence of the protein synthesis inhibitor and could last throughout a 5-hour recording period. Their standard for NELLP in this paper and subsequent papers, however, was potentiation 30 min after washout. They estimated the NE concentration for the half-maximal response of both forms of NE-potentiation with NELLP requiring a somewhat higher concentration than NEP. They then showed that a priming concentration of an adenylate cyclase activator, forskolin, shifted the NE dose-response curve to the left by an order of magnitude for both forms of NE potentiation. This established a key role for adenylate cyclase in both types of NE-potentiation. The nonselective β-adrenergic antagonist, propranolol, and the selective β1-adrenergic antagonist, metoprolol, blocked NEP and NELLP. NE-potentiation was not seen in CA1. Protein synthesis inhibition 15 min after the application of NE did not block NELLP, establishing an early time window for the protein synthesis requirement. Stanton and Sarvey also compared – and + isomers of NE as both had been reported to produce NELLP in an *in vivo* study {Winson, 1985). *In vitro* only the biologically active – NE was effective.

In a second paper (Stanton and Sarvey, 1985b), they followed up an *in vivo* report that perforant path homosynaptic LTP was reduced in rats with 6-hydroxydopamine (6-OHDA) depletion of NE (Bliss et al., 1983). Using the depletion method of the *in vivo* study they took slices from normal and NE-depleted hippocampi and evaluated LTP. Sarvey's group was one of the few who were able to successfully induce LTP (100 Hz train 2 sec at intensity eliciting 40% maximum spike) in the dentate gyrus *in vitro* without using disinhibition (Wigstrom and Gustafsson, 1983b; Wigstrom and Gustafsson, 1983a) or lowered magnesium (Nguyen and Kandel, 1996) to increase cell excitability. With NE-depletion, LTP of perforant path population spike amplitude could not be induced in dentate gyrus. As rats recovered from 6-OHDA-induced NE depletion their NE levels increased in hippocampus, as did the ability to elicit LTP from their brain slices, further supporting the dependence of LTP on NE. β-adrenoceptor antagonists that blocked NELLP also blocked LTP in the dentate gyrus in normal slices. Forskolin, the adenylate cyclase activator, at a priming concentration, could restore LTP in NE-depleted slices. CA1 LTP was not affected by NE manipulations, suggesting separate roles in the two hippocampal subregions. Further, only NE, not 5-HT, depletion prevented perforant path

LTP *in vitro*, in contrast to the earlier *in vivo* study, suggesting a direct action of NE in the dentate gyrus and an indirect mediation of 5-HT depletion effects, possibly via modulation of NE activity.

In the final paper in this set Stanton and Sarvey reported that LTP in the dentate gyrus and NELLP were both associated with increased cAMP in the dentate gyrus (Stanton and Sarvey, 1985c). With LTP, cAMP rose 3-fold in the first min after the tetanus and was at control levels at 30 min. NE-depletion prevented the normal tetanus-induced increase in cAMP. With NE, cAMP levels increased with a 3- to 4-fold peak at 1 min and a sustained elevation even during washout. NE-depletion did not prevent the cAMP response to NE. Whole slices from NE-depleted rats showed reduced basal cAMP suggesting a contribution of NE to basal cAMP. Earlier, Segal et al. had shown that NE and histamine at 100 μM elevated cAMP in dentate gyrus, but glutamate, GABA, serotonin and acetylcholine did not (Segal et al., 1981). The pharmacological profile was that of a β1 receptor, although glial cells also showed an increase in cAMP to NE stimulation that had a β2 profile. Basal levels of cAMP were higher in dentate gyrus than CA3 or CA1.

Taken together, these studies confirmed NE heterosynaptic long-lasting potentiation of the perforant path population spike in the dentate gyrus *in vitro*, showed that it depended on activation of β-adrenergic receptors, the activation of adenylate cyclase and rapid protein synthesis and demonstrated that a predicted increase in cAMP was associated with NE application. Frequency-induced homosynaptic LTP in dentate gyrus also required the presence of NE and β-adrenoceptor activation and was associated with the elevation of cAMP. NE enhanced the somatic calcium current during the LTP-inducing tetanus (Stanton and Heinemann, 1986). These parallels argued for converging mechanisms underlying both NELLP and LTP in dentate gyrus.

3. FURTHER INVESTIGATIONS OF NEP AND NELLP *IN VITRO*

Our laboratory published similar *in vitro* results (Lacaille and Harley, 1985). Using a 10 min NE application both perforant path EPSP slope (measured at the soma) and population spike amplitude increased, although long-lasting effects were more common with the population spike. With NE, 47% of spike amplitude potentiation could be accounted for by the increase in EPSP slope, suggesting an additional increase in EPSP/spike coupling. EPSP slope potentiation and spike potentiation occurred in all slices exposed to the β-adrenoceptor agonist isoproterenol. The β-adrenoeptor antagonist, timolol, blocked NE potentiation. An α-receptor agonist (phenylephrine) and antagonist (phentolamine) produced only weak or partial effects, in contrast to the β-adrenoceptor agents. When perforant path stimulation was omitted during NE application potentiation was still seen. While this appears to argue against a co-activation requirement for glutamatergic and noradrenergic input to obtain NE-potentiation effects, Stanton and Sarvey's results suggest cAMP would have been elevated during NE washout when stimulation was resumed (Stanton and Sarvey, 1985c). Prolonged cAMP elevation may provide the co-activation requirement. In the co-activation tests there were no long-lasting effects of NE, so the co-activation requirement of NELLP was not probed.

Antidromic activation of the population spike was not altered by NE (Lacaille and Harley, 1985) suggesting the effect required synaptically-elicited action potentials.

Antidromic activation also did not result in enhanced granule cell calcium entry in the presence of NE, while the same stimulation given orthodromically was effective (Stanton and Heinemann, 1986).

In 1987, Stanton and Sarvey revisited the phenomena of NELLP, and of NE depletion effects on LTP, in experiments in which the dendritic EPSP, as well as the population spike amplitude, were examined in response to perforant path stimulation *in vitro* (Stanton and Sarvey, 1987). Examination of the dendritic EPSP was prompted by reports that NE depletion *in vivo* affected the population spike, but not the somatic EPSP (Robinson and Racine, 1985) and that NE iontophoresis decreased the dendritic EPSP, while not altering the somatic EPSP (Winson and Dahl, 1985). Stanton and Sarvey found LTP of both the dendritic EPSP, and the population spike amplitude, were prevented by NE depletion. NE application alone initiated long-lasting potentiation of the dendritic EPSP as well as population spike amplitude. The finding of NE-induced dendritic EPSP potentiation complimented the earlier finding of somatic EPSP potentiation (Lacaille and Harley, 1985) and was reinforced by the observation that NE application produced a long-lasting increase of glutamate release in the dentate gyrus (Lynch and Bliss, 1986;. see also other evidence for presynaptic release increases Chen and Roper, 2003; Kohara et al., 2001). This increase also appears to be β-adrenoceptor dependent.

The earlier finding of differing roles of NE in LTP in dentate gyrus and CA1 was also later replicated measuring EPSP slope. The β-adrenoceptor antagonist propranolol reduced both induction and maintenance of EPSP slope LTP in dentate gyrus, but not in CA1 (Swanson-Park et al., 1999) A dopamine antagonist was effective in CA1, but ineffective in dentate gyrus.

Membrane effects of NE on *in vitro* dentate granule cells were reported in 1987. Haas and Rose replicated the spike potentiating effect of NE, and of a β-adrenoceptor agonist, although only the β-adrenoceptor agonist, isoproterenol, but not NE, potentiated the EPSP (Haas and Rose, 1987). Long-lasting effects on either spike amplitude or EPSP were seen in a minority of experiments. Intracellular recording revealed an NE suppression of the pronounced afterhyperpolarization in dentate granule cells (Haas and Rose, 1987). NE modulation of the afterhyperpolarization was related to reduction of a calcium-mediated potassium current. Evidence was also obtained for inhibition of a potassium A current by the β-adrenoceptor agonist in some granule cells (Haas and Rose, 1987).. The afterhyperpolarization suppression could contribute to enhanced cell excitability and appeared enduring in some recordings. Reduction of the A current would also increase cell excitability. Gray and Johnston reported that the voltage dependent L channel currents were enhanced in granule cells by activation of β-adrenoceptors (Gray and Johnston, 1987), which would promote calcium-mediated plasticity effects upon NE application.

In summary, the *in vitro* studies to this point were consistent with the hypothesis that NE induces a heterosynaptic long-lasting potentiation of the perforant path evoked EPSP and population spike in the dentate gyrus, which is mediated by activation of β-adrenoceptors and the elevation of cAMP. The increase in population spike amplitude is not wholly accounted for by EPSP slope increases and thus implicates an increase in the coupling of the EPSP to spike generation. The latter effect is likely related to postsynaptic changes in membrane channels. The differences and similarities to frequency-induced LTP are examined in the next section.

4. NE INTERACTIONS WITH NON-ADRENERGIC RECEPTORS AND PATHWAY SELECTIVITY

Burgard et al. examined the effects of NMDA antagonists on NELLP and LTP. In their study, the medial perforant path was selectively stimulated and both spike amplitude and EPSP effects of NE were examined (Burgard et al., 1989). LTP spike potentiation was compared to NELLP with the same concentrations of NMDA antagonists. LTP was effective somewhat more often than NELLP (18/20 slices versus 19/27 for NELLP) in inducing spike amplitude potentiation. NMDA antagonists blocked LTP and NELLP, with more complete antagonism of NELLP. Dendritic EPSP potentiation by NE occurred in 7/14 slices and was also antagonized by NMDA receptor blockade. These results suggested both LTP and NELLP depended on NMDA receptors.

In the same year, Dahl and Sarvey reported a pathway selective action of NE in the dentate gyrus. NE in the presence of an α-adrenoceptor antagonist induced long-lasting potentiation of both the medial perforant path population spike and the medial perforant path dendritic EPSP slope (Dahl and Sarvey, 1989). Concurrently, the lateral perforant path-evoked population spike and EPSP slope were depressed. The β-adrenoceptor agonist, isoproterenol, mimicked these effects and the β-adrenoceptor antagonist, propranolol, blocked them. While LTP of either pathway in the dentate gyrus is associated with depression of the other, LTP effects are not preferential for the medial perforant path as the β-adrenoceptor activation effects were.

In further examinations of the pharmacology of NELLP and its relationship to LTP, Burgard and Sarvey reported that 1 μM muscarine facilitated LTP induction while 10 μmol depressed the population spike, and dendritic EPSP, but did not alter LTP induction (Burgard and Sarvey, 1990). By contrast (Burgard et al., 1993) the same concentrations of muscarine reduced (1 μmol muscarine) or blocked (10 μmol muscarine) NELLP induced by the β-adrenoceptor agonist isoproterenol. No facilitation of NELLP was seen at the lower dose. The ability of isoproterenol to enhance cAMP in the dentate gyrus was unaffected by muscarine however. M1/M3 muscarine receptor antagonists blocked the effects of muscarine in both studies. The authors concluded that LTP and NELLP differed in response to muscarine receptor activation and, in particular, they suggested a presynaptic action of muscarine: the reduction of glutamate release prevents a co-activation of NMDA receptors *required* to induce NELLP, while the post-synaptic depolarization effect of LTP is enhanced by muscarine's post-synaptic increase of cell excitability.

The $GABA_B$ agonist, baclofen, had a different action in dentate gyrus relative to other regions of the hippocampus, and its net effect was disinhibition, as described in detail by Burgard and Sarvey. They demonstrated that the $GABA_B$ agonist alone induces a long-lasting potentiation of the population spike, but not the EPSP, while at a subthreshold dose a $GABA_B$ agonist synergizes with subthreshold isoproterenol to produce NELLP of both the spike and dendritic EPSP (Burgard and Sarvey, 1991). NELLP does not depend on sustained NE release, as shown in this study by application of a β-adrenoceptor antagonist after its initiation. Burgard and Sarvey suggest the synergistic action of $GABA_B$ receptor activation is a function of postsynaptic disinhibition, evidenced by a decrease in paired pulse inhibition even at doses subthreshold for direct potentiation.

Dahl and Sarvey investigated the role of NMDA receptors in the pathway selective dendritic EPSP responses elicited by isoproterenol. Both long-lasting potentiation and long-lasting depression were blocked by prior application of an NMDA antagonist (Dahl et al., 1990). After washout the LLP of the medial perforant pathway to isoproterenol recovered, however the isoproterenol-induced LLD with lateral perforant pathway stimulation did not reappear. The data suggest there are enduring effects of NMDA receptor antagonism on the lateral, but not the medial, perforant path effects of adrenergic receptor activation. The authors also showed that it was not necessary to electrically stimulate perforant path fibers during the 30 min period of drug application and the 30 min period of washout in order to obtain LLP and LLD. Again the earlier evidence that cAMP levels are elevated even after NE application and washout (Stanton and Sarvey, 1985c) may account for the ability to elicit LLP and LLD without perforant path stimulation. More difficult to explain is the loss of lateral perforant path LLD after a single exposure to an NMDA antagonist.

In a later study Pelletier et al. revisited the pathway specificity of noradrenergic plasticity in the dentate gyrus. They replicated the long-lasting potentiation of medial perforant path EPSP and the long-lasting depression of the lateral perforant path EPSP with 1 μM isoproterenol reported by Dahl and Sarvey (1989) and evaluated the interaction with LTP. LTP potentiated both pathways with larger potentiation at medial perforant path synapses (Pelletier et al., 1994). The EPSP returned to baseline values when LTP was applied to the lateral perforant path. If LTP was induced prior to isoproterenol application there was further potentiation with isoproterenol of the medial perforant path, but no change in the level of potentiation on the lateral perforant path. Again prior NMDA receptor activation apparently prevented LLD effects of isoproterenol. The selective β1-antagonist, metoprolol, was tested against LTP. Metoprolol blocked the effect of medial perforant pathway LTP, but on the lateral perforant path significantly greater LTP occurred in the presence of metoprolol. Bath application of 20 μM (but not 1 μM) metoprolol for 30 min produced a depression of medial perforant path responses and a potentiation of lateral perforant path responses. These effects continued to increase during the 60 min washout of metoprolol. It was not possible to conclude that metoprolol antagonized the LTP effect in the medial perforant path since it had its own depressing effect that may have masked LTP.

Bramham et al. evaluated propranolol and timolol, two nonselective β-adrenoceptor antagonists, against the medial and lateral perforant path dendritic EPSP LTP. Propranolol blocked LTP in both the medial and lateral pathways, but timolol had no effect at the time of LTP evaluation (Bramham et al., 1997). At later time points the timolol-treated slices showed a greater decline in LTP consistent with evidence for β-adrenoceptor involvement in late phase LTP. The failure of timolol to antagonize early LTP led Bramham et al. to speculate that there was a subpopulation of timolol insensitive β2-receptors in dentate gyrus.

Nguyen and Kandel, using reduced extracellular magnesium, investigated early LTP and late (3 hr) LTP of the medial perforant path dendritic EPSP. Both forms were NMDA-dependent. Late, but not the early, LTP required protein synthesis and activation of cAMP-dependent protein kinase (Nguyen and Kandel, 1996). Segal's early investigation of neurotransmitters in the dentate gyrus that elevate cAMP suggest NE is the most likely candidate to promote elevation of cAMP and consequent activation of cAMP-dependent protein kinase. Sarvey's laboratory showed that spike LTP depended on NE activation of β-adrenoceptors and that tetani in the slice induced elevation of cAMP when NE was intact. Thus, all data available converge on the hypothesis that enduring LTP of EPSP and spike in the medial perforant path requires NE heterosynaptic facilitation of this homosynaptic mechanism.

While NMDA receptors participate in both NELLP and LTP examined *in vitro*, the two forms of potentiation differ in pathway selectivity with NELLP showing selectivity for the medial perforant path only. The response to other pharmacological tests also suggests NELLP and LTP are likely distinct forms of connectivity modulation despite a common dependence on the cAMP cascade.

5. PRESYNAPTIC EFFECTS OF NE *IN VITRO* AND THE QUESTION OF THE PHYSIOLOGICAL ACTIVATION OF β-ADRENOCEPTORS

Sarvey drew attention to the evidence that there were both pre- and post-synaptic effects of β-adrenoceptor activation (Sarvey et al., 1989). Parfitt et al. (1991, 1992) investigated the presynaptic effect, enhanced release of glutamate, by examining phosphorylation of Synapsin I and Synapsin II in the dentate gyrus. Isoproterenol (250 nM) or norepinephrine (10 μmol) induced phosphorylation of Synapsin I and II. Both the calcium/calmodulin protein kinase II phosphorylation site and the PKA phosphorylation site on Synapsin I were phosphorylated following exposure to these adrenoceptor agonists (Parfitt et al., 1992). Previous work had established that phosphorylation of the calcium/calmodulin protein kinase II site activated enhanced transmitter release and hence could account for NE's effects on glutamate release. Phosphorylation of the calcium/calmodulin protein kinase II site by NE application underscores the likely importance of NE's modulation of calcium as well as cAMP in mediating cellular plasticity in dentate gyrus. In aged rats NE did not induce phosphorylation of the synapsins and the basal phosphorylation of these dentate gyrus proteins was higher than that of younger rats (Parfitt et al., 1991). However, slices from aged rats would show increased phosphorylation with NE if a phosphodiesterase inhibitor were added. Such inhibitors have become popular as memory enhancing candidates for both rodent and human aging populations. Parfitt et al. (1992) found that the time course of dentate gyrus EPSP potentiation and of synapsin phosphorylation in dentate gyrus in response to a lower dose of isoproterenol (250 nM) were similar in their hands (less than 30 min) suggesting NE-induced EPSP potentiation may be dependent on phosphorylation of these presynaptic proteins.

Dahl speculated that the 1 μM concentration of isoproterenol normally used to induced NELLP of the EPSP and population spike in dentate gyrus might be an unrealistically high level of β-adrenoceptor activation *in vivo* that would be unlikely to be sustained given the efficiency of NE uptake (Dahl and Li, 1994a). Dahl undertook a series of studies of the effects of bath applied isoproterenol in the 50-100 nM range. He found short-term increases in population spike amplitude during the 15 min application of these low concentrations of isoproterenol, but no long-term increase and no change in the EPSP slope. With a 30 min washout followed by a repeated application of the same low dose of isoproterenol NELLP of the population spike was observed to the second application of isoproterenol. This demonstrated that a repeated, spaced activation of β-adrenoceptors *in vitro* could engage a long-term potentiation of population spike amplitude, but not of EPSP slope (Dahl and Li, 1994a). A non-selective β-adrenoceptor antagonist and a selective β1-adrenoceptor antagonist blocked these effects, but an NMDA antagonist did not.

Dahl also examined coapplications of isoproterenol and cholecystokinin, which he predicted would have effects similar to that of the $GABA_B$ agonist, baclofen, previously studied (Burgard and Sarvey, 1991). Co-application of cholecystokinin and a low dose of isoproterenol produced long-lasting depression of the population spike while isoproterenol followed by washout and then cholecystokinin produced long-lasting potentiation of the population spike similar to the sequential effect of low dose isoproterenol alone (Dahl and Li, 1994c). The effect occurred even when the isoproterenol concentration was too low (50 nM) to produce any change when applied alone. β-antagonists prevent the potentiation seen with sequential isoproterenol and cholecystokinin applications, but again an NMDA antagonist does not (Dahl and Li, 1994b).

Dahl concluded that there may be two levels of β-adrenergic modulation in the dentate gyrus: one that requires NMDA receptor activation (the higher 1 μM concentration effect) and one that is NMDA receptor-independent (the lower 50-100 nM concentration effect). Only the higher concentration effect modulates EPSP as well as spike amplitude, the lower concentration effect is specific to spike amplitude. Dahl suggested that the effects of lower concentrations may represent the physiological effects of β-adrenoceptor activation (Dahl and Li, 1994a).

6. *IN VIVO* STUDIES OF NE RELEASE

The question Dahl raises "Is the NMDA-receptor independent effect of β-adrenergic modulation the physiological effect of NE at these receptors?" seems to have been answered in the affirmative when we turn our attention to the results of *in vivo* studies.

The source of dentate gyrus NE is the terminals of the locus coeruleus (LC) (Loy et al., 1980) and activation of the LC has been used in a number of studies to characterize the physiological actions of NE release. In one of the first studies to use LC electrical stimulation to examine modulation of the perforant path evoked potential in the dentate gyrus, Dahl and Winson reported potentiation of population spike amplitude and no change in EPSP slope at the cell body level, although they observed a depression of the EPSP recorded at the dendritic level (Dahl and Winson, 1985).

We used glutamatergic activation of LC and found a potentiation of population spike amplitude in all rats with variable durations, ~5-20 min (Harley and Milway, 1986). The population EPSP slope increased briefly (less than 3 min) in about 50% of the experiments, and decreased briefly, or was unchanged, in the remainder. The different time courses suggested EPSP and spike effects were largely uncorrelated. In a later study (Harley and Sara, 1992), EPSP slope increases were more common (>60%), but again were uncorrelated with spike increases. EPSP slope increases occurred on less than 30% of the evoked potentials with increased population spikes.

Most recently, we have used the neuroactive peptide, orexin, to activate the LC in the urethane-anesthetized rat (Walling et al., 2004). Orexin produces an enduring (more than 3 hr) and gradually increasing β-adrenoceptor-dependent potentiation of population spike amplitude without potentiating the EPSP (see Figure 1). The gradual increase in spike

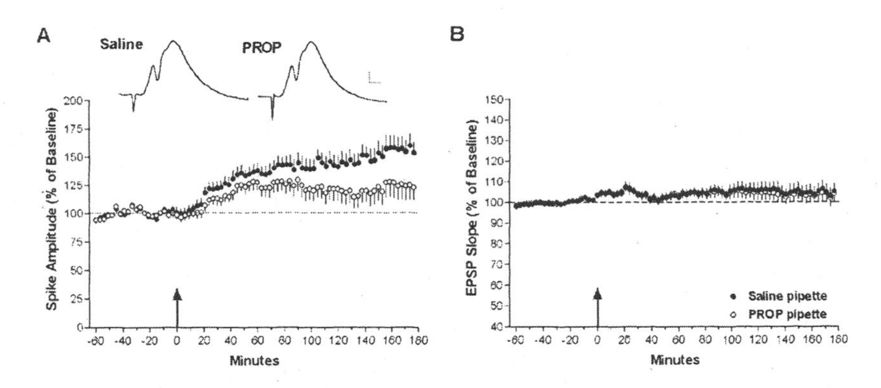

Figure 1. Orexin infused in the locus coeruleus at the arrow produces a significant potentiation of population spike amplitude over the 3 hr recording period (**A**). EPSP slope is not significantly altered (**B**). The β-adrenoceptor antagonist propranolol (PROP) in one of two micropipettes recording the perforant path evoked potential attenuates the spike increase. From Walline et al. (2004) Reprinted with permission from The Journal of Neuroscience.

amplitude is consistent with reports of E-S potentiation recruited by LTP (Jester et al., 1995).

In one study we compared glutamatergic and electrical stimulation of LC (Harley et al., 1989). With electrical stimulation, the potentiating LC effects on population spike amplitude occurred selectively with a preceding ~40 msec interstimulus interval (ISI) as reported previously for LC train stimulation, but repeated LC-perforant path pairings produced 30 min potentiation in about 50% of the experiments. Repeated pairings were associated with enduring population spike increases, but EPSP slope decreased in 60% of the experiments. Long-lasting potentiation occurred more often when potentiation was greater than 140% during the acute stimulation. Ten-Hz stimulation also produced an increase in spike amplitude and no effects on EPSP slope. The β-adrenoceptor antagonist, propranolol, blocked the effects of glutamatergic activation-induced potentiation, as reported previously, but did not alter the potentiating effects of acute LC electrical stimulation-induced potentiation in the same rats. The long-lasting effects of repeated electrical stimulation were not assessed in the presence of a β-adrenoceptor antagonist.

In contrast, Washburn and Moises, using the same stimulation train in the LC or in its efferent dorsal bundle and a 35 msec ISI reported that stimulation giving 50% of the maximal spike amplitude potentiation produced a β-adrenoceptor sensitive potentiation that could also be reduced by clonidine (to activate the terminal NE autoreceptors) or by 6-OHDA depletion (Washburn and Moises, 1989). Sites slightly outside the LC were ineffective in producing spike potentiation. EPSP slope potentiation was not seen (<20% of experiments) at either the somatic or the dendritic level. Repeated pairings were not explored. Thus, LC electrical stimulation may be appropriately selective for NE release under some conditions.

In pharmacological approaches to NE release modulation, Richter-Levin et al. and Sara and Bergis monitored the perforant path evoked potential in the dentate gyrus in anesthetized (Richter-Levin et al., 1991) and awake rats (Sara and Bergis, 1991) receiving the α2 receptor antagonist, idazoxan. Idazoxan increases LC firing and the release of NE. Population spike amplitude was potentiated in both studies. In anesthetized rats, EPSP slope was depressed at the somatic level. Idazoxan was ineffective in 6-OHDA NE-depleted rats confirming NE's role. In the awake rats no changes were seen in EPSP slope. In awake rats paired pulse inhibition was also tested and found to be enhanced contrary to the expectation that there would be decreased inhibition. Segal et al. using amphetamine, also reported potentiation of population spike amplitude in anesthetized rat without consistent effects on EPSP slope (Segal et al., 1991). Amphetamine-associated spike potentiation was sensitive to both β-adrenoceptor and α1-adrenoceptor blockade.

Electrical stimulation of the paragiganocellularis nucleus that provides the major source of glutamatergic excitation to the LC is another method to activate NE release. This produces potentiation of population spike amplitude with EPSP slope increases in <20% of the experiments (Babstock and Harley, 1992). Peak potentiation occurs in a 35-40 msec window, as reported for direct LC stimulation, and is blocked by the β-adrenoceptor antagonist propranolol.

LC stimulation in awake rats has been examined using glutamatergic activation, and using natural events, to recruit activation of LC neurons. We found that glutamatergic LC stimulation in awake rats (Klukowski and Harley, 1994; Walling and Harley, 2004) produces spike potentiation, which can be transient (less than 20 min) or longer term (+20 min to +3 hr depending on the study). EPSP slope potentiation is rare, again occurring in less than 20% of the experiments. With exploration of novel objects in a hole board or exposure to a novel environment, two manipulations previously shown to activate LC, spike potentiation is seen, but not slope potentiation (Kitchigina, 1997). The spike potentiation effects are transient and are blocked by the β-adrenoceptor antagonist, propranolol.

Taken together, a variety of methods for producing physiological increases in synaptic NE, in both anesthetized and awake rats, produces a modulation of the perforant path evoked potential in the dentate gyrus which resembles that reported by Dahl with low dose isoproterenol activation of β-adrenoceptors. The population spike is potentiated, but EPSP slope is rarely affected.

A second parallel between low dose *in vitro* isoproterenol effects and *in vivo* NE release effects is their response to NMDA antagonism. An NMDA antagonist cannot block lasting or long-term low dose isoproterenol potentiation of spike amplitude {Dahl, 1994}. *In vivo,* using glutamatergic activation of the LC, we showed that intraventricular application of the NMDA antagonist ketamine, sufficient to attenuate LTP, did not attenuate, but enhanced, spike potentiation produced by glutamatergic LC activation (Frizzell and Harley, 1994). The enhancement was ascribed to ketamine's ability to reduce NE reuptake. In these experiments transient EPSP slope and spike amplitude increases were seen with all activations. While transient, the potentiations lasted for many minutes, while LC cellular activation by glutamate produces a burst of activity lasting less than .5 sec (Harley and Sara, 1992). Ketamine significantly increased only spike amplitude and also selectively extended the duration of the spike increase. This supports the notion of an NMDA-independent component of NE potentiation, which increases cell excitability, similar to that seen by Dahl and Li (1994a).

Finally, using microdialysis with concomitant intracerebroventricular (ICV) NE *in vivo* we estimated the synaptic concentration of NE required to produce NELLP *in vivo* (Harley et al., 1996). ICV NE in these studies potentiated spike amplitude with no consistent EPSP slope change. The estimated synaptic concentration for long-term spike potentiation (more than 2 hr) was 750 nM NE while transient short-term potentiation was seen at ~1/2 that value. While 750 nM NE is 10X the concentration of isoproterenol used by Dahl and Li to produce long-term spike potentiation with spaced repeated applications, NE concentrations used to evoke NELLP *in vitro* are typically at least 10X (10 μM) higher than isoproterenol (1 μM) concentrations. Thus, 750 nM NE is a low concentration in this context, and would be similar to 75 nM of isoproterenol β-adrenoceptor activation.

These calculations support Dahl and Li's hypothesis that physiological release of NE likely produces low concentration β-adrenoceptor activation rather than high concentration β-adrenoceptor activation (Dahl and Li, 1994a). The general findings that population spike increases occur independently of slope increases after NE release *in vivo* (using a variety of release paradigms) and that such physiological release can induce a potentiation that is NMDA receptor independent, also support the hypothesis.

An alternative explanation of the failure to see consistent EPSP slope potentiation *in vivo* could be the problem of combined medial and lateral perforant pathway contributions to the EPSP slope. In all of the *in vivo* studies described, stimulation was sufficient to evoke a medial perforant pathway population spike. The EPSP slope components would likely represent mixed contributions from both pathways. Enhancement of the medial, and depression of the lateral, EPSP components might result, coincidentally, in a low percentage of apparent EPSP increases. Using paragigantocellularis stimulation to activate NE release we found depression of the lateral olfactory tract EPSP, which is mediated by the lateral perforant pathway, in dentate gyrus when preceded by paragigantocellularis stimulation (Babstock and Harley, 1993). The depression was sensitive to β-adrenoceptor blockade by propranolol. The same stimulation potentiated perforant path spike amplitude. Systematic investigations of medial and lateral perforant path components *in vivo* are warranted and have implications for the cognitive role of hippocampal NE.

7. NE IN INTERACTIONS WITH OTHER NEUROTRANSMITTERS

While it may be suggested that, acting on its own, release of NE from LC terminals primarily modulates population spike amplitude *in vivo*, as hypothesized by Dahl in his model in which physiological NE is at a lower level than that typically used in slice experiments, NE in interaction with the release of other neurotransmitters, may have actions more typical of those reported for the 'higher NE concentration' effects. LTP associated glutamate release, in particular, may alter both the level of NE release locally and the intracellular machinery available for influence by NE. NE release in hippocampal slices is increased by activation of NMDA (e.g, Pittaluga and Raiteri, 1992) and AMPA (e.g., Pittaluga et al., 1994) receptors. This glutamate enhancement of NE release is strongest in the dentate gyrus (Andres et al., 1993), is selective for NE over other transmitters (Fink et al., 1989) and is enhanced by memory promoters (e.g,, Desai et al., 1995). Thus, if brain activity promotes the activation of NMDA and AMPA receptors, NE levels should concomitantly be elevated. NE can enhance glutamate release in dentate gyrus also, which may create further positive feedback (Lynch and Bliss, 1986). LTP tetani do increase NE release in dentate gyrus *in vivo* (Bronzino et al., 2001) likely mediated in part by direct stimulation of NE fibers, but also likely mediated by glutamatergic modulation of local NE release. Elevated levels were sustained for 2 hr following the LTP trains, while spike amplitude was elevated for 24 hr. In a methodological study, levels of NE release varied with the intensity of the LTP manipulation, with stronger manipulations giving rise to more potentiation and higher levels of NE (Bronzino et al., 1999). Thus LTP and NMDA receptor activation in the dentate gyrus are both linked to enhanced NE release.

The ability of NE depletion and β-adrenoceptor blockade to prevent the induction of LTP in dentate gyrus also argues for an key role of NE in LTP plasticity in dentate gyrus, including potentiation of EPSP slope. In the initial NE depletion study *in vivo* Bliss showed that EPSP slope, but not spike potentiation, was impaired (Bliss et al., 1983). We showed that local β-adrenoceptor blockade reduces LTP of EPSP slope, but does not reduce LTP of the population spike *in vivo* (Munro et al., 2001). The reduction of slope potentiation by β-adrenoceptor blockade is largest at later time periods as predicted from brain slice experiments which showed that late phase LTP slope potentiation requires cAMP-dependent protein kinase activation (Nguyen and Kandel, 1996). Thus, physiological circumstances that lead to NMDA receptor activation may always recruit enhanced NE release leading to activation of the 2nd higher concentration NE influence on EPSP slope potentiation described by Sarvey and others.

Decaying LTP of EPSP slope can be restored by electrical stimulation of the LC that elicits exploratory behavior (Ezrokhi et al., 1999). The LC stimulation alone produced a brief EPSP depression followed by mild, but usually nonsignificant potentiation, but when given within minutes or hours of the return to baseline of an LTP-induced EPSP potentiation LC stimulation produced a return to a robust LTP level of EPSP potentiation. Prior LC stimulation, by contrast, did not seem to alter the effectiveness of the LTP train. The authors suggest that either through reduced dephosphorylation, or increased phosphorylation, of the calcium/calmodulin kinase II activation initially set up by LTP NE's β-adrenoceptor recruitment of cAMP restores the activity of the calcium/calmodulin kinase II to return the EPSP to its potentiated level.

The role of β-adrenoceptors in LTP of perforant path spike amplitude *in vivo* varies with the induction protocol. β-adrenoceptors are required for weaker and intermediate induction protocols to generate late phase LTP, but do not appear needed if very strong induction protocols are employed (Straube and Frey, 2003). While EPSP slopes were not measured in these studies, it is likely that they would show similar increases.

In vivo Seidenbecher et al showed that natural reinforcers such as water for water deprived rats or footshock can convert early phase to late phase medial perforant path spike amplitude LTP if given at the same time as, or within 30 min after, the LTP train delivery (Seidenbecher et al., 1997). This effect is blocked by a β-adrenoceptor antagonist. Bergado et al. have similar results with spike amplitude and EPSP slope LTP (Bergado et al., 2001).

Straube et al. (2003) have shown that exposure to a novel environment 15-30 min prior to a weak LTP stimulus converts early phase LTP to late phase LTP (Straube et al., 2003). This conversion is β-adrenoceptor-dependent as well as protein synthesis-dependent. In this study a preference for spike amplitude measures over EPSP measures was ascribed to instability of the latter in recordings in awake rats, however LTP was shown with both measures. The authors note that prior studies of dentate gyrus LTP that only use spike amplitude measures likely also have EPSP potentiation. Isoproterenol administered ICV, at a dose with no effect on its own, mimicked the ability of novelty to convert weak LTP to late phase LTP and supported the hypothesis that β-adrenoceptors were critical. These data suggest sequential interactions occur with initial β-adrenoceptor priming and subsequent LTP events as well as with concurrent and the reverse sequence effects seen in other studies e.g., Seidenbecher et al., 1997.

Our most recent findings suggest there is a 3rd long-term modulatory effect of synaptically released NE (Walling and Harley, 2004). Glutamatergic activation of LC in awake rats produced the typical potentiation of spike amplitude in the dentate gyrus in the majority of rats tested with no significant, or only a very transient, increase in EPSP slope. See Figure 2. This spike potentiation was maintained over a 3 hr period in some rats, an effect consistently observed previously with orexinergic activation of LC (Figure 1). Thus this study initially found a typical low concentration NE effect on spike, but not slope. Twenty-four hr later the EPSP slope of all rats was significantly potentiated, as was spike amplitude (Figure 2). Input output data suggested the increase in EPSP slope at 24 hr accounts for the spike potentiation. Thus, at 3 hr there is E-S potentiation, but at 24 hr there is synaptic potentiation (Figure 3).

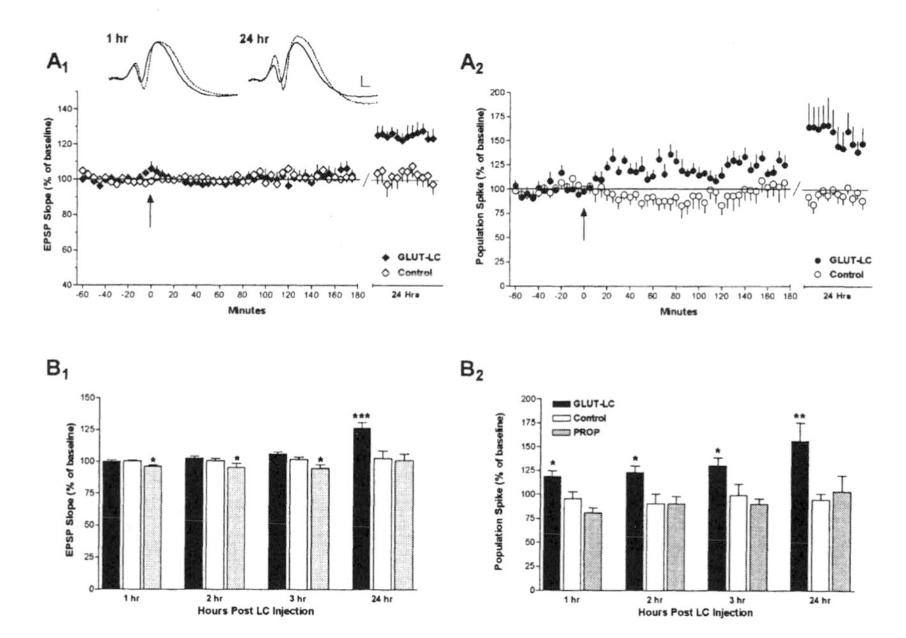

Figure 2. Glutamate infusion in the locus coeruleus at the arrow produces an initial spike potentiation (A_2) with no effect on the EPSP slope (A_1). Twenty-four hours later both EPSP slope and population spike are significantly potentiated. Propranolol prevents the initial and 24 hour effects on EPSP slope (B_1) and spike amplitude (B_2). N=7. From Walling and Harley (2004). Reprinted with permission from The Journal of Neuroscience.

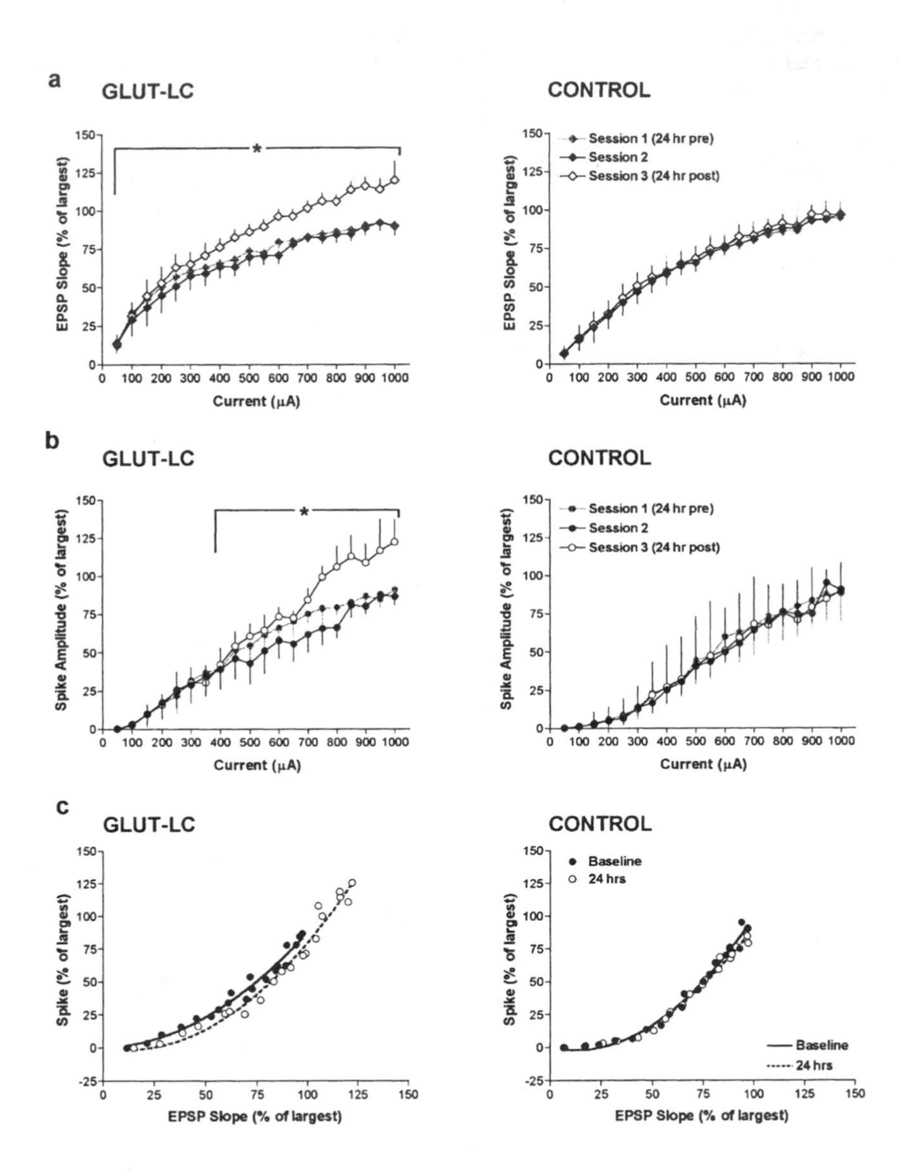

Figure 3. Potentiation of EPSP slope (**a**) and spike amplitude (**b**) at a range of currents prior to (Sessions 1 and 2), and 24 hr following, LC activation by glutamate (N=7). The EPSP-spike relationship is shown in **c**. EPSP slope predicts population spike amplitude both prior to, and 24 hr following, LC-induced potentiation. *indicates significant differences in spike amplitude and EPSP slope 24 hr after LC glutamate infusion relative to baseline. From Walling and Harley (2004). Reprinted with permission from The Journal of Neuroscience.

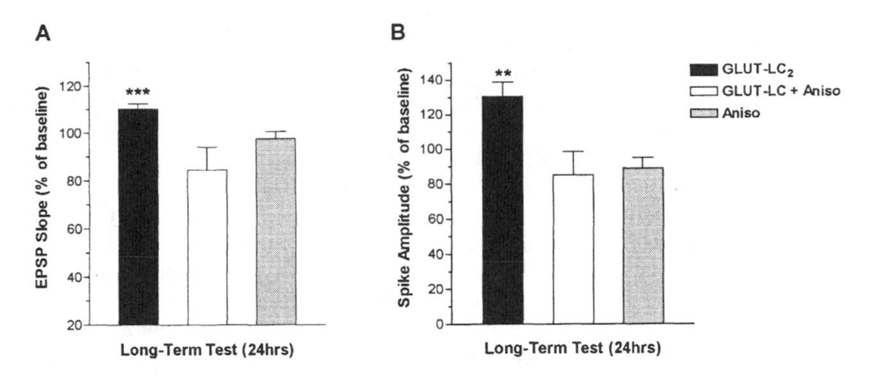

Figure 4. Intracerebroventricular anisomycin (Aniso) given prior to activation of the locus coeruleus (LC) by glutamate (GLUT) prevented the delayed 24 hr potentiation of EPSP slope (**A**) and spike amplitude (**B**) seen in a 2nd group of rats (N=7). Anisomycin alone did not alter recording levels from baseline when taken at the same time point. From Walling and Harley (2004). Reprinted with permission from The Journal of Neuroscience.

The E-S potentiation, characteristic of NE applications in other *in vivo* studies and of the initial effect of LC activation here was no longer seen. The E-S effect had been replaced by a delayed long-term potentiation of EPSP slope and population spike amplitude. The long-term effects were blocked by intraventricular infusion of either a protein synthesis inhibitor (Figure 4) or a β-adrenoceptor blocker (Figure 2). This result parallels a delayed long-term facilitation of synaptic strength described in invertebrates, which is also mediated by cAMP elevation (Emptage and Carew, 1993; Mauelshagen et al., 1998).

8. NE AND PLASTICITY: A SUMMARY AND SUGGESTED MECHANISMS

The present retrospective look at NE effects on the perforant path evoked potential in the dentate gyrus since the early 1980s provides evidence for three distinct effects of noradrenegic modulation on the evoked potential, although only two distinct effects can be described following the initial release of NE.

Low and, possibly physiological, synaptic concentrations of NE produce E-S potentiation. Thus, a given EPSP is associated with a larger spike amplitude than would be predicted by the change in EPSP amplitude alone. This effect can be long-lasting and is typically seen for minutes to hours following a brief period of NE release. Increased NE release associated with LC activation by either glutamatergic or orexinergic activation of LC occurs only in the first microdialysis sample taken at 20 min (Walling et al., 2004).

While E-S potentiation may be related to disinhibition (Staff and Spruston, 2003), our recent measurements of NE suppression of inhibition in dentate gyrus produced by glutamatergic LC activation suggest disinhibitory effects last less than 3 min. The prolonged increase in population spike amplitude is more likely related to changes in intrinsic cell excitability as reported in other systems that exhibit E-S potentiation

(Schrader et al., 2002; Zhang and Linden, 2003; Daoudal and Debanne, 2003; Frick et al., 2004). These changes appear to be post-synaptic changes in voltage-gated channels, particularly potassium channels, that can alter excitability locally in the dendritic arbor. Such channels are known to be modulated by phosphorylation and early evidence for NE modulation of the A type potassium channel was reported by Haas et al. (Haas and Rose, 1987). Recent theoretical models identify the increase in cell excitability as the other half of the Hebb synapse "story", and argue it has an important role in the engram (Schrader et al., 2002; Zhang and Linden, 2003). There is some evidence for an associative component to these changes, although they appear less specific than synaptic changes. The basis of these effects in the dentate gyrus remains to be elucidated.

Higher levels of NE, or low NE levels interacting with other plasticity-promoting neurotransmitters, produce EPSP slope potentiation. In the dentate gyrus, these effects appear intertwined with NMDA-dependent plasticity. NMDA receptor activation enhances NE release and activation of β-adrenoceptor pathways through NE release can enhance NMDA post-synaptic currents (Xie and Lewis, 1997). NE also promotes an increase in evoked presynaptic glutamate release and both the pre- and postsynaptic effects are β-adrenoceptor dependent. NE or β-adrenoceptor agonists also play a role in the conversion of early to late LTP and the recovery of decaying LTP possibly through their enhancement of the L channel calcium current, and promotion of calcium/calmodulin translocation to the nucleus (Mermelstein et al., 2001).

The third effect of NE is the most recently discovered and may be a result of physiological release of NE interacting with glutamatergic input in the dentate gyrus. The conditions for the development of the 24 hr synaptic and spike potentiation effect remain to be described with precision. It is possible that a period of sleep or some other state change must intervene to convert the glutamate-induced burst activation of LC concomitant with the activation of perforant path input into a delayed potentiation of the perforant path connection.

The associative requirements for the E-S potentiation effects characteristic of lower NE levels have not been clarified. However, it is likely that the effects of higher concentrations of NE in promoting synaptic potentiation have an associative component since NMDA antagonists block them. Thus, EPSP slope potentiation by NE appears to require concomitant glutamate release. Examining the associative properties of NE effects in dentate gyrus *in vivo* is complicated by the differential modulation of input from the medial and lateral perforant pathways. The depression observed in the lateral perforant pathway may be related to the disparate effects of opioid and glutamate release in that pathway. The opioids released depress NMDA post-synaptic currents, while NE may enhance the same currents (Xie and Lewis, 1997). LTP stimulation of lateral perforant path fibers releases sufficient opioids for the disinhibitory effects of the opioids to offset their reduction of NMDA currents. If this balance is altered with single pulse stimulation than the competitive reduction of NMDA currents might reduce calcium entry to the level required for LTD. In the delayed EPSP potentiation study (Walling and Harley, 2004), the failure to see EPSP slope potentiation initially could be related to stimulation of a consistent mix of medial and lateral fibers, although this seems unlikely. If it occurred, the later appearance of significant EPSP slope potentiation argues that lateral perforant path depression is transient. *In vitro* low levels of β-receptor activation failed to modulate EPSP slope and a similar lack of modulation seems more likely to account for the *in vivo* observations.

NE is normally released when new events happen in the world (Aston-Jones and Bloom, 1981; Sara and Segal, 1991; Sara et al., 1994; Vankov et al., 1995; Bouret and Sara, 2004). These are clearly moments when the ability to encode new information is most critical. Dahl and Sarvey's early observation still seems pertinent to our present views of NE's role in the dentate gyrus. "Selective persistent NE-induced effects upon neocortical input to the hippocampal formation via the entorhinal cortex may underlie mechanisms of attention, learning and memory (Dahl and Sarvey, 1989)." Our survey of studies on NE's effects in the dentate gyrus suggest NE alters information processing in the dentate gyrus through three β-adrenergic mechanisms: (1) a postsynaptic increase in cell excitability that may be linked to the phosphorylation and closing of voltage-dependent potassium channels in the dendrites of granule cells, (2) an increase in EPSP slope that may be associated with a presynaptic increase in transmitter release through phosphorylation of release proteins like synapsin and/or with an increase in LTP effects through enhancement of NMDA currents, increases in calcium entry via L channels with enhancement of calcium/calmodulin activity, and (3) a delayed potentiation of EPSP slope that resembles that seen in invertebrates, but the mechanism of which has not been identified.

9. ACKNOWLEDGEMENTS

We thank the Natural Sciences Research Council for funding to support our research into the noradrenergic role in plasticity in the dentate gyrus.

10. REFERENCES

Andres ME, Bustos G, Gysling K (1993) Regulation of [3H]norepinephrine release by N-methyl-D-aspartate receptors in minislices from the dentate gyrus and the CA1-CA3 area of the rat hippocampus. Biochem Pharmacol 46: 1983-1987.

Aston-Jones G, Bloom FE (1981) Nonrepinephrine-containing locus coeruleus neurons in behaving rats exhibit pronounced responses to non-noxious environmental stimuli. J Neurosci 1: 887-900.

Babstock DM, Harley CW (1992) Paragigantocellularis stimulation induces beta-adrenergic hippocampal potentiation. Brain Res Bull 28: 709-714.

Babstock DM, Harley CW (1993) Lateral olfactory tract input to dentate gyrus is depressed by prior noradrenergic activation using nucleus paragigantocellularis stimulation. Brain Res 629: 149-154.

Bergado JA, Almaguer W, Ravelo J, Rosillo JC, Frey JU (2001) Behavioral reinforcement of long-term potentiation is impaired in aged rats with cognitive deficiencies. Neuroscience 108: 1-5.

Bliss TV, Gardner-Medwin AR (1973) Long-lasting potentiation of synaptic transmission in the dentate area of the unanaestetized rabbit following stimulation of the perforant path. J Physiol 232: 357-374.

Bliss TV, Goddard GV, Riives M (1983) Reduction of long-term potentiation in the dentate gyrus of the rat following selective depletion of monoamines. J Physiol 334: 475-491.

Bliss TV, Lømo T (1973) Long-lasting potentiation of synaptic transmission in the dentate area of the anaesthetized rabbit following stimulation of the perforant path. J Physiol 232: 331-356.

Bouret S, Sara SJ (2004) Reward expectation, orientation of attention and locus coeruleus-medial frontal cortex interplay during learning. Eur J Neurosci 20: 791-802.

Bramham CR, Bacher-Svendsen K, Sarvey JM (1997) LTP in the lateral perforant path is beta-adrenergic receptor-dependent. Neuroreport 8: 719-724.

Bronzino JD, Kehoe P, Hendriks R, Vita L, Golas B, Vivona C, Morgane PJ (1999) Hippocampal neurochemical and electrophysiological measures from freely moving rats. Exp Neurol 155: 150-155.

Bronzino JD, Kehoe P, Mallinson K, Fortin DA (2001) Increased extracellular release of hippocampal NE is associated with tetanization of the medial perforant pathway in the freely moving adult male rat. Hippocampus 11: 423-429.

Burgard EC, Cote TE, Sarvey JM (1993) Muscarinic depression of synaptic transmission and blockade of norepinephrine-induced long-lasting potentiation in the dentate gyrus. Neuroscience 54: 377-389.

Burgard EC, Decker G, Sarvey JM (1989) NMDA receptor antagonists block norepinephrine-induced long-lasting potentiation and long-term potentiation in rat dentate gyrus. Brain Res 482: 351-355.

Burgard EC, Sarvey JM (1990) Muscarinic receptor activation facilitates the induction of long-term potentiation (LTP) in the rat dentate gyrus. Neurosci Lett 116: 34-39.

Burgard EC, Sarvey JM (1991) Long-lasting potentiation and epileptiform activity produced by $GABA_B$ receptor activation in the dentate gyrus of rat hippocampal slice. J Neurosci 11: 1198-1209.

Chen HX, Roper SN (2003) PKA and PKC enhance excitatory synaptic transmission in human dentate gyrus. J Neurophysiol 89: 2482-2488.

Dahl D, Burgard EC, Sarvey JM (1990) NMDA receptor antagonists reduce medial, but not lateral, perforant path-evoked EPSPs in dentate gyrus of rat hippocampal slice. Exp Brain Res 83: 172-177.

Dahl D, Li J (1994a) Induction of long-lasting potentiation by sequenced applications of isoproterenol. Neuroreport 5: 657-660.

Dahl D, Li J (1994b) Interactive and discontinuous long-lasting potentiation in the dentate gyrus. Neuroreport 5: 1769-1772.

Dahl D, Li J (1994c) Long-lasting potentiation and depression by novel isoproterenol and cholecystokinin 8-S interactions in the dentate gyrus. Exp Brain Res 100: 155-159.

Dahl D, Sarvey JM (1989) Norepinephrine induces pathway-specific long-lasting potentiation and depression in the hippocampal dentate gyrus. Proc Natl Acad Sci U S A 86: 4776-4780.

Dahl D, Winson J (1985) Action of norepinephrine in the dentate gyrus. I. Stimulation of locus coeruleus. Exp Brain Res 59: 491-496.

Daoudal G, Debanne D (2003) Long-term plasticity of intrinsic excitability: learning rules and mechanisms. Learn Mem 10: 456-465.

Desai MA, Valli MJ, Monn JA, Schoepp DD (1995) 1-BCP, a memory-enhancing agent, selectively potentiates AMPA-induced [3H]norepinephrine release in rat hippocampal slices. Neuropharmacology 34: 141-147.

Emptage NJ, Carew TJ (1993) Long-term synaptic facilitation in the absence of short-term facilitation in Aplysia neurons. Science 262: 253-256.

Ezrokhi VL, Zosimovskii VA, Korshunov VA, Markevich VA (1999) Restoration of decaying long-term potentiation in the hippocampal formation by stimulation of neuromodulatory nuclei in freely moving rats. Neuroscience 88: 741-753.

Fink K, Gothert M, Molderings G, Schlicker E (1989) N-methyl-D-aspartate (NMDA) receptor-mediated stimulation of noradrenaline release, but not release of other neurotransmitters, in the rat brain cortex: receptor location, characterization and desensitization. Naunyn Schmiedebergs Arch Pharmacol 339: 514-521.

Frick A, Magee J, Johnston D (2004) LTP is accompanied by an enhanced local excitability of pyramidal neuron dendrites. Nat Neurosci 7: 126-135.

Frizzell LM, Harley CW (1994) The N-methyl-D-aspartate channel blocker ketamine does not attenuate, but enhances, locus coeruleus-induced potentiation in rat dentate gyrus. Brain Res 663: 173-178.

Gray R, Johnston D (1987) Noradrenaline and beta-adrenoceptor agonists increase activity of voltage-dependent calcium channels in hippocampal neurons. Nature 327: 620-622.

Haas HL, Rose GM (1987) Noradrenaline blocks potassium conductance in rat dentate granule cells in vitro. Neurosci Lett 78: 171-174.

Harley C, Milway JS, Lacaille JC (1989) Locus coeruleus potentiation of dentate gyrus responses: evidence for two systems. Brain Res Bull 22: 643-650.

Harley CW, Lalies MD, Nutt DJ (1996) Estimating the synaptic concentration of norepinephrine in dentate gyrus which produces beta-receptor mediated long-lasting potentiation in vivo using microdialysis and intracerebroventricular norepinephrine. Brain Res 710: 293-298.

Harley CW, Milway JS (1986) Glutamate ejection in the locus coeruleus enhances the perforant path-evoked population spike in the dentate gyrus. Exp Brain Res 63: 143-150.

Harley CW, Sara SJ (1992) Locus coeruleus bursts induced by glutamate trigger delayed perforant path spike amplitude potentiation in the dentate gyrus. Exp Brain Res 89: 581-587.

Jester JM, Campbell LW, Sejnowski TJ (1995) Associative EPSP--spike potentiation induced by pairing orthodromic and antidromic stimulation in rat hippocampal slices. J Physiol 484 (Pt 3): 689-705.

Kitchigina V., Vankov A., Harley C., Sara S.J. (1997) Novelty-elicited, noradrenaline-dependent enhancement of excitability in the dentate gyrus. Eur.J.Neurosci. 9: 41-47.

Klukowski G, Harley CW (1994) Locus coeruleus activation induces perforant path-evoked population spike potentiation in the dentate gyrus of awake rat. Exp Brain Res 102: 165-170.

Kohara K, Ogura A, Akagawa K, Yamaguchi K (2001) Increase in number of functional release sites by cyclic AMP-dependent protein kinase in cultured neurons isolated from hippocampal dentate gyrus. Neurosci Res 41: 79-88.

Lacaille JC, Harley CW (1985) The action of norepinephrine in the dentate gyrus: beta-mediated facilitation of evoked potentials in vitro. Brain Res 358: 210-220.

Loy R, Koziell DA, Lindsey JD, Moore RY (1980) Noradrenergic innervation of the adult rat hippocampal formation. J Comp Neurol 189: 699-710.

Lynch MA, Bliss TV (1986) Noradrenaline modulates the release of [14C]glutamate from dentate but not from CA1/CA3 slices of rat hippocampus. Neuropharmacology 25: 493-498.

Mauelshagen J, Sherff CM, Carew TJ (1998) Differential induction of long-term synaptic facilitation by spaced and massed applications of serotonin at sensory neuron synapses of Aplysia californica. Learn Mem 5: 246-256.

Mermelstein PG, Deisseroth K, Dasgupta N, Isaksen AL, Tsien RW (2001) Calmodulin priming: nuclear translocation of a calmodulin complex and the memory of prior neuronal activity. Proc Natl Acad Sci U S A 98: 15342-15347.

Munro CA, Walling SG, Evans JH, Harley CW (2001) Beta-adrenergic blockade in the dentate gyrus in vivo prevents high frequency-induced long-term potentiation of EPSP slope, but not long-term potentiation of population spike amplitude. Hippocampus 11: 322-328.

Neuman RS, Harley CW (1983) Long-lasting potentiation of the dentate gyrus population spike by norepinephrine. Brain Res 273: 162-165.

Nguyen PV, Kandel ER (1996) A macromolecular synthesis-dependent late phase of long-term potentiation requiring cAMP in the medial perforant pathway of rat hippocampal slices. J Neurosci 16: 3189-3198.

Parfitt KD, Doze VA, Madison DV, Browning MD (1992) Isoproterenol increases the phosphorylation of the synapsins and increases synaptic transmission in dentate gyrus, but not in area CA1, of the hippocampus. Hippocampus 2: 59-64.

Parfitt KD, Hoffer BJ, Browning MD (1991) Norepinephrine and isoproterenol increase the phosphorylation of synapsin I and synapsin II in dentate slices of young but not aged Fisher 344 rats. Proc Natl Acad Sci U S A 88: 2361-2365.

Pelletier MR, Kirkby RD, Jones SJ, Corcoran ME (1994) Pathway specificity of noradrenergic plasticity in the dentate gyrus. Hippocampus 4: 181-188.

Pittaluga A, Raiteri M (1992) N-methyl-D-aspartic acid (NMDA) and non-NMDA receptors regulating hippocampal norepinephrine release. III. Changes in the NMDA receptor complex induced by their functional cooperation. J Pharmacol Exp Ther 263: 327-333.

Pittaluga A, Thellung S, Maura G, Raiteri M (1994) Characterization of two central AMPA-preferring receptors having distinct location, function and pharmacology. Naunyn Schmiedebergs Arch Pharmacol 349: 555-558.

Richter-Levin G, Segal M, Sara S (1991) An alpha 2 antagonist, idazoxan, enhances EPSP-spike coupling in the rat dentate gyrus. Brain Res 540: 291-294.

Robinson GB, Racine RJ (1985) Long-term potentiation in the dentate gyrus: effects of noradrenaline depletion in the awake rat. Brain Res 325: 71-78.

Sara SJ, Bergis O (1991) Enhancement of excitability and inhibitory processes in hippocampal dentate gyrus by noradrenaline: a pharmacological study in awake, freely moving rats. Neurosci Lett 126: 1-5.

Sara SJ, Segal M (1991) Plasticity of sensory responses of locus coeruleus neurons in the behaving rat: implications for cognition. Prog Brain Res 88: 571-585.

Sara SJ, Vankov A, Herve A (1994) Locus coeruleus-evoked responses in behaving rats: a clue to the role of noradrenaline in memory. Brain Res Bull 35: 457-465.

Sarvey JM, Burgard EC, Decker G (1989) Long-term potentiation: studies in the hippocampal slice. J Neurosci Methods 28: 109-124.

Schrader LA, Anderson AE, Varga AW, Levy M, Sweatt JD (2002) The other half of Hebb: K+ channels and the regulation of neuronal excitability in the hippocampus. Mol Neurobiol 25: 51-66.

Segal M, Greenberger V, Hofstein R (1981) Cyclic AMP-generating systems in rat hippocampal slices. Brain Res 213: 351-364.

Segal M, Markram H, Richter-Levin G (1991) Actions of norepinephrine in the rat hippocampus. Prog Brain Res 88: 323-330.

Seidenbecher T, Reymann KG, Balschun D (1997) A post-tetanic time window for the reinforcement of long-term potentiation by appetitive and aversive stimuli. Proc Natl Acad Sci U S A 94: 1494-1499.

Staff NP, Spruston N (2003) Intracellular correlate of EPSP-spike potentiation in CA1 pyramidal neurons is controlled by GABAergic modulation. Hippocampus 13: 801-805.

Stanton PK, Heinemann U (1986) Norepinephrine enhances stimulus-evoked calcium and potassium concentration changes in dentate granule cell layer. Neurosci Lett 67: 233-238.

Stanton PK, Sarvey JM (1985a) Blockade of norepinephrine-induced long-lasting potentiation in the hippocampal dentate gyrus by an inhibitor of protein synthesis. Brain Res 361: 276-283.

Stanton PK, Sarvey JM (1985b) Depletion of norepinephrine, but not serotonin, reduces long-term potentiation in the dentate gyrus of rat hippocampal slices. J. Neurosci. 5: 2169-2176.

Stanton PK, Sarvey JM (1985c) The effect of high-frequency electrical stimulation and norepinephrine on cyclic AMP levels in normal versus norepinephrine-depleted rat hippocampal slices. Brain Res 358: 343-348.

Stanton PK, Sarvey JM (1987) Norepinephrine regulates long-term potentiation of both the population spike and dendritic EPSP in hippocampal dentate gyrus. Brain Res Bull 18: 115-119.

Straube T, Frey JU (2003) Involvement of beta-adrenergic receptors in protein synthesis-dependent late long-term potentiation (LTP) in the dentate gyrus of freely moving rats: the critical role of the LTP induction strength. Neuroscience 119: 473-479.

Straube T, Korz V, Balschun D, Frey JU (2003) Requirement of beta-adrenergic receptor activation and protein synthesis for LTP-reinforcement by novelty in rat dentate gyrus. J Physiol 552: 953-960.

Swanson-Park JL, Coussens CM, Mason-Parker SE, Raymond CR, Hargreaves EL, Dragunow M, Cohen AS, Abraham WC (1999) A double dissociation within the hippocampus of dopamine D1/D5 receptor and beta-adrenergic receptor contributions to the persistence of long-term potentiation. Neuroscience 92: 485-497.

Vankov A, Herve-Minvielle A, Sara SJ (1995) Response to novelty and its rapid habituation in locus coeruleus neurons of the freely exploring rat. Eur J Neurosci 7: 1180-1187.

Walling SG, Harley CW (2004) Locus ceruleus activation initiates delayed synaptic potentiation of perforant path input to the dentate gyrus in awake rats: a novel beta-adrenergic- and protein synthesis-dependent mammalian plasticity mechanism. J Neurosci 24: 598-604.

Walling, S. G., Nutt, D. J., Lalies, M. D., and Harley, C. W. Orexin-A infusion in the locus coeruleus triggers norepinephrine (NE) release and NE-induced long-term potentiation in the dentate gyrus. J.Neurosci. 24: in press..

Washburn M, Moises HC (1989) Electrophysiological correlates of presynaptic alpha 2-receptor-mediated inhibition of norepinephrine release at locus coeruleus synapses in dentate gyrus. J Neurosci 9: 2131-2140.

Wigstrom H, Gustafsson B (1983a) Facilitated induction of hippocampal long-lasting potentiation during blockade of inhibition. Nature 301: 603-604.

Wigstrom H, Gustafsson B (1983b) Large long-lasting potentiation in the dentate gyrus in vitro during blockade of inhibition. Brain Res 275: 153-158.

Winson J, Dahl D (1985) Action of norepinephrine in the dentate gyrus. II. Iontophoretic studies. Exp Brain Res 59: 497-506.

Xie CW, Lewis DV (1997) Involvement of cAMP-dependent protein kinase in mu-opioid modulation of NMDA-mediated synaptic currents. J Neurophysiol 78: 759-766.

Zhang W, Linden DJ (2003) The other side of the engram: experience-driven changes in neuronal intrinsic excitability. Nat Rev Neurosci 4: 885-900.

THE HISTAMINERGIC SYSTEM IN BRAIN: MEMORY AND SYNAPTIC PLASTICITY

Oliver Selbach, Olga Sergeeva and Helmut L. Haas*

1. INTRODUCTION

Among the aminergic systems in the brain the histaminergic system has received the least attention although it is equally important. The relatively small groups of neurons containing acetylcholine, noradrenaline, dopamine, serotonin and histamine display comparable electrophysiological properties and morphological features. They all project with multifold arborization to most regions of the central nervous system with some notable distinctions between them. They form mutual connections and act in concert to maintain the homoeostasis of many basic body functions and behavioral state. They switch and modulate higher brain functions and hormonal states during sleep and waking, during stress and contemplative life, during dosing and attention, during reproduction and cognition. Memory formation and retrieval relies on synchronous activities in selected groups of neurons located in the hippocampus, the amygdala, the cerebral cortex, cerebellum and the basal ganglia. Aminergic systems can enable such synchronous discharges through modulation of the cortical excitability by using a number of sophisticated mechanisms. The histaminergic neurons are unique with respect to their localisation in the tuberomamillary nucleus of the posterior hypothalamus. They establish a particularly close partnership to the orexinergic neurons in the perifornical area which serve as the conductor in the aminergic orchestra. The involvement of histamine in long lasting changes of synaptic transmission and memory functions is reviewed.

2. THE HISTAMINERGIC SYSTEM

The reason for the relative neglect of the histaminergic system was its late morphological characterisation (Panula et al., 1984; Takeda et al., 1984) by immunohistochemistry. When all other aminergic neurons had been visualized with phtalaldehyde fluorescence in the 1960s attempts to stain histamine neurons failed because the ubiquitous polyamines sper-mine and spermidine were also stained (Green, 1970). Early studies using lesions as well as biochemical and electrophysiological methods had revealed convincing evidence for the existence and the approximate location of the histaminergic neurons in the tuberomamillary nucleus (TMN) (Dismukes et al., 1974; Garbarg et al., 1974; Haas and Wolf, 1977; Haas et al., 1978). The histaminergic nucleus is subdivided in up to 5 groups which are

* O. Selbach, O. Sergeeva, H. Haas, Department of Physiology, Heinrich-Heine University, Dusseldorf, Germany D-4000. E-mail: haas@uni-duesseldorf.de

considered one functional group in the absence of evidence for different specific projections or functions (Ericson et al., 1987; Inagaki et al., 1988). Two ascending and one descending pathway of histaminergic fibers have been described by (Panula et al., 1990), the ventral pathway innervates the hypothalamus, the diagonal band, the septum and the olfactory bulb, the dorsal pathway reaches the thalamus, hippocampus, amygdala, striatum and nucleus accumbens and the cerebral cortex. Thus all structures involved in memory processing receive histaminergic input. All aminergic nuclei are mutually connected with the TMN and a number of brainstem nuclei are strongly innervated, including the nucleus tractus solitarii. There is a considerable mismatch between the density of histaminergic innervation and the number of histamine receptors in certain regions. The hippocampus for instance displays only a low density of fibers but quite a reasonable density of histamine receptors and marked electrophysiologically measured effects of histamine. Like with other aminergic systems histaminergic fibers rarely form synaptic structures (Diewald et al., 1997), varicosities with the histamine-containing vesicles are found at some distance from the target receptors.

The enzyme histidinedecarboxylase (HDC) specifically serves the histamine formation from histidine, and histamine is metabolized by methylation to tele-methylhistamine. No high affinity uptake system is known. Histamine synthesis and release are inhibited by activation of histamine H3-autoreceptors (Arrang et al., 1987). The release from varicosities is also under control of muscarinic, alpha-adrenergic, opioid and galanin receptors as well as NO (Schwartz et al., 1991; Brown et al., 2001b).

Histaminergic neurons fire at high rates during waking and attention, but rarely or not at all during sleep. The histamine system is now recognized to maintain wakefulness rather than terminating sleep (Parmentier et al. 2002). Histamine release in the brain follows the sleep-wake cycle (Philippu and Prast, 1991; Mochizuki et al., 1992). Histamine neurons display a set of very typical membrane currents and fire rather regularly in vivo and in vitro at about 50 mV membrane potential in a pacemaker-like fashion at about 2–3 Hz. They have broad action potentials with a distinct Ca^{2+}-component. The firing seems to arise from dendritic Ca^{2+}-prepotentials and a slow non-inactivating Na^{+}-current (Uteshev et al., 1995; Stevens and Haas, 1996). A fast and a slow transient outward current and a hyperpolarization activated inward current give the recordings from histamine neurons a fingerprint that allows identification in electrophysiological recordings (Haas and Reiner, 1988; Greene et al., 1990). The autoreceptor feedback inhibition of histamine neurone firing, as well as the release from axons (and possibly dendrites) seems to occur exclusively through block of Ca^{2+}-currents (Takeshita et al., 1998). TM neurons receive GABAergic input mainly from the ventrolateral preoptic area (VLPO) which is active during slow wave sleep and inhibits them via GABAA-receptors. This input is presynaptically inhibited by GABAB, H3, adenosine A_1, opioid and α_2 receptors (Flagmeyer et al., 1997; Stevens et al., 2004). Galanin is expressed in some TM neurons and in GABAergic VLPO-neurons, and inhibits TM neuron firing. Excitatory influences on TM neurons include glutamate, ACh (nicotinic), ATP, serotonin and orexins, the latter two by activating an electrogenic Na^{+}-Ca^{2+}-exchanger (Eriksson et al., 2001).

Four G-protein coupled histamine receptors are known, H1, H2 and H3 are well established in the brain and a direct effect of histamine on NMDA-receptors has been shown. The classical antihistaminics with their sedative properties are H1-antagonists and have prompted early suggestions for the waking action of histamine. H1-receptors are coupled through Gq/11 to a potassium conductance (excitatory) and to phospholipase C to inositol phosphate breakdown delivering IP3 and DAG. H2-receptors are coupled through a Gs Protein to adenylyl cyclase, cyclic AMP, PKA regulating hyperpolarisation activated inward currents and a slow afterhyperpolarisation (AHP). H3-receptors are negatively coupled through a Gi/o protein to adenylyl cyclase and to high voltage activated Ca^{2+}-currents (HVACCs) (Schwartz et al., 1991; Haas and Panula, 2003).

3. HISTAMINE AND SYNAPTIC PLASTICITY

Histamine (and hence activity of histaminergic neurons) can cause long lasting changes in synaptic transmission in several structures by itself and it can prime or modulate other forms of synaptic plasticity such as long-term potentiation (LTP) and long-term depression (LTD). These phenomena represent coincidence detection and are associated with learning and memory. An important signal transduction pathway is the transient rise of intracellular Ca^{2+} and its binding to a Ca^{2+}/calmodulin dependent protein kinase. After describing a direct interaction of histamine with the glutamate NMDA-receptor and a brief account on high frequency oscillations, other mechanisms of action potentially relevant for memory processes will be treated consecutively for the H1, H2 and H3 receptor mediated effects. Most of this work was done on slices from the hippocampus, a structure that is indispensable for declarative memory.

3.1. Potentiation of NMDA Receptor Mediated Currents

The interference of the polyamine spermidine with early attempts of staining the histamine neurons came to mind when we detected a direct interaction of the diamine histamine with the NMDA receptor on hippo-campal pyramidal cells (Haas, 1984; Vorobjev et al., 1993; Bekkers, 1993), specifically the NR1 / NR2B subunit containing variant (Williams, 1994) which is particularly involved in the production of LTP. This potentiation of NMDA-receptor mediated currents is independent on any of the known histamine receptors or the glycine site but akin to the known potentiation by polyamines, it is occluded by spermidine. It occurs at concentrations of histamine within the possible physiological range and is highly pH dependent. Protons reduce NMDA-mediated currents, poly-amines (Traynelis et al., 1995) and histamine counteract this inhibition. The effect has also been demonstrated on autaptic transmission in hippocampal cultures (Bekkers, 1993) and on excitatory synaptic potentials (EPSPs) or currents (EPSCs) in hippocampal slices (Brown et al., 1995; Saybasili et al., 1995; Yanovsky et al., 1995). Under conditions of decreased pH as it occurs during intense and synchronized discharge of neighboring neurons histamine is expected to be most effective in potentiating NMDA-receptor-mediated currents. Such discharges occur for instance in the cornu ammonis 3 (CA3) pyramidal cell population during sharp waves which are probably a decisive element in memory trace formation and represent a natural trigger for LTP production (Buzsaki et al., 1987; Selbach et al., 2004). In most experiments LTP is elicited by the electrical stimulation of a large number of Schaffer collaterals emanating from CA3 and innervating CA1 dendritic regions. Sharp waves cause high frequency oscillations ("ripples", 200 Hz) in CA1.

3.2. Histamine and High Frequency Oscillations

Sharp waves and the ensuing ripples are state dependent: they occur when subcortical inputs to the hippocampus are suppressed, the excitatory drive to interneurons is diminished leading to disinhibition of a primed population (by the input from the entorhinal cortex–dentate area) of CA3 pyramidal cells. While histamine strongly promotes burst firing in the CA3 area (Yanovsky and Haas, 1998) its injection in the lateral ventri-cles of freely moving rats caused rather a reduction of ripple occurrence. This injection mainly affected the septum pellucidum where cholinergic neurons are excited (H1-receptor) that are known to switch the hippocampus from the bursting mode to theta activity. More importantly, hista-mine H1-receptor antagonists strongly increased ripple occurrence indica-ting the intrinsic activity of the histaminergic system. Ripples occur most frequently during quiet waking or sleep (Buzsaki, 1998). The replay of ripples during sleep has been proposed as critical for memory consolidation (Kudrimoti et al., 1999; Ponomarenko et al., 2003).

3.3. Histamine H1-Receptor

High vigilance and attention are no doubt a basis for the perception of sensory information. Excitatory effects of H1-receptor activation in the basal forebrain, thalamus and cortex are held responsible for the waking action of histamine (McCormick and Williamson, 1991; Reiner and Kamondi, 1994). They include block of a potassium conductance, inositol phosphate breakdown to inositoltrisphosphate (IP3) and diacylglycerol (DAG). IP3 releases Ca^{2+} from internal stores: an important signal for synaptic plasticity. DAG enhances the proteinkinase C (PKC) activity, which has been shown to weaken the Mg^{2+} block that normally keeps NMDA channels shut for Ca^{2+}. The raised intracellular Ca^{2+} can also activate nitric oxide synthase (NOS) to produce NO, that diffuses freely to presynaptic varicosities where it furthers glutamate release and induction of presynaptic LTP through activation of guanylyl cyclase, which can also enhance the opening of gap junctions (Hatton and Yang, 1996). Furthermore, Ca^{2+} can activate an electrogenic Na^{+}/Ca^{2+} exchanger (NCX) or a cation channel of the TRPC type (Brown et al., 2001a; Sergeeva et al., 2003). In summary a large number of mechanisms with positive influence on synaptic transmission and plasticity are mediated by H1-receptor activation. Interestingly, in the hippocampus (of rodents) H1-receptor activation causes a hyperpolarization of pyramidal cells and long lasting reductions in firing rate. This is the result of the intracellular increase of Ca^{2+} which leads to opening of Ca^{2+}-dependent K^{+}-channels. Finally H1-receptors support (trough PKC) any H2-receptor mediated actions: histamine alone is more potent in this respect than an H2-receptor agonist (impromidine).

3.4. Histamine H2-Receptor

H2-receptor activation stimulates gastric acid secretion through cyclic AMP and protein kinase A (PKA). The H2-receptor was identified by Black et al. (Black et al., 1972) using pharmacological characterization of this histamine action that was not blocked by the classic antihistamines. The development of H2-receptor antagonists revolutionized the therapy of stomac ulcers. It soon became clear that H2-receptors are also widespread in the brain, in particular in the basal ganglia, the hippocampus and the amygdala (Traiffort et al., 1992). Jim Black had previously developed the adrenergic beta-blockers. Noradrenaline (β-receptor) and histamine (H2-receptor) as well as some dopamine-, serotonin-, CRH- and VIP-receptors have very similar actions on hippocampal excitability: they block the Ca^{2+}-activated potassium conductance (small K) which is responsible for the accommodation of firing (Madison and Nicoll, 1982; Haas and Konnerth, 1983; Pedarzani and Storm, 1993) and the long lasting (seconds) afterhyperpolarization following action potentials, more remarkably groups of action potentials which lead to a more substantial Ca^{2+} inflow. The accommodation of firing determines the number of action potentials in response to a longer (ca 100 ms) depolarizing stimulus, i.e. the number of action potentials in a burst, and the amount of Ca^{2+} inflow through HVACCs or NMDA-channels. These actions are mimicked by cyclic AMP and lead to activation of PKA; they are of obvious importance for plasticity. The catalytic subunit of PKA phosphorylates the transcription factor CREB, a pathway involved in synaptic plasticity (Sheng et al., 1991). The cyclic AMP pathway is necessary for the late phase of NMDA receptor dependent LTP (Stanton and Sarvey, 1985; Dahl and Sarvey, 1990; Frey et al., 1993) and for the NMDA-independent mossy fiber LTP in the CA3 area.

Cyclic AMP can also interact direcly with another current, the hyperpolarisation activated cation current (Ih), by shifting its activation curve to more positive levels, leading to depolarisation in thalamic neurons (McCormick and Williamson, 1991) and hippocampal pyramidal cells (Pedarzani and Storm, 1995).

Brief exposure of hippocampal slices to histamine, noradrenaline or cyclic AMP causes long lasting (at least several hours) enhancements of CA1 population spikes and pyramidal neurone firing in the absence of any electrical stimulation (Kostopoulos et al., 1988; Dahl and Sarvey, 1989; Selbach et al., 1997). A weak tetanus that normally causes a short lasting potentiation of less than 1 hour duration elicits long lasting LTP if histamine was present during the tetanus (Brown et al., 1995). We have not yet tested the dependence on protein synthesis but circumstantial evidence for this is quite high (Stanton and Sarvey, 1985).

Adenosine A1 receptors are negatively coupled to adenylyl cyclase and have opposite effects to histamine H2-receptor activation: increasing the long lasting afterhyperpolarisation and accommodation of firing. Adenosin is sedative (in contrast to the waking action of histamine), it is a major factor in sleep pressure (Huston et al., 1996; Porkka-Heiskanen et al., 1997). Adenosine can antagonize LTP and is probably responsible for some forms of LTD.

3.5. Histamine H3-Receptor

The histamine autoreceptor was termed H3 by Arrang et al. (1983) who first described it. H3-receptors are present on TM neuron somata, dendrites and axons. They are coupled directly through a Gi/o protein to Ca^{2+} channels causing inhibition of TM neuron firing, histamine release from varicosities. On the other hand adenylyl cyclase is depressed by H3-receptor activation too. Different from histaminergic neurons, other known autoreceptors are often also coupled to inwardly rectifying potassium channels (GIRKs). Importantly H3-receptors are found on many non-histaminergic varicosities such as those containing glutamate, ACh, dopamine, noradrenaline, serotonin, GABA and various peptides (Schlicker et al., 1994). Excitatory transmission in the glutamatergic perforant pathway from the entorhinal cortex to the dentate area is depressed by about 30 % by histamine in rats (but not in mice) (Brown and Reymann, 1996; Brown and Haas, 1999). The glutamatergic cortico-striatal pathway (Doreulee et al., 2001) and the dopaminergic nigro-striatal pathway (Schlicker et al., 1993; Hill et al., 1997) are also suppressed. In the substantia nigra we have observed a reduction of GABAergic potentials (Brown and Haas, 1999; Brown et al., 2001b)

4. HISTAMINE AND MEMORY

4.1. The links between histamine related synaptic plasticity and memory

Almost every paper on long-term potentiation mentions the relevance of synaptic plasticity for learning and memory and there is quite a number of papers on the involvement of histamine in memory. Work correlating synaptic plasticity directly with memory mechanisms is much rarer and such data are much more difficult to obtain. The effects of histamine H1- and H2-receptor activation on principal cells in hippo-campus, cortex, amygdala or striatum are quite compatible with a positive influence on synaptic plasticity and memory. The excitatory action and the intracellular increase of Ca^{2+} ions (H1) are thought to be most important mechanisms for maintaining wakefulness and attention, a prerequisite for effective learning. The block of firing-accommodation (H2) shifts target neurons towards a state of quiet readiness – not much difference in membrane potential and single action potential firing but a dramatically changed response to longer lasting (< 100 ms) depolarizing signals. This is a response to compartmentalized Ca^{2+}-signals. Classical conditioning (eye-blink reflex in rabbits) can reduce the amplitude and duration of the Ca^{2+}-dependent afterhyperpolarization in hippocampal pyramidal cells (Coulter et al., 1989).

These mechanisms are also considered with respect to patho-physiological events leading to exaggerated activity and epilepsy. Contrary to expectation histamine H1-antagonists (the classical anti-histamines) are slightly epileptogenic (Iinuma et al., 1993). Two clues can be offered for this from our experimentation: 1. H1-receptor activation leads to hyperpolarization and inhibition of principal neurone firing in the (rodent) hippocampus through activation of Ca^{2+}-dependent potassium channel opening. 2. Interneurones in the hippocampus are strongly excited by H2-receptor activation (Yanovsky et al., 1997). The same interneurons suffer a cutoff at high frequency firing (Atzori et al., 2000). Thus the final action on a whole structure and its output cannot be predicted by studying principal cells alone and in vitro. H2- and H3-receptor blocking has revealed ambiguous results on learning paradigms (Onodera et al., 1994; Flood et al., 1998; Blandina et al., 1996; Rubio et al., 2002).

In the light of multiple mechanisms of action on multiple types of neurons it is not surprising that seemingly contradictory results were obtained for histaminergic activation or inactivation in a variety of tests for learning and memory. Different species or strains and different methods of drug application may well be responsible for some of the "opposite" results that have been described; e.g. blocking histamine synthesis incompletely by α-fluoromethylhistidine, impaired (Chen et al., 1999) or improved (Sakai et al., 1998) spatial memory in the radial arm maze. Active avoidance conditioning has also been affected in this or the other way (Alvarez and Banzan, 1996).

Dere et al (2003) have proposed that histamine may indeed affect memory through at least two independent mechanisms, directly on the synaptic plasticity in hippocampus, striatum and amygdala and by an indirect inhibitory effect through the brain's reinforcement system (Huston et al., 1997). H1-receptor antagonists can indeed act as reinforcers, alone or together with opiates, cocaine or amphetamine. Most antihistamines have antimuscarinic actions and can block dopamine uptake (McKearney, 1982).

4.2. H3-receptor involvement in learning and memory

H3-receptor activation causes autoinhibition of histamine neurons and their axons (release from varicosities) and inhibition of the release of other transmitters from their presynaptic structures. A number of H3-agonists and antagonists that pass the blood-brain barrier have been tested in learning tasks. Agonists were mostly inhibitory, presumably through the inhibition of release from histaminergic, glutamatergic and cholinergic varicosities. In the striatum and in the area dentata of the hippocampus for instance, glutamate release is reduced for prolonged periods after a brief application leading to a long-term depression (Brown and Haas, 1999; Doreulee et al., 2001). Blandina et al. (1996) found an inhibition of cortical acetylcholine release and cognitive performance in rats. H3-receptor antagonists have facilitatory actions on learning and memory presumably by the removal of a tonic inhibition of the release of these and other transmitters (Bacciottini et al., 1999; Bacciottini et al., 2001; Miyazaki et al., 1997; Miyazaki et al., 1995; Flood et al., 1998). Thus H3-receptor antagonism is a prime target for the development of drugs for the treatment of memory disorders.

4.3. Evidence from knock-out mice

Acute selective, complete and possibly reversible removal of the histamine system or its receptors should give the salient information on its physiological role in memory trace formation. The pharmacological approach has been particularly studied with H1-antagonists, which improved water-maze but reduced radial-maze performance (Hasenohrl et al., 1999). Locomotor activity and exploratory behavior are reduced (Inoue et al., 1996) but learning and memory has been found unaffected in H1-receptor knockout mice by (Yanai et al., 1998). H3-receptor -/- mice displayed enhanced spatial learning and reduced anxiety (Rizk et al., 2004).

Lesions of the histaminergic nucleus lead to facilitation of learning and pharmacological inactivation by post-trial injection of lidocaine to memory improvement (Frisch et al., 1998; Frisch et al., 1999). Such lesions or stimulations of the tuberomamillary nucleus are very difficult or impossible to perform without affecting neighboring structures or bypassing fibers (Weiler et al., 1998).

Histidinecarboxylase KO mice cannot synthesize histamine. Parmentier et al. (2002) have described an important defect in $HDC^{-/-}$ mice that is quite relevant for memory formation: their response to novelty is strongly impaired, they hardly explore a new environment but rather fall asleep. Dere and colleagues have revealed a number of rather specific changes in HDC -/- mice: an improvement in negatively reinforced water-maze performance but a defect in (non-reinforced) episodic object memory (Dere et al., 2003). Furthermore these mice displayed reduced exploratory activity but normal habituation to a novel environment, more hight-fear and a superior motor coordination. Higher acetylcholine concentrations in the cortex (not in striatum) and an increased serotonin-turnover may represent compensatory measures to overcome the lack of histamine (Dere et al., 2004).

5. CONCLUSION

The histamine system is an important player in the orchestra of aminergic neurones, including the cholinergic nuclei, the noradrenergic locus coeruleus, the serotonergic raphe and the dopaminergic ventral tegmental area and substantia nigra, pars compacta. Their mutual interactions at somatic and axonal (varicosities) levels allow playing without a conductor. Nevertheless the orexinergic/hypocretinergic neurones fulfill such a role of coordination. The histaminergic neurones possess a special relationship with these neurones whose degeneration leads to narcolepsy. The program is the regulation of basic body functions like sleep-waking, food and water intake, energy administration, the endocrinium and - last not least - setting cortical excitability and synaptic plasticity.

Memory trace formation occurs in the hippocampus where a selected population of CA3 pyramidal cells produces a synchronous burst that is relayed to the CA1 area and acts as the natural trigger for long-term potentiation. Whether or not and at which intensity this burst is fired depends on the subcortical control by the abovementioned orchestra. Histamine interacts at all levels of this process and plays a prominent role. It is involved with intracellular signalling, phosphorylations by protein kinases A and C, and the Ca^{2+} level, mechanisms deeply concerned with learning and memory.

Direct measurements of learning and memory under the influence of histaminergic manipulation have revealed many and sometimes seemingly contradictory results. Some of these may be resolved by the observation that histamine (and the histaminergic system) can act by at least two separate ways: indirectly through modulation of subcortical reinforcement systems and directly through mechanisms of synaptic plasticity, such as long-term potentiation and depression in cortical structures.

6. REFERENCES

Alvarez EO, Banzan AM (1996) Hippocampus and learning: possible role of histamine receptors. Medicina (B Aires) 56:155-160.

Arrang JM, Garbarg M, Schwartz, JC (1983) Autoinhibition of brain histamine release mediated by a novel class (H3) of histamine receptor. Nature 302:832-837.

Arrang JM, Garbarg M, Schwartz, JC (1987) Autoinhibition of histamine synthesis mediated by presynaptic H3-receptors. Neuroscience 23:149-157.

Atzori M, Lau D, Tansey EP, Chow A, Ozaita A, Rudy B, McBain CJ (2000) H2 histamine receptor-phosphorylation of Kv3.2 modulates interneuron fast spiking. Nat.Neurosci. 3:791-798.

Bacciottini L, Mannaioni PF, Chiappetta M, Giovannini MG, Blandina P (1999) Acetylcholine release from hippocampus of freely moving rats is modulated by thioperamide and cimetidine. Inflamm.Res. 48:S63-S64.

Bacciottini L, Passani MB, Mannaioni PF, Blandina P (2001) Interactions between histaminergic and cholinergic systems in learning and memory. Behav.Brain Res. 124:183-194.

Bekkers JM (1993) Enhancement by histamine of NMDA-mediated synaptic transmission in the hippocampus. Science 261:104-106.

Black JW, Duncan WA, Durant CJ, Ganellin CR, Parsons EM (1972) Definition and antagonism of histamine H 2 -receptors. Nature 236:385-390.

Blandina P, Giorgetti M, Bartolini L, Cecchi M, Timmerman H, Leurs R, Pepeu G, Giovannini MG (1996) Inhibition of cortical acetylcholine release and cognitive performance by histamine H3 receptor activation in rats. Br.J Pharmacol. 119:1656-1664.

Brown RE, Fedorov NB, Haas HL, Reymann KG (1995) Histaminergic modulation of synaptic plasticity in area CA1 of rat hippocampal slices. Neuropharmacology 34:181-190.

Brown RE, Haas HL (1999) On the mechanism of histaminergic inhibition of glutamate release in the rat dentate gyrus. J Physiol 515 , 777-786.

Brown RE, Reymann KG (1996) Histamine H3 receptor-mediated depression of synaptic transmission in the dentate gyrus of the rat in vitro. J Physiol 496:175-184.

Brown RE, Sergeeva O, Eriksson KS, Haas HL (2001a) Orexin A excites serotonergic neurons in the dorsal raphe nucleus of the rat. Neuropharmacology 40:457-459.

Brown RE, Stevens DR, Haas, HL (2001b) The physiology of brain histamine. Prog.Neurobiol. 63:637-672.

Buzsaki G (1998) Memory consolidation during sleep: a neurophysiological perspective. J Sleep Res. 7 Suppl 1:17-23.

Buzsaki G, Haas HL, Anderson EG (1987) Long-term potentiation induced by physiologically relevant stimulus patterns. Brain Res. 435:331-333.

Chen Z, Sugimoto Y, Kamei C (1999) Effects of intracerebroventricular injection of alpha-fluoromethylhistidine on radial maze performance in rats. Pharmacol.Biochem.Behav. 64:513-518.

Coulter DA, Lo Turco JJ, Kubota M, Disterhoft JF, Moore JW, Alkon DL (1989) Classical conditioning reduces amplitude and duration of calcium-dependent afterhyperpolarization in rabbit hippocampal pyramidal cells. J Neurophysiol. 61:971-981.

Dahl D, Sarvey JM (1989) Norepinephrine induces pathway-specific long-lasting potentiation and depression in the hippocampal dentate gyrus. Proc.Natl.Acad.Sci.U.S.A 86:4776-4780.

Dahl D, Sarvey JM (1990) Beta-adrenergic agonist-induced long-lasting synaptic modifications in hippocampal dentate gyrus require activation of NMDA receptors, but not electrical activation of afferents. Brain Res. 526:347-350.

Dere E, Souza-Silva MA, Spieler RE, Lin JS, Ohtsu H, Haas HL, Huston JP (2004) Changes in motoric, exploratory and emotional behaviours and neuronal acetylcholine content and 5-HT turnover in histidine decarboxylase-KO mice. Eur.J Neurosci. 20:1051-1058.

Dere E, Souza-Silva MA, Topic B, Spieler RE, Haas HL, Huston JP (2003) Histidine-decarboxylase knockout mice show deficient nonreinforced episodic object memory, improved negatively reinforced water-maze performance, and increased neo- and ventro-striatal dopamine turnover. Learn.Mem. 10:510-519.

Diewald L, Heimrich B, Busselberg D, Watanabe T, Haas HL (1997) Histaminergic system in co-cultures of hippocampus and posterior hypothalamus: a morphological and electrophysiological study in the rat. Eur.J Neurosci. 9:2406-2413.

Dismukes K, Kuhar MJ, Snyder SH (1974) Brain histamine alterations after hypothalamic isolation. Brain Res. 78:144-151.

Doreulee N, Yanovsky Y, Flagmeyer I, Stevens DR, Haas HL, Brown RE (2001) Histamine H(3) receptors depress synaptic transmission in the corticostriatal pathway. Neuropharmacology 40:106-113.

Ericson H, Watanabe T, Kohler C (1987) Morphological analysis of the tuberomammillary nucleus in the rat brain: delineation of subgroups with antibody against L-histidine decarboxylase as a marker. J Comp Neurol. 263:1-24.

Eriksson KS, Stevens DR, Haas HL (2001) Serotonin excites tuberomammillary neurons by activation of Na(+)/Ca(2+)-exchange. Neuropharmacology 40:345-351.

Flagmeyer I, Haas HL, Stevens DR (1997) Adenosine A1 receptor-mediated depression of corticostriatal and thalamostriatal glutamatergic synaptic potentials in vitro. Brain Res. 778:178-185.

Flood JF, Uezu K, Morley JE (1998) Effect of histamine H2 and H3 receptor modulation in the septum on post-training memory processing. Psychopharmacology (Berl) 140:279-284.

Frey U, Huang YY, Kandel ER (1993) Effects of cAMP simulate a late stage of LTP in hippocampal CA1 neurons. Science 260:1661-1664.

Frisch C, Hasenohrl RU, Haas HL, Weiler HT, Steinbusch HW, Huston JP (1998) Facilitation of learning after lesions of the tuberomammillary nucleus region in adult and aged rats. Exp.Brain Res. 118:447-456.

Frisch C, Hasenohrl RU, Huston JP (1999) Memory improvement by post-trial injection of lidocaine into the tuberomammillary nucleus, the source of neuronal histamine. Neurobiol.Learn.Mem. 72:69-77.

Garbarg M, Barbin G, Feger J, Schwartz JC (1974) Histaminergic pathway in rat brain evidenced by lesions of the medial forebrain bundle. Science 186:833-835.

Green JP (1970) HISTAMINE. In Handbook of Neurochemistry, ed. Lajtha, A., pp. 221-250. Plenum, New York - London.

Greene RW, Haas HL, Reiner PB (1990) Two transient outward currents in histamine neurones of the rat hypothalamus in vitro. J Physiol 420:149-163.

Haas H, Panula P (2003) The role of histamine and the tuberomamillary nucleus in the nervous system. Nat.Rev.Neurosci. 4:121-130.

Haas HL (1984) Histamine potentiates neuronal excitation by blocking a calcium-dependent potassium conductance. Agents Actions 14:534-537.

Haas HL, Konnerth A (1983) Histamine and noradrenaline decrease calcium-activated potassium conductance in hippocampal pyramidal cells. Nature 302:432-434.

Haas HL, Reiner PB (1988) Membrane properties of histaminergic tuberomammillary neurones of the rat hypothalamus in vitro. J Physiol 399:633-646.

Haas HL, Wolf P (1977) Central actions of histamine: microelectrophoretic studies. Brain Res. 122:269-279.

Haas HL, Wolf P, Palacios JM, Garbarg M, Barbin G, Schwartz JC (1978) Hypersensitivity to histamine in the guinea-pig brain: microiontophoretic and biochemical studies. Brain Res. 156:275-291.

Hasenohrl RU, Weth K, Huston JP (1999) Intraventricular infusion of the histamine H(1) receptor antagonist chlorpheniramine improves maze performance and has anxiolytic-like effects in aged hybrid Fischer 344xBrown Norway rats. Exp.Brain Res. 128:435-440.

Hatton GI, Yang QZ (1996) Synaptically released histamine increases dye coupling among vasopressinergic neurons of the supraoptic nucleus: mediation by H1 receptors and cyclic nucleotides. J Neurosci. 16:123-129.

Hill SJ, Ganellin CR, Timmerman H, Schwartz JC, Shankley NP, Young JM, Schunack W, Levi R, Haas HL (1997) International Union of Pharmacology. XIII. Classification of histamine receptors. Pharmacol.Rev. 49:253-278.

Huston JP, Haas HL, Boix F, Pfister M, Decking U, Schrader J, Schwarting RK (1996) Extracellular adenosine levels in neostriatum and hippocampus during rest and activity periods of rats. Neuroscience 73:99-107.

Huston JP, Wagner U, Hasenohrl RU (1997) The tuberomammillary nucleus projections in the control of learning, memory and reinforcement processes: evidence for an inhibitory role. Behav.Brain Res. 83:97-105.

Iinuma K, Yokoyama H, Otsuki T, Yanai K, Watanabe T, Ido T, Itoh M (1993) Histamine H1 receptors in complex partial seizures. Lancet 341:238.

Inagaki N, Yamatodani A, Ando-Yamamoto M, Tohyama M, Watanabe T, Wada H (1988) Organization of histaminergic fibers in the rat brain. J Comp Neurol. 273:283-300.

Inoue I, Yanai K, Kitamura D, Taniuchi I, Kobayashi T, Niimura K, Watanabe T, Watanabe T (1996) Impaired locomotor activity and exploratory behavior in mice lacking histamine H1 receptors. Proc.Natl.Acad.Sci.U.S A 93:13316-13320.

Kostopoulos G, Psarropoulou C, Haas HL (1988) Membrane properties, response to amines and to tetanic stimulation of hippocampal neurons in the genetically epileptic mutant mouse tottering. Exp.Brain Res. 72:45-50.

Kudrimoti HS, Barnes CA, McNaughton BL (1999) Reactivation of hippocampal cell assemblies: effects of behavioral state, experience, and EEG dynamics. J.Neurosci. 19:4090-4101.

Madison DV, Nicoll RA (1982) Noradrenaline blocks accommodation of pyramidal cell discharge in the hippocampus. Nature 299:636-638.

McCormick DA, Williamson A (1991) Modulation of neuronal firing mode in cat and guinea pig LGNd by histamine: possible cellular mechanisms of histaminergic control of arousal. J Neurosci. 11:3188-3199.

McKearney JW (1982) Effects of tricyclic antidepressant and anticholinergic drugs on fixed-interval responding in the squirrel monkey. J Pharmacol.Exp.Ther. 222:215-219.

Miyazaki S, Imaizumi M, Onodera K (1995) Effects of thioperamide on the cholinergic system and the step-through passive avoidance test in mice. Methods Find.Exp.Clin.Pharmacol. 17:653-658.

Miyazaki S, Onodera K, Imaizumi M, Timmerman H (1997) Effects of clobenpropit (VUF-9153), a histamine H3-receptor antagonist, on learning and memory, and on cholinergic and monoaminergic systems in mice. Life Sciences 61:355-361.

Mochizuki T, Yamatodani A, Okakura K, Horii A, Inagaki N, Wada H (1992) Circadian rhythm of histamine release from the hypothalamus of freely moving rats. Physiol Behav. 51:391-394.

Onodera K, Yamatodani A, Watanabe T, Wada H (1994) Neuropharmacology of the histaminergic neuron system in the brain and its relationship with behavioral disorders. Prog.Neurobiol. 42:685-702.

Panula P, Airaksinen MS, Pirvola U, Kotilainen E (1990) A histamine-containing neuronal system in human brain. Neuroscience 34:127-132.

Panula P, Yang HY, Costa E (1984) Histamine-containing neurons in the rat hypothalamus. Proc.Natl.Acad.Sci.U.S A 81:2572-2576.

Parmentier R, Ohtsu H, Djebbara-Hannas Z, Valatx JL, Watanabe T, Lin JS (2002) Anatomical, physiological, and pharmacological characteristics of histidine decarboxylase knock-out mice: evidence for the role of brain histamine in behavioral and sleep-wake control. J Neurosci. 22:7695-7711.

Pedarzani P, Storm JF (1993) PKA mediates the effects of monoamine transmitters on the K+ current underlying the slow spike frequency adaptation in hippocampal neurons. Neuron 11:1023-1035.

Pedarzani P, Storm JF (1995) Protein kinase A-independent modulation of ion channels in the brain by cyclic AMP. Proc.Natl.Acad.Sci.U.S.A 92:11716-11720.

Philippu A, Prast H (1991) Patterns of histamine release in the brain. Agents Actions 33:124-125.

Ponomarenko AA, Lin JS, Selbach O, Haas HL (2003) Temporal pattern of hippocampal high-frequency oscillations during sleep after stimulant-evoked waking. Neuroscience 121:759-769.

Porkka-Heiskanen T, Strecker RE, Thakkar M, Bjorkum AA, Greene RW, McCarley RW (1997) Adenosine: a mediator of the sleep-inducing effects of prolonged wakefulness. Science 276:1265-1268.

Reiner PB, Kamondi A (1994) Mechanisms of antihistamine-induced sedation in the human brain: H1 receptor activation reduces a background leakage potassium current. Neuroscience 59:579-588.

Rizk A, Curley J, Robertson J, Raber J (2004) Anxiety and cognition in histamine H3 receptor-/- mice. Eur.J Neurosci. 19:1992-1996.

Rubio S, Begega A, Santin LJ, Arias JL (2002) Improvement of spatial memory by (R)-alpha-methylhistamine, a histamine H(3)-receptor agonist, on the Morris water-maze in rat
94. Behav.Brain Res. 129:77-82.

Sakai N, Sakurai E, Sakurai E, Yanai K, Mirua Y, Watanabe T (1998) Depletion of brain histamine induced by alpha-fluoromethylhistidine enhances radial maze performance in rats with modulation of brain amino acid levels. Life Sciences 62:989-994.

Saybasili H, Stevens DR, Haas HL (1995) pH-dependent modulation of N-methyl-D-aspartate receptor-mediated synaptic currents by histamine in rat hippocampus in vitro. Neurosci.Lett. 199:225-227.

Schlicker E, Fink K, Detzner M, Gothert M (1993) Histamine inhibits dopamine release in the mouse striatum via presynaptic H3 receptors. J Neural Transm.Gen.Sect. 93:1-10.

Schlicker E, Malinowska B, Kathmann M, Gothert M (1994) Modulation of neurotransmitter release via histamine H3 heteroreceptors. Fundam.Clin.Pharmacol. 8:128-137.

Schwartz JC, Arrang JM, Garbarg M, Pollard H, Ruat M (1991) Histaminergic transmission in the mammalian brain. Physiol Rev. 71:1-51.

Selbach O, Brown RE, Haas HL (1997) Long-term increase of hippocampal excitability by histamine and cyclic AMP. Neuropharmacology 36:1539-1548.

Selbach O, Doreulee N, Bohla C, Eriksson KS, Sergeeva OA, Poelchen W, Brown RE, Haas HL (2004) Orexins/hypocretins cause sharp wave- and theta-related synaptic plasticity in the hippocampus via glutamatergic, gabaergic, noradrenergic, and cholinergic signaling. Neuroscience 127:519-528.

Sergeeva OA, Korotkova TM, Scherer A, Brown RE, Haas HL (2003) Co-expression of non-selective cation channels of the transient receptor potential canonical family in central aminergic neurones. J Neurochem. 85:1547-1552.

Sheng M, Thompson MA, Greenberg ME (1991) CREB: a Ca(2+)-regulated transcription factor phosphorylated by calmodulin-dependent kinases. Science 252:1427-1430.

Stanton PK, Sarvey JM (1985) Blockade of norepinephrine-induced long-lasting potentiation in the hippocampal dentate gyrus by an inhibitor of protein synthesis. Brain Res. 361:276-283.

Stevens DR, Haas HL (1996) Calcium-dependent prepotentials contribute to spontaneous activity in rat tuberomammillary neurons. J Physiol 493:747-754.

Stevens DR, Kuramasu A, Eriksson KS, Selbach O, Haas HL (2004) α_2-Adrenergic receptor-mediated presynaptic inhibition of GABAergic IPSPs in rat histaminergic neurons. Neuropharmacology 46:1018-1022.

Takeda N, Inagaki S, Taguchi Y, Tohyama M, Watanabe T, Wada H (1984) Origins of histamine-containing fibers in the cerebral cortex of rats studied by immunohistochemistry with histidine decarboxylase as a marker and transection. Brain Res. 323:55-63.

Takeshita Y, Watanabe T, Sakata T, Munakata M, Ishibashi H, Akaike N (1998) Histamine modulates high-voltage-activated calcium channels in neurons dissociated from the rat tuberomammillary nucleus. Neuroscience 87:797-805.

Traiffort E, Pollard H, Moreau J, Ruat M, Schwartz JC, Martinez-Mir MI, Palacios JM (1992) Pharmacological characterization and autoradiographic localization of histamine H2 receptors in human brain identified with [125I]iodoaminopotentidine. J Neurochem. 59:290-299.

Traynelis SF, Hartley M, Heinemann SF (1995) Control of proton sensitivity of the NMDA receptor by RNA splicing and polyamines. Science 268:873-876.

Uteshev V, Stevens DR, Haas HL (1995) A persistent sodium current in acutely isolated histaminergic neurons from rat hypothalamus. Neuroscience 66:143-149.

Vorobjev VS, Sharonova IN, Walsh IB, Haas HL (1993) Histamine potentiates N-methyl-D-aspartate responses in acutely isolated hippocampal neurons. Neuron 11:837-844.

Weiler HT, Hasenohrl RU, van Landeghem AA, van Landeghem M, Brankack J, Huston JP, Haas HL (1998) Differential modulation of hippocampal signal transfer by tuberomammillary nucleus stimulation in freely moving rats dependent on behavioral state. Synapse 28:294-301.

Williams K (1994) Subunit-specific potentiation of recombinant N-methyl-D-aspartate receptors by histamine. Mol.Pharmacol. 46:531-541.

Yanai K, Son LZ, Endou M, Sakurai E, Watanabe T (1998) Targeting disruption of histamine H1 receptors in mice: behavioral and neurochemical characterization. Life Sciences 62:1607-1610.

Yanovsky Y, Haas HL (1998) Histamine increases the bursting activity of pyramidal cells in the CA3 region of mouse hippocampus. Neurosci.Lett. 240:110-112.

Yanovsky Y, Reymann K, Haas HL (1995) pH-dependent facilitation of synaptic transmission by histamine in the CA1 region of mouse hippocampus. Eur.J Neurosci. 7:2017-2020.

Yanovsky Y, Sergeeva OA, Freund TF, Haas HL (1997) Activation of interneurons at the stratum oriens/alveus border suppresses excitatory transmission to apical dendrites in the CA1 area of the mouse hippocampus. Neuroscience 77:87-96.

SPIKE TIMING DEPENDENT PLASTICITY OF RAT HIPPOCAMPAL AND CORTICAL SYNAPSES AND CONTROL BY MUSCARINIC TRANSMISSION

Wolfgang Müller, Jochen Winterer, and Patric K. Stanton*

1. INTRODUCTION

Synaptic long-term potentiation (LTP) was suggested in the late 1940s by Donald Hebb as a candidate mechanism for learning and memory in neuronal networks like those found in the central nervous system of mammals, but has been demonstrated experimentally only in the early 1970s (Bliss and Lømo, 1973). LTP of excitatory postsynaptic potentials (EPSPs) in hippocampal dentate gyrus granule cells was observed after high frequency stimulation (12-20 Hz) of perforant path fibers in anaesthetized rabbits. It was recognized quite early that coincident pre- and postsynaptic activity was crucial for successful induction of homosynaptic LTP (Gustafsson, et al., 1987). That LTP induction generally required activation of NMDA receptors became particularly intriguing after the characterization of the NMDA receptor channel as a coincidence detector of the transmitter glutamate and of postsynaptic depolarization that is required to remove a blocking Mg^{2+} ion from the outer channel mouth. This coincidence detection allows association of different synaptic inputs by charge integration on the postsynaptic membrane after rapid spread of charge. It is now generally accepted that Ca^{2+} influx through the NMDA receptor channel is crucial for induction of most forms of LTP, presumably because of a close functional coupling of NMDA receptor channels on dendritic spines to Ca^{2+}-dependent processes that are required for the expression of LTP (Regehr and Tank, 1990, Müller and Connor, 1991, Perkel, et al., 1993, Yuste and Denk, 1995). Ca^{2+}-influx through voltage activated Ca^{2+}-channels appears to contribute only to a much lesser extent (Koester and Sakmann, 1998, Sabatini, et al., 2002), but can be sufficient to induce LTP by high frequency stimulation (Grover and Teyler, 1990).

* Wolfgang Müller and Jochen Winterer, Neuroscience Research Institute, Charité, Humboldt University, Berlin, D-10117, Germany; Patric K. Stanton, Dept Cell Biology & Anatomy, New York Medical College, Valhalla, NY 10595, USA. E-mail: wolfgang.mueller@charite.de

Shortly after the demonstration of LTP, an opposite synaptic modification, long-term depression (LTD) of EPSPs, was also described. After induction of LTP, a simultaneous expression of long-term depression (LTD) of inputs inactive during induction was observed (heterosynaptic LTD, Lynch, et al., 1977). It was also demonstrated that low frequency stimulation (LFS) reversed stable LTP, a phenomenon called depotentiation (Barrionuevo, et al., 1980). Interestingly, it took almost another ten years to show that LFS can also induce LTD without the requirement for prior induction and expression of LTP (Stanton and Sejnowski, 1989, Mulkey and Malenka, 1992, Dudek and Bear, 1993).

Although these forms of LTP and LTD together form a theoretical basis of synaptic memory storage and clearance with homeostasis of excitability of neuronal networks, the induction protocols using high and low frequency tetanic trains always had a sense of being highly artificial, even though brief bursts of high frequency action potential firing has been observed in *in vivo* recordings. Moreover, there was apparently little information capacity connected to such activity as compared to complex temporal patterns of "natural" spike trains. A pair of papers reemphasized not only the importance of timing, but also demonstrated the induction of LTP and LTD by the pairing of single action potentials in pre- and postsynaptic neurons, plasticity where the polarity of the change depended on the temporal order of pre- and postsynaptic action potentials (Magee and Johnston, 1997; Markram, et al., 1997). This "spike timing dependent plasticity" was accepted immediately as a breakthrough in connecting LTP and LTD to natural neuronal activity patterns. Most synapses express LTP after repeated pairing of presynaptic firing *followed* within 10-15 ms by postsynaptic firing (pre-before-post), and LTD after repeated pairing of presynaptic firing 10-15 ms *after* postsynaptic firing (post-before-pre). These plasticity laws enhance synaptic contacts contributing to postsynaptic firing while simultaneously weakening those synaptic contacts that have a distinct temporal relationship that is not consistent with the possibility of having made a contribution to the postsynaptic firing.

2. SPIKE TIMING DEPENDENT PLASTICITY

Spike timing dependent synaptic plasticity (STDP), as described in the previous paragraph, is often represented by a fairly symmetric graph of synaptic change versus inter-spike interval (ISI = time of postsynaptic spike - time of presynaptic spike), resembling a plot of the simple hyperbolic function $y=1/x$ modified by bridging the singularity at 0 with a steep linear function (Fig. 1). It is a widely held view that repeated pairings of action potentials at a particular inter-spike interval will result in asymptotic approximation of synaptic strength to the value given by this relation as the number of pairings increases. This view is based, although, on a rather limited variety of spike pairing protocols. Further, functional expression of the synaptic changes elicited develops gradually over a time of 5-40 min, following the repeated pairings (cf. Fig. 3).

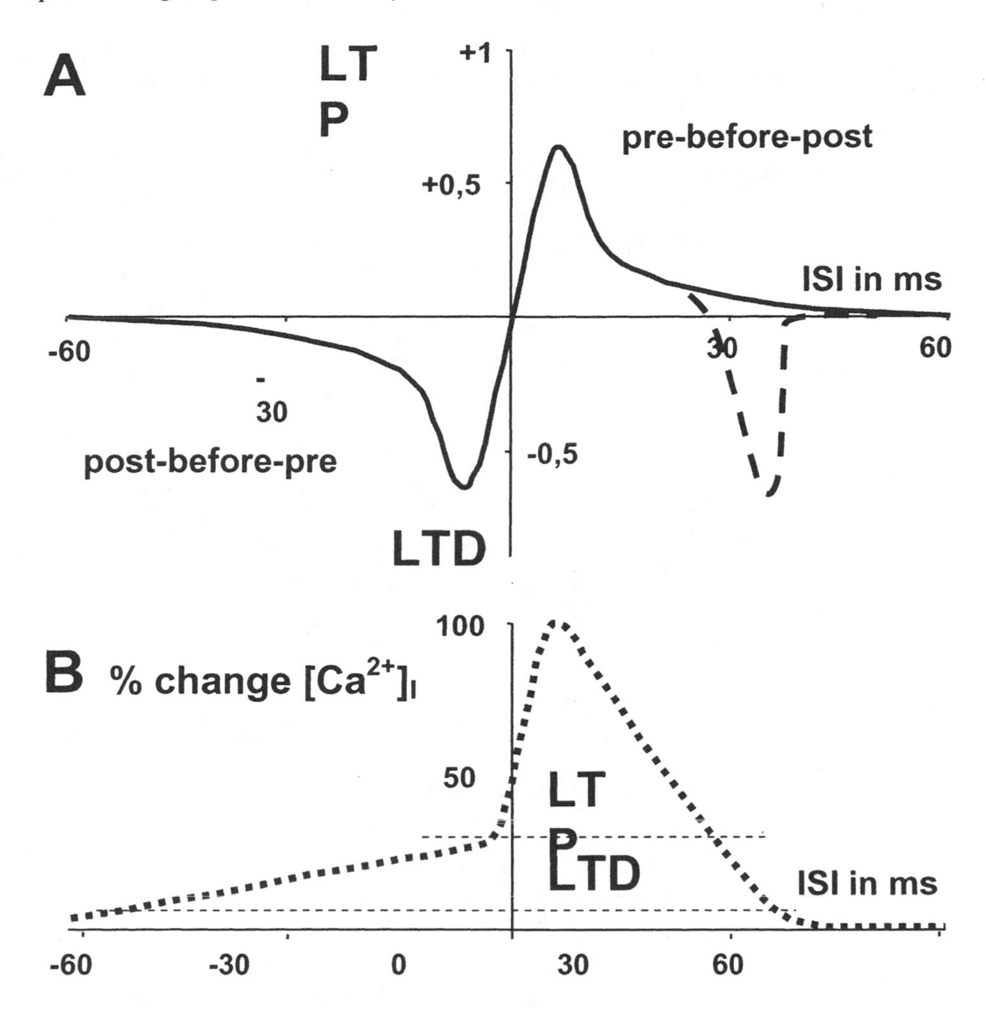

Figure 1. Dependence of relative synaptic change (A) and intracellular change of $[Ca^{2+}]_i$ (B) in STDP on inter-spike interval (ISI = time of postsynaptic spike - time of presynaptic spike). The dashed line in A shows a second LTD window at long positive ISIs due to weak Ca^{2+}-increases (see B and text).

STDP crucially depends on the back-propagation of the postsynaptic action potential to the dendritic site of synaptic contact (see below) to unblock NMDA receptors by depolarization-dependent unbinding of Mg^{2+}. The STDP relation "LTD/LTP versus ISI" can be understood in the context of the Bienenstock-Cooper-Munro (BCM) model of synaptic plasticity derived from theoretical issues of network stability, and experimental properties of low frequency stimulus-induced LTD and high frequency stimulation inducing LTP (Bienenstock, et al., 1982). In this model, LTD increases with the relevant Ca^{2+}-concentration in the dendritic spine above a certain Ca^{2+}-concentration threshold, followed by a second threshold (LTP threshold 1 in Fig. 2), where LTD apparently decreases and/or LTP increases with a further increase of $[Ca^{2+}]$. LTP appears to reach a

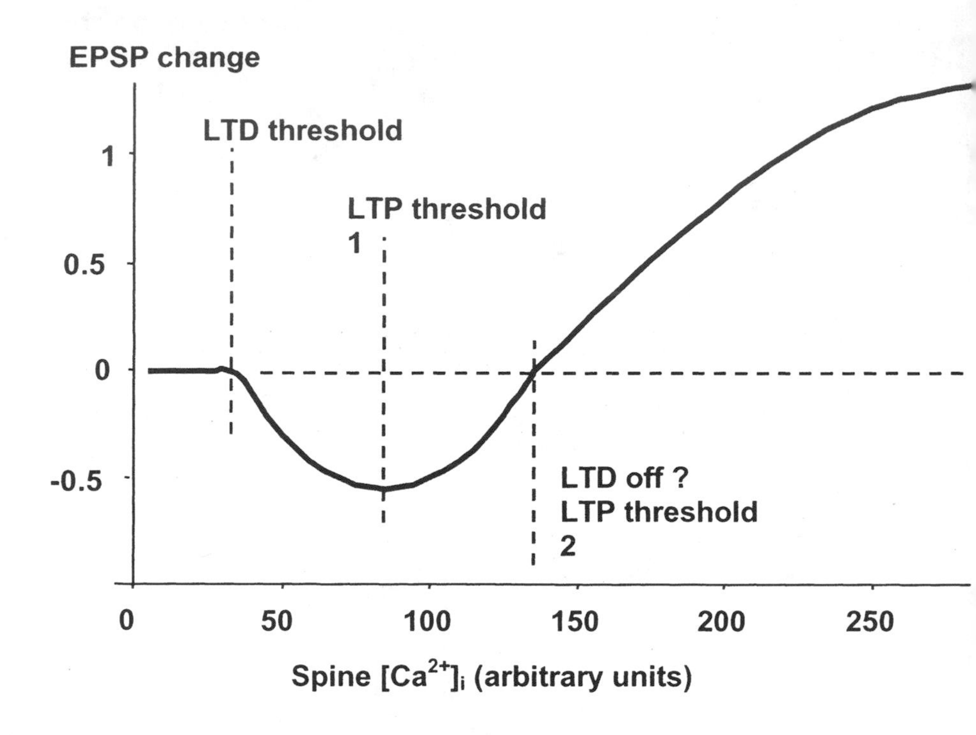

Figure 2. Dependence of synaptic change on peak Ca^{2+}-concentration (arbitrary units) at the sensor(s) in the dendritic spine. Peak Ca^{2+}-concentration depends on the precise timing of paired pre- and postsynaptic spikes. Several thresholds for synaptic modification processes are hypothesized.

saturating level for very high Ca^{2+}-concentrations (Fig. 2; (Bienenstock, et al., 1982, Abarbanel, et al., 2002, Karmarkar, et al., 2002, Lisman, 1989). Classically, either a high or low frequency train stimulation that produces a long lasting sustained increase of free Ca^{2+} concentration, a weak one lasting several minutes for LTD and a strong one lasting 500 ms or longer for LTP, has been considered essential for the induction of long-term plasticity. In contrast, STDP induction protocols are associated with very brief rise times of free Ca^{2+} concentration in spines of ≤4.5 ms for back-propagating action potentials and 9-20 ms for EPSPs (20 to 80 % rise time, Koester and Sakmann, 1998).

Whether LTD can still be induced at higher Ca^{2+} concentrations above LTP threshold 2 is unclear, as it would be masked by simultaneous strong LTP. The timing of action potentials determines the time course and amplitude of Ca^{2+}-influx into the spine. This amplitude time course is determined approximately by the time course of Mg^{2+} unblock of the NMDA receptor by the postsynaptic dendritic spike, multiplied by the time course of glutamate activated NMDA receptors. The latter time course has been recorded as the slow NMDA receptor mediated EPSC in the absence of Mg^{2+} or with strong membrane depolarization. This time course is the result of the presynaptic spike, vesicular release of glutamate into the synaptic cleft, diffusion to postsynaptic receptors, including binding and unbinding, and removal of glutamate due to uptake by neuronal and glial transporters. Coincidence of maximal spike membrane depolarization at the spine, filtered by back-propagation of the postsynaptic spike from the axon hillock to the synaptic site, with maximal activation of NMDA receptors, should result in maximal Ca^{2+}-influx and Ca^{2+}dependent potentiation, assuming that the Ca^{2+}-influx integration time sensed is rather short ($\leq$ 5 ms; cf. Koester and Sakmann, 1998). With further delay of the postsynaptic spike, the associated Ca^{2+}-signal should vary from this maximum in a time course similar to that of the NMDA-EPSP, just slightly shifted in time for the removal of Mg^{2+} from the pore. With the decay of the NMDA-EPSC, the dendritic spike-associated Ca^{2+}-increase will then eventually fall into the range of intermediate levels associated with LTD induction (Shouval, et al., 2002). This has been actually observed (Nishiyama, et al., 2000; Winterer et al., in preparation, see dashed line in fig. 1A). For zero and negative time, delayed coincidence detection could be due to the spike afterdepolarization caused by the membrane capacity and, *in vivo*, by synaptic currents from the other inputs driving the cell to fire an action potential. This afterdepolarization determines a time course of a weak Mg^{2+}-unblock, resulting together with the dendritic spike in weak $[Ca^{2+}]$-increases that cause proportional stronger and weaker LTD, balancing close to no change for an inter-spike interval of zero (Karmarkar, et al., 2002). For a zero ISI, the fast afterhyperpolarization could stabilize the Mg^{2+}-block of the NMDA channel pore so that the $[Ca^{2+}]$-increase could fall below the threshold for LTD induction (cf. Fig. 2). More likely, the $[Ca^{2+}]$-increase comes close to LTP threshold 2 (Fig.2) where, again, no synaptic change has been observed. The quite distinct mechanistic procedures described above actually result in a weak and distorted symmetry of the "LTD/LTP vs. ISI" relation for negative versus positive spike timing intervals (Fig. 1 dashed line).

3. MECHANISMS

3.1. Back Propagation of Action Potentials into Dendrites

Dendritic Na^{+}-, Ca^{2+}- and K^{+}- conductances allow active back-propagation of action potentials generated in the axon hillock. Because of the voltage dependence of NMDA receptor mediated currents, back-propagation of the action potential strongly facilitates or is required for induction of synaptic long-term changes. This back-propagation depends crucially on the current density for Na^{+} and Ca^{2+} in relation to leak and outward voltage- and/or Ca^{2+}-dependent conductances, as well as synaptic K^{+} and Cl^{-} conductances (e.g. during AMPA receptor-mediated EPSPs and $GABA_A$ and $GABA_B$ receptor-mediated IPSPs, cf. (Mott and Lewis, 1991, Davies, et al., 1991, Tsubokawa and Ross, 1996). Dendritic excitability varies therefore with sub-threshold depolarization-induced inactivation and hyperpolarization-evoked recovery from inactivation of Na^{+}-, Ca^{2+}-, and A-type K^{+}-channels (I_A). Initial studies on hippocampal CA1 pyramidal neurons demonstrated an important role for the transient K^{+}-current I_A, with I_A-channel density increasing along the primary apical dendrite with distance from the soma (Hoffman, et al., 1997). I_A strongly attenuates back-propagation of action potentials into dendrites.

Properly timed synaptic input can inactivate I_A by depolarization to 'boost' dendritic action potential propagation (Magee and Johnston, 1997). This is effective because I_A channels inactivate faster than do Na^+ channels, thereby transiently increasing the Na^+/I_A current ratio (Pan and Colbert, 2001).

In cortical layer 5 (L5) neurons, dendritic I_A channel density is lower than in hippocampal CA1 pyramidal neurons and does not increase as one moves towards distal sites (Bekkers, 2000; Korngreen and Sakmann, 2000). Still, in these neurons EPSPs can also boost dendritic spike propagation (Larkum, et al., 2001) here by activation of Na^+-channels, rather than inactivation of I_A channels (Stuart and Hausser, 2001). The time dependence of AP amplification by an EPSP is in good agreement with that for induction of spike timing-dependent LTP in these neurons (Markram, et al., 1997), indicating the importance of AP amplification in long-term plasticity induced by single APs or brief bursts in these neurons.

On the other hand, synaptic inhibition of dendritic excitability by hyperpolarization and/or shunting can block back-propagation (Tsubokawa and Ross, 1996, Pare, et al., 1998). Moreover, learning and memory performance is well known to depend on modulatory neurotransmission, particularly cholinergic-muscarinic modulation. Cholinergic-muscarinic activation induces theta rhythmic oscillations that favor synchronized firing activity across cell populations and that presumably depolarize dendrites rhythmically by local synaptic input (Gloveli, et al., 1999). In hippocampal CA1 pyramidal neurons, cholinergic agonists can enhance dendritic slow depolarization during repetitive synaptic firing by potentiation of NMDA receptor mediated synaptic currents (Egorov, et al., 1999) and increase dendritic action-potential amplitude (Hoffman and Johnston, 1999) or inhibit decrement of later dendritic action potentials in a train (Tsubokawa and Ross, 1997).

Depending on the local dendritic density of Na^+- and Ca^{2+}-channels, action potentials can also be initiated directly in a dendritic segment in layer V cortical neurons (Larkum, et al., 2001). Varying interaction of the two initiation zones, depending on proximal and distal currents, can generate complex firing patterns that are likely to be important for more complex neuronal information processing.

Higher firing frequencies during natural activity result in APs that could interact with two or more spikes from the other side of the synapse. In this situation, the efficacy of each spike to induce synaptic modification is suppressed by the preceding spike in the same cell for several tens of ms. Therefore, for natural spike trains, the first AP in a burst is dominant in synaptic modification. These results are in good agreement with synaptic modifications induced by firing activity in response to visual stimuli (Froemke and Dan, 2003).

The concept of STDP emphasizes timing of spikes over rate. Still, rate is important, too. LTP induction at low frequency requires stronger somatic depolarization by multiple inputs (Sjostrom, et al., 2001). This kind of cooperativity and associativity may reflect a requirement for sufficient dendritic excitability to enhance back-propagating spikes (see above). At high frequencies, residual depolarization is provided by preceding spikes and, in natural scenarios, by synaptic currents within a burst input (Egorov and Müller, 1999). Hyperpolarizing the neuron between APs blocks synaptic plasticity even when presynaptic and postsynaptic firing occurred with optimal timing and frequency (Sjostrom, et al., 2001).

Spike-timing dependent LTD depends differently on dendritic voltage and frequency. Low frequency post-before-pre pairing reliably induces robust LTD with either a weak or strong input. In contrast, at high frequency ($\geq$ 40 Hz) post-before-pre pairing produces LTP, probably because although each postsynaptic spike occurs 10 ms prior to the presynaptic spike, it also follows 15 ms after the preceding presynaptic spike. In this way timing requirements for both LTD and LTP are met, consistent with the BCM model of distinct [Ca^{2+}] thresholds for LTD and LTP (Sjostrom, et al., 2001). It remains possible, indeed likely, that stimulus patterns exist where conditions for presynaptic changes in one direction (i.e. LTD) are met *simultaneously* with conditions for postsynaptic changes in the opposite direction (i.e. LTP), while only the arithmetic sum of changes in synaptic efficacy is observed.

3.2. Coincidence Detection of Negative Inter-spike Intervals

Induction of LTD at negative ISIs (and long positive ISIs) is NMDA receptor-dependent and requires a detection mechanism for these time intervals. Negative ISI of 10-15 ms could be detected due to a slow afterpotential, e.g. an afterdepolarization lasting at least 15 ms and producing significant relief of the Mg^{2+}-block at the NMDA receptor during the EPSP (Karmarkar, et al., 2002). This afterdepolarization is primarily determined by the membrane time constant and kinetic behavior of multiple potassium channels. When postsynaptic firing is caused by synaptic activity from other inputs, these excitatory synaptic currents, including NMDA receptor mediated currents, will also contribute, with their time courses, to depolarization following the postsynaptic spike.

3.3. Augmentation and Spread of Ca^{2+}-signals by Release from Intracellular Stores

Ca^{2+}-influx through NMDA receptors can evoke Ca^{2+}-induced Ca^{2+}-release from intracellular stores (Segal and Manor, 1992; Liang, et al., 2002; Emptage, et al., 1999). IP_3 dependent Ca^{2+}-release has been demonstrated directly in proximal and somatic regions of CA1 pyramidal neurons, using Ca^{2+}-imaging, to be triggered by a train of back-propagating action potentials (Nakamura, et al., 2000). Presynaptic Ca^{2+}-dependent Ca^{2+}-release via ryanodine receptors, and postsynaptic IP_3 receptor mediated Ca^{2+}-release, play important roles in the induction of homosynaptic LTD (Reyes and Stanton, 1996). Propagating waves of type 1 IP_3 receptor mediated Ca^{2+}-release have been suggested to be essential in the induction of homo- as well as heterosynaptic LTD by a spike timing induction protocol applied to the hippocampal Schaffer collateral-CA1 synapse (Nishiyama, et al., 2000). An LTD induction protocol applied to CA1 pyramidal neurons in slices from IP_3R1 knockout mice or with intracellular application of specific, blocking antibodies resulted not in LTD but in LTP. Interestingly, this is not hypothesized to be caused by a reduction of the Ca^{2+}-signals but by an enhancement of these signals due to enhanced excitability, e.g. because of. insufficient activation of Ca^{2+}-

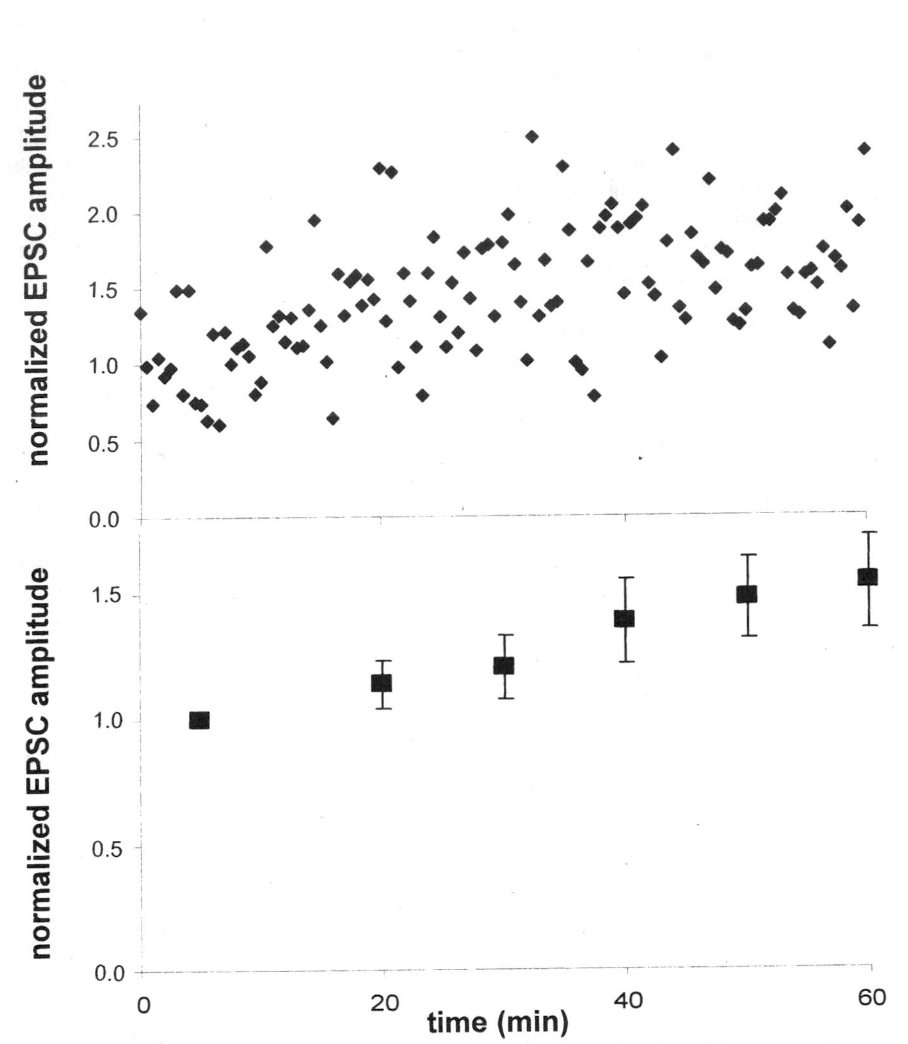

Figure 3. Time courses of EPSC amplitudes in an hippocampal CA1 pyramidal neuron and averages ± SEM from 8 cells (bottom) before and after spike-pairing induction of long-term potentiation (vertical bar, Schaffer collateral stimulation), in control conditions, demonstrate slow but robust expression of LTP (75 pairings at a frequency of 0.33 Hz using fiber stimulation with an extracellular electrode in combination with voltage/current clamp recording from a CA1 pyramidal neuron; adapted from (Adams, et al., 2004).

activated K^+-channels or reduced NMDA receptor activity (but see Markram and Segal (1992).In this way, the results could still be in agreement with the BCM model. In line with this argumentation, elimination of IP_3R1 mediated signaling enhances LTP in response to the spike timing LTP induction and produces, in addition, heterosynaptic LTP, i.e. specificity of LTP is lost (Nishiyama, et al., 2000).

In contrast, in intercalated neurons between the basolateral amygdala and the nucleus centralis, low frequency stimulation induces homosynaptic LTD associated with a heterosynaptic LTP (Royer and Pare, 2003). This balancing mechanism maintains total synaptic weight. Although heterosynaptic LTP requires Ca^{2+}-release from intracellular stores, the $[Ca^{2+}]$-threshold for LTP in the heterosynaptic inputs, as suggested by the BCM model, apparently has not been met (Royer and Pare, 2003).

3.4. Stability of Synaptic Changes and Cholinergic-Muscarinic Modulation

One of the most striking features of mammalian memory is the recognition of sensory information that has not been experienced for many years or even decades. Apparently, there must be synaptic memory patterns that are extremely stable. In contrast, models of plastic networks using the basic rules of synaptic plasticity exhibit ongoing run-up of strong, and run-down of weak, synaptic connections. Another problem is that basic rules of synaptic plasticity would suggest that ongoing activity would constantly shift synaptic weights up and down, depending on spike intervals that would be different during processing of different sensory information. How does the brain control information loss in a synaptic network during different behavioral states? One good candidate mechanism is cholinergic neuromodulation. Disruption of cholinergic or muscarinic transmission by neurotoxins does impair mammalian memory and early loss of cholinergic neurons in Alzheimer's disease appears to cause functional deficits typical of this disease. Hippocampus and neocortex both receive strong and widespread cholinergic input. Repetitive stimulation of fibers in the hippocampal slice evokes a slow cholinergic muscarinic excitation consisting of blockade of adaptation of firing and of the slow afterhyperpolarization (sAHP) following repetitive firing and, at stronger stimulation intensity, a slow depolarization due to closure of K^+-channels (Cole and Nicoll, 1984). Apparently, blockade of the sAHP contributes to reduced adaptation of firing and is mediated by a muscarinic receptor subtype distinct from that mediating the slow depolarization (Müller and Misgeld, 1986). Blockade of the sAHP can be mimicked by application of low concentrations of muscarinic agonists; this effect does not desensitize and is not sensitive to Li^+, but is blocked by inhibition of Ca^{2+}/calmodulin-dependent protein kinase II (Müller, et al., 1988, Misgeld, et al., 1989, Müller, et al., 1989, Müller, et al., 1992, Pedarzani and Storm, 1996, Alberi, et al., 2000). In contrast, the slow depolarization requires higher concentrations of agonist, shows strong desensitization, and is blocked easily by the M1 receptor-selective antagonist pirenzepine or rapidly by intracellular Li^+ application (Müller and Misgeld, 1986, Müller, et al., 1988, Misgeld, et al., 1989, Müller, et al., 1989) but not by inhibition of Ca^{2+}/calmodulin-dependent protein kinase II (Müller, et al., 1992). Muscarinic activation enhances the dendritic slow depolarization during repetitive synaptic input by potentiation of NMDA receptor mediated currents (Egorov, et al., 1999). It can further inhibit I_A, thereby slowing AP repolarization (Nakajima, et al., 1986; Müller and Connor, 1991), enhancing dendritic AP amplitude and inhibiting decrement of later dendritic action potentials during repetitive firing (Egorov, et al., 1999, Hoffman and Johnston, 1999, Tsubokawa and Ross, 1997, Tsubokawa, et al., 2000).

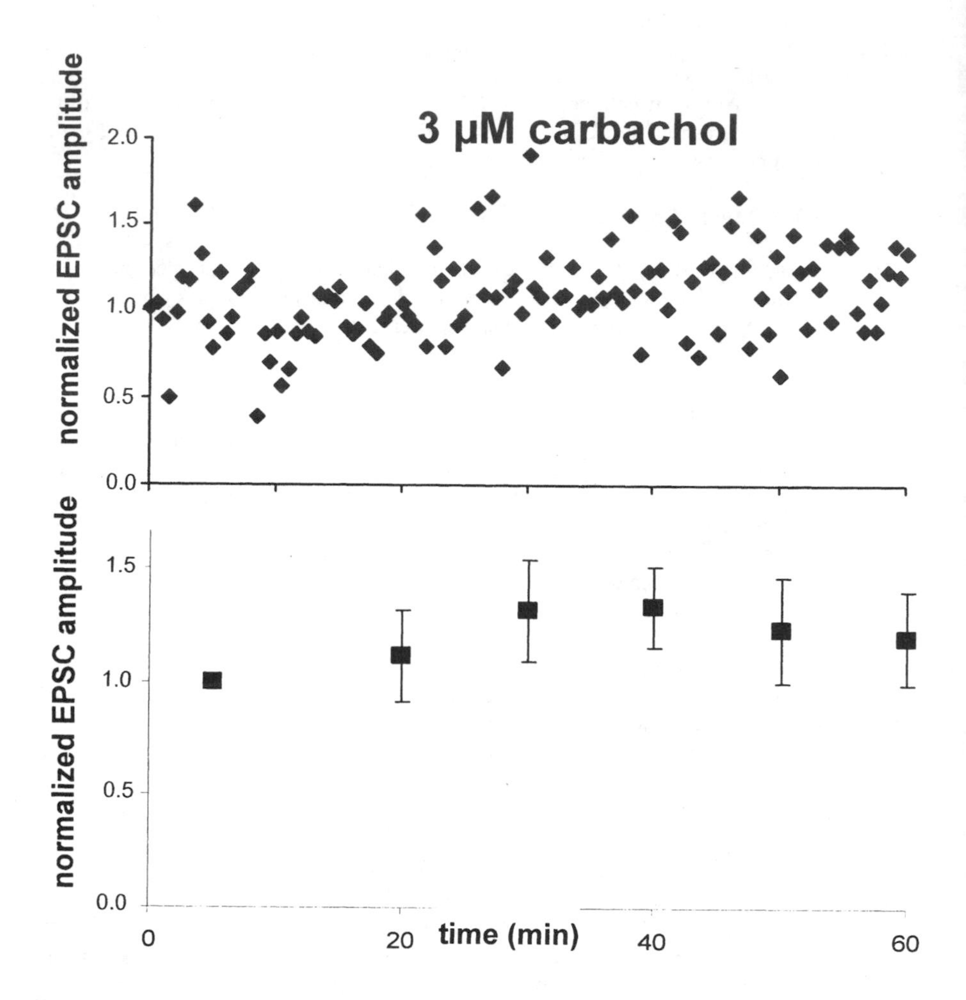

Figure 4. Time courses of EPSC amplitudes in an example neuron and averages ± SEM from five cells (bottom) before and after spike-pairing induction of long-term potentiation (vertical bar) in the presence of the stable cholinergic agonist carbachol (3 μM) demonstrate only transient potentiation of EPSCs (75 pairings at a frequency of 0.33 Hz; adapted from (Adams, et al., 2004).

All these conductance changes may contribute to cholinergic-muscarinic activation of theta rhythmic oscillations. These oscillations favor synchronized firing activity at theta frequencies across cell populations. In addition they presumably depolarize dendrites rhythmically by local synaptic input, thereby enhancing dendritic excitability rhythmically.

During cholinergic induction of a theta rhythm oscillatory state, requirements for synaptic plasticity are dramatically altered; a single burst given at the peak of theta induces homosynaptic LTP, but induces homosynaptic *LTD* of previously potentiated synapses when given at a trough. Heterosynaptic LTD is produced at inactive synapses when others undergo LTP. All these synaptic changes require activation of both NMDA and muscarinic receptors. The enhancement is cooperative and occludes with standard classic LTP (Huerta and Lisman, 1995, Huerta and Lisman, 1993).

What about STDP? Surprisingly, the muscarinic agonist carbachol strongly inhibits spike timing dependent LTP (Figs.3, 4; Adams, et al., 2004). However, blocking muscarinic receptors with atropine completely blocks spike timing dependent LTP (Fig. 5). Apparently, firing one action potential in cholinergic fibers does not evoke visible changes in postsynaptic somatic excitability, but is effective in gating induction of spike timing dependent LTP. Why did carbachol not mimic this effect? Probably a mechanism is required that easily desensitizes. In this way, it can operate properly only when transiently activated by pulsatile release and removal of acetylcholine. In this respect spike timing dependent LTP may differ substantially from burst firing or high frequency induction of LTP that is facilitated by carbachol (Huerta and Lisman, 1995, Huerta and Lisman, 1993, Burgard and Sarvey, 1990, Blitzer, et al., 1990).

4. CONCLUSIONS

Spike timing-based induction of synaptic plasticity is attractive because it links natural neuronal activity patterns, that have higher information capacity, with the most attractive classic neuronal-synaptic memory mechanism comprising specificity, associativity and cooperativity of synaptic plasticity. It enhances computational power by bringing the temporal scale of synaptic memory mechanisms from hundreds of ms for classical induction protocols down to the low ms range. In addition, it very elegantly addresses simultaneous input-specific up- and down-regulation of tens of thousands of synaptic inputs to a single cortical or hippocampal neuron. For long-term synaptic plasticity, induction frequency has not been recognized to be of primary importance once a threshold postsynaptic depolarization has been reached that achieves the cooperativity required for induction of LTP. Of course, average frequency affects timing for regular as well as for random firing activity. In STDP, this results, at 40-50 Hz, in a postsynaptic spike falling not only into the LTD window for the following presynaptic spike but also falling into the LTP window for the preceding presynaptic spike. Therefore, disappearance of LTD at $f \geq 40$ Hz is not surprising (Sjostrom, et al., 2001).

STDP depends on coincidence detection of a back-propagating action potential, plus some slower depolarization and synaptic input necessary to induce LTP. For LTD, coincidence detection is required of a 5-30 ms after-event following the back-propagating action potential, consisting most likely of the spike afterdepolarization and those other synaptic currents that have been required for firing the postsynaptic cell. Therefore, events affecting threshold, amplitude, propagation, repolarization and afterpotentials of postsynaptic dendritic spikes all affect information processing and storage. Postsynaptic Ca^{2+}-dependent Ca^{2+}-release via ryanodine receptors and inositol triphosphate ($InsP_3$)

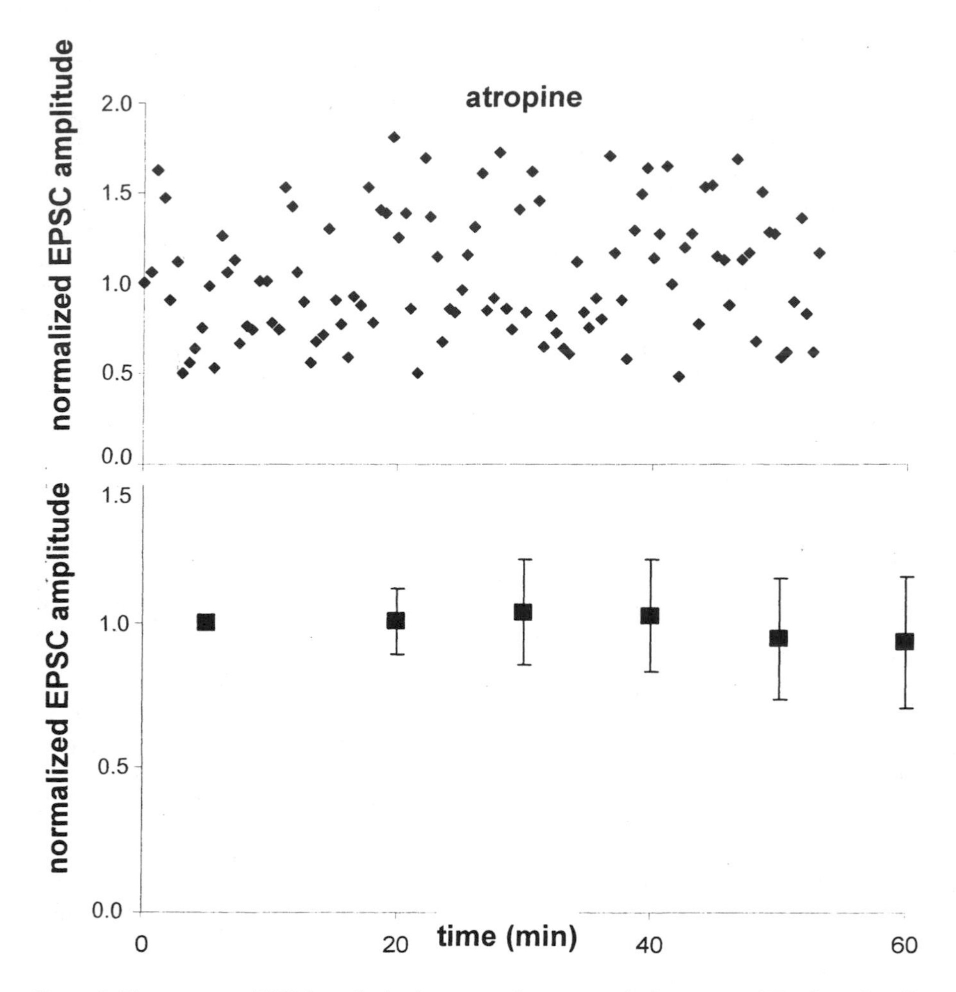

Figure 5. Time courses of EPSC amplitudes in an example neuron and of averages ± SEM from six cells (bottom) before and after spike-pairing induction of long-term potentiation (vertical bar) in the presence of the muscarinic receptor antagonist atropine (10 μM) show a clear failure to express LTP (75 pairings at a frequency of 0.33 Hz; adapted from (Adams, et al., 2004).

receptors type 1 (IP_3R1) has an important role for homo- and heterosynaptic spike timing-dependent LTD, respectively (Nishiyama, et al., 2000).

With long-term plasticity being induced by single spike pairings, stability of synaptic weight changes will be disturbed by ongoing spontaneous activity, processing of novel information by the same neuronal networks and even by activity being employed for recalling previously stored information. Artificial neuronal network simulations usually switch off synaptic long-term plasticity after a training phase by "deus-ex-machina" intervention. Cholinergic-muscarinic modulation of cortex/hippocampus appears to be an important mechanism in controlling synaptic stability in mammalian neuronal networks. Besides affecting neuronal activity patterns by induction of theta rhythm, cholinergic transmission is also required for induction of STDP-LTP in CA1 pyramidal neurons in hippocampal brain slices (Adams, et al., 2004). The mechanism behind this muscarinic effect is rather unclear. Simple muscarinic phospholipase C – $InsP_3$-Ca^{2+}-release signaling does not seem sufficient to explain it.

Induction signaling for classic and spike timing-dependent LTP and LTD seem to share many of the same basic mechanisms, even though temporal requirements for activity differ significantly. Virtually nothing is known about the expression mechanisms of STDP in cortex/hippocampus, with the exception that presynaptic changes have been suggested for connected pairs of neocortical pyramidal neurons (Markram and Tsodyks, 1996) and layer 4 spiny stellate neurons in the barrel field of rat somatosensory cortex (Egger, et al., 1999). Postsynaptic molecular changes are likely to occur as well, but have not, so far, been characterized. Characterization of expression relations for classic LTP and LTD and spike timing dependent LTP and LTD should be only a matter of a little more time.

4. ACKNOWLEDGEMENTS

We thank the Bundesministerium für Bildung und Forschung (BMBF) for support through grant "Virtual Brain - 0311562Y" to WM, the DFG for support through grant SFB515/B9 to WM and GK 238 03 to JW and NINDS for grant NS44421 to PKS.

5. REFERENCES

Abarbanel HD, Huerta R, Rabinovich MI (2002) Dynamical model of long-term synaptic plasticity, Proc Natl Acad Sci U S A 99:10132-7.

Adams SV, Winterer J, Müller W (2004) Muscarinic signaling is required for spike-pairing induction of long-term potentiation at rat Schaffer collateral-CA1 synapses, Hippocampus 14: 413-6.

Alberi S, Boeijinga PH, Raggenbass M, and Boddeke HW (2000) Involvement of calmodulin-dependent protein kinase II in carbachol-induced rhythmic activity in the hippocampus of the rat, Brain Res 872: 11-19.

Barrionuevo G, Schottler F, Lynch G (1980) The effects of repetitive low frequency stimulation on control and "potentiated" synaptic responses in the hippocampus, Life Sci 27: 2385-91.

Bekkers JM (2000) Properties of voltage-gated potassium currents in nucleated patches from large layer 5 cortical pyramidal neurons of the rat, J Physiol 525: 593-609.

Bienenstock EL, Cooper LN, Munro PW (1982) Theory for the development of neuron selectivity: orientation specificity and binocular interaction in visual cortex, J Neurosci 2: 32-48.

Bliss TV, Lømo T (1973) Long-lasting potentiation of synaptic transmission in the dentate area of the anaesthetized rabbit following stimulation of the perforant path, J Physiol 232: 331-56.

Blitzer RD, Gil O, Landau EM (1990) Cholinergic stimulation enhances long-term potentiation in the CA1 region of rat hippocampus, Neurosci Lett 119: 207-210.

Burgard EC, Sarvey JM (1990) Muscarinic receptor activation facilitates the induction of long-term potentiation (LTP) in the rat dentate gyrus, Neurosci Lett 116: 34-39.

Cole AE, Nicoll RA (1984) Characterization of a slow cholinergic post-synaptic potential recorded in vitro from rat hippocampal pyramidal cells, J Physiol Lond 352: 173-188.

Davies CH, Starkey SJ, Pozza MF, Collingridge GL (1991) GABA autoreceptors regulate the induction of LTP, Nature 349: 609-11.

Dudek SM, Bear MF (1993) Bidirectional long-term modification of synaptic effectiveness in the adult and immature hippocampus, J Neurosci 13: 2910-8.

Egger V, Feldmeyer D, Sakmann B (1999) Coincidence detection and changes of synaptic efficacy in spiny stellate neurons in rat barrel cortex, Nat Neurosci 2: 1098-105.

Egorov AV, Gloveli T, Müller W (1999) Muscarinic Control of Dendritic Excitability and Ca(2+) Signaling in CA1 Pyramidal Neurons in Rat Hippocampal Slice., J. Neurophysiol. 82: 1909-1915.

Egorov AV, Müller W (1999) Subcellular muscarinic enhancement of excitability and Ca2+-signals in CA1-dendrites in rat hippocampal slice, Neurosci Lett 261: 77-80.

Emptage N, Bliss TV, Fine A (1999) Single synaptic events evoke NMDA receptor-mediated release of calcium from internal stores in hippocampal dendritic spines, Neuron 22: 115-124.

Froemke RC, Dan Y (2003) Spike-timing-dependent synaptic modification induced by natural spike trains, Nature 416: 433-8.

Gloveli T, Egorov AV, Schmitz D, Heinemann U, Müller W (1999) Carbachol-induced changes in excitability and [Ca2+]i signalling in projection cells of medial entorhinal cortex layers II and III, Eur. J. Neurosci. 11: 3626-3636.

Grover LM, Teyler TJ (1990) Two components of long-term potentiation induced by different patterns of afferent activation, Nature 347: 477-9.

Gustafsson B, Wigstrom H, Abraham WC, Huang YY (1987) Long-term potentiation in the hippocampus using depolarizing current pulses as the conditioning stimulus to single volley synaptic potentials, J Neurosci 7: 774-80.

Hoffman DA, Johnston D (1999) Neuromodulation of dendritic action potentials, J Neurophysiol 81: 408-411.

Hoffman DA, Magee JC, Colbert CM, Johnston D (1997) K+ channel regulation of signal propagation in dendrites of hippocampal pyramidal neurons, Nature 387: 869-875.

Huerta PT, Lisman JE (1993) Heightened synaptic plasticity of hippocampal CA1 neurons during a cholinergically induced rhythmic state, Nature 364: 723-725.

Huerta PT, Lisman JE (1995) Bidirectional synaptic plasticity induced by a single burst during cholinergic theta oscillation in CA1 in vitro, Neuron 15: 1053-1063.

Karmarkar UR, Najarian MT, Buonomano DV (2002) Mechanisms and significance of spike-timing dependent plasticity, Biol Cybern 87: 373-82.

Koester HJ, Sakmann B (1998) Calcium dynamics in single spines during coincident pre- and postsynaptic activity depend on relative timing of back-propagating action potentials and subthreshold excitatory postsynaptic potentials, Proc Natl Acad Sci U S A 95: 9596-601.

Korngreen A, Sakmann B (2000) Voltage-gated K+ channels in layer 5 neocortical pyramidal neurones from young rats: subtypes and gradients, J Physiol 525: 621-39.

Larkum ME, Zhu JJ, Sakmann B (2001) Dendritic mechanisms underlying the coupling of the dendritic with the axonal action potential initiation zone of adult rat layer 5 pyramidal neurons, J Physiol 533: 447-66.

Levy WB, Steward O (1983) Temporal contiguity requirements for long-term associative potentiation/depression in the hippocampus, Neuroscience 8: 791-7.

Liang Y, Yuan LL, Johnston D, Gray R (2002) Calcium signaling at single mossy fiber presynaptic terminals in the rat hippocampus, J Neurophysiol 87: 1132-7.

Lisman JE (1989) A mechanism for the Hebb and the anti-Hebb processes underlying learning and memory, Proc Natl Acad Sci U S A 86: 9574-9578.

Lynch GS, Dunwiddie T, Gribkoff V (1977) Heterosynaptic depression: a postsynaptic correlate of long-term potentiation, Nature 266: 737-9.

Magee JC, Johnston D (1997) A synaptically controlled, associative signal for Hebbian plasticity in hippocampal neurons, Science 275: 209-13.

Markram H, Lubke J, Frotscher M, Sakmann B (1997) Regulation of synaptic efficacy by coincidence of postsynaptic APs and EPSPs, Science 275: 213-215.

Markram H, Segal M (1992) The inositol 1,4,5-trisphosphate pathway mediates cholinergic potentiation of rat hippocampal neuronal responses to NMDA, J Physiol Lond 447: 513-533.

Markram H, Tsodyks M (1996) Redistribution of synaptic efficacy between neocortical pyramidal neurons, Nature 382: 807-810.

Misgeld U, Müller W, Polder HR (1989) Potentiation and suppression by eserine of muscarinic synaptic transmission in the guinea-pig hippocampal slice, J Physiol Lond 409: 191-206.

Mott DD, Lewis DV (1991) Facilitation of the induction of long-term potentiation by GABAB receptors, Science 252: 1718-20.

Mulkey RM, Malenka RC (1992) Mechanisms underlying induction of homosynaptic long-term depression in area CA1 of the hippocampus, Neuron 9: 967-975.

Müller W, Brunner H, Misgeld U (1989) Lithium discriminates between muscarinic receptor subtypes on guinea pig hippocampal neurons in vitro, Neurosci Lett 100: 135-140.

Müller W, Connor JA (1991a) Cholinergic input uncouples Ca2+ changes from K+ conductance activation and amplifies intradendritic Ca2+ changes in hippocampal neurons, Neuron 6: 901-905.

Müller W, Connor JA (1991b) Dendritic spines as individual neuronal compartments for synaptic Ca2+ responses, Nature 354: 73-76.

Müller W, Misgeld U (1986) Slow cholinergic excitation of guinea pig hippocampal neurons is mediated by two muscarinic receptor subtypes, Neurosci Lett 67: 107-112.

Müller W, Misgeld U, Heinemann U (1988) Carbachol effects on hippocampal neurons in vitro: dependence on the rate of rise of carbachol tissue concentration, Exp Brain Res 72: 287-298.

Müller W, Petrozzino JJ, Griffith LC, Danho W, Connor JA (1992) Specific involvement of Ca(2+)-calmodulin kinase II in cholinergic modulation of neuronal responsiveness, J Neurophysiol 68: 2264-2269.

Nakajima Y, Nakajima S, Leonard RJ, Yamaguchi Y (1986) Acetylcholine raises excitability by inhibiting the fast transient potassium current in cultured hippocampal neurons, Proc Natl Acad Sci U S A 83: 3022-3026.

Nakamura T, Nakamura K, Lasser RN, Barbara JG, Sandler VM, Ross WN (2000) Inositol 1,4,5-trisphosphate (IP3)-mediated Ca2+ release evoked by metabotropic agonists and backpropagating action potentials in hippocampal CA1 pyramidal neurons, J Neurosci 20: 8365-8376.

Nishiyama M, Hong K, Mikoshiba K, Poo MM, Kato K (2000) Calcium stores regulate the polarity and input specificity of synaptic modification, Nature 408: 584-8.

Pan E, Colbert CM (2001) Subthreshold inactivation of Na+ and K+ channels supports activity-dependent enhancement of back-propagating action potentials in hippocampal CA1, J Neurophysiol 85: 1013-6.

Pare D, Shink E, Gaudreau H, Destexhe A, Lang EJ (1998) Impact of spontaneous synaptic activity on the resting properties of cat neocortical pyramidal neurons In vivo, J Neurophysiol 79: 1450-60.

Pedarzani P, Storm JF (1996) Evidence that Ca/calmodulin-dependent protein kinase mediates the modulation of the Ca2+-dependent K+ current, IAHP, by acetylcholine, but not by glutamate, in hippocampal neurons, Pflugers Arch 431: 723-728.

Perkel DJ, Petrozzino JJ, Nicoll RA, Connor JA (1993) The role of Ca2+ entry via synaptically activated NMDA receptors in the induction of long-term potentiation, Neuron 11: 817-823.

Regehr WG, Tank DW (1990) Postsynaptic NMDA receptor-mediated calcium accumulation in hippocampal CA1 pyramidal cell dendrites, Nature 345: 807-810.60

Reyes M, Stanton PK (1996) Induction of hippocampal long-term depression requires release of Ca2+ from separate presynaptic and postsynaptic intracellular stores. J Neurosci 16:5951-5960.

Royer S, Pare D (2003) Conservation of total synaptic weight through balanced synaptic depression and potentiation, Nature 422: 518-22.

Sabatini, BL, Oertner TG, Svoboda K (2002) The life cycle of Ca(2+) ions in dendritic spines, Neuron 33: 439-452.

Segal M, Manor D (1992) Confocal microscopic imaging of [Ca2+]i in cultured rat hippocampal neurons following exposure to N-methyl-D-aspartate, J Physiol (Lond) 448: 655-676.

Shouval HZ, Bear MF, Cooper LN (2002) A unified model of NMDA receptor-dependent bidirectional synaptic plasticity, Proc Natl Acad Sci U S A 99: 10831-6. Epub 2002 Jul 22.

Sjostrom PJ, Turrigiano GG, Nelson SB (2001) Rate, timing, and cooperativity jointly determine cortical synaptic plasticity, Neuron 32: 1149-64.

Stanton PK, Sejnowski TJ (1989) Associative long-term depression in the hippocampus induced by hebbian covariance, Nature 339: 215-8.

Stuart GJ, Hausser M (2001) Dendritic coincidence detection of EPSPs and action potentials, Nat Neurosci 4: 63-71.

Tsubokawa H, Offermanns S, Simon M, Kano M (2000) Calcium-dependent persistent facilitation of spike backpropagation in the CA1 pyramidal neurons, J Neurosci 20: 4878-4884.

Tsubokawa H, Ross WN (1996) IPSPs modulate spike backpropagation and associated [Ca2+]i changes in the dendrites of hippocampal CA1 pyramidal neurons, J Neurophysiol 76: 2896-2906.

Tsubokawa H, Ross WN (1997) Muscarinic modulation of spike backpropagation in the apical dendrites of hippocampal CA1 pyramidal neurons, J Neurosci 17: 5782-5791.

Yuste R, Denk W (1995) Dendritic spines as basic functional units of neuronal integration, Nature 375: 682-684.

HYDROGEN PEROXIDE REGULATES METAPLASTICITY IN THE HIPPOCAMPUS

A. Kamsler and M. Segal*

1. INTRODUCTION

The brain ties a thread between temporally distinct events through the use of memory. The biological substrate of memory, the memory "trace" or the "engram" has been extensively searched for over the past century and much of what we know about the cellular and molecular machinery of memory comes from studies conducted on the hippocampus. The Schaffer collateral synapse in region CA1 which can be potentiated by a train of pulses has been most instrumental in demonstrating the different phases of memory formation and the mechanisms that control them. In their seminal paper, Stanton and Sarvey (1984), demonstrated the need for protein synthesis in the expression of hippocampal long term potentiation (LTP). They also showed that short term potentiation did not require new protein synthesis. These findings offered a molecular basis for behavioral phenomena described earlier, namely, that memory itself can be divided into temporally distinct phases. In the present chapter we will discuss the results of a series of experiments that were also triggered by a seminal work of John Sarvey (Pellmar et al. 1991). We describe how the ability to produce LTP is controlled by different concentrations of hydrogen peroxide (H_2O_2), a phenomena of meta-plasticity that is altered throughout the life of the organism and may control the ability of some brain circuits to remember.

A continuing rise in the incidence of age-associated neurodegenerative diseases has spurred extensive research in an attempt to discover their mechanisms, etiology, and prospects for therapy. A leading hypothesis proposes the involvement of reactive oxygen species (ROS) including superoxide radical (O_2^-), H_2O_2 and hydroxyl radical ($OH^·$) in neurodegenerative diseases. ROS have been proposed to be involved in molecular processes leading to neurodegeneration through the adverse effects of oxidative stress – a condition in which more ROS are produced than the cellular defense mechanisms can handle, leading to eventual neuronal apoptosis. In this regard, the oxidative stress that induces apoptosis is believed to be the underlying mechanism of decline in neuronal

*A. Kamsler and M. Segal, Department of Neurobiology, Weizmann Institute of Science, Rehovot, Israel 76100. E-mail: menahem.segal@weizmann.ac.il

efficacy. This mechanism has been proposed for Alzheimer's disease (AD), Parkinson's disease (PD) and amyotrophic lateral sclerosis (ALS) (Halliwell 1992), diseases of the nervous system involving death of specific neurons, and an impairment of neurological systems in mostly aged patients. However, support for this hypothesis comes mostly from in-vitro studies, many of which employ high concentrations of ROS that rarely exist in-vivo (Kanno et al 1999, Burlacu et al. 2001, Datta et al. 2002) . Nevertheless, and despite the inability to directly measure ROS reliably in living brains, there is a growing body of evidence indicating that aged brains are indeed exposed to higher concentrations of ROS than young ones (O'Donnell et al 2000). The cause for this rise remains unknown but a breakdown in mitochondrial regulation of intermediate oxidation products has been shown to occur in senescent individuals (Lopez-Torres et al (2002). Likewise, the cellular consequences of the putative rise in ROS are not clearly understood; which of the many molecular targets of ROS are the ones responsible for the loss of neuronal functions and how is this loss actually expressed (Beckman and Ames 1998).
We propose a mechanism by which reduced control over ROS production in aged brains leads to impaired neuronal plasticity manifested as cognitive decline irrespective of, and prior to ultimate cell death. We submit that the role of ROS is more complex than simple mediation of neuronal death, and we provide evidence to suggest that ROS play a facilitatory action towards neuronal plasticity. Such a role has also been suggested recently (Kamsler and Segal 2003a). We will review studies on bimodal action of ROS on neuronal properties, proposing a role for H_2O_2 as a specific diffusible messenger molecule that modulates the activity of protein phosphatases, resulting in modulation of neuronal plasticity. Thus, when H_2O_2 levels are not under optimal regulation, cells may lose the ability to utilize H_2O_2, leading to changes in neuronal functions that could be expressed as cognitive or motor impairments before and irrespective of the eventual cell death evident in neurodegenerative diseases.

2. REACTIVE OXYGEN SPECIES

Since the beginning of aerobic life the high energy yield obtained by reducing oxygen atoms has served as a double edged sword (Halliwell 1992). Eukaryotic cells have quarantined these processes in the mitochondria, intracellular membranous organelles that create a metabolically dependent proton gradient that can be utilized for producing ATP with superoxide anions as a by-product. Although the mitochondria are probably the major source for superoxide anions, they can also be produced by NMDA receptor activation (Lafon-Cazal et al. 1993), which could be important for neuronal signaling. Superoxide anions contain an unpaired electron, which is highly reactive (Beckman and Ames 1998, Benzi and Moretti, 1995). If left unchecked, it can oxidize proton-rich cellular components such as lipids, proteins, and nucleic acids causing reduction in membrane fluidity, disturbance of cellular metabolism, or mutations, respectively. The enzyme superoxide dismutase (SOD) catalyses the reaction that converts superoxide to H_2O_2. H_2O_2 in itself is much less toxic than superoxide, however, it can be converted via the Fenton reaction in the presence of iron ions to hydroxyl radicals that are more reactive than superoxide (figure 1). The *in vivo* occurrence of this reaction depends on the availability of free H_2O_2 and free iron (Halliwell, 1992) and has been regarded as the mechanism by which H_2O_2 can become toxic. H_2O_2 is normally converted to H_2O and O_2 by cellular antioxidants including catalase and glutathione peroxidase, however, under conditions termed "oxidative stress" more ROS are produced than can be handled and the overall redox state of a cell can be altered.

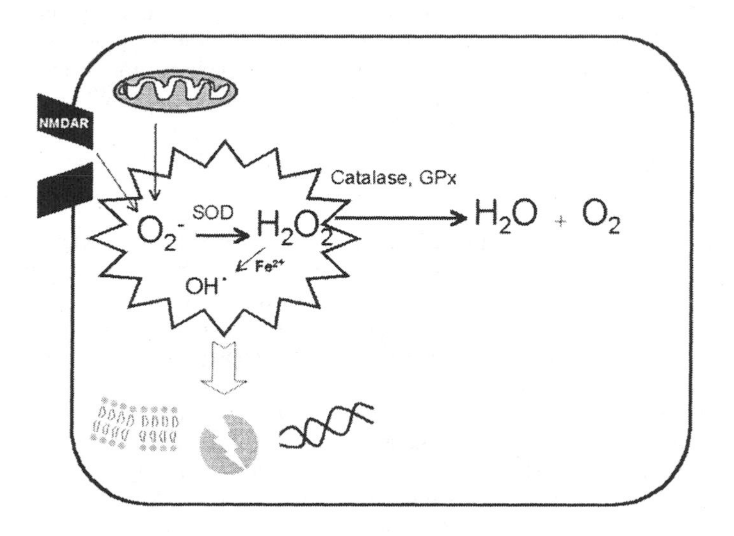

Figure 1. Schematic view of cellular ROS management. Superoxide radicals are produced by mitochondria and NMDA receptors. This highly active radical can undergo dismutation by the enzyme SOD to form hydrogen peroxide, which in turn can form hydroxyl radicals via the Fenton reaction in the presence of free iron cations. These ROS can cause damage to lipids, proteins and nucleic acids thus causing a disruption of cellular activities. The antioxidative enzymes catalase and glutathione peroxidase can facilitate the conversion of H_2O_2 to the benign water and oxygen molecules. (Modified from Kamsler and Segal, 2003a).

This view of ROS as agents of destruction wreaking havoc on lipids, proteins and DNA promotes studies that use concentrations of ROS that are several orders of magnitude higher than those expected to be present in living cells, in an attempt to accelerate processes that are perceived to occur *in vivo*. But, what are the concentrations of ROS *in vivo*? A study by Hyslop et al.(1995), found that the highest concentration of H_2O_2 in the striatum of a reperfused brain after an ischemic insult was estimated to be 100μM. In another microdialysis study, Lei et al.(1998) determined basal H_2O_2 level in gerbil hippocampus to be about 1μM. These studies demonstrated a sub-millimolar concentration of H_2O_2 even under extreme acute pathological conditions that are known to generate ROS. Despite these low estimates there are virtually hundreds of studies showing that millimolar concentrations of H_2O_2 can produce apoptosis in different cell types including neurons (Kano et al.1999, Jang and Surh, 2001, Crossthwaite et al2002, Bhat and Zhang 1999, Herson et al, 1999). These studies aim at understanding mechanisms of neurodegeneration by studying oxidative stress-induced apoptosis in different cell types. But what is the validity of these studies as models for neurodegeneration in the brain if the concentrations of H_2O_2 employed are at least 10-100 times higher than those assumed to be present in vivo? While it may be argued that the high concentrations of extracellular antioxidants present in the brain mask the true intracellular concentrations of ROS, this argument is not relevant to studies conducted with cell cultures that are grown in serum containing medium that is rich in these same antioxidants. Furthermore, neuronal apoptosis has not been proven to be the cause of the functional deficits seen in early stages of AD. In fact, early stages of AD and other neurodegenerative disorders are characterized by episodes of decline in neuronal functions followed by remission. A mechanism of neurodegeneration involving cell death, an irreversible process, would intuitively fail to explain these remissions. We must therefore search for mechanisms that affect the function of living neurons and not their viability. We assume that such mechanisms are involved in neuronal plasticity.

3. NEURONAL PLASTICITY

A cortical structure that has been employed extensively in the study of neurodegeneration is the hippocampus. The hippocampus is part of the temporal lobe, long known for its involvement in various forms of short term memory processes in the brain. In humans, damaged hippocampus has been associated with loss of short term memory, and these observations have been confirmed in numerous animal studies (Manns et al 2003). The hippocampus undergoes massive degeneration in AD and as such has been a convenient target for many studies attempting to dissect the cellular and molecular mechanisms underlying age and disease related cell death, and their functional implications.
LTP and long term depression (LTD) especially in the Schaffer collateral -CA1 synapses of hippocampal slices and in perforant path-dentate gyrus synapse in vivo are extensively studied models of synaptic plasticity. LTP is a long lasting increase in synaptic efficacy following a potentiating event. Potentiating trains of pulses cause a persistent depolarization of post synaptic membrane resulting in calcium ion flow through NMDA receptors and/or voltage gated calcium channels (VGCCs) both in vivo and in vitro Morgan and Teyler, 1999, Bliss and Collingridge 1993). This calcium flux results in well-characterized signal transduction cascades leading to a change in AMPA receptor permeability (Bliss and Collingridge 1993, Dudek and Bear 1992) which is manifested in larger (LTP) or smaller (LTD) excitatory post synaptic potentials (EPSPs). This persistent change in EPSPs is regarded as a model for synaptic plasticity sharing mechanisms that are assumed to operate in learning and memory. Many of the transducing processes during and immediately after the potentiating stimulus are carried out by kinases and phosphatases (reviewed in Soderling and Derkach 2000).

4. CALCINEURIN IS A KEY PARTICIPANT IN NEURONAL PLASTICITY

One protein phosphatase that has been implicated in LTP modulation is calcineurin, a calcium dependent serine/threonine phosphatase that can dephosphorylate protein kinase A (PKA) substrates (such as calcium calmodulin kinase II, CamKII). The cellular shift in phosphorylation/ dephosphorylation ratio of PKA substrates is enhanced by the fact that calcineurin can also activate protein phosphatase 1 (PP1) by dephosphorylating inhibitor-1. It has been hypothesized (Bito et al, 1996) that this route is active in calcineurin inhibition of pCREB activation in response to depolarizing stimuli in hippocampal neurons demonstrating lower levels of activity downstream to potentiating events when calcineurin activity is high. Genetic models of mice overexpressing calcineurin have yielded conflicting results, while enhanced activity of calcineurin in a genetic model (Winder et al1998) caused a decrease in some forms of LTP, inhibiting brain calcineurin in another genetic model enhanced memory and LTP (Malleret et al, 2001). Interestingly, forebrain specific calcineurin knockout resulted mostly in impaired LTD and working memory (Zeng et al, 2001) without affecting LTP. Inhibiting calcineurin activity in CA1 cells caused potentiation of EPSPs in slices from adult but not young rats (Wang and Kelly 1997), while in another study Onuma et al.(1998) show that blocking calcineurin inhibited VGCC LTP in CA1 of 7-10 week old mice. These studies establish a role for calcineurin in synaptic plasticity with the direction of this effect (enhancing or inhibiting LTP) depending on the background conditions. Interestingly, a calcineurin inhibitory gene *DSCR1* is expressed in brains of AD patients three fold over that of controls. The expression of this gene could be induced by the amyloid β 1-42 peptide associated with senile plaques (Ermak et al 2001) which may indicate that pathologically low calcineurin activity in AD brains is a contributing factor in mental decline. Conversely, it may indicate a response to pathologically high activity of calcineurin, which may be induced by ROS.

5. CALCINEURIN AND AGE DEPENDENT HIPPOCAMPAL DECLINE

Calcineurin is a good candidate for linking age dependent deficiencies in hippocampal functions and calcium dependant cellular mechanisms, as suggested by Foster et al. (2001). They have shown an elevation of calcineurin activity as well as the activity of calcineurin-regulated PP1 in aged rats that were impaired in hippocampus dependent memory tasks.

Indirect evidence from cell cultures shows that calcineurin can be activated by H_2O_2. Nuclear factor of activated T cells (NFAT) is a transcription factor that is activated by calcineurin. NFAT activity can be induced by asbestos (Li et al, 2002) or vanadium (Huang et al. 2001) in a manner that is dependent on H_2O_2 production. It can be enhanced by the H_2O_2 producing enzyme superoxide dismutase (SOD) and can be blocked by the calcineurin inhibitor Cyclosporin A. Inactivation of calcineurin by H_2O_2 has been shown (Bogumil et al2000), however this study used 1mM H_2O_2 for time dependence of the reaction showing 75% activity decrease over 30 minutes and only as little as 0.3mM H_2O_2 for dose dependence, under which an activity plot was extrapolated. Accordingly, activation of calcineurin by micromolar concentrations of H_2O_2 cannot be ruled out in that study (Bogumil et al 2000).

CamKII activates Calcineurin after being exposed to a rise in intracellular calcium ($[Ca^{2+}]_{in}$). Several studies have shown that H_2O_2 can induce an increase in $[Ca^{2+}]_{in}$ using various cell types and 0.01-10mM of H_2O_2 (Herson et al 1999, Roveri et al 1992, Volk et al 1997, Nakazaki et al, 2000,Lee et al 1998, Yermolaiva et al 2000, Yang et al 1999, Nam et al 2002, Gen et al. 2001). In some of these studies the increase in $[Ca^{2+}]_i$ was sensitive to thapsigargin, a blocker of endoplasmic reticulum calcium uptake channels indicating its origin in intracellular calcium stores. These studies show that a change in ROS can induce a change in $[Ca^{2+}]_i$, which can then have an effect on calcineurin.

6. ROS IN BRAIN PLASTICITY STUDIES

As mentioned above, aging of the brain is accompanied by an increase in ROS production. Concomitantly, hippocampal slices taken from aged and young adult rats exhibit different responses to similar trains of stimulation. Slices from aged rats are impaired in LTP and they exhibit LTD in response to frequencies of stimulation that do not affect young rats. Nifedipine – an L-type VGCC blocker can reverse some of these changes (Norris et al 1998). O'Donnell et al. (2000) also reported an age dependent decrease in LTP in rats; a decline thought to be due to oxidative stress. They also demonstrated an age dependent increase in SOD activity. Moreover, stress activated genes that are upregulated in aged individuals could be activated in synaptosomes by applying H_2O_2 (albeit at a high concentration of 5 mM). An interesting study by Vereker et al.(2001) demonstrated a connection between stress, SOD activity and LTP decline. Stress induced a decrease in LTP, an increase in SOD activity and an increase in IL-1β concentration. An increase in IL-1β concentration could increase SOD activity without the stress. IL-1β alone also inhibited LTP in a manner that was reversible by adding anti oxidants, demonstrating that the IL-1β mediated stress induced LTP decline was associated with a rise in ROS. Moreover, they showed a similar decline in LTP in the presence of 200μM H_2O_2. Interestingly, dietary manipulation with anti oxidants could restore LTP in aged individuals that were otherwise impaired. McGahon et al (1999), demonstrated a reversal of age dependent LTP decline by dietary supplementation with omega-3 fatty acids as did Liu et al (2002) by feeding other anti oxidant fatty acids, acetyl-L-carnitine and/or R-α-lipoic acid.

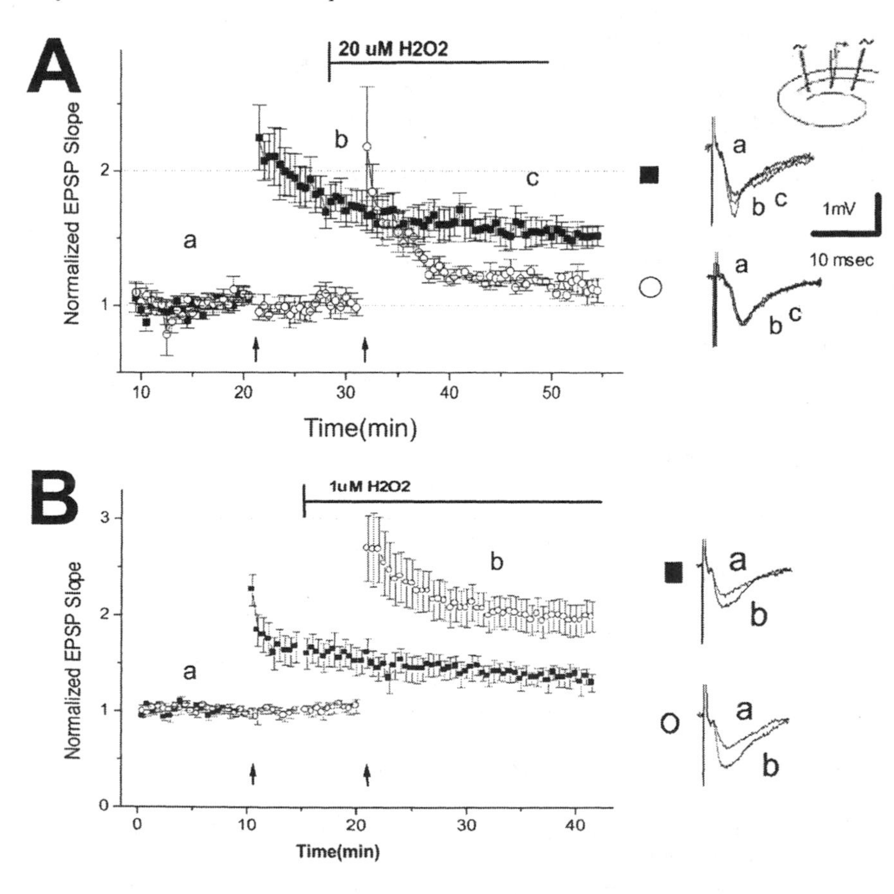

Figure 2. Long term potentiation is modulated by hydrogen peroxide in a dose-dependent manner. (A) Application of 20 μM of H_2O_2 did not affect a previously potentiated channel. It did however inhibit new LTP. (Schematic representation of the two channels is in the inset.) Average normalized EPSP slope is plotted

versus time. Inset, a and b, are representative tracings at indicated times. The arrows indicate applied HFS (100 Hz for 1 sec at twice the test stimulus intensity). (B) Application of 1μM of H_2O_2 (bar) increases LTP twofold as compared to control. (Modified from Kamsler et al 2003a.)

7. DIRECT APPLICATION OF H_2O_2 TO BRAIN SLICES

Several groups have applied H_2O_2 to brain slices for the purpose of studying the effects of ROS on synaptic plasticity. Pellmar et al.(1991) have shown that 0.72mM H_2O_2 can inhibit long term potentiation (LTP) in guinea pig hippocampal slices. Avshalumov et al.(2000) showed a dramatic effect of 1.2-2 mM H_2O_2 in reducing population spikes in hippocampal slices with a consequent epileptiform activity when H_2O_2 was washed out (Avshalumov and Rice, 2002). These concentrations of H_2O_2.are likely to produce non-specific effects, with respect to LTP induction mechanisms. In contrast, some of the LTP related phenomena seen in aged animals could be mimicked in younger animals exposed to physiologically relevant concentrations of H_2O_2; a low concentration of H_2O_2 (29μM) inhibits muscarinic as well as tetanically induced LTP in hippocampal slices (Auerbach and Segal, 1997).

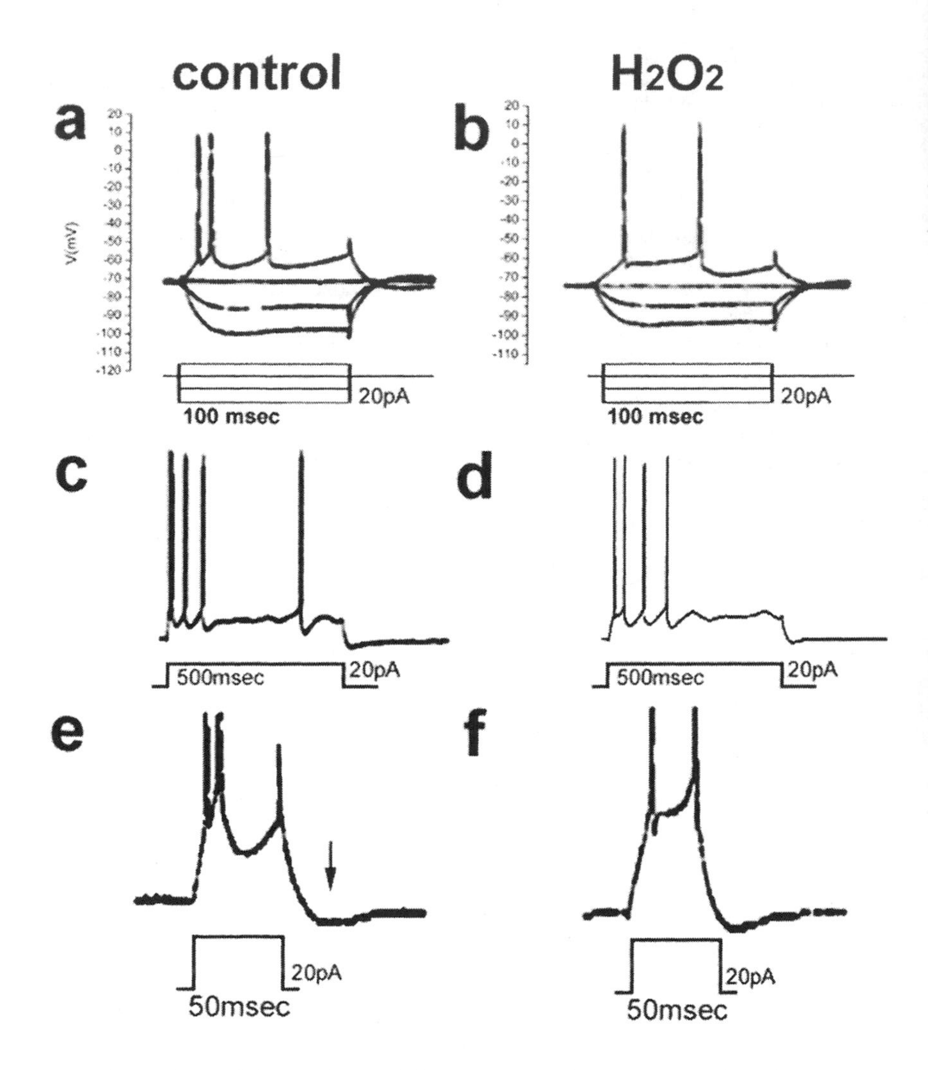

Figure 3. Traces of inracellular recordings made in a rat hippocampal slice. Representative voltage recordings obtained from CA1 hippocampal pyramidal neurons in response to current pulse stimulation before (a, c, e) and 15 minutes after (b, d, f) onset of perfusion with H_2O_2. In a-b, the traces are generated by several 100 msec voltage steps showing healthy current-voltage relationships of the cells and the ability to generate spikes in response to depolarizing stimuli. In c-d, are recordings made in response to a 500 msec depolarizing step exhibiting attenuation of the action potentials. In e-f, are responses to a 50msec current step, demonstrating a fast after hyper polarization (arrow).

We have shown recently (Kamsler and Segal, 2003a) *(*figure 2), that applying 20μM of H_2O_2 to hippocampal slices taken from rats resulted in reduction of LTP and an increase in LTD without affecting baseline properties. Table 1 and figure 3 show the stability of electrical properties of CA1 neurons recorded in a slice after treatment with this concentration of H_2O_2, indicating that H_2O_2 did not affect LTP by changing intrinsic neuronal properties.

Table 1. Intracellularly recorded parameters of rat hippocampal pyramidal neurons exposed to 20μM H_2O_2.

Parameter	Control	15' H_2O_2
Spike amplitude(mV)	60.4 ± 1.3	59 ± 3.5
Spike rise time(ms)	0.675 ± 0.033	0.733 ± 0.087
AHP size(mV)	3.33 ± 0.54	3.74 ± 0.54
AHP dur. (ms)	8.1 ± 1.15	9.98 ± 1.99

The table represents recordings made in 6 hippocampal pyramidal neurons. The cells selected had a resting potential of −60 ± 2 mV throughout the recording and input resistance of about 40 MOhm. The control measurements were made after 10 minutes of stable recording and were immediately followed by 20μM H_2O_2.

Blocking calcineurin could antagonize this effect of H_2O_2. Calcineurin was also shown to be more active in hippocampal slices exposed to 20μM H_2O_2. Interestingly, 1μM of H_2O_2 had an opposite effect, enhancing LTP to double that of control (Fig 2). This effect of H_2O_2 was also blocked by calcineurin inhibitors but also by rapamycin – an inhibitor of FKBP-12, which does not interact with calcineurin. When bound by rapamycin, FKBP-12 dissociates from ryanodine receptors that control calcium flow from internal stores. H_2O_2 at a concentration of 1μM also decreased the range of low stimulation frequencies that could elicit LTD, 600 pulses at 1 Hz, which depressed control EPSP's by 20%, did not have an effect under these conditions.
While these studies demonstrate the acute effects of H_2O_2 on synaptic plasticity, aged individuals are exposed to chronically altered levels of ROS, which may affect synaptic plasticity in different ways.

8. TRANSGENIC SOD OVEREXPRESSING MICE

Under normal conditions, superoxide radicals that are produced by the mitochondria but also by the activity of ion channels, are scavenged by the enzyme SOD. SOD converts superoxide to H_2O_2, which is a less reactive, membrane permeable intermediate. Interestingly, the gene encoding the SOD message resides in humans on chromosome 21 in a region that is triplicated in Down's syndrome – a genetic form of mental retardation that shares pathological hallmarks with AD. Transgenic mice overexpressing human SOD (tg-SOD) were generated (Epstein et al 1987) and have been extensively studied as tentative models of neurodegeneration. These mice express several copies of SOD resulting in 6 fold enhanced activity of the enzyme over controls. The neuromuscular junction in the tongue of tg-SOD mice is degenerated[53] and they also exhibit thymic abnormalities (Peled Kamar et al 1995). On the other hand, tg-SOD mice were found to be less susceptible to focal cerebral ischemic injury (Kinouch et al 1991, Saito et al 2003) and SOD over expressing rats were also protected against ischemia (Sugawara et al 2002). While kainic acid induced apoptosis is exacerbated in cultured neurons from tg-SOD mice Peled-Kamar et al, 1995), whole animals injected with kainic acid are protected from seizure in comparison to controls (Levkovitz et al. 1999). Thymocytes from tg-SOD mice have been shown to produce more H_2O_2 than controls (Peled-Kamar et al 1995). We have shown (Gahtan et al 1998) that transgenic mice overexpressing SOD were impaired in spatial memory tasks. Hippocampal slices taken from these mice were impaired in LTP in a manner that was reversed by catalase, an enzyme that breaks down H_2O_2, and also by the spin trapping agent N-t-butyl-phenylnitrone. We also found impairment in perforant path LTP measured in vivo in SOD transgenic mice (Levkovitz et al 1999), as well as resistance to kainic acid which induced seizures in wild-type controls. Hippocampal cells in the transgenic mice were under a high level of GABAergic inhibition manifested in over-activity of interneurons as measured in slices from SOD mice. Consequently, bicuculline, a GABAergic antagonist, could restore LTP in the dentate gyrus.

In our recent studies conducted with these mice (Kamsler and Segal 2003b) we were able to restore LTP in hippocampal slices of tg-SOD mice by perfusing them with 50 μM H_2O_2 in a calcineurin dependent manner. This concentration of H_2O_2 inhibited LTP in the wt controls. We were intrigued to find that aged (2yr old) tg-SOD slices had larger LTP than controls. Aged wt slices were impaired in LTP in a manner reversible by 50 μM H_2O_2, as was the case with young tg-SOD mice, and they also had high levels of endogenous ROS and phosphatase activity (Figure 4).

Interestingly, mutations leading to the neurodegenerative disease ALS have been mapped to the SOD gene, however it is believed that these mutations do not inhibit the dismutase activity, rather, they constitute a "gain of function" which may also lead to oxidative stress through a decrease in the enzymes affinity to zinc (Beckman et al 2001).

It has been suggested that the LTP impairment as well as memory deficiencies seen in tg-SOD mice are due to a rapid elimination of superoxide radicals produced by the activation of NMDA receptors and not as a consequence of high levels of H_2O_2. This can be deduced from the elimination of LTP with the addition of superoxide scavengers (Klann 1998, Thiels et al 2000) and is supported by the fact that catalase added to hippocampal slices exposed to SOD as well as to slices from extracellular SOD overexpressing mice (Thiels et al 2000), could not restore LTP. These results were interpreted as an indication that the overproduction of H_2O_2 by itself is not the cause for the reduction in efficacy of LTP, but perhaps the fast removal of superoxides by SOD. While this is an interesting interpretation of the data, the biochemical nomenclature of unit definitions of different enzymes is such that enzymatic activity of different enzymes cannot be compared on a scale of units. Furthermore, externally perfused catalase would have to permeate the slice and be active at the synaptic cleft, and with negative results, and lack of independent confirmation that the enzyme did work, such data is hard to interpret. Taken together, these studies further support the involvement of ROS in the regulation of synaptic plasticity.

It has been hypothesized that superoxide can activate protein kinases and inhibit protein phosphatases, such as calcineurin (Klann et al 1998, Klann and Thiels 1999), however most of the evidence suggesting phosphatase inhibition is based on experiments which use high levels of oxidants. Interestingly, Knapp and Klann (2002) have demonstrated that LTP could be induced in rat hippocampal slices via a PKC dependent pathway by generating superoxide radicals with the xanthine/ xanthine oxidase system.

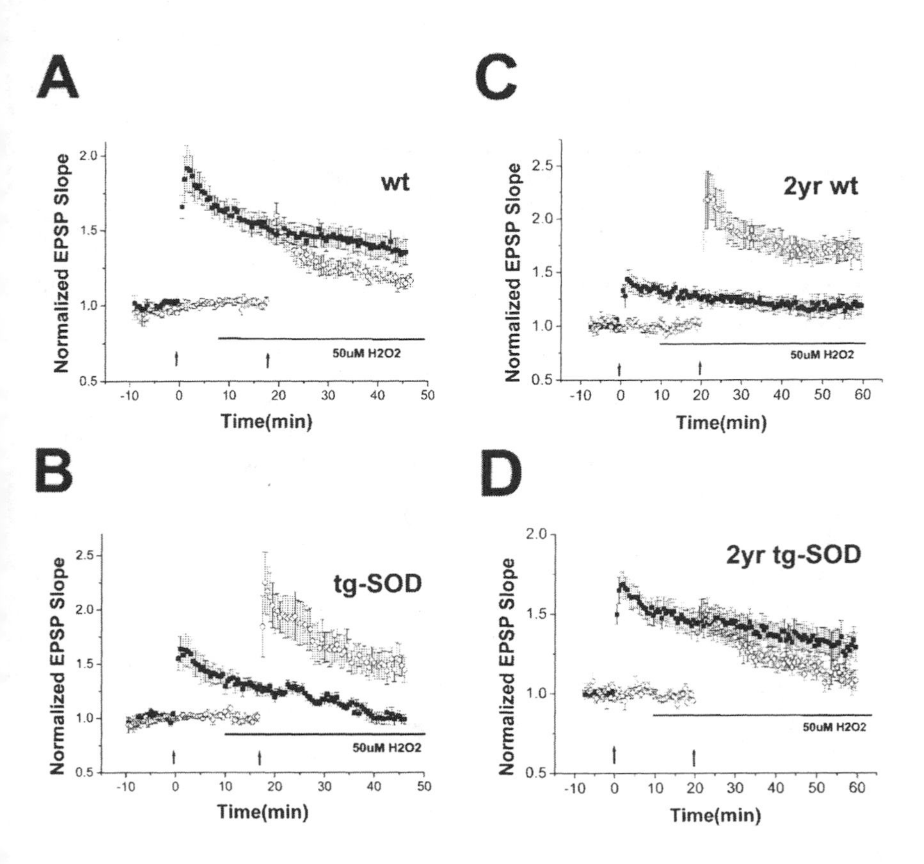

Figure 4. H_2O_2 reverses impaired LTP in young Tg-SOD mice and has opposite effects in aged mice. (A) LTP in hippocampal slices from 2 month old wt mice was induced by Theta Burst Stimulation (TBS) in the first pathway (arrow). Prior to TBS in the second pathway the slice was perfused with 50 μM H_2O_2 which inhibited LTP. (B) A similar experiment shows that H_2O_2 markedly enhances potentiation in slices taken from tg-SOD mice, which otherwise express a low level of LTP. (C)TBS (arrows), applied to aged wt mice before (squares) and after (circles) perfusion with 50μM H2O2 resulted in a large enhancement of both short and long term potentiation. (D) In tg-SOD slices, under the same conditions, 50μM H_2O_2 markedly attenuated LTP. (Modified from Kamsler and Segal, 2003b).

9. H_2O_2 AND CALCIUM CHANNELS

Aging, SOD overexpression and direct application of H_2O_2, can increase the tissue level of H_2O_2. Elevated H_2O_2 can induce the release of calcium from intracellular stores and this excess calcium can induce calcineurin activity via CamKII. Active calcineurin dephosphorylates I-1 resulting in higher activity of PP1 (figure 5). Which substrate of PP1 could induce changes in LTP?

Norris et al.(1998) found that blocking PP1 and PP2A enhanced EPSPs in aged but not in adult rats whereas H-7, a serine / threonine kinase inhibitor decreased EPSPs in adult but not aged rats. Blocking potassium currents with apamin was also effective in restoring LTP in aged slices (Norris et al. 1998b), suggesting a role for calcium dependent potassium channels in age dependent LTP decline. These data are compatible with

Campbell et al. (1996) who measured an age dependent increase in L-type VGCC current. Furthermore, Mermelstein et al. (2000) have shown that the state of activity of L-type VGCCs can differentially regulate signal transduction cascades. Nifedipine, an L-type VGCC blocker was effective in reversing the effects of 1μM H_2O_2 on LTP (Kamsler and Segal, 2003). A direct link between VGCCs and calcineurin was demonstrated by Norris et al. (2002) who showed that specific blocking of calcineurin decreased VGCC currents in cultured hippocampal neurons. This effect also had age dependent characteristics since the block was increased in cultures that were 4 weeks old over cultures that were 2 weeks old. Interestingly, VGCC permeability can be controlled by PKA phosphorylation of several sites on the channel protein (Hell et al 1995, Davare et al 1999) these sites are potential candidates for PP1 dephosphorylation.

A calcium imaging study conducted in hippocampal slices taken from young adult and aged rats measured responses to 7 Hz stimulation, aged slices were found to be impaired in frequency facilitation in a manner that could be mimicked in young slices by adding the VGCC agonist Bay K8644 (Thibault et al. 2001) demonstrating an age dependent change in VGCC permeability.

When neurons are depolarized by sustained stimulation, VGCCs remain open long enough to allow an inward calcium current that acts as a second messenger molecule. One of the targets of this calcium current is a calcium dependent potassium channel that affects the cells ability to undergo further depolarization. Changing the calcium permeability of VGCCs by H_2O_2 can thus change the message that a post synaptic cell is receiving. Such increases in the after-hyperpolarization (AHP) has been demonstrated in slices from aged rats. Interestingly, we have not found such changes in the AHP when perfusing 20μM H_2O_2. This may indicate that the effects of H_2O_2 on LTP are mediated by biochemical events that are downstream of the potentiating train and not by a change in intrinsic neuronal properties.

10. CONCLUSIONS

Physiologically relevant concentrations of H_2O_2, within the 1 to 50 micromolar range, can induce changes in synaptic plasticity. These concentrations of H_2O_2 have been shown to release calcium from intracellular stores. The release of calcium may result from a redox sensitive domain on proteins controlling this release such as ryanodine receptors. Once a shift in redox state has occurred, either by external addition of ROS or by genetic manipulation of cellular anti oxidants or by normal aging, a change in calcium may follow. This excess calcium can activate calcium dependent proteins such as CamKII, calcineurin and PP1, which in turn can alter the calcium permeability of L-type VGCCs. An altered calcium flux at the time of synaptic potentiating events can change the neuronal meaning of that event by inducing changes in calcium dependent potassium currents (figure 5). While our data make a strong case for calcineurin as the main transducer of H_2O_2 mediated signaling in hippocampal slices, this signaling may involve other, as yet undiscovered transduction cascades. H_2O_2 is a short-lived, membrane permeable, oxidant that is well suited for the role of a diffusible messenger. This messenger can induce the release of calcium on both sides of the synapse triggering concerted activity. Accordingly, H_2O_2 acting as an acute messenger molecule produced by the activity of ion channels depends on existing levels of H_2O_2 prior to generation of plastic events. A high background level of H_2O_2 can induce higher activity of antioxidants, it can also alter the redox sensitivity of target molecules. In this way, a high

ambient H_2O_2 level will dampen the effect of an H_2O_2 flux that results from synaptic activity.
H_2O_2 acting as a messenger molecule that is sensitive to changes such as age, sickness or stress is a good candidate for controlling the ability of the hippocampus to remember. In this way, the synaptic "engram" that is driven by sodium spikes and modulated by calcium fluxes exhibits a higher level of control by which the size of the calcium flux is controlled by phosphorylation in a way dependent on redox conditions and H_2O_2 concentration. This hypothesis represents a modulator of memory machinery adding to our understanding of the complexity of the "simple" hippocampal synapse.

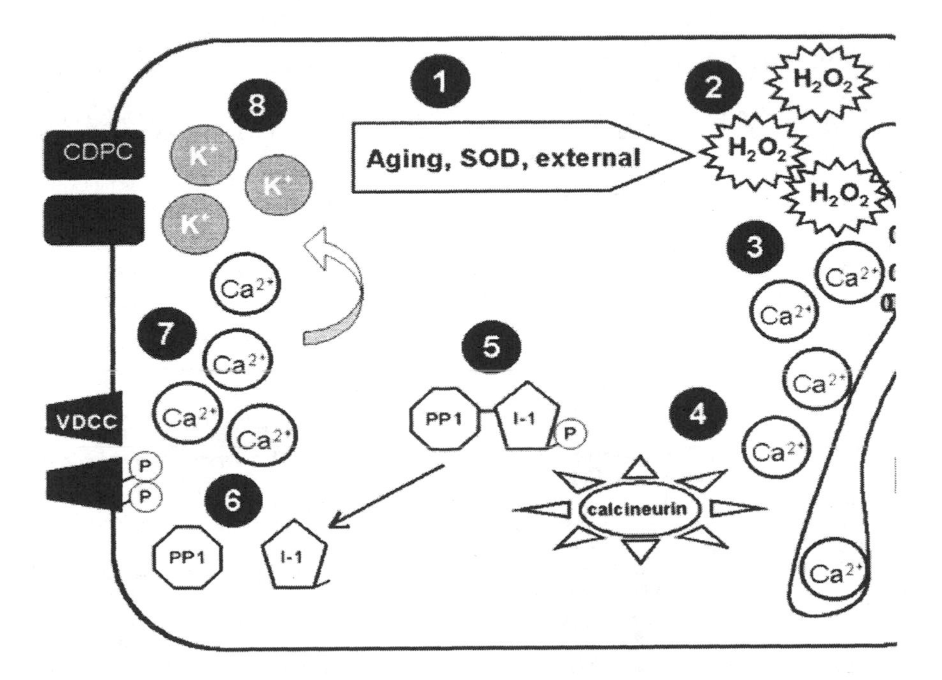

Figure 5. Representation of the sequence of events that may link redox changes with alterations in synaptic plasticity. Aging, transgenic intervention or exogenous addition of H2O2 (1) can increase the intracellular concentration of H_2O_2 (2) which can then cause the release of calcium from internal stores (3) activating calcineurin (4), calcineurin mediated dephosphorylation of Inhibitor-1 (5) allows protein phosphatase 1 to de-phosphorylate PKA substrates on VGCCs (6) altering the permeability of these to calcium (7) which may alter the opening time of calcium dependant potassium channels (8) leading to a change in synaptic plasticity. (Modified from Kamsler and Segal 2004).

11. ACKNOWLEDGEMENTS

This work was supported by a grant from the Azheimer's Association.

12. REFERENCES

Avshalumov MV, Chen BT, Rice ME. (2000) Mechanisms underlying H(2)O(2)-mediated inhibition of synaptic transmission in rat hippocampal slices. Brain Res. 882:86-94.

Avshalumov MV, Rice ME. (2002) NMDA receptor activation mediates hydrogen peroxide induced pathophysiology in rat hippocampal slices. J Neurophysiol. 87:2896-903.
Auerbach JM, Segal M. (1997) Peroxide modulation of slow onset potentiation in rat hippocampus. J Neurosci. 17:8695-701.
Beckman KB, Ames BN. (1998) The free radical theory of aging matures. Physiol Rev. 78:547-81.
Beckman JS, Estevez AG, Crow JP, Barbeito L. (2001) Superoxide dismutase and the death of motoneurons in ALS. Trends Neurosci. 24:S15-20.
Benzi G, Moretti A. (1995) Are reactive oxygen species involved in Alzheimer's disease? Neurobiol Aging. 16:661-74.
Bliss TV, Collingridge GL (1993) A synaptic model of memory: long-term potentiation in the hippocampus. Nature. 361:31-9.
Bhat NR, Zhang P. (1999) Hydrogen peroxide activation of multiple mitogen-activated protein kinases in an oligodendrocyte cell line: role of extracellular signal-regulated kinase in hydrogen peroxide-induced cell death. J Neurochem. 72:112-9.
Bito H, Deisseroth K, Tsien RW. (1996) CREB phosphorylation and dephosphorylation: a Ca^{2+}- and stimulus duration-dependent switch for hippocampal gene expression. Cell. 87:1203-14.
Bogumil R, Namgaladze D, Schaarschmidt D, Schmachtel T, Hellstern S, Mutzel R, Ullrich V. (2000) Inactivation of calcineurin by hydrogen peroxide and phenylarsine oxide. Evidence for a dithiol-disulfide equilibrium and implications for redox regulation. Eur J Biochem. 267:1407-15.
Borroni AM, Fichtenholtz H, Woodside BL, Teyler TJ. (2000) Role of voltage-dependent calcium channel long-term potentiation (LTP) and NMDA LTP in spatial memory. J Neurosci. 20:9272-6.
Burlacu A, Jinga V, Gafencu AV, Simionescu M. (2001) Severity of oxidative stress generates different mechanisms of endothelial cell death. Cell Tissue Res. 306:409-16.
Campbell LW, Hao SY, Thibault O, Blalock EM, Landfield PW. (1996) Aging changes in voltage-gated calcium currents in hippocampal CA1 neurons. J Neurosci. 16:6286-95.
Crossthwaite AJ, Hasan S, Williams RJ. (2002) Hydrogen peroxide-mediated phosphorylation of ERK1/2, Akt/PKB and JNK in cortical neurones: dependence on $Ca(^{2+})$ and PI3-kinase. J Neurochem. 80:24-35 .
Datta K, Babbar P, Srivastava T, Sinha S, Chattopadhyay P. (2002) p53 dependent apoptosis in glioma cell lines in response to hydrogen peroxide induced oxidative stress. Int J Biochem Cell Biol. 34:148-57.
Davare MA, Dong F, Rubin CS, Hell JW. (1999) The A-kinase anchor protein MAP2B and cAMP-dependent protein kinase are associated with class C L-type calcium channels in neurons. J Biol Chem. 274:30280-7
Dudek SM, Bear MF. (1992) Homosynaptic long-term depression in area CA1 of hippocampus and effects of N-methyl-D-aspartate receptor blockade. Proc Natl Acad Sci U S A. 89:4363-7.
Epstein CJ, Avraham KB, Lovett M, Smith S, Elroy-Stein O, Rotman G, Bry C, Groner Y. (1987) Transgenic mice with increased Cu/Zn-superoxide dismutase activity: animal model of dosage effects in Down syndrome.Proc Natl Acad Sci U S A. 84:8044-8.
Ermak G, Morgan TE, Davies KJ. (2001) Chronic overexpression of the calcineurin inhibitory gene DSCR1 (Adapt78) is associated with Alzheimer's disease. J Biol Chem. 276:38787-94.
Foster TC, Sharrow KM, Masse JR, Norris CM, Kumar A. (2001) Calcineurin links Ca^{2+} dysregulation with brain aging. J Neurosci. 21:4066-73.
Gahtan E, Auerbach JM, Groner Y, Segal M. (1998) Reversible impairment of long-term potentiation in transgenic Cu/Zn-SOD mice. Eur J Neurosci. 10:538-44 .
Gen W, Tani M, Takeshita J, Ebihara Y, Tamaki K. (2001) Mechanisms of Ca^{2+} overload induced by extracellular H2O2 in quiescent isolated rat cardiomyocytes. Basic Res Cardiol. 96:623-9 .
Halliwell B. (1992) Reactive oxygen species and the central nervous system. J Neurochem. 59,1609-23.
Hell JW, Yokoyama CT, Breeze LJ, Chavkin C, Catterall WA. (1995) Phosphorylation of presynaptic and postsynaptic calcium channels by cAMP-dependent protein kinase in hippocampal neurons. EMBO J. 14:3036-44.
Herson PS, Lee K, Pinnock RD, Hughes J, Ashford ML. (1999) Hydrogen peroxide induces intracellular calcium overload by activation of a non-selective cation channel in an insulin-secreting cell line. J Biol Chem. 274:833-41.
Huang C, Ding M, Li J, Leonard SS, Rojanasakul Y, Castranova V, Vallyathan V, Ju G, Shi X. (2001) Vanadium-induced nuclear factor of activated T cells activation through hydrogen peroxide. J Biol Chem. 276:22397-403.
Hyslop PA, Zhang Z, Pearson DV, Phebus LA. (1995) Measurement of striatal H_2O_2 by microdialysis following global forebrain ischemia and reperfusion in the rat: correlation with the cytotoxic potential of H_2O_2 in vitro. Brain Res. 671:181-6.
Jang JH, Surh YJ. (2001) Protective effects of resveratrol on hydrogen peroxide-induced apoptosis in rat pheochromocytoma (PC12) cells. Mutat Res. 496:181-90.
Kamsler A, Segal M. (2003a) Hydrogen peroxide modulation of synaptic plasticity. J Neurosci. 23:269-76.

Kamsler A, Segal M. (2003b) Paradoxical actions of hydrogen peroxide on LTP in transgenic SOD-1 mice. J. Neurosci. 23:10359-10367.
Kamsler A., Segal. M. (2004) Hydrogen peroxide as a diffusible signal molecule in synaptic plasticity. Mol Neurobiol 29:167-78.
Kanno S, Ishikawa M, Takayanagi M, Takayanagi Y, Sasaki K. (1999) Exposure to hydrogen peroxide induces cell death via apoptosis in primary cultured mouse hepatocytes. Biol Pharm Bull. 22:1296-300.
Kinouchi H, Epstein CJ, Mizui T, Carlson E, Chen SF, Chan PH. (1991) Attenuation of focal cerebral ischemic injury in transgenic mice overexpressing CuZn superoxide dismutase. Proc Natl Acad Sci U S A. 88:11158-62.
Klann E. (1998) Cell-permeable scavengers of superoxide prevent long-term potentiation in hippocampal area CA1. J Neurophysiol. 80:452-7 .
Klann E, Roberson ED, Knapp LT, Sweatt JD. (1998) A role for superoxide in protein kinase C activation and induction of long-term potentiation. J Biol Chem. 273:4516-22.
Klann E, Thiels E (1999) Modulation of protein kinases and protein phosphatases by reactive oxygen species: implications for hippocampal synaptic plasticity. Prog Neuropsychopharmacol Biol Psychiat. 23:359-76.
Knapp LT, Klann E. (2002) Potentiation of hippocampal synaptic transmission by superoxide requires the oxidative activation of protein kinase C. J Neurosci. 22:674-83.
Lafon-Cazal M, Pietri S, Culcasi M, Bockaert J. (1993) NMDA-dependent superoxide production and neurotoxicity. Nature. 364:535-7.
Lee ZW, Kweon SM, Kim BC, Leem SH, Shin I, Kim JH, Ha KS. (1998) Phosphatidic acid-induced elevation of intracellular Ca2+ is mediated by RhoA and H2O2 in Rat-2 fibroblasts. J Biol Chem. 273:12710-5.
Lei B, Adachi N, Arai T. (1998) Measurement of the extracellular H_2O_2 in the brain by microdialysis. Brain Res Brain Res Protoc. 3:33-6.
Levkovitz Y, Avignone E, Groner Y, Segal M. (1999) Upregulation of GABA neurotransmission suppresses hippocampal excitability and prevents long-term potentiation in transgenic superoxide dismutase-overexpressing mice. Neurosci. 19:10977-84.
Li J, Huang B, Shi X, Castranova V, Vallyathan V, Huang C. (2002) Involvement of hydrogen peroxide in asbestos-induced NFAT activation. Mol Cell Biochem. 234-235:161-8.
Liu J, Head E, Gharib AM, Yuan W, Ingersoll RT, Hagen TM, Cotman CW, Ames BN. (2002) Memory loss in old rats is associated with brain mitochondrial decay and RNA/DNA oxidation: partial reversal by feeding acetyl-L carnitine and/or R-alpha-lipoic acid. Proc Natl Acad Sci U S A. 99:2356-61.
Lopez-Torres M, Gredilla R, Sanz A, Barja G. (2002) Influence of aging and long-term caloric restriction on oxygen radical generation and oxidative DNA damage in rat liver mitochondria. Free Radic Biol Med. 32:882-9.
Malleret G, Haditsch U, Genoux D, Jones MW, Bliss TV, Vanhoose AM, Weitlauf C, Kandel ER, Winder DG, Mansuy IM. (2001) Inducible and reversible enhancement of learning, memory, and long-term potentiation by genetic inhibition of calcineurin. Cell. 104:675-86.
Manns JR, Hopkins RO, Squire LR. (2003) Semantic memory and the human hippocampus. Neuron. 38:127-33
McGahon BM, Martin DS, Horrobin DF, Lynch MA. (1999) Age-related changes in synaptic function: analysis of the effect of dietary supplementation with omega-3 fatty acids. Neuroscience. 94:305-14.
Mermelstein PG, Bito H, Deisseroth K, Tsien RW. (2000) Critical dependence of cAMP response element-binding protein phosphorylation on L-type calcium channels supports a selective response to EPSPs in preference to action potentials. J Neurosci. 20:266-73.
Morgan SL, Teyler TJ. (1999) VDCCs and NMDARs underlie two forms of LTP in CA1 hippocampus in vivo. J Neurophysiol. 82:736-40.
Nakazaki M, Kakei M, Yaekura K, Koriyama N, Morimitsu S, Ichinari K, Yada T, Tei C. (2000) Diverse effects of hydrogen peroxide on cytosolic Ca^{2+} homeostasis in rat pancreatic beta-cells. Cell Struct Funct. 25:187-93.
Nam SH, Jung SY, Yoo CM, Ahn EH, Suh CK. (2002) H2O2 enhances Ca^{2+} release from osteoblast internal stores. Yonsei Med J. 43:229-35 .
Norris CM, Halpain S, Foster TC. (1998a) Alterations in the balance of protein kinase/phosphatase activities parallel reduced synaptic strength during aging. J Neurophysiol. 80:1567-70 .
Norris CM, Halpain S, Foster TC. (1998b) Reversal of age-related alterations in synaptic plasticity by blockade of L-type Ca2+ channels. J Neurosci. 18:3171-9.
Norris CM, Blalock EM, Chen KC, Porter NM, Landfield PW. (2002) Calcineurin enhances L-type Ca^{2+} channel activity in hippocampal neurons: increased effect with age in culture. Neuroscience. 110:213-25.
O'Donnell E, Vereker E, Lynch MA. (2000) Age-related impairment in LTP is accompanied by enhanced activity of stress-activated protein kinases: analysis of underlying mechanisms. Eur J Neurosci. 12:345-52
Onuma H, Lu YF, Tomizawa K, Moriwaki A, Tokuda M, Hatase O, Matsui H. (1998) A calcineurin inhibitor, FK506 blocks voltage-gated calcium channel-dependent LTP in the hippocampus. Neurosci Res 30:313-9

Peled-Kamar M, Lotem J, Okon E, Sachs L, Groner Y. (1995) Thymic abnormalities and enhanced apoptosis of thymocytes and bone marrow cells in transgenic mice overexpressing Cu/Zn-superoxide dismutase: implications for Down syndrome. EMBO J. 14:4985-93.
Pellmar TC, Hollinden GE, Sarvey JM. (1991) Free radicals accelerate the decay of long-term potentiation in field CA1 of guinea-pig hippocampus. Neuroscience. 44:353-9.
Roveri A, Coassin M, Maiorino M, Zamburlini A, van Amsterdam FT, Ratti E, Ursini F. (1992) Effect of hydrogen peroxide on calcium homeostasis in smooth muscle cells. Arch Biochem Biophys. 297:265-70.
Saito A, Hayashi T, Okuno S, Ferrand-Drake M, Chan PH. (2003) Overexpression of copper/zinc superoxide dismutase in transgenic mice protects against neuronal cell death after transient focal ischemia by blocking activation of the Bad cell death signaling pathway. J Neurosci. 23:1710-8.
Soderling TR, Derkach VA. (2000) Postsynaptic protein phosphorylation and LTP. Trends Neurosci. 23:75-80.
Stanton PK, Sarvey JM. (1984) Blockade of long-term potentiation in rat hippocampal CA1 region by inhibitors of protein synthesis. J Neurosci. 4:3080-8.
Sugawara T, Noshita N, Lewen A, Gasche Y, Ferrand-Drake M, Fujimura M, Morita-Fujimura Y, Chan PH. (2002) Overexpression of copper/zinc superoxide dismutase in transgenic rats protects vulnerable neurons against ischemic damage by blocking the mitochondrial pathway of caspase activation. J Neurosci. 22:209-17
Thibault O, Hadley R, Landfield PW. (2001) Elevated postsynaptic $[Ca^{2+}]_i$ and L-type calcium channel activity in aged hippocampal neurons: relationship to impaired synaptic plasticity. J Neurosci. 21:9744-56.
Thiels E, Urban NN, Gonzalez-Burgos GR, Kanterewicz BI, Barrionuevo G, Chu CT, Oury TD, Klann E. (2000) Impairment of long-term potentiation and associative memory in mice that overexpress extracellular superoxide dismutase. J Neurosci. 20:7631-9.
Vereker E, O'Donnell E, Lynch A, Kelly A, Nolan Y, Lynch MA. (2001) Evidence that interleukin-1beta and reactive oxygen species production play a pivotal role in stress-induced impairment of LTP in the rat dentate gyrus. Eur J Neurosci. 14:1809-19.
Volk T, Hensel M, Kox WJ. (1997) Transient Ca^{2+} changes in endothelial cells induced by low doses of reactive oxygen species: role of hydrogen peroxide. Mol Cell Biochem. 171:11-21.
Wang JH:Kelly PT. (1997) Postsynaptic calcineurin activity downregulates synaptic transmission by weakening intracellular Ca2+ signaling mechanisms in hippocampal CA1 neurons. J Neurosci. 17:4600-11
Winder DG, Mansuy IM, Osman M, Moallem TM, Kandel ER. (1998) Genetic and pharmacological evidence for a novel, intermediate phase of long-term potentiation suppressed by calcineurin. Cell. 92:25-37.
Yang ZW, Zheng T, Wang J, Zhang A, Altura BT, Altura BM. (1999) Hydrogen peroxide induces contraction and raises $[Ca^{2+}]_i$ in canine cerebral arterial smooth muscle: participation of cellular signaling pathways. Naunyn Schmiedebergs Arch Pharmacol. 360:646-53.
Yarom R, Sapoznikov D, Havivi Y, Avraham KB, Schickler M, Groner Y. (1988) Premature aging changes in neuromuscular junctions of transgenic mice with an extra human CuZnSOD gene: a model for tongue pathology in Down's syndrome. J Neurol Sci. 88:41-53.
Yermolaieva O, Brot N, Weissbach H, Heinemann SH, Hoshi T. (2000) Reactive oxygen species and nitric oxide mediate plasticity of neuronal calcium signaling. Proc Natl Acad Sci U S A. 97:448-53.
Zeng H, Chattarji S, Barbarosie M, Rondi-Reig L, Philpot BD, Miyakawa T, Bear MF, Tonegawa S. (2001) Forebrain-specific calcineurin knockout selectively impairs bi-directional synaptic plasticity and working/episodic-like memory. Cell. 107:617-29.

NEURONAL PLASTICITY AND SEIZURE SPREAD IN THE ENTORHINAL CORTEX AND HIPPOCAMPUS OF AMYGDALA KINDLED RATS

U. Heinemann, D. Albrecht, A. Behr, D. von Haebler, and T. Gloveli[1]

1. INTRODUCTION

The entorhinal cortex is the main input and output region for the hippocampus and forms with the DG, the cornu ammonis and the subicular regions a functional complex which is involved in learning and memory of explicit information, stress adaptation and sexually oriented cognitive tasks. With respect to learning and memory the entorhinal cortex- hippocampus complex displays different working modes. For storage of information the structure has to decide whether stored information is novel; before a memory trace is formed. Once information is stored the memory traces are still labile and therefore require consolidation. Stored information must be retrieved. Apart from the novelty detection mode, the storage mode, the consolidation mode and the retrieval mode the entorhinal cortex has also a distributor task as not all information reaching this complex is transferred to the hippocampus. Thus temporally structured information is usually thought to be processed in frontal cortex areas. Indeed a number of such behavioral tasks depend on intact information processing in the entorhinal and perirhinal cortex areas (Zola-Morgan et al., 1989). This requires that there is also a distributor mode which decides where information is stored.

The richness of different working modes and the different functions in which the entorhinal cortex and hippocampus are involved suggests that maldevelopment and lesions in these structures contribute to a wide range of neurological and psychiatric disorders. Indeed, in Alzheimer's disease the first lesions are seen in superficial layers of the perirhinal, postrhinal and entorhinal cortex (Braak and Braak, 1991). Likewise, the entorhinal cortex is also involved in the cognitive deficits which develop in the transition from Parkinson's disease to the Parkinson plus syndrome (Schousboe et al., 1993). In this case the first cells which are affected are located in the deep layers of the entorhinal cortex. Cognitive disturbances are also characteristic for temporal lobe epilepsy where cells in the middle layers of the entorhinal cortex apart from the well known cell death in the hippocampus are known to be affected.

[1] Johannes Müller Institut für Physiologie, Charité Universitätsmedizin Berlin, Humboldt Universität, Tucholskystr. 2, D 10117 Berlin, Germany E-mail: uwe.heinemann@charite.de

The entorhinal cortex seems also to be involved in psychiatric disorders such as anxiety and panic attacks, depression and schizophrenia. In the latter case developmental abnormalities have been noted in a subgroup of patients with clusters of layer II stellate cells being located in middle layers of the entorhinal cortex (Bogerts, 2002).

In this review we summarize some of the principal functions of the entorhinal cortex and hippocampal complex. One of the most devasting diseases involving the entorhinal cortex and hippocampus are the mesial temporal lobe epilepsies. These come in two forms: Mesial temporal lobe epilepsy is frequently associated with cell loss in the hippocampal formation leading to functional alterations on the cellular synaptic and network level (hippocampal sclerosis). By contrast a minority of temporal lobe epilepsies have a lesion outside the hippocampus (non-hippocampal sclerosis). In these cases cell loss and network reorganization is minimal and functional alterations determine the seizure susceptibility. This condition is mimicked by the kindling epilepsy. This form of epilepsy is induced by once daily recurrent stimulation of the amygdala, the perforant path or the hippocampus. In this review we therefore discuss some functional properties of the entorhinal cortex and hippocampus, then describe alterations in synaptic plasticity in the entorhinal cortex hippocampus formation following kindling and alterations which lead to facilitated seizure spread.

2. PROPERTIES OF THE ENTORHINAL CORTEX HIPPOCAMPAL FORMATION

The perirhinal and postrhinal cortex receive in still segregated subfields input from almost any of the association cortex areas (**Fig.1**). Perirhinal and postrhinal cells send their axons into superficial layers of the EC where they interact with projection cells in layer II and III of the entorhinal cortex but also with apical dendrites of the deep layer entorhinal cortex cells (Witter et al., 1989). These receive input from olfactory cortex, the subiculum and cholinergic input from the basal ganglia and the nucleus basalis Meynert as well as from the septum. Layer II stellate cells form the major portion of the perforant path which projects to the dentate gyrus while layer III cells project to the subiculum and in addition to the stratum moleculare of the CA1 area and perhaps also to area CA3 (Dugladze et al., 2001). Deep layer cells in the EC project to thalamic structures, the subiculum, to superficial layers of the EC and surprisingly also to the dentate gyrus. The subiculum has apart from output into the EC also output to a number of different structures in the CNS including the n. accumbens, basal ganglia and thalamic nuclei.

The question arises which of these and additional connectivities in the hippocampus are used for the different working modes. For novelty detection it can be postulated that incoming information is compared with stored information. Structures involved in novelty detection should therefore receive input and output of the hippocampus. The first structure to be considered is layer II in the EC where input from the perirhinal cortex and postrhinal cortex converge with input from the deep layers of the EC which are activated by output information from the hippocampus and subiculum. Likewise the dentate gyrus receives input from layer II cells and output from deep layer cells in the EC. Finally the subiculum might be in a position to do comparison between incoming information and stored information. While at present there is no clear evidence for a function of layer II in novelty detection behavioral and lesion studies point to an important role of the DG (Straube et al., 2003). In vivo recordings in the subiculum have suggested that this structure is also important in novelty detection(Eichenbaum, 1999).

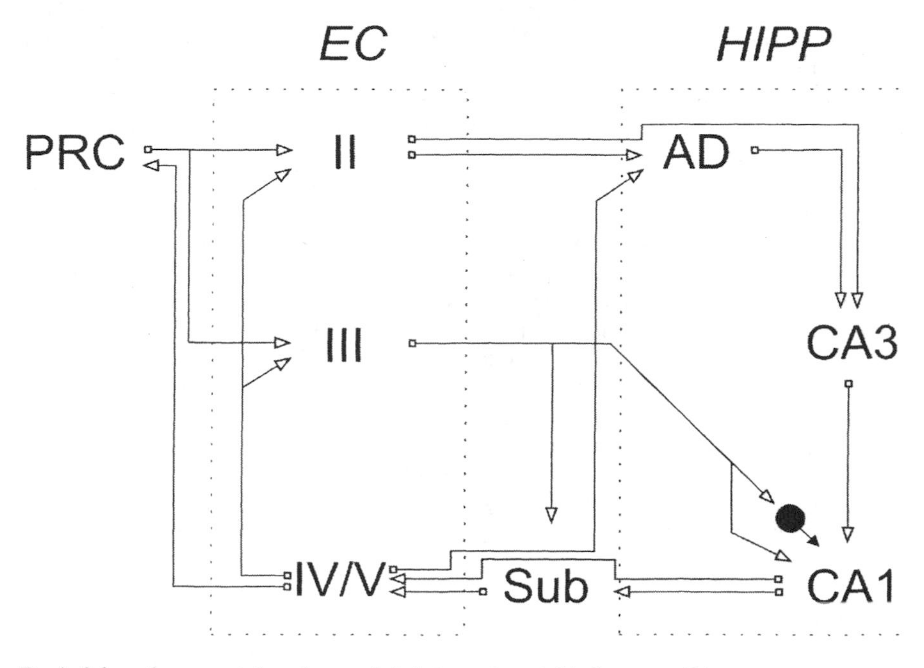

Fig. 1 Schematic representation of connectivity between the entorhinal cortex and hippocampus

For the storage mode most evidence points to an important role for the EC layer II- DG-CA3-CA1 pathway for memory storage. This trisynaptic loop uses glutamate as a fast excitatory neurotransmitter and induction of LTP in the DG and associational synapses within area CA3 and in area CA1 is NMDA receptor dependent (Nicoll, 2003). These prominently postsynaptic forms of LTP are contrasted with a presynaptic form of LTP in the mossy fiber CA3 pyramidal cell synapse where induction and expression of LTP depends on presynaptic mechanisms involving high affinity kainate receptors (Schmitz et al., 2003). The retrieval mode was suggested to depend on activation of layer III cells and activation of neurons in area CA1 and the subiculum but perhaps also in area CA3. Finally consolidation of memory requires activation of stored information either during sleep or during resting states of the mind and requires silencing of disturbing input into the hippocampus proper from the peri-, post- and entorhinal cortex. It might also be postulated that frequent recall serves for memory consolidation. Memory consolidation may under some conditions require transfer of information from the hippocampus into other structures of the CNS while other memory traces can be consolidated within the hippocampus (Squire and Alvarez, 1995).

If the use of connectivities in the entorhinal cortex hippocampal formation are dependent on different activity modes then the question arises which types of neuronal activities correlate with the different working modes. Studies in freely behaving animals have suggested that during explorative behaviors theta and gamma oscillation dominate the ensemble activity as revealed by depth recordings of local field potential recordings (Buzsaki, 2002). The generation of such rhythmic activities requires pacemaking and network properties. Theta oscillations can be induced by near threshold depolarization in perirhinal cortex cells, in deep layer cells projecting to the dentate gyrus and to superficial cells in the EC and in pyramidal cells and some types of interneurons in hippocampal area CA3 and CA1 (Heinemann et al., 2000). At least in stellate cells and pyramidal cells application of carbachole can induce theta oscillations as these agents depolarize the neurons into the membrane potential range required for generation of such oscillations (Weiss et al., 2003; Klink and Alonso, 1997). The interneuronal synchronization of such membrane oscillations is likely dependent on synaptic reset mechanisms involving transient hyperpolarisation of the cells. Gamma oscillations by contrast are dependent on network interactions involving interneuron principal cell interactions and interactions between interneurons (Jefferys et al., 1996). Pharmacologically gamma oscillations can be induced by application of carbachole, kainate, metabotropic glutamate receptor agonists and elevations in potassium concentration. Also short high frequency stimulus trains induce gamma oscillations. While carbachole and kainate induced gamma oscillations have a dominant frequency near 40 Hz (Buhl et al., 2003; Fisahn et al., 1998; Weiss et al., 2003) stimulus and glutamate induced gamma oscillation come usually with higher frequencies (Poschel et al., 2003; Poschel et al., 2002). At present it is an open question whether the same network is accountable for different forms of gamma oscillations or whether subsets of oscillating networks exist. Interestingly, carbachole induced gamma oscillations are sensitive to dopamine and serotonin which were shown to suppress this activity (Weiss et al., 2003) while the effects of these agents on stimulus and glutamate induced gamma oscillations is still unclear.

It is usually assumed that during explorative behavior and therefore theta gamma oscillation storage of information within the hippocampus is facilitated. This requires that at these frequencies stellate cells become readily activated. Indeed, stellate cells present with synaptically induced gamma oscillations when ACH or carbachole is applied focally to deep layers (Gloveli et al., 1999). Stellate cells show a flat synaptic input output curve. However, when repetitive stimulation is used to activate these cells firing probability strongly increases when frequencies above 10 Hz are used (Gloveli et al., 1997a). The facilitated information transfer may depend on resonance properties of these cells (Erchova et al., 2004; Schreiber et al., 2004). Resonance describes the input impedance of a given cell and indicates that at some frequencies the input impedance is higher. The resonance frequency of stellate cells is in the range of 5 to 15 Hz. Whether the resonance frequency is synaptically modifiable and changes over long times is not yet known.

Layer III entorhinal cortex cells projecting to the subiculum, area CA1 and potentially also to area CA3 display low pass filter properties (Erchova et al., 2004) and in response to repetitive stimulation fail to transmit their action potentials to the hippocampus (Gloveli et al., 1997a). They possess a steep input output curve and activate within area CA1 only a small number of pyramidal cells (Empson and Heinemann, 1995). These properties make them very suited for recall of information.

Memory consolidation is thought to be facilitated during resting states of the mind and during sleep. REM sleep is characterized by theta and gamma oscillations and evidence has been presented that neuronal activity patterns observed during explorative behavior are played during REM sleep in a phase dependent manner as during explorative behavior (Wilson and McNaughton, 1994). It is at present not clear though whether phase related stimulation has long lasting effects on synaptic plasticity. On the other hand theta burst stimulation mimicking this physiological activity is known to be well suitable to induce LTP in different subregions of the hippocampal formation.

Resting states of the mind and slow wave sleep are known to be characterized by low frequency EEG activity as well as low discharge frequencies of neurons in the entorhinal cortex hippocampal formation. On the background of this activity sharp wave ripple complexes are observed (Buzsaki, 1986). These are 30 – 80 ms long field potential transients which occur in the hippocampus, the subiculum and the entorhinal cortex. They are superimposed by high frequency oscillations of around 200 Hz (Chrobak and Buzsáki, 1996). During such events pyramidal cells fire in synchrony with the extracellular field potential transients (Maier et al., 2003). Since 200 Hz stimulus trains are very efficient in inducing LTP, it was suggested that they are potentially involved in memory consolidation (Buzsáki et al., 1987). Sharp wave ripple complexes can also be observed in hippocampal slices of rodents. They are prominently generated in ventral portions of the hippocampus. The underlying physiological mechanisms could therefore be investigated. They are generated in area CA3, propagate to area CA1 and the subiculum (Maier et al., 2003). Both inhibitory and excitatory synaptic potential accompany these events. Interestingly in slices which present with sharp wave ripple complexes it is difficult to induce LTP presumably as a consequence of saturating LTP during generation of such events (Colgin et al., 2004).

Long-term potentiation has for a long time been considered as one of the cellular equivalents of learning and memory. Different patterns of repetitive neuronal activation were shown to induce LTP. These include high frequency stimulation with 100 and 200 Hz, theta burst stimulation mimicking the behavioral situation during REM sleep and explorative behavior. It was later recognized that appropriate timing of EPSPs with retrogradely propagated action potentials can also induce LTP (Paulsen and Sejnowski, 2000). As many neuromodulatory systems are involved in learning and memory their function on LTP induction was also studied. John Sarvey's lab (Dahl et al., 1990; Stanton and Sarvey, 1985a; Stanton and Sarvey, 1985b) was important in showing that norepinephrine (NE) is required for induction of LTP in the DG and that this effect is mediated by ß receptors for norepinephrine. It is also of interest that NE induced long-lasting potentiation is dependent on NMDA receptor activation (Stanton et al., 1989). The role of dopamine seems to be limited to area CA1, where norepinephrine no longer facilitates generation of LTP.

3. SEIZURE SUSCEPTIBILITY OF THE ENTORHINAL CORTEX AND SEIZURE SPREAD

In vitro studies on the normal interaction between the entorhinal cortex and hippocampus from adult rats have revealed that seizure susceptibility is larger in the EC than in the hippocampus which tends to develop only interictal discharges and short ictal events. In adult rats seizure like events spread from the EC to the subiculum but rarely fully recruit area CA1 and the DG (Buchheim et al., 1999; Dreier and Heinemann, 1991; Stanton et al., 1987). This limitation of spread seems to relate to the extensive feedforward and feedback inhibitory network in the DG (Buckmaster and Schwartzkroin, 1995) and the prominent activation of inhibitory interneurons in area CA1 (Empson and Heinemann, 1995). The DG was therefore assumed to play a gating role in transfer of information from layer II of the EC to the DG (Heinemann et al., 1992). Which properties of the DG are responsible for the limitation of seizure spread? Studies with J. Sarvey revealed powerful paired pulse inhibition in the DG (Rausche et al., 1989). Intracellular recordings revealed a prominent GABA A mediated IPSP followed by a slow GABA B mediated IPSP. Upon repetitive stimulation this GABA B mediated IPSP became reduced. Apart from the general accepted concept that GABA B autoreceptors lead to the decline of slow IPSP's we were able to show that potassium released from granule cells during the late GABA B mediated IPSP shifts its reversal potential into depolarizing direction. These findings suggest that part of the synaptic depolarization during repetitive stimulation is mediated by activation of GABA B receptors. In order to determine which inhibitory mechanism is involved in controlling seizure spread through the DG to area CA3 we determined whether application of the GABA A receptor antagonist bicuculline would facilitate seizure spread from the EC through the DG to area CA3. Surprisingly this was not the case. By contrast when baclofen was applied seizure spread was facilitated suggesting that inhibitory cells in the hippocampus are innervated by other inhibitory cells and that by disinhibition seizure spread could be initiated or that potassium mediated loss in efficacy of slow synaptic inhibition contributes to seizure spread.

4. KINDLING AND SEIZURE SPREAD FROM THE ENTORHINAL CORTEX TO THE HIPPOCAMPUS

Kindling affects the interaction between the entorhinal cortex and hippocampus. In amygdala kindled animals we found that after kindling low Mg^{2+} induced seizure like events propagated more readily through the dentate gyrus to the hippocampus while also readily recruiting the subiculum into seizure activity (Behr et al., 1996; Behr et al., 1998). Even when seizure activity was locally induced in the entorhinal cortex by focal application of bicuculline and elevated K^+ seizure spread in kindled animals was facilitated through the dentate gyrus.

Interestingly, data from different labs indicates upregulation of slow IPSP´s after kindling (Oliver and Miller, 1985; Otis and Mody, 1993) and a subunit change in synaptic GABA receptors rendering them more sensitive to Zn which can be released from mossy fiber terminals during seizure like events (Buhl et al., 1996). Further alterations in granule cells are the seizure induced upregulation of the GABA synthesizing enzyme GAD67 (Schwarzer and Sperk, 1995). This enzyme is constitutively expressed in granule cells but becomes strongly upregulated after seizures (Sloviter et al., 1996). Present evidence suggests that GABA is co-released with glutamate after a seizure at mossy fiber terminals onto CA3 pyramidal cells (Gutierrez and Heinemann, 2001; Gutierrez, 2002; Lamas et al., 2001). We therefore stimulated the DG and recorded synaptic potentials from CA3 pyramidal cells in brain slices from kindled and control rats. In both preparations, DG stimulation caused excitatory postsynaptic potential (EPSP)/inhibitory postsynaptic potential (IPSP) sequences. These potentials could be completely blocked by glutamate receptor antagonists in control rats, while in the kindled rats, a bicuculline-sensitive fast IPSP remained, with onset latency similar to that of the control EPSP. Interestingly, this IPSP disappeared 1 month after the last seizure. When synaptic responses were evoked by high-frequency stimulation, EPSPs in normal rats readily summate to evoke action potentials. In slices from kindled rats, a summation of IPSPs overrides that of the EPSPs and reduces the probability of evoking action potentials. However, at frequencies above 50 Hz this inhibitory effect fades presumably due to K^+ dependent reduced efficacy of inhibition. Our data show that kindling induces functionally relevant activity-dependent expression of fast inhibition onto pyramidal cells, coming from the DG that can limit CA3 excitation in a frequency-dependent manner. Interestingly, contribution of GABA to synaptic transmission in this pathway is characteristic for juvenile animals and it can be provoked by a single seizure and even under *in vitro* conditions by protocols which induce LTP (Gutierrez and Heinemann, 2001; Gutierrez, 2002). This transient inhibitory effect on CA3 cells may be in contrast to activation of networks in the dentate gyrus. Increased release of GABA from mossy fiber recurrent axon collaterals will lead to disinhibition by reducing activation of interneurons and mossy cells and could thereby contribute to disinhibition.

The granule cell excitability is not only controlled by phasic inhibition. In these cells very prominent tonic GABAergic currents have been reported. These are extrasynaptic receptors which have a high affinity for GABA but low affinity for the GABA A antagonist GABAzine and are thought to contain delta subunits (Jensen et al., 2003; Nusser and Mody, 2002; Soltesz et al., 1995; Stell et al., 2003; Stell and Mody, 2002). Whether tonic inhibition is altered in dentate granule cells or other cells after kindling is not yet known. In the kindled hippocampus an upregulation of kainate receptors was shown to have a strong depressing effect on GABA mediated inhibition. This would effectively remove inhibition and thus contribute to the facilitated seizure spread.

The kindling induced alteration of gating in the DG may, however, not solely depend on changes in the inhibitory network. We therefore tested for the relative contribution of NMDA, AMPA and kainate receptors to frequency facilitation in transmission from the perforant path to granule cells. This may in part be due to alterations in presynaptic release properties. Indeed we have observed an upregulation of the synaptobrevin-synaptophysin complex following kindling both in cortex and hippocampus (Hinz et al., 2001). This may provide for a reserve pool of synaptobrevin, promoting enhanced synaptic transmission in the kindled state. However, there was no evidence for altered synaptically evoked AMPA currents in kindled animals (Behr et al., 2000a). Several lines of evidence indicate a substantial contribution of kainate receptors to temporal lobe seizures. The activation of kainate receptors located on hippocampal inhibitory interneurons was shown to reduce GABA release (Kullmann, 2001). A reduced GABA release secondary to kainate receptor activation could contribute to enhanced seizure susceptibility. As the dentate gyrus serves a pivotal gating function in the spread of limbic seizures, we (Behr et al., 2002) tested the role of kainate receptors in the regulation of GABA release in the dentate gyrus of control and kindled animals. Application of glutamate (100 μM) in the presence of the NMDA receptor antagonist d-APV and the AMPA receptor antagonist, SYM 2206 caused a slight depression of evoked monosynaptic inhibitory postsynaptic currents (IPSCs) in control, but a substantial decrease in kindled dentate granule cells. The observation that kainate receptor activation altered paired-pulse depression and reduced the frequency of TTX-insensitive miniature IPSCs without affecting their amplitude is consistent with a presynaptic action of glutamate via kainate receptors on the inhibitory terminal to reduce GABA release. In kindled preparations, neither glutamate (100 μM) nor kainate (10 μM) applied in a concentration known to depolarize hippocampal interneurons led to an increase of the TTX-sensitive spontaneous IPSC frequency nor to changes of the postsynaptic membrane properties. Consistently, the inhibitory effect on evoked IPSCs was not affected by presence of the GABA-B receptor antagonist, CGP55845A, excluding a depression by enhanced release of GABA acting on presynaptic GABA-B receptors. The enhanced inhibition of GABA release following presynaptic kainate receptor activation favors a use-dependent hyperexcitability in the epileptic dentate gyrus.

Apart from this effect on use-dependent disinhibition there are two transient effects which may also contribute to facilitated kindling. One is the transient upregulation of NMDA receptors (Behr et al., 2000b; Mody and Heinemann, 1987; Mody et al., 1988) which contributes to frequency potentiation of EPSP´s. In the DG medial entorhinal cortex afferents innervate the middle portion of the granule cell dendrites while input to the outer third of the granule cell dendrites comes mostly from the lateral entorhinal cortex. Activation of the MEC input to DG granule cells induces already at resting potential an NMDA dependent postsynaptic potential while lateral entorhinal cortex input does not. After kindling NMDA receptors contribute more effectively to synaptic transmission and NMDA dependent components are evoked at resting membrane potential also when the lateral part of the perforant path was stimulated. NMDA receptors depend on expression of NMDA NR1 subunits and expression of NMDA 2A-D subunits. The NR2C and 2D subunits convey a lower channel conductance (Fleidervish et al., 1998). Importantly they can already be activated at resting membrane potential. This suggests that the increased NMDA component in the EPSP evoked by perforant path activation is due to an upregulation of NR2C or 2D.

We (Behr et al., 2000b) further investigated dentate gyrus field potentials and granule cell excitatory postsynaptic potentials (EPSPs) following high-frequency stimulation (10-100 Hz) of the lateral perforant path in kindled rats. Although control slices showed steady EPSP depression at frequencies greater than 20 Hz, slices taken from animals 48 h after the last seizure presented pronounced EPSP facilitation at 50 and 100 Hz, followed by steady depression. However, 28 days after kindling, the EPSP facilitation was no longer detectable. Using the specific N-methyl-D-aspartate (NMDA) and RS-alpha-amino-3-hydroxy-5-methyl-4-isoxazolepropionic acid (AMPA) receptor antagonists 2-amino-5-phosphonovaleric acid and SYM 2206, we examined the time course of alterations in glutamate receptor-dependent synaptic currents that parallel transient EPSP facilitation. Forty-eight hours after kindling, the fractional AMPA and NMDA receptor-mediated excitatory postsynaptic current (EPSC) components shifted dramatically in favor of the NMDA receptor-mediated response. Four weeks after kindling, however, AMPA and NMDA receptor-mediated EPSCs reverted to control-like values. Although the granule cells of the dentate gyrus contain mRNA-encoding for kainate receptors, neither single nor repetitive perforant path stimuli evoked kainate receptor-mediated EPSCs in control or in kindled rats. The enhanced excitability of the kindled dentate gyrus 48 h after the last seizure, as well as the breakdown of its gating function, appear to result from transiently enhanced NMDA receptor activation that provides significantly slower EPSC kinetics than those observed in control slices and in slices from kindled animals with a 28-day seizure-free interval. Therefore, NMDA receptors seem to play a critical role in the acute throughput of seizure activity and in the induction of the kindled state but not in the persistence of enhanced seizure susceptibility. Changes in metabotropic glutamate receptor expression may also contribute to facilitated seizure spread (Friedl et al., 1999).

5. ALTERATIONS IN NOREPINEPHRINE INDUCED LTP AFTER KINDLING

Mechanisms of action of norepinephrine (NE) on dentate gyrus granule cells were studied in rat hippocampal slices using extra- and intracellular recordings and measurements of stimulus and amino acid-induced changes in extracellular Ca^{2+} and K^+ concentration. Bath application of NE (10-50 μM) induced long-lasting potentiation of perforant path evoked potentials, and markedly enhanced high-frequency stimulus-induced Ca^{2+} influx and K^+ efflux, actions blocked by beta-receptor antagonists and mimicked by beta agonists. Enhanced Ca^{2+} influx was primarily postsynaptic, since presynaptic delta $[Ca^{2+}]_o$ in the stratum molecularé synaptic field was not altered by NE. Interestingly, the potentiation of both ionic fluxes and evoked population potentials were antagonized by the N-methyl-D-aspartate (NMDA) receptor antagonist 2-amino-5-phosphonovalerate (APV). Furthermore, NE selectively enhanced the ionic signals and slow field potentials elicited by iontophoretically applied NMDA, but not those induced by the excitatory amino acid quisqualate. These results suggest that granule cell influx of Ca^{2+} through NMDA ionophores is enhanced by NE via beta-receptor activation. In intracellular recordings, NE depolarized granule cells (4.8 ± 1.1 mV), and increased input resistance (RN) by 34 ± 6.5%. These actions were also blocked by either the beta-antagonist propranolol or specific beta 1-blocker metoprolol. Moreover, the depolarization and RN increase persisted for long periods (93 ± 12 min) after NE washout. In contrast, while NE, in the presence of APV, still depolarized granule cells and increased RN, APV made these actions quickly reversible upon NE washout (16 ± 9 min). This suggested that NE induction of long-term, but not short-term, plasticity in the dentate gyrus requires NMDA receptor activation. NE may be enhancing granule cell firing by some combination of blockade on the late Ca^{2+}-activated K^{+} conductance and depolarization of granule cells, both actions that can bring granule cells into a voltage range where NMDA receptors are more easily activated. Furthermore, NE also elicited activity-independent long-lasting depolarization and R_N increases, which required functional NMDA receptors to persist. Whether this involves NR2A/2B or NR2C/2D receptors is unknown.

Following these studies we tested for the actions of norepinephrine (NE) in the dentate gyrus before and after kindling-induced epilepsy. NE, acting on beta 1-receptors, depolarized granule cells, increased input resistance, firing and influx of Ca^{2+} in response to repetitive stimulation, and elicited long-lasting potentiation of synaptic potentials. In addition, NE acting via alpha 1-receptors, attenuated Ca^{2+}-dependent regenerative potentials. After kindling-induced plasticity, there were marked reductions in all these effects of NE on granule cells, changes likely to influence kindling-induced seizures, protecting against further enhancement of excitability once plasticity is in place.

6. KINDLING INDUCED ALTERATIONS OF ENTORHINAL CORTEX PROPERTIES

Alterations in kindling do not only affect the hippocampus but also alters the amygdala and input of the amygdala to the entorhinal cortex. This was found in a recent study where we analyzed the neuropathological consequences of kindling with a sensitive silver-staining method for the visualization of damaged neurons and Nissl staining for the estimation of the neuronal densities in different limbic areas (Bohlen et al., 2004).

Amygdala-kindled animals had reduced cell density in the amygdala and increased density of fragments of degenerated axons both ipsi- and contralateral to the stimulation site. Reduced neuronal density and the occurrence of degenerated axons in kindled animals were more prominent in the ipsilateral than in the contra lateral hemisphere. In addition, more degenerated axons were found in entorhinal and perirhinal cortical structures of kindled than sham-operated animals. These results indicate that kindling induced morphological alterations that were not restricted to either the ipsilateral hemisphere or the stimulated region. With respect to the entorhinal cortex we found a particularly interesting loss of innervation from the amygdala and a somewhat reduced cell number in layer III of the entorhinal cortex which was less marked than the findings of cell loss in the pilocarpine model and kainate model of temporal lobe epilepsy (Du et al., 1993; Du et al., 1995).

We (Gloveli et al., 1998; Gloveli et al., 2003) therefore began to study alterations in entorhinal cortex layer III cells which project to the subiculum and to area CA1. We studied the effect of kindling on the frequency-dependent information transfer from the entorhinal cortex layer III to area CA1 in vitro. In control rats repetitive synaptic activation of layer III projection cells resulted in a frequency dependent depression of the synaptic transfer of action potentials to the hippocampus. One-to-two-days after kindling this effect was strongly reduced. Although no substantial change in synaptic inhibition upon single electrical stimulation was detected in kindled rats, there was a significant depression in the prolonged inhibition (Gloveli et al., 1997b) following high frequency stimulation.

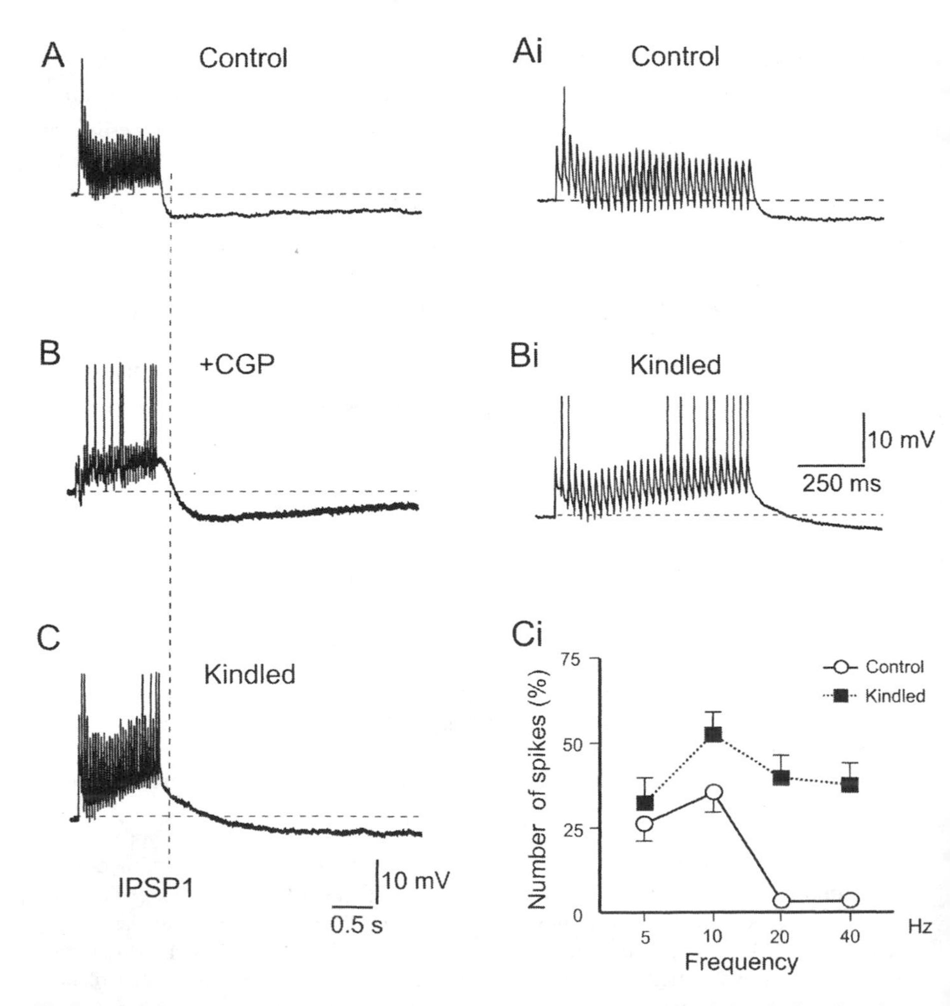

Fig.2 GABA B dependent alterations in frequency habituation and kindling induced alterations in frequency habituations in entorhinal cortex layer III pyramidal cells. Note the early part of stimulus induced after-hyperpolarisation (A) is missing after application of a specific GABA B blocker and also in a similar neuron in kindled entorhinal cortex. Note also change in frequency dependent generation of action potentials in normal and kindled entorhinal cortex cells.

In kindled animals, paired-pulse depression (PPD) of stimulus-evoked IPSCs in layer III neurons was significantly stronger than in control rats. The increase of PPD is most likely caused by an increased presynaptic GABA(B) receptor-mediated auto inhibition. In kindled animals activation of presynaptic GABA(B) receptors by baclofen (10 μM) suppressed monosynaptic IPSCs significantly more than in control rats. In contrast, activation of postsynaptic GABA(B) receptors by baclofen was accompanied by comparable changes of the membrane conductance in both animal groups. Thus, in kindled animals activation of the layer III-CA1 pathway is facilitated by an increased GABA(B) receptor-mediated auto inhibition leading to an enhanced activation of the monosynaptic EC-CA1 pathway.

We therefore investigated further the effects which high frequency activation of the perforant path portion from EC layer III to CA1 would have on the plasticity in this area. We found that stimulation of the PP would induce a potentiation of evoked responses in area CA1. Interestingly high frequency stimulation had a suppressant effect on Schaffer collateral evoked field EPSP's and intracellularly recorded EPSP's. This activation protocol could in part reverse LTP induced by high frequency stimulation of the stratum radiatum. This suggests that memory deficits in patients with temporal lobe epilepsy may have an active component mediated by altered transmission properties of the entorhinal cortex layer III cells to the hippocampal field CA1 (Heinemann et al, in preparation).

Following kindling we also noted enhanced muscarinic synaptic transmission in CA3 pyramidal cells and an increased sensitivity of the entorhinal cortex and hippocampus to acetylcholine and carbachole in induction of epileptiform discharges (Gutierrez and Heinemann, 2001). This raises the question which effects such an upregulation of cholinergic transmission might have on synaptic plasticity. The same applies to the transient downregulation of K conductances underlying slow after hyperpolarisations.

7. CONCLUSION

Our findings add to the widespread observations that kindling affects a wide spectrum of cellular and synaptic functions within mesial temporal lobe structures. Some of these alterations are lasting while many of them are transient in nature. On a network level the alterations in the DG favor seizure spread from the entorhinal cortex into the hippocampus. In addition they have effects on synaptic plasticity which might contribute to cognitive impairments characteristic for patients which suffer from temporal lobe epilepsy.

Kindling is thought to model one condition of human temporal lobe epilepsy where hippocampal and extrahippocampal cell loss is minimal. In these patients the functional alterations in cellular properties seem to be more important than the structural reorganization which is associated with hippocampal sclerosis. However, not all alterations which we know from kindled animals also occur in the tissue of patients suffering from nonAHS. Thus the down regulation of calbindin 28 which is typical for AHS patients and for models of temporal lobe epilepsy induced by a prior status epilepticus and kindling is not observed in tissue from non AHS patients.

Our findings regarding effects of up regulation of GABA B receptors in a model of TLE also explain why GABA B agonists such as baclofen had disappointing effects in the treatment of focal epilepsies. This suggests that preclinical monitoring of potential anticonvulsants should involve studies in chronic epileptic animals and where possible also in human epileptic tissue.

Another interesting aspect in the comparison between LTP and kindling is the fact that kindling is a form of pathogenic plasticity induced by stimuli which are similarly structured as stimuli which induce LTP. The major difference is that kindling stimuli are tested for threshold of induction of local afterdischarges. By contrast this is not the case with in vivo induction of LTP. Recent data suggest that the overall excitability of neurons after LTP induction does not increase (Dragoi et al., 2003). This suggests that homeostatic mechanisms exist which limit progression. This might depend on induction of LTP in subclasses of interneurons (Alle et al., 2001). By contrast local seizure like events may disrupt the blood brain barrier thereby leading to exudation of proteins in the extracellular space, edema. In this context it is important that opening of the blood brain barrier can indeed induce an epileptic focus (Seiffert et al., 2004). Whether this or immunological factors or local cell death and reorganization of neuronal networks at the site of stimulation are involved in the pathogenic plasticity underlying kindling is still an enigma which awaits experimental clarification.

8. REFERENCES

Alle H, Jonas P, Geiger JR (2001) PTP and LTP at a hippocampal mossy fiber-interneuron synapse. Proc Natl Acad Sci U S A 98: 14708-14713.

Behr J, Gebhardt C, Heinemann U, Mody I (2002) Kindling enhances kainate receptor-mediated depression of GABAergic inhibition in rat granule cells. Eur J Neurosci 16: 861-867.

Behr J, Gloveli T, Gutiérrez R, Heinemann U (1996) Spread of low Mg^{2+} induced epileptiform activity from the rat entorhinal cortex to the hippocampus after kindling studied in vitro. Neurosci Lett 216: 41-44.

Behr J, Heinemann U, Mody I (2000a) Glutamate receptor activation in the kindled dentate gyrus. Epilepsia 41: S100-S103.

Behr, J., Heinemann, U., and Mody, I. Kindling induces transient NMDA receptor-mediated facilitation of high frequency input in the rat dentate gyrus. 2000b.

Behr J, Lyson KJ, Mody I (1998) Enhanced propagation of epileptiform activity through the kindled dentate gyrus. J Neurophysiol 79: 1726-1732.

Bogerts, B. The neuropathology of schizophrenic diseases: historical aspects and present knowledge. Eur.Arch.Psychiatry Clin.Neurosci. 249 Suppl 4, 2-13. 2002.

Bohlen HO, Schulze K, Albrecht D (2004) Amygdala-kindling induces alterations in neuronal density and in density of degenerated fibers. Hippocampus 14: 311-318.

Braak H, Braak E (1991) Neuropathological stageing of Alzheimer-related changes. Acta Neuropathol Berl 82: 239-259.

Buchheim K, Schuchmann S, Siegmund H, Gabriel H-J, Heinemann U, Meierkord H (1999) Intrinsic optical signal measurement reveal characteristic features during different forms of spontaneous neuronal hyperactivity associated with ECS shrinkage *in vitro*. Eur J Neurosci 11: 1877-1882.

Buckmaster PS, Schwartzkroin PA (1995) Interneurons and inhibition in the dentate gyrus of the rat *in vivo*. J Neurosci 15: 774-789.

Buhl DL, Harris KD, Hormuzdi SG, Monyer H, Buzsaki G (2003) Selective impairment of hippocampal gamma oscillations in connexin-36 knock-out mouse in vivo. J Neurosci 23: 1013-1018.

Buhl EH, Otis TS, Mody I (1996) Zinc-induced collapse of augmented inhibition by GABA in a temporal lobe epilepsy model. Science 271: 369-373.

Buzsaki G (1986) Hippocampal sharp waves: their origin and significance. Brain Res 398: 242-252.

Buzsaki G (2002) Theta oscillations in the hippocampus. Neuron 33: 325-340.

Buzsáki G, Haas HL, Anderson EG (1987) Long-term potentiation induced by physiologically relevant stimulus patterns. Brain Res 435: 331-333.

Chrobak JJ, Buzsáki G (1996) High-frequency oscillations in the output networks of the hippocampal-entorhinal axis of the freely behaving rat. J Neurosci 16: 3056-3066.

Colgin LL, Kubota D, Jia Y, Rex CS, Lynch G (2004) Long-term potentiation is impaired in rat hippocampal slices that produce spontaneous sharp waves. J Physiol 558: 953-961.

Dahl D, Burgard EC, Sarvey JM (1990) NMDA receptor antagonists reduce medial, but not lateral, perforant path-evoked EPSPs in dentate gyrus of rat hippocampal slice. Exp Brain Res 83: 172-177.

Dragoi G, Harris KD, Buzsaki G (2003) Place representation within hippocampal networks is modified by long-term potentiation. Neuron 39: 843-853.

Dreier JP, Heinemann U (1991) Regional and time dependent variations of low magnesium induced epileptiform activity in rat temporal cortex. Exp Brain Res 87: 581-596.

Du F, Eid T, Lothman EW, Köhler C, Schwarcz R (1995) Preferential neuronal loss in layer III of the medial entorhinal cortex in rat models of temporal lobe epilepsy. J Neurosci 15: 6301-6313.
Du F, Whetsell WO, Jr., Abou-Khalil B, Blumenkopf B, Lothman EW, Schwarcz R (1993) Preferential neuronal loss in layer III of the entorhinal cortex in patients with temporal lobe epilepsy. Epilepsy Res 16: 223-233.
Dugladze T, Heinemann U, Gloveli T (2001) Entorhinal cortex projection cells to the hippocampal formation in vitro. Brain Res 905: 224-231.
Eichenbaum H (1999) The hippocampus: The shock of the new. Curr Biol 9: R482-R484.
Empson RM, Heinemann U (1995) The perforant path projection to hippocampal area CA1 in the rat hippocampal-entorhinal cortex combined slice. J Physiol (Lond) 484: 707-729.
Erchova I, Kreck G, Heinemann U, Herz A (2004) Dynamics of rat entorhinal cortex layer II/III cells: characteristics of membrane potential resonance at rest predict oscillation properties near threshold. J Physiol ..
Fisahn A, Pike FG, Buhl EH, Paulsen O (1998) Cholinergic induction of network oscillations at 40 Hz in the hippocampus *in vitro*. Nature 394: 186-189.
Fleidervish IA, Binshtok AM, Gutnick MJ (1998) Functionally distinct NMDA receptors mediate horizontal connectivity within layer 4 of mouse barrel cortex. Neuron 21: 1055-1065.
Friedl M, Clusmann H, Kral T, Dietrich D, Schramm J (1999) Analysing metabotropic glutamate group III receptor mediated modulation of synaptic transmission in the amygdala-kindled dentate gyrus of the rat. Brain Res 821: 117-123.
Gloveli T, Behr J, Dugladze T, Kokaia Z, Kokaia M, Heinemann U (2003) Kindling alters entorhinal cortex-hippocampal interaction by increased efficacy of presynaptic GABA(B) autoreceptors in layer III of the entorhinal cortex. Neurobiology of Disease 13: 203-212.
Gloveli T, Egorov AV, Schmitz D, Heinemann U, Müller W (1999) Carbachol-induced changes in excitability and $[Ca^{2+}]_i$ signalling in projection cells of medial entorhinal cortex layers II and III. Eur J Neurosci 11: 3626-3636.
Gloveli T, Schmitz D, Empson RM, Heinemann U (1997a) Frequency-dependent information flow from the entorhinal cortex to the hippocampus. J Neurophysiol 78: 3444-3449.
Gloveli T, Schmitz D, Heinemann U (1997b) Prolonged inhibitory potentials in layer III projection cells of the rat medial entorhinal cortex induced by synaptic stimulation *in vitro*. Neuroscience 80: 119-131.
Gloveli T, Schmitz D, Heinemann U (1998) Interaction between superficial layers of the entorhinal cortex and the hippocampus in normal and epileptic temporal lobe. Epilepsy Res 32: 183-193.
Gutierrez R (2002) Activity-dependent expression of simultaneous glutamatergic and GABAergic neurotransmission from the mossy fibers in vitro. J Neurophysiol 87: 2562-2570.
Gutierrez R, Heinemann U (2001) Kindling induces transient fast inhibition in the dentate gyrus-CA3 projection. Eur J Neurosci 13: 1371-1379.
Heinemann U, Beck H, Dreier JP, Ficker E, Stabel J, Zhang CL (1992) The dentate gyrus as a regulated gate for the propagation of epileptiform activity. In: The dentate gyrus and its role in seizures (Ribak CE, Gall CM, Mody I, eds), pp 273-280. Amsterdam: Elsevier Science Publishers BV.
Heinemann U, Schmitz D, Eder C, Gloveli T (2000) Properties of entorhinal cortex projection cells to the hippocampal formation. Ann N Y Acad Sci 911: 112-126.
Hinz B, Becher A, Mitter D, Schulze K, Heinemann U, Draguhn A, Ahnert-Hilger G (2001) Activity-dependent changes of the presynaptic synaptophysin- synaptobrevin complex in adult rat brain. Eur J Cell Biol 80: 615-619.
Jefferys JGR, Traub RD, Whittington MA (1996) Neuronal networks for induced '40 Hz' rhythms. Trends Neurosci 19: 202-208.
Jensen K, Chiu CS, Sokolova I, Lester HA, Mody I (2003) GABA transporter-1 (GAT1)-deficient mice: differential tonic activation of GABAA versus GABAB receptors in the hippocampus. J Neurophysiol 90: 2690-2701.
Klink R, Alonso A (1997) Muscarinic modulation of the oscillatory and repetitive firing properties of entorhinal cortex layer II neurons. J Neurophysiol 77: 1813-1828.
Kullmann DM (2001) Presynaptic kainate receptors in the hippocampus: slowly emerging from obscurity. Neuron 32: 561-564.
Lamas M, Gomez-Lira G, Gutierrez R (2001) Vesicular GABA transporter mRNA expression in the dentate gyrus and in mossy fiber synaptosomes. Brain Res Mol Brain Res 93: 209-214.
Maier N, Nimmrich V, Draguhn A (2003) Cellular and network mechanisms underlying spontaneous sharp wave-ripple complexes in mouse hippocampal slices. J Physiol 550: 873-887.
Mody I, Heinemann U (1987) NMDA receptors of dentate gyrus granule cells participate in synaptic transmission following kindling. Nature 326: 701-704.
Mody I, Stanton PK, Heinemann U (1988) Activation of *N*-methyl-D-aspartate receptors parallels changes in cellular and synaptic properties of dentate gyrus granule cells after kindling. J Neurophysiol 59: 1033-1054.
Nicoll RA (2003) Expression mechanisms underlying long-term potentiation: a postsynaptic view. Philos Trans R Soc Lond [Biol] 358: 721-726.

Nusser Z, Mody I (2002) Selective modulation of tonic and phasic inhibitions in dentate gyrus granule cells. J Neurophysiol 87: 2624-2628.

Oliver MW, Miller JJ (1985) Alterations of inhibitory processes in the dentate gyrus following kindling-induced epilepsy. Exp Brain Res 57: 443-447.

Otis TS, Mody I (1993) Three mechanisms for increased GABAergic inhibition after kindling-induced epilepsy. Soc Neurosci Abstr 19: 1267.

Paulsen O, Sejnowski TJ (2000) Natural patterns of activity and long-term synaptic plasticity. Curr Opin Neurobiol 10: 172-179.

Poschel B, Draguhn A, Heinemann U (2002) Glutamate-induced gamma oscillations in the dentate gyrus of rat hippocampal slices. Brain Res 938: 22-28.

Poschel B, Heinemann U, Draguhn A (2003) High frequency oscillations in the dentate gyrus of rat hippocampal slices induced by tetanic stimulation. Brain Res 959: 320-327.

Rausche G, Sarvey JM, Heinemann U (1989) Slow synaptic inhibition in relation to frequency habituation in dentate granule cells of rat hippocampal slices. Exp Brain Res 78: 233-242.

Schmitz D, Mellor J, Breustedt J, Nicoll RA (2003) Presynaptic kainate receptors impart an associative property to hippocampal mossy fiber long-term potentiation. Nat Neurosci 6: 1058-1063.

Schousboe A, Bachevalier J, Braak H, Heinemann U, Nitsch R, Schröder H, Wetmore C (1993) Structural correlates and cellular mechanisms in entorhinal- hippocampal dysfunction. Hippocampus 3 Suppl.: 293-302.

Schreiber S, Erchova I, Heinemann U, Herz AV (2004) Subthreshold resonance explains the frequency-dependent integration of periodic as well as random stimuli in the entorhinal cortex. J Neurophysiol 92: 408-415.

Schwarzer C, Sperk G (1995) Hippocampal granule cells express glutamic acid decarboxylase- 67 after limbic seizures in the rat. Neuroscience 69: 705-709.

Seiffert E, Dreier JP, Ivens S, Bechmann I, Tomkins O, Heinemann U, Friedman A (2004) Lasting blood-brain barrier disruption induces epileptic focus in the rat somatosensory cortex. J Neurosci 24: 7829-7836.

Sloviter RS, Dichter MA, Rachinsky TL, Dean E, Goodman JH, Sollas AL, Martin DL (1996) Basal expression and induction of glutamate decarboxylase and GABA in excitatory granule cells of the rat and monkey hippocampal dentate gyrus. J Comp Neurol 373: 593-618.

Soltesz I, Smetters DK, Mody I (1995) Tonic inhibition originates from synapses close to the soma. Neuron 14: 1273-1283.

Squire LR, Alvarez P (1995) Retrograde amnesia and memory consolidation: a neurobiological perspective. Curr Opin Neurobiol 5: 169-177.

Stanton PK, Jones RSG, Mody I, Heinemann U (1987) Epileptiform activity induced by lowering extracellular [Mg^{2+}] in combined hippocampal-entorhinal cortex slices: modulation by receptors for norepinephrine and N-methyl-D-aspartate. Epilepsy Res 1: 53-62.

Stanton PK, Mody I, Heinemann U (1989) A role for N-methyl-D-aspartate receptors in norepinephrine-induced long-lasting potentiation in the dentate gyrus. Exp Brain Res 77: 517-530.

Stanton PK, Sarvey JM (1985a) Depletion of norepinephrine, but not serotonin, reduces long-term potentiation in the dentate gyrus of rat hippocampal slices. J Neurosci 5: 2169-2176.

Stanton PK, Sarvey JM (1985b) The effect of high-frequency electrical stimulation and norepinephrine on cyclic AMP levels in normal versus norepinephrine-depleted rat hippocampal slices. Brain Res 358: 343-348.

Stell BM, Brickley SG, Tang CY, Farrant M, Mody I (2003) Neuroactive steroids reduce neuronal excitability by selectively enhancing tonic inhibition mediated by delta subunit-containing GABAA receptors. Proc Natl Acad Sci U S A 100: 14439-14444.

Stell BM, Mody I (2002) Receptors with different affinities mediate phasic and tonic GABA(A) conductances in hippocampal neurons. J Neurosci 22: RC223.

Straube T, Korz V, Balschun D, Frey JU (2003) Requirement of beta-adrenergic receptor activation and protein synthesis for LTP-reinforcement by novelty in rat dentate gyrus. J Physiol 552: 953-960.

Weiss T, Veh RW, Heinemann U (2003) Dopamine depresses cholinergic oscillatory network activity in rat hippocampus. Eur J Neurosci 18: 2573-2580.

Wilson MA, McNaughton BL (1994) Reactivation of hippocampal ensemble memories during sleep. Science 265: 676-679.

Witter MP, Groenewegen HJ, Da Silva FHL, Lohman AHM (1989) Functional organization of the extrinsic and intrinsic circuitry of the parahippocampal region. Prog Neurobiol 33: 161-253.

Zola-Morgan SM, Squire LR, Amaral DG (1989) Lesions of the hippocampal formation but not lesions of the fornix or the mammillary nuclei produce long-lasting memory impairment in monkeys. J Neurosci 9: 898-913.

PRESYNAPTIC IONOTROPIC GABA RECEPTORS

A homeostatic feedback mechanism at axon terminals of inhibitory interneurons

Nikolai Axmacher, Kristin Hartmann, and Andreas Draguhn*

1. INTRODUCTION

Neuronal network activity and -synchrony can vary widely between different functional states of the brain. This makes it necessary to maintain stability by homeostatic mechanisms. One common stabilizing mechanism at most chemical synapses is comprised by the negative feedback of transmitters on further vesicular release through autoreceptors. Most of these presynaptically located receptors are G-protein coupled and act in the time frame of tens to hundreds of milliseconds. The inhibitory transmitter GABA is known to suppress vesicle release via presynaptic $GABA_BR$ which are present at most inhibitory- and even at many excitatory- synapses (Misgeld et al., 1995). Being formed as dimers of proteins with the typical 7-transmembrane domain motive they belong to the family of metabotropic transmitter receptors (Kaupmann et al., 1997). There is, however, increasing evidence for ligand-gated ion channels in presynaptic terminal membranes which could exert much faster feedback than metabolically coupled receptors (MacDermott et al., 1999). Such fast acting channels might be of special importance at inhibitory synapses which can be activated at very high frequencies, depending on the state of the network. Some subtypes of hippocampal interneurons, for example, fire action potentials at frequencies around 200 Hz during fast network oscillations (Csicsvari et al., 1999; Klausberger et al., 2003). In order to cope with inter-spike intervals of few milliseconds, regulatory mechanisms thus have to be similarly fast.

Our recent work has focused on presynaptic GABA-gated ion channels which regulate the release of GABA from hippocampal interneurons onto their (principal) target cells. It should be noted, however, that $GABA_A$- or $GABA_C$-receptors have also been

*Nikolai Axmacher, Klinik für Epileptologie, Universität Bonn, Sigmund-Freud-Str. 25, 53105 Bonn ,Germany; Kristin Hartmann and Andreas Draguhn, Institut für Physiologie und Pathophysiologie, Ruprecht-Karls-Universität Heidelberg, Im Neuenheimer Feld 326, 69120 Heidelberg, Germany.
E-mail: Andreas.Draguhn@urz.uni-heidelberg.de

identified at the endings of non-GABAergic cells, where GABA thus has paracrine, rather than autocrine, functions. In the rodent hippocampus, $GABA_A$ receptors are present at Schaffer collateral terminals (Stasheff et al., 1993) and at mossy fibers (Ruiz et al., 2003), i.e. two types of glutamatergic axons. These and other examples show an enormous functional heterogeneity of GABAergic signaling at axons: at the glutamatergic endings of the CA3-to-CA1 projection GABA triggers action potentials which travel antidromically into CA3 pyramidal cells (Stasheff et al., 1993). At mossy fibers, in contrast, GABA reduces excitability by shunting action potentials, much more similar to its "standard" inhibitory function (Ruiz et al., 2003). At the calyx of Held, glutamate release is facilitated by presynaptic glycine- or GABA-receptors (Turecek and Trussell, 2001; 2002). In the spinal cord, axo-axonal synapses inhibit transmission from muscle spindle afferents onto motor neurons by depolarizing GABA responses which drive sodium channels into inactivation - the classical afferent depolarizing shift by Eccles and coworkers (Eccles, 1964). More recent work by the Akaike-group revealed that presynaptic $GABA_A$ receptors facilitate action potential-independent release of glycine from spinal cord neurons but suppress action potential-dependent release (Jang et al., 2002). These examples show that the heterogeneous effects of presynaptic GABA receptors depend on several factors: axonal membrane potential, transmembrane chloride gradient, the molecular nature, density and functional state of voltage-gated ion channels etc. Unfortunately, these parameters are difficult to assess at small synaptic endings. Most urgently, one would need reliable data on $[Cl^-]$ within the axon and its endings (see below). This experimental program, however, has to await refinement of available imaging or electrical recording techniques.

Besides the examples given above, presynaptic ionotropic GABA receptors have been identified at cultured hippocampal neurons (Vautrin et al., 1994). A direct observation of $GABA_A$- or $GABA_C$-receptor mediated effects of GABA at interneurons *in situ* has, however, not been reported prior to our work. We did therefore look for such receptors in acutely prepared rat hippocampal brain slices, using inhibitory postsynaptic currents (IPSCs) and destaining of fluorescence-labeled vesicles as functional assays (Axmacher and Draguhn, 2004a, Axmacher et al., 2004b). Our observations suggest that GABAergic inhibition is indeed regulated by feedback of GABA through axonally expressed autoreceptors at the endings of CA3 hippocampal interneurons. These receptors exert massive effects on action potential-dependent and -independent release of GABA and it is therefore likely that they provide a mechanism for fast, frequency-dependent regulation of GABA release.

2. EXPERIMENTAL METHODS

Horizontal hippocampal slices were prepared from juvenile (second postnatal week) rats using standard methods. We recorded from CA3 pyramidal cells with patch clamp techniques in the whole cell configuration (Hamill et al., 1981). Our electrode solution contained a high chloride concentration (140 mM CsCl) leading to depolarizing GABA-induced potentials or inward currents, respectively. It should be noted that IPSCs were recorded as an indicator of GABA release from presynaptic terminals. We did not aim at analyzing the native postsynaptic effects of GABA which would require recording techniques which do not interfere with the internal chloride concentration (Ebihara et al.,

1995) or imaging of intracellular [Cl⁻] (Kuner et al., 2000). We isolated GABAergic IPSCs pharmacologically by addition of CNQX and APV (each 30 μM) and suppressed effects of $GABA_B$ receptors by adding 2 μM CGP 55845A. Measurements were taken at room temperature (20-25°C) under visual control with an upright high-magnification microscope. We did carefully monitor input and access resistance throughout the experiments, especially when comparing miniature IPSCs before and after application of GABAergic drugs. Evoked IPSCs were elicited by stimulation (200 μs, < 50 V) with a locally positioned patch electrode. The software package developed by John Dempster (University of Glasgow, Scotland) was used for analysis of miniature IPSCs (mIPSCs) which were detected using a threshold detection algorithm. Parameters were adjusted by comparison with hand-evaluated data.

In a previous series of experiments we had observed changes in the frequency of mIPSCs after manipulating GABA metabolism in cultured hippocampal slices (Engel et al., 2001) (see below). Slice cultures were prepared and maintained as described by Yamamoto and coworkers (Yamamoto et al., 1989, 1992) following the protocol by Stoppini and coworkers (Stoppini et al., 1991).

For imaging, we prepared horizontal hippocampal slices from 9-14 days old Wistar rats. Slices were transferred into a recording chamber under submerged conditions and at room temperature. In order to monitor vesicular release we loaded the readily releasable pool of vesicles with the fluorescent dye FM1-43 by brief (25 sec) superfusion with hypertonic solution (ACSF supplemented with sucrose to 800 mOsmol) according to the protocol of Stanton et al. (2001, 2003). We blocked action potentials by addition of 1 μM TTX and suppressed glutamate-evoked GABA release by adding 30 μM CNQX and 10 μM APV. Modulation of transmitter release by $GABA_BR$ was excluded by 2 μM CGP 55845A. Two-photon confocal imaging was performed using a Leica microscope and a Ti:sapphire laser (Spectra-Physics, Freemont, USA). Regions of interest were selected at bright fluorescent spots (putative synaptic boutons) in the perisomatic region of CA3 pyramidal cells. Control regions showing no puncta were selected in each slice and were used for subtraction of background and bleaching. As a control, we also monitored destaining of putative excitatory synapses located further outward in the dendritic layers. Destaining was monitored over 40 minutes.

Our model of presynaptic vesicle dynamics (Axmacher et al., 2004c) was based on a reduced version of the vesicle cycle described by Sudhof (1995). We divided vesicles into three pools; the readily releasable pool (RRP), the reserve pool and the pool of fused vesicles. Transitions between the pools were modeled by a set of ordinary differential equations. Filling of vesicles was assumed to be a bi-directional flux of transmitter into and out of vesicles with saturation when influx and efflux are equal. Numerical simulation was performed using the Matlab program (Mathworks, Natick, USA).

3. RESULTS AND DISCUSSION

We examined the GABAergic modulation of GABA-release onto CA3 pyramidal neurons using both electrophysiological and imaging techniques. GABAergic postsynaptic currents were pharmacologically isolated and $GABA_BR$ were blocked to exclude confounding effects by this well-established negative feedback loop (see Methods). In our electrophysiological approach we recorded synaptic currents from CA3 pyramidal cells in whole cell configuration. IPSCs were evoked by local electrical stimulation with an extracellularly positioned patch pipette, adjusting parameters such that about 75% of our stimuli caused postsynaptic responses. Miniature IPSCs were recorded as spontaneous GABAergic events in the presence of TTX.

After recording 20-30 stimulus-triggered traces under baseline conditions, we applied muscimol at 1 μM via bath perfusion. Our aim was to tonically stimulate presynaptic $GABA_AR$, if present, and to test for their effects on stimulus-evoked IPSCs. It is obvious that perfusion of the recording chamber with a GABA receptor agonist will not selectively activate presynaptic GABA receptors but will also open GABA-gated ion channels in the postsynaptic membrane. This can massively distort the results by desensitizing postsynaptic receptors or by changing the recording conditions (increased membrane conductance). Such effects could, however, be attenuated by recording at 5 minutes after washout of muscimol. While, at this time, the acute effects of the substance on membrane conductance and noise were eliminated, the presynaptic effects were still present. This prolonged action of muscimol at presynaptic terminals is an interesting finding in itself and has not yet been understood (see below). Furthermore, muscimol did not change the responses to iontophoretically applied GABA nor did it suppress amplitudes of mIPSCs. These findings do largely exclude that any effects of muscimol on stimulus-induced or miniature IPSCs are confounded by receptor desensitization or by deterioration of the experimental conditions.

Evoked IPSCs were massively reduced by muscimol, both in amplitude and in response rate (i.e., failure rate increased; see Figure 1). We suggest that this effect reflects a reduced probability of transmitter release from endings of inhibitory interneurons in the presence of muscimol. The experiment does, however, not clarify whether the underlying mechanism is located at the bouton itself or whether more indirect mechanisms, e.g. shunting of the axonal membrane, cause the suppression of evoked release. Recent work by Ruiz and colleagues has indeed shown that glutamatergic transmission from granule cells to CA3 pyramids is regulated by $GABA_AR$ located at the

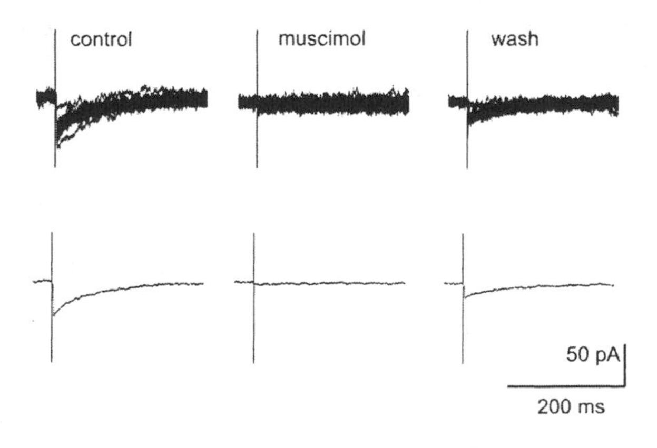

Figure 1. Stimulus-induced IPSCs are suppressed by muscimol. Middle traces were recorded 5 min after washout of muscimol, right traces after 25 min of wash. Bottom traces are averages from the single events shown above. Modified from Axmacher and Draguhn, 2004a.

axons, rather than the effects are restricted to the endings we recorded action potential-independent IPSCs in the presence of TTX. Similar to the evoked responses these mIPSCs were reduced in frequency after application of muscimol. Their amplitude, however, remained unaltered, confirming the presynaptic origin of the action of muscimol. Similar effects were obtained by bath-application of isoguvacine, an agonist with specificity for $GABA_AR$ over $GABA_CR$ (Figure 2). Thus, we assume that presynaptically located $GABA_A$ receptors mediate a negative feedback on vesicular release of GABA.

Our approach, however, did rely on the application of GABAergic drugs while measuring postsynaptic GABAergic currents as an indicator of the presynaptic modulatory effects. This experimental paradigm impedes a clear separation between the hypothesized mechanism and the test signal. Therefore, we searched for an alternative approach which would make us independent from the postsynaptic IPSCs. We recruited to the method developed by W. Müller and P. Stanton who had used two-photon confocal video-microscopy for FM1-43 imaging of vesicular release in hippocampal slices (Stanton et al., 2001, 2003). We focused on perisomatic synapses of CA3 pyramidal cells which can be assumed to be preferentially GABAergic, as compared to more distal dendritic synapses. Indeed, basal destaining in the presence of TTX was faster in the former, compatible with the known higher frequency of mIPSCs, compared to miniature excitatory postsynaptic currents. In the somatic region, synapses destained at a rate of roughly 10% per 10 min. Muscimol was applied after 12 min of spontaneous destaining in control solution. In these experiments, we left muscimol in the bath solution for 10 min (compared to 1 min in our electrophysiological studies), since postsynaptic receptor desensitization was no concern. Muscimol clearly slowedthe rate of destaining of perisomatic synapses (Figure 3). This effect was irreversible within the time of observation, fitting well with the long-lasting effects of muscimol in our previous study on mIPSCs. In contrast to perisomatic boutons, distal (dendritic) synapses showed a slower destaining which was resistant to modulation by muscimol.

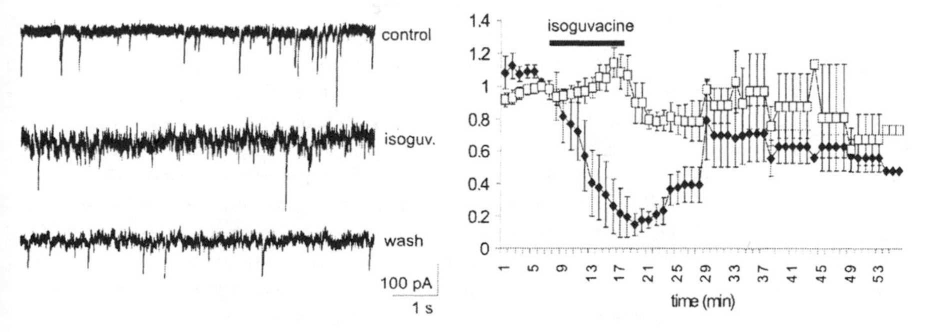

Figure 2. Reduction of mIPSC frequency by the $GABA_AR$ agonist isoguvacine (10 μM). Left part of the figure shows original recordings of mIPSCs from a CA3 pyramidal cell. Summary graph (right) shows mean normalized frequency (solid diamonds) and amplitude (open squares) from 5 cells. Note stability of amplitude while frequency is reduced. Modified from Axmacher and Draguhn, 2004a.

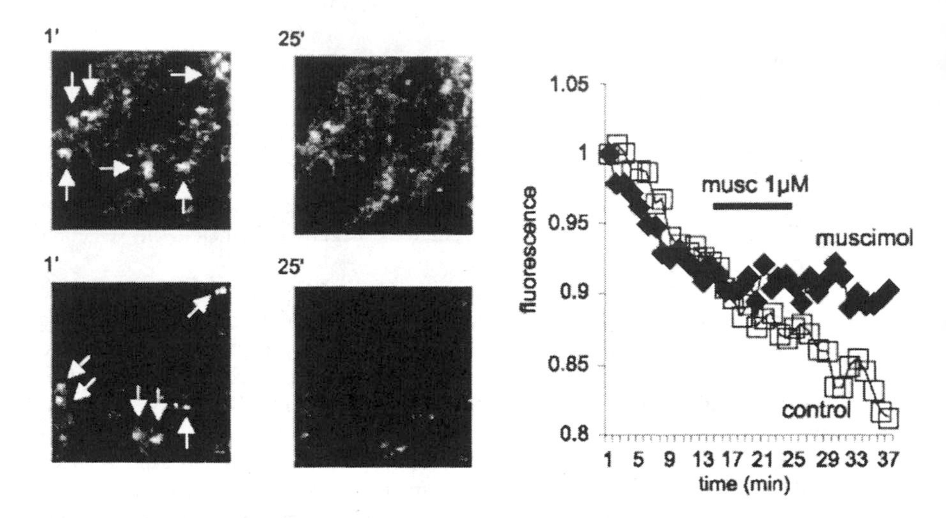

Figure 3. Fluorescence signals of FM1-43 stained putative inhibitory synapses in the perisomatic region of CA3 pyramidal cells. Top images show the slow decay observed when muscimol was added after 12 min, bottom images show faster decay in the absence of muscimol. Quantitative data (right panel) show continuous decrease of vesicle-staining in the absence of muscimol (open squares) and plateau formation after addition of muscimol (solid diamonds). Modified from Axmacher et al., 2004c.

Together, our data show that perisomatic inhibition in rat hippocampal CA3 pyramidal cells is negatively regulated by presynaptic feedback of GABA on $GABA_AR$. This constitutes a potential mechanism of short-term plasticity and is of obvious importance for the dynamics of synaptic inhibition and for the function of interneuron-networks, especially during oscillations (Whittington and Traub, 2003). Ionotropic receptors can act faster than the metabotropically coupled $GABA_BR$ and therefore might influence release at a millisecond time-scale, although this has not yet been directly proven. A *caveat* is given by the observation that the effects of muscimol outlasted the presence of the substance by several minutes. This suggests that plastic effects of presynaptic ionotropic GABA receptors last longer than the typical time course of postsynaptic IPSCs, constituting a more sustained modulation of synaptic inhibition. The underlying mechanisms are, as to yet, unclear. An intriguing possibility would be that ionic fluxes induce volume changes at the presynaptic terminal which are known to influence release probability, hence the use of hypertonic solution for massive release of the RRP (Stanton et al., 2003). The tiny volume of presynaptic boutons can be heavily affected even by a rather small ionic load, e.g. influx of chloride through $GABA_AR$. Direct observation of the volume of presynaptic boutons in living tissue is beyond present technical possibilities. We are also lacking information about the expression of volume-regulatory proteins at the presynaptic boutons of interneurons. It would be of special importance to know whether the chloride extruding transporter KCC-2 is present at these boutons. This would also help to clarify the chloride reversal potential which, in conjunction with resting potential, defines the direction and size of the GABA-induced chloride flux.

Another important information at the molecular level concerns the type of receptors expressed. Many extrasynaptic GABA receptors are composed of subunits which confer a specially high affinity towards GABA, making them ideal detectors of low levels of the transmitter in the extracellular space (Stell and Mody, 2002). In the retina, this sensitive subtype of GABA receptors is the $GABA_CR$ (Shields et al., 2000), assembled from rho-subunits which belong to the same gene family as $GABA_AR$ subunits (Bormann and Feigenspan, 1995). While the effects of isoguvacine argue against the expression of $GABA_CR$ at presynaptic endings in the CA3 region, it is feasible that other receptor subunits with high affinity are sorted to the axons of hippocampal interneurons (Stell and Mody, 2002).

Experimental (Engel et al., 2001; Overstreet and Westbrook, 2001; Wu et al., 2003) and theoretical (Axmacher et al., 2004b) evidence indicates that presynaptic ionotropic GABA receptors play a crucial role for the regulation of inhibition in a special situation of inhibitory synaptic plasticity: the adaptation of GABA-metabolism to changing activity within local networks. Inhibitory synapses appear to generate more GABA when network activity is increased and down-regulate transmitter production when network activity is lowered. Evidence for this "metabolic plasticity" comes from several studies in tissue with increased or decreased network activity, respectively. The GABA-synthesizing enzyme glutamate decarboxylase (GAD) is up-regulated in the hippocampus of epileptic rats (Esclapez and Houser, 1999; Feldblum et al., 1990). In contrast, GABA-production is down-regulated when network activity is reduced, e.g. in deafferentiated areas of somatosensory cortex (Garraghty et al., 1991; Gierdalski et al., 1999; Hendry and Carder, 1992). Thus, it appears that the production of GABA follows a homeostatic principle (Turrigiano, 1999), adapting inhibition to the "needs" of the network. The cellular and subcellular mechanisms of this regulation are, however, far from trivial: several complex, non-linear steps lie between the modulation of GABA production and inhibitory synaptic efficacy (Axmacher et al., 2004b). In order to yield enhanced filling of synaptic vesicles upon increasing cytosolic GABA-concentration, normally filled vesicles must have a reserve for additional uptake of the transmitter. Given that such vesicles contain more GABA, the postsynaptic responses will only be increased if normal IPSPs are non-saturating, i.e. there is a receptor reserve (Frerking et al., 1995; Nusser et al., 1997; Yee et al., 1998). In addition, an increased cytosolic GABA-concentration may enhance non-vesicular release of GABA via reverse operation of GABA transporters. The latter mechanism might explain the observed increase in GABA-induced membrane noise (Engel et al., 2001; Overstreet and Westbrook, 2001; Wu et al., 2003) and GABA-release (Yee et al., 1998) in experimental situations of increased cellular GABA content.

We have increased cellular GABA content in two different experimental systems, namely in acutely prepared (Axmacher and Draguhn, 2004a) and in cultured (Engel et al., 2001) rat hippocampal slices. Chronic (~4 days) incubation of cultured slices with a blocker of the GABA-degrading enzyme GABA transaminase (γ-vinyl-GABA) led to an increase in the amplitude of mIPSCs as well as to increased membrane noise. These findings indicate increased vesicle filling, a postsynaptic reserve pool of GABA receptors and increased tonic inhibition, possibly via non-vesicular release. Surprisingly, the treatment did also increase the frequency of mIPSCs. Moreover, the events tended to occur in bursts, rather than isolated (Engel et al., 2001). In contrast, incubation of acutely prepared slices with γ-vinyl-GABA for 3-9 hours led to a dramatic (> 90%) decrease in the frequency of mIPSCs in CA3 pyramidal cells (Axmacher and Draguhn, 2004a).

Similar observations have been previously reported by Overstreet and Westbrook (2001). The reduction in mIPSC frequency may be a direct consequence of the negative feedback of GABA on its own release via ionotropic $GABA_AR$ (see Figure 4; $GABA_BR$ had been blocked in these experiments).

The feedback may be mediated by GABA released from vesicles with enhanced transmitter content or, alternatively, by a tonically increased ambient GABA concentration, as indicated by the enhanced GABAergic membrane noise in these experiments (Engel et al., 2001; Overstreet and Westbrook, 2001; Wu et al., 2003). Why did we observe an opposite effect (increase) on mIPSC frequency in cultured hippocampal slices? One possible explanation for these apparently contradictory results from two different preparations is that the axonal chloride gradient differs between acutely prepared and chronically stored hippocampal slices. Depolarizing GABA responses in the cultured tissue might increase presynaptic calcium levels and thereby facilitate spontaneous vesicle release while hyperpolarizing responses in acutely prepared tissue would inhibit vesicle release. Again, it would be important to know the chloride equilibrium and resting potentials at presynaptic boutons. Functionally, the negative feedback observed in acutely prepared slices will shift GABAergic inhibition from phasic to tonic mode, i.e. postsynaptic cells receive less GABA from vesicular release but are tonically inhibited by an increased ambient GABA concentration. While this may still

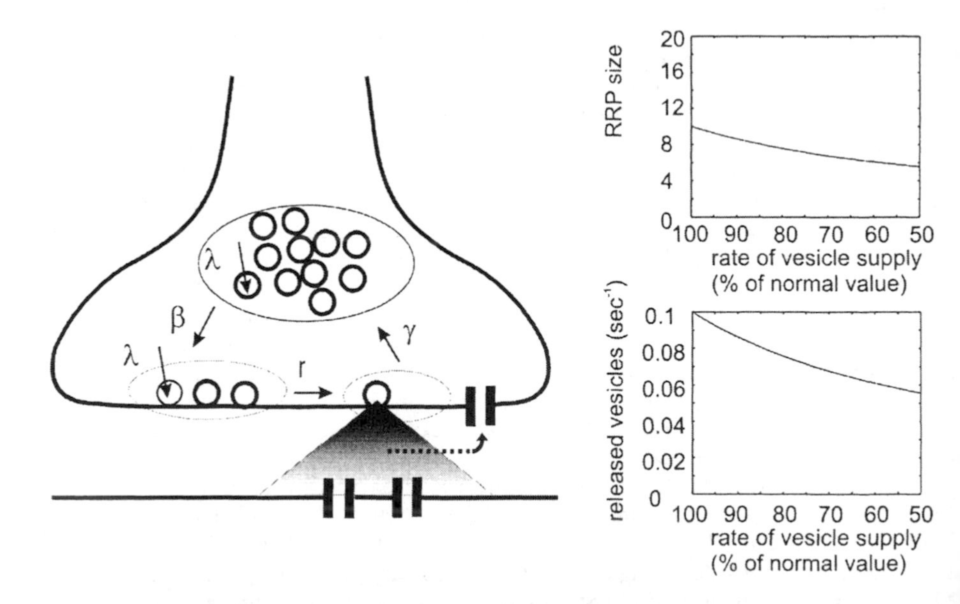

Figure 4. Simplified model of the presynaptic terminal. Vesicles are divided into three pools with different transition rates from the reserve pool to the RRP (membrane-attached vesicles, left) and to the released state. Rate of filling is labeled λ. Released GABA feeds back onto presynaptic GABA-A receptors. Right panels show the resulting simulations assuming that GABAergic feedback inhibits the transition from the reserve pool to the RRP. Note decrease release rate of vesicles with decreasing supply of vesicles into the RRP, consistent with experiments.

cause highly efficient inhibition, it might induce changes in interneuron-driven network activity like gamma oscillations (Whittington and Traub, 2003). Possibly, side-effects of GABAergic drugs may be associated with such specific disturbances of spike timing in neuronal networks which are organized by interneurons.

Modeling the effects of altered GABA metabolism on vesicular release did indeed reveal that positive or negative feedback of GABA on further release of GABAergic vesicles is a likely mechanism underlying the observed changes in mIPSC frequency (Axmacher et al., 2004b). It further turned out that the interaction between released GABA and presynaptic vesicle dynamics takes most likely place at the transition from the reserve pool to the RRP, providing more or less release-ready vesicles. Our model shows that this indirect mechanism allows for sustained changes of the frequency of release, while a direct effect of GABA release would tend to deplete the RRP.

4. CONCLUSIONS

Presynaptic autoreceptors modulate the release of GABA from terminals of inhibitory interneurons. We have shown that in rodent hippocampal CA3 this feedback involves the activation of ionotropic $GABA_AR$ which suppress further release in acutely prepared tissue. Under conditions of different E_{Cl^-}, the same feedback might be positively coupled to release, giving rise to bursts of IPSCs after an initial (large) event. Our further experimental and theoretical studies show that presynaptic ionotropic $GABA_AR$ are specifically important during one mechanism of inhibitory synaptic plasticity, namely changes in GABA metabolism. Besides alterations of the amount of released GABA this homeostatic mechanism induces changes in the frequency of vesicle release which are likely to be mediated by presynaptic ionotropic $GABA_AR$. This new feedback mechanism will be important to understand the dynamics of inhibitory synaptic signaling and the ratio between tonic and phasic GABAergic inhibition.

5. ACKNOWLEDGEMENTS

This work was supported by the DFG (Deutsche Forschungsgemeinschaft) grant SFB 515/B1.

6. REFERENCES

Axmacher N, Draguhn A (2004a) Inhibition of GABA release by presynaptic ionotropic GABA receptors in hippocampal CA3. NeuroReport 15:329-334.
Axmacher N, Stemmler M, Engel D, Draguhn A, Ritz R (2004b) Transmitter metabolism as a mechanism of synaptic plasticity: a modeling study. J Neurophysiol 91:25-39.
Axmacher N, Winterer J, Stanton PK, Draguhn A, Muller W (2004c) Two-photon imaging of spontaneous vesicular release in acute brain slices and its modulation by presynaptic GABAA receptors. Neuroimage 22:1014-1021.
Bormann J, Feigenspan A (1995) GABAC receptors. Trends Neurosci 18:515-519.
Csicsvari J, Hirase H, Czurkó A, Mamiya A Buzsáki G (1999) Oscillatory coupling of hippocampal pyramidal cells and interneurons in the behaving Rat. J Neurosci 19:274-287.
Ebihara S, Shirato K, Harata N, Akaike N (1995) Gramicidin-perforated patch recording: GABA response in mammalian neurons with intact intracellular chloride. J Physiol 484:77-86.

Eccles JC (1964) Presynaptic inhibition in the spinal cord. Prog Brain Res 12:65-91.
Engel D, Pahner I, Schulze K, Frahm C, Jarry H, Ahnert-Hilger G, Draguhn A (2001) Plasticity of rat central inhibitory synapses through GABA metabolism. J Physiol 535:473-82.
Esclapez M, Houser CR (1999) Up-regulation of GAD65 and GAD67 in remaining hippocampal GABA neurons in a model of temporal lobe epilepsy. J Comp Neurol 412:488-505.
Feldblum S, Ackermann RF, Tobin AJ (1990) Long-term increase of glutamate decarboxylase mRNA in a rat model of temporal lobe epilepsy. Neuron 5:361-371.
Frerking M, Borges S, Wilson M (1995) Variation in GABA mini amplitude is the consequence of variation in transmitter concentration. Neuron 15:885-895.
Frerking M, Wilson M (1996) Saturation of postsynaptic receptors at central synapses? Curr Opin Neurobiol 6:395-403.
Garraghty PE, LaChica AE, Kaas JH (1991) Injury-induced reorganization of somatosensory cortex is accompanied by reductions in GABA staining. Somatosens Mot Res 8:347-354.
Gierdalski M, Jablonska B, Smith A, Skangiel-Kramska J, Kossut M (1999) Deafferentation induced changes in GAD67 and GluR2 mRNA expression in mouse somatosensory cortex. Brain Res Mol Brain Res 71:111-119.
Hamill OP, Marty A, Neher E, Sakmann B, Sigworth FJ (1981) Improved patch-clamp techniques for high-resolution current recording from cells and cell-free membrane patches. Pflugers Arch 391:85-100.
Hendry S, Carder RK (1992) Organization and plasticity of GABA neurons and receptors in monkey visual cortex. Prog Brain Res 90:477-502.
Jang IS, Jeong HJ, Katsurabayashi S, Akaike N (2002) Functional roles of presynaptic GABA(A) receptors on glycinergic nerve terminals in the rat spinal cord. J Physiol 541:423-434.
Kaupmann K, Huggel K, Heid J, Flor PJ, Bischoff S, Mickel SJ, McMaster G, Angst C, Bittiger H, Froestl W Bettler B (1997) Expression cloning of GABA(B) receptors uncovers similarity to metabotropic glutamate receptors. Nature 386:239-246.
Klausberger T, Magill PJ, Marton LF, Roberts JD, Cobden PM, Buzsáki G, Somogyi P (2003) Brain-state- and cell-type-specific firing of hippocampal interneurons in vivo. Nature 421:844-848.
Kuner T, Augustine GJ (2000) A genetically encoded ratiometric indicator for chloride: capturing chloride transients in cultured hippocampal neurons. Neuron 27:447-459.
MacDermott AB, Role LW, Siegelbaum SA. (1999) Presynaptic ionotropic receptors and the control of transmitter release. Annu Rev Neurosci 22:443-85.
Misgeld U, Bijak M, Jarolimek W (1995) A physiological role for GABAB receptors and the effects of baclofen in the mammalian central nervous system. Prog Neurobiol 46:423-462.
Nusser Z, Cull-Candy S, Farrant M (1997) Differences in synaptic GABA(A) receptor number underlie variation in GABA mini amplitude. Neuron 19:697-709.
Overstreet LS, Westbrook GL (2001) Paradoxical reduction of synaptic inhibition by vigabatrin. J Neurophysiol 86:596-603.
Ruiz A, Fabian-Fine A, Scott A, Walker MC, Rusakov AD, Kullmann AD (2003) GABAA receptors at hippocampal mossy fibers. Neuron 39:961-973.
Shields CR, Tran MN, Wong RO, Lukasiewicz PD (2000) Distinct ionotropic GABA receptors mediate presynaptic and postsynaptic inhibition in retinal bipolar cells. J Neurosci 20:2673-2682.
Stanton P, Heinemann U, Muller W (2001) FM1-43 imaging reveals cGMP-dependent long-term depression of presynaptic transmitter release. J Neurosci 21:167-173.
Stanton PK, Winterer J, Bailey CP, Kyrozis A, Raginov I, Laube G, Veh RW, Nguyen CQ, Muller W (2003) Long-term depression of presynaptic release from the readily releasable vesicle pool induced by NMDA receptor-dependent retrograde nitric oxide. J Neurosci 23:5936-44.
Stasheff SF, Mott DD, Wilson WA (1993) Axon terminal hyperexcitability associated with epileptogenesis in vitro. II. Pharmacological regulation by NMDA and GABAA receptors. J Neurophysiol 70:976-984.
Stell BM, Mody I (2002) Receptors with different affinities mediate phasic and tonic GABA(A) conductances in hippocampal neurons. J Neurosci 22:223-227.
Stoppini L, Buchs PA, Muller D (1991) A simple method for organotypic cultures of nervous tissue. J Neurosci Methods 37:173-182.
Sudhof TC (1995) The synaptic vesicle cycle: a cascade of protein-protein interactions. Nature 375:645-653.
Turecek R, Trussell LO (2001) Presynaptic glycine receptors enhance transmitter release at a mammalian central synapse. Nature 411:587-590.
Turecek R, Trussell LO (2002) Reciprocal developmental regulation of presynaptic ionotropic receptors. Proc Natl Acad Sci U S A 99:13884-13889.
Turrigiano GG (1999) Homeostatic plasticity in neuronal networks: the more things change, the more they stay the same. Trends Neurosci 22:221-227.

Vautrin J, Schaffner AE, Barker JL (1994) Fast presynaptic GABAA receptor-mediated Cl- conductance in cultured rat hippocampal neurons. J Physiol 479:53-63.
Whittington MA, Traub RD (2003) Interneuron diversity series: inhibitory interneurons and network oscillations in vitro. Trends Neurosci 26:676-682.
Wu Y, Wang W, Richerson GB (2003) Vigabatrin induces tonic inhibition via GABA transporter reversal without increasing vesicular GABA release. J Neurophysiol 89:2021-2034.
Yamamoto N, Kurotani T, Toyama K (1989) Neural connections between the lateral geniculate nucleus and visual cortex in vitro. Science 245:192-194.
Yamamoto N, Yamada K, Kurotani T, Toyama K (1992) Laminar specificity of extrinsic cortical connections studied in coculture preparations. Neuron 9:217-228.
Yee JM, Agulian S, Kocsis JD (1998) Vigabatrin enhances promoted release of GABA in neonatal rat optic nerve. Epilepsy Res 29:195-200.

ACTIVITY DEPENDENT REGULATION OF THE Cl^- TRANSPORTING SYSTEM IN NEURONS

Stefan Titz and Ulrich Misgeld*

1. PLASTICITY IN PATHWAYS RESPONSIBLE FOR POSTSYNAPTIC INHIBITORY ACTION

Long term potentiation (LTP) became a topic of intense research as soon as the hippocampal slice preparation had proven its usefulness for studies with intracellular recording techniques (Schwartzkroin, 1975). The early studies addressed predominantly two questions: **1)** Is LTP specific to the activated input? **2)** What kinds of changes in postsynaptic potentials underlie LTP (Andersen et al., 1977; Lynch et al., 1977; Yamamoto and Chujo, 1978)? In a study on synaptic plasticity in the hippocampal field CA3, John Sarvey, Manfred Klee and myself (UM), at that time working together in the Max-Planck-Institute for Brain Research in Frankfurt/M, Germany, found long lasting changes that were not specific to the stimulated input (Misgeld et al., 1979). We used the term 'heterosynaptic postactivation potentiation' to describe this phenomenon (**Fig. 1A, B**). Potentiation is defined as an enhancement of synaptic transmission and appears in extracellular recording as an increased probability of firing of synaptically excited neurons. Therefore, on the cellular level, an increased EPSP amplitude was to be expected. What we observed, however, was an increased IPSP amplitude in the majority of CA3 neurons and a decreased IPSP amplitude in a minority, in which indeed the EPSP amplitude was increased (**Fig. 1C**). To explain these data we adopted the following model: Whereas a potentiation of only a few cells in the field would suffice to generate an enhanced field potential amplitude, the potentiated output from these neurons would, via the recurrent inhibitory pathway, strengthen the inhibition in the neighboring 'non-potentiated' neurons. This proposal could account for the enhanced IPSP amplitudes, but did not provide an explanation for the decrease in inhibition. In any case, our major conclusion was that plasticity of inhibitory circuits is an important asset of plasticity in brain function. Although the possible mechanisms for changes in pathways conveying postsynaptic inhibitory action are manifold and comprise pre- and postsynaptic sites, the

* Stefan Titz and Ulrich Misgeld, Institute of Physiology and Pathophysiology and IZN, University of Heidelberg, D-69120 Heidelberg, Germany. E-mail: ulrich.misgeld@pio1.uni-heidelberg.de

interest of our group in the years to come focused on the role of chloride and chloride transport in postsynaptic $GABA_A$ receptor mediated inhibition.

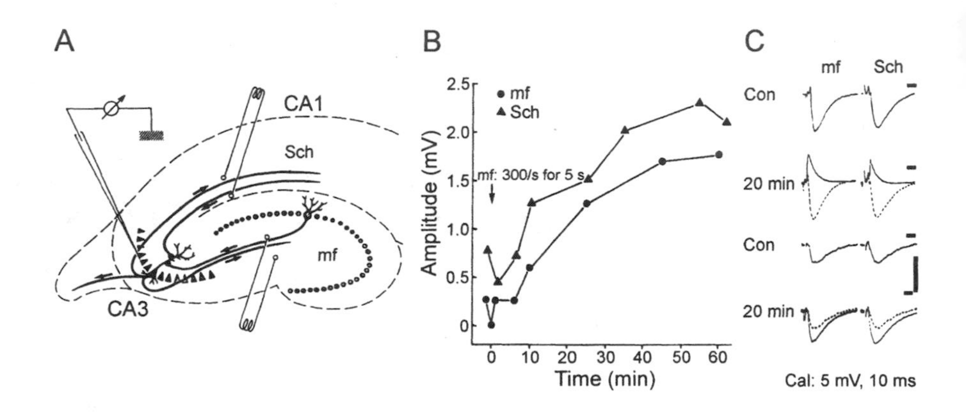

Figure 1. Heterosynaptic postactivation potentiation. Schematic representation of the position of the electrodes to test for heterosynaptic activation. In a hippocampal slice of the guinea-pig brain, bipolar stimulation electrodes to evoke the test stimuli were positioned in the hilar region and in the CA1 region. Intracellular recording with sharp microelectrodes was obtained from CA3 pyramidal cells. Tetanic stimulation was applied via the electrode in the hilar region. B) The plot of the population spike amplitude in the responses from the two electrodes revealed a long lasting increase of both responses after a tetanic stimulation of 300 Hz for 5 s. C) Typical postsynaptic potentials recorded from CA3 neurons in control and 20 min after tetanic stimulation, the latter superimposed on the control response (broken lines). In the upper example the response evoked from both stimulation sites changed from a predominant IPSP to a predominant EPSP, in the lower example the amplitude of the IPSP increased (modified from Misgeld et al., 1979).

2. ROLE OF CHLORIDE TRANSPORT IN POSTSYNAPTIC INHIBITION

A balance of synaptic excitation and inhibition is a basic requirement for a normal function of neuronal networks. A striking example of an excitation – inhibition imbalance is the occurrence of 'epileptiform discharge activity' upon pharmacological blockade of inhibition. The classical studies on postsynaptic inhibition in cat motoneurons (Eccles, 1964) already revealed two important principles of operation, hyperpolarizing inhibition and shunting inhibition. The fact that the long known primary afferent depolarization of sensory fibers was the correlate of presynaptic inhibition in the spinal cord (Eccles et al., 1962) was clear evidence that the inhibitory capacity does not depend solely on hyperpolarizing membrane potential changes. Instead, the necessity for a membrane potential change remained somewhat unsettled for quite some time although it is obvious that a membrane hyperpolarization increases the gap between membrane and spike threshold potentials.

Cation-Cl^- cotransporters are ubiquitous in eukaryotic cells and involved in volume regulation corresponding to the role of Cl^- as the main osmotically effective anion in the body fluids. In neurons, Cl^- has an additional role in that it is the charge carrier for currents that inhibit neurons by dampening their discharge activity. The most common inhibitory neurotransmitters in the CNS, GABA and glycine, act through receptors that are ligand gated unselective anion channels with a, by far highest, permeability for Cl^-. If a conductance increase for Cl^- ions generates a current which hyperpolarizes a membrane, an electromotor force must exist which drives Cl^- into the cell. An inwardly directed driving force for Cl^- ions (which generate outward currents because of the negative charge they carry) can only be maintained if, over time, Cl^- extrusion by an active transporter exceeds passive Cl^- uptake through conducting pathways. Such a task can be accomplished without too much of an energetic cost by a secondarily active transport with constitutive activity operating near its thermodynamic equilibrium on the one hand and a low Cl^- leak conductance on the other hand. Both pre-conditions seem to be fulfilled in most mature neurons.

H.D. Lux (1971) was the first to demonstrate that hyperpolarizing IPSPs are generated by the influx of Cl^- ions down a gradient that is maintained by an outwardly directed chloride transport mechanism in cat motoneurons. We used the slice preparation to investigate the role of Cl^- transport in inhibitory responses of CA3 neurons and granule cells of the guinea pig hippocampus. In both cell types, increasing the intracellular Cl^- concentration by Cl^- injection shifted the reversal potential of the inhibitory responses in a positive direction. Blocking the outward transport of Cl^- by furosemide dramatically slowed their recovery from the injection. In addition, hyperpolarizing and depolarizing responses to the inhibitory neurotransmitter GABA decreased in the presence of furosemide due to a shift in their reversal potential towards a passive distribution. We concluded that $[Cl^-]_i$ of adult pyramidal neurons was regulated by furosemide-sensitive cation-Cl^- cotransport (Misgeld et al., 1986). Further, our electrophysiological studies indicated the co-existence of two Cl^- transport directions in neurons of the hippocampus, one extruding Cl^- and another accumulating it, and suggested a higher $[Cl^-]_i$ in the dendrites than in the soma (Misgeld et al., 1986; Müller et al., 1989).

It took another 10 years until the molecular structure of the presumed Cl^- regulating systems in neurons were identified. It turned out that, among the family of cation-Cl^- cotransporters (CCC, **Fig. 2A**) which are widely distributed and play an important physiological role in the recovery of cell volume after swelling or shrinkage, one member is not expressed ubiquitously, but is found exclusively in neurons (Payne et al., 1996). The CCC isoform that is specific to neurons belongs to the group of K^+-Cl^- cotransporters (KCC) and is termed KCC2. The KCC2 gene contains a neuronal restrictive silencing element which provides the molecular basis for its specific expression in neurons (Karadsheh and Delpire, 2001). The primary function of the transporter does not appear to be volume regulation but regulation of $[Cl^-]_i$ and, possibly, $[K^+]_o$. It differs from other members of the KCC family in that it has a somewhat higher affinity for extracellular K^+. KCC2 is an electroneutral transport and its driving force is provided by the transmembranal chemical gradients for K^+ and Cl^-. Therefore, it operates near its equilibrium under physiological ion concentrations (Payne, 1997). There are no specific inhibitors available for KCC2, although a drug with notorious toxicity, DIOA (Pond et al., 2004), was considered specific for some time. The loop diuretics furosemide and bumetanide inhibit KCC2 and provide useful pharmacological tools for most studies ($k_i \approx$ 50 µM; Payne et al., 2003) as long as the inhibition of other CCCs and other effects (blockade of Cl^- channels and of $GABA_A$ receptor subtypes etc.) are taken into consideration.

The properties described above were those of KCC2 expressed in a heterologous expression system. We were interested to examine the Cl^- regulating system in neurons and to compare its properties to those of KCC2 in heterologous expression systems. In cultured neurons we established an assay that allowed us to determine the direction of KCC2 transport as a function of $[K^+]_o$ (Jarolimek et al., 1999). A major advantage of the assay is that the driving force for the transporter is set by defined intracellular and extracellular ion concentrations and that $GABA_A$ currents are not confounded by the influence of HCO_3^- ions (Kaila et al., 1993) and, hence, can be used to calculate $[Cl^-]_i$ from their reversal potentials.

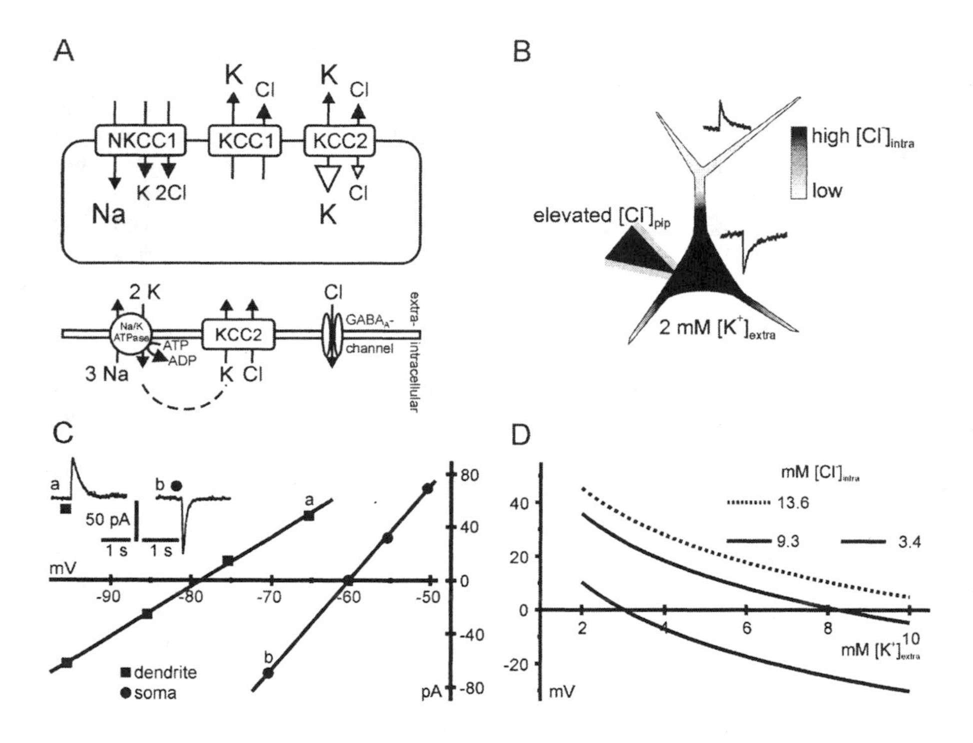

Figure 2. Properties and thermodynamics of KCC2. A) Upper scheme: The main representatives of the family of cation-Cl^- cotransporters (CCCs) are shown, comprising the ubiquitous Na^+-K^+-$2Cl^-$ cotransporter NKCC1 and K^+-Cl^- cotransporters, of which KCC1 is ubiquitous and KCC2 neuron-specific. Under physiological conditions NKCCs accumulate and KCCs extrude Cl^-. KCC2 accumulates Cl^-, however, if $[K^+]_o$ increases during intense neuronal activity. Lower scheme: KCC2 is an electroneutral transport which takes its energy from the transmembranal gradients for K^+ and Cl^-, $[K^+]_i/[K^+]_o$ being established by the activity of the Na^+-K^+ ATPase. The electrochemical gradient established for Cl^- in this way allows Cl^- influx and, hence, membrane hyperpolarization as soon as Cl^- channels open, e.g. if GABA or glycine bind to their receptors. B, C) Principles of an assay for KCC2 activity. A neuron is loaded with Cl^- by the patch pipette at its soma. If $[K^+]_o$ is substantially lower than $[Cl^-]_i$, the capacity of the neuron to extrude Cl^- will reduce $[Cl^-]_i$ of dendrites below $[Cl^-]_i$ of the soma (B). As a consequence, reversal potential of GABA currents induced by a focal application of GABA to a dendrite have a reversal potential, which is more negative than the reversal potential of currents induced by focal GABA applications to the soma (C). D) Plot of the driving force for KCC as a function of $[K^+]_o$ for different $[Cl^-]_i$. $[Cl^-]_i$ of the neurons was calculated with the Nernst equation using measured E_{GABA} values in the dendrites of neurons as shown in (C). Positive values indicate outward transport (modified from Jarolimek et al., 1999Based on our work in slices we expected that the neuronal Cl^- regulation system can establish a difference between $[Cl^-]_i$ in dendrites and soma

provided one of the two sites is loaded with Cl^- locally (**Fig. 2A**). If the regulating system is inhibited pharmacologically or there is no driving force through which it can operate, Cl^- diffusion is expected to prevent any intracellular Cl^- gradient. Under whole cell recording configuration with open access to the cell, the $[Cl^-]$ in the cell soma is determined by the $[Cl^-]$ in the solution of the pipette. This allows to load the cell soma with Cl^- by the pipette and, thereby, to create a gradient with $[Cl^-]_i$ being higher in soma than in dendrites in this case (**Fig. 2B, C**). Furosemide indeed reduced the difference in $[Cl^-]_i$ between soma and dendrites created in such a way. The same effect had varying $[K^+]_o$ and $[Cl^-]_i$ to values for which it could be calculated that no driving force for KCC existed (**Fig. 2D**). The assay allowed to study KCC mediated outward transport of Cl^- only. Using this assay we could establish that, at appropriate concentrations of cations and anions, transport can reverse its direction. Hence, $[Cl^-]_i$ is tightly coupled to $[K^+]_o$ (Jarolimek et al., 1999).

As we had reported it for hippocampal neurons in slices, the neuronal Cl^- regulating system could be inhibited by furosemide in cultured cells. The comparison of various blockers (furosemide, DIDs and bumetanide) revealed a pharmacological profile very much like that reported for KCC2 in heterologous systems (Kelsch et al., 2001). It turned out, however, that the efficacy of furosemide depended strongly on the extracellular composition and concentration of monovalent cations (Misgeld et al., 2004). Because the extracellular binding/transporting site is not selective for K^+, other cations can be taken up into the neurons through this Cl^- transporting pathway. A prominent example is provided by NH_4^+ ions with the effect that, under conditions encountered in hepatic encephalopathy, NH_4^+ transport through the neuronal Cl^- regulating system imposes a continuous acid load (Liu et al., 2003).

The molecular identification of the neuronal Cl^- extruding K^+-Cl^- cotransporter KCC2 renewed the interest in Cl^- regulation in neurons and its consequences for postsynaptic inhibition. Strong evidence was provided that KCC2 is the main Cl^- extruder to establish hyperpolarizing postsynaptic inhibition in neurons of the mammalian CNS. Pivotal was the demonstration of a tight link between the expression of KCC2 and the presence of $GABA_A$ receptor mediated hyperpolarizing responses in hippocampal pyramidal cells. Pyramidal cells of hippocampal slice cultures exposed to antisense oligodeoxynucleotides against KCC2 mRNA had an E_{GABA} which was close to resting membrane potential in sharp contrast to control cells in which E_{GABA} was negative to the resting potential (Rivera et al., 1999). KCC2 knockout experiments supported an important role of the hyperpolarization accompanying inhibition. KCC2 knockout mice died immediately after birth due to an inability to breathe and, in brainstem preparations from these mice, spontaneous rhythmic activity of respiratory output neurons was absent (Hübner et al., 2001). Mice with only 5-10 % of KCC2 displayed spontaneous, generalized seizures, a severe loss of inhibitory interneurons in the cortex and hippocampus and died shortly after birth. Adult heterozygote animals showed increased susceptibility for epileptogenic drugs (Woo et al., 2002).

Antibodies raised against the intracellular C-terminus (Williams et al., 1999) or the intracellular N-terminus (Hübner et al., 2001) allowed to study the location of KCC2 using immunocytochemical techniques. Main issues were the question for a co-localization with inhibitory synapses and/or receptors and for a discriminate distribution across neurons. A co-localization with the β_2/β_3 subunits of $GABA_A$ receptors in the cerebellum (Williams et al., 1999) and gephyrin in the spinal cord (Hübner et al., 2001) was noted, but the findings did not indicate that KCC2 was co-localized exclusively with inhibitory synapses. In fact, in rat hippocampus KCC2 is highly expressed in the vicinity of excitatory synapses (Gulyás et al., 2001), whereas there is substantial expression in the vicinity of inhibitory synapses on GABAergic neurons in the substantia nigra (Gulácsi et al., 2003). The co-localization with receptors but not synapses can be taken as evidence for a co-localization with extrasynaptic $GABA_A$ or glycine receptors, but except for the cerebellum direct evidence to this point is missing.

It is, however, established that the protein can be expressed in membranes in a distinct spatial fashion with considerable differences between cells or even between different cell compartments. KCC2 mRNA expression appears to be less abundant in dorsal vagal motoneurons innervating smooth muscles than in other motoneurons which innervate striated muscles. This expression difference was associated with a higher $[Cl^-]_i$, a slower Cl^- extrusion after prolonged GABA applications and a stronger activity dependent depression of IPSPs in dorsal vagal motoneurons than found in the other motoneurons (Ueno et al., 2002). In rat substantia nigra, KCC2 is expressed in GABAergic but not in dopaminergic neurons, and the $GABA_A$ IPSP reversal potential is significantly less negative in dopaminergic than in GABAergic neurons (Gulácsi et al., 2003). In the rat hippocampus, parvalbumine-immunoreactive GABAergic cells in the subfields CA1 and CA3 show the highest level of KCC2 mRNA and protein expression (Gulyás et al., 2001). A differential expression of KCC2 was also reported for the retina, the expression being high in cells and cell compartments in which $[Cl^-]_i$ was expected to be low. For bipolar cells, according to their spatial expression pattern, the expectation was that $[Cl^-]_i$ would be significantly lower in their synaptic terminals than in their dendrites (Vardi et al., 2000). Electrophysiological experiments, however, revealed only a minor difference between $[Cl^-]_i$ in dendrites and synaptic terminals of rat retinal ON bipolar cells (Billups and Attwell, 2002).

There are further transport proteins expressed in neurons which may participate in neuronal Cl^- regulation. In particular, a Na^+-K^+-$2Cl^-$ cotransporter (NKCC1) of which the function is to accumulate Cl^- is expressed in certain neurons. A popular assumption is that $[Cl^-]_i$ is high where NKCC1 is strongly expressed and $[Cl^-]_i$ is low where KCC2 expression dominates (Lu et al., 1999; Vardi et al., 2000; Shimizu-Okabe et al., 2002). In dorsal root ganglion cells in which NKCC1 and KCC2 are co-expressed, GABA responses are depolarizing (Lu et al., 1999). Other reports describe a lack of a significant amount or an absence of KCC2 expression in dorsal root ganglia (Boettger et al., 2003; Coull et al., 2003). There is a negative shift in E_{GABA} of dorsal root ganglion cells upon disruption of the gene encoding NKCC1 (Sung et al., 2000), but no obvious disturbance of central nervous system function. In hippocampal CA3 pyramidal neurons in which NKCC1 and KCC2 are co-expressed, somatic GABA responses are hyperpolarizing (Misgeld et al., 1986). In the adult hippocampus, NKCC expression is maintained but the immunoreactivity changes from a somatic to a predominantly dendritic localization (Marty et al., 2002). The physiological meaning of higher NKCC expression on the dendrites remains to be determined. The co-expression of these two transporters might allow a tight regulation of $[Cl^-]_i$ in dendrites and cell bodies, and their differential topographic distribution could be the basis for putative intracellular Cl^- gradients (Misgeld et al., 1986; Hara et al., 1992; Kuner and Augustine, 2000). The scenario, however, can be much more complex than indicated here, because the properties and demands of cellular volume regulation are not considered.

Other pathways including passive Cl^- conductance will also play a role. They could be the reason for the finding that $[Cl^-]_i$ in bipolar ON cells is not strongly influenced by the topographic distribution of transporter proteins (Vardi et al., 2000; Billups and Attwell, 2002). Further depolarizing GABA responses can be found in neurons lacking expression of NKCC1 mRNA (Balakrishnan et al., 2003) or in neurons lacking evidence for a major contribution of a functional Cl^- inward transport to $[Cl^-]_i$ regulation even under conditions under which a contribution of HCO_3^- ions can be excluded (Titz et al., 2003; Misgeld et al., 2004; Titz et al., 2004). It has been suggested that co-expression of NKCC1 and KCC2 implies that these two proteins are functionally coupled in the control of neuronal $[Cl^-]_i$ (Payne et al., 2003). Any experimental hints as to the mechanism of this 'coupling' are missing to date, but if NKCC1 is regulated by $[Cl^-]_i$ or osmotic challenge, as it is in non-neuronal systems (Russell, 2000), such a functional coupling is not easy to reconcile on the mere basis of expression levels.

The KCC isoform KCC3 has also been suggested to participate in neuronal Cl^- homeostasis. The disruption of its gene produced a phenotype with functional and morphological changes in the brain (Boettger et al., 2003) possibly resulting from cellular dysfunction. The changes were tentatively explained by a disturbed volume regulation in conjunction with a hyperexcitability resulting from impaired inhibition because of an elevated $[Cl^-]_i$. It was suggested that $[Cl^-]_i$ increased from 5.6 mM to 8.3 mM if the main Cl^- regulating system KCC2 was not supported by KCC3 in Purkinje cells. It was further argued that KCC3 transport activity was low under isotonic conditions and hence, KCC2 activity would prevail. However, if a 'thermodynamic regulation' governs KCC2 (Payne et al., 2003), any change in $[Cl^-]_i$ resulting from KCC3 deletion in presence of an intact KCC2 transporter is difficult to understand. A considerable body of evidence suggests that degeneration of neurons is associated with an impaired $[Cl^-]_i$ regulation through KCC2 (see below), therefore the cause for the increased $[Cl^-]_i$ could be well sought in the process leading to neuronal degeneration initiated in neurons by the KCC3 deletion.

In conclusion, all available evidence suggests that KCC2 is the main Cl^- regulating system in neurons. The $GABA_A$ and glycine channel permeability assign a specific role for Cl^- in postsynaptic inhibition. The polarity of the membrane potential change resulting from a change in Cl^- conductance depends on $[Cl^-]_i$, possibly with a minor contribution of HCO_3^- ions (Kaila, 1994). Under varying activities of KCC2, $[Cl^-]_i$ could change and, hence, the driving force for postsynaptic inhibitory currents. This could provide a considerable plasticity and activity dependence to postsynaptic inhibition and, thereby, also play a role in long term plasticity. In this scenario the Cl^- conductance activated by the neurotransmitters GABA and glycine covers a voltage range which allows a perfect and flexible tuning of neuronal activity within a voltage range delineated by K^+ conductances activated during K^+ dependent synaptic inhibition on the one hand and unspecific cationic conductance activated during synaptic excitation on the other hand.

3. REGULATION OF CHLORIDE TRANSPORT AND PLASTICITY OF POSTSYNAPTIC INHIBITION

In this review we would like to explore whether KCC2 up- and/or downregulation can play a role in the generation of synaptic plasticity including LTP. The claim for a role would require that KCC2 is regulated by neuronal activity. Processes which could involve a regulation of KCC2 by activity are: **1)** The developmental switch in the response to inhibitory transmitters provided $[Ca^{2+}]_i$ increases are an important trigger for the upregulation of KCC2 function. **2)** The downregulation of KCC2 following deafferentation. **3)** The downregulation of KCC2 mediated by the release of BDNF during epileptic activity. **4)** Changes in Cl^- transporter activity by coincident pre- and postsynaptic activity. Although these are diverse processes with very different regulations of Cl^- transport they might share some common mechanisms of which a better understanding has to await future studies.

3.1 Development

There is a profound change in the response to the inhibitory transmitters, GABA and glycine, during the ontogenetic development of the brain: The response switches from depolarizing, eventually excitatory, to hyperpolarizing. Many studies have addressed the question which molecular factors might control the transformation of the response. The major change held responsible for the developmental switch is an upregulation of KCC2 (Rivera et al., 1999) possibly in conjunction with a downregulation of NKCC1 (Payne et al., 2003). It was argued that the developmental switch in the GABA response depends on GABA itself (see below) which suggests that the developmental switch depends on synaptic and, possibly, electrical activity. The developmental switch may even provide insights in factors regulating KCC2 activity throughout life.

In immature hippocampal neurons, the synergistic action of depolarizing GABA and glutamate responses drives giant depolarizing potentials (GDPs) which are network-driven membrane oscillations due to recurrent membrane depolarization with superimposed fast action potentials (Ben-Ari et al., 1989). They have been recorded in acute slices and also in the intact brain of rat pups (Leinekugel et al., 2002). The depolarizing action of GABA results in calcium influx through the activation of NMDA receptors and of voltage dependent calcium channels. Thereby GABA may mediate several developmentally important functions such as cell proliferation, migration, neurite outgrowth (Borodinsky et al., 2003) and enhancement of synaptic efficacy at emerging synapses (Meier et al., 2003; Kasyanov et al., 2004). GDPs disappear towards the end of the first postnatal week, when GABA becomes inhibitory because of a developmental shift of E_{GABA} towards more hyperpolarized potentials (Ben-Ari et al., 1989).

Because KCC2 establishes the hyperpolarizing inhibition (Rivera et al., 1999), the developmental upregulation of KCC2 function limits the trophic action of GABA. It was argued that the limitation by upregulation of KCC2 expression was promoted by GABA itself in cultured hippocampal neurons (Ganguly et al., 2001). Culturing in the presence of the $GABA_A$ receptor antagonist bicuculline prevented the developmental disappearance of GABA induced $[Ca^{2+}]_i$ increases and the upregulation in the expression level of KCC2. However, electrical activity was not needed for the change in the GABA response, hence the authors suggested that action potential independent GABA release was sufficient to change the GABA response. Pharmacological experiments suggested an involvement of L-type Ca^{2+} channels. Thus, the suggestion was that GABA itself through $GABA_A$ receptors would regulate its own trophic action in a self limited fashion by increasing $[Ca^{2+}]_i$ which then promoted the upregulation of KCC2.

Unfortunately, a toxic effect of the pharmacological manipulations was not excluded which is important, because KCC2 is downregulated in traumatized neurons (see below). This may be the reason, why we and others were not able to confirm this attractive hypothesis. In a variety of preparations, neither electrical activity nor synaptic activity nor an activation of ionotropic receptors were a necessary requirement for the developmental change in the GABA response (Titz et al., 2003) and the upregulation of KCC2 protein expression (Ludwig et al., 2003). This is not to say that ambient GABA if surrounding developing neurons would not have an effect. Ambient GABA through the activation of $GABA_A$ receptors will dissipate existing Cl^- gradients, unless they are maintained by transport mechanisms. We could show that, in the nominal absence of HCO_3^- ions, there is not much constitutive capacity for Cl^- accumulation in cultured neurons even if they display depolarizing GABA responses (Titz et al., 2003). Therefore, it is likely that the absence or the loss of constitutive KCC2 activity allows a variety of factors including membrane depolarization and HCO_3^- dependent and independent Cl^- inward transport to elevate $[Cl^-]_i$. A contribution of NKCCs to the Cl^- accumulating capacity of immature neurons (Payne et al., 2003) is also possible, but requires regulated activity (Russell, 2000) in the absence of a constitutive activity which obscures a simple correlation to expression levels. Summarizing these data, the developmental switch is probably not a good example for an activity dependent regulation of KCC2 expression.

3.2 Trauma

Traumatic insults may restore an immature state in neurons including the reappearance of depolarizing GABA responses. Several reports have shown that direct damage of a neuron reduces the expression of KCC2 (van den Pol et al., 1996; Nabekura et al., 2002; Malek et al., 2003; Toyoda et al., 2003; Pond et al., 2004). In contrast, there is only few data available indicating that altered synaptic activity after traumatic insults also changes KCC2 expression and the strength of inhibition. Recently it has been shown that peripheral neuropathy induced by transient compression of the sciatic nerve results in a trans-synaptic (downstream from the injured neuron) positive shift in E_{GABA} and downregulation of KCC2 in neurons of lamina I of the superficial dorsal horn. In some neurons the change in E_{GABA} even resulted in GABA induced $[Ca^{2+}]_i$ increases or GABA evoked action potentials. Selective knock-down of the exporter using spinal administration of an antisense oligodeoxynucleotide against KCC2 messenger RNA reduced the nociceptive threshold (Coull et al., 2003). Taken together these data suggest that the chronic pain resulting from the transient compression of the nerve may be fostered by the impediment of inhibition by the increased $[Cl^-]_i$. Interestingly in parallel with the collapsed $[Cl^-]_i$, peripheral neuropathy caused a reorganization at lamina I synapses, thereby unmasking a $GABA_A$ receptor mediated component in addition to the glycinergic component. The expression of two transmitter systems resembles the synaptic organization observed in immature laminae I–II neurons (Jonas et al., 1998). Together with the observed depolarizing GABA responses this suggests that peripheral neuropathy induced changes in the synaptic activity converging on neurons in lamina I of the superficial dorsal horn can revert them to an immature state.

Another example for a trans-synaptic and, hence, activity dependent regulation of KCC2 is provided by deafness induced changes in E_{IPSC} in gerbil inferior colliculus neurons (Vale and Sanes, 2000; Vale and Sanes, 2002; Vale et al., 2003). Bilateral cochlea ablation at P7 resulted in a functional disruption of chloride extrusion mechanisms 1–7 d after surgery while the expression level of KCC2 and also NKCC1 remained unchanged (Vale et al., 2003). Pharmacological blockade of cotransport by bumetanide and inhibition of protein tyrosine kinase by genistein which inhibits KCC2 mediated Cl^- extrusion (Kelsch et al., 2001) resulted in a positive shift of E_{IPSC} in control neurons, but in a significantly smaller shift in neurons from animals with cochlear ablation. This indicates that normal auditory activity regulates inhibitory synaptic strength through the functional status of the chloride cotransporter. The authors suggested that the activity of KCC2 was regulated by its phosphorylation status at the tyrosine residue mediated by a putative release from afferent terminals of neurotrophins that elicited postsynaptic KCC2 phosphorylation (Vale et al., 2003).

3.3 Epileptiform Activity Dependent Downregulation of KCC2

A causal link between neuronal hyperactivity and KCC2 activity was demonstrated in studies on experimental models of the epilepsies (Rivera et al., 2002; Rivera et al., 2004). The relationship appears to work in both directions, epileptiform activity impairing KCC2 function and impaired Cl^- extrusion capacity producing epileptiform activity. Depolarizing GABA actions could provide the hallmark of an increased seizure susceptibility. Indeed, in slices obtained from the resection material of patients submitted to surgery because of temporal lobe epilepsy, some subicular neurons displayed depolarizing GABA responses. These depolarizations appeared to be involved in the generation of activity reminiscent of interictal discharges (Köhling et al., 1998; Cohen et al., 2002).

Some indication for the importance of KCC2 in the generation of epileptiform activity resulted from KCC2 knock-down experiments. While KCC2 knockout mice die immediately after birth (Hübner et al., 2001), 'hypomorphic' KCC2 gene-targeted mice with only 5-10% of KCC2 expression display spontaneous, generalized seizures and die shortly after birth (Woo et al., 2002). KCC2 mRNA levels covary with seizure susceptibility in the inferior colliculus of the post-ischemic audiogenic seizure-prone rat (Reid et al., 2001).

Studies on BDNF induced tyrosine receptor kinase B (TrkB) activation provided a closer link between KCC2 and epileptiform activity. Kindling induced seizures *in vivo* downregulated KCC2 mRNA and protein expression in the mouse hippocampus (Rivera et al., 2002). The spatiotemporal profile of this downregulation was complementary to an upregulation of TrkB and BDNF and mediated by endogenous BDNF acting on TrkB. *In vitro*, a downregulation of KCC2 was induced by exogenous BDNF or neurotrophin-4 or by several forms of pharmacologically induced epileptiform activity in the hippocampal CA1 region. The fast removal of cell surface bound KCC2 which surprisingly occurred within minutes resulted in an impairment of neuronal Cl^- extrusion capacity. Direct evidence for the involvement of TrkB receptors was provided by pharmacological blockade and knock-down technology. It was suggested that cascades down-stream of the TrkB receptor involve Shc/FRS-2 (src homology 2 domain containing transforming protein/FGF receptor substrate 2) and PLCγ (phospholipase Cγ)-cAMP response element-binding protein signaling. Interestingly, these pathways are also involved in the upregulation of KCC2 expression. While activation of the PLCγ and Shc/Frs2 pathways together mediate the downregulation of KCC2, activation of the Shc/Frs2 pathway alone upregulates KCC2 expression (Rivera et al., 2004). The dualism of signaling pathways may explain the different effect of BDNF on KCC2 expression observed early in development (Aguado et al., 2003) where BDNF upregulates the expression of KCC2.

3.4 Downregulation by Coincident Activity

Because there is a necessity for new synthesis or rapid turnover of proteins for hippocampal LTP (Stanton and Sarvey, 1984), it is very likely that a protein with a turnover rate as high as it was reported for KCC2 (Rivera et al., 2004) can undergo alterations in the time window assigned to LTP. The question we would like to pose, however, is whether KCC2 activity can even be adjusted with a rapidity which, probably, does not allow for *de novo* synthesis of KCC2 protein. The adjustment should be initiated by signals which are or can be generated by either pre- or postsynaptic discharge activity. Here we explore examples which fulfill one or both these requirements.

The strongest evidence for a possible contribution of KCC2 to synaptic plasticity comes from a study reporting that coincident pre- and postsynaptic spiking activity in hippocampal neurons in culture and slices produced a positive shift of E_{IPSC} with immediate onset. Precondition was that an inhibitory presynaptic neuron produced a postsynaptic current mediated by $GABA_A$ receptors and that the postsynaptic spike fell into a narrow time window of ± 20 ms before and after the onset of the inhibitory current. The resulting shift in E_{IPSC} was long lasting on the one hand, but restricted to the synapse used. E_{IPSC}s for postsynaptic currents generated by inhibitory synapses converging on the same target neuron but not taking part in the coincident activation were not changed. The postsynaptic action potential was needed to activate Ca^{2+} influx through L-type Ca^{2+} channels. The shift of E_{IPSC} did not occur after pharmacological blockade of CCCs or in neurons lacking transport activity and was highly dependent on $[K^+]_o$ which findings all suggested that the change in E_{IPSC} was due to a local decrease in K^+-Cl^- cotransport activity, most likely KCC2 (Woodin et al., 2003).

However, this interesting proposal raises many questions for which an answer is needed to come to a better understanding. The questions concern the assumption that the downregulation of KCC2 activity and, hence, the increase in $[Cl^-]_i$ is restricted to very small dendritic compartments near the activated GABA synapse. A requirement for such a concept is an impediment of the Cl^- mobility in the cytosol guaranteeing that the change in E_{IPSC} remains stable over tens of minutes. Further, as all available immunohistochemical evidence indicates that KCC2 is not exclusively restricted to GABA synapses, there is a necessity for a close proximity of L-type Ca^{2+} channels and/or the regulatory network modulating KCC2. Such an organization into functional complexes was hitherto not proven for GABA synapses.

An attractive hypothesis is that endogenous BDNF might provide a signal in activity dependent postsynaptic modulation of $GABA_A$ receptor mediated inhibition (Rivera et al., 2004). BDNF decreases the efficacy of inhibitory transmission within minutes in cultured GABAergic but not glutamatergic cells. The apparent specificity may have resulted from the fact that the cultured glutamatergic cells did not reach sufficient maturity to express functional KCC2 (Wardle and Poo, 2003). An acute downregulation of K^+-Cl^- cotransporter activity by BDNF could result from a rapid degradation of membrane bound KCC2 which also underlies the activity dependent reduction of the capacity for Cl^- extrusion of CA1 pyramidal neurons in hippocampal slices after a prolonged period of epileptiform activity (Rivera et al., 2004; see above).

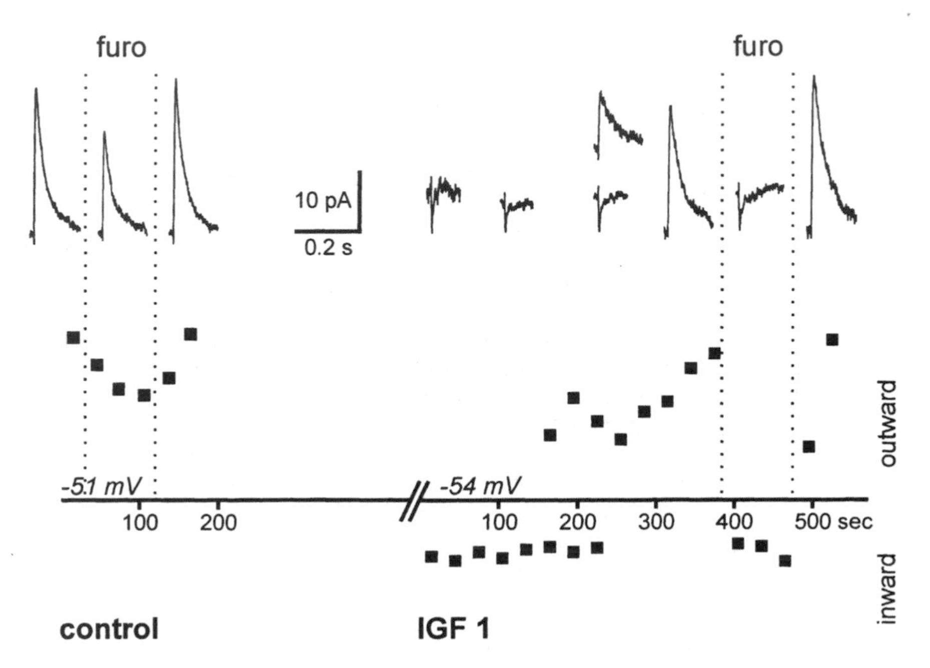

Figure 3. Rapid increase of Cl⁻ outward transport activity upon stimulation of IGF-1 receptors in a cultured hippocampal neuron perfused intracellularily with c-Src-kinase (60 U/ml). Before application of IGF-1 the inhibitor of cation-Cl⁻ cotransport furosemide (furo, 100 μM) has little effect on the spontaneous outward current mediated by $GABA_A$ receptors. Resetting the holding potential from -51mV to -54 mV reversed the inhibitory synaptic currents to inward currents which re-reversed to outward currents because Cl⁻ was increasingly extruded from the cell. Under IGF-1 receptor activation, the furosemide effect became pronounced indicating that the cell had developed considerable Cl⁻ outward transport activity. Such a change was neither observed in cells perfused internally with c-Src alone, nor externally with IGF-1 alone. Dots represent averaged spontaneous IPSCs for which sample recordings are shown on top collected and averaged within time bins of 20 s.

Most available evidence so far points to the possibility of a rapid downregulation of KCC2 function. We provided an example for an activation of which the onset occurred very rapidly, possibly before *de novo* synthesis of the protein could take place. Insulin-like growth factor 1 (IGF-1) and a cytosolic tyrosine kinase activated chloride outward transport in immature cultured hippocampal neurons within min (**Fig. 3**) and, inhibitors of tyrosine kinase dependent phosphorylation reduced KCC2 function (Kelsch et al., 2001). However, the regulation of KCC2 activity through protein tyrosine kinases may be by far more complex than anticipated. A glimpse to this complexity is provided by the observation that, depending on the activation of distinct TrkB-linked pathways, BDNF has opposite effects on KCC2, in one case triggering an upregulation, in the other a downregulation of KCC2 expression (Rivera et al., 2004). In any case, neuronal Cl⁻ regulation through KCC2 is an attractive target for a postsynaptic modulation and, hence, the plasticity of postsynaptic inhibition through ionotropic receptors.

4. CONCLUSION

Mounting evidence suggests that regulation of the neuronal Cl^- transport capacity contributes to the plasticity of Cl^- dependent postsynaptic inhibition in the brain. Shifts of $[Cl^-]_i$ in neurons allow synaptic tuning across a voltage domain that covers the range between the membrane potentials set by the other main ligand gated conductances, K^+ conductance on the one hand and unspecific cation conductance on the other hand. So far the existing literature has allowed not much more but a glimpse at a network of regulating factors by which activity dependent signals could adjust KCC2 and, hence, $[Cl^-]_i$ to changing demands for inhibition in neuronal networks. However, also irritating or deteriorating influences may lead to $[Cl^-]_i$ elevations because of reduced KCC2 function and, hence, be reinforced by an impaired inhibition, thereby contributing to neuronal vulnerability.

5. ACKNOWLEDGEMENTS

Our own studies included in this review were supported by the Deutsche Forschungsgemeinschaft (TP D9, Collaborative Research Center 488: Molecular and Cellular Bases of Neural Development, Heidelberg). The authors thank Dr. W. Kelsch for giving access to some of his unpublished work. The excellent editorial assistance of A. Lewen is gratefully acknowledged.

6. REFERENCES

Aguado F, Carmona MA, Pozas E, Aguiló A, Martínez-Guijarro FJ, Alcantara S, Borrell V, Yuste R, Ibañez CF, Soriano E (2003) BDNF regulates spontaneous correlated activity at early developmental stages by increasing synaptogenesis and expression of the K^+/Cl^- co-transporter KCC2, Development 130:1267-1280.

Andersen P, Sundberg SH, Sveen O, Wigström H (1977) Specific long-lasting potentiation of synaptic transmission in hippocampal slices, Nature 266:736-737.

Balakrishnan V, Becker M, Löhrke S, Nothwang HG, Guresir E, Friauf E (2003) Expression and function of chloride transporters during development of inhibitory neurotransmission in the auditory brainstem, J Neurosci 23:4134-4145.

Ben-Ari Y, Cherubini E, Corradetti R, Gaiarsa J-L (1989) Giant synaptic potentials in immature rat CA3 hippocampal neurones, J Physiol (Lond) 416:303-325.

Billups D, Attwell D (2002) Control of intracellular chloride concentrations and GABA response polarity in rat retinal ON bipolar cells, J Physiol (Lond) 545:183-198.

Boettger T, Rust MB, Maier H, Seidenbecher T, Schweizer M, Keating DJ, Faulhaber J, Ehmke H, Pfeffer C, Scheel O, Lemcke B, Horst J, Leuwer R, Pape H-C, Völkl H, Hübner CA, Jentsch TJ (2003) Loss of K-Cl co-transporter KCC3 causes deafness, neurodegeneration and reduced seizure threshold, EMBO J 22:5422-5434.

Borodinsky LN, O'Leary D, Neale JH, Vicini S, Coso OA, Fiszman ML (2003) GABA-induced neurite outgrowth of cerebellar granule cells is mediated by $GABA_A$ receptor activation, calcium influx and CaMKII and erk1/2 pathways, J Neurochem 84:1411-1420.

Cohen I, Navarro V, Clemenceau S, Baulac M, Miles R (2002) On the origin of interictal activity in human temporal lobe epilepsy in vitro, Science 298:1418-1421.

Coull JAM, Boudreau D, Bachand K, Prescott SA, Nault F, Sík A, De Koninck P, De Koninck Y (2003) Trans-synaptic shift in anion gradient in spinal lamina I neurons as a mechanism of neuropathic pain, Nature 424:938-942.

Eccles JC(1964) The Physiology of Synapses. Springer, Berlin.

Eccles JC, Kostyuk PG, Schmidt RF (1962) Central pathways responsible for depolarization of primary afferent fibres, J Physiol (Paris) 161:237-257.

Ganguly K, Schinder AF, Wong ST, Poo M-m (2001) GABA itself promotes the developmental switch of neuronal GABAergic responses from excitation to inhibition, Cell 105:521-532.

Gulácsi A, Lee CR, Sík A, Viitanen T, Kaila K, Tepper JM, Freund TF (2003) Cell type-specific differences in chloride-regulatory mechanisms and $GABA_A$ receptor-mediated inhibition in rat substantia nigra, J Neurosci 23:8237-8246.

Gulyás AI, Sík A, Payne JA, Kaila K, Freund TF (2001) The KCl cotransporter, KCC2, is highly expressed in the vicinity of excitatory synapses in the rat hippocampus, Eur J Neurosci 13:2205-2217.

Hara M, Inoue M, Yasukura T, Ohnishi S, Mikami Y, Inagaki C (1992) Uneven distribution of intracellular Cl^- in rat hippocampal neurons, Neurosci Lett 143:135-138.

Hübner CA, Stein V, Hermans-Borgmeyer I, Meyer T, Ballanyi K, Jentsch TJ (2001) Disruption of KCC2 reveals an essential role of K-Cl cotransport already in early synaptic inhibition, Neuron 30:515-524.

Jarolimek W, Lewen A, Misgeld U (1999) A furosemide-sensitive K^+-Cl^- cotransporter counteracts intracellular Cl^- accumulation and depletion in cultured rat midbrain neurons, J Neurosci 19:4695-4704.

Jonas P, Bischofberger J, Sandkühler J (1998) Corelease of two fast neurotransmitters at a central synapse, Science 281:419-424.

Kaila K (1994) Ionic basis of $GABA_A$ receptor channel function in the nervous system, Prog Neurobiol 42:489-537.

Kaila K, Voipio J, Paalasmaa P, Pasternack M, Deisz RA (1993) The role of bicarbonate in $GABA_A$ receptor-mediated IPSPs of rat neocortical neurones, J Physiol (Lond) 464:273-289.

Karadsheh MF, Delpire E (2001) Neuronal restrictive silencing element is found in the KCC2 gene: molecular basis for KCC2-specific expression in neurons, J Neurophysiol 85:995-997.

Kasyanov AM, Safiulina VF, Voronin LL, Cherubini E (2004) GABA-mediated giant depolarizing potentials as coincidence detectors for enhancing synaptic efficacy in the developing hippocampus, Proc Natl Acad Sci USA 101:3967-3972.

Kelsch W, Hormuzdi S, Straube E, Lewen A, Monyer H, Misgeld U (2001) Insulin-like growth factor 1 and a cytosolic tyrosine kinase activate chloride outward transport during maturation of hippocampal neurons, J Neurosci 21:8339-8347.

Köhling R, Lücke A, Straub H, Speckmann E-J, Tuxhorn I, Wolf P, Pannek H, Oppel F (1998) Spontaneous sharp waves in human neocortical slices excised from epileptic patients, Brain 121:1073-1087.

Kuner T, Augustine GJ (2000) A genetically encoded ratiometric neurotechnique indicator for chloride: capturing chloride transients in cultured hippocampal neurons, Neuron 27:447-459.

Leinekugel X, Khazipov R, Cannon R, Hirase H, Ben-Ari Y, Buzsáki G (2002) Correlated bursts of activity in the neonatal hippocampus in vivo, Science 296:2049-2052.

Liu X, Titz S, Lewen A, Misgeld U (2003) KCC2 mediates NH_4^+ uptake in cultured rat brain neurons, J Neurophysiol 90:2785-2790.

Lu J, Karadsheh M, Delpire E (1999) Developmental regulation of the neuronal-specific isoform of K-Cl cotransporter KCC2 in postnatal rat brains, J Neurobiol 39:558-568.

Ludwig A, Li H, Saarma M, Kaila K, Rivera C (2003) Developmental up-regulation of KCC2 in the absence of GABAergic and glutamatergic transmission, Eur J Neurosci 18:3199-3206.

Lux HD (1971) Ammonium and chloride extrusion: hyperpolarizing synaptic inhibition in spinal motoneurons, Science 173:555-557.

Lynch GS, Dunwiddie T, Gribkoff V (1977) Heterosynaptic depression: a postsynaptic correlate of long-term potentiation, Nature 266:737-739.

Malek SA, Coderre E, Stys PK (2003) Aberrant chloride transport contributes to anoxic/ischemic white matter injury, J Neurosci 23:3826-3836.

Marty S, Wehrlé R, Alvarez-Leefmans FJ, Gasnier B, Sotelo C (2002) Postnatal maturation of Na^+, K^+, $2Cl^-$ cotransporter expression and inhibitory synaptogenesis in the rat hippocampus: an immunocytochemical analysis, Eur J Neurosci 15:233-245.

Meier J, Akyeli J, Kirischuk S, Grantyn R (2003) $GABA_A$ receptor activity and PKC control inhibitory synaptogenesis in CNS tissue slices, Mol Cell Neurosci 23:600-613.

Misgeld U, Sarvey JM, Klee MR (1979) Heterosynaptic postactivation potentiation in hippocampal CA3 neurons: long-term changes of the postsynaptic potentials, Exp Brain Res 37:217-229.

Misgeld U, Deisz RA, Dodt HU, Lux HD (1986) The role of chloride transport in postsynaptic inhibition of hippocampal neurons, Science 232:1413-1415.

Misgeld U, Liu X, Kelsch W, Lewen A, Titz S (2004) Dependence of neuronal K-Cl cotransport on monovalent cations, Pflügers Arch Suppl 447:S87.

Müller W,Misgeld U, Lux HD (1989) γ-Aminobutyric acid-induced ion movements in the guinea pig hippocampal slice, Brain Res 484:184-191.

Nabekura J, Ueno T, Okabe A, Furuta A, Iwaki T, Shimizu-Okabe C, Fukuda A, Akaike N (2002) Reduction of KCC2 expression and $GABA_A$ receptor-mediated excitation after *in vivo* axonal injury, J Neurosci 22:4412-4417.

Payne JA (1997) Functional characterization of the neuronal-specific K-Cl cotransporter: implications for $[K^+]_o$ regulation, Am J Physiol Cell Physiol 273:C1516-C1525.

Payne JA, Stevenson TJ, Donaldson LF (1996) Molecular characterization of a putative K-Cl cotransporter in rat brain. A neuronal-specific isoform, J Biol Chem 271:16245-16252.

Payne JA, Rivera C, Voipio J, Kaila K (2003) Cation-chloride co-transporters in neuronal communication, development and trauma, Trends Neurosci 26:199-206.

Pond BB, Galeffi F, Ahrens R, Schwartz-Bloom RD (2004) Chloride transport inhibitors influence recovery from oxygen-glucose deprivation-induced cellular injury in adult hippocampus, Neuropharmacology 47:253-262.

Reid KH, Li GY, Payne RS, Schurr A, Cooper NG (2001) The mRNA level of the potassium-chloride cotransporter KCC2 covaries with seizure susceptibility in inferior colliculus of the post-ischemic audiogenic seizure-prone rat, Neurosci Lett 308:29-32.

Rivera C, Voipio J, Payne JA, Ruusuvuori E, Lahtinen H, Lamsa K, Pirvola U, Saarma M, Kaila K (1999) The K^+/Cl^- co-transporter KCC2 renders GABA hyperpolarizing during neuronal maturation, Nature 397:251-255.

Rivera C, Li H, Thomas-Crusells J, Lahtinen H, Viitanen T, Nanobashvili A, Kokaia Z, Airaksinen MS, Voipio J, Kaila K, Saarma M (2002) BDNF-induced TrkB activation down-regulates the K^+-Cl^- cotransporter KCC2 and impairs neuronal Cl^- extrusion, J Cell Biol 159:747-752.

Rivera C, Voipio J, Thomas-Crusells J, Li H, Emri Z, Sipilä S, Payne JA, Minichiello L, Saarma M, Kaila K (2004) Mechanism of activity-dependent downregulation of the neuron-specific K-Cl cotransporter KCC2, J Neurosci 24:4683-4691.

Russell JM (2000) Sodium-potassium-chloride cotransport, Physiol Rev 80:211-276.

Schwartzkroin PA (1975) Characteristics of CA1 neurons recorded intracellularly in the hippocampal in vitro slice preparation, Brain Res 85:423-436.

Shimizu-Okabe C, Yokokura M, Okabe A, Ikeda M, Sato K, Kilb W, Luhmann HJ, Fukuda A (2002) Layer-specific expression of Cl^- transporters and differential $[Cl^-]_i$ in newborn rat cortex, NeuroReport 13:2433-2437.

Stanton PK, Sarvey JM (1984) Blockade of long-term potentiation in rat hippocampal CA1 region by inhibitors of protein synthesis, J Neurosci 4:3080-3088.

Sung K-W, Kirby M, McDonald MP, Lovinger DM, Delpire E (2000) Abnormal $GABA_A$ receptor-mediated currents in dorsal root ganglion neurons isolated from Na-K-2Cl cotransporter null mice, J Neurosci 20:7531-7538.

Titz S, Hans M, Kelsch W, Lewen A, Swandulla D, Misgeld U (2003) Hyperpolarizing inhibition develops without trophic support by GABA in cultured rat midbrain neurons, J Physiol (Lond) 550:719-730.

Titz S, Hormuzdi S, Lewen A, Monyer H, Misgeld U (2004) Regulation of the developmental switch in the GABA response of cultured rat hippocampal neurons, Pflügers Arch Suppl 447:S63.

Toyoda H, Ohno K, Yamada J, Ikeda M, Okabe A, Sato K, Hashimoto K, Fukuda A (2003) Induction of NMDA and $GABA_A$ receptor-mediated Ca^{2+} oscillations with KCC2 mRNA downregulation in injured facial motoneurons, J Neurophysiol 89:1353-1362.

Ueno T, Okabe A, Akaike N, Fukuda A, Nabekura J (2002) Diversity of neuron-specific K^+-Cl^- cotransporter expression and inhibitory postsynaptic potential depression in rat motoneurons, J Biol Chem 277:4945-4950.

Vale C, Sanes DH (2000) Afferent regulation of inhibitory synaptic transmission in the developing auditory midbrain, J Neurosci 20:1912-1921.

Vale C, Sanes DH (2002) The effect of bilateral deafness on excitatory and inhibitory synaptic strength in the inferior colliculus, Eur J Neurosci 16:2394-2404.

Vale C, Schoorlemmer J, Sanes DH (2003) Deafness disrupts chloride transporter function and inhibitory synaptic transmission, J Neurosci 23:7516-7524.

van den Pol AN, Obrietan K, Chen G (1996) Excitatory actions of GABA after neuronal trauma, J Neurosci 16:4283-4292.

Vardi N, Zhang L-L, Payne JA, Sterling P (2000) Evidence that different cation chloride cotransporters in retinal neurons allow opposite responses to GABA, J Neurosci 20:7657-7663.

Wardle RA, Poo M-m (2003) Brain-derived neurotrophic factor modulation of GABAergic synapses by postsynaptic regulation of chloride transport, J Neurosci 23:8722-8732.

Williams JR, Sharp JW, Kumari VG, Wilson M, Payne JA (1999) The neuron-specific K-Cl cotransporter, KCC2. Antibody development and initial characterization of the protein, J Biol Chem 274:12656-12664.

Woo N-S, Lu J, England R, McClellan R, Dufour S, Mount DB, Deutch AY, Lovinger DM, Delpire E (2002) Hyperexcitability and epilepsy associated with disruption of the mouse neuronal-specific K-Cl cotransporter gene, Hippocampus 12:258-268.

Woodin MA, Ganguly K, Poo M-m (2003) Coincident pre- and postsynaptic activity modifies GABAergic synapses by postsynaptic changes in Cl^- transporter activity, Neuron 39:807-820.

Yamamoto C, Chujo T (1978) Long-term potentiation in thin hippocampal sections studied by intracellular and extracellular recordings, Exp Neurol 58:242-250.

THE TRUTH ABOUT MOSSY FIBER LONG-TERM POTENTIATION

Joe L. Martinez, Jr., William J. Meilandt, and Kenira J. Thompson*

1. INTRODUCTION

The mossy fiber-CA3 projection is a hippocampal pathway that is well characterized anatomically (Claiborne et al., 1986; Amaral and Witter, 1989; Chicurel and Harris, 1992; Henze et al., 2000), yet the functional involvement of this pathway in synaptic plasticity and learning remain controversial. In this chapter we address several of these issues and review current findings regarding the role of the mossy fibers in hippocampal function.

2. AUTOASSOCIATIVE CA3 NETWORK MODELS

The hippocampus is a component of the limbic system that is found within the medial temporal lobe of the brain. It is comprised of two characteristic interlocking C-shaped layers of cells, which include pyramidal cells of the hippocampus proper (fields CA1, CA2, and CA3) and granule cells of the dentate gyrus. As seen in Figure 1, the hippocampal CA3 region is uniquely situated to receive direct cortical (e.g., entorhinal cortex) and subcortical (e.g., medial septum) information via the perforant path and septal afferents, respectively, in addition to intrinsic information from the dentate gyrus via the mossy fiber pathway (Steward, 1976; Steward et al., 1976; Steward and Scoville, 1976; Amaral and Witter, 1989). Once information is processed within the CA3 region, it is then sent to the CA1 region via the Schaffer collaterals, to the contralateral CA3 region via the commissural fibers, and back to itself and neighboring CA3 cells via the recurrent/associative fibers. The CA3 pyramidal cells also send connections back to subcortical structures (e.g., lateral septum). The extensive afferent and efferent connectivity within the CA3 region suggests that this region is likely to play an important role in information processing. In fact, several computational models of the CA3 region have been developed, based on the unique circuitry of the CA3, to predict how this region processes and stores information.

*Joe L. Martinez, Jr., William J. Meilandt, Kenira J. Thompson, Cajal Neuroscience Institute, University of Texas at San Antonio, San Antonio, Texas, 78249. E-mail: JMartinez@utsa.edu

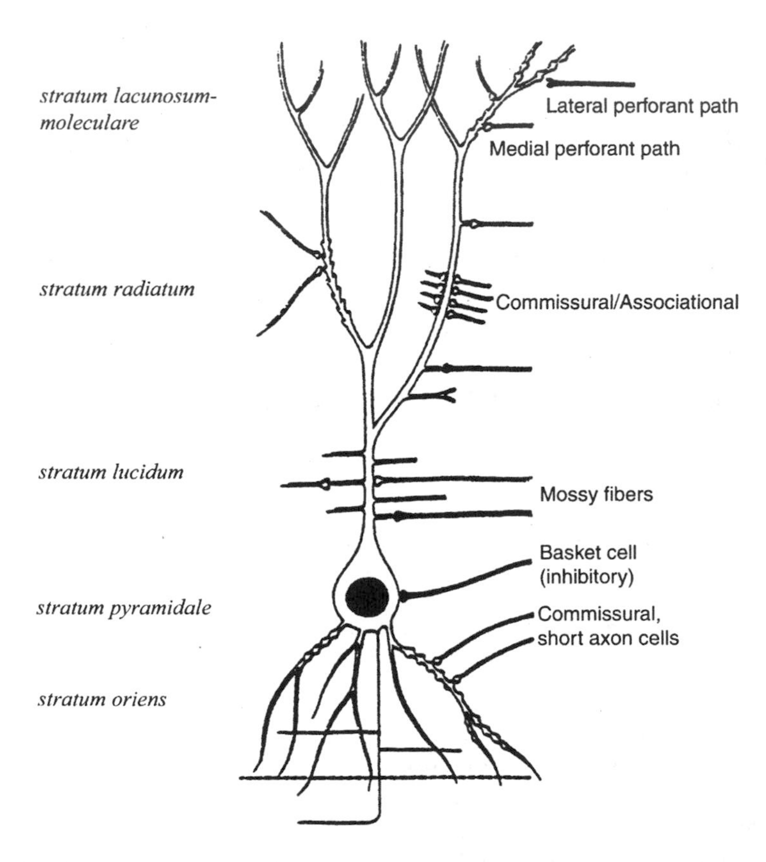

Figure 1. A hippocampal CA3 pyramidal neuron receives afferent projections from the entorhinal cortex via the medial and lateral perforant path, the dentate gyrus via the mossy fibers, and from neighboring and contralateral CA3 pyramidal cells via the Commissural/Associational fibers. Adapted from (Martinez and Barea-Rodriguez, 1997).

In 1971, David Marr proposed one of the earliest and most influential computational models of hippocampal function. In fact, portions of his model continue to be incorporated into current computational models, now called Hebb-Marr models. Marr believed that the neocortex was important for storing event memories or episodic memories. He proposed that the neocortex could store patterns of information within the strengthened connections (or synapses) of the network of cells activated during the pattern presentation. Marr believed that when the neocortex was presented with a portion of a previously learned pattern, it would be able to retrieve the full pattern through a process known as pattern completion. Marr found that the neocortex was not able to support rapid memory storage and proposed that the hippocampus may act as a short-term episodic memory storage devise that could gradually transfer patterns of information to the neocortex where it would then be incorporated with existing knowledge. He believed that this transfer occurred during sleep, a concept that remains popular (Buzsaki, 1998; Shen et al., 1998).

Marr also developed the concept of an autoassociative network that could learn to associate an input pattern with an identical output pattern through the "collateral effect" (Marr, 1971). An autoassociative network must have three basic elements: it must have 1) sparse, yet strong synapses from afferents that act as forcing synapses, 2) a high degree of recurrent connections, and 3) the synapses between coactive cells must become associatively modified or strengthened (plasticity). Although the autoassociative network proposed by Marr did not specifically identify regions of the hippocampus involved in this process, properties of CA3 pyramidal cells satisfy all of the requirements. CA3 pyramidal cells receive approximately 12000 synaptic contacts from their extensive recurrent collaterals, creating a high degree of connectivity (2%) between CA3 pyramidal cells (Ishizuka et al., 1990) that is sufficient to allow for autoassociation (Rolls, 1996). The ability of the hippocampus to store large amounts of information is facilitated by sparse encoding. For instance, information entering the hippocampus from the entorhinal cortex is first distributed over 1×10^6 dentate granule cells, which allows for pattern separation (Treves and Rolls, 1994). Pyramidal cells in the CA3 (3×10^5) then receive sparse connections from the granule cells, with approximately 50 mossy fiber inputs per pyramidal cell, creating a sparseness of 0.005% (Amaral et al., 1990; Chicurel and Harris, 1992; Claiborne et al., 1993; Rolls, 1996). The mossy fiber synapses are very large and are located proximal to the soma of CA3 pyramidal cells (Claiborne et al., 1986). In fact, activation of a single dentate granule cell is sufficient to fire CA3 pyramidal cells *in vivo* (Henze et al., 2002b; Henze et al., 2002a), and is considered to act as a "conditional-detonator". Finally, the recurrent CA3 synapses display Hebbian forms of synaptic plasticity (Bliss and Lømo, 1973).

Several computational models have been developed based on Marr's original autoassociative network model to identify the involvement of the hippocampus in various forms of memory. These include spatial memory (McNaughton and Morris, 1987b; Levy, 1996; Recce and Harris, 1996), episodic memory (Steward et al., 1976; McNaughton and Morris, 1987b; Treves and Rolls, 1994; Hasselmo, 1995), and sequence learning (Levy, 1996).

Given that the CA3 contains the architectural and synaptic properties necessary for an autoassociative network, the CA3 is believed to play an important role in the storage and retrieval of episodic memories. First, information or patterns of activity (sequences) are presented to the CA3 pyramidal cells via the sparse, yet powerful, mossy fiber synapses that can force CA3 pyramidal cells to fire (Henze et al., 2002b; Henze et al., 2002a). Associative LTP occurring in the recurrent collaterals between coactive cells then store the patterns of activity. Later during recall, when a portion of the original pattern is presented to the CA3 via the weaker perforant path, the CA3 pyramidal cells become activated and fire (Treves and Rolls, 1994). Activity in the recurrent collaterals then activates the previously modified synapses on neighboring CA3 cells until the complete pattern is retrieved.

McNaughton and Morris (1987b) proposed an alternative associative memory model, based on Marr's original model, that could associate patterns of input from sparse non-modifiable detonator synapses (corresponding to the mossy fiber input to CA3) with a second modifiable input (corresponding to the recurrent associative/commissural inputs to CA3). Patterns stored in this network could latter be recalled even when presented with a partial sequence (or cue). Their model proposed that the CA3 region might contain both a heteroassociative network (between the medial and lateral perforant path synapses) embedded within an autoassociative network of the CA3 recurrents. This model focused on the importance of synaptic plasticity (LTP) within the hippocampus for memory storage and suggested that if synapses became completely saturated, then memory retrieval and new learning would be impaired (McNaughton et al., 1986; Barnes et al., 1994; Moser et al., 1998).

Hasselmo and colleagues (Hasselmo, 1999; Hasselmo and McClelland, 1999) proposed a two-stage model of hippocampal memory storage that is modulated by acetylcholine. Their model suggests that during active waking (stage one) when acetylcholine levels are high (Kametani and Kawamura, 1990), synaptic activity in the CA3 recurrents, CA1 Schaffer collaterals, entorhinal cortex, and neocortex are suppressed. The suppression of activity in these pathways allows for information entering the CA3 to be strengthened and stored as an intermediate-term memory. Acetylcholine suppression during learning also prevents active recall of previously stored memories, which could interfere with the formation of new memories. During slow-wave sleep and quite waking (stage two) acetylcholine levels are low, which relieves the suppression and increases the activity of CA3 recurrents often leading to sharp wave activity (Buzsaki, 1998). The increased recurrent activity during low acetylcholine levels allows for the network to recall previously stored patterns while preventing new learning from occurring. Recent evidence shows that $GABA_B$ receptor modulation by GABA is also important for memory storage (Hasselmo et al., 1996; Sohal and Hasselmo, 1998a, b).

Treves and Rolls proposed a model of hippocampal function based on the anatomy of the hippocampus and from the results of lesion studies (Treves and Rolls, 1994; Rolls, 1996). Their model suggest that the hippocampus acts as an intermediate-term memory storage device for episodic events (both spatial and non-spatial), and that the CA3 region, acting as an autoassociative network, is critical for the storage and retrieval of episodic memories. They particularly stress the importance of the mossy fiber to CA3 input for learning and suggest that the mossy fiber input may act as a "teaching synapse" necessary to drive CA3 firing. Information reaching the CA3 via the perforant path is not strong enough to support learning but synaptic modifications occurring at these synapses during learning are necessary for the successful storage and retrieval of information (Treves and Rolls, 1992). Once CA3 cells are activated, the memory representation is believed to be stored within the associatively modified synapses of the recurrent collaterals. Retrieval of episodic memory representations is achieved by presenting the CA3 with partial cues via the perforant path.

In summary, these models suggest that the mossy fibers are necessary for learning new information and for the encoding of hippocampal dependent memory, whereas the perforant path to CA3 inputs are necessary for the retrieval of previously stored information. The unique forms of receptor dependent plasticity found in these pathways may therefore contribute differently to either the encoding or retrieval of certain types of memory.

3. ASSOCIATIVITY AT THE MOSSY FIBER-CA3 PATHWAY

> *"The idea is an old one, that any two cells or systems of cells that are repeatedly active at the same time will tend to become associated, so that activity in one facilitates activity in the other."*

This statement, postulated by D.O. Hebb in 1949, established the basis for associative networks in memory storage. As mentioned earlier in the chapter, in 1971 Marr proposed a similar associative network in area CA3 of the hippocampus in which distributed patterns of activity were imposed on principal cells. According to Marr, the memory trace became established as the synaptic connections were strengthened, thus supporting Hebb's Postulate. Recent work indicates that this principle is still valid: "Inputs that fire together, wire together" (Bear, 1996). Since the discovery of LTP (Bliss and Lømo, 1973) and the work of Hebb, associative connections between neurons that strengthen as a result of activity are referred to as Hebbian Synapses.

The mossy fiber projection to area CA3 of the hippocampus displays an associative (Hebbian) form of LTP that is not as well-known and studied as the typical NMDA receptor dependent forms of LTP (Bliss and Lømo, 1973; Martin, 1983; Ishihara et al., 1990; Jaffe and Johnston, 1990; Bramham et al., 1991a, b; Derrick et al., 1991; Urban and Barrionuevo, 1996). LTP induction at the mossy fiber CA3 synapse depends on the activation of μ-opioid receptors (Derrick et al., 1992) and on repetitive mossy fiber activity (Jaffe and Johnston, 1990; Zalutsky and Nicoll, 1990, 1992; Derrick and Martinez, 1994a). The time course for LTP at the mossy fiber-CA3 pathway also differs from NMDA receptor dependent LTP. Whereas NMDA-dependent LTP reaches its maximum almost immediately and can then begin to decay, mossy fiber-CA3 LTP induced *in vivo* takes about 1 hr to reach its maximum and does not decay (Derrick et al., 1991; Derrick et al., 1992; Breindl et al., 1994; Derrick and Martinez, 1994a, b). These differences are important in understanding the relevant mechanisms underlying each form of LTP.

Associative mossy fiber LTP can be elicited by high-frequency stimulation of non-opioidergic commissural afferents in conjunction with low-intensity mossy fiber stimulation (Derrick and Martinez, 1994b). The commissural-CA3 pathway expresses NMDA-receptor dependent LTP (Derrick and Martinez, 1994a, b). Figure 2 shows an example of associative mossy fiber LTP. In this experiment, the commissural fibers are first stimulated at a suprathreshold intensity, resulting in LTP in this pathway but not in the mossy fiber pathway, thus indicating that the projection of the two pathways to CA3 is independent. Upon delivery of stimulation at 25% of the intensity required to elicit LTP in the mossy fiber pathway administered concurrently with a suprathreshold stimulation to the commissural pathway, LTP is observed at both pathways. Induction of associative mossy fiber and commissural LTP is blocked by both μ-opioid and NMDA receptor antagonists, and it occludes the induction of homosynaptic mossy fiber LTP (Derrick et al., 1991; Derrick et al., 1992; Derrick and Martinez, 1994a). This associative form of mossy fiber LTP displays a time course and changes in paired pulse facilitation that are typical of homosynaptic mossy fiber LTP (Zalutsky and Nicoll, 1990; Derrick et al., 1991; Derrick et al., 1992; Derrick and Martinez, 1994a, b), suggesting that co-

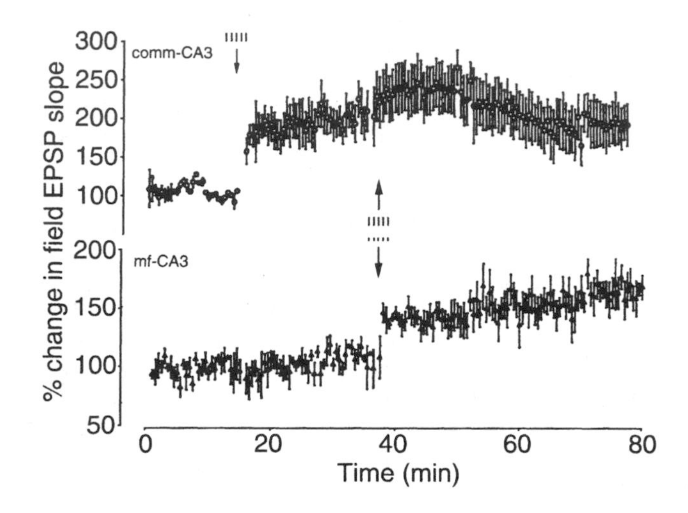

Figure 2. Associative mossy fiber LTP is observed *in vivo* when low intensity mossy fiber stimulation (mf-CA3) is paired with high intensity commissural-CA3 stimulation (comm-CA3). Adapted from (Derrick and Martinez, 1994b).

activation of mossy fiber synapses with commissural-CA3 afferents induces mossy fiber LTP.

Despite the mounting evidence describing associativity at the mossy fiber synapse, there is some controversy in the literature regarding its locus and underlying mechanisms. Some have suggested that mossy fiber LTP is independent of the postsynaptic element, and therefore non-associative (Zalutsky and Nicoll, 1992; Langdon et al., 1995). One group found that associative LTP could be induced at the commissural synapses, but not at the mossy fiber synapses (Chattarji et al., 1989). Also, presynaptic kainate receptors were found to impart associativity to mossy fiber LTP (Schmitz et al., 2003). More recent work indicates that proponents of this "presynaptic" associativity cannot rule out a postsynaptic mechanism, because kainate receptors are found both pre and post-synaptically (Darstein et al., 2003). More details on the controversy regarding the locus of LTP induction at the mossy fiber synapse are discussed below.

4. OPIOID RECEPTOR DEPENDENT LTP

The induction of LTP in the mossy fiber-CA3 projection is independent of NMDA receptor activation (Harris and Cotman, 1986). In fact, there are few NMDA receptors in the region where mossy fibers contact the CA3 pyramidal cells (Wedzony and Czyrak, 1997; Fritschy et al., 1998). However, the mossy fibers contain the densest concentration of opioids in the entire CNS. Thus, it seemed reasonable to suggest that opioids might contribute to synaptic plasticity in this region. Martin (1983) was the first to show that mossy fiber LTP induction was prevented by the opioid receptor antagonist, naloxone, in an *in vitro* hippocampal slice preparation. Since Martin's early findings, a number of different laboratories (Williams and Johnston, 1989; Ishihara et al., 1990; Derrick et al.,

1991; Derrick et al., 1992; Derrick and Martinez, 1994a, b; Jin and Chavkin, 1999) have shown that opioid receptor antagonists prevent the induction of either *in vivo* or *in vitro* mossy fiber LTP in guinea pigs and rats. However, Weiskoff et al., (1993) failed to find an effect of naloxone on LTP induction in the mossy fibers. This discrepancy is likely due to differences in stimulation frequencies used by different investigators. For example, Mata et al., (2000) showed that mossy fiber LTP induced by a one 5 second, 25 Hz train is not affected by naloxone. By contrast, mossy fiber LTP induced by two, 1 second, 100 Hz trains separated by 20 seconds was blocked by naloxone. Thus, high frequency activity of the mossy fibers is necessary to activate opioid activity. This conclusion agrees with results showing that displacement of μ-opioid receptor ligands was observed in the *stratum lucidum* of area CA3 following high frequency stimulation (Wagner et al., 1990).

Interestingly, both opioid receptor-dependent and opioid receptor-independent induction of LTP in the mossy fibers is blocked by AIDA, a class 1 metabotropic glutamate receptor (mGluR) antagonist, showing that the opioidergic modulation at the mossy fiber pathway is potentially mediated by glutamate. However, the site of action of opioids in the hippocampus is not fully understood. Most opioid receptors in the hippocampus are found on interneurons (Drake and Milner, 1999), and Jin and Chavkin (1999) suggests that effects of opioids on synaptic plasticity are mediated through a reduction in $GABA_B$ receptor activation. Others report opioid receptors exist on CA3 pyramidal cells (Stengaard-Pedersen, 1983; Stengaard-Pedersen et al., 1983; Onodera and Kogure, 1988) and that opioids have direct actions on the M-current (Moore et al., 1988; Moore et al., 1994; Madamba et al., 1999). Future research will clarify where the site of action of opioids is located, and it may turn out that opioids act at multiple sites to affect synaptic plasticity.

The fact that opioids only affect synaptic plasticity under conditions of high frequency stimulation suggests that their actions have behavioral relevance. Granule cells are relatively quiescent and are under tight inhibitory control (Cohen et al., 2003). They become active, at frequencies that will release opioids, only when the animal is exploring (Buzsaki and Czeh, 1992; Moser et al., 1994; Moser, 1996). It is during such times that animals will make associations that involve space. Thus, as discussed above, the dentate gyrus, which is likely responsible for the initial establishment of memory traces, utilizes opioid systems in a unique way, that enables learning.

5. MOSSY FIBER-CA3 LTP INDUCTION: PRE-POST?

The mechanisms underlying mossy fiber-CA3 LTP are particularly important because of the pathway's strategic localization in the hippocampal circuitry. Evidence suggests that Ca^{2+} entry is necessary for mossy fiber LTP induction (Ito and Sugiyama, 1991; Bashir et al., 1993b; Bashir et al., 1993a; Williams and Johnston, 1996; Yeckel et al., 1999). Multiple gene expression changes involved in synaptic plasticity, neurotransmission, transcription factors, cell survival, trafficking, and ion channels occur in the hippocampus following LTP induction at the mossy fiber pathway (Thompson et al., 2003). These gene alterations have enhanced our understanding of the potential pre- and postsynaptic mechanisms underlying opioid-dependent synaptic plasticity in the hippocampus. However, little is known regarding the specific receptor systems that interact with opioid receptors to result in LTP induction at the mossy fiber pathway.

Some of the most recent electrophysiological work focuses on elucidating the role of metabotropic (mGluR) and kainate glutamate receptors on LTP at the mossy fiber-CA3 synapse. Recently, kainate 1 (KA1) and kainate 2 (KA2) receptors were found both presynaptically and postsynaptically at the mossy fiber synapses and are thought to coassemble with mGluR6 subunits to modulate mossy fiber activity (Darstein et al., 2003). Also, presynaptic kainate receptors at mossy fiber synapses are suggested to initiate a cascade involving Ca^{2+} release from intracellular stores that is important in both short-term and long-term plasticity at this synapse (Lauri et al., 2001; Lauri et al., 2003). As with other systems involved in mossy fiber transmission, the results of mGluRs activation/inhibition on mossy fiber LTP have been quite contradictory. For example, the mGluR antagonist MCPG blocked mossy fiber LTP *in vitro* (Bashir et al., 1993a) in one preparation, but not in another (Manzoni et al., 1994; Yeckel et al., 1999). Yeckel et al., (1999) reported that *in vitro* mossy fiber LTP induction was dependent on the activation of mGluRs, whereas Mellor and Nicoll (2001) found that mGluRs antagonist had no effect on mossy fiber LTP *in vitro*. Additional work using mGluR1-knockout mice and mGluR antagonist did not identify a role for mGluRs in mossy fiber LTP (Hsia et al., 1995), whereas a different study using mGluR1-knockout mice reported impaired mossy fiber LTP (Conquet et al., 1994).

Class I mGluRs have been localized postsynaptically to CA3 dendritic spines (Lujan et al., 1996) and when activated evoke intracellular Ca^{2+} release in CA3 neurons (Kapur et al., 2001). We found that the mGluR antagonist AIDA blocks mossy fiber LTP *in vivo* both by systemic and intrahippocampal administration (Thompson, K.J. et al., unpublished work). Our findings reiterate the importance of postsynaptic Ca^{2+} in LTP induction at the mossy fiber synapse. Various studies show that any experimental procedure that impedes an increase in postsynaptic Ca^{2+} consequently disrupts LTP at this synapse (Yeckel et al., 1999; Kapur et al., 2001). Our work *in vivo* with AIDA shows an mGluR-mediated blockade of LTP induction at the mossy fiber synapse that is potentially due to an mGluR effect on Ca^{2+}-mediated transmission at the mossy fiber-CA3 synapse.

Despite the obvious involvement of multiple receptor systems in mossy fiber LTP induction, there is still some controversy regarding the underlying mechanisms for this induction. Most would agree that LTP at this pathway depends on trains of pulses and extracellular Ca^{2+}. However, the site of Ca^{2+} entry is still being disputed (Williams and Johnston, 1989; Zalutsky and Nicoll, 1990). Some studies show that LTP induction at the mossy fiber pathway is entirely presynaptic (Staubli et al., 1990; Zalutsky and Nicoll, 1990; Ito and Sugiyama, 1991; Nicoll et al., 1994; Langdon et al., 1995). Others indicate that LTP induction at this pathway displays clear Hebbian characteristics, and requires both pre and post-synaptic activity (Jaffe and Johnston, 1990; Johnston et al., 1992; Derrick and Martinez, 1994b). A possible resolution to this controversy lies in the induction protocol used. Urban and Barrionuevo (1996) found that different trains of stimulation result in LTP induction at this pathway. One form of LTP at the mossy fiber pathway is induced by long-lasting high frequency stimulation, and requires at least 30 pulses during a single train (Derrick and Martinez, 1994a, b). On the contrary, mossy fiber LTP induction by brief tetanic train requires both pre and postsynaptic mechanisms.

Mossy fiber terminals also contain high concentrations of chelatable zinc (Zn^{2+}), which is released in a stimulus-dependent manner at the hilar and CA3 regions (Amaral and Dent, 1981; Ueno et al., 2002). Translocation of synaptically released Zn^{2+} is required for LTP induction at the mossy fiber-CA3 pathway (Li et al., 2001a; Li et al., 2001b). More recently, an imaging analysis of neurotransmitter release at the mossy fiber synapse using the zinc sensitive dye, FluoZin-3, revealed a clear postsynaptic expression of LTP at this synapse (Quian and Noebels, 2004). There is also some evidence to suggest that Zn^{2+} interacts with opioid receptors (Stengaard-Pedersen et al., 1981), which is particularly important due to the high concentration of opioid peptides at the mossy fiber synapse. The definitive role of Zn^{2+} in mossy fiber LTP has not been established, yet much evidence suggests that Zn^{2+} activity is widely implicated in synaptic plasticity at the mossy fiber synapse.

Part of the debate may be related to alterations in the mossy fiber system due to *in vitro* preparations of hippocampal slices (Fiala et al., 2003; Danzer et al., 2004). Also, there are differential expression patterns of opioid peptides and receptors between rats, mice, and guinea pigs, that could potentially account for some of the observed differences (McLean et al., 1986; McLean et al., 1987).

6. MOSSY FIBERS AND BEHAVIOR

LTP is one of the leading synaptic models for information storage in the brain (Martinez and Derrick, 1996). Numerous studies have shown that treatments that impair NMDA receptor-dependent forms of LTP often lead to impairments in hippocampally dependent learning tasks, such as the Morris water maze (Morris et al., 1986). As for the mossy fibers, which display an NMDA-independent form of LTP, there are several conflicting reports describing the importance of mossy fiber LTP for spatial learning. For example, treatments that attenuate mossy fiber LTP, such as deleting the gene for the type I adenylyl cyclase or the metabotropic glutamate receptor 1 (mGluR1) by homologous recombination in mice, cause spatial memory impairments during probe trials when tested in a Morris water maze (Conquet et al., 1994; Wu et al., 1995; Villacres et al., 1998). In contrast, others found that targeted deletions of tissue-type plasminogen activator (Huang and Kandel, 1996) or the regulatory or catalytic subunits of protein kinase A (Huang et al., 1995) block mossy fiber LTP but does not impair spatial memory in a water maze. Activation of μ-opioid receptors has been shown to be important for the induction of mossy fiber and lateral perforant path LTP (Derrick et al., 1992; Breindl et al., 1994; Bramham and Sarvey, 1996). Recent studies (Matthies et al., 2000; Jamot et al., 2003; Jang et al., 2003) found that deletions of the μ-opioid receptor gene not only impaired mossy fiber and lateral perforant path LTP, but also impaired the acquisition of spatial water maze learning. In contrast, we recently found that a different strain of μ-opioid receptor knockout mice (Sora et al., 1997) did not display deficits in mossy fiber LTP and instead, exhibited an enhanced spatial memory retention and long-term recognition memory in the homozygous knockout mice (Meilandt et al., 2004b), suggesting that the behavioral and physiological differences between these different knockout strains may be dependent on the genetic background of the knockout mice or perhaps which exon (exon 1-3) was targeted for deletion. It is important to note that these studies were performed in mice lacking the gene throughout development, therefore one cannot rule out the possibility that compensatory gene expression changes may account for the discrepancies.

Regardless of discrepancies between mossy fiber LTP and spatial learning, there is evidence that shows that superior spatial learning is correlated with large intra/infrapyramidal (IIF) mossy fiber projections in mice, rats, and certain strains of voles (Schopke et al., 1991; Pleskacheva et al., 2000). Animals that perform worse in the Morris water maze tend to have smaller IIF-mossy fiber projections. In addition, extensive Morris water maze training induces mossy fiber synaptogenesis (Ramirez-Amaya et al., 1999) within the IIF-mossy fiber projections. Synaptogenesis may be a mechanism by which the hippocampus is able to create more storage space for information. In agreement with this idea, opioid antagonists (Escobar et al., 1997) and treatments that impair water maze learning (Ramirez-Amaya et al., 2001) block mossy fiber synaptogenesis.

As mentioned previously, one of the unique features of the CA3 recurrent/associative fibers is that there is a high degree of synaptic connectivity between neighboring CA3 pyramidal cells along the transverse and longitudinal hippocampal planes, that are believed to be necessary for information storage in autoassociative networks (McNaughton and Morris, 1987a; Treves and Rolls, 1994). The mossy fibers inputs to area CA3 are believed to be necessary for learning or encoding of new information, whereas the perforant path inputs to CA3 are necessary for memory retrieval (Treves and Rolls, 1992, 1994; Rolls, 1996). A study by Lee and Kesner (2004) recently aimed to differentiate the mossy fiber input from the perforant path input to area CA3 for spatial learning and retrieval by selectively lesioning the dentate gyrus with colchicine (indirectly destroying the mossy fibers) and using electrolytic lesions in the CA3 region to destroy the perforant path inputs. Following training in a modified spatial Hebb-Williams maze, animals with dentate lesions were selectively impaired in a learning index, (i.e., encoding of spatial information), compared with control and perforant path lesioned animals. Perforant path lesioned animals, however, were selectively impaired in a retrieval index, but not a learning index, when compared with control and mossy fiber lesioned animals. Lassalle et al., (2000) reported that blocking mossy fiber input to CA3 caused impairments in the acquisition of spatial learning but not retrieval in mice trained in a water maze. Steffenach et al., (2002) also found that knife cuts severing the longitudinal hippocampal CA3 fibers significantly impaired the retention of previously learned spatial information and impaired new learning in a water maze.

Studies are now beginning to address the extent to which different forms of receptor-dependent synaptic plasticity (LTP) within the CA3 region contributes to learning and memory formation. Since the recurrent/associative fibers and medial perforant path to CA3 fibers display NMDA receptor-dependent forms of synaptic plasticity (LTP), studies have been performed to address the role of CA3 NMDA receptors in spatial learning and memory. Although pharmacological blockade (Lee and Kesner, 2002) or genetic ablation (Nakazawa et al., 2002) of hippocampal CA3 NMDA receptors block recurrent/associational LTP but not mossy fiber LTP, these treatments do not impair the acquisition or retention of spatial learning in a reference memory version of the water maze when tested in a familiar environment. NMDA receptors may, however, be necessary for recalling memories in the presence of limited cues (Nakazawa et al., 2002) or when animals are trained in either a new spatial environment (Lee and Kesner, 2002) or in a working memory version of the Morris water maze (Nakazawa et al., 2003).

Consistent with these studies, we have recently reported that blocking μ-opioid receptors within the dorsal CA3 region of the hippocampus significantly impaired the acquisition and retrieval of spatial memory in a reference memory version of the water maze (Meilandt et al., 2004a). Building on the findings from Lee and Kesner (2004), our results suggest that μ-opioid receptor activation in the mossy fiber pathway, and perhaps the lateral perforant pathway, to area CA3 are particularly important for the encoding of new spatial memories, whereas spatial memory retrieval is most likely dependent on μ-opioid receptor activation in the lateral perforant pathway to area CA3. Our findings provide the first evidence that μ-opioid receptor dependent plasticity in the mossy fiber and lateral perforant pathways to area CA3 are necessary for the encoding and retrieval of spatial memory as predicted by the computational models of Treves and Rolls (Treves and Rolls, 1994; Rolls, 1996).

7. ACKNOWLEDGMENTS

We thank Brian E. Derrick for comments on this manuscript. A grant from the National Institute on Drug Abuse, DA 04195 (JLM), the Texas Consortium for Behavioral Neuroscience, 1T32 MH65728-2 (WJM and KJT), and the Ewing Halsell Foundation (JLM) supported this work.

8. REFERENCES

Amaral DG, Dent JA (1981) Development of the mossy fibers of the dentate gyrus: I. A light and electron microscopic study of the mossy fibers and their expansions, J Comp Neurol 195:51-86.

Amaral DG, Witter MP (1989) The three-dimensional organization of the hippocampal formation: a review of anatomical data, Neuroscience 31:571-591.

Amaral DG, Ishizuka N, Claiborne B (1990) Neurons, numbers and the hippocampal network, Prog Brain Res 83:1-11.

Barnes C A, Jung M W, McNaughton BL, Korol DL, Andreasson K, Worley PF (1994) LTP saturation and spatial learning disruption: effects of task variables and saturation levels, J Neurosci 14:5793-5806.

Bashir ZI, Jane DE, Sunter DC, Watkins JC, Collingridge GL (1993a) Metabotropic glutamate receptors contribute to the induction of long-term depression in the CA1 region of the hippocampus, Eur J Pharmacol 239:265-266.

Bashir ZI, Bortolotto ZA, Davies CH, Berretta N, Irving AJ, Seal AJ, Henley JM, Jane DE, Watkins JC, Collingridge GL (1993b) Induction of LTP in the hippocampus needs synaptic activation of glutamate metabotropic receptors, Nature 363:347-350.

Bear MF (1996) A synaptic basis for memory storage in the cerebral cortex, Proc Natl Acad Sci U S A 93:13453-13459.

Bliss TV, Lømo T (1973) Long-lasting potentiation of synaptic transmission in the dentate area of the anaesthetized rabbit following stimulation of the perforant path, J Physiol 232:331-356.

Bramham CR, Sarvey JM (1996) Endogenous activation of mu and delta-1 opioid receptors is required for long-term potentiation induction in the lateral perforant path: dependence on GABAergic inhibition, J Neurosci 16:8123-8131.

Bramham CR, Milgram NW, Srebro B (1991a) Activation of AP5-sensitive NMDA Receptors is Not Required to Induce LTP of Synaptic Transmission in the Lateral Perforant Path, Eur J Neurosci 3:1300-1308.

Bramham CR, Milgram NW, Srebro B (1991b) Delta opioid receptor activation is required to induce LTP of synaptic transmission in the lateral perforant path in vivo, Brain Res 567:42-50.

Breindl A, Derrick BE, Rodriguez SB, Martinez JL, Jr. (1994) Opioid receptor-dependent long-term potentiation at the lateral perforant path-CA3 synapse in rat hippocampus, Brain Res Bull 33:17-24.

Buzsaki G (1998) Memory consolidation during sleep: a neurophysiological perspective, J Sleep Res 7 Suppl 1:17-23.

Buzsaki G, Czeh G (1992) Physiological function of granule cells: a hypothesis, Epilepsy Res Suppl 7:281-290.

Chattarji S, Stanton PK, Sejnowski TJ (1989) Commissural synapses, but not mossy fiber synapses, in hippocampal field CA3 exhibit associative long-term potentiation and depression, Brain Res 495:145-150.

Chicurel ME, Harris KM (1992) Three-dimensional analysis of the structure and composition of CA3 branched dendritic spines and their synaptic relationships with mossy fiber boutons in the rat hippocampus, J Comp Neurol 325:169-182.

Claiborne BJ, Amaral DG, Cowan WM (1986) A light and electron microscopic analysis of the mossy fibers of the rat dentate gyrus, J Comp Neurol 246:435-458.

Claiborne BJ, Xiang Z, Brown TH (1993) Hippocampal circuitry complicates analysis of long-term potentiation in mossy fiber synapses, Hippocampus 3:115-121.

Cohen AS, Lin DD, Quirk GL, Coulter DA (2003) Dentate granule cell GABA(A) receptors in epileptic hippocampus: enhanced synaptic efficacy and altered pharmacology, Eur J Neurosci 17:1607-1616.

Conquet F, Bashir ZI, Davies CH, Daniel H, Ferraguti F, Bordi F, Franz-Bacon K, Reggiani A, Matarese V, Conde F, et al. (1994) Motor deficit and impairment of synaptic plasticity in mice lacking mGluR1, Nature 372:237-243.

Danzer SC, Pan E, Nef S, Parada LF, McNamara JO (2004) Altered regulation of brain-derived neurotrophic factor protein in hippocampus following slice preparation, Neuroscience 126:859-869.

Darstein M, Petralia RS, Swanson GT, Wenthold RJ, Heinemann SF (2003) Distribution of kainate receptor subunits at hippocampal mossy fiber synapses, J Neurosci 23:8013-8019.

Derrick BE, Martinez JL, Jr. (1994a) Opioid receptor activation is one factor underlying the frequency dependence of mossy fiber LTP induction, J Neurosci 14:4359-4367.

Derrick BE, Martinez JL, Jr. (1994b) Frequency-dependent associative long-term potentiation at the hippocampal mossy fiber-CA3 synapse, Proc Natl Acad Sci U S A 91:10290-10294.

Derrick BE, Weinberger SB, Martinez JL, Jr. (1991) Opioid receptors are involved in an NMDA receptor-independent mechanism of LTP induction at hippocampal mossy fiber-CA3 synapses, Brain Res Bull 27:219-223.

Derrick BE, Rodriguez SB, Lieberman DN, Martinez JL, Jr. (1992) Mu opioid receptors are associated with the induction of hippocampal mossy fiber long-term potentiation, J Pharmacol Exp Ther 263:725-733.

Drake CT, Milner TA (1999) Mu opioid receptors are in somatodendritic and axonal compartments of GABAergic neurons in rat hippocampal formation, Brain Res 849:203-215.

Escobar ML, Barea-Rodriguez EJ, Derrick BE, Reyes JA, Martinez JL, Jr. (1997) Opioid receptor modulation of mossy fiber synaptogenesis: independence from long-term potentiation, Brain Res 751:330-335.

Fiala JC, Kirov SA, Feinberg MD, Petrak LJ, George P, Goddard CA, Harris KM (2003) Timing of neuronal and glial ultrastructure disruption during brain slice preparation and recovery in vitro, J Comp Neurol 465:90-103.

Fritschy JM, Weinmann O, Wenzel A, Benke D (1998) Synapse-specific localization of NMDA and GABA(A) receptor subunits revealed by antigen-retrieval immunohistochemistry, J Comp Neurol 390:194-210.

Harris EW, Cotman CW (1986) Long-term potentiation of guinea pig mossy fiber responses is not blocked by N-methyl D-aspartate antagonists, Neurosci Lett 70:132-137.

Hasselmo ME (1995) Neuromodulation and cortical function: modeling the physiological basis of behavior, Behav Brain Res 67:1-27.

Hasselmo ME (1999) Neuromodulation and the hippocampus: memory function and dysfunction in a network simulation, Prog Brain Res 121:3-18.

Hasselmo ME, McClelland JL (1999) Neural models of memory, Curr Opin Neurobiol 9:184-188.

Hasselmo ME, Wyble BP, Wallenstein GV (1996) Encoding and retrieval of episodic memories: role of cholinergic and GABAergic modulation in the hippocampus, Hippocampus 6:693-708.

Henze DA, Urban NN, Barrionuevo G (2000) The multifarious hippocampal mossy fiber pathway: a review, Neuroscience 98:407-427.

Henze DA, Wittner L, Buzsaki G (2002a) Single granule cells reliably discharge targets in the hippocampal CA3 network in vivo, Nat Neurosci 5:790-795.

Henze DA, McMahon DB, Harris KM, Barrionuevo G (2002b) Giant miniature EPSCs at the hippocampal mossy fiber to CA3 pyramidal cell synapse are monoquantal, J Neurophysiol 87:15-29.

Hsia AY, Salin PA, Castillo PE, Aiba A, Abeliovich A, Tonegawa S, Nicoll RA (1995) Evidence against a role for metabotropic glutamate receptors in mossy fiber LTP: the use of mutant mice and pharmacological antagonists, Neuropharmacology 34:1567-1572.

Huang YY, Kandel ER (1996) Modulation of both the early and the late phase of mossy fiber LTP by the activation of beta-adrenergic receptors, Neuron 16:611-617.

Huang YY, Kandel ER, Varshavsky L, Brandon EP, Qi M, Idzerda RL, McKnight GS, Bourtchouladze R (1995) A genetic test of the effects of mutations in PKA on mossy fiber LTP and its relation to spatial and contextual learning, Cell 83:1211-1222.

Ishihara K, Katsuki H, Sugimura M, Kaneko S, Satoh M (1990) Different drug-susceptibilities of long-term potentiation in three input systems to the CA3 region of the guinea pig hippocampus in vitro, Neuropharmacology 29:487-492.

Ishizuka N, Weber J, Amaral DG (1990) Organization of intrahippocampal projections originating from CA3 pyramidal cells in the rat, J Comp Neurol 295:580-623.

Ito I, Sugiyama H (1991) Roles of glutamate receptors in long-term potentiation at hippocampal mossy fiber synapses, Neuroreport 2:333-336.

Jaffe D, Johnston D (1990) Induction of long-term potentiation at hippocampal mossy-fiber synapses follows a Hebbian rule, J Neurophysiol 64:948-960.

Jamot L, Matthes HW, Simonin F, Kieffer BL, Roder JC (2003) Differential involvement of the mu and kappa opioid receptors in spatial learning, Genes Brain Behav 2:80-92.

Jang CG, Lee SY, Yoo JH, Yan JJ, Song DK, Loh HH, Ho IK (2003) Impaired water maze learning performance in mu-opioid receptor knockout mice, Brain Res Mol Brain Res 117:68-72.

Jin W, Chavkin C (1999) Mu opioids enhance mossy fiber synaptic transmission indirectly by reducing GABAB receptor activation, Brain Res 821:286-293.

Johnston D, Williams S, Jaffe D, Gray R (1992) NMDA-receptor-independent long-term potentiation, Annu Rev Physiol 54:489-505.

Kametani H, Kawamura H (1990) Alterations in acetylcholine release in the rat hippocampus during sleep-wakefulness detected by intracerebral dialysis, Life Sci 47:421-426.

Kapur A, Yeckel M, Johnston D (2001) Hippocampal mossy fiber activity evokes Ca2+ release in CA3 pyramidal neurons via a metabotropic glutamate receptor pathway, Neuroscience 107:59-69.

Langdon RB, Johnson JW, Barrionuevo G (1995) Posttetanic potentiation and presynaptically induced long-term potentiation at the mossy fiber synapse in rat hippocampus, J Neurobiol 26:370-385.

Lassalle JM, Bataille T, Halley H (2000) Reversible inactivation of the hippocampal mossy fiber synapses in mice impairs spatial learning, but neither consolidation nor memory retrieval, in the Morris navigation task, Neurobiol Learn Mem 73:243-257.

Lauri SE, Bortolotto ZA, Bleakman D, Ornstein PL, Lodge D, Isaac JT, Collingridge GL (2001) A critical role of a facilitatory presynaptic kainate receptor in mossy fiber LTP, Neuron 32:697-709.

Lauri SE, Bortolotto ZA, Nistico R, Bleakman D, Ornstein PL, Lodge D, Isaac JT, Collingridge GL (2003) A role for Ca2+ stores in kainate receptor-dependent synaptic facilitation and LTP at mossy fiber synapses in the hippocampus, Neuron 39:327-341.

Lee I, Kesner RP (2002) Differential contribution of NMDA receptors in hippocampal subregions to spatial working memory, Nat Neurosci 5:162-168.

Lee I, Kesner RP (2004) Encoding versus retrieval of spatial memory: double dissociation between the dentate gyrus and the perforant path inputs into CA3 in the dorsal hippocampus, Hippocampus 14:66-76.

Levy WB (1996) A sequence predicting CA3 is a flexible associator that learns and uses context to solve hippocampal-like tasks, Hippocampus 6:579-590.

Li Y, Hough CJ, Frederickson CJ, Sarvey JM (2001a) Induction of mossy fiber --> Ca3 long-term potentiation requires translocation of synaptically released Zn2+, J Neurosci 21:8015-8025.

Li Y, Hough CJ, Suh SW, Sarvey JM, Frederickson CJ (2001b) Rapid translocation of Zn(2+) from presynaptic terminals into postsynaptic hippocampal neurons after physiological stimulation, J Neurophysiol 86:2597-2604.

Lujan R, Nusser Z, Roberts JD, Shigemoto R, Somogyi P (1996) Perisynaptic location of metabotropic glutamate receptors mGluR1 and mGluR5 on dendrites and dendritic spines in the rat hippocampus, Eur J Neurosci 8:1488-1500.

Madamba SG, Schweitzer P, Siggins GR (1999) Dynorphin selectively augments the M-current in hippocampal CA1 neurons by an opiate receptor mechanism, J Neurophysiol 82:1768-1775.

Manzoni OJ, Manabe T, Nicoll RA (1994) Release of adenosine by activation of NMDA receptors in the hippocampus, Science 265:2098-2101.

Marr D (1971) Simple memory: a theory for archicortex, Philos Trans R Soc Lond B Biol Sci 262:23-81.

Martin MR (1983) Naloxone and long term potentiation of hippocampal CA3 field potentials in vitro, Neuropeptides 4:45-50.

Martinez JL, Jr., Derrick BE (1996) Long-term potentiation and learning, Annu Rev Psychol 47:173-203.

Martinez JL, Jr., Barea-Rodriguez EJ (1997) How the brain stores information: Hebbian mechanisms. In: Erinnern und Behalten: Wege zur Erforschung des menschlichen Gedachtnisses (Luer G, Lass L, eds), pp 39-59. Gottingen: Vandenhoeck and Ruprecht.

Mata ML, Barea-Rodriguez E, Martinez JL, Jr. (2000) The involvement of mGluR1 in two forms of mossy fiber long term potentiation in vivo. In: Society for Neuroscience, p 362. New Orleans, LA.

Matthies H, Schroeder H, Becker A, Loh H, Hollt V, Krug M (2000) Lack of expression of long-term potentiation in the dentate gyrus but not in the CA1 region of the hippocampus of mu-opioid receptor-deficient mice, Neuropharmacology 39:952-960.

McLean S, Rothman RB, Herkenham M (1986) Autoradiographic localization of mu- and delta-opiate receptors in the forebrain of the rat, Brain Res 378:49-60.

McLean S, Rothman RB, Jacobson AE, Rice KC, Herkenham M (1987) Distribution of opiate receptor subtypes and enkephalin and dynorphin immunoreactivity in the hippocampus of squirrel, guinea pig, rat, and hamster, J Comp Neurol 255:497-510.
McNaughton BL, Morris RGM (1987a) Hippocampal synaptic enhancement and information storage within a distributed memory system, Trends in Neurosciences 10:408-415.
McNaughton BL, Barnes CA, Rao G, Baldwin J, Rasmussen M (1986) Long-term enhancement of hippocampal synaptic transmission and the acquisition of spatial information, J Neurosci 6:563-571.
McNaughton N, Morris RG (1987b) Chlordiazepoxide, an anxiolytic benzodiazepine, impairs place navigation in rats, Behav Brain Res 24:39-46.
Meilandt WJ, Barea-Rodriguez E, Harvey SA, Martinez JL, Jr. (2004a) Role of hippocampal CA3 mu-opioid receptors in spatial learning and memory, J Neurosci 24:2953-2962.
Meilandt WJ, Barea-Rodriguez EJ, Jaffe DB, Sora I, Hall FS, Uhl GR, Martinez JL, Jr. (2004b) Effects of exon 1 mu-opioid receptor gene deletions on spatial memory, object recognition memory, and mossy fiber synaptic plasticity, unpublished data.
Mellor J, Nicoll RA (2001) Hippocampal mossy fiber LTP is independent of postsynaptic calcium, Nat Neurosci 4:125-126.
Moore SD, Madamba SG, Joels M, Siggins GR (1988) Somatostatin augments the M-current in hippocampal neurons, Science 239:278-280.
Moore SD, Madamba SG, Schweitzer P, Siggins GR (1994) Voltage-dependent effects of opioid peptides on hippocampal CA3 pyramidal neurons in vitro, J Neurosci 14:809-820.
Morris RG, Hagan JJ, Rawlins JN (1986) Allocentric spatial learning by hippocampectomised rats: a further test of the "spatial mapping" and "working memory" theories of hippocampal function, Q J Exp Psychol B 38:365-395.
Moser EI (1996) Altered inhibition of dentate granule cells during spatial learning in an exploration task, J Neurosci 16:1247-1259.
Moser EI, Moser MB, Andersen P (1994) Potentiation of dentate synapses initiated by exploratory learning in rats: dissociation from brain temperature, motor activity, and arousal, Learn Mem 1:55-73.
Moser EI, Krobert KA, Moser MB, Morris RG (1998) Impaired spatial learning after saturation of long-term potentiation, Science 281:2038-2042.
Nakazawa K, Sun LD, Quirk MC, Rondi-Reig L, Wilson MA, Tonegawa S (2003) Hippocampal CA3 NMDA receptors are crucial for memory acquisition of one-time experience, Neuron 38:305-315.
Nakazawa K, Quirk MC, Chitwood RA, Watanabe M, Yeckel MF, Sun LD, Kato A, Carr CA, Johnston D, Wilson MA, Tonegawa S (2002) Requirement for Hippocampal CA3 NMDA Receptors in Associative Memory Recall, Science 30:30.
Nicoll RA, Castillo PE, Weisskopf MG (1994) The role of Ca2+ in transmitter release and long-term potentiation at hippocampal mossy fiber synapses, Adv Second Messenger Phosphoprotein Res 29:497-505.
Onodera H, Kogure K (1988) Autoradiographic localization of opioid and spirodecanone receptors in the gerbil hippocampus as compared with the rat hippocampus, J Cereb Blood Flow Metab 8:568-574.
Pleskacheva MG, Wolfer DP, Kupriyanova IF, Nikolenko DL, Scheffrahn H, Dell'Omo G, Lipp HP (2000) Hippocampal mossy fibers and swimming navigation learning in two vole species occupying different habitats [In Process Citation], Hippocampus 10:17-30.
Quian J, Noebels JL (2004) Postsynaptic expression of LTP at the mossy fiber synapse revealed by imaging neurotransmitter release with a zinc sensitive dye. In: Society for Neuroscience.
Ramirez-Amaya V, Escobar ML, Chao V, Bermudez-Rattoni F (1999) Synaptogenesis of mossy fibers induced by spatial water maze overtraining, Hippocampus 9:631-636.
Ramirez-Amaya V, Balderas I, Sandoval J, Escobar ML, Bermudez-Rattoni F (2001) Spatial long-term memory is related to mossy fiber synaptogenesis, J Neurosci 21:7340-7348.
Recce M, Harris KD (1996) Memory for places: a navigational model in support of Marr's theory of hippocampal function, Hippocampus 6:735-748.
Rolls ET (1996) A theory of hippocampal function in memory, Hippocampus 6:601-620.
Schmitz D, Mellor J, Breustedt J, Nicoll RA (2003) Presynaptic kainate receptors impart an associative property to hippocampal mossy fiber long-term potentiation, Nat Neurosci 6:1058-1063.
Schopke R, Wolfer DP, Lipp HP, Leisinger-Trigona MC (1991) Swimming navigation and structural variations of the infrapyramidal mossy fibers in the hippocampus of the mouse, Hippocampus 1:315-328.
Shen J, Kudrimoti HS, McNaughton BL, Barnes CA (1998) Reactivation of neuronal ensembles in hippocampal dentate gyrus during sleep after spatial experience, J Sleep Res 7 Suppl 1:6-16.
Sohal VS, Hasselmo ME (1998a) Changes in GABAB modulation during a theta cycle may be analogous to the fall of temperature during annealing, Neural Comput 10:869-882.

Sohal VS, Hasselmo ME (1998b) GABA(B) modulation improves sequence disambiguation in computational models of hippocampal region CA3, Hippocampus 8:171-193.

Sora I, Takahashi N, Funada M, Ujike H, Revay RS, Donovan DM, Miner LL, Uhl GR (1997) Opiate receptor knockout mice define mu receptor roles in endogenous nociceptive responses and morphine-induced analgesia, Proc Natl Acad Sci U S A 94:1544-1549.

Staubli U, Larson J, Lynch G (1990) Mossy fiber potentiation and long-term potentiation involve different expression mechanisms, Synapse 5:333-335.

Steffenach HA, Sloviter RS, Moser EI, Moser MB (2002) Impaired retention of spatial memory after transection of longitudinally oriented axons of hippocampal CA3 pyramidal cells, Proc Natl Acad Sci U S A 99:3194-3198.

Stengaard-Pedersen K (1983) Comparative mapping of opioid receptors and enkephalin immunoreactive nerve terminals in the rat hippocampus. A radiohistochemical and immunocytochemical study, Histochemistry 79:311-333.

Stengaard-Pedersen K, Fredens K, Larsson LI (1981) Enkephalin and zinc in the hippocampal mossy fiber System, Brain Res 212:230-233.

Stengaard-Pedersen K, Fredens K, Larsson LI (1983) Comparative localization of enkephalin and cholecystokinin immunoreactivities and heavy metals in the hippocampus, Brain Res 273:81-96.

Steward O (1976) Topographic organization of the projections from the entorhinal area to the hippocampal formation of the rat, J Comp Neurol 167:285-314.

Steward O, Scoville SA (1976) Cells of origin of entorhinal cortical afferents to the hippocampus and fascia dentata of the rat, J Comp Neurol 169:347-370.

Steward O, Cotman C, Lynch G (1976) A quantitative autoradiographic and electrophysiological study of the reinnervation of the dentate gyrus by the contralateral entorhinal cortex following ipsilateral entorhinal lesions, Brain Res 114:181-200.

Thompson KJ, Orfila JE, Achanta P, Martinez JL, Jr. (2003) Gene expression associated with in vivo induction of early phase-long-term potentiation (LTP) in the hippocampal mossy fiber-Cornus Ammonis (CA)3 pathway, Cell Mol Biol (Noisy-le-grand) 49:1281-1287.

Treves A, Rolls ET (1992) Computational constraints suggest the need for two distinct input systems to the hippocampal CA3 network, Hippocampus 2:189-199.

Treves A, Rolls ET (1994) Computational analysis of the role of the hippocampus in memory, Hippocampus 4:374-391.

Ueno S, Tsukamoto M, Hirano T, Kikuchi K, Yamada MK, Nishiyama N, Nagano T, Matsuki N, Ikegaya Y (2002) Mossy fiber Zn2+ spillover modulates heterosynaptic N-methyl-D-aspartate receptor activity in hippocampal CA3 circuits, J Cell Biol 158:215-220.

Urban NN, Barrionuevo G (1996) Induction of hebbian and non-hebbian mossy fiber long-term potentiation by distinct patterns of high-frequency stimulation, J Neurosci 16:4293-4299.

Villacres EC, Wong ST, Chavkin C, Storm DR (1998) Type I adenylyl cyclase mutant mice have impaired mossy fiber long-term potentiation, J Neurosci 18:3186-3194.

Wagner JJ, Caudle RM, Neumaier JF, Chavkin C (1990) Stimulation of endogenous opioid release displaces mu receptor binding in rat hippocampus, Neuroscience 37:45-53.

Wedzony K, Czyrak A (1997) The distribution of the NMDA R1 subunit in the rat hippocampus--an immunocytohistochemical study, Brain Res 768:333-337.

Weisskopf MG, Zalutsky RA, Nicoll RA (1993) The opioid peptide dynorphin mediates heterosynaptic depression of hippocampal mossy fibre synapses and modulates long-term potentiation, Nature 365:188.

Williams S, Johnston D (1989) Long-term potentiation of hippocampal mossy fiber synapses is blocked by postsynaptic injection of calcium chelators, Neuron 3:583-588.

Williams SH, Johnston D (1996) Actions of endogenous opioids on NMDA receptor-independent long-term potentiation in area CA3 of the hippocampus, J Neurosci 16:3652-3660.

Wu ZL, Thomas SA, Villacres EC, Xia Z, Simmons ML, Chavkin C, Palmiter RD, Storm DR (1995) Altered behavior and long-term potentiation in type I adenylyl cyclase mutant mice, Proc Natl Acad Sci U S A 92:220-224.

Yeckel MF, Kapur A, Johnston D (1999) Multiple forms of LTP in hippocampal CA3 neurons use a common postsynaptic mechanism, Nat Neurosci 2:625-633.

Zalutsky RA, Nicoll RA (1990) Comparison of two forms of long-term potentiation in single hippocampal neurons, Science 248:1619-1624.

Zalutsky RA, Nicoll RA (1992) Mossy fiber long-term potentiation shows specificity but no apparent cooperativity, Neurosci Lett 138:193-197.

THE GLUZINERGIC SYNAPSE: WHO'S TALKING AND WHO'S LISTENING?

Christopher J. Frederickson, Michal Hershfinkel, and Leonard J. Giblin*

1. INTRODUCTION

It is now almost half a century since Maske (1955) discovered the bright red band of zinc:dithizone staining in the brain that marks the gluzinergic synapses in stratum lucidum of the hippocampal formation. These 50 years have witnessed a dramatic, now exponential, growth in our understanding of these gluzinergic synapses.

Gluzinergic neurons are the neurons that sequester both zinc and glutamate in their presynaptic vesicles and release both in a calcium- and impulse-dependent fashion. These neurons are the private "voice" of cerebrocortical and "limbic" systems, located predominantly in the limbic and cortical regions of the forebrain, and projecting almost exclusively to other cerebrocortical and limbic targets. Here we present a brief survey of the basic descriptive phenomenology of the gluzinergic system, and then focus on the possible role of synaptic zinc in synaptic plasticity, the problem on which John Sarvey was working in his last years.

2. ANATOMY

The anatomy of the gluzinergic system has now been described in broad strokes, with some intriguing details filled in as well (Fig. 1). The gluzinergic pathways (which are those with both glutamate and zinc in the terminals) are found almost exclusively in the cerebrum, with the somata located in either cerebral cortex or "limbic" nuclei of the septal and amygdalar regions (Casanovas-Aguilar et al., 2002; Franco-Pons et al., 2000; Garrett et al., 1992; Howell et al., 1991; Slomianka et al., 1990; Sorensen et al., 1993; Sorensen et al., 1995). The projections of these gluzinergic neurons are exclusively amongst these limbic and cortical regions, and thus form a vast cerebrocortical and limbo-cortical associational network. In some cortical regions, almost all of the axospinous synapses are gluzinergic (Sindreu et al., 2003). In the hippocampal formation, it appears that *all* of the principal, pyramidal, excitatory glutamatergic neurons

* C.J. Frederickson, M. Hershfinkel, L.J. Giblin, Institute for Zinc in Biology, NeuroBiotex Inc., 101 Christopher Columbus Blvd, Galveston, TX 77550-2607. e-mail: chris@neurobiotex.com

are gluzinergic: granule cell∧CA3, CA3∧CA1, and CA1∧Subiculum. In stark contrast, none of the subicular pyramidal neurons are gluzinergic (Slomianka, 1992). This hippocampal anatomy mirrors the overall gluzinergic organization in the brain, inasmuch as virtually all of the pyramids of Ammon's horn have telencephalic targets, whereas the subicular neurons project subcortically (Witter et al., 1990).

3. BIOLOGY AND CHEMISTRY OF THE GLUZINERGIC VESICLE

The life-cycle of zinc in the vesicles of the gluzinergic neurons is still poorly understood. A specific protein (znt3) decorates gluzinergic vesicles, and because mice lacking the protein znt-3 show no zinc staining in the vesicles, it is clear that the znt3 protein is somehow involved in vesicular zinc (Palmiter et al., 1996; Wenzel et al., 1997).

Figure 1. Gluzinergic terminals of the human isocortex are shown stained for zinc by the post-morten silver method of Danscher. Note the lack of terminals in layer IV, where thalamo-cortical afferents terminate. Courtesy of Jeus Perez-Clausell. (WM=white matter)

The notion that znt-3 is a vesicular transmembrane zinc pump is the most parsimonious explanation of the data so far. However, one odd fact is that the staining for zinc in vesicles is not seen while the vesicles are in the somata of neurons, but only after the vesicles have traveled some distance down the axons; moreover, even in the terminals, not all vesicles are stained (Frederickson and Danscher, 1990; Perez-Clausell and Danscher, 1985). Whether this means that zinc is being pumped into the vesicle in transit or has some other interpretation is not yet known.

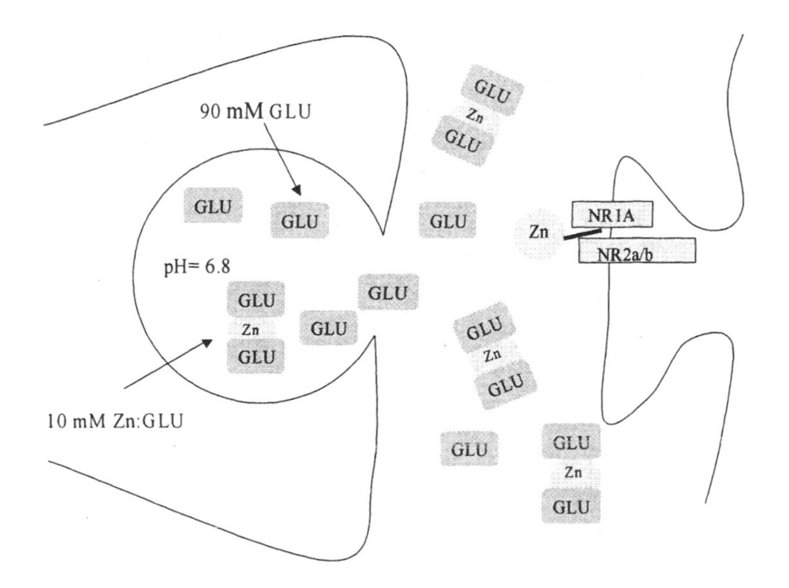

Figure 2. Exocytosis probably releases Zn:Glu complexes into the cleft, where they will dissociate, releasing zinc ions and glutamate ions. Zinc is shown binding to the high-affinity site on the NMDAR, between two specific subunits.

Regardless of how (and when) the zinc gets into the vesicles, the final disposition of the gluzinergic vesicle is that it resides in gluzinergic terminals with about 100 mM of glutamate (Riveros et al., 1986), about 1-10 mM of zinc (Frederickson, 1989), and a pH of about 6.8 (Shioi et al., 1989; Shioi and Ueda, 1990; Tabb et al., 1992). At pH 6.8 the glutamate zwitterions would be –1 anions (glu^-). The size of these vesicles ranges from 50-60 nm to the larger size (found, for example, in mossy boutons) of up to 130 nm. The smaller, 50 nm vesicles hold an estimated 2000 molecules of glutamate (Ventriglia and Di Maio, 2003), which would imply about 200 zinc ions per vesicle. Glutamate binds zinc with an affinity in the mid ($Zn:Glu_2^{2-)}$ to low (Zn:Glu) micromolar range. As a rough approximation, the interior of the vesicle would therefore be filled with about 10 mM of zinc that would be a mixture of $Zn:glu_2^{2-}$ (~ 62%), Zn:Glu (~35%), and 3 % (about 6 ions) of free Zn^{2+}. At pH 6.8, the free proton concentration is ~10 µM. What else is in the vesicle is unknown, and whether the Zn:Glu complexes combine into a polymeric form (as does for example the Zn:insulin complex (Prabhu et al., 2001)) is also unknown (Fig. 2).

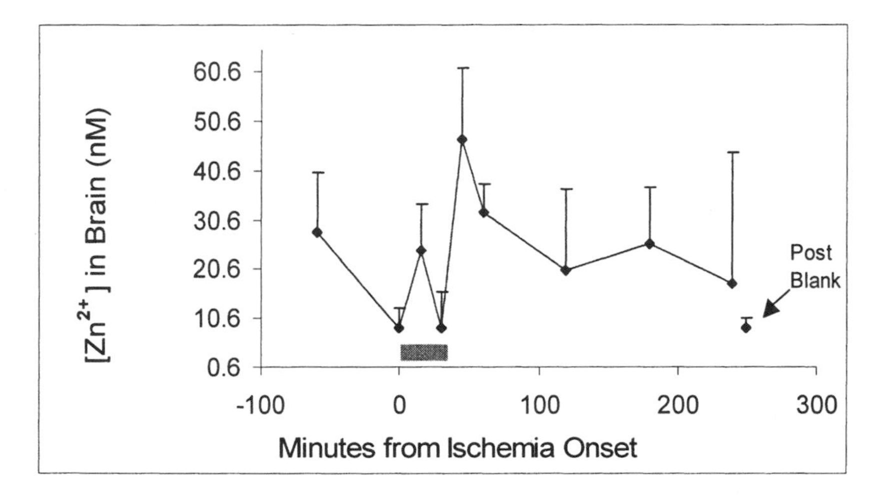

Figure 3. Release of free zinc during ischemia and reperfusion. Fluorimetry was used to measure the free zinc concentration in brain dialysates collected from four rabbits. The zinc level falls after recovery from probe insertion (to below 10 nM), then rises upon ischemia, and especially upon reperfusion. The reperfusion release persists for hours.

4. SYNAPTIC RELEASE

The fact that various types of cells secrete zinc into bodily fluids was discovered over 100 years ago, and the discoverer was nominated repeatedly (though never selected) for a Nobel Prize for his discoveries (Dr. Delezenne, nominated in 1925 & 1926; www.nobelprize.org). That work concerned zinc secretion from salivary (venom) glands and pancreas cells.

Concerning zinc secretion from neurons, fourteen different laboratories have demonstrated the stimulation-induced release of zinc from mossy-fiber axons in the hippocampal formation by various methods (Assaf and Chung, 1984; Charton et al., 1985; Haug et al., 1971; Howell et al., 1984; Itoh et al., 1993; Johansen et al., 1993; Li et al., 2001b; Perez-Clausell and Danscher, 1986; Quinta-Ferreira and Matias, 2004; Takeda et al., 1999; Ueno et al., 2002; Varea et al., 2001; Wei et al., 2004). The first group to show this was the Aarhus group led by Haug and Blackstadt, who transected the mossy fibers and observed that the zinc disappeared from the terminals within 24 hours; long before the terminals showed typical degeneration, which occurred at 2-3 days post transection (Haug et al., 1971). Others have repeated this type of before-and-after examination of zinc in the terminals and observed that the zinc also disappears after 24

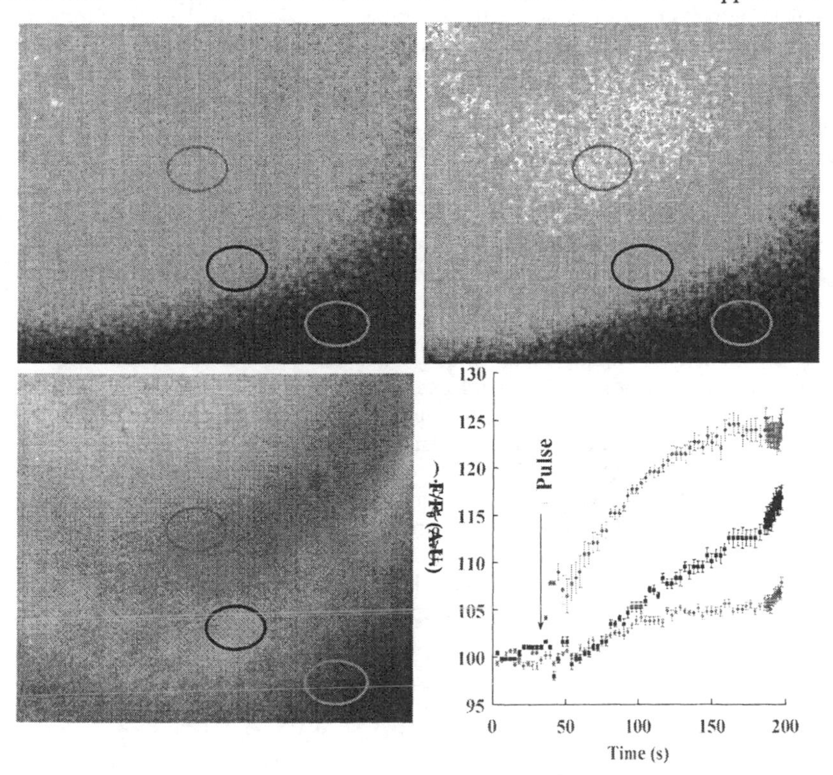

Figure 4. Synaptic release of zinc. The upper two psuedocolor images show the CA3 region of the hippocampus before (left) and after (right) a one-second train of stimulus pulses (vertical line) delivered to the granule cell stratum, outside the field of view. The rise in free zinc concentration in the extracellular fluid (recorded by zinc:ZnAF2 fluorescence) is shown for the stratum lucidum (red oval and graph), stratum pyramidale (blue) and stratum oriens (green). Static-pool slice, brightfield shown lower left.

hours of slow electrical stimulation (Sloviter, 1985), after 90 min or so of status epilepticus (Frederickson et al., 1988), and after ischemia/reperfusion (Wei et al., 2004) or brain trauma (Suh et al., 2000). In short, the zinc has been seen to disappear after essentially any "excitotoxic" brain insult.

Several groups have monitored the release of zinc from brain tissue by collecting superfusates or dialysates before, during, and after brain stimulation by high K+ or electrical stimulation. Calcium- and stimulation-dependent release has been observed by all (Aniksztejn et al., 1987; Assaf and Chung, 1984; Charton et al., 1985; Howell et al.,

1984). The most recent of these experiments has employed a fluorimetric assay of the effluent in order to directly measure free zinc (as opposed to total zinc) in the effluent fluid. In this case microdialysis of the brain of the anesthetized rabbit was used to measure the release of free zinc during the sequence of ischemia and reperfusion (Fig 3).

The definitive method of measuring the release of free zinc from gluzinergic neurons is direct imaging of the terminals in the fluorescence microscope. This was first done using the biosensor for zinc developed by Thompson and Fierke (Thompson et al., 2002) who found that up to 30 μM of free zinc was released into the medium overlying a cultured hippocampal slice (Thompson et al., 2000). Subsequently, John Sarvey's group performed a particularly thorough and elegant version of the release experiment, using acutely prepared hippocampal slices. In that work, the zinc release was detectable after just 4 single electrical pulses (one video frame during stimulation at 100 Hz), under maximum release conditions (100 HZ 5 sec), and reached a peak of about 20 μM of free zinc in the immediate vicinity of the mossy-fiber axons (Li et al., 2001b). Ueda later expanded our understanding of this release by showing that the released zinc could diffuse from the stratum lucidum and actually reach sufficient concentration in the suprajacent stratum radiatum to down-regulate the NMDA receptors in the region (Ueno et al., 2002). The findings of Ueda have recently been confirmed by the Ben Gurion group (Hershfinkel and Sekler), who have also verified the rapid diffusion of the zinc signals into the strata adjacent to stratum lucidum (Fig 4). Interestingly, although the various groups have used probes with radically-different affinities for zinc (K_D from 4 pM (Thompson et al., 2000) to 1 μM (Li et al., 2001b)), they have all reached essentially the same conclusion regarding the magnitude of the free zinc signal, i.e., in the 10-30 μM range.

5. SYNAPTIC EFFECTS OF THE ZINC SIGNAL AT THE CLEFT

There are only two tests for "zinoceptive" sites at this time: the investigator can apply an agonist, the only one of which is zinc *per se,* or one can use "antagonists," the only type of which (so far) are the zinc-binding ligands or "zinc-traps," that might catch zinc in the synapse or in the cleft, thus blocking zinc access to the zinc receptor. History shows that the first approach has tended to produce what appear to be numerous false positives, while the second has tended more toward false negatives.

Thus, to consider the extreme cases, when enough zinc salt is added to a biological solution, virtually everything will change. At the low mM and high μM concentrations that are all too often used, zinc is a Lewis acid, an osmotic factor, an electromotive force, and will stick non-specifically to membranes as well as proteins. Indeed, zinc salts are routinely used as histological *fixatives* (Mugnaini and Dahl, 1983; Wester et al., 2003). Accordingly, unless care is used to keep the concentration of the free (active, rapidly-exchangeable) ion in the low nM range, it is unclear how one might interpret the effects of added zinc.

On the other hand, trying to block endogenous zinc signals *in situ* is also difficult because of the kinetics. It is estimated that a single glutamatergic vesicle releases its contents in about 100 μsec, with comparable intervals (20-100 μsec) required for diffusion across the 20 nM cleft to a receptor (Rusakov and Kullmann, 1998). Presumably, most of that glutamate leaves the vesicle as $ZnGlu_2^{2-}$ and ZnGlu (see above).

This means that in the <100 μsec or so of diffusion, the Zn:Glu complexes must dissociate, freeing the two separate ions to bind to their respective targets. Any zinc-binding ligand or "trap" to be used as an antagonist must therefore have fast enough kinetics to bind substantial amounts of the zinc in these few microseconds of zinc's travels. This has been discussed previously by Nicoll (Vogt et al., 2000) and Li (Li et al., 2001a), both of whom have recognized that the "zinc-specific" chelator CaEDTA has such slow kinetics, that a relatively huge concentration (10 mM) is needed to have any hope of chelating the brief, transient zinc signals that occur in the cleft.

These problems notwithstanding, a consensus has emerged that synaptically-released zinc has important effects upon glutamate receptors and upon GABA receptors. On the glutamate side, both the NMDA and the AMPA/KA receptors are zinc-sensitive, but they are modulated in opposite ways: zinc reduces NMDA currents, and increases the AMPA/KA response (Hollmann et al., 1993). The NMDA receptor has a zinc-sensitive site between segment NR1 and NR2a/b that has a K_i of ~9 nM (Paoletti et al., 1997) Given that the resting pZn of the extracellular fluid of the brain is in the 0.5 to 5 nM range (Thompson, Frederickson, Fierke, unpublished), this zinc site would be 60-80% tonically occupied in a normal brain. This means that the NMDA receptor-channel is most probably tonically inhibited by zinc, as Vogt et al have observed (Vogt et al., 2000).

The kinetics of this site are noteworthy (Tovar et al., 2000). The on rate is in the mid-msec range, and the off rate is correspondingly longer, with $t_{1/2}$ ~ 3 sec (Paoletti et al., 1997). This means that there is a tonic occupancy of this receptor, with the percent occupancy moving up and down relatively slowly compared to the normal μsec regime of synaptic events. Indeed, one might envision that the $Zn{:}Glu_2^{2-}$ and Zn:Glu complexes that are dissociating in the cleft would provide transient buffering of zinc there to near their μM affinities, such that the zinc site on the NMDA receptor would enjoy a slow, sustained input.

The AMPA/KA receptor is also zinc-sensitive, although it is tuned to a zinc concentration about 10,000-fold higher than that of the NMDA receptor. Moreover, the AMPA/KA receptor is potentiated by zinc, not inhibited (Rassendren et al., 1990). With a K_o of 30 μM or so, the AMPA/KA receptor would not be tonically activated, but could easily be activated by the brief "puffs" of zinc which occur during synaptic release. As mentioned, several groups have observed zinc release in the 10-30 μM range using fluorescent imaging studies (Li et al., 2001b; Ueno et al., 2002). Moreover, if the estimate of 10 mM of zinc in a vesicle is correct, then the dilution of a single 50 nm vesicle (~ .06 zL) into a 20 nm wide synaptic cleft, would easily produce local "puffs" or "clouds" of zinc ion of hundreds of micromolar in concentration.

In one recent study of brains made epileptic by prior *status epilepticus*, it was found that newly-sprouted axons releasing zinc actually made the brain more susceptible to further seizures because the zinc was activating (up regulating) the AMPA/KA receptors (Timofeeva and Nadler, 2004).

GABAR's are consistently inhibited by zinc, and the K_i is in the low micromolar range (Barberis et al., 2000; Casagrande et al., 2003; Xie et al., 1994). Thus, like the AMPA/KA receptors, the GABA receptors would be inhibited by zinc only during the release of substantial "puffs" of zinc reaching micromolar concentrations.

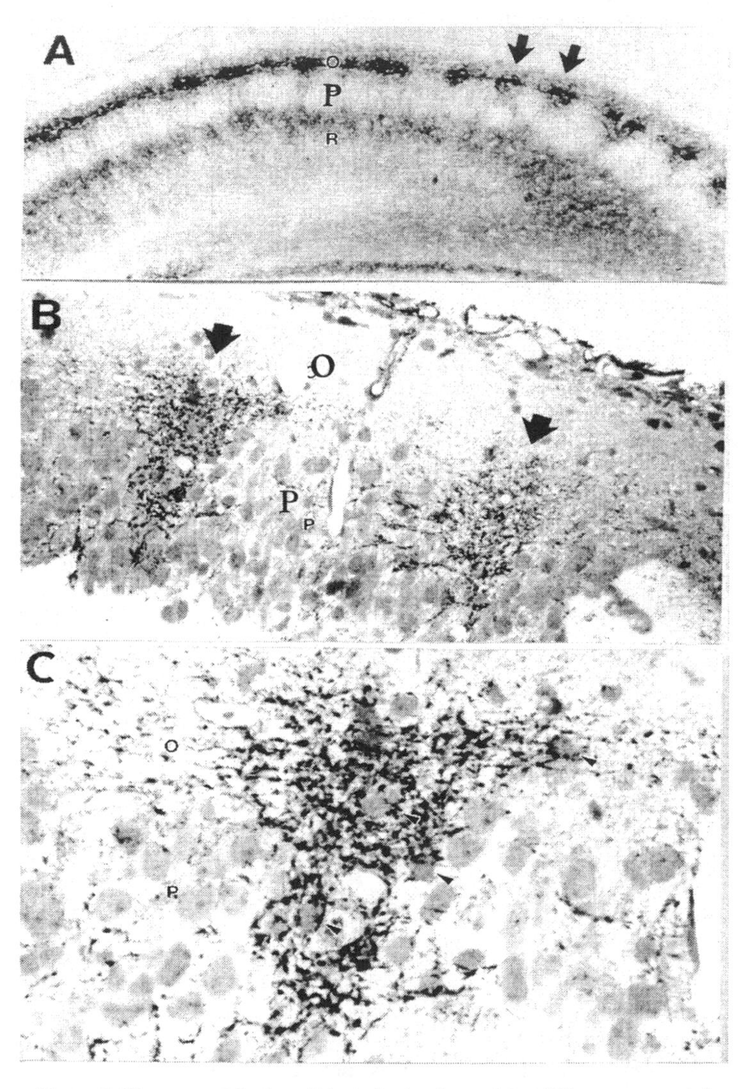

Figure 5. Hippocampal "columns" in the brain of a newborn (P0) rat pup revealed by staining of zinc with the selenium method of Danscher. Progressively-higher magnification is shown (A,B,C) to illustrate the global pattern (in CA1-3; A) as well as the intimate relationship between the gluzinergic fibers and the pyramids (B,C).

6. PLASTICITY: DEVELOPMENTAL

As mentioned, gluzinergic terminals are located almost entirely in the limbic nuclei of the amygdala and septum and in the iso- and allo-cortex of the cerebrum. Thus it is not surprising that theorists have sought to link zinc to some neuronal activity that is uniquely or preferentially associated with cerebrocortical and "limbic" neuroarchitecture. One salient possibility has always been that the gluzinergic system is involved in the neuroplasticity of the cerebral cortex, especially hippocampal-dependent "episodic" type mnemonic functions which are presumably the unique domain of cerebrocortical circuitry. In fact, there is evidence implicating gluzinergic neuronal systems in both experiential plasticity (learning) and developmental plasticity.

Suggesting an involvement of gluzinergic circuitry in developmental plasticity are data that show some intriguing, transient patterns of gluzinergic innervation during early brain development. For example, in the caudate-putamen (striatum) of the young rat pup, the so-called striosomes of the tissue can be seen to stain vividly for zinc in the first few post-natal days. Thereafter, other biochemical markers that distinguish the striosomes from the background or "matrix" tissue can be seen to develop, as if following the "pioneering" of the gluzinergic fibers. Finally, the gluzinergic innervation of the matrix appears, and thereafter the entire striatum is nearly (though not absolutely) homogenous for the gluzinergic terminal staining (Vincent and Semba, 1989). Both the cortex and the amygdala send gluzinergic collaterals to the striatum (Sorensen et al., 1995); which sources are the early, pioneer fibers to the striosomes is not yet determined.

A parallel example is seen in the septa of the barrel fields of the whisker "barrels" of the mouse isocortex. One of the first markers to delineate the barrel fields is the gluzinergic innervation (Dyck and Cynader, 1993; Land and Akhtar, 1999). Even more provocative is the fact that reduced stimulation of a single whisker, or over-stimulation of a single whisker causes, respectively, an increase or a decrease in the intensity of the zinc staining in the gluzinergic terminals of the corresponding whisker barrel (Brown and Dyck, 2002;Tsirka et al., 1996).

As a final example, the CA1-2-3 regions of the hippocampal formation in the rat show a striking "columnar" cytologic organization in gluzinergic innervation (Fig 5). These "columns" are vivid for about 2 days postnatally (day 0 and day1) when brains have been stained by Danscher's intravital selenium method for gluzinergic boutons. Interestingly, theses columnar patches of boutons are most densely distributed through the pyramidal layer (which at that age is many cells thick) and almost seem to encircle pyramids (Fig 5c). Within a few days (by P4-5) these columns disappear, as boutons in the pyramidal stratum thin, and the rest of stratum oriens and radiatum fill in with gluzinergic boutons, leaving only an even, homogeneous gluzinergic innervation. Upon what postsynaptic targets (if any) these boutons synapse during this 2 day postnatal period has not been established, nor is the cellular origin of these fibers known. Finally, it is has not been established whether there is any residual "columnar" organization in the adult hippocampal formation that follows the initial demarcation of these "pioneer" fibers in the very young hippocampal formation. Taken together, these examples suggest the possibility that gluzinergic fibers may serve as pioneer fibers, demarcating focal "patches" or "columns" and triggering induction of specialization within those regions.

Use-disuse effects on gluzinergic terminals have also been explored in the visual system (Dyck et al., 1993). Most recently, it was shown that lesions forcing reorganization of the thalamo-cortical projection system by denervation caused a

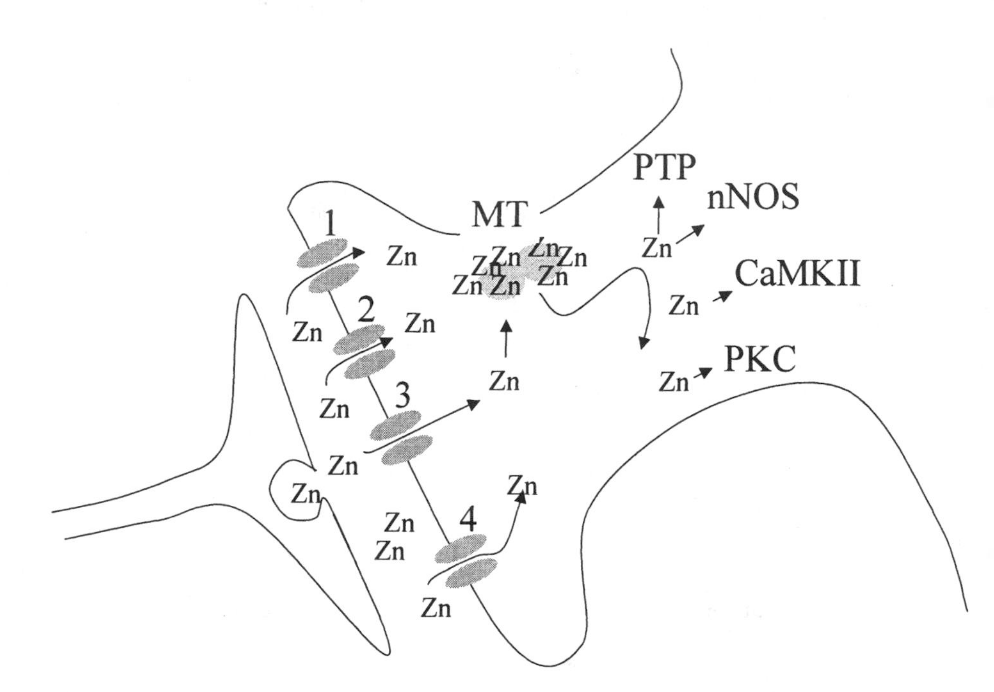

Figure 6. Zinc enters neurons through calcium-permeable AMPA/Kainate-gated channels (1), NMDARs (2), and 2 types of voltage-gated calcium channels (3,4).

contemporaneous change in the gluzinergic innervation pattern (Land and Shamalla-Hannah, 2001). This too raises the possibility of gluzinergic involvement in synaptic remodeling and reorganization. One enzymatic link that is noteworthy is that the matrix metallo-proteinases (MMP's), which act to break down and degrade extracellular matrix structure in the service of regrowth or repair, are all zinc metalloproteins (Pei, 1999).

7. PLASTICITY: EXPERIENTIAL (LTP)

Ever since John Weiss suggested that synaptically-released zinc might be important in long-term-potentiation (Weiss et al., 1989), the hunt has been on for conclusive evidence of a zinc role in LTP at the premier gluzinergic synapse – the mossy fiber to CA3 synapse. At this time, five major papers have been published on the effects of blockade of endogenous zinc signals (by chelation) upon LTP in rat brain slices. The score is running 3 to 2 against. In 1994, Simon Xie and Trevor Smart found that chelating zinc did not block LTP in their hippocampal slice preparation (Xie and Smart, 1994), and later Roger Nicoll's group found the same thing (Vogt et al., 2000). A quite recent paper by Quinta-Ferriera (Quinta-Ferreira and Matias, 2004) also found no effect of chelation upon LTP. On the other hand, both Lu et al (Lu et al., 2000) and Li et al (Li et al., 2001a) were able to block LTP by chelation strategies aimed at chelating synaptically-released zinc.

Comparison among all of these papers indicates that the type and concentration of chelator used to block the synaptically-released zinc signal may be the pivotal independent variable explaining why some groups see effects and others do not. Thus, Li et al confirmed the lack of effect reported by Nicoll with a low dose of chelator (1 mM), but were able to demonstrate an effect with a higher dose (10 mM). As Nicoll pointed out, to be effective, the chelator must have a fast "capture rate" (on rate) so that the zinc released in the few dozen microseconds of synaptic transmission can be captured. Li et al showed, by calculation, that raising the concentration (10X) raises the net effective zinc capture rate. Concerning the other 3 papers, all used different chelators and doses, so it is possible that those who saw effects (Lu et al.) simply used more effective types and doses of chelators than those who did not.

Another dimension to the complexity of this literature is that the two papers that agreed on zinc's importance actually disagreed about where the zinc might act. Li et al marshaled considerable evidence that the pivotal action of zinc actually takes place within the postsynaptic CA3 neurons, into which the zinc appears to be translocated during the LTP-inducing tetanization. Lu et al., on the other hand, offered some evidence that CA3 intracellular zinc was not an important factor in LTP induction, even though their most robust effects were obtained with zinc chelators that were freely permeable through membranes (dithizone and DEDTC).

8. PLASTICITY: EPISODIC MEMORY

It was also reported some years ago that infusion of zinc-binding drugs into the hippocampal formation caused a selective loss of episodic-type place memory, as observed in the Morris Water Maze, without any discernable impairment of procedural memory (Frederickson et al., 1990). Interestingly, the duration of the reversible impairment was directly related to the duration of histochemically-verifiable interference with the bio-availability of vesicular zinc (Lu et al., 2000). Two more recent studies, using mice tested in different behavioral tasks, have replicated the earlier work in all major aspects (Daumas et al., 2004; Lassalle et al., 2000). The one noteworthy difference in the findings of these two studies was that CaEDTA did disrupt memory trace deposition in the more recent but not in the earlier study. This discrepancy is reminiscent of the similar one in the LTP literature, where it is now known that the key difference is the use of a sufficiently large dose of CaEDTA to give effectively fast kinetics. Presumably this small difference would explain the difference in the behavioral literature.

Because virtually all tests of zinc chelation upon single, monosynaptic evoked potentials have found no effect (rev in Frederickson (Frederickson, 1989) and Li et al. (Li et al., 2001a)), one can assume that the effects of chelators upon mnemonic performance involve an effect on some higher-order aspect of synaptic function, perhaps representing the physiological version of the function that LTP is intended to mimic. Thus, the LTP and the behavioral data are loosely concordant. One additional control test that remains to be done is to test the chelators against synapses that are not gluzinergic. For example, if it should turn out that synaptic plasticity in the neocerebellar cortex (essentially without gluzinergic synapses) is not disrupted by any chelator infusion, then stronger conclusions could be drawn.

9. POSSIBLE MOLECULAR MECHANISMS OF ZINC-INDUCED PLASTICITY

Li et al (Li et al., 2001a) adduced substantial evidence that zinc modulates synaptic activity from within the postsynaptic neuron, and the data from Lu et al, as well as from all three behavioral studies of zinc and episodic memory (Daumas et al., 2004; Frederickson et al., 1990; Lassalle et al., 2000) are, by and large, consistent with that proposal. Thus, it is worth considering by what signal pathways an intracellular zinc signal in the postsynaptic pyramidal neuron could lead to a potentiation of the mossy-fiber-to-pyramid synaptic input.

In this regard, Sarvey and coworkers listed several signal cascades that are known to be modulated by zinc and to contribute to modifications in synaptic strength (Li et al., 2003). There are, in fact, a number of prominent pathways via which LTP could be produced by free zinc ions entering the postsynaptic thorny excresences of CA3 pyramidal neurons. For example, zinc modulates the activity of protein kinases, including protein kinase C (Murakami et al., 1987), mitogen-activated protein kinase (Park and Koh, 1999), and calcium-calmodulin activated kinase II (Lengyel et al., 2000). Zinc can also alter the activity of nNOS, though both up- and down-regulation have been observed in different conditions (Koh, 2001; Persechini et al., 1995). Recently, an especially versatile family of messenger sequences have been identified as zinc-modulated, namely, the protein tyrosine phosphatases (PTPs). Because zinc inhibits the PTPs, the influx of zinc would shift the intracellular balance in the direction of increased phosphorylation (Haase and Maret, 2003; Maret et al., 1999). Ultimately, these myriad signal cascades could generate either postsynaptic changes or (via NO*, for example) presynaptic changes; either of which could provide the final step of producing LTP.

10. ACKNOWLEDGEMENTS

Supported by NINDS and NeuroBioTex, Inc. We thank Dennis Dahl and John Sarvey for teaching by example, the meaning of character, humor, and integrity in both science and in life.

11. REFERENCES

Aniksztejn L, Charton G, Ben Ari Y (1987) Selective release of endogenous zinc from the hippocampal mossy fibers in situ. Brain Res 404:58-64.

Assaf SY, Chung SH (1984) Release of endogenous Zn2+ from brain tissue during activity. Nature 308:734-736.

Barberis A, Cherubini E, Mozrzymas JW (2000) Zinc inhibits miniature GABAergic currents by allosteric modulation of GABAA receptor gating. J Neurosci 20:8618-8627.

Brown CE, Dyck RH (2002) Rapid, experience-dependent changes in levels of synaptic zinc in primary somatosensory cortex of the adult mouse. J Neurosci 22:2617-2625.

Casagrande S, Valle L, Cupello A, Robello M (2003) Modulation by Zn(2+) and Cd(2+) of GABA(A) receptors of rat cerebellum granule cells in culture. Eur Biophys J 32:40-46.

Casanovas-Aguilar C, Christensen MK, Reblet C, Martinez-Garcia F, Perez-Clausell J, Bueno-Lopez JL (1995) Callosal neurones give rise to zinc-rich boutons in the rat visual cortex. Neuroreport 6:497-500.

Casanovas-Aguilar C, Miro-Bernie N, Perez-Clausell J (2002) Zinc-rich neurones in the rat visual cortex give rise to two laminar segregated systems of connections. Neuroscience 110:445-458.

Charton G, Rovira C, Ben Ari Y, Leviel V (1985) Spontaneous and evoked release of endogenous Zn2+ in the hippocampal mossy fiber zone of the rat in situ. Exp Brain Res 58:202-205.

Daumas S, Halley H, Lassalle JM (2004) Disruption of hippocampal CA3 network: effects on episodic-like memory processing in C57BL/6J mice. Eur J Neurosci 20:597-600.

Dyck R, Beaulieu C, Cynader M (1993) Histochemical localization of synaptic zinc in the developing cat visual cortex. J Comp Neurol 329:53-67.

Dyck RH, Cynader MS (1993) An interdigitated columnar mosaic of cytochrome oxidase, zinc, and neurotransmitter-related molecules in cat and monkey visual cortex. Proc Natl Acad Sci U S A 90:9066-9069.

Franco-Pons N, Casanovas-Aguilar C, Arroyo S, Rumia J, Perez-Clausell J, Danscher G (2000) Zinc-rich synaptic boutons in human temporal cortex biopsies. Neuroscience 98:429-435.

Frederickson CJ (1989) Neurobiology of zinc and zinc-containing neurons. Int Rev Neurobiol 31:145-238.

Frederickson CJ, Danscher G (1990) Zinc-containing neurons in hippocampus and related CNS structures. Prog Brain Res 83:71-84.

Frederickson CJ, Hernandez MD, Goik SA, Morton JD, McGinty JF (1988) Loss of zinc staining from hippocampal mossy fibers during kainic acid induced seizures: a histofluorescence study. Brain Res 446:383-386.

Frederickson RE, Frederickson CJ, Danscher G (1990) In situ binding of bouton zinc reversibly disrupts performance on a spatial memory task. Behav Brain Res 38:25-33.

Garrett B, Sorensen JC, Slomianka L (1992) Fluoro-Gold tracing of zinc-containing afferent connections in the mouse visual cortices. Anat Embryol (Berl) 185:451-459.

Haase H, Maret W (2003) Intracellular zinc fluctuations modulate protein tyrosine phosphatase activity in insulin/insulin-like growth factor-1 signaling. Exp Cell Res 291:289-298.

Haug FM, Blackstad TW, Simonsen AH, Zimmer J (1971) Timm's sulfide silver reaction for zinc during experimental anterograde degeneration of hippocampal mossy fibers. J Comp Neurol 142:23-31.

Hollmann M, Boulter J, Maron C, Beasley L, Sullivan J, Pecht G, Heinemann S (1993) Zinc potentiates agonist-induced currents at certain splice variants of the NMDA receptor. Neuron 10:943-954.

Howell GA, Perez-Clausell J, Frederickson CJ (1991) Zinc containing projections to the bed nucleus of the stria terminalis. Brain Res 562:181-189.

Howell GA, Welch MG, Frederickson CJ (1984) Stimulation-induced uptake and release of zinc in hippocampal slices. Nature 308:736-738.

Itoh T, Saito T, Fujimura M, Watanabe S, Saito K (1993) Restraint stress-induced changes in endogenous zinc release from the rat hippocampus. Brain Res 618:318-322.

Johansen FF, Tonder N, Berg M, Zimmer J, Diemer NH (1993) Hypothermia protects somatostatinergic neurons in rat dentate hilus from zinc accumulation and cell death after cerebral ischemia. Mol Chem Neuropathol 18:161-172.

Koh JY (2001) Zinc and disease of the brain. Mol Neurobiol 24:99-106.

Land PW, Akhtar ND (1999) Experience-dependent alteration of synaptic zinc in rat somatosensory barrel cortex. Somatosens Mot Res 16:139-150.

Land PW, Shamalla-Hannah L (2001) Transient expression of synaptic zinc during development of uncrossed retinogeniculate projections. J Comp Neurol 433:515-525.

Lassalle JM, Bataille T, Halley H (2000) Reversible inactivation of the hippocampal mossy fiber synapses in mice impairs spatial learning, but neither consolidation nor memory retrieval, in the Morris navigation task. Neurobiol Learn Mem 73:243-257.

Lengyel I, Fieuw-Makaroff S, Hall AL, Sim AT, Rostas JA, Dunkley PR (2000) Modulation of the phosphorylation and activity of calcium/calmodulin-dependent protein kinase II by zinc. J Neurochem 75:594-605.

Li Y, Hough CJ, Frederickson CJ, Sarvey JM (2001a) Induction of mossy fiber --> Ca3 long-term potentiation requires translocation of synaptically released Zn2+. J Neurosci 21:8015-8025.

Li Y, Hough CJ, Suh SW, Sarvey JM, Frederickson CJ (2001b) Rapid translocation of Zn(2+) from presynaptic terminals into postsynaptic hippocampal neurons after physiological stimulation. J Neurophysiol 86:2597-2604.

Li YV, Hough CJ, Sarvey JM (2003) Do we need zinc to think? Sci STKE 2003:e19.

Lu YM, Taverna FA, Tu R, Ackerley CA, Wang YT, Roder J (2000) Endogenous Zn(2+) is required for the induction of long-term potentiation at rat hippocampal mossy fiber-CA3 synapses. Synapse 38:187-197.

Maret W, Jacob C, Vallee BL, Fischer EH (1999) Inhibitory sites in enzymes: zinc removal and reactivation by thionein. Proc Natl Acad Sci U S A 96:1936-1940.

Maske H (1955) Uber den topochemischen Nachweis von Zink im Ammonshorn verschiedener. Saugetiere Naturwissenshaften 42:424.

Mugnaini E, Dahl AL (1983) Zinc-aldehyde fixation for light-microscopic immunocytochemistry of nervous tissues. J Histochem Cytochem 31:1435-1438.

Murakami K, Whiteley MK, Routtenberg A (1987) Regulation of protein kinase C activity by cooperative interaction of Zn2+ and Ca2+. J Biol Chem 262:13902-13906.

Palmiter RD, Cole TB, Quaife CJ, Findley SD (1996) ZnT-3, a putative transporter of zinc into synaptic vesicles. Proc Natl Acad Sci U S A 93:14934-14939.

Paoletti P, Ascher P, Neyton J (1997) High-affinity zinc inhibition of NMDA NR1-NR2A receptors. J Neurosci 17:5711-5725.

Park JA, Koh JY (1999) Induction of an immediate early gene egr-1 by zinc through extracellular signal-regulated kinase activation in cortical culture: its role in zinc-induced neuronal death. J Neurochem 73:450-456.

Pei D (1999) Identification and characterization of the fifth membrane-type matrix metalloproteinase MT5-MMP. J Biol Chem 274:8925-8932.

Perez-Clausell J (1996) Distribution of terminal fields stained for zinc in the neocortex of the rat. J Chem Neuroanat 11:99-111.

Perez-Clausell J, Danscher G (1985) Intravesicular localization of zinc in rat telencephalic boutons. A histochemical study. Brain Res 337:91-98.

Perez-Clausell J, Danscher G (1986) Release of zinc sulphide accumulations into synaptic clefts after in vivo injection of sodium sulphide. Brain Res 362:358-361.

Persechini A, McMillan K, Masters BS (1995) Inhibition of nitric oxide synthase activity by Zn2+ ion. Biochemistry 34:15091-15095.

Prabhu S, Jacknowitz AI, Stout PJ (2001) A study of factors controlling dissolution kinetics of zinc complexed protein suspensions in various ionic species. Int J Pharm 217:71-78.

Quinta-Ferreira ME, Matias CM (2004) Hippocampal mossy fiber calcium transients are maintained during long-term potentiation and are inhibited by endogenous zinc. Brain Res 1004:52-60.

Rassendren FA, Lory P, Pin JP, Nargeot J (1990) Zinc has opposite effects on NMDA and non-NMDA receptors expressed in Xenopus oocytes. Neuron 4:733-740.

Riveros N, Fiedler J, Lagos N, Munoz C, Orrego F (1986) Glutamate in rat brain cortex synaptic vesicles: influence of the vesicle isolation procedure. Brain Res 386:405-408.

Rusakov DA, Kullmann DM (1998) Extrasynaptic glutamate diffusion in the hippocampus: ultrastructural constraints, uptake, and receptor activation. J Neurosci 18:3158-3170.

Shioi J, Naito S, Ueda T (1989) Glutamate uptake into synaptic vesicles of bovine cerebral cortex and electrochemical potential difference of proton across the membrane. Biochem J 258:499-504.

Shioi J, Ueda T (1990) Artificially imposed electrical potentials drive L-glutamate uptake into synaptic vesicles of bovine cerebral cortex. Biochem J 267:63-68.

Sindreu CB, Varoqui H, Erickson JD, Perez-Clausell J (2003) Boutons containing vesicular zinc define a subpopulation of synapses with low AMPAR content in rat hippocampus. Cereb Cortex 13:823-829.

Slomianka L (1992) Neurons of origin of zinc-containing pathways and the distribution of zinc-containing boutons in the hippocampal region of the rat. Neuroscience 48:325-352.

Slomianka L, Danscher G, Frederickson CJ (1990) Labeling of the neurons of origin of zinc-containing pathways by intraperitoneal injections of sodium selenite. Neuroscience 38:843-854.

Sloviter RS (1985) A selective loss of hippocampal mossy fiber Timm stain accompanies granule cell seizure activity induced by perforant path stimulation. Brain Res 330:150-153.

Sorensen JC, Slomianka L, Christensen J, Zimmer J (1995) Zinc-containing telencephalic connections to the rat striatum: a combined Fluoro-Gold tracing and histochemical study. Exp Brain Res 105:370-382.

Sorensen JC, Tonder N, Slomianka L (1993) Zinc-positive afferents to the rat septum originate from distinct subpopulations of zinc-containing neurons in the hippocampal areas and layers. A combined fluoro-gold tracing and histochemical study. Anat Embryol (Berl) 188:107-115.

Suh SW, Chen JW, Motamedi M, Bell B, Listiak K, Pons NF, Danscher G, Frederickson CJ (2000) Evidence that synaptically-released zinc contributes to neuronal injury after traumatic brain injury. Brain Res 852:268-273.

Tabb JS, Kish PE, Van DR, Ueda T (1992) Glutamate transport into synaptic vesicles. Roles of membrane potential, pH gradient, and intravesicular pH. J Biol Chem 267:15412-15418.

Takeda A, Hanajima T, Ijiro H, Ishige A, Iizuka S, Okada S, Oku N (1999) Release of zinc from the brain of El (epilepsy) mice during seizure induction. Brain Res 828:174-178.

Thompson RB, Peterson D, Mahoney W, Cramer M, Maliwal BP, Suh SW, Frederickson C, Fierke C, Herman P (2002) Fluorescent zinc indicators for neurobiology. J Neurosci Methods 118:63-75.

Thompson RB, Whetsell WO, Jr., Maliwal BP, Fierke CA, Frederickson CJ (2000) Fluorescence microscopy of stimulated Zn(II) release from organotypic cultures of mammalian hippocampus using a carbonic anhydrase-based biosensor system. J Neurosci Methods 96:35-45.

Timofeeva OA, Nadler JV (2004) Net facilitation of granule cell epileptiform activity by mossy fiber zinc. Soc Neurosci. Abs (in press)..

Tovar KR, Sprouffske K, Westbrook GL (2000) Fast NMDA receptor-mediated synaptic currents in neurons from mice lacking the epsilon2 (NR2B) subunit. J Neurophysiol 83:616-620.

Tsirka SE, Rogove AD, Strickland S (1996) Neuronal cell death and tPA. Nature 384:123-124.

Ueno S, Tsukamoto M, Hirano T, Kikuchi K, Yamada MK, Nishiyama N, Nagano T, Matsuki N, Ikegaya Y (2002) Mossy fiber Zn2+ spillover modulates heterosynaptic N-methyl-D-aspartate receptor activity in hippocampal CA3 circuits. J Cell Biol 158:215-220.

Varea E, Ponsoda X, Molowny A, Danscher G, Lopez-Garcia C (2001) Imaging synaptic zinc release in living nervous tissue. J Neurosci Methods 110:57-63.

Ventriglia F, Di Maio V (2003) Stochastic fluctuations of the quantal EPSC amplitude in computer simulated excitatory synapses of hippocampus. Biosystems 71:195-204.

Vincent SR, Semba K (1989) A heavy metal marker of the developing striatal mosaic. Brain Res Dev Brain Res 45:155-159.

Vogt K, Mellor J, Tong G, Nicoll R (2000) The actions of synaptically released zinc at hippocampal mossy fiber synapses. Neuron 26:187-196.

Wei G, Hough CJ, Li Y, Sarvey JM (2004) Characterization of extracellular accumulation of Zn2+ during ischemia and reperfusion of hippocampus slices in rat. Neuroscience 125:867-877.

Weiss JH, Koh JY, Christine CW, Choi DW (1989) Zinc and LTP. Nature 338:212.

Wenzel HJ, Cole TB, Born DE, Schwartzkroin PA, Palmiter RD (1997) Ultrastructural localization of zinc transporter-3 (ZnT-3) to synaptic vesicle membranes within mossy fiber boutons in the hippocampus of mouse and monkey. Proc Natl Acad Sci U S A 94:12676-12681.

Wester K, Asplund A, Backvall H, Micke P, Derveniece A, Hartmane I, Malmstrom PU, Ponten F (2003) Zinc-based fixative improves preservation of genomic DNA and proteins in histoprocessing of human tissues. Lab Invest 83:889-899.

Witter MP, Ostendorf RH, Groenewegen HJ (1990) Heterogeneity in the Dorsal Subiculum of the Rat. Distinct Neuronal Zones Project to Different Cortical and Subcortical Targets. Eur J Neurosci 2:718-725.

Xie X, Hider RC, Smart TG (1994) Modulation of GABA-mediated synaptic transmission by endogenous zinc in the immature rat hippocampus in vitro. J Physiol 478 (Pt 1):75-86.

Xie X, Smart TG (1994) Modulation of long-term potentiation in rat hippocampal pyramidal neurons by zinc. Pflugers Arch 427:481-486.

ZINC DYSHOMEOSTASIS IN NEURONAL INJURY

Jade-Ming Jeng and Stefano L. Sensi*

1. INTRODUCTION

Zn^{2+} plays a vital role in diverse physiological processes, yet in excess amounts, is potently neurotoxic. In the cerebral cortex, Zn^{2+} co-localizes with glutamate in the synaptic vesicles of a subset of excitatory neurons, and is released in a Ca^{2+}-dependent fashion. *In vivo* trans-synaptic movement of Zn^{2+} and its subsequent postsynaptic intracellular accumulation contributes to the neuronal injury observed in several excitotoxic conditions, such as cerebral ischemia, epilepsy, and head trauma. Zn^{2+} may enter postsynaptic neurons through NMDA receptor-mediated channels (NMDAR), voltage-sensitive calcium channels (VSCC), Ca^{2+}-permeable AMPA/kainate receptor-mediated (Ca-A/K) channels, or Zn^{2+}-sensitive membrane transporters. The cation is also released from intracellular sites of sequestration such as metallothioneins and mitochondria. The mechanisms by which Zn^{2+} wields its potent toxic effects involve several cytosolic signaling pathways, including mitochondrial and extra-mitochondrial generation of reactive oxygen species (ROS) and disruption of metabolic enzyme activity, which ultimately lead to the activation of apoptotic and/or necrotic processes.†

As is the case with Ca^{2+}, neuronal mitochondria take up Zn^{2+} as a way of modulating cytosolic Zn^{2+} ($[Zn^{2+}]_i$) levels, yet at the same time, excessive mitochondrial Zn^{2+} sequestration leads to a marked dysfunction of these organelles characterized by prolonged ROS generation. However, Zn^{2+} appears to induce these changes with a considerably greater degree of potency than Ca^{2+}. Different intensities of $[Zn^{2+}]_i$ load may activate distinct pathways of injury, as large (i.e., micromolar) rises in $[Zn^{2+}]_i$ favor necrotic neuronal death, while submicromolar $[Zn^{2+}]_i$ increases promote release of pro-apoptotic factors. Moreover, Zn^{2+} homeostasis appears to be particularly sensitive to the environmental changes observed in ischemia, such as acidosis and oxidative stress,

---

* JM Jeng and SL Sensi, Department of Neurology, University of California, Irvine, Irvine, California, 92697-4292 and Department of Neurology, CESI-Center for Excellence on Aging, University 'G. d'Annunzio', Chieti, 66013, Italy. E-mail: ssensi@uci.edu

† Abbreviations: Ca-A/K, *calcium-permeable AMPA/kainate receptor*; Ca-A/K(+), *strongly Ca-A/K-expressing*; mPTP, *mitochondrial permeability transition pore*; MTs, *metallothioneins*; NO, *nitric oxide*; ROS, *reactive oxygen species*; TGI, *transient global ischemia*; VSCC, *voltage-sensitive calcium channel*; Zn^{2+}, *"free" ionic zinc*; $[Zn^{2+}]_i$, *intracellular* Zn^{2+} *concentration*; $\Delta\psi_m$, *mitochondrial membrane potential*

indicating that alterations in $[Zn^{2+}]_i$ may play a very significant role in the development of ischemic neuronal damage. In this chapter, we summarize the current state of knowledge regarding cellular Zn^{2+} homeostasis and pathophysiology as illuminated by investigations into the role of this ion in neuronal injury.

2. Zn^{2+} HOMEOSTASIS

As Zn^{2+} may contribute to both normal and pathologic cellular functioning, depending on intracellular concentration, it is useful to review the presently known mechanisms of cellular Zn^{2+} homeostasis. Like Ca^{2+} homeostasis, regulation of $[Zn^{2+}]_i$ levels consists of a balance between ion sequestration into subcellular compartments, intracellular buffering by macromolecules, and extrusion from the cytosol into the extracellular milieu. Sequestration and buffering appear to be largely controlled by a family of proteins called metallothioneins, while membrane-associated Zn^{2+} transporters mediate Zn^{2+} extrusion (Kägi, 1993; Hidalgo et al., 2001). Mitochondria also play a particularly important and perhaps even watershed role in the sequestration and buffering of Zn^{2+}.

2.1. Metallothioneins

Metallothioneins (MTs) are low molecular weight, cysteine-rich proteins ubiquitously expressed throughout the body, though varying levels of isoforms have been found in specific tissues. Three such isoforms (MT-1, -2, -3) are present in the central nervous system with distinct patterns of expression. MT-1 and MT-2 are found largely in astrocytes and spinal glia but conspicuously absent in neurons, while MT-3 is complementarily expressed predominantly in neurons and sparingly in glial cells. MT-3 is also particularly abundant in the hippocampal glutamatergic terminals known to be rich in vesicular Zn^{2+} (Hidalgo et al., 2001).

All three MT isoforms have a similar structure consisting of a single polypeptide chain, 61–68 amino acids in size, with a highly conserved sequence of 20 cysteine residues (Kägi, 1993). These cysteines are grouped into two domains for Zn^{2+} binding, resulting in a dumbbell-shaped physical conformation (Robbins et al., 1991; Arseniev et al., 1988). A total of seven Zn^{2+} ions bind to the Cys cluster regions with very high affinity ($K_d = 2 \times 10^{12}\ M^{-1}$ at pH 7.0; Kägi, 1993). These two Zn^{2+}/Cys cluster regions are critically important in the regulation of Zn^{2+} binding, which can be readily modulated by shifts in acid-base equilibrium or, more significantly, by changes in the redox state of the two clusters (Kägi, 1993; Maret and Vallee, 1998; Jacob et al., 1998; Maret, 1994; Jiang et al., 2000).

For example, cellular oxidants have been found to promote Zn^{2+} release from MTs while shifting to a more reduced intracellular environment facilitates Zn^{2+} binding (Maret and Vallee, 1998; Jiang et al., 1998; Aizenman et al., 2000). Changes in the glutathione redox state (i.e., the ratio between GSH (glutathione) and GSSG (glutathione disulfide)) may serve a principal role in regulating Zn^{2+}-MT interactions, as GSH binds directly to MT and is thought to "activate" the protein in order to facilitate GSSG-mediated Zn^{2+} release (Maret, 1994; Jiang et al., 1998).

Nitric oxide (NO), an important cellular signaling molecule, has also been described to promote Zn^{2+} release from MTs *in vitro* and *in vivo* (Kroncke et al., 1994; Cuajungco and Lees, 1998; Frederickson et al., 2002). This is of particular interest because NO interacts preferentially with MT-3 (Chen et al., 2002), suggesting that MT-3 is uniquely positioned to translate NO signaling into Zn^{2+} signaling and resulting in potentially significant implications for the role of MT-3 in neuronal injury mediated by oxidative or nitrosative stress (Bossy-Wetzel et al., 2004).

2.2. Membrane-Associated Zn^{2+} Transport Proteins

The proteins directly involved in active movement of Zn^{2+} across cellular membranes generally belong to two families of transporters: the CDF (Cation Diffusion Facilitator) and ZIP ("Zn^{2+}-regulated metal transporter, Iron-regulated metal transporter-like Protein") families. Those most directly associated with Zn^{2+} transport in humans are the ZnT (Zn^{2+} Transporter) proteins, members of the CDF family which favor Zn^{2+} movement out of the cytosol, either by extrusion or sequestration of the cation into intracellular compartments.

Nine human CDF genes (also called SLC30 genes) have been described, with seven ZnTs identified (ZnT-1-7) *in vitro* (Palmiter and Huang, 2004). Of these, ZnT-1 and ZnT-3 are most relevant to synaptic physiology. ZnT-1 is largely expressed on neuronal plasma membranes and plays an important role in modulating Zn^{2+} homeostasis in the brain, although the exact nature of this role appears to be complex and has not been fully elucidated. ZnT-3 is strongly expressed in brain regions that are rich in histochemically reactive Zn^{2+}, such as the entorhinal cortex, the amygdala, and the hippocampus. This transporter is particularly present in the mossy fiber tract and localizes to the membranes of Zn^{2+}-containing vesicles in the mossy fiber synaptic boutons. ZnT-3 knockout mice display a conspicuous lack of Zn^{2+} in their hippocampi, and ultrastructural examination demonstrates the absence of Zn^{2+} in their mossy fiber boutons, thus confirming that this transporter is essential for Zn^{2+} influx into synaptic vesicles (Cole et al., 1999). The remaining ZnT proteins have variable levels of expression in the brain, and much less is known about their function in general.

In contrast to the direction of Zn^{2+} movement mediated by ZnTs, the ZIP family of transporters controls Zn^{2+} transport into the cytosol. ZIP proteins generally possess eight transmembrane domains, and those which have been characterized appear to transport metal ions from either the lumen of cytoplasmic organelles, or from extracellular compartments into the cytoplasm, by as-yet unknown mechanism(s) of substrate transport (Eide, 2003). However, as ZIP transporters to date have been described largely in eukaryotic and plant systems, the extent to which any of their human isoforms participate in Zn^{2+} homeostasis in the brain remains at this time unknown.

2.3. Mitochondria As Homeostatic Zn^{2+} Buffers

Mitochondria are known to be crucial in the buffering of intracellular Ca^{2+}, and several lines of evidence suggest that these organelles also constitute a site of cytosolic Zn^{2+} uptake. Zn^{2+} sequestration by isolated mitochondria is a long-established phenomenon (Brierley and Knight, 1967) which may be mediated by the Ca^{2+} uniporter (Saris and Niva, 1994; Jiang et al., 2001), and mitochondrial Zn^{2+} uptake has been directly visualized in intact neurons using Zn^{2+}-sensitive mitochondrial fluorophores. Such uptake is blocked by treatment with the protonophore FCCP, which induces strong mitochondrial depolarization and thus eliminates the driving force for cation entry (Sensi et al., 2000, 2003). Moreover, inhibition of mitochondrial Zn^{2+} uptake in neurons leads to elevation or prolongation of experimentally-induced cytosolic $[Zn^{2+}]_i$ rises, suggesting that these organelles have a high Zn^{2+} uptake capacity and as such might serve a key role in the clearance of cytosolic Zn^{2+} loads (Sensi et al., 2000).

Induced $[Zn^{2+}]_i$ elevations which are buffered by sequestration of the cation into mitochondria may subsequently be re-released into the cytosol in a Ca^{2+}-dependent fashion (Sensi et al., 2002), and recent studies suggest that Ca^{2+}-dependent release of Zn^{2+} from mitochondria may occur even under resting conditions (Sensi et al., 2003). While the physiological purpose of this mitochondrial Zn^{2+} is currently unexplored, the cation's presence suggests that these organelles could act as important sources for Zn^{2+} release under physiological and/or pathological conditions.

3. TRANSSYNAPTIC MOVEMENT OF Zn^{2+} IN NEURONAL INJURY

A great deal of evidence exists indicating that Zn^{2+} is a potently neurotoxic agent that plays a critical role in the neuronal loss observed in a variety of excitotoxic neurological conditions (Frederickson and Bush, 2001). In fact, much of our current knowledge regarding the physiology of Zn^{2+} at the synapse has evolved from studies of the cation's role in mechanisms of neuronal injury. To begin with, exposure to pathologically relevant (e.g., several hundred micromolar) concentrations of Zn^{2+} *in vitro* triggers both neuronal and glial injury (Yokoyama et al., 1986; Choi et al., 1988), and direct injection of Zn^{2+} into the brain *in vivo* promotes strong neurotoxicity (Itoh and Ebadi, 1982). In addition, excessive Zn^{2+} exposure and $[Zn^{2+}]_i$ rises are potently toxic to neurons in culture (Choi et al., 1988; Lobner et al., 2000; Sensi et al., 2003). *In vivo*, $[Zn^{2+}]_i$ increases are observed in excitotoxic conditions such as ischemia, epilepsy, and brain trauma (Sloviter, 1985; Frederickson et al., 1989; Tonder et al., 1990; Suh et al., 2000). In particular, $[Zn^{2+}]_i$ increases precede neuronal degeneration, and application of an extracellular Zn^{2+} chelator has proved neuroprotective in both transient global and focal ischemia (Koh et al., 1996; Lee et al., 2002).

But more germane to the mechanism of neuronal excitotoxicity is the finding that Zn^{2+} is released from pre-synaptic terminals upon sustained synaptic activity in a Ca^{2+}-dependent fashion (Assaf and Chung, 1984; Howell et al., 1943; Aniksztejn et al., 1987). Furthermore, strong stimulation of the Zn^{2+}-rich perforant path in the hippocampus induces a loss of Zn^{2+} in pre-synaptic mossy fiber terminals in tandem with concurrent neuronal death of hilar interneurons and CA3 pyramidal cells, the post-synaptic targets of the perforant path (Sloviter, 1985). Zn^{2+}'s involvement in a common mechanism of excitotoxic injury was confirmed through the demonstration that status epilepticus also induces both Zn^{2+} depletion from mossy fiber boutons and injurious intracellular Zn^{2+} ($[Zn^{2+}]_i$) accumulation in adjacent post-synaptic hilar neurons (Frederickson, et al., 1988,1989).

This trans-synaptic Zn^{2+} movement, dubbed "Zn^{2+} translocation," has been shown to play a key role in ischemic injury (Tonder et al., 1990; Koh et al., 1996); such translocation occurs following transient global ischemia (TGI), and the neurons enriched in cytosolic Zn^{2+} are the same neurons which exhibit cellular markers of injury after ischemia. The role played by Zn^{2+} in TGI has been of particular interest due to the selective and delayed degeneration of certain hippocampal pyramidal neurons (Pulsinelli et al., 1982). The administration of the extracellular Zn^{2+} chelator Ca-EDTA before and during TGI is able to block $[Zn^{2+}]_i$ accumulation as well as protect against damage in the subpopulation of neurons most vulnerable to TGI, the pyramidal neurons of the CA1 hippocampal subregion (Koh et al., 1996; Pulsinelli et al., 1982). These studies were conducted in the 1990's, thus cementing our current concept of Zn^{2+} involvement in ischemic neuronal injury for the past decade.

The correlation between post-synaptic $[Zn^{2+}]_i$ rises and subsequent neuronal injury has raised two central scientific questions: what is the source of the $[Zn^{2+}]_i$ increase, and by what mechanism might it mediate cellular injury? For the past decade, these rises were thought to be exclusively the result of Zn^{2+} translocation, i.e., extracellular and presumably synaptically-released Zn^{2+} crossing the postsynaptic membrane via some type of ion channel or transporter. Within the last three years, however, studies of transgenic or mutant mice lacking vesicular Zn^{2+} at presynaptic terminal have challenged the assumption that translocation is the only source of increased $[Zn^{2+}]_i$, and several recent reports suggest that a pool of intracellular Zn^{2+} exists which may significantly contribute to "post-synaptic" $[Zn^{2+}]_i$ rises during ischemia.

3.1. Routes Of Postsynaptic Zn^{2+} Entry: The Translocation Model

Since Zn^{2+} translocation is observed in models of ischemia and epilepsy, both of which are excitotoxic conditions, and that Zn^{2+} is co-released with glutamate at excitatory synapses, it is reasonable to consider that this phenomenon involves glutamate receptor activation. In order to identify routes of post-synaptic Zn^{2+} entry, it is useful to first consider the permeability of a similar endogenous divalent cation, Ca^{2+}.

Microfluorimetric and electrophysiological studies of cultured neurons have revealed that Zn^{2+} may enter neurons through NMDA receptor-associated channels and VSCC, both of which are ubiquitously expressed on neurons throughout the brain. However, Zn^{2+} may also flux through Ca^{2+}-permeable AMPA/kainate receptor-mediated channels (Ca-A/K channels), an atypical subtype of AMPAR distinctive for its high Ca^{2+} permeability as well as its selectively increased expression in minority subpopulations of forebrain and spinal cord neurons (Weiss et al., 1993; Yin and Weiss, 1995; Cheng and Reynolds, 1998; Sensi et al., 1997; Canzoniero et al., 1999; Sensi et al., 1999a,b). Ca-A/K channels lack the GluR2 subunit, whose presence in the typical heterotetrameric AMPAR assembly blocks Ca^{2+} entry. In addition, Zn^{2+} can serve as a substrate for the Na^+-Ca^{2+} exchanger in place of Ca^{2+}, which may bring Zn^{2+} into neurons and/or contribute to Zn^{2+} extrusion following intracellular accumulation (Sensi et al., 1997). More recent evidence suggests the presence of a putative Na^+/Zn^{2+} exchanger, likely a separate molecule from the Na^+/Ca^{2+} exchanger (Ohana et al., 2004).

However, there are significant differences in the relative permeabilities of these channels to Zn^{2+} and Ca^{2+}. *In vitro* studies indicate that while Ca^{2+} entry through either NMDAR-associated channels or Ca-A/K channels results in similarly high (many micromolar) $[Ca^{2+}]_i$, Ca-A/K channels have the greatest permeability to Zn^{2+} (Yin and Weiss, 1995; Sensi et al., 1999a), with VSCC and NMDAR showing intermediate and minimal permeability, respectively. In fact, simultaneous activation of all three entry routes (NMDAR, AMPAR and VSCC) using glutamate results in preferential $[Zn^{2+}]_i$ rises in strongly Ca-A/K channel-expressing (Ca-A/K(+)) neurons (Sensi et al., 1999b). Moreover, selective pharmacological inhibition of Ca-A/K channels has been found to be highly neuroprotective against CA1 pyramidal neuron loss in an *in vitro* model of global ischemia, while NMDAR and VSCC blockade were each found to be only marginally beneficial (Yin et al., 2002). These findings strongly support a model of excitotoxic injury in which upon excessive glutamate and Zn^{2+} co-release *in vivo,* Ca-A/K channels act as the main route for injurious Zn^{2+} influx (Sensi et al., 1999b; Yin et al., 2002). In addition, the differential permeability of Ca-A/K channels to Zn^{2+} may be pathologically relevant in TGI, as these channels are both concentrated on post-synaptic membranes where the highest levels of synaptically-released Zn^{2+} are likely to be achieved, and selectively expressed in subpopulations of neurons such as TGI vulnerable CA1 pyramidal neurons. Although CA1 pyramidal neurons lack Ca-A/K channels at their soma, they do appear to express some of these receptors in their dendritic tree, where they would likely play important roles in neurotransmission and injury induction (Lerma et al., 1994; Yin et al., 1999).

Interestingly, not only are these channels present in the dendrites of these most vulnerable neurons, but in the context of TGI, they may also be subject to dynamic, injury-driven up-regulation. Several studies have demonstrated selective decreases in GluR2 subunit expression (and thus increases in the number of functional Ca-A/K channels) in CA1 pyramidal neurons after TGI (Gorter et al., 1997; Opitz et al., 2000). The "GluR2 hypothesis," which postulates that some forms of neuronal insult selectively trigger an increase in the number of Ca-A/K channels present on the plasma membrane of certain neurons, is based on these observations, and proposes that such a phenomenon likely underlies their selective vulnerability to injury in these conditions (Benett et al., 1996; Pellegrini-Giampietro et al., 1997; Opitz et al., 2000).

3.2. Emerging Models of Intracellular Zn^{2+} Release

The conventional thinking regarding the role of Zn^{2+} translocation in neuronal injury has been challenged by recent unexpected observations in ZnT-3 knockout mice (ZnT-3 KO). Despite having virtually no histochemically reactive Zn^{2+} in their pre-synaptic terminals, the hippocampal neurons of these animals undergo intracellular Zn^{2+} accumulation and injury following an excitotoxic insult (Lee et al., 2000). Even more intriguingly, ZnT-3 KO mice appear to have no significant differences from wild-type mice across a range of electrophysiological and behavioral parameters (Lee et al., 2000; Cole et al., 2001; Lopantsev et al., 2003).

These data may seem at first glance to contradict the concept that Zn^{2+} exerts a critical role in modulating cell physiology, suggesting instead that the cation is simply not as essential a modulator as previously thought. However, as with any experimental evidence gathered from transgenic animals, one should bear in mind that ZnT-3 KO mice may have adapted to the lack of vesicular Zn^{2+} during gestation by over-expressing other, non-vesicular sources of Zn^{2+}. Thus, one might argue instead that Zn^{2+} is in fact so vital to cell physiology that multiple systems are in place to regulate Zn^{2+} homeostasis, such that organisms are able to adapt to the loss of any single system (e.g. vesicular Zn^{2+}) with a powerful compensatory up-regulation of other(s).

For instance, one possible substitute for vesicular Zn^{2+} release could be release from non-vesicular, presynaptic sites; i.e., Zn^{2+} mobilized from histochemically invisible intracellular stores in the presynaptic terminal, such as mitochondria or MTs, might in theory be released in the synaptic cleft via plasma membrane systems such as the ZnT-1 transporter or the Na^{+}-Zn^{2+} and Zn^{2+}-H^{+} exchangers. While this specific mechanism is largely speculative, the concept that non-vesicular, presynaptic Zn^{2+} release occurs is supported by the otherwise puzzling observation that Ca-EDTA is still able to protect against Zn^{2+} accumulation and injury in both CA1 and CA3 regions in ZnT-3 KO mice (Lee et al., 2000).

An alternative explanation for the injurious Zn^{2+} accumulation seen in ZnT-3 KO mice is that some or all of this Zn^{2+} originates from one or more sites in the post-synaptic neuron itself. This scenario does not preclude the neuroprotective effects of Ca-EDTA observed in ZnT-3 KO mice, as extracellular Zn^{2+} chelators have been shown to act as Zn^{2+} "sponges" which promote removal of the cation from intracellular compartments (Frederickson et al., 2002). Given that neurons do possess intracellular sources of the cation such as MTs and mitochondria, this possibility of deleterious Zn^{2+} release from within the postsynaptic cell may be a more appealing model.

In the case of MTs, a number of recent findings strongly imply that these Zn^{2+}-binding proteins play a part in modulating excitotoxic and ischemic injury, although the exact nature of this role is not entirely clear. Several factors suggest that in some instances, MTs serve a protective purpose in neurons, particularly in focal ischemia. MT-1 and MT-2 mRNA expression is rapidly upregulated following transient focal ischemic insult, and increased expression of MT-1 has been shown to be neuroprotective in focal ischemia (Van Lookeren et al., 1999; Tredelenburg et al., 2002). In addition, MT-1 and MT-2 KO mice were observed to develop three times larger infarcts in focal ischemia than wild-type mice, and a separate study found MT-3 KO mice were more sensitive to excitotoxic injury as well (Erickson et al., 1997; Van Lookeren et al., 1999; Tredelenburg et al., 2002). In interpreting these data, one should bear in mind that MTs possess intrinsic antioxidant properties which might also contribute to the overall beneficial role exerted by these proteins (Ibadi et al., 1996). Together, these studies support the concept of MTs as a "passive" cellular defense mechanism against toxic $[Zn^{2+}]_i$ elevations.

However, emerging evidence suggests that in other instances, MTs may in fact act mainly as a source of injurious Zn^{2+} release. In principle, the ability of MTs to release Zn^{2+} upon changes in the cellular redox state renders these proteins a reservoir of readily-available Zn^{2+} under conditions of oxidative stress, which does occur in ischemia. Indeed, the additional knockout of MT-3 in ZnT-3 KO mice results in substantial protection from the excitotoxic injury otherwise observed in CA1 (Cole et al., 2000). Furthermore, recent data indicate that NO-triggered $[Zn^{2+}]_i$ rises and subsequent neuronal loss in the CA1 region are substantially reduced in MT-3 KO mice compared to wild type animals (Lee et al., 2003). MTs may thus ultimately act not only as Zn^{2+} buffers, but also as sources for potentially deleterious cation release.

Another site of intracellular Zn^{2+} release is offered by mitochondria (Sensi et al., 2002, 2003), and the implications of this mitochondrial Zn^{2+} release will be discussed in further detail below. Overall, recent findings seem to indicate that the mechanisms underlying injurious Zn^{2+} accumulation may be more multifaceted than previously thought and encompass pre-synaptic (vesicular/non-vesicular) as well as post-synaptic sources, as summarized in **Fig. 1**.

4. INTRACELLULAR MECHANISMS OF Zn^{2+} TOXICITY

The multiple mechanisms by which intracellular Zn^{2+} promotes cell death are just beginning to be understood. Given the cation's multidirectional effects on cell physiology, Zn^{2+}-dependent injury pathways are likely to be complex, with some overlap or interaction between mechanisms. Significantly, the two classical models of cell death,

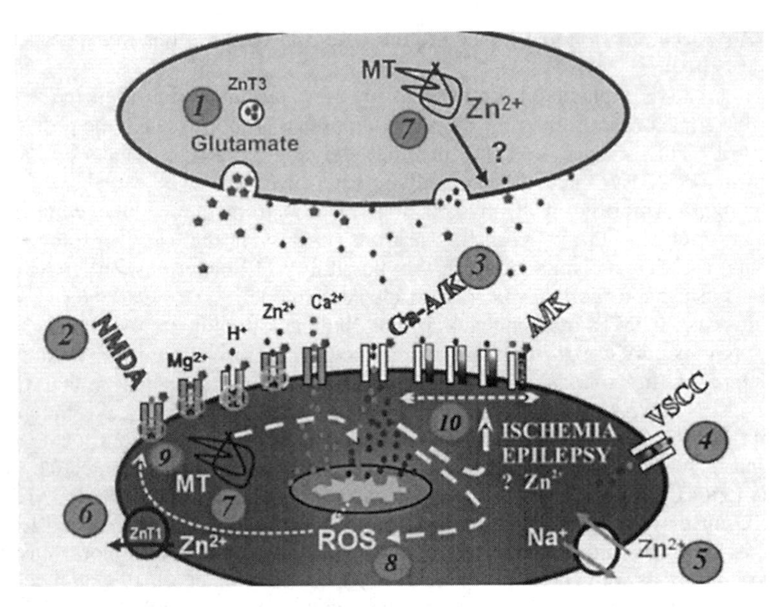

Figure 1. Sources of postsynaptic Zn^{2+} accumulation. Zn^{2+} is transported into pre-synaptic vesicles of certain glutamatergic neurons by the vesicular transporter, ZnT-3, (**1**) and is co-released with glutamate into the synaptic cleft following sustained synaptic activation. Synaptic Zn^{2+}, extracellular H^+, and physiological Mg^{2+} substantially block NMDAR (**2**), likely resulting in a net inhibition of Ca^{2+} influx during ischemia. Some neurons (i.e., the highly TGI-vulnerable CA1 pyramidal neurons) express Ca-A/K channels, a subtype of AMPAR highly permeable to Zn^{2+} as well as to Ca^{2+} (**3**). Zn^{2+} may also flux across the plasma membrane through largely extra-synaptic routes such as VSCC (**4**) and the putative Na^+/Zn^+ exchanger (**5**). The Zn^{2+} transporter, ZnT-1 (**6**) promotes Zn^{2+} efflux from neurons and likely helps to maintain low $[Zn^{2+}]_i$ under physiological conditions. Metallothioneins may act either as a Zn^{2+}-buffering system or as a source for Zn^{2+} release into the cytosol under oxidizing conditions (**7**). Rapid Zn^{2+} entry through Ca-A/K-receptor channels can cause mitochondrial dysfunction and reactive oxygen species (ROS) generation (**8**), which in turn may further inhibit NMDAR activity (**9**), and/or induce release of Zn^{2+} from MT (**7**). Zn^{2+} might also interfere with transcription factors, leading to GluR2 subunit downregulation and thus promoting an increase in the number of Ca-A/K channels (**10**). Note that acidosis blocks NMDAR activity (also blocked by Zn^{2+}), yet increases Zn^{2+} (but not Ca^{2+}) permeability of both Ca-A/K channels and VSCC.

apoptosis and necrosis, are coexisting phenomena which may occur in parallel in cells exposed to the same stimulus. The intensity of the insult and its effect on the status of cellular energy levels determines which process is more prominent in a given cell; by controlling ATP production, mitochondria appear to function as the fulcrum between a more predominantly necrotic or apoptotic demise. For instance, a subacute ischemic insult may leave affected neurons with injured but still partially-functioning mitochondria, therefore generating sufficient ATP levels to allow apoptosis to fully progress. In contrast, a fulminant insult would result in the rapid compromise of mitochondria and intracellular energy levels, forcing cells to abandon the apoptotic sequence in favor of necrosis (Nicotera et al., 1999) As Zn^{2+} has well-known effects on mitochondrial function, it is not surprising that the cation has been implicated in the induction of both necrotic and apoptotic processes (Lobner et al., 2000). In particular, metabolic inhibition and reactive oxygen species (ROS) generation may be crucial to both rapid and slow Zn^{2+} neurotoxicity (Kim et al., 1999; Nicholls and Budd, 2000).

4.1. Zn^{2+} in Necrosis

4.1.1. Generation of Mitochondrial ROS

Several studies have suggested that mitochondria are important sites for the toxic effects of elevated intracellular Zn^{2+} as well as Ca^{2+} (Kim et al., 1999; Sensi et al., 1999a, 1999b, 2000, 2002, 2003; Lobner et al., 2000). Mitochondria are the major cellular source of ROS, which are routinely produced and rapidly utilized as recyclable co-factors in the electron transport chain. Under normal physiological conditions, ROS are secured within the mitochondrial membranes, where they cannot adversely affect cellular function. Following rapid and massive cytosolic Zn^{2+} loads, however, a profound disruption of mitochondrial function occurs which manifests as a dramatic loss of mitochondrial membrane potential ($\Delta\Psi_m$) and increased ROS generation (Sensi et al., 1999a, 1999b, 2000). As a result, ROS are able to diffuse out of these organelles and disrupt plasma membrane lipids, ultimately promoting neuronal death.

Zn^{2+} induces mitochondrial ROS production largely through its inhibition of cellular respiration, which may additionally interfere with homeostatic maintenance of $\Delta\Psi_m$ (Skulachev et al., 1967; Nicholls and Malviya, 1968; Kleiner and von Jagow, 1972; Kleiner, 1974]. The cation likely acts at several sites within the electron transport chain, including cytochrome bc_1 in complex III (Link and von Jagow, 1995) and KGDHG (α-ketoglutarate dehydrogenase) in complex I (Lorusso et al., 1991; Brown et al., 2000). Zn^{2+} also affects multiple activities of LADH (lipoamide dehydrogenase); this enzyme is a component of the KDGHC complex in mitochondria, but also catalyzes NADH oxidation and produces ROS as a by-product. Zn^{2+} strongly inhibits the LADH reaction in mitochondria in both directions, on one hand interfering with respiration (Gazaryan et al., 2002), and on the other decreasing the formation of a potent ROS scavenger and antioxidant regenerator (Packer et al., 1997). With respect to NADH oxidation, Zn^{2+} accelerates LADH catalysis of the oxidative reaction 5-fold (Gazaryan et al., 2002), resulting in an overall increase in cellular oxidative burden by both direct and indirect means.

4.1.2. Cytosolic ROS Generation

Milder cytosolic $[Zn^{2+}]_i$ rises may also induce Zn^{2+}-dependent oxidative stress independent of mitochondria, which would favor necrotic cell death as well. Zn^{2+} is known to modulate a number of cytosolic enzymes which generate ROS secondary to their main physiological activity. Similar to its effect on LADH in NADH oxidation described above, Zn^{2+} appears to induce the activity of NADPH oxidase (a multi-subunit enzyme widely expressed in central neurons) via activation of protein kinase C (Noh et al., 1999; Kim and Koh, 2002), subsequently producing ROS. Another possible effector of Zn^{2+}-mediated free radical production in the cytosol is nNOS (neuronal nitric oxide synthase), which together with superoxide produces peroxynitrite. Although a biochemical assay of nNOS activity *in vitro* demonstrates inhibition by high concentrations of Zn^{2+}, cortical neurons exposed to Zn^{2+} show nNOS activation and increased levels of NO and nitrites (Persechini et al., 1995; Kim et al., 1999). The effect of Zn^{2+} on this enzyme *in vivo* is of particular interest because the number of neurons expressing nNOS may increase following focal cerebral ischemia (Holtz et al., 2001).

In addition to the lipid membrane degradation described above, intracellular oxidative stress can also result in DNA strand breakage. Such DNA damage triggers a cascade of events including the activation of PARP (poly(ADP ribose) polymerase), consumption of NAD^+ during the formation of PAR polymers, and eventually, death due to ATP depletion. PARP is an enzyme which regulates its own activity in conjunction with PARG (poly(ADP ribose) glycohydrolase); the PARP/PARG cycle appears to be necessary for the persistent PARP-dependent activity which leads to NAD^+ and ATP depletion and eventual cell death. Pharmacological inhibition of both these enzymes in cortical neurons results in significant neuroprotection against Zn^{2+} neurotoxicity (Kim et al., 1999; Kim and Koh, 2002).

A final consideration is that oxidative stress may also trigger additional intracellular Zn^{2+} release. As noted earlier, cellular oxidation promotes the release of Zn^{2+} from MTs, and given the ROS-generating activity of Zn^{2+} summarized above, it is not difficult to imagine that a dangerous feed-forward cycle of cellular injury might develop as a consequence (Kroncke et al., 1994; Cuajungco et al., 1998; Frederickson et al., 2002; Gazaryan et al., 2002; Kim et al., 1999; Kim and Koh, 2002; **Fig. 2**).

4.1.3. Disruption Of Cellular Metabolism

Zn^{2+} can also contribute to necrotic cell death through direct modulation of key enzymes in neuronal glycolysis. *In vitro* biochemical assays have demonstrated Zn^{2+} inhibition of GAPDH (glyceraldehyde-3-phosphate dehydrogenase; Krotiewska et al., 1992), phosphofructokinase (Ikeda et al., 1980), and NAD^+ glycohydrolase (Kukimoto et al., 1996). In intact cortical neurons, moderate (submicromolar) $[Zn^{2+}]_i$ rises are sufficient to trigger a powerful inhibition of GAPDH, which leads to ATP depletion and neuronal death (**Fig. 2**). This inhibition by Zn^{2+} involves decreased levels of cytosolic NAD^+, mediated by an as-yet unknown mechanism, since restoring NAD^+ by the addition of pyruvate results in strong neuroprotection specifically against Zn^{2+}-dependent toxicity (Sheline et al., 2000). Interestingly, pyruvate has also been found to dramatically decrease both ischemic $[Zn^{2+}]_i$ rises and injury in an animal model of TGI (Lee et al., 2001).

4.2. Zn^{2+} Mediation of Apoptosis

4.2.1 Mitochondrial Release Of Pro-Apoptotic Factors

The selective neuronal injury observed in TGI is likely a necrotic process, resulting from an abrupt decline in intracellular energy levels and mitochondrial function triggered by robust Ca-A/K channel-mediated increase of $[Zn^{2+}]_i$. By contrast, milder cytosolic Zn^{2+} loads may reasonably be expected to be seen in neurons lacking Ca-A/K channels and in the penumbral areas of TGI as a consequence of cation entry through less permeable but more ubiquitously-expressed routes such as VSCC (Lee et al., 2002). These lower $[Zn^{2+}]_i$ rises may bring about less intense disruption of mitochondrial function, allowing neurons to activate apoptotic machinery instead.

Whereas high (micromolar) levels of $[Zn^{2+}]_i$ inhibit the electron transport chain and decrease cellular respiration, lower levels of Zn^{2+} have been shown to elicit the opposite effect. Submicromolar Zn^{2+} levels lead to mitochondrial swelling and increased respiration in isolated mitochondria, effects consistent with the induction of mPTP (mitochondrial permeability transition pore) (Wudarczyk et al., 1999; Jiang et al., 2001; Sensi et al., 2003). This milder Zn^{2+} burden has also been observed to promote the release of pro-apoptotic mitochondrial proteins such as Cyt-C (cytochrome C) and AIF (apoptosis inducing factor), both associated with mPTP opening (Jiang et al., 2001). In intact cortical neurons, similar submicromolar $[Zn^{2+}]_i$ rises mediated by VSCC are sufficient to trigger substantial mitochondrial swelling and release of Cyt-C and AIF, while inhibition of mPTP opening attenuates both Zn^{2+}-triggered release of these factors as well as subsequent neuronal loss (Jiang et al., 2001).

4.2.2. Modulation Of Apoptotic Signaling

Zn^{2+} may also mediate cell death through other, parallel pathways resulting in apoptosis. For example, oxidation-induced $[Zn^{2+}]_i$ rises have been demonstrated to alter intracellular K^+ content, which is known to play a key role in neuronal apoptosis (Yu et al., 2001). Experiments in neuronal culture have demonstrated that the addition of exogenous oxidizing agents such as DTDP (2,2'-dithiodipyridine) or NO (resulting in production of peroxynitrite) mobilizes Zn^{2+} release from MTs, initiating an injury cascade with Zn^{2+}-induced p38 MAP kinase activation leading to caspase-independent K+ efflux, cell volume loss, and apoptotic neuronal death (Aizenman et al., 2000; McLaughlin et al., 2001; Bossy-Wetzel et al., 2004).

Zn^{2+} specifically modulates two other mediators of apoptotic signaling, p75(NTR)/NADE and Egr-1, both of which have been implicated in models of TGI injury *in vivo*. Zn^{2+} exposure stimulates activity of the low-affinity neurotrophin receptor p75(NTR) and its associated death executor protein NADE in cortical neurons while various other neurotoxic conditions do not, and the resultant neuronal loss is blocked by caspase inhibitors. In an *in vivo* model of TGI, both p75(NTR) and NADE were induced in degenerating CA1 neurons, an effect entirely suppressed by extracellular Zn^{2+} chelation, further supporting a positive correlation between Zn^{2+} accumulation and their induction (Lee et al., 1995; Kokaia et al., 1998). Egr-1 is an immediate-early gene transcription factor that is induced after cerebral ischemia (Beckmann and Wilce, 1997), and brief exposure to Zn^{2+} but not Ca^{2+} has been reported to trigger sustained Erk 1/2 activation, an event upstream of Egr-1 induction. Pharmacological inhibition of Erk 1/2 blocks both Egr-1 activity and Zn^{2+}-dependent neurotoxicity (Park and Koh, 1999).

4.3. Interaction Between Pathways of Injury

As we have described, Zn^{2+} exerts its toxic effects at mitochondrial and extra-mitochondrial sites of action, effecting both rapid, necrotic injury and slower, apoptotic processes. It is quite possible, however, that these pathways interact in a synergistic manner. For instance, recent findings demonstrate how cytosolic PARP activation leads to decreased NAD^+ levels, prompting the release of AIF from mitochondria. Cytosolic AIF eventually promotes the collapse of $\Delta\Psi_m$, triggering the release of Cyt-C and initiating the apoptotic cascade (Yu et al., 2002; Du et al., 2003).

The mobilization of intracellular Zn^{2+} which occurs upon oxidative stress might also serve as a link between pathway. Recent observations in intact neurons suggest that the level of oxidation-induced Zn^{2+} release from protein-bound stores is capable of causing a partial loss of $\Delta\psi_m$; comparable $[Zn^{2+}]$ rises in isolated mitochondria elicit a multi-conductance ion channel activity consistent with mPTP opening (Jiang et al., 2001 ; Sensi et al., 2003). Conversely, Zn^{2+}-induced mitochondrial ROS generation might promote yet more Zn^{2+} release from the protein-bound pool. Thus, rather than acting separately, these two processes could together form a self-perpetuating, vicious cycle of neuronal injury (**Fig. 2**).

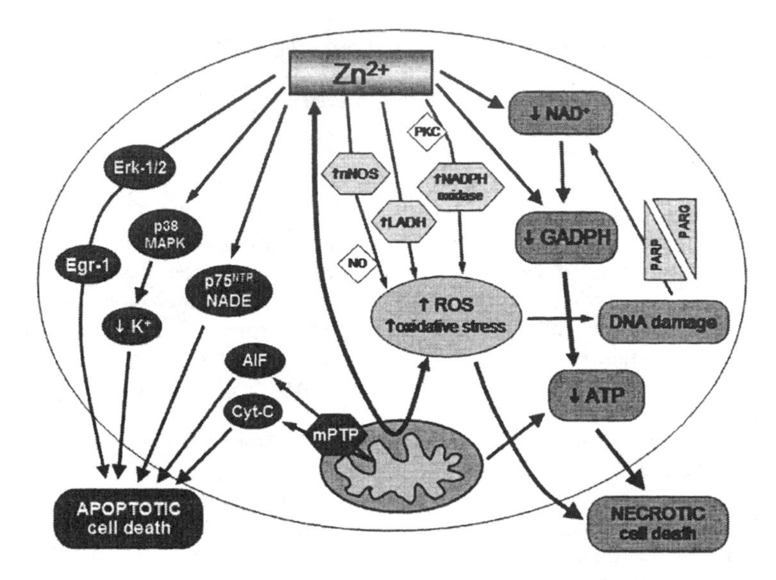

Figure 2. Zn^{2+} activates multiple intracellular pathways leading to both necrotic and apoptotic neuronal death. Zn^{2+} can trigger necrosis (right) by increasing cellular oxidative stress, promoting ATP depletion through direct inhibition of glycolytic pathways, or indirectly via Zn^{2+}-induced mitochondrial dysfunction and/or another as yet unknown mechanism. Zn^{2+} induces oxidative stress either by triggering release of mitochondrial ROS or by activating cytosolic enzymes favoring ROS generation (nNOS, NADPH oxidase, LADH). Oxidative stress can also contribute to NAD^+ depletion upon DNA damage, and promote further Zn^{2+} release from MTs.

Apoptotic neuronal death (left) may be triggered by Zn^{2+}-triggered release of apoptotic factors (Cyt-C, AIF) from mitochondria, Zn^{2+} induction of p38-dependent K^+ efflux resulting in depletion of intracellular K^+, or induction of apoptotic signaling factors p75/NADE or Egr-1. Note the potential for interaction between pathways; for example, Zn^{2+}-triggered mitochondrial dysfunction can also favor the release of pro-apoptotic factors via induction of mPTP.

It is also worth noting that PARP and other poly(ADP-ribosyl) transferases are localized within mitochondria as well as in the cytosol, and activation of mitochondrial PARP may be able to bring about loss of $\Delta\Psi_m$ and NAD^+ depletion as well (Du et al., 2003). As discussed above, Zn^{2+}-induced reduction of NAD^+ levels is instrumental in inhibiting the glycolytic pathway and triggering neuronal death (Sheline et al., 2000), and thus Zn^{2+}-dependent reduction of mitochondrial NAD^+ may further enhance the injury sequence (**Fig. 2**).

5. MODULATION OF ZN^{2+} EFFECTS IN PATHOLOGIC ENVIRONMENTS

5.1. Zn^{2+} Influx And Homeostasis Under Ischemic Conditions

To date, the majority of studies of Zn^{2+} transmembrane movement, including those identifying routes of extracellular Zn^{2+} entry or elucidating homeostatic mechanisms, have been conducted under "normal" physiological conditions in vitro. In ischemia, however, several characteristics of the extra- and intra-cellular milieu are significantly altered, most notably the characteristic acidosis which develops in ischemic tissue (cf. Lipton, 1999). Thus, it is pertinent to consider how the systems involved in each model might respond in the specific environment of ischemia, and accordingly, potentially contribute to the development of ischemic pathology.

Acidosis appears to modulate several routes of intracellular Zn^{2+} entry. Firstly, while the relative permeability of Zn^{2+} is less than that of Ca^{2+} for both VSCC and Ca-A/K channels (Sensi et al., 1997, 1999a; Kerchner et al., 2000; Jia et al., 2002), this is interestingly reversed under acidic conditions: Zn^{2+} permeability increases while Ca^{2+} permeability of these channels (and NMDA receptor-mediated channels) decreases (Tang et al., 1990; Traynelis and Cull-Candy, 1991; Canzoniero et al., 1999; Kerchner et al., 2000).

Decreased pH also affects other Zn^{2+} transporter systems involved with Zn^{2+} fluxes. A recent report demonstrated that under basal conditions, Zn^{2+} accumulated in cortical neurons in cytosolic organelles tentatively identified as mitochondria, but that this intracellular Zn^{2+} uptake was blocked by extracellular acidification. The same study found that extracellular acidosis inhibited overall neuronal accumulation of $^{65}Zn^{2+}$ by (pH 6.0), suggesting the presence of a putative plasma membrane Zn^{2+}/proton antiporter (Colvin et al., 2002, 2003).

Additionally, acidic shifts rapidly destabilize the interaction between MTs and Zn^{2+} on a biochemical level, favoring an overall release of Zn^{2+} (Kägi, 1993; Jiang et al., 2000). This phenomenon has been observed in intact neurons, where intracellular acidification markedly increases $[Zn^{2+}]_i$ rises generated by oxidation of MTs. These acid-induced $[Zn^{2+}]_i$ rises also interfere with mitochondrial function by triggering partial depolarization of these organelles (Sensi et al., 2003).

Finally, a recent study reports that Zn^{2+} is able to both induce intracellular acidification and/or prolong recovery from intracellular acidosis (Dineley et al., 2002). The Zn^{2+}-induced acidification is dependent on extracellular Ca^{2+} entry via the Na^+/Ca^{2+} exchanger, whereas the delayed recovery from intracellular acidosis is attributed to Zn^{2+} inhibition of proton efflux by the Cl^-/HCO_3^- exchanger, one of the major systems involved in maintaining neuronal acid-base equilibrium (Schwiening and Boron, 1994; Canzoniero et al., 1996; Dineley et al., 2002).

Furthermore, persistent intracellular acidosis may affect $[Zn^{2+}]_i$ and subsequently promote neuronal injury by interfering with the overall neuronal redox state, either by reducing the activity of cellular antioxidant enzymes or increasing hydroxyl radical formation (Ying et al., 1999). An acidotic environment may thus influence Zn^{2+} homeostasis through modulation of Zn^{2+} influx, transport, and/or intracellular mobilization. Zn^{2+} dyshomeostasis may in turn disrupt neuronal acid-base equilibrium, thereby creating a yet another potentially injurious feed-forward loop.

5.2 Relative Strength Of Zn^{2+} Effects On Mitochondria

Several mechanisms of Zn^{2+}-dependent neurotoxicity seem to share common targets with Ca^{2+}-dependent injury cascades. Both Zn^{2+} and Ca^{2+} can effect different degrees of mitochondrial dysfunction and thus favor either acute necrotic or slower apoptotic forms of neuronal injury. These two divalent cations differ significantly, however, in the potency of their disruptive effects. With respect to acute necrotic processes induced by large cytosolic rises of either cation, comparing the achieved intracellular concentration and degree of subsequent toxicity indicates, surprisingly, that cytosolic Zn^{2+} acts far more robustly than Ca^{2+} in producing neuronal dysfunction and cell death (Turetsky et al., 1994; Yin and Weiss, 1995; Sensi et al., 1999a).

In Ca-A/K(+) neurons, recovery from Ca-A/K channel-mediated $[Zn^{2+}]_i$ rises is much slower than recovery of similarly-induced $[Ca^{2+}]_i$ increases, despite the lower absolute $[cation]_i$ observed with Zn^{2+} (Sensi et al., 1999a, 2000). As might be expected with these different temporal profiles of cytosolic clearance, Zn^{2+}-triggered disruption of mitochondrial function (i.e., loss of $\Delta\Psi_m$ and ROS generation) persists longer than that elicited with Ca^{2+} (Sensi et al., 2000). Zn^{2+} has been found to have a much higher potency than Ca^{2+} in regard to apoptotic processes as well, as assessed by cation induction of mitochondrial swelling, mPTP opening, and release of pro-apoptotic factors (Jiang et al., 2001).

Zn^{2+}'s interactions with mitochondria in ischemia and injury should also take into account that the ischemic brain is more often than not an aging brain as well, and that the capability of these organelles to cope with stressors and death signaling dramatically decreases with maturity. Thus, any Zn^{2+}-triggered disruption of mitochondrial function would almost certainly have a greater impact on cell viability in older neurons than what the above models, largely based on experimental findings from young, healthy neuronal preparations, have proposed. In addition, while individual mitochondria have a relatively rapid rate of turnover, in general neurons do not, and therefore any potential mutations in mitochondrial DNA brought about by increased oxidative stress would be perpetuated and subsequently accumulate in neurons over time. Such mutations could result in the expression of defective mitochondrial proteins, generating less-efficient metabolism and disrupting critical recycling of ROS, resulting in further oxidative stress in a self-propagating, feed-forward cycle (Wallace, 1999).

Finally, while glutamate-induced intracellular Ca^{2+} overload is key factor in the development of ischemic neuronal injury, a recent study has reported the intriguing finding that Zn^{2+} may have modulatory effects on Ca^{2+} homeostasis. The study demonstrates that extracellular Zn^{2+} can specifically induce Ca^{2+} mobilization from the endoplasmic reticulum of non-neuronal cells, via a G-protein/phospholipase C-mediated mechanism. In addition, these Zn^{2+}-triggered $[Ca^{2+}]_i$ rises are able to increase the activity of the Na^+-H^+ exchanger, thus potentially interfering with the intracellular acid-base equilibrium as well (Hershfinkel et al., 2001). Given that Zn^{2+} shows far greater potency than Ca^{2+} in promoting ROS generation, these findings taken together may support a model which privileges the role played by Zn^{2+} in the evolution of ischemic injury.

6. CONCLUDING REMARKS: NEW PARADIGMS IN EXCITOTOXICITY

Given the emerging pathogenic role played by intracellular Zn^{2+} accumulation in neuronal death and the fact that the ion seems to promote injury with greater potency compared to Ca^{2+}, Zn^{2+} may be an underappreciated mediator of excitotoxicity, which has for the most part been thought of as a purely Ca^{2+}-dependent phenomenon. Moreover, the deleterious effects of cytosolic $[Ca^{2+}]_i$ increases in ischemia should perhaps be re-evaluated in light of emerging data regarding the Ca^{2+} dependence of intracellular mobilization of Zn^{2+}. The observation that intracellular Zn^{2+} release from mitochondria is particularly prominent in the case of large, glutamate-evoked $[Ca^{2+}]_i$ rises, coupled with the likely probability that Ca^{2+}-induced mitochondrial ROS generation would also promote Zn^{2+} release from MTs, offers the possibility of a more complex injury paradigm than previously imagined. In such a model, glutamate-driven $[Ca^{2+}]_i$ rises might actually serve as an "accomplice" to spark the release of the true ionic mediator of neuronal damage: Zn^{2+}.

7. ACKNOWLEDGEMENTS

Supported by NIH grant AG00919 (SLS). The authors are indebted to Dr. Patric Stanton for his patience and perseverance as both editor and friend. We thank Dr. John Weiss, Dr. Hong Z. Yin and Mr. Dien Ton-That for essential experimental contribution to several of the results discussed. Finally, we would like to think that our friend and colleague John Sarvey would find the hypotheses proposed above of interest, and worthy of continued dialogue.

8. REFERENCES

Aizenman E, Stout AK, Hartnett KA, Dineley KE, Mclaughlin B, Reynolds IJ (2000) Induction of neuronal apoptosis by thiol oxidation: putative role of intracellular zinc release, J Neurochem 75:1878-1888.

Aniksztejn L, Charton G, Ben-Ari Y (1987) Selective release of endogenous zinc from the hippocampal mossy fibers in situ, Brain Res 404:58-64.

Aravindakumar CT, Ceulemans J, De Ley M (1999) Nitric oxide induces Zn2+ release from metallothionein by destroying zinc–sulphur clusters without concomitant formation of S-nitrosothiol, Biochem J 344:253-258.

Arseniev A, Schultze P, Worgotter E, Braun W, Wagner G, Vasak M, Kagi JH, Wuthrich K (1988) Three-dimensional structure of rabbit liver [Cd7]metallothionein-2a in aqueous solution determined by nuclear magnetic resonance, J Mol Biol 201:637-657.

Assaf SY, Chung SH (1984) Release of endogenous Zn2+ from brain tissue during activity, Nature 308:734-736.

Beckmann AM, Wilce PA (1997) Egr transcription factors in the nervous system, Neurochem Int 31:477-510.

Bennett MVL, Pellegrini-Giampietro DE, Gorter JA, Aronica E, Connor JA, Zukin RS (1996) The GluR2 hypothesis: Ca2+-permeable AMPA receptors in delayed neurodegeneration, Cold Spring Harbor Symposia Quantitative Biology 61:373-384.

Bossy-Wetzel E, Talantova MV, Lee WD, Scholzke MN, Harrop A, Mathews E, Gotz T, Han J, Ellisman MH, Perkins GA, Lipton SA (2004) Crosstalk between nitric oxide and zinc pathways to neuronal cell death involving mitochondrial dysfunction and p38-activated K+ channels, Neuron 41:351-365.

Brierley GP, Knight VA (1967) Ion transport by heart mitochondria. X. The uptake and release of Zn2+ and its relation to the energy-linked accumulation of magnesium, Biochemistry 6:3892-3901.

Brown AM, Kristal BS, Effron MS, Shestopalov AI, Ullucci PA, Sheu KFR, Blass JP, Cooper AJL (2000) Zn2+ inhibits alpha-ketoglutarate-stimulated mitochondrial respiration and the isolated alpha-ketoglutarate dehydrogenase complex., J Biol Chem 275:13441–13447.

Canzoniero LM, Sensi SL, Choi DW (1996) Recovery from NMDA-induced intracellular acidification is delayed and dependent on extracellular bicarbonate, Am J Physiol 270:593–599.

Canzoniero LM, Turetsky DM, Choi DW (1999) Measurement of intracellular free zinc concentrations accompanying zinc-induced neuronal death, J Neurosci 19(RC31):1-6.

Chen Y, Irie Y, Keung WM, Maret W (2002) S-Nitrosothiols React Preferentially with Zinc Thiolate Clusters of Metallothionein III through Transnitrosation, Biochemistry 41:8360 -8367.

Cheng C, Reynolds IJ (1998) Calcium-sensitive fluorescent dyes can report increases in intracellular free zinc concentration in cultured forebrain neurons, J Neurochem 71:2401-2410.

Choi DW, Yokoyama M, Koh J (1988) Zinc neurotoxicity in cortical cell culture, Neurosci 24:67-79.

Cole TB, Wenzel HJ, Kafer KE, Schwartzkroin PA, Palmiter RD (1999) Elimination of zinc from synaptic vesicles in the intact mouse brain by disruption of the ZnT3 gene, Proc Natl Acad Sci USA 96:1716-1721.

Cole TB, Robbins CA, Wenzel HJ, Schwartzkroin PA, Palmiter RD (2000) Seizures and neuronal damage in mice lacking vesicular zinc, Epilepsy Res 39:153-169.

Cole TB, Martyanova A, Palmiter RD (2001) Removing zinc from synaptic vesicles does not impair spatial learning, memory, or sensorimotor functions in the mouse, Brain Res 891:253-265.

Colvin RA (2002) pH dependence and compartmentalization of zinc transported across plasma membrane of rat cortical neurons, Am J Physiol Cell Physiol 282:C317-329.

Colvin RA, Fontaine CP, Laskowski M, Thomas D (2003) Zn2+ transporters and Zn2+ homeostasis in neurons,

Eur J Pharmacol 479:171-185.

Cuajungco MP, Lees GJ (1998) Nitric oxide generators produce accumulation of chelatable zinc in hippocampal neuronal perikarya, Brain Res 799:118-129.

Dineley KE, Brocard JB, Reynolds IJ (2002) Elevated intracellular zinc and altered proton homeostasis in forebrain neurons, Neurosci 114:439-449.

Dineley KE, Votyakova TV, Reynolds IJ (2003) Zinc inhibition of cellular energy production: implications for mitochondria and neurodegeneration, J Neurochem 85:563-570.

Du L, Zhang X, Han YY, Burke NA, Kochanek PM, Watkins SC, Graham SH, Carcillo JA, Szabo C, Clark RSB (2003) Intra-mitochondrial Poly(ADP-ribosylation) Contributes to NAD+ Depletion and Cell Death Induced by Oxidative Stress, J Biol Chem 278:18426-18433.

Ebadi M, Leuschen MP, El Refaey H, Hamada FM, Rojas P (1996) The antioxidant properties of zinc and metallothionein, Neurochem Int 29:159-166.

Eide DJ (2003) Multiple regulatory mechanisms maintain zinc homeostasis in Saccharomyces cerevisiae, J Nutr 133:1532S-1535.

Erickson JC, Hollopeter G, Thomas SA, Froelick GJ, Palmiter RD (1997) Disruption of the metallothionein-III gene in mice: analysis of brain zinc, behavior, and neuron vulnerability to metals, aging, and seizures, J Neurosci 17:1271-1281.

Frederickson CJ, Hernandez MD, Goik SA, Morton JD, Mcginty JF (1988) Loss of zinc staining from hippocampal mossy fibers during kainic acid induced seizures: a histofluorescence study, Brain Res 446:383-386.

Frederickson CJ, Hernandez MD, Mcginty JF (1989) Translocation of zinc may contribute to seizure-induced death of neurons, Brain Res 480:317-321.

Frederickson CJ, Bush AI (2001) Synaptically released zinc: Physiological functions and pathological effects, BioMetals 14:353-366.

Frederickson CJ, Cuajungco MP, Labuda CJ, Suh SW (2002) Nitric oxide causes apparent release of zinc from presynaptic boutons, Neuroscience 115:471-474.

Frederickson CJ, Suh SW, Koh JY, Cha YK, Thompson RB, Labuda CJ, Balaji RV, Cuajungco MP (2002) Depletion of intracellular zinc from neurons by use of an extracellular chelator in vivo and in vitro, J Histochem Cytochem 50:1659-1662.

Gazaryan IG, Krasnikov BF, G. A. Ashby GA, Thorneley RN, Kristal BS, Brown AM (2002) Zinc is a potent inhibitor of thiol oxidoreductase activity and stimulates reactive oxygen species production by lipoamide dehydrogenase, J Biol Chem 277:10064-10072.

Gorter JA, Petrozzino JJ, Aronica EM, Rosenbaum DM, Opitz T, Bennett MVL, Connor JA, Zukin RS (1997) Global ischemia induces downregulation of Glur2 mRNA and increases AMPA receptor-mediated Ca2+ influx in hippocampal CA1 neurons of gerbil, J Neurosci 17:6179-6188.

Hershfinkel M, Moran A, Grossman N, Sekler I (2001) A zinc-sensing receptor triggers the release of intracellular Ca2+ and regulates ion transport., Proc Natl Acad Sci USA 98:11749-1175.

Hidalgo J, Aschner M, Zatta P, Vasak M (2001) Roles of the metallothionein family of proteins in the central nervous system, Brain Res Bull 55:133-145.

Holtz ML, Craddock SD, Pettigrew LC (2001) Rapid expression of neuronal and inducible nitric oxide synthases during post-ischemic reperfusion in rat brain., Brain Res 898:49-60.

Howell GA, Welch G, and Frederickson CJ (1984) Stimulation-induced uptake and release of zinc in hippocampal slices, Nature 308:736-738.

Ikeda T, Kimura K, Morioka S, Tamaki N (1980) Inhibitory effects of Zn2+ on muscle glycolysis and their reversal by histidine, J Nutr Sci Vitaminol (Tokyo) 26:357-366.

Itoh M, Ebadi M (1982) The selective inhibition of hippocampal glutamic acid decarboxylase in zinc-induced epileptic seizures, Neurochem Res 7:1287-1298.

Jacob C, Maret W, Vallee BL (1998) Control of zinc transfer between thionein, metallothionein, and zinc proteins, Proc Natl Acad Sci USA 95:3489-3494.

Jia Y, Jeng JM, Sensi SL, Weiss JH (2002) Zn2+ currents are mediated by calcium-permeable AMPA/kainate channels in cultured murine hippocampal neurones, J Physiol (Lond) 543:35-48.

Jeng JM, Jia Y, Bonanni L, Weiss JH (2002) Divergent effects of pH on Zn2+ and Ca2+ flux through Ca2+-permeable AMPA/kainate channels (CAKR), Soc Neurosci Abstr 29:539.

Jiang LJ, Maret W, Vallee BL (1998) The glutathione redox couple modulates zinc transfer from metallothionein to zinc-depleted sorbitol dehydrogenase, Proc Natl Acad Sci USA 95:3483–3488.

Jiang LJ, Vasak M, Vallee BL, Maret W (2000) Zinc transfer potentials of the alpha - and beta-clusters of metallothionein are affected by domain interactions in the whole molecule, Proc Natl Acad Sci USA 97:2503-2508.

Jiang D, Sullivan PG, Sensi SL, Steward O, Weiss JH (2001) Zn(2+) induces permeability transition pore opening and release of pro-apoptotic peptides from neuronal mitochondria, J Biol Chem 276:47524-47529.

Kägi JHR (1993) In: Metallothionein III, Biological Roles and Medical Implications, Suzuki KT, Imura N, Kimura M, Eds. (Birkhäuser Verlag, Basel), pp. 29-56.

Kerchner G, Canzoniero L, Yu S, Ling C, Choi DW (2000) Zn2+ current is mediated by voltage-gated Ca2+ channels and enhanced by extracellular acidity in mouse cortical neurones, J Physiol 528:39-52.

Kim YH, Kim EY, Gwag BJ, Sohn S, Koh JY (1999) Zinc-induced cortical neuronal death with features of apoptosis and necrosis: mediation by free radicals, Neurosci 89:175-182.

Kim TY, Hwang JJ, Yun SH, Jung MW, Koh JY (2002) Augmentation by zinc of NMDA receptor-mediated synaptic responses in CA1 of rat hippocampal slices: mediation by Src family tyrosine kinases, Synapse 46:49-56.

Kim Y-H, Koh J-Y (2002) The role of NADPH oxidase and neuronal nitric oxide synthase in zinc-induced poly(ADP-ribose) polymerase activation and cell death in cortical culture, Exp Neurol 177:407-418.

Kleiner D, Von Jagow G (1972) On the inhibition of mitochondrial electron transport by Zn(2+) ions, FEBS Lett 20:229-232.

Kleiner D (1974) The effect of Zn2+ ions on mitochondrial electron transport, Arch Biochem Biophys 165:121-125.

Koh JY, Suh SW, Gwag BJ, He YY, Hsu CY, Choi DW (1996) The role of zinc in selective neuronal death after transient global cerebral ischemia, Science 272:1013-1016.

Kokaia Z, Andsberg G, Martinez-Serrano A, Lindvall O (1998) Focal cerebral ischemia in rats induces expression of P75 neurotrophin receptor in resistant striatal cholinergic neurons, Neurosci 84:1113-1125.

Korichneva I, Hoyos B, Chua R, Levi E, Hammerling U (2002) Zinc release from protein kinase C as the common event during activation by lipid second messenger or reactive oxygen, J Biol Chem 277:44327-44331.

Kroncke KD, Fehsel K, Schmidt T, Zenke FT, Dasting I, Wesener JR, Bettermann H, Breunig KD, Kolb-Bachofen V (1994) Nitric oxide destroys zinc-sulfur clusters inducing zinc release from metallothionein and inhibition of the zinc finger-type yeast transcription activator LAC9, Biochem Biophys Res Commun 200:1105-1110.

Krotkiewska B, Banas T (1992) Interaction of Zn2+ and Cu2+ ions with glyceraldehyde-3-phosphate dehydrogenase from bovine heart and rabbit muscle, Int J Biochem 24:1501-1505.

Kukimoto I, Hoshino S, Kontani K, Inageda K, Nishina H, Takahashi K, Katada T (1996) Stimulation of ADP-ribosyl cyclase activity of the cell surface antigen CD38 by zinc ions resulting from inhibition of its NAD+ glycohydrolase activity, Eur J Biochem 239:177-182.

Lee TH, Abe K, Kogure K, Itoyama Y (1995) Expressions of nerve growth factor and p75 low affinity receptor after transient forebrain ischemia in gerbil hippocampal CA1 neurons, J Neurosci Res 41:684-695.

Lee JY, Cole TB, Palmiter RD, Koh JY (2000) Accumulation of zinc in degenerating hippocampal neurons of ZnT3-null mice after seizures: evidence against synaptic vesicle origin, J Neurosci 20:RC79.

Lee JY, Kim YH, Koh JY (2001) Protection by pyruvate against transient forebrain ischemia in rats, J Neurosci 21:RC171,1-6.

Lee JM, Zipfel GJ, Park KH, He YY, Hsu CY, Choi DW (2002) Zinc translocation accelerates infarction after mild transient focal ischemia, Neurosci 115:871-878.

Lee JY, Kim JH, Palmiter RD, Koh JY (2003) Zinc released from metallothionein-III may contribute to hippocampal CA1 and thalamic neuronal death following acute brain injury, Exp Neurol 184:337-347.

Lerma J, Morales M, Ibarz JM, Somohano F (1994) Rectification properties and Ca2+ permeability of glutamate receptor channels in hippocampal cells, Eur J Neurosci 6:1080-1088.

Link TA, Von Jagow G (1995) Zinc ions inhibit the QP center of bovine heart mitochondrial bc1 complex by blocking a protonatable group, J Biol Chem 270:25001-25006.

Lipton P (1999) Ischemic Cell Death in Brain Neurons, Physiol Rev 79:1431-1568.

Lobner D, Canzoniero LM, Manzerra P, Gottron F, Ying H, Knudson M, Tian M, Dugan LL, Kerchner GA, Sheline CT, Korsmeyer SJ, Choi DW (2000) Zinc-induced neuronal death in cortical neurons, Cell Mol Biol (Noisy-le-grand) 46:797-806.

Lopantsev V, Wenzel HJ, Cole TB, Palmiter RD, Schwartzkroin PA (2003) Lack of vesicular zinc in mossy fibers does not affect synaptic excitability of CA3 pyramidal cells in zinc transporter 3 knockout mice, Neurosci 116:237-248.

Lorusso M, Cocco T, Sardanelli AM, Minuto M, Bonomi F, Papa S. (1991) Interaction of Zn2+ with the bovine-heart mitochondrial bc1 complex, Eur J Biochem 197:555-561.

Maret W (1994) Oxidative metal release from metallothionein via zinc-thiol/disulfide interchange, Proc Natl Acad Sci USA 91:237-241.

Maret W, Vallee BL (1998) Thiolate ligands in metallothionein confer redox activity on zinc clusters, Proc Natl Acad Sci USA 95:3478-3482.

McLaughlin B, Pal S, Tran MP, Parsons AA, Barone FC, Erhardt JA, Aizenman E (2001) p38 Activation Is Required Upstream of Potassium Current Enhancement and Caspase Cleavage in Thiol Oxidant-Induced

Neuronal Apoptosis, J. Neurosci. 21:3303-3311.
Nicholls DG, Budd SL (2000) Mitochondria and Neuronal Survival Physiol Rev. 80:315-360.
Nicholls P, Malviya AN (1968) Inhibition of nonphosphorylating electron transfer by zinc. The problem of delineating interaction sites, Biochemistry 7:305-310.
Nicotera P, Leist M, Ferrando-May E (1999) Apoptosis and necrosis: different execution of the same death, Biochem Soc Symp 66.
Noh KM, Kim YH, Koh JY (1999) Mediation by membrane protein kinase C of zinc-induced oxidative neuronal injury in mouse cortical cultures, J Neurochem 72:1609-1616.
Ohana E, Segal D, Palty R, Ton-That D, Moran A, Sensi SL, Weiss JH, Hershfinkel M, Sekler I (2004) A sodium zinc exchange mechanism is mediating extrusion of zinc in mammalian cells, J Biol Chem 279:4278-4284.
Opitz T, Grooms SY, Bennett MVL, Zukin RS (2000) Remodeling of alpha -amino-3-hydroxy-5-methyl-4-isoxazole-propionic acid receptor subunit composition in hippocampal neurons after global ischemia, Proc Natl Acad Sci USA 97:13360-13365.
Packer L, Tritschler HJ, Wessel K (1997) Neuroprotection by the Metabolic Antioxidant Lipoic Acid, Free Radical Biology and Medicine 22:359-378.
Palmiter RD, Huang L (2004) Efflux and compartmentalization of zinc by members of the SLC30 family of solute carriers, Pflugers Archiv 447:744-751.
Park JA, Koh JY (1999) Induction of an Immediate Early Gene egr-1 by Zinc Through Extracellular Signal-Regulated Kinase Activation in Cortical Culture, J Neurochem 73:450-456.
Park JA, Lee JY, Sato TA, Koh JY (2000) Co-induction of p75NTR and p75NTR-associated death executor in neurons after zinc exposure in cortical culture or transient ischemia in the rat, J Neurosci 20:9096-9103.
Pellegrini-Giampietro DE, Gorter JA, Bennett MV, Zukin RS (1997) The GluR2 (GluR-B) hypothesis: Ca(2+)-permeable AMPA receptors in neurological disorders, Trends Neurosci 20:464-470.
Persechini A, McMillan K, Masters BS (1995) Inhibition of nitric oxide synthase activity by Zn2+ ion, Biochemistry 34:15091-15095.
Pulsinelli WA, Brierley JB, Plum F (1982) Temporal profile of neuronal damage in a model of transient forebrain ischemia, Ann Neurol 11:491-498.
Robbins AH, Mcree DE, Williamson M, Collett SA, Xuong NH, Furey WF, Wang BC, Stout CD (1991) Refined crystal structure of Cd, Zn metallothionein at 2.0 A resolution, J Mol Biol 221:1269-1293.
Saris NE, Niva K (1994) Is Zn2+ transported by the mitochondrial calcium uniporter?, FEBS Lett 356:195-198.
Schwiening CJ, Boron WF (1994) Regulation of intracellular pH in pyramidal neurones from the rat hippocampus by Na(+)-dependent Cl(-)-HCO3- exchange, J Physiol 475:59–67.
Sensi SL, Canzoniero LM, Yu SP, Ying HS, Koh JY, Kerchner GA, Choi DW (1997) Measurement of intracellular free zinc in living cortical neurons: routes of entry, J Neurosci 17:9554-9564.
Sensi SL, Yin HZ, Carriedo SG, Rao SS, Weiss JH (1999) Preferential Zn2+ influx through Ca2+-permeable AMPA/kainate channels triggers prolonged mitochondrial superoxide production, Proc Natl Acad Sci U S A 96:2414-2419.
Sensi SL, Yin HZ, Weiss JH (1999) Glutamate triggers preferential Zn2+ flux through Ca2+ permeable AMPA channels and consequent ROS production, Neuroreport 10:1723-1727.
Sensi SL, Yin HZ, Weiss JH (2000) AMPA/kainate receptor-triggered Zn2+ entry into cortical neurons induces mitochondrial Zn2+ uptake and persistent mitochondrial dysfunction, Eur J Neurosci 12:3813-3818.
Sensi SL, Ton-That D, Weiss JH (2002) Mitochondrial sequestration and Ca(2+)-dependent release of cytosolic Zn(2+) loads in cortical neurons, Neurobiol Dis 10:100-108.
Sensi SL, Ton-That D, Sullivan PG, Jonas EA, Gee KR, Kaczmarek LK, Weiss JH (2003) Modulation of mitochondrial function by endogenous Zn2+ pools, Proc Natl Acad Sci USA 100:6157-6162.
Sheline CT, Behrens MM, Choi DW (2000) Zinc-induced cortical neuronal death: contribution of energy failure attributable to loss of NAD(+) and inhibition of glycolysis, J Neurosci 20:3139-3146.
Skulachev VP, Chistyakov VV, Jasaitis AA, Smirnova EG (1967) Inhibition of the respiratory chain by zinc ions, Biochem Biophys Res Commun 26:1-6.
Sloviter RS (1985) A selective loss of hippocampal mossy fiber Timm stain accompanies granule cell seizure activity induced by perforant path stimulation, Brain Res 330:150-153.
Suh SW, Chen JW, Motamedi M, Bell B, Listiak K, Pons NF, Danscher G, Frederickson CJ (2000) Evidence that synaptically-released zinc contributes to neuronal injury after traumatic brain injury, Brain Res 852:268-273.
Tang C, Dichter M, Morad M (1990) Modulation of the N-Methyl-D-Aspartate Channel by Extracellular H+, Proc Natl Acad Sci USA 87:6445-6449.
Tonder N, Johansen FF, Frederickson CJ, Zimmer J, Diemer NH (1990) Possible role of zinc in the selective degeneration of dentate hilar neurons after cerebral ischemia in the adult rat, Neurosci Lett 109:247-252.
Traynelis S, Cull-Candy S (1991) Pharmacological properties and H+ sensitivity of excitatory amino acid

receptor channels in rat cerebellar granule neurones, J Physiol (Lond) 433:727-763.

Trendelenburg G, Prass GK, Priller J, Kapinya K, Polley A, Muselmann C, Ruscher K, Kannbley U, Schmitt AO, Castell S, Wiegand F (2002) A. Meisel, A. Rosenthal, and U. Dirnagl, Serial Analysis of Gene Expression Identifies Metallothionein-II as Major Neuroprotective Gene in Mouse Focal Cerebral Ischemia, J Neurosci 22:5879-5888.

Turetsky DM, Canzoniero LMT, Sensi SL, Weiss JH, Goldberg MP, Choi DW (1994) Cortical neurones exhibiting kainate-activated Co2+uptake are selectively vulnerable to AMPA/kainate receptor-mediated toxicity, Neurobiology of Disease 1:101-110.

Van Lookeren Campagne M, Thibodeaux H, Van Bruggen N, Cairns B, Gerlai R, Palmer JT, Williams SP, Lowe DG (1999) Evidence for a protective role of metallothionein-1 in focal cerebral ischemia, Proc Natl Acad Sci USA 96:12870-12875.

Wallace DC (1999) Mitochondrial Diseases in Man and Mouse, Science 283:1482-1488.

Weiss JH, Hartley DM, Koh JY, Choi DW (1993) AMPA receptor activation potentiates zinc neurotoxicity, Neuron 10:43-49.

Wudarczyk J, Debska G, Lenartowicz E (1999) Zinc as an inducer of the membrane permeability transition in rat liver mitochondria, Arch Biochem Biophys 363:1-8.

Yin HZ, Weiss JH (1995) Zn(2+) permeates Ca(2+) permeable AMPA/kainate channels and triggers selective neural injury, Neuroreport 6:2553-2556.

Yin HZ, Sensi SL, Carriedo SG, Weiss JH (1999) Dendritic localization of Ca2+-permeable AMPA/kainate channels in hippocampal pyramidal neurons, J Comp Neurol 409:250-260.

Yin HZ, Sensi SL, Ogoshi F, Weiss JH (2002) Blockade of Ca2+-permeable AMPA/kainate channels decreases oxygen-glucose deprivation-induced Zn2+ accumulation and neuronal loss in hippocampal pyramidal neurons, J Neurosci 22:1273-1279.

Ying W, Han SK, Miller JW, Swanson RA (1999) Acidosis potentiates oxidative neuronal death by multiple mechanisms, J Neurochem 73:1549-1556.

Yokoyama M, Koh J, Choi DW (1986) Brief exposure to zinc is toxic to cortical neurons, Neurosci Lett 71:351-355.

Yu SP, Canzoniero LM, Choi DW (2001) Ion homeostasis and apoptosis, Curr Opin Cell Biol 13:405-411.

Yu SW, Wang H, Poitras MF, Coombs C, Bowers WJ, Federoff HJ, Poirier GG, Dawson TM, Dawson VL (2002) Mediation of Poly(ADP-Ribose) Polymerase-1-Dependent Cell Death by Apoptosis-Inducing Factor, Science 297:259-263.

BDNF AS A TRIGGER FOR TRANSSYNAPTIC CONSOLIDATION IN THE ADULT BRAIN

Clive R. Bramham and Elhoucine Messaoudi*

1. INTRODUCTION

The neurotrophin family of signaling proteins, including nerve growth factor (NGF), brain-derived neurotrophic factor (BDNF), neurotrophin-3 (NT-3), and NT-4/5 are crucially involved in regulating the survival and differentiation of neuronal populations during development (Levi Montalcini, 1987; Davies, 1994; Lewin and Barde, 1996). In addition to these well-established functions in development, a large body of work suggests that neurotrophins continue to shape the structure and function of neuronal connections throughout life (Schnell et al., 1994; Thoenen, 1995; Bonhoeffer, 1996; Prakash et al., 1996; Cabelli et al., 1997; Alsina et al., 2001; Maffei, 2002; Bolanos and Nestler, 2004; Duman, 2004; Tuszynski and Blesch, 2004). While neurotrophins traditionally were thought to operate on a time scale of days and weeks, extremely rapid effects have now been demonstrated on ion channels, neurotransmitter release, axon pathfinding, gene expression and mRNA translation (Song and Poo, 1999; Desai et al., 1999; Schinder and Poo, 2000).

It has nevertheless been difficult to pin down precise functions for specific neurotrophins in adulthood. One of the most contested areas is the contribution of neurotrophins to activity-dependent synaptic plasticity. In a series of recent advances several lines of evidence have converged to specficially implicate BDNF in long-term potentiation (LTP), the most widely studied form of synaptic plasticity in the adult brain. BDNF is uniquely positioned to regulate synaptic efficacy through bidirectional effects at the glutamate synapse. The complexity and versatility of BDNF signaling is reflected in the multiple roles of this neurotrophin not only in LTP, but also in modulation of long-term depression (LTD), various forms of short-term synaptic plasticity, and homeostatic regulation of intrinsic neuronal excitability (Desai et al., 1999; Sermasi et al., 2000; Asztely et al., 2000; Kumura et al., 2000; Ikegaya et al., 2002; Jiang et al., 2003).

In this chapter we will briefly review key evidence for permissive and instructive actions of BDNF in hippocampal LTP. We will further elaborate on new evidence suggesting that BDNF drives the formation of stable, protein synthesis–dependent LTP—a process we refer to as synaptic consolidation. A working model for synaptic consolidation based on induction of the immediate early gene, Arc/Arg, and local enhancement of dendritic protein synthesis, is proposed.

---

* Department of Biomedicine and Bergen Mental Health Research Center, University of Bergen, Jonas Lies vei 91, N-5009 Bergen, Norway.

2. SYNAPTIC CONSOLIDATION

Experience-dependent changes in behavior are thought to derive from lasting changes in synaptic strength and neuronal excitability, remodeling of synapses, and, in the dentate gyrus, neurogenesis (Gould and Gross, 2002; Morris et al., 2003). All of these mechanisms are encompassed in LTP evoked by high-frequency stimulation (HFS) of excitatory inputs. LTP induction requires a rapid rise in postsynaptic calcium; at most glutamatergic synapses this is critically provided by NMDA receptor (NMDAR) activation. The ensuing maintenance of LTP consists of at least two phases, early and late. Early LTP (lasting some 1-2 hours) requires covalent modification of existing proteins and protein trafficking at synapses, but not new protein synthesis (Bliss and Collingridge, 1993; Lisman et al., 2002; Malinow and Malenka, 2002). Development of late LTP, like long-term memory, depends on *de novo* mRNA and protein synthesis (Frey et al., 1988; Otani and Abraham, 1989; Matthies et al., 1990; Nguyen et al., 1994; Nguyen and Kandel, 1996; Frey et al., 1996; Raymond et al., 2000; Davis et al., 2000; Kandel, 2001). LTP is associated with both rapid and more delayed changes in gene expression. The early window of gene expression occurring during the first 30 minutes or so after HFS is associated with activation of several constitutively expressed transcription factors, including cyclic-AMP/calcium responsive-element binding protein (CREB) and Elk-1, leading to induction of a functionally diverse group of immediate early genes (IEGs). Not surprisingly, a long list of protein kinases is implicated in transcriptional regulation underlying late LTP. Critical roles of cyclic-AMP dependent protein kinase (PKA) and extracellular-signal regulated kinase (ERK) acting through phosphorylation of CREB have been described. The notion of protein synthesis–dependent consolidation is borne out in various forms of long-term synaptic plasticity and memory consolidation from flies to man (Kandel, 2001). By analogy to memory consolidation, synaptic consolidation refers to protein synthesis-dependent strengthening of synaptic transmission.

The NMDA receptor is a calcium gate exquisitely designed to detect coincident pre- and postsynaptic activity, thereby triggering LTP (or LTD). LTP is initially unstable and reversible through depotentiation (O'Dell and Kandel, 1994; Staubli and Chun, 1996). Constrasting with this molecular switch idea, the macromolecular synthesis underlying synaptic consolidation places greater demands on cellular resources and is not immediately reversible. From a systems level standpoint stable changes in synapses represent a commitment of the network in which the altered synapses are embedded. Rather than being dictated slavishly by the LTP induction event, synaptic consolidation is likely be to a highly regulated process with its own set of controls. Control mechanisms may exist from the molecular level to the neural systems level. Modulatory transmitters such as norepinephrine, serotonin, dopamine, and acetylcholine are all implicated in modulation of LTP induction or maintenance (Stanton and Sarvey, 1985a; Stanton and Sarvey, 1985b; Stanton and Sarvey, 1985c; Frey et al., 1991; Bramham et al., 1997; Swanson-Park et al., 1999; Graves et al., 2001; Kulla and Manahan-Vaughan, 2002; Straube and Frey, 2003; Harley et al., 2004). These extrinsic inputs have characteristically diffuse patterns of innervation. Activity in these systems is typically a function of the animal's behavioral or attentional state, with changes in activity dictating the functional modes of networks (i.e. local rhythmic activity, population discharges and synchronization, timing of synaptic events, frequency and duration of action potential firing) while setting the biochemical tone of target neurons. Some of the molecular targets for modulation of late LTP have already been identified, convergent regulation of PKA and CREB being eminent examples. However, these extrinsic systems are not designed for high-fidelity, synapse-specific control of excitatory synapses. Such effects would be better served by activity-dependent signaling occurring at the glutamate synapse itself. As discussed below, the BDNF/TrkB system has several features which fit the bill for a transsynaptic consolidation factor acting in tandem with glutamate.

3. BDNF: THE TIES THAT BIND

Neurotrophins activate one or more receptor tyrosine kinases of the tropomyosin-related kinase (Trk) family (Kaplan and Miller, 2000; Patapoutian and Reichardt, 2001). NGF binds preferentially to TrkA, BDNF and NT-4 to TrkB, and NT-3 to Trk C. In addition to Trk receptors, all neurotrophins bind to the p75 neurotrophin receptor ($p75^{NTR}$), a member of the tumor necrosis factor superfamily. The role of $p75^{NTR}$ is slowly beginning to emerge. One important function may be facilitation of Trk activation, either by presenting the neurotrophin to Trks or by inducing a favorable conformational change in the receptor (Chao and Bothwell, 2002). There is also evidence that pro-neurotrophins, including pro-BDNF, is released and preferentially activates $p75^{NTR}$ (Lu, 2003). Ligand binding to Trk leads to autophosphorylation of tyrosine residues within the intracellular domains of the receptor, creating docking sites for second messengers. The adaptor proteins Shc and FRS-2 bind to a common docking site coupling to activation of the Ras-raf-ERK cascade and the phosphatidylinositol-3-OH kinase (PI3K)/Akt pathway. Docking of phopholipase Cγ (PLCγ) to a separate site leads to production of diacylglyerol, a transient activator of protein kinase C (PKC), and inositol trisphosphate (IP3), which mobilizes intracellular calcium (see by chapter Amaral and Pozzo-Miller for more on calcium regulation). Signaling through these pathways, particularly pathways affecting gene expression, underlie the ability of neurotrophins to regulate neuronal differentiation, survival and outgrowth during development.

Functional diversity within the neurotrophin family is suggested by the distinct anatomical distributions of each neurotrophin/Trk receptor pair (Kokaia et al., 1993; Miranda et al., 1993; Schmidt Kastner et al., 1996; Yan et al., 1997; Tanaka et al., 1997; Conner et al., 1997). In the hippocampus, NGF is expressed in populations of principal neurons (granule cells and pyramidal cells) while TrkA receptors are located on cholinergic fibers projecting from the medial septum/diagonal band, consistent with a specialized function for NGF in modulating septal-hippocampal function (Blesch *et al.*, 2001). In contrast, BDNF and NT-3 and their respective Trk receptors are expressed on principal neurons and certain types of interneurons, implying extensive signaling within the intrinsic hippocampal network. NT-3 has a patchy distribution in the hippocampus, being expressed mainly in granule cells and CA2 pyramidal cells. Of all the neurotrophins, BDNF/TrkB is the only signaling system exhibiting widespread distribution across the subregions of the hippocampus and the adult forebrain.

BDNF is synthesized, stored and released from glutamatergic neurons. Storage and activity-dependent release has been demonstrated in dendrites and axon terminals, though not at the same synaptic sites. In principal neurons of the hippocampus, BDNF appears to be stored in dendritic processes in dense-core vesicles from which it is released in response to HFS (Blochl and Thoenen, 1996; Hartmann et al., 2001; Kohara et al., 2001; Balkowiec and Katz, 2002). Catalytic, signal transducing TrkB receptors have been localized to pre- and postsynaptically elements of glutamatergic synapses by immuno-electronmicroscopy (Drake *et al.*, 1999). Catalytic TrkB receptors are found in the PSD; TrkB is enriched in the postsynaptic density (PSD) fraction and co-immunprecipitates with NMDAR complex proteins (Wu et al., 1996; Aoki et al., 2000; Husi et al., 2000). These properties make BDNF attractive as a bidirectional modulator of excitatory transmission and plasticity. Two addition features make BDNF-TrkB particularly attractive as a mediator of synaptic consolidation: 1) BDNF regulates protein synthesis through both transcriptional and post-transcriptional mechanisms, and 2) BDNF is capable of stimulating its own release, possibly allowing sustained, regenerative signaling at synaptic sites. These features will be discussed more later.

4. BDNF HAS MULTIPLE, DISTINCT FUNCTIONS IN LTP

A combination of genetic and pharmacological approaches has revealed multiple, distinct contributions of BDNF signaling to LTP. These actions may be classified as *permissive* or *instructive*. Permissive refers to effects of BDNF that make synapses capable of LTP in the first place, but which are not causally involved in generating LTP. For example, basal (non-evoked) release of BDNF maintains the presynaptic release machinery, allowing sustained presynaptic transmission (and LTP induction) during HFS. In this way, the prior history of BDNF transmission determines whether LTP can be induced. In contrast, instructive refers to BDNF signaling that is initiated in response to HFS and causally involved in the development of LTP. Immediate and more delayed instructive roles have both been reported. Key evidence supporting these roles is described below. The predicted time course of BDNF release and mechanism of action is shown in Table 1. For a more extensive account of the literature see (Bramham and Messaoudi, 2005).

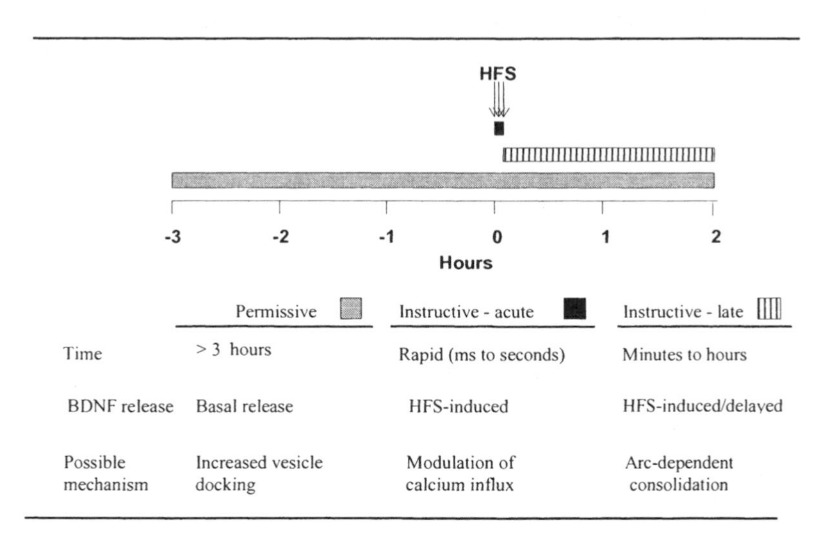

	Permissive	Instructive - acute	Instructive - late
Time	> 3 hours	Rapid (ms to seconds)	Minutes to hours
BDNF release	Basal release	HFS-induced	HFS-induced/delayed
Possible mechanism	Increased vesicle docking	Modulation of calcium influx	Arc-dependent consolidation

Table 1. Multiple roles of BDNF in hippocampal long-term potentiation.

4.1. Permissive: Setting the Stage for Activity-Dependent Synaptic Plasticity

Synaptic fatigue is a reduction in EPSP amplitude observed in response to consecutive stimuli in a stimulus train. Evidence suggests that BDNF modulates LTP indirectly by inhibiting synaptic fatigue. In the first study to address this issue Figurov et al. (1996) used a BDNF scavenger fusion protein, TrkB-Fc, to sequester released BDNF in adult rat hippocampal slices. Inhibition of BDNF signaling enhanced synaptic fatigue and impaired the induction and early maintenance of LTP at CA3-CA1 synapses. Conversely, application of BDNF to early postnatal hippocampal slices, in which endogenous BDNF levels are low, attenuated synaptic fatigue and facilitated LTP induction. The same BDNF treatment of adult hippocampal slices had no effect. Subsequent electrophysiological and biochemical studies identified a presynaptic locus for BDNF regulation of synaptic fatigue (Gottschalk *et al.*, 1998). Pozzo-Miller et al. (1999) demonstrated enhanced synaptic fatigue and impairment of LTP in slices obtained from BDNF knockout mice. The effects in BDNF mutants correlates with reduced expression of the synaptic vesicle-associated proteins synaptobrevin and synaptophysin and a reduction in the proportion of docked (readily releasable) vesicles in the active zone. A similar enhancement in synaptic fatigue and reduction in the expression of synaptic vesicle-associated proteins is seen in TrkB mutant mice (Martinez et al., 1998; Xu et al., 2000). However, mice in which TrkB receptors are selectively deleted from postsynaptic neurons at CA3-CA1 synapses exhibit normal synaptic fatigue and intact early LTP (Xu *et al.*, 2000).

BDNF incubation of slices obtained from BDNF knockouts restores expression of presynaptic proteins and reverses the effects on synaptic fatigue and LTP. The effects require at least 3-4 hours of BDNF incubation and involves transcription-dependent and transcription-independent mechanisms (Bradley and Sporns, 1999; Thakker-Varia et al., 2001; Tartaglia et al., 2001). Significantly, work in primary hippocampal cultures shows that BDNF enhances transcription of Rab3a, a small GTP-binding protein important for trafficking transmitter vesicles to the active zone (Thakker-Varia *et al.*, 2001). Taken together these studies suggest that BDNF facilitates frequency-dependent transmission and LTP induction through regulated synthesis of proteins involved in vesicle trafficking and neurotransmitter exocytosis. For review of the presynaptic actions of BDNF see (Schinder and Poo, 2000; Tyler et al., 2002).

4.2. Instructive: Acute Modulation

Endogenous BDNF is clearly capable of modulating LTP through mechanisms that do not involve suppression of synaptic fatigue. Thus, acute application of inhibitors shortly before HFS impairs LTP without affecting synaptic fatigue. This has been demonstrated using TrkB-Fc as well as function blocking antibodies to TrkB and BDNF (Figurov et al., 1996; Kang et al., 1997; Chen et al., 1999; Patterson et al., 2001). In an intriguing twist the effect of acutely applied inhibitors on early LTP were shown to depend on the stimulation pattern used for LTP induction. LTP induced by short high-frequency bursts applied at theta rhythm frequency (theta burst stimulation) was inhibited while LTP induced by a single, continuous train of stimuli was not impaired (Kang et al., 1997; Chen et al., 1999). The reason for this difference is still unclear but recent work suggests that theta burst stimulation at CA3-CA1 synapses recruits a presynaptic component to LTP expression that requires BDNF signaling (Zakharenko et al., 2003). See Bramham and Messoudi (2005) for further discussion.

In an elegant study employing photo-induced release of caged BDNF antibody, (Kossel et al., 2001) sought to resolve rapid actions of BDNF during HFS. Hippocampal slices were incubated in medium containing caged antibody and flashes of UV light were applied from 2 minutes before until 2 minutes after HFS. LTP was impaired in the period from immediately after HFS to approximately 30 minutes thereafter. Although the effects were modest, this work provided direct evidence for immediate instructive actions of endogenous BDNF in LTP.

BDNF is one of the most potent neuroexcitants in the brain. Using focal, puff application in acute hippocampal slices, extremely rapid (millisecond) depolarizations are evoked by nanomolar concentrations of BDNF (Kafitz et al., 1999). This rapid depolarization is mediated by a tetrodotoxin-insensitive voltage-gated sodium channel (Na_V 1.9), which is thought to couple directly to TrkB, independently of second messenger signaling (Blum *et al.*, 2002). Kovalchuk et al., (2002) explored the ability of BDNF application to modulate LTP at medial perforant path-granule cell synapse. When puffed into onto the synaptic zone, BDNF induced a sharp rise in calcium levels in the spines and shafts of granule cells dendrites and a burst of action potentials in the cell body. BDNF had no effect on efficacy when given alone. However, LTP could be induced by pairing the BDNF puff with HFS that by itself was too weak to induced LTP. The pairing effect was obtained when stimulation was applied within 1 second of BDNF application, a time window exactly corresponding to the time course of the BDNF-induced depolarization. The effect was abolished by blocking NMDA receptors, activation of voltage-dependent calcium channels, and by chelation of postsynaptic calcium. Thus, puffs of BDNF can directly gate LTP induction through rapid postsynaptic calcium influx. In future studies it will be important to determine whether endogenous BDNF modulates LTP through activation of the postsynaptic Na_v 1.9 channels.

4.3. Instructive: Late LTP

Genetic and pharmacological studies suggest an important function for BDNF and TrkB activation in late LTP (Minichiello et al., 1999; Patterson et al., 2001; Minichiello et al., 2002). The first evidence came from studies of BDNF germline knockout mice (Korte et al., 1995; Patterson et al., 1996). By focusing on slices with significant early LTP and monitoring the responses over longer time periods, a deficit in long-term maintenance of the response was apparent (Korte et al., 1998). Kandel and colleagues have shown that spaced, but not single HFS, produces transcription-dependent LTP (Huang et al., 1996). Using this spaced HFS protocol Kang et al. (1997) showed that TrkB antibody prevented development of late LTP while leaving early LTP almost completely intact. The authors were able to resolve the time point of BDNF's effects by perfusing hippocampal slices with TrkB-Fc at different time points during LTP maintenance. Remarkably, LTP was reversed when the scavenger was applied from 30-60 min, but not 70- 100 min, after HFS. These results suggested that formation of late LTP depends on a critical period of TrkB signaling *after* HFS. As elaborated in the next section, this evidence is complimented by studies showing that exogenously applied BDNF directly activates late LTP.

5. INSIGHTS FROM BDNF-INDUCED LTP

(Lohof et al., 1993) were the first to show neurotrophin-evoked increases in synaptic transmission. This original observation at the frog nerve-muscle synapse was followed by flurry of studies on the effects of exogenously applied neurotrophins on hippocampal synaptic transmission. The response to exogenous neurotrophins in the hippocampus appears to be a function of the preparation used (cell culture, slice, whole animal) as well as the method and duration of application. For example, application of BDNF to cultured hippocampal neurons from embryonic or early postnatal hippocampus results in a transient potentiation of excitatory synaptic transmission lasting some 10 to 30 minutes upon switching to normal medium (Lessmann et al., 1994; Knipper et al., 1994; Levine et al., 1995). In contrast, BDNF application in the adult hippocampus can trigger a long-lasting increase in synaptic efficacy dubbed BDNF-induced LTP (or simply BDNF-LTP). Persistent potentiation was first shown at CA3-CA1 synapses in response to bath perfusion of hippocampal slices with BDNF (Kang and Schuman, 1995b; Kang and Schuman, 1996). BDNF-LTP was subsequently shown at medial perforant path-granule cell synapses of the dentate gyrus *in vivo*, in the insular cortex *in vivo*, and in the visual cortex *in vitro* and *in vivo* (Messaoudi et al., 1998; Jiang et al., 2001; Escobar et al., 2003). Although exogenous NT-3 (but not NGF) elicits long-lasting potentiation in the CA1 region, this effect appears to be mediated by TrkB activation (Ma *et al.*, 1999). Studies of exogenous BDNF-LTP have helped to elucidate cellular mechanisms of synaptic plasticity specifically mediated by this neurotrophin. The discussion below elaborates on our recent findings in the dentate gyrus *in vivo*.

5.1 Some Basic Properties

BDNF-LTP is induced at medial perforant path-granule cell synapses in the dentate gyrus by a single, brief (25 min) infusion of BDNF. The infusion site is located 300 μm above the medial perforant path synapse. Field EPSPs are significantly elevated 15 min after infusion and climb to a stable plateau within 2-3 hours. The full duration of BDNF-LTP has not been determined, but it lasts at least 15 hours in anesthetized rats (Messaoudi et al., 1998; Messaoudi et al., 2002; Ying et al., 2002), and 24 hours in freely moving rats (Messaoudi and Bramham, unpublished). Like LTP, BDNF-LTP is associated with enhanced EPSP-spike coupling in addition to enhanced synaptic efficacy (Bliss and Lomo, 1973; Abraham et al., 1987; Lu et al., 2000; Messaoudi et al., 2002).

BDNF activation of TrkB receptors is implicated in epileptogenesis (Binder et al., 2001; He et al., 2004). BDNF and other neurotrophins can also modulate GABAergic transmission (Tanaka et al., 1997; Frerking et al., 1998) Recently, (Scharfman et al., 2003) showed that chronic (2 week) application of BDNF is associated with spontaneous seizures and mossy fiber sprouting in the dentate hilar region. In hippocampal slice cultures picrotoxin-induced seizures induces release of endogenous BDNF leading to axonal branching of mossy fibers and development of hyperexcitable reentrant circuits in the dentate gyrus (Koyama et al., 2004). However, acute application of BDNF leading to BDNF-LTP does not lead to changes in recurrent GABAergic inhibition onto dentate granule cells, hyperexcitability (e.g. multiple population spikes) or epileptiform spiking (Messaoudi *et al.*, 1998). It will be important to determine whether BDNF-induced changes in glutamatergic transmission contribute to seizure pathogenesis.

The fact that BDNF is capable of acutely increasing glutamate release raises the possibility that BDNF indirectly induces NMDAR-dependent potentiation. To address this issue, BDNF was infused into the dentate gyrus following systemic administration of the a competitive NMDAR antagonist (Messaoudi et al., 2002). While HFS-LTP was abolished, BDNF infusion induced robust potentiation during NMDAR blockade. BDNF-LTP at CA3-CA1 synapses in hippocampal slices is similarly NMDAR-independent (Kang and Schuman, 1995a). While information on BDNF release is not yet available from LTP studies in the adult brain, BDNF-GFP is released from dendrites of cultured hippocampal neurons following HFS and this release depends on NMDAR and AMPA-type glutamate receptor activation (Hartmann et al., 2001; Lever et al., 2001). Exogenous BDNF application may bypass this initial release event to trigger synaptic consolidation.

5.2. BDNF-LTP Occludes with Late Phase LTP.

A crucial issue is whether exogenous BDNF reflects the physiological actions of endogenous BDNF. If two forms of LTP utilize a common mechanism of expression the generation of one should occlude (inhibit) the other. Messaoudi et al. (2002) examined the effect of BDNF infusion at time points corresponding to early and late LTP (Fig 1). BDNF applied during early LTP induced robust potentiation indicating a distinct mechanisms of expression. Strikingly, complete occlusion was observed when BDNF was applied during late LTP. Conversely, at CA3-CA1 synapses *in vitro*, Kang et al (1997) showed that prior induction of BDNF-LTP occludes expression of late, but not early, HFS-LTP. This time-dependent pattern of occlusion suggests that exogenous BDNF specifically activates mechanisms common to late LTP. Consistent with previous work (Frey et al., 1995), it also suggests a rapid switch in the mechanism of expression between early and late phase LTP.

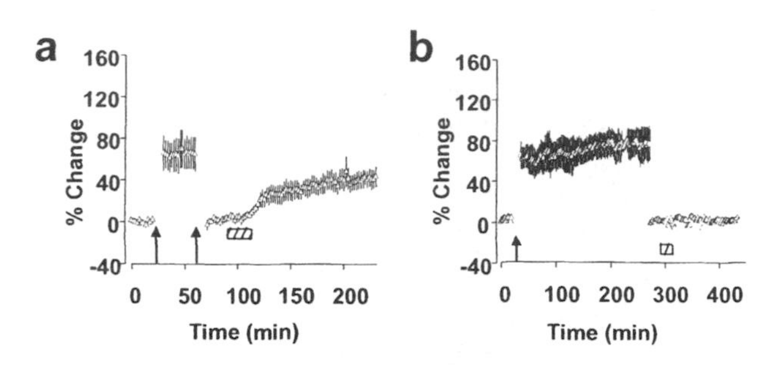

Figure 1. BDNF-LTP is occluded by late phase, but not early phase, HFS-LTP. *A*, LTP was induced by three sessions of HFS (400 Hz) and recorded for 30 min. The stimulus intensity was then lowered to reset the fEPSP slope to baseline. A second session of HFS produced no further increase, demonstrating saturation of HFS-LTP. BDNF infusion (hatched bar) induced normal BDNF-LTP. *B*, HFS-LTP was induced and recorded for 240 min, then reset to baseline. BDNF-LTP was occluded at this time. Values are group means (± SEM) expressed in percent of baseline. Adapted from (Messaoudi et al., 2002).

5.3. BDNF-LTP Induction Requires Rapid ERK Activation and *De Novo* Gene Expression

ERK signaling leading to CREB activation is required for late LTP and hippocampal-dependent memory formation. Ying et al. (2002) examined the role of ERK signaling in BDNF-induced LTP. Local infusion of the MEK (MAPK, or ERK, kinase) inhibitors PD98059 and U0126 completely abolished BDNF-LTP induction but had no effect on established BDNF-LTP (Fig. 2). Immunoblot analysis performed in homogenates obtained from microdissected dentate gyrus confirmed rapid phosphorylation of ERK. Treatment with MEK inhibitors blocked this activation in parallel with BDNF-LTP. Thus, MEK-ERK activation is required for the induction, but not the maintenance of, BDNF-LTP. Furthermore, BDNF-LTP induction is

5.3. BDNF-LTP Induction Requires Rapid ERK Activation and *De Novo* Gene Expression

ERK signaling leading to CREB activation is required for late LTP and hippocampal-dependent memory formation. Ying et al. (2002) examined the role of ERK signaling in BDNF-induced LTP. Local infusion of the MEK (MAPK, or ERK, kinase) inhibitors PD98059 and U0126 completely abolished BDNF-LTP induction but had no effect on established BDNF-LTP (Fig. 2). Immunoblot analysis performed in homogenates obtained from microdissected dentate gyrus confirmed rapid phosphorylation of ERK. Treatment with MEK inhibitors blocked this activation in parallel with BDNF-LTP. Thus, MEK-ERK activation is required for the induction, but not the maintenance of, BDNF-LTP. Furthermore, BDNF-LTP induction is transcription-dependent and associated with ERK-dependent phosphorylation of CREB on serine-133, which is required for CRE-driven gene expression. Taken together these experiments suggest that BDNF triggers synaptic consolidation through rapid activation of MEK-ERK and *de novo* gene expression. At CA3-CA1 synapses, TrkB-coupled PLC may be dominant to TrkB-ERK in regulating late LTP formation (Minichiello et al., 2002; Ernfors and Bramham, 2003). In these synapses, TrkB signaling is required for nuclear translocation of ERK, but may not be directly involved in modulating ERK activity (Patterson et al., 2001).

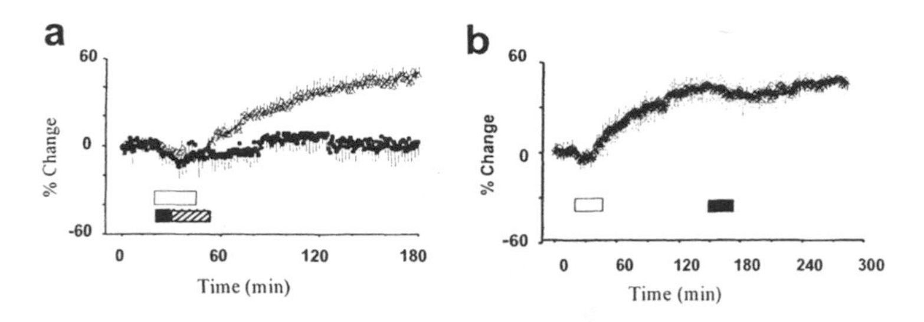

Figure 2. BDNF-LTP induction, but not maintenance, requires ERK signaling. A, BDNF infusion (white bar; 2μg in 2μl, 25min) induces LTP of medial perforant path-evoked fEPSPs. Infusion of the MEK inhibitor U0126 (30μM; black and hatched bar) blocked BDNF-LTP. B, U0126 applied 2 hours after BDNF had no effect. Adapted from Ying et al. (2002).

5.4. BDNF Triggers Arc-Dependent Synaptic Consolidation

IEGs induced following LTP induction encode a miscellany of transcription factor and non-transcription factor proteins (Cole et al., 1989; Wisden et al., 1990; Qian et al., 1993; Abraham et al., 1993; Meberg et al., 1993; Link et al., 1995; Williams et al., 1995; Lyford et al., 1995; Tsui et al., 1996; Lanahan et al., 1997). Transcription factor IEGs such as zif268 (a.k.a. egr-1, ngfi-a, krox24) and nurr1 regulate late response genes, although the targets of these genes have not been identified. The non-transcription factor genes encode synaptic proteins such as activity-regulated cytoskeleton-associated protein (Arc, a.k.a. Arg3.1) and Homer1a, as well as secreted proteins such as tissue plasminogen activator (tPA), neuronal activity-regulated pentraxin (NARP), and BDNF itself.

Arc and zif268 are both implicated in LTP maintenance and memory consolidation. Mice with germline knockout of zif268, a zinc-finger transcriptional activator, have impaired late LTP (1 day post-HFS) and impaired hippocampal-dependent memory (Jones et al., 2001). Arc mRNA is rapidly induced and trafficked to dendritic processes following LTP induction. Using intrahippocampal injection of Arc antisense (AS) oligodeoxynucleotides, Guzowski et al. (2000) showed that Arc is required for consolidation, but not acquisition of, hippocampal-dependent learning tasks. These effects of AS treatment on LTP were less conclusive. Rats treated with Arc AS prior to HFS exhibited LTP lasting several days, but the potentiation was smaller and decayed to baseline sooner than in control-infused rats. Ying et al. (2002) examined expression of Arc and zif268 following BDNF-LTP (Fig. 4). Arc mRNA and protein were both sharply enhanced, whereas zif268 expression was unaffected at the same time points. *In situ* hybridization of brains collected 2 hours after BDNF infusion revealed enhanced Arc expression across the granule cell layer and molecular layer of the dendrite gyrus, indicating delivery of transcripts to dendritic processes. Upregulation of Arc protein depended on ERK activation and new transcription (Messaoudi et al., 2002; Ying et al., 2002). The requirement for new transcription indicated that Arc synthesis stems from translation of newly induced, rather than preexisting, mRNA.

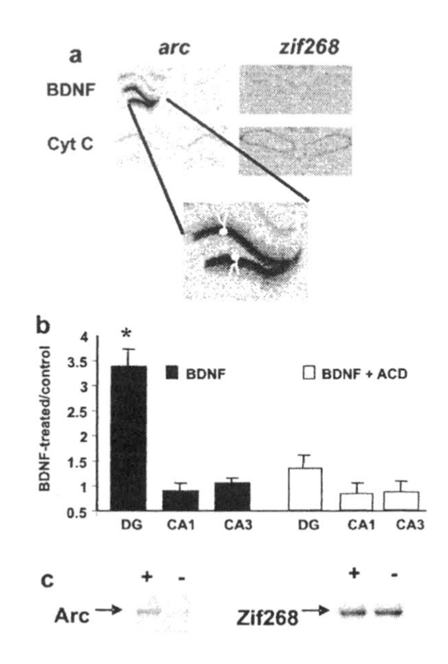

Figure 3. BDNF induces selective upregulation of Arc mRNA and protein expression. *A*, Autoradiographs of *in situ* hybridization signals showing upregulation of *Arc* mRNA levels in granule cells somata and dendrites 2 hours after BDNF-LTP induction. *B*, Enhancement in Arc protein expression is specific to the infused dentate gyrus and blocked by the transcription-inhibitor actinomycin D (ACD), showing that Arc protein derives from new mRNA synthesis. *Significantly different from BDNF control. Zif268 mRNA and protein levels were unchanged. *C*, Representative Western blots. Adapted from Messaoudi et al (2002) and Ying et al. (2002).

BDNF is expected to modulate many genes having no role or only a subsidiary role in LTP. A possible causal role for Arc in BDNF-LTP was recently investigated using local infusion of Arc AS oligodeoxynucleotides (Messaoudi et al., 2003). Treatment with Arc AS prior to BDNF infusion had no effect on baseline synaptic transmission but abolished BDNF-LTP and the associated upregulation of Arc protein, indicating a requirement for Arc induction (Fig. 4). Remarkably, Arc antisense applied 2 hours after BDNF rapidly reversed ongoing BDNF-LTP. The same treatment 4 hours after BDNF infusion had no effect, demonstrating a sharply defined time-window for Arc function. The same critical time-window of Arc translation was found during the maintenance of HFS-LTP. The reversal of ongoing potentiation was coupled to a rapid reduction in Arc protein expression, while levels of β-actin, PSD-95, and αCaMKII were unchanged. These results suggested that Arc translation defines a window of consolidation in long-term synaptic plasticity. Sustained translation during this narrow time-window is required for stable LTP. Furthermore, BDNF directly activates Arc-dependent synaptic strengthening and its time-dependent consolidation.

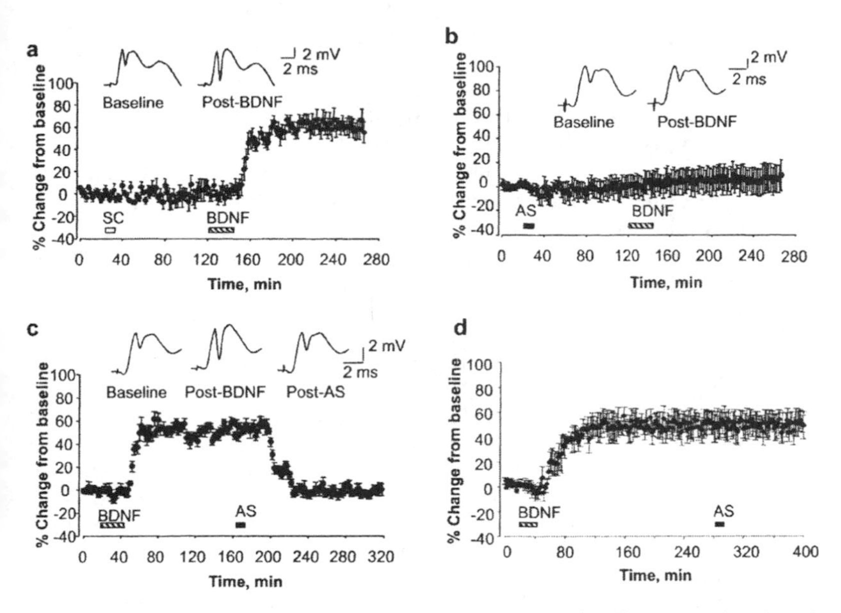

Figure 4. BDNF-LTP induction and consolidation requires Arc translation. Time course plots show changes (means ± SEM) in the medial perforant path-evoked fEPSP slope expressed in percent of baseline. Arc AS oligodeoxynucleotide or scrambled (SC) Arc sequence were infused 90 minutes before BDNF. A, Robust BDNF-LTP was induced in SC-treat rats. B, Arc AS pre-treatment abolished BDNF-LTP . C, Arc AS infusion at 2 hours rapidly reverses ongoing BDNF-LTP. D, Arc AS infusion at 4 hours had no effect. N=5-7 in all groups. Adapted from Messaoudi et al. (2003).

Changes in Arc protein levels following LTP induction have recently been examined by immuno-electronmicroscopy (Rodriguez et al., 2003; Moga et al., 2004). Interestingly, the time course of Arc protein elevation in dendritic spines of medial perforant path-granule cell synapses (up at 2 hours, down at 4 hours) (Rodriguez *et al.*, 2003), matches the critical period of synaptic consolidation shown using Arc antisense. Arc localizes to the PSD, co-precipitates with F-actin, and contains a spectrin homology domain suggesting a structural role (Lyford et al., 1995; Husi et al., 2000). Stable LTP is associated with insertion of glutamate receptors at postsynaptic membranes, thickening of the PSD, and changes in spine shape (Geinisman, 2000; Weeks et al., 2001; Harris et al., 2003). The critical, yet transient, action of Arc suggests that it mediates a coordinated cell biological process leading to persistent changes in synaptic strength.

Taken together the work in the dentate gyrus suggests that BDNF induces late LTP through TrkB-coupled activation of MEK-ERK leading to synthesis of Arc. Although Arc mRNA is transported throughout the dendritic tree following LTP induction, the protein is expressed more specifically in activated dendritic spines, emphasizing the importance of local translation mechanisms (Steward and Worley, 2001; Rodriguez et al., 2003; Moga et al., 2004).

6. BDNF, DENDRITIC PROTEIN SYNTHESIS, AND TRANSLATION CONTROL

A recent surge of studies has demonstrated the existence of protein synthesis in dendrites (Feig and Lipton, 1993; Wu et al., 1998; Kacharmina et al., 2000; Pierce et al., 2000; Steward and Schuman, 2001; Eberwine et al., 2001; Aakalu et al., 2001). The foundations of compartmental protein synthesis have been elegantly elaborated in oocyte maturation, early embryogenesis, and myelinization in oligodendrocytes (Carson et al., 1998; Bashirullah et al., 1999; Richter, 2001; de Moor and Richter, 2001; Bashirullah et al., 2001; Johnstone and Lasko, 2001). As in these biological systems, protein synthesis in dendrites depends on regulated mRNA transport, localization, and translation (Wells and Fallon, 2000; Steward and Schuman, 2003). Activity-dependent translation of dendritically localized mRNA provides a means for local, perhaps even synapse-specific, modulation of structure and function.

Whereas Arc mRNA is activity-induced and routed to dendrites, other mRNA species are stored locally in dendrites and translated in response to synaptic input (Steward and Levy, 1982; Steward et al., 1996; Steward, 1997). The mRNA coding for the α-subunit of CaMKII (αCaMKII) is one of a handful of mRNA species exhibiting abundant constitutive expression in dendrites of adult hippocampal principal neurons. CaMKII is a major component of the PSD and a pivotal molecule in the early stages of LTP expression involving membrane delivery of AMPA receptors (Lisman et al., 2002; Malinow and Malenka, 2002; Lisman et al., 2004). Activity-dependent translation of αCaMKII mRNA has been demonstrated in synaptic plasticity of the developing visual cortex (Wu et al., 1998). In the hippocampus, CaMKII protein expression is rapidly (5 min) increased in CA1 pyramidal cell dendrites following LTP induction (Ouyang et al., 1999). In an electron microscopic study, Ostroff et al., (2002) et al. used 3-D reconstruction of serial sections to examine the distrubtion of polyribosomes following LTP induction in the CA1 region of hippocampal slices. LTP was associated with an increase in the number of polyribosomes in dendritic spines commensurate with translocation of ribosomes from the shaft into the spine head. Havik et al. (2003) showed that αCaMKII mRNA levels are rapidly elevated in synaptodendrosomes following LTP induction in awake rats. The synaptodendrosomes, obtained by subcellular fractionation of dentate gyrus homogenates, are enriched in excitatory terminals attached to pinched-off resealed dendritic spines. No changes in mRNA levels occurred in whole dentate gyrus homogenates, suggesting trafficking of stored, pre-existing αCaMKII mRNA into dendritic spines. This increase of αCaMKII mRNA was paralleled by enhanced expression of CaMKII protein, again specific to the synaptodendrosome fraction. Routing of αCaMKII mRNA from cell body to dendrites depends on cis-acting elements in the 3' untranslated region (UTR). Mice in which the 3'UTR of the gene is removed by targeted mutagenesis do not express dendritic αCaMKII mRNA. These mice have normal early LTP and memory acquisition but impaired late LTP and memory consolidation (Miller et al., 2002). While the evidence is still fragmentary it clearly implicates dendritic protein synthesis in LTP. Arguably the most compelling demonstration of local protein synthesis-dependent synaptic plasticity comes from studies of long-term facilitation in *Aplysia* (Casadio et al., 1999; Sherff and Carew, 1999). Studying connections between single sensorimotor synapses, Casadio et al. (1999) showed that local application of five puffs of serotonin induced synapses-specific facilitation requiring local protein synthesis.

Dendritic protein synthesis in mammalian neurons is modulated by BDNF, glutamate, acetylcholine, and no doubt a number of other neurotransmitters (Feig and Lipton, 1993; Kacharmina et al., 2000; Aakalu et al., 2001). In the first direct visualization of dendritic protein synthesis, Aakalu et al., (2001) showed that BDNF induces hotspots of reporter GFP synthesis in isolated dendrites from cultured hippocampal neurons. Dendritic localization was obtained by flanking the reporter with the 5' and 3'UTRs of αCaMKII. Likewise, BDNF-LTP at adult CA3-CA1 synapses appears to require dendritic protein synthesis; hippocampal slices in which the CA1 dendrites and the CA3 axons were severed from the cell body exhibited protein synthesis-dependent BDNF-LTP (Kang et al., 1996).

Recent work has begun to explore the signal transduction pathways by which BDNF modulates dendritic protein synthesis. The rate-limiting step in translation of most mammalian mRNAs is phosphorylation of eukaryotic initiation factor 4E (eIF-4E) (Gingras *et al.*, 2004). eIF4E binds to the 7-methyl-guanosine cap structure at the 5' end of target mRNAs. Phosphorylation of eIF4E on Ser209 is correlated with enhanced rates of translation, whereas hypophosphorylation is associated with decreased translation (Flynn et al., 1997; Takei et al., 2001; Gingras et al., 2004). eIF4E is phosphorylated by mitogen-activating integrating kinase (MNK1), whose activity is regulated by ERK and p38 MAPK. The availability of eIF4E is controlled by binding proteins (4E-BPs). Phosphorylation of 4E-BP releases eIF4E and promotes cap-dependent translation (Gingras et al., 2004). Trk-coupled PI3K is thought to stimulate translation through activation of mammalian target of rapamycin (mTOR or FRAP), a multifunctional serine/threonine kinase that leads to phosphorylation of 4E-BPs. Importantly, the immunosuppresent drug rapamycin, which inhibits mTOR, blocks both late LTP and BDNF-LTP at CA3-CA1 synapses (Tang et al., 2002; Cammalleri et al., 2003). Recent work suggests a critical role for ERK signaling in translation control underlying late LTP in the hippocampus. Using both dominant negative MEK mice and pharmacological inhibitors, Kelleher, et al., (2004) showed that eIF4E is phosphorylated through an ERK signaling pathway in hippocampal neurons. In MEK1 mutant mice impaired eIF4E phosphorylation was associated with specific deficits in translation-dependent late LTP at CA3-CA1 synapses and impaired hippocampal-dependent memory formation. In the dentate gyrus *in vivo*, (Kanhema et al., 2005) showed that BDNF-LTP is coupled to an ERK-dependent phosphorylation of eIF4E. The synaptic actions of BDNF were examined *in vitro* using synaptodendrosomes. BDNF treatment of synaptodendrosomes led to rapid (5 min) phosphorylation of eIF4E and enhanced expression of αCaMKII, consistent with a cap-dependent translation of αCaMKII mRNA in dendritic spines. Finally, in addition to increasing eIF4E phosphorylation, BDNF also induces a redistribution of this translation factor to an mRNA granule-rich cytoskeletal fraction (Smart *et al.*, 2003).

Polyadenylation of the 3'UTR of target transcripts is another means of modulating translation. Polyadenylation induced activation of αCaMKII translation has been demonstrated in the developing visual cortex in vivo and in hippocampal neurons in vitro (Wu et al., 1998; Wells et al., 2001; Huang et al., 2002). Polyadenylation of αCaMKII is triggered by phosphorylation of the cytoplasmic polyadenylation binding protein (CPEB), which binds to recognition elements in the αCaMKII 3'UTR (Wells et al., 2000; Huang and Richter, 2004). The phosphorylation of CPEB in neurons appears to require NMDAR signaling through an aurora kinase. Current evidence suggests that eIF4E attaches to CPEB through the protein maskin, which releases eIF4E upon CPEB phosphorylation. Thus, both CPEB and mTOR regulate the availability of eIF4E. Interestingly, local upregulation of CPEB in sensory neurites is required for the stable maintenance of long-term facilitation in *Aplysia*. The N-terminus of the *Aplysia* CPEB protein has a prion-like switch which may provide stable, self-perpetuating enhancement of CPEB-regulated translation (Si et al., 2003a; Si et al., 2003b)

Protein synthesis is also regulated at the level of peptide chain elongation. Eukaryotic elongation factor-2 (eEF2) is a GTP-binding protein that mediates translocation of peptidyl-tRNAs from the A-site to the P site on the ribosome. Phosphorylation of eEF2 on Thr56 inhibits ribosome binding and arrests peptide chain elongation (Nairn and Palfrey, 1987; Ryazanov et al., 1988; Nairn et al., 2001). *In vivo*

BDNF-LTP is associated with a transient ERK-dependent phosphorylation of eEF2 in whole dentate gyrus (Kanhema et al., 2005). However, net eEF2 phosphorylation is unchanged in BDNF-treated synaptodendrosomes, suggesting that BDNF may have compartmental (synaptic and non-synaptic) effects on eEF2 phosphorylation. Immunocytochemical localization of phosphorylated translation factors is needed to resolve this issue. Peptide chain elongation is highly energy consuming, and decreases in ATP levels lead to phosphorylation of eEF2 and arrest of global protein synthesis (Marin et al., 1997; Browne and Proud, 2002; Chotiner et al., 2003). Decreases as well as increases in protein synthesis are seen during LTP (Fazeli et al., 1993; Chotiner et al., 2003). Perhaps inhibition of eEF2 serves to conserve metabolic energy during periods of intensive protein synthesis at synapses. As eEF2 only arrests ongoing translation, synthesis can be rapidly resumed upon dephosphorylation.

To summarize, changes in local dendritic synthesis depend on the availability of message for translation, the positioning of the translational apparatus, and biochemical regulation of translation factors. Recent evidence shows that BDNF can stimulate Arc synthesis in isolated synaptoneurosomes (Yin et al., 2002). BDNF may therefore play a dual role in inducing Arc transcription and modulating the local translation of newly synthesized (Arc) and locally resident (i.e. αCaMKII) mRNAs.

7. PRESYNAPTIC MECHANISMS AND RETROGRADE NUCLEAR SIGNALING

LTP is thought to involve coordinate pre- and postsynaptic modifications, as synapses increase in size. The discussion of Arc and dendritic protein synthesis emphasizes postsynaptic mechanisms of BDNF-TrkB signaling in the induction and expression of LTP. However, BDNF also acutely enhances glutamate release from synaptosomes and transiently enhances presynaptic transmission (Lessmann and Heumann, 1998; Jovanovic et al., 2000; Gooney and Lynch, 2001). Does BDNF also have an instructive presynaptic role in LTP?

At perforant path-granule cell synapses, both quantal analysis and biochemical studies support a contribution of enhanced glutamate release to LTP expression (Min et al., 1998; Errington et al., 2003). Evidence supporting a role for BDNF in enhanced glutamate transmitter release during LTP includes the following: 1) The maintenance phase of BDNF-LTP, like HFS-LTP, is associated with a lasting increase in potassium-evoked glutamate release from synaptosomes (Gooney et al., 2004), 2) TrkB receptors are autophosphorylated in synaptosomes collected during the maintenance phase of both HFS- and BDNF-induced LTP (Gooney et al., 2002; Gooney et al., 2004), 3) The Trk inhibitor, K252a, blocks presynaptic Trk activation and the sustained enhancement in neurotransmitter release. Finally, LTP maintenance is associated with enhanced, depolarization-evoked release of BDNF from dentate gyrus tissue (Gooney and Lynch, 2001). The relationship between enhanced BDNF release and glutamate release is far from clear. BDNF may act transiently to trigger a persistent presynaptic modification. Alternatively, the persistent component may be the release of BDNF itself. There is no data discriminating between these scenarios at the moment. The fact that acute enhancement of glutamate release by BDNF requires ERK signaling, while expression of BDNF-LTP and HFS-LTP does not, argues against a mechanism based on acute modulation of glutamate release. The exact sites and dynamics of endogenous BDNF release and TrkB activation in LTP remain to be determined. BDNF is released postsynaptically following HFS in hippocampal cultures (Hartmann et al., 2001). In an elegant study employing two-photon imaging of FM 1-43 in mice with conditional deletion of BDNF, (Zakharenko et al., 2003) found that presynaptic BDNF is required for the expression of presynaptic components of early LTP at CA3-CA1 synapses.

The classic hypothesis of target-derived trophic support involves signaling from the nerve terminal to the nucleus. Insights into the underlying molecular mechanisms has come from studies of sympathetic and sensory neurons (Riccio et al., 1997; Watson et al., 1999; Watson et al., 2001; Ginty and Segal, 2002; Delcroix et al., 2003; Campenot and MacInnis, 2004). Neurotrophin binding to presynaptic Trk receptors activates retrograde signaling pathways in axons leading to activation of nuclear substrates, such as CREB, and modulation of gene expression. The possible contribution of retrograde nuclear signaling to LTP has not been explored in any detail. However, recent evidence suggests that such mechanisms may play a role. HFS of the perforant pathway leads to CREB phosphorylation in the entorhinal cortex and this effect is blocked by intracerebroventricular application of the Trk inhibitor K252a (Kelly et al., 2000; Gooney and Lynch, 2001). Simliar effects are seen with BDNF-LTP indicating that local signaling in the dentate gyrus leads to retrograde activation of CREB in parent cell bodies located some 4 millimeters away (Gooney et al., 2004).

8. BDNF AS A SYNAPTIC CONSOLIDATION FACTOR: A MODEL

Figure 5 collates recent findings into a summary model we will call the BDNF hypothesis of synaptic consolidation. This is meant as a working model for future investigation. Postsynaptic NMDA receptor activation triggers the initial release of BDNF during and shortly after HFS (lightning bolt in Figure 5). Postsynaptic TrkB receptors leads to ERK-dependent phosphorylation eIF4E and local enhancement of cap-dependent translation in dendritic spines. ERK signaling simultaneously leads to rapid CREB activation and Arc gene expression[1]. A population of Arc mRNA is trafficked to dendritic processes of granule cells. By mechanisms still unknown, HFS also triggers delivery of αCaMKII (and probably other mRNAs) from local sites of storage to sites of translation in or near the spine head[2]. In this model, activation of the translation apparatus coincides with the presentation of mRNA from sites of local storage (mRNA granules) as well as Arc mRNA en route from the soma. Arc mRNA distributes throughout the dendritic tree following standard LTP induction protocols, but Arc protein expression is concentrated in the zone of activated medial perforant path-granule cell synapse within dendritic spines. We suggest that BDNF-induced enhancement of transcript initiation enables capture of incoming mRNA at specific synapses or dendritic domains. In the terminology of Frey and Morris (1997), BDNF sets a synaptic tag (discussed in Section 9).

The function of Arc in late phase LTP expression is unknown. Given evidence of synaptic growth and PSD expansion, roles for Arc in co-translational assembly or trafficking of PSD proteins should be considered. It will likewise be important to determine the relationship between Arc, zif268 and other LTP-regulated genes. Arc synthesis would be expected to precede, yet overlap with, transport of late gene products from the cell soma. Presynaptically, BDNF appears to trigger a sustained increase in evoked glutamate release, although the contribution of this effect to LTP expression remains to be determined.

---

[1] Although ERK signaling is required from CREB activation and Arc induction, the role of CREB is unclear as there is no CRE in expected 5' region of the Arc promoter (Waltereit et al., 2001).

[2] The mechanisms of CaMKII redistribution is NMDAR-independent (Havik et al., 2003).

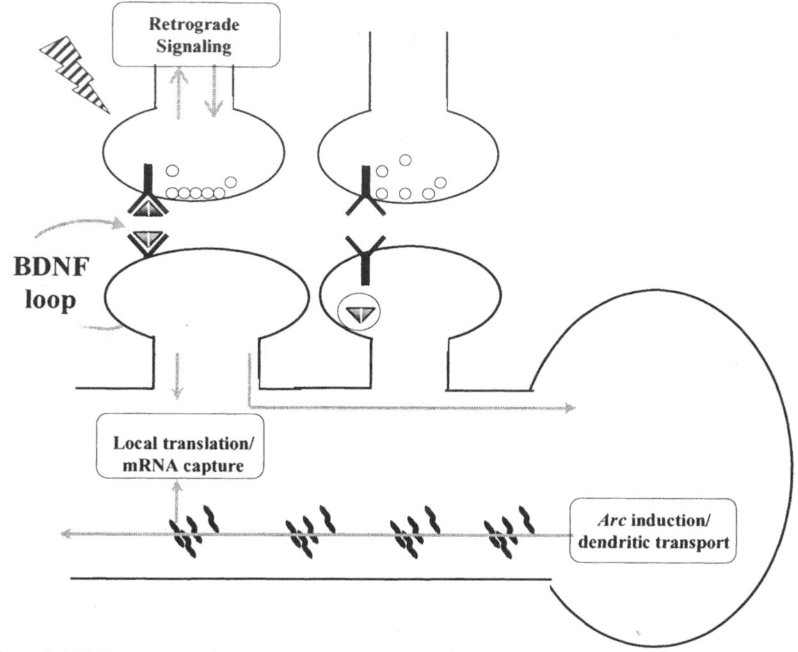

Figure 5. BDNF hypothesis of synaptic consolidation.

Synaptic consolidation appears to require a critical period of BDNF signaling after LTP induction (Kang et al., 1997). One possible mechanism for this is self-sustaining BDNF release. Regenerative autocrine loops of neurotrophin-induced neurotrophin release are implicated in the maintenance of sensory neurons (Davies and Wright, 1995). In the hippocampus, BDNF has been shown to induce BDNF release through TrkB-coupled PLC activation and mobilization of intracellular calcium (Canossa et al., 1997; Canossa et al., 2001). Mizoguchi and Nabekura (2003) have reported long-lasting (90 min) increases in neuronal intracellular calcium concentration following and 1-2 minute BDNF application. Amaral and Pozzo-Miller (this volume) have provided evidence of lasting (minutes) increases in dendritic calcium levels following puff application of BDNF. Their data also raises the intriguing possibility that BDNF-induced mobilization of intracellular calcium is amplified by calcium entry from the extracellular space, possibly through a plasma membrane non-selective cationic channel such as TRPC. A loop of this sort may drive the formation of late LTP in an activity-dependent manner. Interestingly, recent work suggests that coupling of TRPC family channels to IP3 receptors may be regulated by another LTP-regulated immediate early gene, homer1a (Yuan et al., 2003).

9. BDNF AND SYNAPTIC TAGGING

(Frey and Morris, 1997) have suggested that HFS sets a synaptic tag that allows the capture of proteins involved in late LTP. In their experimental paradigm two convergent inputs to CA1 pyramidal cell dendrites were stimulated. Input 1 received strong HFS leading to protein synthesis-dependent late LTP. They found that weak HFS applied to input 2, which normal gives only early LTP, induced late LTP when applied within the first 3 hours after stimulation of input 1. Importantly, development of stable LTP on the weak input was insensitive to protein synthesis inhibition. This suggests that weak stimulation leads to a hijacking of proteins (or secondary effects of these proteins) produced following strong stimulation on input 1. The proteins are thought to come partly from dendritic synthesis. BDNF could be involved simply by stimulating protein synthesis for subsequent capture at tagged synapses.

The critical question, however, is whether BDNF sets the tag. The synaptic consolidation model proposes that synapses are tagged by activation of the translational machinery, and translation of Arc is required for readout of the tag. The problem is that readout of such a tag (in response to input 2) would be protein synthesis-dependent and therefore at odds with the data of Frey and Morris. Perhaps this reflects differences in the kinetics of the tag in the CA1 region and dentate gyrus. The tag was demonstrated in the CA1 region, whereas Arc-dependent consolidation has thus far only been shown in the dentate gyrus. The duration of activity-induced Arc mRNA expression varies greatly between brain regions (Kelly and Deadwyler, 2003). Recent work suggests a more rapid decline in Arc protein levels in the CA1 region compared to the dentate gyrus following spatial exploration (Ramirez-Amaya et al., 2004). If Arc-dependent consolidation exists in CA1 it is likely to be of much shorter duration than in the dentate gyrus. In that case, a protein synthesis-dependent readout of the tag would not be seen in the paradigm used by Frey and Morris. Work on long-term facilitation in *Aplysia* sensory neurons has already suggested the existence of protein synthesis-dependent and independent tags occurring at different time points in the same cell (Martin et al., 1997; Casadio et al., 1999; Martin and Kosik, 2002)

10. TRUNCATED TRKB AND SPATIALLY RESTRICTED SIGNALING: SOURCE OF CONTROVERSY?

In organotypic visual cortex slices release of BDNF from a point source (single-cell) produces spatially restricted (within 4.5 μm) effects on dendritic outgrowth, suggesting very limited diffusion (Horch and Katz, 2002). Diffusion of BDNF appears to be restricted by binding to non-catalytic, truncated TrkB receptors. These receptors are expressed on dendritic shafts and glial processes and highly upregulated during development (Anderson et al., 1995; Biffo et al., 1995; Eide et al., 1996; Drake et al., 1999). Truncated TrkB could serve to concentrate BDNF to sites of release. By the same token, truncated TrkB may curtail access of exogenously applied BDNF to full-length TrkB receptors within excitatory synapses. Several authors have failed to observe BDNF-LTP in hippocampal slices (Figurov et al., 1996; Patterson et al., 1996; Scharfman, 1997; Frerking et al., 1998). Consistent with the notion of a diffusion barrier, BDNF-LTP in the hippocampal slice preparation has only been observed with use of high rates of bath application, correlating with increased penetration of BDNF into the slice (Kang et al., 1996). *In vivo*, BDNF is applied as a concentrated bolus which may be a more effective means of saturating the diffusion barrier.

11. FUTURE PERSPECTIVES AND IMPLICATIONS

BDNF can signal bidirectionally at glutamate synapses where it triggers events on a time scale from milliseconds to hours. Many basic issues such as the exact sites of neurotrophin release, the spatial distribution and dynamics of receptor (TrkB and $p75^{NTR}$) activation are still unclear, particularly in the context of adult synaptic signaling. In the dentate gyrus, BDNF appears to drive synaptic consolidation through dual effects on gene expression (Arc) and local protein synthesis. A major goal is to delineate the cell biological function of Arc in the consolidation process. A better understanding of Arc regulation could lead to very specific ways of contracting or expanding the window of synaptic consolidation with potential implications for the management of memory disorders and unipolar depression. BDNF is increasingly implicated in the pathogenesis of depression and the action of antidepressant drugs (Nestler et al., 2002; Monteggia et al., 2004). While we have found that BDNF induces Arc-dependent synaptic strengthening, similar infusion of BDNF into the dentate gyrus of behaving rats has antidepressant-like effects (Shirayama *et al.*, 2002). Given evidence that Arc mRNA elevation in dendrites outlasts the increase in dendritic spines (Lyford et al., 1995; Steward and Worley, 2001), synaptic consolidation might be enhanced or extended at the level of mRNA translation.

A variety of mechanisms have been proposed to contribute to formation of late LTP. There is no reason to suspect that Arc is acting alone. In addition to zif268 and other IEGs, the list of critical players includes a constitutively active form of a protein kinase C isozyme, PKM-zeta, N-cadherin, and members of the integrin receptor family (Bahr et al., 1997; Bozdagi et al., 2000; Ling et al., 2002; Chan et al., 2003). Putting together this mosaic of interactions on both sides of the synapses (and in-between) is one of the most exciting challenges for the future.

12. FURTHER DETAILS

Address correspondence to: Clive Bramham, MD, PhD. Department of Biomedicine, University of Bergen, Jonas Lies vei 91, N-5009 Bergen, Norway. Phone: 47 55 58 60 32. Fax: 47 55 58 64 10. Email: clive.bramham@biomed.uib.no

13. ACKNOWLEDGEMENTS

Funded by the European Union Biotechnology program (BIO4-CT98-0333) and the Norwegian Research Council. BDNF was kindly provided by Amgen-Regeneron partners.

14. REFERENCES

Aakalu G, Smith WB, Nguyen N, Jiang C, Schuman EM (2001) Dynamic visualization of local protein synthesis in hippocampal neurons. Neuron 30: 489-502.

Abraham WC, Gustafsson B, Wigstrom H (1987) Long-term potentiation involves enhanced synaptic excitation relative to synaptic inhibition in guinea-pig hippocampus. J Physiol Lond 394: 367-380.

Abraham WC, Mason SE, Demmer J, Williams JM, Richardson CL, Tate WP, Lawlor PA, Dragunow M (1993) Correlations between immediate early gene induction and the persistence of long-term potentiation. Neuroscience 56: 717-727.

Alsina B, Vu T, Cohen-Cory S (2001) Visualizing synapse formation in arborizing optic axons in vivo: dynamics and modulation by BDNF. Nat Neurosci 4: 1093-1101.

Amaral MD, Pozzo-Miller L (2005) On the role of neurotrophins in dendritic calcium signaling: implications for transsynaptic plasticity. In: Synaptic plasticity and transsynaptic signaling.

Anderson KD, Alderson RF, Altar CA, DiStefano PS, Corcoran TL, Lindsay RM, Wiegand SJ (1995) Differential distribution of exogenous BDNF, NGF, and NT-3 in the brain corresponds to the relative abundance and distribution of high-affinity and low-affinity neurotrophin receptors. J Comp Neurol 357: 296-317.

Aoki C, Wu K, Elste A, Len G, Lin S, McAuliffe G, Black IB (2000) Localization of brain-derived neurotrophic factor and TrkB receptors to postsynaptic densities of adult rat cerebral cortex. J Neurosci Res 59: 454-463.

Asztely F, Kokaia M, Olofsdotter K, Ortegren U, Lindvall O (2000) Afferent-specific modulation of short-term

synaptic plasticity by neurotrophins in dentate gyrus. Eur J Neurosci 12: 662-669.
Bahr BA, Staubli U, Xiao P, Chun D, Ji ZX, Esteban ET, Lynch G (1997) Arg-Gly-Asp-Ser-selective adhesion and the stabilization of long- term potentiation: pharmacological studies and the characterization of a candidate matrix receptor. J Neurosci 17: 1320-1329.
Balkowiec A, Katz DM (2002) Cellular mechanisms regulating activity-dependent release of native brain-derived neurotrophic factor from hippocampal neurons. J Neurosci 22: 10399-10407.
Bashirullah A, Cooperstock RL, Lipshitz HD (2001) Spatial and temporal control of RNA stability. Proc Natl Acad Sci U S A 98: 7025-7028.
Bashirullah A, Halsell SR, Cooperstock RL, Kloc M, Karaiskakis A, Fisher WW, Fu W, Hamilton JK, Etkin LD, Lipshitz HD (1999) Joint action of two RNA degradation pathways controls the timing of maternal transcript elimination at the midblastula transition in Drosophila melanogaster. EMBO J 18: 2610-2620.
Biffo S, Offenhauser N, Carter BD, Barde YA (1995) Selective binding and internalisation by truncated receptors restrict the availability of BDNF during development. Development 121: 2461-2470.
Binder DK, Croll SD, Gall CM, Scharfman HE (2001) BDNF and epilepsy: too much of a good thing? Trends Neurosci 24: 47-53.
Blesch A, Conner JM, Tuszynski MH (2001) Modulation of neuronal survival and axonal growth in vivo by tetracycline-regulated neurotrophin expression. Gene Ther 8: 954-960.
Bliss TV, Collingridge GL (1993) A synaptic model of memory: long-term potentiation in the hippocampus. Nature 361: 31-39.
Bliss TV, Lomo T (1973) Long-lasting potentiation of synaptic transmission in the dentate area of the anaesthetized rabbit following stimulation of the perforant path. J Physiol 232: 331-356.
Blochl A, Thoenen H (1996) Localization of cellular storage compartments and sites of constitutive and activity-dependent release of nerve growth factor (NGF) in primary cultures of hippocampal neurons. Mol Cell Neurosci 7: 173-190.
Blum R, Kafitz KW, Konnerth A (2002) Neurotrophin-evoked depolarization requires the sodium channel Na(V)1.9. Nature 419: 687-693.
Bolanos CA, Nestler EJ (2004) Neurotrophic mechanisms in drug addiction. Neuromolecular Med 5: 69-83.
Bonhoeffer T (1996) Neurotrophins and activity-dependent development of the neocortex. Curr Opin Neurobiol 6: 119-126.
Bozdagi O, Shan W, Tanaka H, Benson DL, Huntley GW (2000) Increasing numbers of synaptic puncta during late-phase LTP: N-cadherin is synthesized, recruited to synaptic sites, and required for potentiation. Neuron 28: 245-259.
Bradley J, Sporns O (1999) BDNF-dependent enhancement of exocytosis in cultured cortical neurons requires translation but not transcription. Brain Res 815: 140-149.
Bramham CR, Bacher-Svendsen K, Sarvey JM (1997) LTP in the lateral perforant path is beta-adrenergic receptor-dependent. Neuroreport 8: 719-724.
Bramham CR, Messaoudi E (2005) BDNF mechanisms and functions in adult synaptic plasticity: the synaptic consolidation hypothesis. Prog. Neurobiol.
Browne GJ, Proud CG (2002) Regulation of peptide-chain elongation in mammalian cells. Eur J Biochem 269: 5360-5368.
Cabelli RJ, Shelton DL, Segal RA, Shatz CJ (1997) Blockade of endogenous ligands of trkB inhibits formation of ocular dominance columns. Neuron 19: 63-76.
Cammalleri M, Lutjens R, Berton F, King AR, Simpson C, Francesconi W, Sanna PP (2003) Time-restricted role for dendritic activation of the mTOR-p70S6K pathway in the induction of late-phase long-term potentiation in the CA1. Proc Natl Acad Sci U S A 100: 14368-14373.
Campenot RB, MacInnis BL (2004) Retrograde transport of neurotrophins: fact and function. J Neurobiol 58: 217-229.
Canossa M, Gartner A, Campana G, Inagaki N, Thoenen H (2001) Regulated secretion of neurotrophins by metabotropic glutamate group I (mGluRI) and Trk receptor activation is mediated via phospholipase C signalling pathways. EMBO J 20: 1640-1650.
Canossa M, Griesbeck O, Berninger B, Campana G, Kolbeck R, Thoenen H (1997) Neurotrophin release by neurotrophins: implications for activity- dependent neuronal plasticity. Proc Natl Acad Sci U S A 94: 13279-13286.
Carson JH, Kwon S, Barbarese E (1998) RNA trafficking in myelinating cells. Curr Opin Neurobiol 8: 607-612.
Casadio A, Martin KC, Giustetto M, Zhu H, Chen M, Bartsch D, Bailey CH, Kandel ER (1999) A transient, neuron-wide form of CREB-mediated long-term facilitation can be stabilized at specific synapses by local protein synthesis. Cell 99: 221-237.
Chan CS, Weeber EJ, Kurup S, Sweatt JD, Davis RL (2003) Integrin requirement for hippocampal synaptic plasticity and spatial memory. J Neurosci 23: 7107-7116.
Chao MV, Bothwell M (2002) Neurotrophins: to cleave or not to cleave. Neuron 33: 9-12.
Chen G, Kolbeck R, Barde YA, Bonhoeffer T, Kossel A (1999) Relative contribution of endogenous neurotrophins in hippocampal long-term potentiation. J Neurosci 19: 7983-7990.
Chotiner JK, Khorasani H, Nairn AC, O'Dell TJ, Watson JB (2003) Adenylyl cyclase-dependent form of chemical long-term potentiation triggers translational regulation at the elongation step. Neuroscience 116: 743-752.

Cole AJ, Saffen DW, Baraban JM, Worley PF (1989) Rapid increase of an immediate early gene messenger RNA in hippocampal neurons by synaptic NMDA receptor activation. Nature 340: 474-476.

Conner JM, Lauterborn JC, Yan Q, Gall CM, Varon S (1997) Distribution of brain-derived neurotrophic factor (BDNF) protein and mRNA in the normal adult rat CNS: evidence for anterograde axonal transport. J Neurosci 17: 2295-2313.

Davies AM (1994) The role of neurotrophins in the developing nervous system. J Neurobiol 25: 1334-1348.

Davies AM, Wright EM (1995) Neurotrophic factors: Neurotrophin autocrine loops. Curr Biol 5: 723-726.

Davis S, Vanhoutte P, Pages C, Caboche J, Laroche S (2000) The MAPK/ERK cascade targets both Elk-1 and cAMP response element-binding protein to control long-term potentiation-dependent gene expression in the dentate gyrus in vivo. J Neurosci 20: 4563-4572.

de Moor CH, Richter JD (2001) Translational control in vertebrate development. Int Rev Cytol 203: 567-608.

Delcroix JD, Valletta JS, Wu C, Hunt SJ, Kowal AS, Mobley WC (2003) NGF signaling in sensory neurons: evidence that early endosomes carry NGF retrograde signals. Neuron 39: 69-84.

Desai NS, Rutherford LC, Turrigiano GG (1999) BDNF regulates the intrinsic excitability of cortical neurons. Learn Mem 6: 284-291.

Drake CT, Milner TA, Patterson SL (1999) Ultrastructural localization of full-length trkB immunoreactivity in rat hippocampus suggests multiple roles in modulating activity-dependent synaptic plasticity. J Neurosci 19: 8009-8026.

Duman RS (2004) Role of neurotrophic factors in the etiology and treatment of mood disorders. Neuromolecular Med 5: 11-25.

Eberwine J, Miyashiro K, Kacharmina JE, Job C (2001) Local translation of classes of mRNAs that are targeted to neuronal dendrites. Proc Natl Acad Sci U S A 98: 7080-7085.

Eide FF, Vining ER, Eide BL, Zang K, Wang XY, Reichardt LF (1996) Naturally occurring truncated trkB receptors have dominant inhibitory effects on brain-derived neurotrophic factor signaling. J Neurosci 16: 3123-3129.

Ernfors P, Bramham CR (2003) The coupling of a trkB tyrosine residue to LTP. Trends Neurosci 26: 171-173.

Errington ML, Galley PT, Bliss TV (2003) Long-term potentiation in the dentate gyrus of the anaesthetized rat is accompanied by an increase in extracellular glutamate: real-time measurements using a novel dialysis electrode. Philos Trans R Soc Lond B Biol Sci 358: 675-687.

Escobar ML, Figueroa-Guzman Y, Gomez-Palacio-Schjetnan A (2003) In vivo insular cortex LTP induced by brain-derived neurotrophic factor. Brain Res 991: 274-279.

Fazeli MS, Corbet J, Dunn MJ, Dolphin AC, Bliss TV (1993) Changes in protein synthesis accompanying long-term potentiation in the dentate gyrus in vivo. J Neurosci 13: 1346-1353.

Feig S, Lipton P (1993) Pairing the cholinergic agonist carbachol with patterned Schaffer collateral stimulation initiates protein synthesis in hippocampal CA1 pyramidal cell dendrites via a muscarinic, NMDA-dependent mechanism. J Neurosci 13: 1010-1021.

Figurov A, Pozzo Miller LD, Olafsson P, Wang T, Lu B (1996) Regulation of synaptic responses to high-frequency stimulation and LTP by neurotrophins in the hippocampus. Nature 381: 706-709.

Flynn A, Vries RG, Proud CG (1997) Signalling pathways which regulate eIF4E. Biochem Soc Trans 25: 192S.

Frerking M, Malenka RC, Nicoll RA (1998) Brain-derived neurotrophic factor (BDNF) modulates inhibitory, but not excitatory, transmission in the CA1 region of the hippocampus. J Neurophysiol 80: 3383-3386.

Frey U, Frey S, Schollmeier F, Krug M (1996) Influence of actinomycin D, a RNA synthesis inhibitor, on long-term potentiation in rat hippocampal neurons in vivo and in vitro. J Physiol Lond 490: 703-711.

Frey U, Krug M, Reymann KG, Matthies H (1988) Anisomycin, an inhibitor of protein synthesis, blocks late phases of LTP phenomena in the hippocampal CA1 region in vitro. Brain Res 452: 57-65.

Frey U, Matthies H, Reymann KG (1991) The effect of dopaminergic D1 receptor blockade during tetanization on the expression of long-term potentiation in the rat CA1 region in vitro. Neurosci Lett 129: 111-114.

Frey U, Morris RG (1997) Synaptic tagging and long-term potentiation [see comments]. Nature 385: 533-536.

Frey U, Schollmeier K, Reymann KG, Seidenbecher T (1995) Asymptotic hippocampal long-term potentiation in rats does not preclude additional potentiation at later phases. Neuroscience 67: 799-807.

Geinisman Y (2000) Structural synaptic modifications associated with hippocampal LTP and behavioral learning. Cereb Cortex 10: 952-962.

Gingras AC, Raught B, Sonenberg N (2004) mTOR signaling to translation. Curr Top Microbiol Immunol 279: 169-197.

Ginty DD, Segal RA (2002) Retrograde neurotrophin signaling: Trk-ing along the axon. Curr Opin Neurobiol 12: 268-274.

Gooney M, Lynch MA (2001) Long-term potentiation in the dentate gyrus of the rat hippocampus is accompanied by brain-derived neurotrophic factor-induced activation of TrkB. J Neurochem 77: 1198-1207.

Gooney M, Messaoudi E, Maher FO, Bramham CR, Lynch MA (2004) BDNF-induced LTP in dentate gyurs is impaired with age: analysis of changes in cell signaling events. Neurobiol Aging In press, published online Apr 17.

Gooney M, Shaw K, Kelly A, O'Mara SM, Lynch MA (2002) Long-term potentiation and spatial learning are associated with increased phosphorylation of TrkB and extracellular signal-regulated kinase (ERK) in the dentate gyrus: evidence for a role for brain-derived neurotrophic factor. Behav Neurosci 116: 455-463.

Gottschalk W, Pozzo-Miller LD, Figurov A, Lu B (1998) Presynaptic modulation of synaptic transmission and plasticity by brain-derived neurotrophic factor in the developing hippocampus. J Neurosci 18: 6830-6839.
Gould E, Gross CG (2002) Neurogenesis in adult mammals: some progress and problems. J Neurosci 22: 619-623.
Graves L, Pack A, Abel T (2001) Sleep and memory: a molecular perspective. Trends Neurosci 24: 237-243.
Harley CW, Walling SG, Brown RAM (2004) The three faces of norepinephrine: plasticity at the perforant path-dentate gyrus synapse. In: Synaptic plasticity and transsynaptic signaling.
Harris KM, Fiala JC, Ostroff L (2003) Structural changes at dendritic spine synapses during long-term potentiation. Philos Trans R Soc Lond B Biol Sci 358: 745-748.
Hartmann M, Heumann R, Lessmann V (2001) Synaptic secretion of BDNF after high-frequency stimulation of glutamatergic synapses. EMBO J 20: 5887-5897.
Havik B, Rokke H, Bardsen K, Davanger S, Bramham CR (2003) Bursts of high-frequency stimulation trigger rapid delivery of pre-existing alpha-CaMKII mRNA to synapses: a mechanism in dendritic protein synthesis during long-term potentiation in adult awake rats. Eur J Neurosci 17: 2679-2689.
He XP, Kotloski R, Nef S, Luikart BW, Parada LF, McNamara JO (2004) Conditional deletion of TrkB but not BDNF prevents epileptogenesis in the kindling model. Neuron 43: 31-42.
Horch HW, Katz LC (2002) BDNF release from single cells elicits local dendritic growth in nearby neurons. Nat Neurosci 5: 1177-1184.
Huang YS, Jung MY, Sarkissian M, Richter JD (2002) N-methyl-D-aspartate receptor signaling results in Aurora kinase-catalyzed CPEB phosphorylation and alpha-CaMKII mRNA polyadenylation at synapses. EMBO J 21: in press.
Huang YS, Richter JD (2004) Regulation of local mRNA translation. Curr Opin Cell Biol 16: 308-313.
Huang YY, Nguyen PV, Abel T, Kandel ER (1996) Long-lasting forms of synaptic potentiation in the mammalian hippocampus. Learn Mem 3: 74-85.
Husi H, Ward MA, Choudhary JS, Blackstock WP, Grant SG (2000) Proteomic analysis of NMDA receptor-adhesion protein signaling complexes [see comments]. Nat Neurosci 3: 661-669.
Ikegaya Y, Ishizaka Y, Matsuki N (2002) BDNF attenuates hippocampal LTD via activation of phospholipase C: implications for a vertical shift in the frequency-response curve of synaptic plasticity. Eur J Neurosci 16: 145-148.
Jiang B, Akaneya Y, Hata Y, Tsumoto T (2003) Long-term depression is not induced by low-frequency stimulation in rat visual cortex in vivo: a possible preventing role of endogenous brain-derived neurotrophic factor. J Neurosci 23: 3761-3770.
Jiang B, Akaneya Y, Ohshima M, Ichisaka S, Hata Y, Tsumoto T (2001) Brain-derived neurotrophic factor induces long-lasting potentiation of synaptic transmission in visual cortex in vivo in young rats, but not in the adult. Eur J Neurosci 14: 1219-1228.
Johnstone O, Lasko P (2001) Translational regulation and RNA localization in Drosophila oocytes and embryos. Annu Rev Genet 35: 365-406.
Jones MW, Errington ML, French PJ, Fine A, Bliss TV, Garel S, Charnay P, Bozon B, Laroche S, Davis S (2001) A requirement for the immediate early gene Zif268 in the expression of late LTP and long-term memories. Nat Neurosci 4: 289-296.
Jovanovic JN, Czernik AJ, Fienberg AA, Greengard P, Sihra TS (2000) Synapsins as mediators of BDNF-enhanced neurotransmitter release. Nat Neurosci 3: 323-329.
Kacharmina JE, Job C, Crino P, Eberwine J (2000) Stimulation of glutamate receptor protein synthesis and membrane insertion within isolated neuronal dendrites. Proc Natl Acad Sci U S A 97: 11545-11550.
Kafitz KW, Rose CR, Thoenen H, Konnerth A (1999) Neurotrophin-evoked rapid excitation through TrkB receptors. Nature 401: 918-921.
Kandel ER (2001) The molecular biology of memory storage: a dialogue between genes and synapses. Science 294: 1030-1038.
Kang H, Schuman EM (1995a) Long-lasting neurotrophin-induced enhancement of synaptic transmission in the adult hippocampus. Science 267: 1658-1662.
Kang H, Schuman EM (1996) A requirement for local protein synthesis in neurotrophin- induced hippocampal synaptic plasticity. Science 273: 1402-1406.
Kang H, Welcher AA, Shelton D, Schuman EM (1997) Neurotrophins and time: different roles for TrkB signaling in hippocampal long-term potentiation. Neuron 19: 653-664.
Kang HJ, Jia LZ, Suh K-Y, Tang L, Schuman EM (1996) Determinants of BDNF-induced hippocampal synaptic plasticity: role of the Trk B receptor and the kinetics of neurotrophin delivery. Learning and memory 3: 188-196.
Kang HJ, Schuman EM (1995b) Long-lasting neurotrophin-induced enhancement of synaptic transmission in the adult hippocampus. Science 267: 1658-1662.
Kanhema T, Dagestad G, Havik B, Ying SW, Nairn AC, Sonenberg N, Bramham CR (2004) BDNF regulates translation initiation and elongation in long-term synaptic plasticity and enhances dendritic alpha-CaMKII synthesis. Submitted.
Kaplan DR, Miller FD (2000) Neurotrophin signal transduction in the nervous system. Curr Opin Neurobiol 10: 381-391.
Kelleher RJ, III, Govindarajan A, Jung HY, Kang H, Tonegawa S (2004) Translational control by MAPK

signaling in long-term synaptic plasticity and memory. Cell 116: 467-479.
Kelly A, Mullany PM, Lynch MA (2000) Protein synthesis in entorhinal cortex and long-term potentiation in dentate gyrus. Hippocampus 10: 431-437.
Kelly MP, Deadwyler SA (2003) Experience-dependent regulation of the immediate-early gene arc differs across brain regions. J Neurosci 23: 6443-6451.
Knipper M, Leung LS, Zhao D, Rylett RJ (1994) Short-term modulation of glutamatergic synapses in adult rat hippocampus by NGF. Neuroreport 5: 2433-2436.
Kohara K, Kitamura A, Morishima M, Tsumoto T (2001) Activity-dependent transfer of brain-derived neurotrophic factor to postsynaptic neurons. Science 291: 2419-2423.
Kokaia Z, Bengzon J, Metsis M, Kokaia M, Persson H, Lindvall O (1993) Coexpression of neurotrophins and their receptors in neurons of the central nervous system. Proc Natl Acad Sci U S A 90: 6711-6715.
Korte M, Carroll P, Wolf E, Brem G, Thoenen H, Bonhoeffer T (1995) Hippocampal long-term potentiation is impaired in mice lacking brain-derived neurotrophic factor. Proc Natl Acad Sci U S A 92: 8856-8860.
Korte M, Kang H, Bonhoeffer T, Schuman EM (1998) A role for BDNF in the late-phase of hippocampal long-term potentiation. Neuropharmacology 37: 553-559.
Kossel AH, Cambridge SB, Wagner U, Bonhoeffer T (2001) A caged Ab reveals an immediate/instructive effect of BDNF during hippocampal synaptic potentiation. Proc Natl Acad Sci U S A 98: 14702-14707.
Kovalchuk Y, Hanse E, Kafitz KW, Konnerth A (2002) Postsynaptic induction of BDNF-mediated long-term potentiation. Science 295: 1729-1734.
Koyama R, Yamada MK, Fujisawa S, Katoh-Semba R, Matsuki N, Ikegaya Y (2004) Brain-derived neurotrophic factor induces hyperexcitable reentrant circuits in the dentate gyrus. J Neurosci 24: 9215-9224.
Kulla A, Manahan-Vaughan D (2002) Modulation by Serotonin 5-HT(4) Receptors of Long-term Potentiation and Depotentiation in the Dentate Gyrus of Freely Moving Rats. Cereb Cortex 12: 150-162.
Kumura E, Kimura F, Taniguchi N, Tsumoto T (2000) Brain-derived neurotrophic factor blocks long-term depression in solitary neurones cultured from rat visual cortex. J Physiol Lond 524 Pt 1: 195-204.
Lanahan A, Lyford G, Stevenson GS, Worley PF, Barnes CA (1997) Selective alteration of long-term potentiation-induced transcriptional response in hippocampus of aged, memory-impaired rats. J Neurosci 17: 2876-2885.
Lessmann V, Gottmann K, Heumann R (1994) BDNF and NT-4/5 enhance glutamatergic synaptic transmission in cultured hippocampal neurones. Neuroreport 6: 21-25.
Lessmann V, Heumann R (1998) Modulation of unitary glutamatergic synapses by neurotrophin-4/5 or brain-derived neurotrophic factor in hippocampal microcultures: presynaptic enhancement depends on pre-established paired-pulse facilitation. Neuroscience 86: 399-413.
Lever IJ, Bradbury EJ, Cunningham JR, Adelson DW, Jones MG, McMahon SB, Marvizon JC, Malcangio M (2001) Brain-derived neurotrophic factor is released in the dorsal horn by distinctive patterns of afferent fiber stimulation. J Neurosci 21: 4469-4477.
Levi Montalcini R (1987) The nerve growth factor 35 years later. Science 237: 1154-1162.
Levine ES, Dreyfus CF, Black IB, Plummer MR (1995) Brain-derived neurotrophic factor rapidly enhances synaptic transmission in hippocampal neurons via postsynaptic tyrosine kinase receptors. Proc Natl Acad Sci USA 92: 8074-8077.
Lewin GR, Barde YA (1996) Physiology of the neurotrophins. Ann Rev Neurosci 19: 289-317.
Ling DS, Benardo LS, Serrano PA, Blace N, Kelly MT, Crary JF, Sacktor TC (2002) Protein kinase Mzeta is necessary and sufficient for LTP maintenance. Nat Neurosci 5: 295-296.
Link W, Konietzko U, Kauselmann G, Krug M, Schwanke B, Frey U, Kuhl D (1995) Somatodendritic expression of an immediate early gene is regulated by synaptic activity. Proc Natl Acad Sci U S A 92: 5734-5738.
Lisman J, Schulman H, Cline H (2002) The molecular basis of CaMKII function in synaptic and behavioural memory. Nat Rev Neurosci 3: 175-190.
Lisman JE, Raghavachari S, Otmakhov N, Otmakhova NA (2004) The phases of LTP: the new complexities. In: Synaptic plasticity and transsynaptic signaling.
Lohof AM, Ip NY, Poo MM (1993) Potentiation of developing neuromuscular synapses by the neurotrophins NT-3 and BDNF. Nature 363: 350-353.
Lu B (2003) Pro-region of neurotrophins: role in synaptic modulation. Neuron 39: 735-738.
Lu YM, Mansuy IM, Kandel ER, Roder J (2000) Calcineurin-mediated LTD of GABAergic inhibition underlies the increased excitability of CA1 neurons associated with LTP. Neuron 26: 197-205.
Lyford GL, Yamagata K, Kaufmann WE, Barnes CA, Sanders LK, Copeland NG, Gilbert DJ, Jenkins NA, Lanahan AA, Worley PF (1995) Arc, a growth factor and activity-regulated gene, encodes a novel cytoskeleton-associated protein that is enriched in neuronal dendrites. Neuron 14: 433-445.
Ma L, Reis G, Parada LF, Schuman EM (1999) Neuronal NT-3 is not required for synaptic transmission or long-term potentiation in area CA1 of the adult rat hippocampus. Learn Mem 6: 267-275.
Maffei L (2002) Plasticity in the visual system: role of neurotrophins and electrical activity. Arch Ital Biol 140: 341-346.
Malinow R, Malenka RC (2002) AMPA receptor trafficking and synaptic plasticity. Annu Rev Neurosci 25: 103-126.

Marin P, Nastiuk KL, Daniel N, Girault JA, Czernik AJ, Glowinski J, Nairn AC, Premont J (1997) Glutamate-dependent phosphorylation of elongation factor-2 and inhibition of protein synthesis in neurons. J Neurosci 17: 3445-3454.

Martin KC, Casadio A, Zhu H, E-Y, Rose JC, Chen M, Bailey CH, Kandel ER (1997) Synapse-specific, long-term facilitation of aplysia sensory to motor synapses: a function for local protein synthesis in memory storage. Cell 91: 927-938.

Martin KC, Kosik KS (2002) Synaptic tagging -- who's it? Nat Rev Neurosci 3: 813-820.

Martinez A, Alcantara S, Borrell V, Del Rio JA, Blasi J, Otal R, Campos N, Boronat A, Barbacid M, Silos-Santiago I, Soriano E (1998) TrkB and TrkC signaling are required for maturation and synaptogenesis of hippocampal connections. J Neurosci 18: 7336-7350.

Matthies H, Frey U, Reymann K, Krug M, Jork R, Schroeder H (1990) Different mechanisms and multiple stages of LTP. Adv Exp Med Biol 268: 359-368.

Meberg PJ, Barnes CA, McNaughton BL, Routtenberg A (1993) Protein kinase C and F1/GAP-43 gene expression in hippocampus inversely related to synaptic enhancement lasting 3 days. Proc Natl Acad Sci U S A 90: 12050-12054.

Messaoudi E, Bardsen K, Srebro B, Bramham CR (1998) Acute intrahippocampal infusion of BDNF induces lasting potentiation of synaptic transmission in the rat dentate gyrus. J Neurophysiol 79: 496-499.

Messaoudi, E., Kanhema, T., and Bramham, C. R. BDNF-LTP requires Arc protein expression in rat dentate gyrus *in vivo*. Soc Neurosci Abstr 583.4. 2003.

Messaoudi E, Ying SW, Kanhema T, Croll SD, Bramham CR (2002) BDNF triggers transcription-dependent, late phase LTP in vivo. J Neurosci 22: 7453-7461.

Miller S, Yasuda M, Coats JK, Jones Y, Martone ME, Mayford M (2002) Disruption of dendritic translation of CaMKIIalpha impairs stabilization of synaptic plasticity and memory consolidation. Neuron 36: 507-519.

Min MY, Asztely F, Kokaia M, Kullmann DM (1998) Long-term potentiation and dual-component quantal signaling in the dentate gyrus. Proc Natl Acad Sci U S A 95: 4702-4707.

Minichiello L, Calella A, Medina D, Bonhoeffer T, Klein R, Korte M (2002) Mechanism of TrkB-Mediated Hippocampal Long-Term Potentiation. Neuron 36: 121.

Minichiello L, Korte M, Wolfer D, Kuhn R, Unsicker K, Cestari V, Rossi AC, Lipp HP, Bonhoeffer T, Klein R (1999) Essential role for TrkB receptors in hippocampus-mediated learning. Neuron 24: 401-414.

Miranda RC, Sohrabji F, Toran-Allerand CD (1993) Neuronal colocalization of mRNAs for neurotrophins and their receptors in the developing central nervous system suggests a potential for autocrine interactions. Proc Natl Acad Sci USA 90: 6439-6443.

Mizoguchi Y, Nabekura J (2003) Sustained intracellular Ca2+ elevation induced by a brief BDNF application in rat visual cortex neurons. Neuroreport 14: 1481-1483.

Moga DE, Calhoun ME, Chowdhury A, Worley P, Morrison JH, Shapiro ML (2004) Activity-regulated cytoskeletal-associated protein is localized to recently activated excitatory synapses. Neuroscience 125: 7-11.

Monteggia LM, Barrot M, Powell CM, Berton O, Galanis V, Gemelli T, Meuth S, Nagy A, Greene RW, Nestler EJ (2004) Essential role of brain-derived neurotrophic factor in adult hippocampal function. Proc Natl Acad Sci U S A 101: 10827-10832.

Morris RG, Moser EI, Riedel G, Martin SJ, Sandin J, Day M, O'Carroll C (2003) Elements of a neurobiological theory of the hippocampus: the role of activity-dependent synaptic plasticity in memory. Philos Trans R Soc Lond B Biol Sci 358: 773-786.

Nairn AC, Matsushita M, Nastiuk K, Horiuchi A, Mitsui K, Shimizu Y, Palfrey HC (2001) Elongation factor-2 phosphorylation and the regulation of protein synthesis by calcium. Prog Mol Subcell Biol 27: 91-129.

Nairn AC, Palfrey HC (1987) Identification of the major Mr 100,000 substrate for calmodulin-dependent protein kinase III in mammalian cells as elongation factor-2. J Biol Chem 262: 17299-17303.

Nestler EJ, Barrot M, DiLeone RJ, Eisch AJ, Gold SJ, Monteggia LM (2002) Neurobiology of depression. Neuron 34: 13-25.

Nguyen PV, Abel T, Kandel ER (1994) Requirement of a critical period of transcription for induction of a late phase of LTP. Science 265: 1104-1107.

Nguyen PV, Kandel ER (1996) A macromolecular synthesis-dependent late phase of long-term potentiation requiring cAMP in the medial perforant pathway of rat hippocampal slices. J Neurosci 16: 3189-3198.

O'Dell TJ, Kandel ER (1994) Low-frequency stimulation erases LTP through an NMDA receptor-mediated activation of protein phosphatases. Learn Mem 1: 129-139.

Ostroff LE, Fiala JC, Allwardt B, Harris KM (2002) Polyribosomes redistribute from dendritic shafts into spines with enlarged synapses during LTP in developing rat hippocampal slices. Neuron 35: 535-545.

Otani S, Abraham WC (1989) Inhibition of protein synthesis in the dentate gyrus, but not the entorhinal cortex, blocks maintenance of long-term potentiation in rats. Neurosci Lett 106: 175-180.

Ouyang Y, Rosenstein A, Kreiman G, Schuman EM, Kennedy MB (1999) Tetanic stimulation leads to increased accumulation of Ca(2+)/calmodulin-dependent protein kinase II via dendritic protein synthesis in hippocampal neurons. J Neurosci 19: 7823-7833.

Patapoutian A, Reichardt LF (2001) Trk receptors: mediators of neurotrophin action. Curr Opin Neurobiol 11: 272-280.

Patterson SL, Abel T, Deuel TA, Martin KC, Rose JC, Kandel ER (1996) Recombinant BDNF rescues deficits

in basal synaptic transmission and hippocampal LTP in BDNF knockout mice. Neuron 16: 1137-1145.
Patterson SL, Pittenger C, Morozov A, Martin KC, Scanlin H, Drake C, Kandel ER (2001) Some forms of cAMP-mediated long-lasting potentiation are associated with release of BDNF and nuclear translocation of phospho-MAP kinase. Neuron 32: 123-140.
Pierce JP, van Leyen K, McCarthy JB (2000) Translocation machinery for synthesis of integral membrane and secretory proteins in dendritic spines. Nat Neurosci 3: 311-313.
Prakash N, Cohen Cory S, Frostig RD (1996) Rapid and opposite effects of BDNF and NGF on the functional organization of the adult cortex in vivo. Nature 381: 702-706.
Qian Z, Gilbert ME, Colicos MA, Kandel ER, Kuhl D (1993) Tissue-plasminogen activator is induced as an immediate-early gene during seizure, kindling and long-term potentiation. Nature 361: 453-457.
Ramirez-Amaya V, Vazdarjanova A, Houston FP, Olson K, Worley PF, Barnes CA (2004) Time course of Arc protein expression in the hippocampus after spatial exploration. Soc Neurosci Abstr 519.6.
Raymond CR, Thompson VL, Tate WP, Abraham WC (2000) Metabotropic glutamate receptors trigger homosynaptic protein synthesis to prolong long-term potentiation. J Neurosci 20: 969-976.
Riccio A, Pierchala BA, Ciarallo CL, Ginty DD (1997) An NGF-TrkA-mediated retrograde signal to transcription factor CREB in sympathetic neurons [see comments]. Science 277: 1097-1100.
Richter JD (2001) Think globally, translate locally: what mitotic spindles and neuronal synapses have in common. Proc Natl Acad Sci U S A 98: 7069-7071.
Rodriguez JJ, Davies HA, Silva AR, DeSouza IEJ, Peddie CJ, Colyer FM, Fine A, Errington ML, Bliss TVP, Stewart MG (2003) Dynamic changes in arc distribution in the dentate gyrus after induction of LTP: an ultrastructural study. Soc Neurosci Abstr 807.6.
Ryazanov AG, Shestakova EA, Natapov PG (1988) Phosphorylation of elongation factor 2 by EF-2 kinase affects rate of translation. Nature 334: 170-173.
Scharfman HE (1997) Hyperexcitability in combined entorhinal/hippocampal slices of adult rat after exposure to brain-derived neurotrophic factor. J Neurophysiol 78: 1082-1095.
Scharfman HE, Mercurio TC, Goodman JH, Wilson MA, MacLusky NJ (2003) Hippocampal excitability increases during the estrous cycle in the rat: a potential role for brain-derived neurotrophic factor. J Neurosci 23: 11641-11652.
Schinder AF, Poo MM (2000) The neurotrophin hypothesis for synaptic plasticity. Trends Neurosci 23: 639-645.
Schmidt Kastner R, Wetmore C, Olson L (1996) Comparative study of brain-derived neurotrophic factor messenger RNA and protein at the cellular level suggests multiple roles in hippocampus, striatum and cortex. Neuroscience 74: 161-183.
Schnell L, Schneider R, Kolbeck R, Barde YA, Schwab ME (1994) Neurotrophin-3 enhances sprouting of corticospinal tract during development and after adult spinal cord lesion [see comments]. Nature 367: 170-173.
Sermasi E, Margotti E, Cattaneo A, Domenici L (2000) Trk B signalling controls LTP but not LTD expression in the developing rat visual cortex. Eur J Neurosci 12: 1411-1419.
Sherff CM, Carew TJ (1999) Coincident induction of long-term facilitation in Aplysia: cooperativity between cell bodies and remote synapses. Science 285: 1911-1914.
Shirayama Y, Chen AC, Nakagawa S, Russell DS, Duman RS (2002) Brain-derived neurotrophic factor produces antidepressant effects in behavioral models of depression. J Neurosci 22: 3251-3261.
Si K, Giustetto M, Etkin A, Hsu R, Janisiewicz AM, Miniaci MC, Kim JH, Zhu H, Kandel ER (2003a) A neuronal isoform of CPEB regulates local protein synthesis and stabilizes synapse-specific long-term facilitation in aplysia. Cell 115: 893-904.
Si K, Lindquist S, Kandel ER (2003b) A neuronal isoform of the aplysia CPEB has prion-like properties. Cell 115: 879-891.
Smart FM, Edelman GM, Vanderklish PW (2003) BDNF induces translocation of initiation factor 4E to mRNA granules: evidence for a role of synaptic microfilaments and integrins. Proc Natl Acad Sci U S A 100: 14403-14408.
Song HJ, Poo MM (1999) Signal transduction underlying growth cone guidance by diffusible factors. Curr Opin Neurobiol 9: 355-363.
Stanton PK, Sarvey JM (1985a) Blockade of norepinephrine-induced long-lasting potentiation in the hippocampal dentate gyrus by an inhibitor of protein synthesis. Brain Res 361: 276-283.
Stanton PK, Sarvey JM (1985b) Depletion of norepinephrine, but not serotonin, reduces long- term potentiation in the dentate gyrus of rat hippocampal slices. J Neurosci 5: 2169-2176.
Stanton PK, Sarvey JM (1985c) The effect of high-frequency electrical stimulation and norepinephrine on cyclic AMP levels in normal versus norepinephrine-depleted rat hippocampal slices. Brain Res 358: 343-348.
Staubli U, Chun D (1996) Factors regulating the reversibility of long-term potentiation. J Neurosci 16: 853-860.
Steward O (1997) mRNA localization in neurons: a multipurpose mechanism? Neuron 18: 9-12.
Steward O, Falk PM, Torre ER (1996) Ultrastructural basis for gene expression at the synapse: synapse-associated polyribosome complexes. J Neurocytol 25: 717-734.
Steward O, Levy WB (1982) Preferential localization of polyribosomes under the base of dendritic spines in granule cells of the dentate gyrus. J Neurosci 2: 284-291.
Steward O, Schuman EM (2001) Protein synthesis at synaptic sites on dendrites. Annu Rev Neurosci 24: 299-

325.
Steward O, Schuman EM (2003) Compartmentalized synthesis and degradation of proteins in neurons. Neuron 40: 347-359.
Steward O, Worley PF (2001) A cellular mechanism for targeting newly synthesized mRNAs to synaptic sites on dendrites. Proc Natl Acad Sci U S A 98: 7062-7068.
Straube T, Frey JU (2003) Involvement of beta-adrenergic receptors in protein synthesis-dependent late long-term potentiation (LTP) in the dentate gyrus of freely moving rats: the critical role of the LTP induction strength. Neuroscience 119: 473-479.
Swanson-Park JL, Coussens CM, Mason-Parker SE, Raymond CR, Hargreaves EL, Dragunow M, Cohen AS, Abraham WC (1999) A double dissociation within the hippocampus of dopamine D1/D5 receptor and beta-adrenergic receptor contributions to the persistence of long-term potentiation. Neuroscience 92: 485-497.
Takei N, Kawamura M, Hara K, Yonezawa K, Nawa H (2001) Brain-derived neurotrophic factor enhances neuronal translation by activating multiple initiation processes: comparison with the effects of insulin. J Biol Chem 276: 42818-42825.
Tanaka T, Saito H, Matsuki N (1997) Inhibition of GABAA synaptic responses by brain-derived neurotrophic factor (BDNF) in rat hippocampus. J Neurosci 17: 2959-2966.
Tang SJ, Reis G, Kang H, Gingras AC, Sonenberg N, Schuman EM (2002) A rapamycin-sensitive signaling pathway contributes to long-term synaptic plasticity in the hippocampus. Proc Natl Acad Sci U S A 99: 467-472.
Tartaglia N, Du J, Tyler WJ, Neale E, Pozzo-Miller L, Lu B (2001) Protein synthesis-dependent and -independent regulation of hippocampal synapses by brain-derived neurotrophic factor. J Biol Chem 276: 37585-37593.
Thakker-Varia S, Alder J, Crozier RA, Plummer MR, Black IB (2001) Rab3A is required for brain-derived neurotrophic factor-induced synaptic plasticity: transcriptional analysis at the population and single-cell levels. J Neurosci 21: 6782-6790.
Thoenen H (1995) Neurotrophins and neuronal plasticity. Science 270: 593-598.
Tsui CC, Copeland NG, Gilbert DJ, Jenkins NA, Barnes C, Worley PF (1996) Narp, a novel member of the pentraxin family, promotes neurite outgrowth and is dynamically regulated by neuronal activity. J Neurosci 16: 2463-2478.
Tuszynski MH, Blesch A (2004) Nerve growth factor: from animal models of cholinergic neuronal degeneration to gene therapy in Alzheimer's disease. Prog Brain Res 146: 441-449.
Tyler WJ, Perrett SP, Pozzo-Miller LD (2002) The role of neurotrophins in neurotransmitter release. Neuroscientist 8: 524-531.
Waltereit R, Dammermann B, Wulff P, Scafidi J, Staubli U, Kauselmann G, Bundman M, Kuhl D (2001) Arg3.1/Arc mRNA induction by Ca2+ and cAMP requires protein kinase A and mitogen-activated protein kinase/extracellular regulated kinase activation. J Neurosci 21: 5484-5493.
Watson FL, Heerssen HM, Bhattacharyya A, Klesse L, Lin MZ, Segal RA (2001) Neurotrophins use the Erk5 pathway to mediate a retrograde survival response. Nat Neurosci 4: 981-988.
Watson FL, Heerssen HM, Moheban DB, Lin MZ, Sauvageot CM, Bhattacharyya A, Pomeroy SL, Segal RA (1999) Rapid nuclear responses to target-derived neurotrophins require retrograde transport of ligand-receptor complex. J Neurosci 19: 7889-7900.
Weeks AC, Ivanco TL, LeBoutillier JC, Racine RJ, Petit TL (2001) Sequential changes in the synaptic structural profile following long-term potentiation in the rat dentate gyrus: III. Long-term maintenance phase. Synapse 40: 74-84.
Wells DG, Dong X, Quinlan EM, Huang YS, Bear MF, Richter JD, Fallon JR (2001) A role for the cytoplasmic polyadenylation element in NMDA receptor-regulated mRNA translation in neurons. J Neurosci 21: 9541-9548.
Wells DG, Fallon JR (2000) Dendritic mRNA translation: deciphering the uncoded. Nat Neurosci 3: 1062-1064.
Wells DG, Richter JD, Fallon JR (2000) Molecular mechanisms for activity-regulated protein synthesis in the synapto-dendritic compartment. Curr Opin Neurobiol 10: 132-137.
Williams J, Dragunow M, Lawlor P, Mason S, Abraham WC, Leah J, Bravo R, Demmer J, Tate W (1995) Krox20 may play a key role in the stabilization of long-term potentiation. Exp Neurol 152: 1-15.
Wisden W, Errington ML, Williams S, Dunnett SB, Waters C, Hitchcock D, Evan G, Bliss TV, Hunt SP (1990) Differential expression of immediate early genes in the hippocampus and spinal cord. Neuron 4: 603-614.
Wu K, Xu JL, Suen PC, Levine E, Huang YY, Mount HT, Lin SY, Black IB (1996) Functional trkB neurotrophin receptors are intrinsic components of the adult brain postsynaptic density. Brain Res Mol Brain Res 43: 286-290.
Wu L, Wells D, Tay J, Mendis D, Abbott MA, Barnitt A, Quinlan E, Heynen A, Fallon JR, Richter JD (1998) CPEB-mediated cytoplasmic polyadenylation and the regulation of experience-dependent translation of alpha-CaMKII mRNA at synapses [see comments]. Neuron 21: 1129-1139.
Xu B, Gottschalk W, Chow A, Wilson RI, Schnell E, Zang K, Wang D, Nicoll RA, Lu B, Reichardt LF (2000) The role of brain-derived neurotrophic factor receptors in the mature hippocampus: modulation of long-term potentiation through a presynaptic mechanism involving TrkB. J Neurosci 20: 6888-6897.
Yan Q, Radeke MJ, Matheson CR, Talvenheimo J, Welcher AA, Feinstein SC (1997) Immunocytochemical

localization of TrkB in the central nervous system of the adult rat. J Comp Neurol 378: 135-157.

Yin Y, Edelman GM, Vanderklish PW (2002) The brain-derived neurotrophic factor enhances synthesis of Arc in synaptoneurosomes. Proc Natl Acad Sci U S A 99: 2368-2373.

Ying SW, Futter M, Rosenblum K, Webber MJ, Hunt SP, Bliss TV, Bramham CR (2002) Brain-derived neurotrophic factor induces long-term potentiation in intact adult hippocampus: requirement for ERK activation coupled to CREB and upregulation of Arc synthesis. J Neurosci 22: 1532-1540.

Yuan JP, Kiselyov K, Shin DM, Chen J, Shcheynikov N, Kang SH, Dehoff MH, Schwarz MK, Seeburg PH, Muallem S, Worley PF (2003) Homer binds TRPC family channels and is required for gating of TRPC1 by IP3 receptors. Cell 114: 777-789.

Zakharenko SS, Patterson SL, Dragatsis I, Zeitlin SO, Siegelbaum SA, Kandel ER, Morozov A (2003) Presynaptic BDNF required for a presynaptic but not postsynaptic component of LTP at hippocampal CA1-CA3 synapses. Neuron 39: 975-990.

ON THE ROLE OF NEUROTROPHINS IN DENDRITIC CALCIUM SIGNALING

Implications for hippocampal transsynaptic plasticity

Michelle D. Amaral and Lucas Pozzo-Miller *

1. INTRODUCTION

Recent advances in neuroscience have provided experimental evidence of the cellular and molecular mechanisms underlying the ability of an organism to translate experience-induced changes of the Central Nervous System (CNS) into the acquisition, consolidation, retention, and subsequent recall of relevant information. Several models of activity-dependent neuronal plasticity focus on the malleability of synapses and propose the existence of extracellular signals that render synaptic activity into enduring structural changes. The family of neurotrophic factors, and brain-derived neurotrophic factor (BDNF) in particular, have emerged as prime candidates to mediate this fundamental role due to their strong actions on synaptic transmission and plasticity, most notably in the hippocampus (Lo, 1995; Thoenen, 1995; Poo, 2001; Tyler et al., 2002a; Lu, 2003). The neurotrophins are secretory proteins involved in neuronal survival and differentiation during early brain development (Barde, 1989). Both BDNF and its receptor TrkB are highly expressed in the pyramidal cells of the CA1 and CA3 regions, as well as in the granule cells of the dentate gyrus in the adult hippocampus, a region well known for its role in learning and memory (Dugich-Djordjevic et al., 1995; Fryer et al., 1996; Schmidt-Kastner et al., 1996; Yan et al., 1997a; Yan et al., 1997b). Four neurotrophins have been identified in mammals, and all are widely expressed in the CNS: nerve growth factor (NGF), BDNF, neurotrophin-3 (NT-3), and NT-4/5 (Lewin and Barde, 1996). They exert their effects by binding to high-affinity tyrosine kinase receptors, members of the *trk* family of protooncogenes related to insulin and epidermal growth factor receptors. NGF binds selectively to TrkA, BDNF and NT-4/5 to TrkB, and NT-3 to TrkC (Chao, 1992). Activation of these Trk receptors by autophosphorylation following neurotrophin-induced dimerization initiates several intracellular signaling cascades (Figure 1) (Segal and Greenberg, 1996). One of these, the Ras-mitogen-activated protein kinase (MAPK) signaling pathway, has been involved in hippocampal synaptic plasticity because MAPK

*M.D. Amaral and L. Pozzo-Miller, Department of Neurobiology & Civitan International Research Center, University of Alabama at Birmingham, Birmingham, AL 35294. E-mail: pozzomiller@nrc.uab.edu

is abundantly expressed in hippocampal pyramidal dendrites and is activated by Ca^{2+} influx via NMDA receptors or voltage-gated L-type Ca^{2+} channels (Bading et al., 1993; Rosen et al., 1994; Segal and Greenberg, 1996). Second, the phosphatidylinositol 3-kinase (PI3-K) cascade is responsible for dendritic targeting of BDNF and TrkB mRNA (Righi et al., 2000) and has been implicated in spatial learning (Mizuno et al., 2003). Lastly, the phospholipase C (PLCγ) cascade also plays a critical role in hippocampal synaptic plasticity, as demonstrated by the lack of long-term potentiation (LTP) in conditional knock-out mice carrying a deletion of the PLCγ adaptor site in Trk receptors (Minichiello et al., 2002).

The hippocampus is one of the brain regions most susceptible to excitotoxic and neurodegenerative injuries. Due to its role in explicit learning and memory, severe impairments in cognitive performance occur in patients suffering ischemic strokes and neurodegenerative diseases involving hippocampal areas innervated by cholinergic systems (i.e. the septum). Neurotrophins have been implicated in the maintenance of

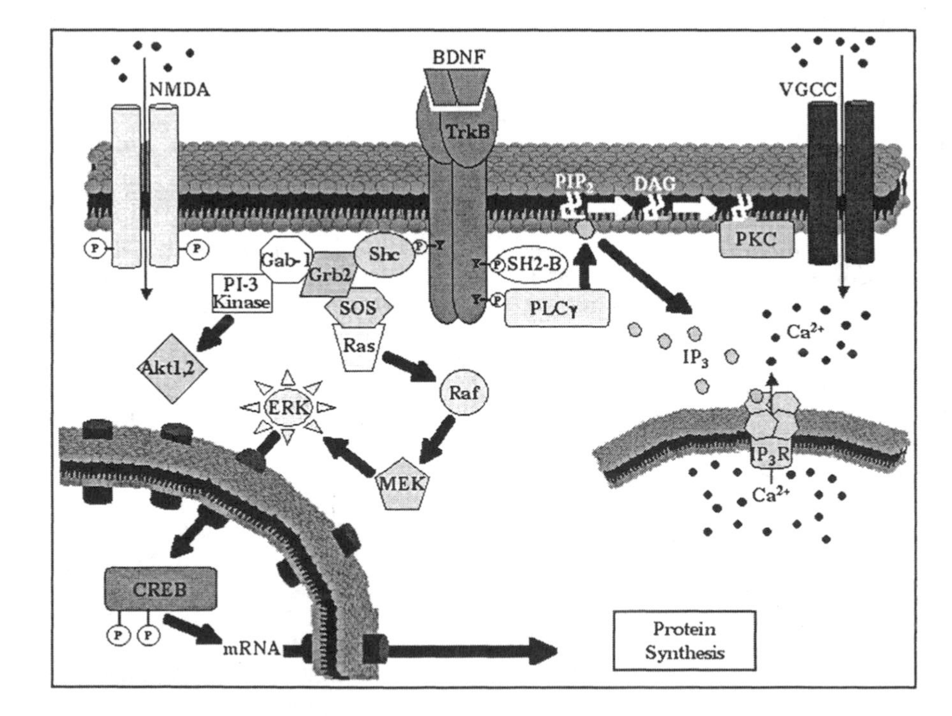

Figure 1. Intracellular signaling cascades activated by BDNF-stimulated autophosphorylation of TrkB receptors may affect prominent pathways responsible for elevations of intracellular Ca^{2+} levels in neurons. Activation of the PLC-γ/IP_3 signaling pathway leads to Ca^{2+} mobilization from IP_3-sensitive intracellular stores. In addition, NMDA receptors and voltage-gated Ca^{2+} channels could be directly phosphorylated by activated TrkB receptors or indirectly through the Ras-MAPK or PI3-K pathways. Combined, these covalent modifications potentially enhance Ca^{2+} signals within spines and dendrites.

neuronal viability in adulthood, possibly underlying the reported neuroprotection and restoration of impaired brain function in neurodegenerative disorders, such as Alzheimer's disease (Lindvall et al., 1994). These neuroprotective effects have prompted significant research on their potential clinical use as therapeutic agents. Intimately tied to neuronal excitotoxic and neurodegenerative injuries is the remarkable sensitivity of CA1 pyramidal neurons to deregulation of intracellular Ca^{2+} homeostasis. Understanding the role of neurotrophins in synapse formation, plasticity, and maintenance through their modulation of Ca^{2+} homeostasis in hippocampal neurons will make fundamental contributions to the development of therapeutic strategies for the improvement of cognitive functions in certain neurodegenerative diseases and after stroke-induced ischemia.

2. CALCIUM SIGNALS EVOKED BY EXCITATORY SYNAPTIC TRANSMISSION

When released from presynaptic terminals, the excitatory neurotransmitter glutamate induces transient elevations of intracellular Ca^{2+} concentration in postsynaptic pyramidal neurons by different mechanisms (Ghosh and Greenberg, 1995; Berridge, 1998). First, activation of AMPA-type ionotropic glutamate receptors causes a local depolarization that opens low-threshold voltage-gated Ca^{2+} channels (VGCC) leading to spatially restricted Ca^{2+} influx (Magee et al., 1995). When these synaptically initiated depolarizations reach action potential threshold, spikes back-propagate into dendrites causing widespread voltage-gated Ca^{2+} influx throughout the dendritic tree (Jaffe et al., 1992; Miyakawa et al., 1992; Pozzo-Miller et al., 1993; Spruston et al., 1995; Magee and Johnston, 1997). Second, activation of NMDA-type ionotropic glutamate receptors elicits localized dendritic Ca^{2+} elevations restricted to activated postsynaptic sites, usually spine heads (Alford et al., 1993; Perkel et al., 1993; Malinow et al., 1994; Mainen et al., 1999; Pozzo-Miller et al., 1999b; Yuste et al., 1999; Nimchinsky et al., 2004), but only when the postsynaptic membrane is sufficiently depolarized to release the NMDA channel block by extracellular Mg^{2+} ions (Mayer et al., 1984; Mayer and Westbrook, 1987; Ascher et al., 1988). Lastly, activation of group-I metabotropic glutamate receptors (mGluR1/5) coupled through G_q/G_{11}-type G-proteins to PLCβ trigger Ca^{2+} mobilization from intracellular stores through inositol-1,4,5-triphosphate (IP_3) receptors (Conn and Pin, 1997; Valenti et al., 2002) and/or via Ca^{2+}-induced Ca^{2+} release (CICR) through ryanodine receptors (RyR) (Tsien and Tsien, 1990; Fagni et al., 2000). Indeed, widespread activation of mGluRs with selective agonists evokes Ca^{2+} mobilization from intracellular stores (Murphy and Miller, 1988; Jaffe and Brown, 1994; Shirasaki et al., 1994; Bianchi et al., 1999; Rae and Irving, 2004). Furthemore, afferent synaptic stimulation elicited delayed and slow Ca^{2+} elevations in apical dendrites of CA3 pyramidal neurons under conditions of minimal voltage- and ligand-gated Ca^{2+} influx (Pozzo-Miller et al., 1996), a signal necessary for the induction of LTP at mossy fiber-CA3 synapses (Yeckel et al., 1999). These delayed Ca^{2+} elevations showed a progressive run-down with successive stimuli, recovered after Ca^{2+} loading of intracellular stores, and required the activity of the smooth endoplasmic reticulum Ca^{2+}-ATPase (SERCA) pump (Pozzo-Miller et al., 1996; Kapur et al., 2001). These observations demonstrate that Ca^{2+} can be released from intracellular stores by a signaling cascade involving $G_{q/11}$/PLCβ/IP_3

initiated at glutamatergic synapses. In addition, IP_3 mediated Ca^{2+} mobilization can be synergistically enhanced by Ca^{2+} influx through NMDA receptors or through VGCC opened during back-propagation of action potentials (Nakamura et al., 1999). Despite these direct observations of Ca^{2+} release and the established role of smooth ER stores as a Ca^{2+} source and sink (Pozzo-Miller et al., 2000), the contribution of Ca^{2+} mobilization from intracellular stores during physiological glutamatergic synaptic transmission, as well as during the induction phase of LTP is still rather controversial (Emptage et al., 1999; Svoboda and Mainen, 1999; Rose and Konnerth, 2001).

3. PRESYNAPTIC AND POSTSYNAPTIC MECHANISMS OF BDNF ACTION

BDNF can contribute to dendritic Ca^{2+} homeostasis through direct or indirect actions on membrane depolarization at both presynaptic and postsynaptic sites. At the presynaptic level, BDNF modulates quantal synaptic transmission. Acute application of BDNF enhances excitatory synaptic transmission in embryonic hippocampal neurons in primary culture (Lessman et al., 1994; Levine et al., 1995a; Li et al., 1998). Similarly, chronic treatment of postnatal hippocampal slice cultures with BDNF increases the density of docked synaptic vesicles, as well as the frequency of miniature excitatory postsynaptic currents (mEPSC) (Tyler and Pozzo-Miller, 2001). The observation that BDNF increases both the frequency of mEPSCs and the packing density of docked vesicles at the active zone reinforces the hypothesis that those vesicles correspond to the readily releasable pool (RRP) of quanta (Harris and Sultan, 1995; Zucker, 1996; Schikorski and Stevens, 1997). Increasing the size of the RRP would allow synapses to sustain longer epochs or higher frequencies of transmitter release before vesicle depletion compromises information transfer across the synapse (Dobrunz and Stevens, 1997; Murthy et al., 1997). These observations provide support for the interpretation that BDNF facilitates the induction of LTP at immature synapses by allowing them to follow high-frequency afferent activity (Figurov et al., 1996; Gottschalk et al., 1998; Pozzo-Miller et al., 1999a). It has also been shown that BDNF enhances glutamate release from synaptosomes *in vitro* by MAPK-dependent phosphorylation of synapsin-I, an effect absent in preparations from synapsin-I and/or -II knock-out mice (Jovanovic et al., 2000). Because one of the major functions of synapsins is to regulate the trafficking of synaptic vesicles between distinct pools within presynaptic terminals (Greengard et al., 1993), it seems likely that BDNF facilitates synaptic vesicle docking via the modulation of presynaptic vesicle proteins (Pozzo-Miller et al., 1999a; Tartaglia et al., 2001).

Additionally, BDNF potentiates voltage-dependent spontaneous Ca^{2+} oscillations in cultured embryonic hippocampal neurons (Sakai et al., 1997; Numakawa et al., 2002), most likely through its enhancement of glutamate release (Berninger and Poo, 1996; Tyler et al., 2002b). However, these effects on Ca^{2+} levels can be accounted for only by enhanced network activity leading to spike-driven voltage-dependent Ca^{2+} influx. In fact, tetrodotoxin (TTX), the AMPA/kainate receptor antagonist CNQX, and the NMDA receptor antagonist APV, all completely blocked spontaneous Ca^{2+} oscillations, as well as the potentiation of their frequency by BDNF (Sakai et al., 1997). Also, BDNF increases Ca^{2+} levels within presynaptic terminals of *Xenopus* neuromuscular junction in culture leading to a transient enhancement of transmitter release, an effect dependent upon extracellular Ca^{2+} levels and presynaptic depolarization (Stoop and Poo, 1996; Boulanger and Poo, 1999).

The potential neurotrophin modulation of postsynaptic mechanisms known to be critical for the induction of enduring changes in synaptic efficacy, such as NMDA-dependent Ca^{2+} elevations, and activation of protein kinases, strongly argues for a postsynaptic site of action as well. The posttranslational modification of ion channels or neurotransmitter receptors by neurotrophins via protein phosphorylation may provide the molecular mechanism of their action at hippocampal synapses. BDNF rapidly and selectively enhanced the phosphorylation of NMDA receptor subunits 1 and 2B in isolated postsynaptic densities (Suen et al., 1997; Lin et al., 1998), and it increased the open channel probability of NMDA receptors (Jarvis et al., 1997; Levine et al., 1998). These covalent modifications of NMDA receptors may have profound effects on hippocampal synaptic plasticity.

In addition to its effects on excitatory synapses, BDNF also modulates GABAergic synapses on hippocampal neurons (Marty et al., 1997; Tanaka et al., 1997; Rico et al., 2002). Acute application of BDNF to cultured neurons caused an initial increase, followed by a prolonged reduction in the amplitude of miniature inhibitory postsynaptic currents (mIPSCs). This effect involved different phosphorylation states of postsynaptic $GABA_A$ receptors, and was modulated by protein kinase C (PKC) and protein phosphatase 2A (PP2A) (Jovanovic et al., 2004). Taken altogether, these observations suggest that the net effect of BDNF signaling in the hippocampus is to enhance excitatory synaptic transmission.

Neurotrophins have also been shown to enhance voltage-gated ionic currents (Lesser et al., 1997), including Ca^{2+} currents (Levine et al., 1995b). Activation of TTX-insensitive $Na_v1.9$ channels underlies a direct membrane depolarization observed when BDNF is rapidly ejected onto hippocampal dendrites (Blum et al., 2002). In addition, it has been found that chronic BDNF treatment of embryonic hippocampal neurons in culture increases the expression and channel density of presynaptic P/Q and N-type channels, without affecting postsynaptic L-type channels (Baldelli et al., 1999; Baldelli et al., 2000). Because of the well-established role of postsynaptic Ca^{2+} elevations in the induction of synaptic plasticity, the neuromodulatory effects of BDNF on voltage-gated Ca^{2+} influx evoked by back-propagating action potentials (bAPs) was investigated in CA1 pyramidal neurons maintained in hippocampal slice cultures. Due to their clustering at the soma and in the base of proximal dendrites, high-threshold L-type VGCCs are the most likely mode of Ca^{2+} entry in those regions during bAPs (Ahlijanian et al., 1990; Westenbroek et al., 1990). The spatio-temporal patterns and amplitude of dendritic and somatic Ca^{2+} transients evoked by trains of bAPs were not significantly different between BDNF-treated and control slices kept in serum-free media (SF) (Figure 2). Furthermore, the L-type channel blocker nifedipine reduced dendritic Ca^{2+} transients during bAP trains by 19% in SF controls and by 26% in BDNF-treated neurons. Similarly, somatic Ca^{2+} transients were reduced by 22% and 20% in SF controls and BDNF-treated neurons, respectively. The lack of significant differences between controls and BDNF-treated neurons in the amount of L-type channel block suggests that BDNF does not affect Ca^{2+} entry in proximal apical dendrites and cell bodies mediated by dihydropyridine-sensitive L-type Ca^{2+} channels during trains of bAPs (McCutchen et al., 2002). Thus, despite the profound effects of BDNF on hippocampal synaptic plasticity, and of L-type Ca^{2+} channels on neuronal gene transcription (Bading et al., 1993; Mermelstein et al., 2000), it appears as though the actions of BDNF do not involve modulation of voltage-gated dendritic Ca^{2+} signaling mediated by L-type channels in the proximal apical dendrites and somata of CA1 pyramidal neurons. It is likely that modulation of other non-L-type Ca^{2+}

channels by BDNF could be related to synaptic plasticity, while L-type Ca^{2+} channel modulation by other neurotrophins may contribute to more classic neurotrophic effects, such as neuronal survival and differentiation through modulation of gene transcription (Ghosh and Greenberg, 1995).

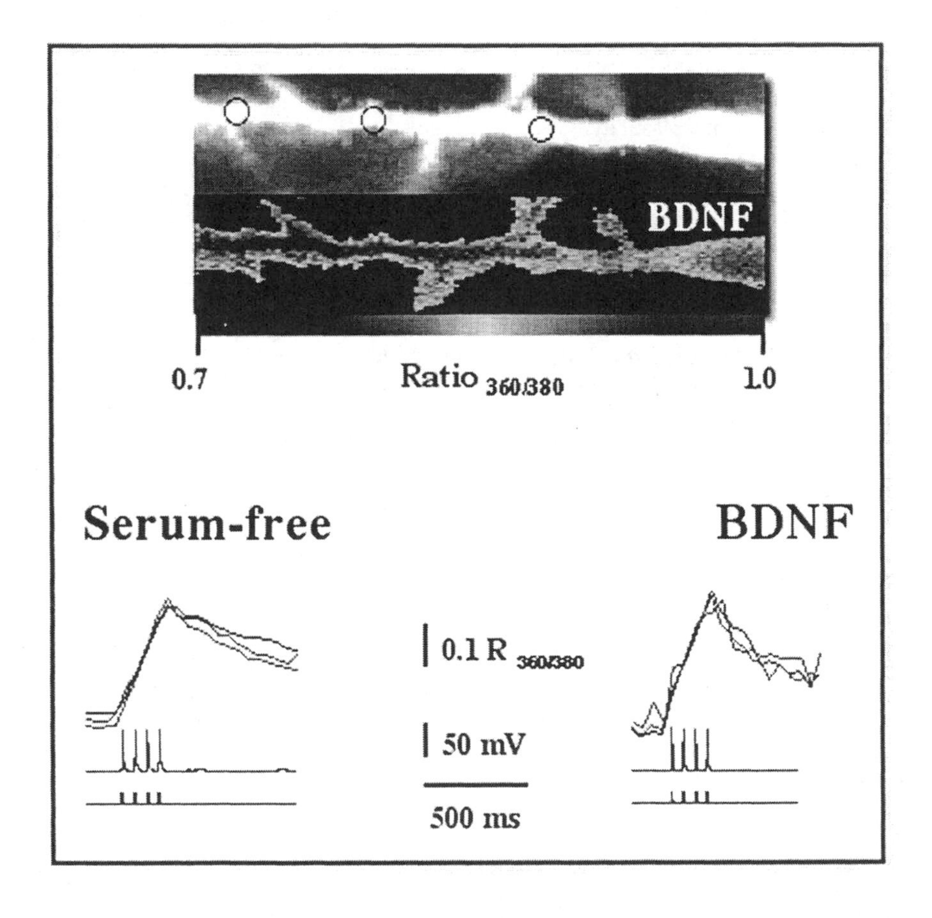

Figure 2. Dendritic Ca^{2+} elevations evoked by trains of bAPs. The fluorescence image (bis-fura-2, 380nm excitation) shows an example of an apical spiny dendrite, in this case from a BDNF-treated CA1 pyramidal neuron. The colors of the individual ROIs correspond to the Ca^{2+} traces shown below. The lower image is a pixel-by-pixel ratio (360/380nm) image displayed using the pseudocolor scale shown below. Top traces show Ca^{2+} levels expressed as bis-fura-2 ratios within the color-coded ROIs shown above. The traces below are from simultaneous whole-cell recordings of membrane voltage (red) and current (blue) in the current-clamp mode. The whole-cell pipette contained a K-gluconate solution and 200μM bis-fura-2. Digital imaging was performed with a cooled CCD camera at 20-33 frames-per-second. Short trains of brief depolarizing current pulses (5ms) delivered by the whole-cell recording electrode elicited action potentials.

Considering that specific Ca^{2+} entry pathways, such as VGCC or NMDA receptors, lead to differential activation of particular effector genes, differential BDNF modulation of dendritic Ca^{2+} signals originating from various sources of Ca^{2+} may ultimately be expressed by differential activation of signaling pathways that control the expression of particular genes (Ghosh and Greenberg, 1995). Thus, enduring changes in synaptic function and structure thought to underlie memory consolidation are ultimately translated into long-term changes in gene expression (Kandel, 2001).

4. BDNF CAUSES MOBILIZATION OF CALCIUM FROM INTRACELLULAR STORES

One of the prominent signaling cascades activated by Trk neurotrophin receptors, the hydrolysis of phosphatidylinositol-4,5-biphosphate (PIP_2) by activated PLC-γ leading to IP_3 formation and subsequent Ca^{2+} mobilization from smooth ER through IP_3 receptors (Berridge et al., 2000), is expected to cause Ca^{2+} elevations in neurons (Segal and Greenberg, 1996). This pathway converges on the same IP_3R-containing Ca^{2+} stores targeted by other metabotropic receptors in hippocampal pyramidal neurons, such as group-I mGluR (Conn and Pin, 1997). The amplification of Ca^{2+} signals by mobilization from intracellular stores allows the enhancement of such processes as transmitter release and gene expression, which are fundamental for LTP induction. Although NMDA receptors are well-known Ca^{2+} entry routes necessary for LTP induction, depletion of intracellular Ca^{2+} stores using thapsigargin, an inhibitor of the SERCA pump, also prevents LTP induction, demonstrating the importance of Ca^{2+} mobilization (Harvey and Collingridge, 1992).

NGF has been reported to increase Ca^{2+} levels in PC12 cells (Pandiella-Alonso et al., 1986), most likely by PLC-γ stimulation (Vetter et al., 1991), leading to IP_3 formation and subsequent release from smooth ER cisterns endowed with IP_3Rs. BDNF induces rapid phosphorylation of PLC-γ1 increasing phosphatidylinositol (PI) turnover and IP_3 levels in embryonic cortical neurons in culture (Widmer et al., 1992; Widmer et al., 1993) causing Ca^{2+} elevations (Zirrgiebel et al., 1995; Behar et al., 1997; Matsumoto et al., 2001). In embryonic hippocampal neurons in culture, BDNF also induces elevations of intracellular Ca^{2+} (Berninger et al., 1993; Marsh and Palfrey, 1996; Canossa et al., 1997; Finkbeiner et al., 1997; Li et al., 1998). In contrast, BDNF fails to increase Ca^{2+} levels in cultured cerebellar granule cells (Gaiddon et al., 1996); but see (Numakawa et al., 2001), and in cultured neurons and acute slices from visual cortex (Pizzorusso et al., 2000); but see (Mizoguchi et al., 2002; Mizoguchi and Nabekura, 2003). Lastly, it has been shown that intracellular Ca^{2+} stores are important for neurotrophin-induced synaptic potentiation in the hippocampus (Kang and Schuman, 2000). Depletion of intracellular Ca^{2+} stores with thapsigargin does not affect basal synaptic transmission but does prevent the neurotrophin-induced enhancement of excitatory synaptic transmission. However, experimental evidence of neurotrophin-initiated Ca^{2+} signaling is sparse, controversial, and limited to cell lines or embryonic cultured neurons.

Unfortunately, most published Ca^{2+} imaging studies regarding the actions of neurotrophins on neuronal Ca^{2+} homeostasis were done without simultaneous membrane voltage control, making it difficult to isolate the contribution of cell depolarization to the observed Ca^{2+} changes. Figure 3 shows a preliminary example of intracellular Ca^{2+}

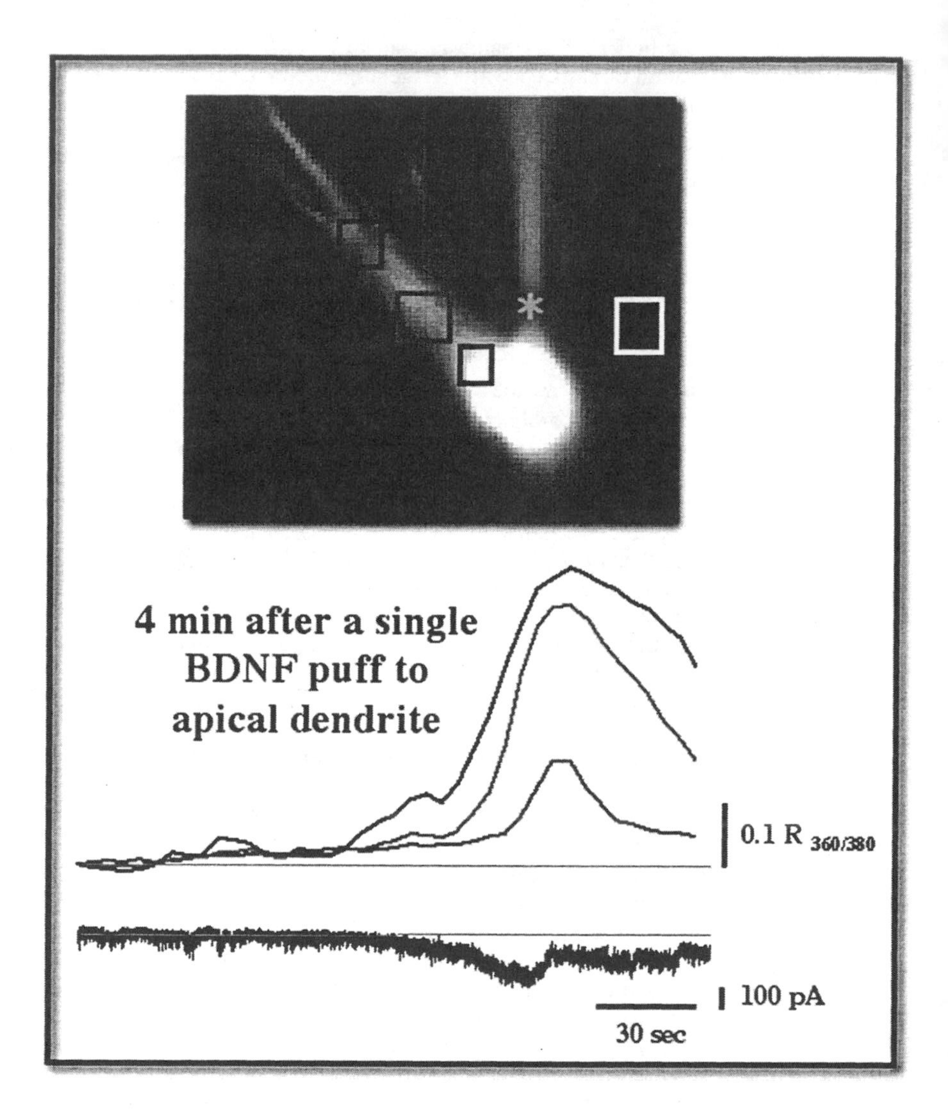

Figure 3. BDNF induces slowly developing intracellular Ca^{2+} elevations, which are associated with inward currents that have similar kinetics. The fluorescence image (380nm excitation) shows a CA1 pyramidal neuron filled with the Ca^{2+} indicator bis-fura-2 through the whole-cell recording electrode (asterisk). The traces represent intracellular Ca^{2+} levels within the color-coded ROIs and the simultaneously recorded membrane current. The whole-cell electrode contained a Cs-gluconate solution and 100μM bis-fura-2; voltage-gated Na^+ channels were blocked by TTX (500nM); the holding voltage was -65mV. BDNF was applied by pressure ejection (25sec, 30psi) onto the apical dendrite of the neuron under recording.

signals evoked by a single BDNF puff application to the apical dendrite of a CA1 pyramidal neuron during simultaneous whole-cell voltage-clamp recording. BDNF evoked intracellular Ca^{2+} elevations that initiated near the soma at the base of the apical dendrite. These Ca^{2+} responses had slow rise times (time to peak ~120sec), and slowly spread back toward the distal apical dendrite before returning to baseline levels. Note that during the initial slow and sustained Ca^{2+} rise there is no noticeable membrane current but when the Ca^{2+} levels reach their peak there is a concomitant slow inward current. The slow kinetics of this inward current, and the lack of faster spike-like membrane currents strongly suggest that BDNF-induced Ca^{2+} signals reflect intracellular mobilization, possibly amplified by a Ca^{2+}-dependent non-selective cationic current.

5. DOES BDNF INDUCE CAPACITATIVE CALCIUM ENTRY?

In addition to the well-established voltage- and ligand-gated mechanisms of Ca^{2+} influx discussed above, Ca^{2+} entry associated with depletion of intracellular Ca^{2+} stores is beginning to emerge as a fundamental component of neuronal Ca^{2+} signaling. The depletion of these stores activates plasma membrane channels allowing Ca^{2+} influx for their replenishment, a process called capacitative Ca^{2+} entry (Berridge, 1995). Cultured hippocampal neurons seem to exhibit capacitative Ca^{2+} entry (Bouron, 2000; Emptage et al., 2001; Baba et al., 2003), and impairments in this Ca^{2+} influx mechanism have been observed in neurons from transgenic mice carrying Alzheimer's disease-associated mutations in presenilin (Yoo et al., 2000).

Recent evidence suggests that members of the transient receptor potential (TRP) ion channel family play a critical role in these responses (Birnbaumer et al., 1996). TRP channels were originally described in *Drosophila* photoreceptors, where *trp* mutants exhibit a transient response to continuous light (Montell et al., 1985). Their mammalian homologues are currently implicated not only in capacitative Ca^{2+} entry but also in a variety of cellular functions, ranging from sensory transduction (e.g. temperature, touch, pain, osmolarity, pheromone, taste) to modulation of the cell cycle (Montell et al., 2002; Clapham, 2003). Mammalian TRP homologues are classified into TRPC (canonical), TRPV (vanilloid), TRPM (melastatin), TRPP (polycystin), and TRPML (mucolipin) subfamilies (Clapham, 2003). All mammalian TRPC proteins (TRPC1 through TRPC7) appear to be homologues of the TRP channels involved in *Drosophila* phototransduction, in that they function as receptor-operated channels. TRPC channels can be activated in neurons by stimulation of G_q/G_{11}-type G protein-coupled receptors (i.e. group-I mGluRs) and receptor tyrosine kinases (i.e. Trk receptors) leading to PLC-dependent hydrolysis of PIP_2 into IP_3 and diacylglycerol (DAG). Figure 4 illustrates our working hypothesis for TRPC activation by BDNF-initiated TrkB signaling. The activation mechanism of mammalian TRPC channels is still highly controversial, mostly because it has been studied exclusively in cell lines expressing specific subunits forming homomeric channels. TRPC channels have been proposed to be gated by a direct interaction with IP_3Rs, by DAG itself, as well as by a soluble Ca^{2+} influx factor produced in response to Ca^{2+} store depletion (Clapham et al., 2001). The fact that most TRPC channels are able to form functional heteromeric channels with activation properties distinct from those of the homomers contributes to the current disparity in the proposed models for the activation and modulation of native TRPC channels expressed by neurons in the CNS.

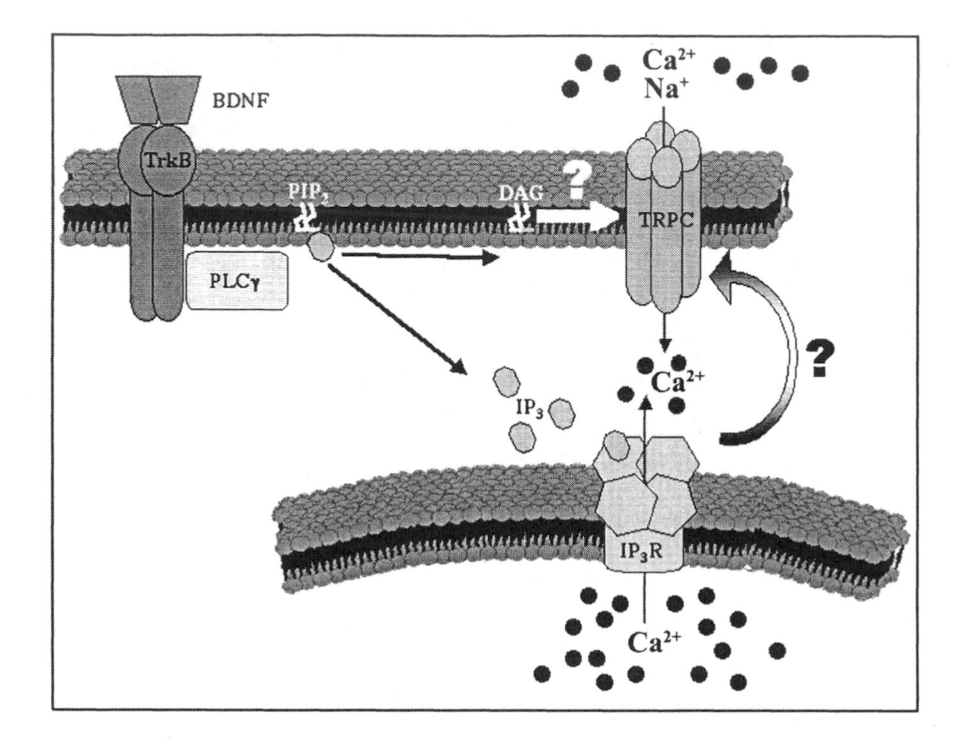

Figure 4. TrkB-initiated stimulation of PLCγ causes PIP_2 hydrolysis and formation of DAG and IP_3. Activation of IP_3Rs leads to Ca^{2+} mobilization from intracellular stores. Native TRPC channels expressed in hippocampal neurons, most likely heteromultimeric, may be gated by a diffusible factor, Ca^{2+} ions released from IP_3-sensitive Ca^{2+} stores, a physical interaction with activated IP_3Rs, or by DAG itself. TRPC channels are known to mediate a non-selective cationic current that requires intact IP_3R signaling, full intracellular Ca^{2+} stores, and extracellular Ca^{2+} ions. Preliminary results exemplified in Figure 3 suggest that BDNF-induced dendritic Ca^{2+} elevations include a Ca^{2+} mobilization component as well as Ca^{2+} entry from the extracellular space.

Intriguingly, when BDNF is applied to cultures in the absence of extracellular Ca^{2+}, the resultant Ca^{2+} elevations are reduced but not completely blocked, suggesting that both mobilization and entry contribute to the effects of BDNF on intracellular Ca^{2+} levels (Marsh and Palfrey, 1996; Canossa et al., 1997; Finkbeiner et al., 1997). It has been assumed that Trk receptor activation does not modulate plasma membrane ion channels, other than by direct or indirect protein phosphorylation (Segal and Greenberg, 1996). However, TrkB receptor-initiated formation of IP_3 and subsequent Ca^{2+} mobilization from IP_3R-containing intracellular Ca^{2+} stores is one of the best-established mechanisms for activation of non-selective cationic currents mediated by plasma membrane TRPC channels (Montell et al., 2002; Clapham, 2003). The intracellular cascades that converge on Ca^{2+} mobilization and IP_3 formation represent intriguing signaling links between TrkB receptor activation, group-I metabotropic glutamate receptor activation, and those prominent Ca^{2+}-dependent processes that are responsible for synaptic plasticity (Berridge, 1998; Zucker, 1999).

6. CONCLUSIONS

In addition to its strong presynaptic actions on quantal neurotransmitter release, the regulation of the spatial and temporal patterns of postsynaptic Ca^{2+} elevations by BDNF is also a likely mechanism for its modulation of synaptic plasticity. The intracellular signaling cascades activated by TrkB receptors include several well-characterized protein kinases that target most of the routes of Ca^{2+} entry into hippocampal neurons. In addition, TrkB activation leads to IP_3 formation, strongly arguing for direct Ca^{2+} mobilization from intracellular Ca^{2+} stores. Lastly, depletion of intracellular Ca^{2+} stores is associated with the activation of plasma membrane non-selective cationic currents thought to mediate Ca^{2+} store refilling. These membrane currents mediated by members of the TRPC family of ion channels not only represent novel downstream effects of neurotrophin action, but also are intriguing points of convergence with other intracellular signaling cascades, such as those triggered by group-I metabotropic glutamate receptors. The information gained from future experiments in this rapidly evolving field will integrate the actions of BDNF at synapses with the requirement of dendritic Ca^{2+} signals necessary for the induction of synaptic plasticity. Ultimately, the challenge ahead is to assimilate the varied functional and structural consequences of BDNF signaling through TrkB receptors at both sides of the synaptic cleft at excitatory synapses in the hippocampus with its intriguing role in the consolidation of hippocampal-dependent learning. Indeed, BDNF represents the prototypical example of a consolidation factor necessary for trans-synaptic plasticity at hippocampal synapses.

7. ACKOWLEDGEMENTS

We would like to thank NIH grants R01-NS40593 (LP-M), T32-GM008111 (Graduate Training for MDA), MRRC P30-HD38985, and PO1-HD38760, as well as AMGEN for the supply of BDNF.

8. REFERENCES

Ahlijanian MK, Westenbroek RE, Catterall WA (1990) Subunit structure and localization of dihydropyridine-sensitive calcium channels in mammalian brain, spinal cord, and retina, Neuron 4:819-832.

Alford S, Frenguelly BG, Schofield JG, Collingridge GL (1993) Characterization of Ca^{2+} signals induced in hippocampal CA1 neurones by the synaptic activation of NMDA receptors, J. Physiol. 469:693-716.

Ascher P, Bregestovsky P, Nowak L (1988) N-methyl-D-aspartate-activated channels of mouse central neurones in magnesium-free solutions, J. Physiol. 399:207-226.

Baba A, Yasui T, Fujisawa S, Yamada RX, Yamada MK, Nishiyama N, Matsuki N, Ikegaya Y (2003) Activity-evoked capacitative Ca^{2+} entry: implications in synaptic plasticity, J. Neurosci. 23:7737-7741.

Bading H, Ginty DD, Greenberg ME (1993) Regulation of gene expression in hippocampal neurnons by distinct calcium signaling pathways, Science 260:181-186.

Baldelli P, Forni PE, Carbone E (2000) BDNF, NT-3 and NGF induce distinct new Ca^{2+} channel synthesis in developing hippocampal neurons, Eur J Neurosci 12:4017-4032.

Baldelli P, Magnelli V, Carbone E (1999) Selective up-regulation of P- and R-type Ca^{2+} channels in rat embryo motoneurons by BDNF, Eur. J. Neurosci. 11:1127-1133.

Barde Y, (1989) Trophic factors and neuronal survival, Neuron 2:1525-1534.

Behar TN, Dugich-Djordjevic MM, Li YX, Ma W, Somogyi R, Wen X, Brown E, Scott C, Mckay RD, Barker JL (1997) Neurotrophins stimulate chemotaxis of embryonic cortical neurons, Eur. J. Neurosci. 9:2561-2570.

Berninger B, Garcia DE, Inagaki N, Hahnel C, Lindholm D (1993) BDNF and NT-3 induce intracellular Ca^{2+} elevation in hippocampal neurones, Neuroreport 4:1303-1306.

Berninger B, Poo M (1996) Fast actions of neurotrophic factors, Curr. Opin. Neurobiol. 6:324-330.

Berridge MJ (1995) Capacitative calcium entry, Biochem. J. 312:1-11.

Berridge MJ (1998) Neuronal calcium signaling, Neuron 21:13-26.

Berridge MJ, Lipp P, Bootman MD (2000) The versatility and universality of calcium signalling, Nat. Rev. Mol. Cell Biol. 1:11-21.

Bianchi R, Young SR, Wong RKS (1999) Group I mGluR activation causes voltage-dependent and -independent Ca^{2+} rises in hippocampal pyramidal cells, J. Neurophysiol. 81:2903-2913.

Birnbaumer L, Zhu X, Jiang M, Boulay G, Peyton M, Vannier B, Brown D, Platano D, Sadeghi H, Stefani E, Birnbaumer M (1996) On the molecular basis and regulation of cellular capacitative calcium entry: roles for Trp proteins, Proc. Natl. Acad. Sci. USA 93:15195-15202.

Blum R, Kafitz KW, Konnerth A (2002) Neurotrophin-evoked depolarization requires the sodium channel $Na_V1.9$, Nature 419:687-693.

Boulanger L, Poo MM (1999) Presynaptic depolarization facilitates neurotrophin-induced synaptic potentiation, Nat. Neurosci. 2:346-351.

Bouron A (2000) Activation of a capacitative Ca^{2+} entry pathway by store depletion in cultured hippocampal neurones, FEBS Lett. 470:269-272.

Canossa M, Griesbeck O, Berninger B, Campana G, Kolbeck R, Thoenen H (1997) Neurotrophin release by neurotrophins: implications for activity-dependent neuronal plasticity, Proc. Natl. Acad. Sci. USA 94:13279-13286.

Chao MV (1992) Neurotrophin receptors: a window into neuronal differentiation, Neuron 9:583-593.

Clapham DE (2003) TRP channels as cellular sensors, Nature 426:517-524.

Clapham DE, Runnels LW, Strubing C (2001) The TRP ion channel family, Nat. Rev. Neurosci. 2:387-396.

Conn PJ, Pin JP (1997) Pharmacology and functions of metabotropic glutamate receptors, Annu. Rev. Pharmacol. Toxicol. 37:205-237.

Dobrunz LE, Stevens CF (1997) Heterogeneity of release probability, facilitation, and depletion at central synapses, Neuron 18:995-1008.

Dugich-Djordjevic MM, Peterson C, Isono F, Ohsawa F, Widmer HR, Denton TL, Bennett GL, Hefti F (1995) Immunohistochemical visualization of brain-derived neurotrophic factor in the rat brain, Eur. J. Neurosci. 7:1831-1839.

Emptage N, Bliss TV, Fine A (1999) Single synaptic events evoke NMDA receptor-mediated release of calcium from internal stores in hippocampal dendritic spines, Neuron 22:115-124.

Emptage NJ, Reid CA, Fine A (2001) Calcium stores in hippocampal synaptic boutons mediate short-term plasticity, store-operated Ca^{2+} entry, and spontaneous transmitter release, Neuron 29:197-208.

Fagni L, Chavis P, Ango F, Bockaert J (2000) Complex interactions between mGluRs, intracellular Ca^{2+} stores and ion channels in neurons, Trends Neurosci. 23:80-88.

Figurov A, Pozzo-Miller LD, Olafsson P, Wang T, Lu B (1996) Regulation of synaptic responses to high-frequency stimulation and LTP by neurotrophins in the hippocampus, Nature 381:706-709.

Finkbeiner S, Tavazoie SF, Maloratsky A, Jacobs KM, Harris KM, Greenberg ME (1997) CREB: a major mediator of neuronal neurotrophin responses, Neuron 19:1031-1047.

Fryer RH, Kaplan DR, Feinstein SC, Radeke MJ, Grayson DR, Kromer LF (1996) Developmental and mature expression of full-length and truncated TrkB receptors in the rat forebrain, J. Comp. Neurol. 374:21-40.

Gaiddon C, Loeffler JP, Larmet Y (1996) Brain-derived neurotrophic factor stimulates AP-1 and cyclic AMP-responsive element dependent transcriptional activity in central nervous system neurons, J. Neurochem. 66:2279-2286.

Ghosh A, Greenberg ME (1995) Calcium signaling in neurons: molecular mechanisms and cellular consequences, Science 268:239-247.

Gottschalk W, Pozzo-Miller LD, Figurov A, Lu B (1998) Presynaptic modulation of synaptic transmission and plasticity by brain- derived neurotrophic factor in the developing hippocampus, J. Neurosci. 18:6830-6839.

Greengard P, Valtorta F, Czernik AJ, Benfenati F (1993) Synaptic vesicle phosphoproteins and regulation of synaptic function, Science 259:780-785.

Harris KM, Sultan P (1995) Variation in the number, location and size of synaptic vesicles provides an anatomical basis for the nonuniform probability of release at hippocampal CA1 synapses, Neuropharmacology 34:1387-1395.

Harvey J, Collingridge GL (1992) Thapsigargin blocks the induction of long-term potentiation in rat hippocampal slices, Neurosci. Lett. 139:197-200.

Jaffe DB, Brown TH (1994) Metabotropic glutamate receptor activation induces calcium waves within hippocampal dendrites, J. Neurophysiol. 72:471-474.

Jaffe DB, Johnston D, Lasser-Ross N, Lisman JE, Miyakawa H, Ross WN (1992) The spread of Na^+ spikes determines the pattern of dendritic Ca^{2+} entry into hippocampal neurons, Nature 357:244-246.

Jarvis CR, Xiong ZG, Plant JR, Churchill D, Lu WY, Macvicar BA, Macdonald JF (1997) Neurotrophin modulation of NMDA receptors in cultured murine and isolated rat neurons, J. Neurophysiol. 78:2363-2371.

Jovanovic JN, Czernik AJ, Fienberg AA, Greengard P, Sihra TS (2000) Synapsins as mediators of BDNF-enhanced neurotransmitter release, Nat. Neurosci. 3:323-329.

Jovanovic JN, Thomas P, Kittler JT, Smart TG, Moss SJ, (2004) Brain-derived neurotrophic factor modulates fast synaptic inhibition by regulating $GABA_A$ receptor phosphorylation, activity, and cell-surface stability, J. Neurosci. 24:522-530.

Kandel ER (2001) The molecular biology of memory storage: a dialogue between genes and synapses, Science 294:1030-1038.

Kang H, Schuman EM (2000) Intracellular Ca^{2+} signaling is required for neurotrophin-induced potentiation in the adult rat hippocampus, Neurosci. Lett. 282:141-144.

Kapur A, Yeckel M, Johnston D (2001) Hippocampal mossy fiber activity evokes Ca^{2+} release in CA3 pyramidal neurons via a metabotropic glutamate receptor pathway, Neuroscience 107:59-69.

Lesser SS, Sherwood NT, Lo DC (1997) Neurotrophins differentially regulate voltage-gated ion channels, Mol. Cell. Neurosci. 10:173-183.

Lessman V, Gottmann K, Heumann R (1994) BDNF and NT-4/5 enhance glutamatergic synaptic transmission in cultured hippocampal neurons, Neuroreport 6:21-25.

Levine ES, Crozier RA, Black IB, Plummer MR (1998) Brain-derived neurotrophic factor modulates hippocampal synaptic transmission by increasing N-methyl-D-aspartic acid receptor activity, Proc. Natl. Acad. Sci. USA 95:10235-10239.

Levine ES, Dreyfus CF, Black IB, Plummer MR (1995a) Brain-derived neurotrophic factor rapidly enhances synaptic transmission in hippocampal neurons via postsynaptic tyrosine kinase receptors, Proc. Natl. Acad. Sci. USA 92:8074-8077.

Levine ES, Dreyfus CF, Black IB, Plummer MR (1995b) Differential effects of NGF and BDNF on voltage-gated calcium currents in embryonic basal forebrain neurons, J. Neurosci. 15:3084-3091.

Lewin GR, Barde Y (1996) Physiology of the neurotrophins, Ann. Rev. Neurosci. 19:289-317.

Li YX, Zhang Y, Lester HA, Schuman EM, Davidson N (1998) Enhancement of neurotransmitter release induced by brain-derived neurotrophic factor in cultured hippocampal neurons, J. Neurosci. 18:10231-10240.

Lin SY, Wu K, Levine ES, Mount HT, Suen PC, Black IB (1998) BDNF acutely increases tyrosine phosphorylation of the NMDA receptor subunit 2B in cortical and hippocampal postsynaptic densities, Brain Res. Mol. Brain Res. 55:20-27.

Lindvall O, Kokaia Z, Bengzon J, Elmer E, Kokaia M (1994) Neurotrophins and brain insults, Trends Neurosci. 17:490-496.

Lo DC (1995) Neurotrophic factors and synaptic plasticity, Neuron 15:979-981.

Lu B (2003) BDNF and activity-dependent synaptic modulation, Learn. Mem. 10:86-98.

Magee JC, Christofi G, Miyakawa H, Christie B, Lasser-Ross N, Johnston D (1995) Subthreshold synaptic activation of voltage-gated Ca^{2+} channels mediates a localized Ca^{2+} influx into the dendrites of hippocampal pyramidal neurons, J. Neurophysiol. 74:1335-1342.

Magee JC, Johnston D (1997) A synaptically controlled, associative signal for Hebbian plasticity in hippocampal neurons, Science 275:209-213.

Mainen ZF, Malinow R, Svoboda K (1999) Synaptic calcium transients in single spines indicate that NMDA receptors are not saturated, Nature 399:151-155.

Malinow R, Otmakhov N, Blum KI, Lisman J (1994) Visualizing hippocampal synaptic function by optical detection of Ca^{2+} entry through the N-methyl-D-aspartate channel, Proc. Natl. Acad. Sci. USA 91:8170-8174.

Marsh HN, Palfrey HC (1996) Neurotrophin-3 and brain-derived neurotrophic factor activate multiple signal transduction events but are not survival factors for hippocampal pyramidal neurons, J. Neurochem. 67:952-963.

Marty S, Berzaghi Mda P, Berninger B (1997) Neurotrophins and activity-dependent plasticity of cortical interneurons, Trends Neurosci. 20:198-202.

Matsumoto T, Numakawa T, Adachi N, Yokomaku D, Yamagishi S, Takei N, Hatanaka H (2001) Brain-derived neurotrophic factor enhances depolarization-evoked glutamate release in cultured cortical neurons, J. Neurochem. 79:522-530.

Mayer ML, Westbrook GL (1987) Permeation and block of N-methyl-D-aspartic acid receptor channels by divalent cations in mouse cultured central neurones, J. Physiol. 394:501-527.

Mayer ML, Westbrook GL, Guthrie PB (1984) Voltage-dependent block by Mg^{+} of NMDA responses in spinal cord neurones, Nature 309:261-263.

McCutchen ME, Bramham CR, Pozzo-Miller LD (2002) Modulation of neuronal calcium signaling by neurotrophic factors, Int. J. Dev. Neurosci. 20:199-207.

Mermelstein PG, Bito H, Deisseroth K, Tsien RW (2000) Critical dependence of cAMP response element-binding protein phosphorylation on L-type calcium channels supports a selective response to EPSPs in preference to action potentials, J. Neurosci. 20:266-273.

Minichiello L, Calella AM, Medina DL, Bonhoeffer T, Klein R, Korte M (2002) Mechanism of TrkB-mediated hippocampal long-term potentiation, Neuron 36:121-137.

Miyakawa H, Ross WN, Jaffe D, Callaway JC, Lasser-Ross N, Lisman JE, Johnston D (1992) Synaptically activated increases in Ca^{2+} concentration in hippocampal CA1 pyramidal cells are primarily due to voltage-gated Ca^{2+} channels, Neuron 9:1163-1173.

Mizoguchi Y, Monji A, Nabekura J (2002) Brain-derived neurotrophic factor induces long-lasting Ca^{2+}-activated K+ currents in rat visual cortex neurons, Eur. J. Neurosci. 16:1417-1424.

Mizoguchi Y, Nabekura J (2003) Sustained intracellular Ca^{2+} elevation induced by a brief BDNF application in rat visual cortex neurons, Neuroreport 14:1481-1483.

Mizuno M, Yamada K, Takei N, Tran MH, He J, Nakajima A, Nawa H, Nabeshima T (2003) Phosphatidylinositol 3-kinase: a molecule mediating BDNF-dependent spatial memory formation, Mol. Psychiatry 8:217-224.

Montell C, Birnbaumer L, Flockerzi V (2002) The TRP channels, a remarkably functional family, Cell 108:595-598.

Montell C, Jones K, Hafen E, Rubin G (1985) Rescue of the Drosophila phototransduction mutation trp by germline transformation, Science 230:1040-1043.

Murphy SN, Miller RJ (1988) A glutamate receptor regulates Ca^{2+} mobilization in hippocampal neurons, Proc.Natl.Acad.Sci. USA 85:8737-8741.

Murthy VN, Sejnowski TJ, Stevens CF (1997) Heterogeneous release properties of visualized individual hippocampal synapses, Neuron 18:599-612.

Nakamura T, Barbara JG, Nakamura K, Ross WN (1999) Synergistic release of Ca^{2+} from IP_3-sensitive stores evoked by synaptic activation of mGluRs paired with backpropagating action potentials, Neuron 24:727-737.

Nimchinsky EA, Yasuda R, Oertner TG, Svoboda K (2004) The number of glutamate receptors opened by synaptic stimulation in single hippocampal spines, J. Neurosci. 24:2054-2064.

Numakawa T, Matsumoto T, Adachi N, Yokomaku D, Kojima M, Takei N, Hatanaka H (2001) Brain-derived neurotrophic factor triggers a rapid glutamate release through increase of intracellular Ca^{2+} and Na^{+} in cultured cerebellar neurons, J. Neurosci. Res. 66:96-108.

Numakawa T, Yamagishi S, Adachi N, Matsumoto T, Yokomaku D, Yamada M, Hatanaka H (2002) Brain-derived neurotrophic factor-induced potentiation of Ca^{2+} oscillations in developing cortical neurons, J. Biol. Chem. 277:6520-6529.

Pandiella-Alonso A, Malgaroli A, Vicentini LM, Meldolesi J (1986) Early rise of cytosolic Ca^{2+} induced by NGF in PC12 and chromaffin cells, FEBS Lett. 208:48-51.

Perkel DJ, Petrozzino JJ, Nicoll RA, Connor JA (1993) The role of Ca^{2+} entry via synaptically activated NMDA receptors in the induction of long-term potentiation, Neuron 11:817-823.

Pizzorusso T, Ratto GM, Putignano E, Maffei L (2000) Brain-derived neurotrophic factor causes cAMP response element-binding protein phosphorylation in absence of calcium increases in slices and cultured neurons from rat visual cortex, J. Neurosci. 20:2809-2816.

Poo MM (2001), Neurotrophins as synaptic modulators, Nat. Rev. Neurosci. 2:24-32.

Pozzo-Miller LD, Connor JA, Andrews SB (2000) Microheterogeneity of calcium signalling in dendrites, J. Physiol. 525:53-61.

Pozzo-Miller LD, Gottschalk W, Zhang L, Mcdermott K, Du J, Gopalakrishnan R, Oho C, Sheng ZH, Lu B (1999a) Impairments in high-frequency transmission, synaptic vesicle docking, and synaptic protein distribution in the hippocampus of BDNF knockout mice, J. Neurosci. 19:4972-4983.

Pozzo-Miller LD, Inoue T, Murphy DD (1999b) Estradiol increases spine density and NMDA-dependent Ca^{2+} transients in spines of CA1 pyramidal neurons from hippocampal slices, J. Neurophysiol. 81:1404-1411.

Pozzo-Miller LD, Petrozzino JJ, Golarai G, Connor JA (1996) Ca^{2+} release from intracellular stores induced by afferent stimulation of CA3 pyramidal neurons in hippocampal slices, J. Neurophysiol. 76:554-562.

Pozzo-Miller LD, Petrozzino JJ, Mahanty NK, Connor JA (1993) Optical imaging of cytosolic calcium, electrophysiology, and ultrastructure in pyramidal neurons of organotypic slice cultures from rat hippocampus, Neuroimage 1:109-120.

Rae MG, Irving AJ (2004) Both mGluR1 and mGluR5 mediate Ca^{2+} release and inward currents in hippocampal CA1 pyramidal neurons, Neuropharmacology 46:1057-1069.

Rico B, Xu B, Reichardt LF (2002) TrkB receptor signaling is required for establishment of GABAergic synapses in the cerebellum, Nat. Neurosci. 5:225-233.

Righi M, Tongiorgi E, Cattaneo A (2000) Brain-derived neurotrophic factor (BDNF) induces dendritic targeting of BDNF and tyrosine kinase B mRNAs in hippocampal neurons through a phosphatidylinositol-3 kinase-dependent pathway, J. Neurosci. 20:3165-3174.

Rose CR, Konnerth A (2001) Stores not just for storage. intracellular calcium release and synaptic plasticity, Neuron 31:519-522.

Rosen LB, Ginty DD, Weber MJ, Greenberg ME (1994) Membrane depolarization and calcium influx stimulate MEK and MAP kinase via activation of Ras, Neuron 12:1207-1221.

Sakai N, Yamada M, Numakawa T, Ogura A, Hatanaka H (1997) BDNF potentiates spontaneous Ca^{2+} oscillations in cultured hippocampal neurons, Brain Res. 778:318-328.

Schikorski T, Stevens CF (1997) Quantitative ultrastructural analysis of hippocampal excitatory synapses, J. Neurosci. 17:5858-5867.

Schmidt-Kastner R, Wetmore C, Olson L (1996) Comparative study of brain-derived neurotrophic factor messenger RNA and protein at the cellular level suggests multiple roles in hippocampus, striatum and cortex, Neuroscience 74:161-183.

Segal RA, Greenberg ME (1996) Intracellular signaling pathways activated by neurotrophic factors, Annu. Rev. Neurosci. 19:463-489.

Shirasaki T, Harata N, Akaike N (1994) Metabotropic glutamate response in acutely dissociated hippocampal CA1 pyramidal neurones of the rat, J. Physiol. 475:439-453.

Spruston N, Schiller Y, Stuart G, Sakmann B (1995) Activity-dependent action potential invasion and calcium influx into hippocampal CA1 dendrites, Science 268:297-300.

Stoop R, Poo MM (1996) Synaptic modulation by neurotrophic factors: differential and synergistic effects of brain-derived neurotrophic factor and ciliary neurotrophic factor, J. Neurosci. 16:3256-3264.

Suen PC, Wu K, Levine ES, Mount HT, Xu JL, Lin SY, Black IB (1997) Brain-derived neurotrophic factor rapidly enhances phosphorylation of the postsynaptic N-methyl-D-aspartate receptor subunit 1, Proc. Natl. Acad. Sci. USA 94:8191-8195.

Svoboda K, Mainen ZF (1999) Synaptic $[Ca^{2+}]$: intracellular stores spill their guts, Neuron 22:427-430.

Tanaka T, Saito H, Matsuki N (1997) Inhibition of $GABA_A$ synaptic responses by brain-derived neurotrophic factor (BDNF) in rat hippocampus, J. Neurosci. 17:2959-2966.

Tartaglia N, Du J, Tyler WJ, Neale E, Pozzo-Miller L, Lu B (2001) Protein synthesis-dependent and -independent regulation of hippocampal synapses by brain-derived neurotrophic factor, J. Biol. Chem. 276:37585-37593.

Thoenen H (1995) Neurotrophins and neuronal plasticity, Science 270:593-598.

Tsien RW, Tsien RY (1990) Calcium channels, stores, and oscillations, Annu. Rev. Cell Biol. 6:715-760.

Tyler WJ, Alonso M, Bramham CR, Pozzo-Miller LD (2002a) From acquisition to consolidation: on the role of brain-derived neurotrophic factor signaling in hippocampal-dependent learning, Learn. Mem. 9:224-237.

Tyler WJ, Perrett SP, Pozzo-Miller LD (2002b) The role of neurotrophins in neurotransmitter release, Neuroscientist 8:524-531.

Tyler WJ, Pozzo-Miller LD (2001) BDNF enhances quantal neurotransmitter release and increases the number of docked vesicles at the active zones of hippocampal excitatory synapses, J. Neurosci. 21:4249-4258.

Valenti O, Conn PJ, Marino MJ (2002) Distinct physiological roles of the Gq-coupled metabotropic glutamate receptors Co-expressed in the same neuronal populations, J. Cell Physiol. 191:125-137.

Vetter ML, Martin-Zanca D, Parada LF, Bishop JM, Kaplan DR (1991) Nerve growth factor rapidly stimulates tyrosine phosphorylation of phospholipase C-γ 1 by a kinase activity associated with the product of the trk protooncogene, Proc. Natl. Acad. Sci. USA 88:5650-5654.

Westenbroek RE, Ahlijanian MK, Catterall WA (1990) Clustering of L-type Ca^{2+} channels at the base of major dendrites in hippocampal pyramidal neurons, Nature 347:281-284.

Widmer HR, Kaplan DR, Rabin SJ, Beck KD, Hefti F, Knusel B (1993) Rapid phosphorylation of phospholipase C gamma 1 by brain-derived neurotrophic factor and neurotrophin-3 in cultures of embryonic rat cortical neurons, J. Neurochem. 60:2111-2123.

Widmer HR, Knusel B, Hefti F (1992) Stimulation of phosphatidylinositol hydrolysis by brain-derived neurotrophic factor and neurotrophin-3 in rat cerebral cortical neurons developing in culture, J. Neurochem. 59:2113-2124.

Yan Q, Radeke MJ, Matheson CR, Talvenheimo J, Welcher AA, Feinstein SC (1997a) Immunocytochemical localization of TrkB in the central nervous system of the adult rat, J. Comp. Neurol. 378:135-157.
Yan Q, Rosenfeld RD, Matheson CR, Hawkins N, Lopez OT, Bennett L, Welcher AA (1997b) Expression of brain-derived neurotrophic factor protein in the adult rat central nervous system, Neuroscience 78:431-448.
Yeckel MF, Kapur A, Johnston D (1999) Multiple forms of LTP in hippocampal CA3 neurons use a common postsynaptic mechanism, Nat. Neurosci. 2:625-633.
Yoo AS, Cheng I, Chung S, Grenfell TZ, Lee H, Pack-Chung E, Handler M, Shen J, Xia W, Tesco G, Saunders AJ, Ding K, Frosch MP, Tanzi RE, Kim TW (2000) Presenilin-mediated modulation of capacitative calcium entry, Neuron 27:561-572.
Yuste R, Majewska A, Cash SS, Denk W (1999) Mechanisms of calcium influx into hippocampal spines: heterogeneity among spines, coincidence detection by NMDA receptors, and optical quantal analysis, J. Neurosci. 19:1976-1987.
Zirrgiebel U, Ohga Y, Carter B, Berninger B, Inagaki N, Thoenen H, Lindholm D (1995) Characterization of TrkB receptor-mediated signaling pathways in rat cerebellar granule neurons: involvement of protein kinase C in neuronal survival, J. Neurochem. 65:2241-2250.
Zucker RS (1996) Exocytosis: a molecular and physiological perspective, Neuron 17:1049-1055.
Zucker RS (1999) Calcium- and activity-dependent synaptic plasticity, Curr. Opin. Neurobiol. 9:305-313.

BRAIN-DERIVED NEUROTROPHIC FACTOR (BDNF) AND THE DENTATE GYRUS MOSSY FIBERS: IMPLICATIONS FOR EPILEPSY

Helen E. Scharfman*

1. INTRODUCTION

In the last decade, one of the most avidly studied compounds in the central nervous system (CNS) has been the neurotrophin BDNF. Although historically it had been studied in the context of development, where it plays numerous critical roles, more recent studies have shown striking actions on a number of pathways and processes that are critical for normal function of the *adult* CNS. In addition, studies of neurological and psychiatric diseases indicate a potential role for BDNF in pathology.

Our interest in BDNF began with the demonstration that one of the regions where BDNF protein expression is among the highest in the adult rat brain is the dentate gyrus (Conner *et al.*, 1997; Yan *et al.*, 1997). Previous studies had not clearly defined where BDNF protein was expressed, although mRNA was definitely in hippocampal neurons, and many other adult brain regions. One of the interesting aspects of BDNF protein expression in the dentate gyrus that came to light after the development of specific antibodies was that its location was extremely specific. Thus, in the normal adult rat, hippocampal BDNF protein expression is much greater in the axons of the dentate granule cells, the mossy fibers, than anywhere else in hippocampus. This led us to study mossy fiber BDNF specifically, and the results showed that mossy fiber BDNF has potent functional effects. These studies, and their implications for epilepsy, are reviewed below.

---

* Helen E. Scharfman, Center for Neural Recovery and Rehabilitation Research (CNRRR), Helen Hayes Hospital, New York State Department of Health, West Haverstraw, NY 10993, and Departments of Pharmacology and Neurology, Columbia University.

2. FUNDAMENTAL STUDIES OF BDNF IN THE CNS

Historically, BDNF has been studied as one of the critical mediators of CNS development. This is no surprise given it is one of the family of neurotrophins, which include nerve growth factor (NGF), BDNF, as well as other potent neurotrophic molecules such as neurotrophin-3 (NT-3) and neurotrophin-4/5 (NT-4/5; Figure 1). All neurotrophins have potent actions at tropomyosin receptor kinases (trk), as shown in Figure 1. BDNF binds with high specificity to trkB, which is also a ligand for NT-4/5, and possibly NT-3 (Figure 1). Neurotrophins also bind to what was originally termed the low affinity neurotrophin receptor (LNTR), now referred to as p75. Both trkB and p75 are coupled to a complex array of signal transduction pathways and mediate numerous actions in different systems (Bibel and Barde, 2000; Vicario-Abejon *et al.*, 2002; Chao, 2003). More recently, it has become clear that proneurotrophins may be functionally active, as well as the mature peptide (Brake *et al.*, 2001).

Our interest in this field began with the studies of (Lohof *et al.*, 1993), who showed that BDNF exposure to xenopus cultures could increase synaptic transmission. Subsequently it became clear that this was a principle that could be generalized well beyond xenopus. Relevance to the hippocampus became evident when it was shown that exposure of adult rat hippocampal slices to recombinant BDNF led to a long lasting potentiation of synaptic transmission in area CA1 (Kang and Schuman, 1995). Further studies at a similar time also showed that BDNF contributes to a long-lasting potentiation, similar to long-term potentiation (Patterson *et al.*, 1996; Korte *et al.*, 1996), and it is now well accepted that BDNF has an important role in synaptic plasticity in area CA1.

Our interest to enter this field arose from the demonstration that BDNF protein is expressed much more strongly in the axons of granule cells, the mossy fibers, than in area CA1. Although it is not always the case that brain regions with relatively greater amounts of protein are associated with enhanced function, this relative difference piqued our interest in the potential functional effects of BDNF in the dentate gyrus and CA3 regions. An additional reason for this interest was the known role of the mossy fibers in the hippocampal trisynaptic circuit. The trisynaptic circuit refers to a sequence of glutamatergic synapses that mediates much of the signal processing from principal cells of the entorhinal cortex to the granule cells, pyramidal cells, and back to cortex. As a key

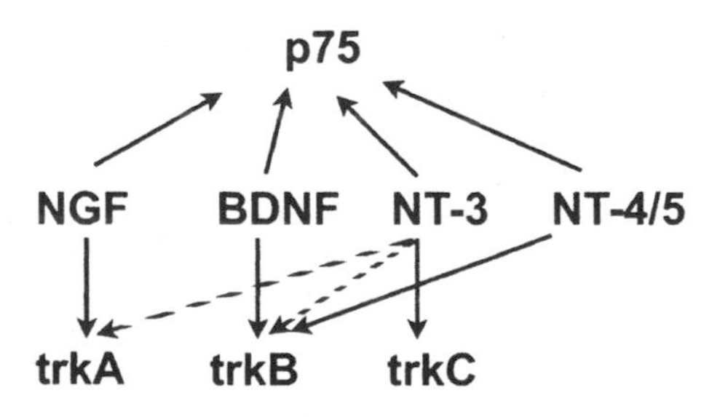

Figure 1. The neurotrophin family and its receptors. Nerve growth factor, NGF; Brain-derived neurotrophic factor, BDNF; Neurotrophin-3, NT-3; Neurotrophin-4/5, NT-4/5.

intermediary synapse in this circuit, i.e., the synapse from granule cells to CA3 pyramidal cells, it is likely that the relative strength of mossy fiber transmission plays an important role in information processing between cortex and hippocampus. Indeed, there are several control points for the trisynaptic circuit, and although much attention has focused on perforant path - to - granule cell synapse as a primary "gate" for entry into hippocampus, the mossy fibers may have an accessory gate function. One reason for this is that recent studies of the mossy fibers have illustrated that these axons primarily inhibit CA3 neurons because they activate GABAergic neurons so well, and the GABAergic neurons innervate the CA3 pyramidal cells (Acsady *et al.*, 1998). Yet the underlying excitation by glutamate released from mossy fibers directly onto pyramidal cells, which GABAergic inhibition normally masks, is extremely powerful; furthermore, it facilitates more strongly than many other hippocampal synapses (Scharfman *et al.*, 1990; von Kitzing *et al.*, 1994). It therefore is likely that normal inhibition of mossy fiber glutamatergic transmission plays a critical role in the control of information flow through the hippocampus.

3. EFFECTS OF BDNF ON MOSSY FIBER TRANSMISSION IN NORMAL ADULT RATS

Our first study was simply to examine the mossy fiber synapses and determine if exogenous BDNF application in adult rat hippocampal slices could influence this pathway. We also questioned whether BDNF might influence other aspects of granule cell function, perhaps by having a retrograde effect, given that there are many examples in the literature of retrograde effects of neurotrophins.

We found striking effects of BDNF on mossy fiber transmission, similar to the studies of others in area CA1. Thus, exposure to 50-250 ng/ml recombinant BDNF (generously provided by Regeneron Pharmaceuticals) for over 30 min led to a potentiation of the extracellularly-recorded response of pyramidal cells to an electrical stimulus to the mossy fibers, i.e. a potentiation of the amplitude of the population spike (Figure 2). Importantly, the fiber volley (an index of mossy fiber action potentials) was unchanged, indicating non-specific excitatory effects could not explain the result. In addition, the effect maximized at approximately 60 min and outlasted BDNF exposure. In fact, we could not reverse the potentiation, even if drug-free buffer was continuously applied for several hours after BDNF had been removed. These robust effects were supported by control experiments showing that there was no effect of vehicle (bovine serum albumin, diluted to the same concentration as was used for BDNF application), or cytochrome C, which has similar physical chemistry as BDNF but does not bind trk receptors. Similarly, there was no effect of heat-inactivated BDNF (Scharfman, 1997). The trk antagonist K252a could block the effect, but the NMDA receptor antagonist D-APV did not. The latter was consistent with the known characteristics of normal mossy fiber transmission, and mossy fiber LTP, which have a very small NMDA receptor-mediated component.

Interestingly, other inputs to the CA3 neurons were not influenced by exposure to BDNF. Thus, stimuli to the fimbria or Schaffer collaterals evoked responses that were unchanged in the same slices with potentiated mossy fiber responses. This was important because it suggested that in fact we were able to stimulate mossy fibers selectively by electrical activation of the deep hilus, which is not trivial. Taken together, these studies

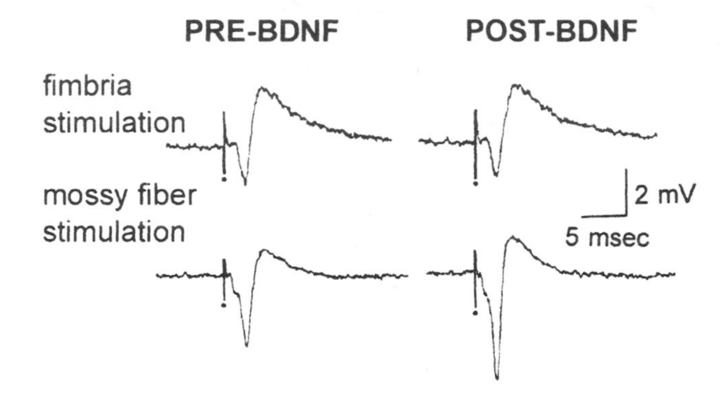

Figure 2. Selective potentiation of mossy fiber-evoked, but not fimbria-evoked, responses of CA3 neurons by BDNF. Simultaneous extracellular recordings in the CA3 cell layer to single stimuli applied to the fimbria or mossy fibers, before and after exposure to BDNF in a slice from an adult male rat (Scharfman, 1997).

suggested that BDNF exposure could lead to a long-lasting and specific potentiation of mossy fiber transmission, a homosynaptic mossy fiber - "LTP - like" effect. The fact that other inputs appeared unaffected suggested a presynaptic mechanism, or at least a mechanism highly restricted to the mossy fiber synapses. Given the large literature that has now accumulated to suggest that BDNF has presynaptic effects (Tyler *et al.*, 2002), and that trkB receptors have been shown on mossy fiber boutons (Drake *et al.*, 1999), we currently favor the hypothesis that BDNF's actions are presynaptic. We hypothesize that BDNF binds to presynaptic trkB receptors, leading to a change in release mechanisms. Indeed, there are data to suggest that BDNF acts to phosphorylate proteins critical to transmitter release (Jovanic *et al.*, 1996; Tartaglia *et al.*, 2001). However, postsynaptic effects of BDNF have been documented in several systems (Black, 1999; Kovalchuk *et al.*, 2002; Tongiorgi *et al.*, 2004), and trkB receptors have been identified postsynaptically in CA3 (Drake *et al.*, 1999). Potential actions on GABAergic neurons and glia also can not be overlooked (Schinder *et al.*, 2000; Rose *et al.*, 2003). Therefore, we continued to examine effects of BDNF in the dentate gyrus and CA3 and indeed are still doing so. Some of these experiments, particularly those conducted towards defining the pre- vs. postsynaptic locus of BDNF's effects, are described below.

It is important to raise the point that the majority of slices, but not all slices, demonstrated effects of exogenous BDNF. This has been an issue that has been noticed by our colleagues also. In fact, some laboratories have found that BDNF exposure to area CA1 does not potentiate synaptic transmission. In our own work, we found no effects of BDNF obtained from some manufacturers, or if BDNF had been frozen and then thawed. Other factors may also contribute to variability, such as the fact that some slices might not preserve all constituents of the signaling cascade initiated by trkB. There is also a significant difficulty in perfusion of tissue by BDNF because it is a large molecule that does not readily diffuse through neuropil. Depending on the preparation, differential access of BDNF to its receptors may lead to variable effects. Another possibility was raised recently by a study which showed that BDNF levels can be

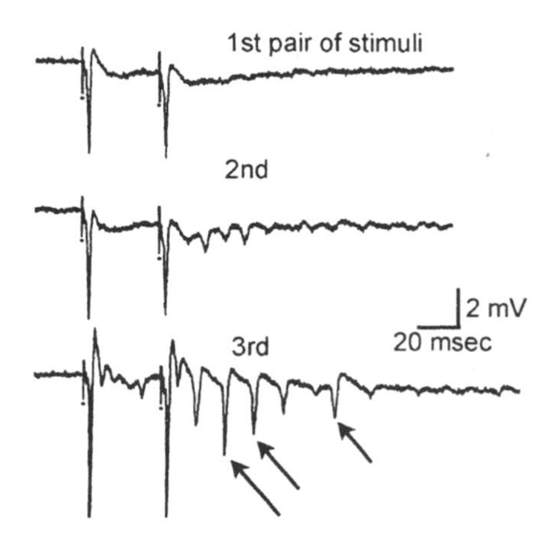

Figure 3. BDNF exposure leads to hyperexcitability in area CA3 upon mossy fiber stimulation. Extracellular recordings from a slice of an adult male rat demonstrate multiple population spikes (defined here as hyperexcitability; arrows) in response to 3 pairs of 1 Hz stimuli to the mossy fibers in the presence of BDNF are shown (Scharfman, 1997).

increased simply by slicing (Danzer *et al.*, 2004); if so, effects of exogenous BDNF thereafter might be occluded, or at the very least altered.

Besides examination of BDNF's influence on responses to single stimuli to the mossy fibers, we tested the effects of 1 Hz stimuli. The paradigm we developed was paired stimuli (40 msec apart, using 50% of the maximum population spike amplitude) at 1 Hz for 5-10 sec (total of 10-20 stimuli). Surprisingly, even after as few as 2-3 pairs of stimuli, BDNF-exposed slices demonstrated multiple population spikes (Figure 3). If stimulation continued at 1 Hz for the full 10 sec period, spreading depression often occurred. Yet even 20 pairs of stimuli at maximum intensity using other inputs to CA3, in the same slices, could not elicit multiple population spikes or spreading depression (Scharfman, 1997). This striking effect of 1 Hz paired stimuli has continued to be useful as a benchmark of BDNF's effects across numerous experimental preparations (see below). It appears to be unique among compounds that induce hyperexcitability, because no other convulsant we have tested has had the same type of effect. For example, we have tried to use very low (0.1 up to 10 μM) concentrations of the $GABA_A$ receptor antagonist bicuculline methiodide, because it is well known that $GABA_A$ receptor antagonism leads to multiple population spikes in response to stimulation in hippocampus. However, bicuculline never mimicked the effects of BDNF at any time during its administration or at the time it reached its final concentration. The effect of low doses of bicuculline (also observed in the initial minutes after adding a higher dose) was to elicit 2-3 population spikes per single stimulus. These occurred immediately after the stimulus, and were independent of stimulus site. If concentration was raised, spontaneous bursts developed. In contrast, BDNF exposure did not lead to multiple population spikes in response to one stimulus. Instead, the paired 1 Hz paradigm was required, and in that case the population spikes did not immediately follow the stimulus.

Instead, slower, longer duration trains of population spikes were evoked. Furthermore, they only were evoked in response to mossy fiber stimuli, and spontaneous bursts never developed. Thus, although hyperexcitability can be invoked by a number of convulsants using the slice preparation, or a number of manipulations of the perfusing buffer, BDNF appears to induce a qualitatively unique pattern of hyperexcitability.

The fact that hyperexcitability was present in area CA3 after exposure to BDNF was surprising to us, because it had been widely considered that BDNF was neuroprotective, critical to normal development, and important to learning and memory (Cirulli *et al.*, 2004; Koponen *et al.*, 2004; Monteggia *et al.*, 2004). However, upon further examination, there have been studies of CA3 neuronal damage after seizures that indicate that BDNF is exactly the opposite, i.e., it can exacerbate damage (Rudge *et al.*, 1998; Lahteinen *et al.*, 2003). It may be that a low level of BDNF is necessary and important to normal function, but excess BDNF (as would occur after administration of exogenous BDNF, or manipulations that induce its expression) leads to precisely the opposite, that BDNF in excess might not necessarily a "good thing" (Binder *et al.*, 2001). Indeed, BDNF overexpressing mice had cognitive deficits and deficits in LTP, as well as increased susceptibility to seizures (Croll *et al.*, 1999). Indeed, others have identified that BDNF has actions that were consistent with a pro-epileptogenic effect (Binder *et al.*, 1999), although others have not always concluded the same (Larmet *et al.*, 1995; Osehobo *et al.*, 1999). Whether or not BDNF is proconvulsant may be dependent on how BDNF is manipulated and the ultimate concentration, as well as other factors, such as compensatory changes. Indeed, prolonged BDNF exposure may downregulate trkB receptors (Xu *et al.*, 2004) and increase the expression of neuropeptide Y, which is anticonvulsant (Reibel *et al.*, 2000). In summary, it may be that hippocampal BDNF is critical to development, and the maintenance of dendritic and synaptic plasticity in the normal adult under most conditions. However, under abnormal conditions, it may be maladaptive for the hippocampus to have such a high concentration and contribute to hippocampal seizures susceptibility (see Figure 8).

To fully examine where BDNF might have effects in the dentate gyrus, we expanded our tests of synaptic and nonsynaptic function outside CA3. Thus, we examined antidromic transmission to granule cells. However, no effects on the antidromic population spike were ever detected, suggesting that axon conduction was an unlikely target. In addition, we tested orthodromic inputs to granule cells. Interestingly, extracellularly-evoked responses to stimulation of the outer molecular layer, to activate the perforant path input to the dentate granule cells, were unaffected by exogenous BDNF, using the same concentrations that had a robust influence on mossy fiber transmission (Scharfman, 1997). Furthermore, paired pulse inhibition of the population spike was unaffected, suggesting that BDNF did not act by depression of inhibition (Scharfman, 1997). However, other studies *in vivo* (Messaoudi *et al.*, 1998) subsequently identified that BDNF can potentiate the perforant path input. Furthermore, studies of granule cell inhibition in heterozygous knockouts suggests a role of BDNF at dentate gyrus inhibitory synapses (Olofsdotter *et al.*, 2000). This is unlikely to be due to alterations in GABAergic inhibition (due to developmental deficiency in the knockouts), because a scavenger of BDNF had similar effects. Others have identified that BDNF can depress IPSPs in area CA1 (Tanaka *et al.*, 1997; Frerking *et al.*, 1998), although the mechanism in the dentate gyrus appeared to be presynaptic, and in CA1 this was not necessarily the case. Additional effects of BDNF have also been reported that further complicate analysis of its actions: there appears to be a very rapid effect of BDNF to depolarize neurons by an action on a specific subtype of sodium channel, Nav 1.9 (Kafitz *et al.*, 1999; Blum *et al.*, 2002).

These differences in effects of BDNF could be due to the different methods that were used, different sources/concentrations of BDNF (possibly due to contaminants of proneurotrophins, binding proteins, etc.), and also that some synapses/regions could simply be influenced differentially. They raise the important point that BDNF's effects may be highly sensitive to the preparation of the neurotrophin, experimental setting, and conditions of an experiment. They also could be explained by the fact that many synapses can be influenced by BDNF, but mossy fiber transmission might simply be one of the most important given protein expression there is highest. This leads naturally to a key question: are the robust effects on mossy fiber transmission reflective of physiological actions of BDNF? What is the physiological effect, in contrast to the pharmacological effect, of BDNF? In other words, what is the effect of endogenous BDNF?

4. BDNF OVEREXPRESSING MICE

To address this issue, we chose to take an approach that would not rely on exogenous application of recombinant BDNF at all, but to find a way to probe actions of BDNF that might occur *in vivo*. We were fortunate to have the opportunity to work with a transgenic mouse line that overexpressed BDNF, provided by Regeneron Pharmaceuticals. This mouse overexpresses BDNF by approximately 30%, and overexpression occurs in the areas of the CNS where the β-actin promoter is present (Croll *et al.*, 1999). Thus, in a blinded study, we examined mossy fiber transmission in slices of transgenics and wild type controls. Exogenous BDNF was not added at all. Our hypothesis was that, without the addition of BDNF, we would be able to examine any endogenous effects of BDNF simply by comparison to the wild type. A critical assumption was that transgenic overexpression would not lead to abnormal effects that would be confounding, but reflect effects of endogenous BDNF. Therefore, an important first step was an analysis of gross morphology and hippocampal structure, and no apparent abnormalities were present in the animals. Indeed, the animals appeared normal, were viable, and were fertile, although there of course could have been fine changes that more specific markers could have potentially detected.

We found that normal mossy fiber transmission was robust in both the transgenic and wild type mouse in response to a single stimulus. Although a comparison of population spike amplitude evoked by selective mossy fiber stimulation is difficult across slices, extensive differences in the amplitude of the population spike that could be evoked using standardized electrode locations were not noticed. However, there was a large difference when we used the paradigm described above that involved repeated stimuli. In the transgenics, we found an abnormality of mossy fiber transmission that was striking in its similarity to the rat slices that were exposed to exogenous BDNF: multiple population spikes and spreading depression could be evoked by pairs of mossy fiber stimuli at 1 Hz (Figure 4). This was not observed in response to other inputs to CA3, and no differences were detected in dentate gyrus recordings of the response to perforant path input, all reminiscent of the effects of exogenous BDNF in normal adult male rat slices.

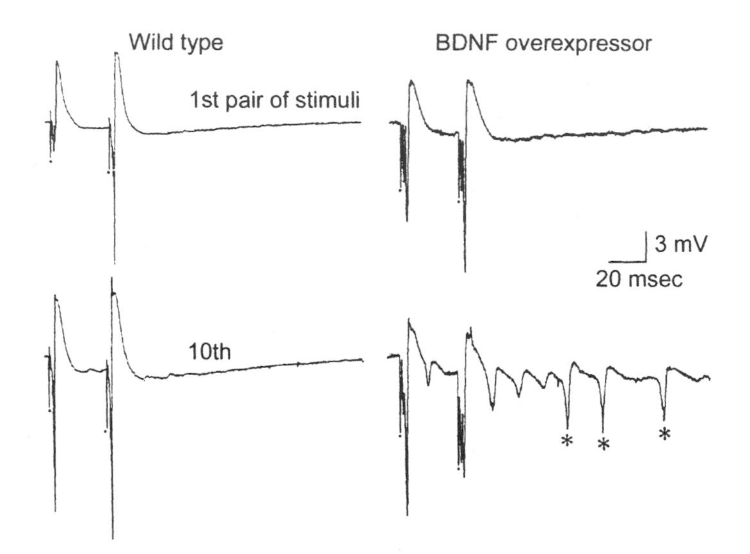

Figure 4. Hyperexcitability after mossy fiber stimuli in BDNF overexpressing mice. Extracellular recordings from the area CA3 cell layer in response to mossy fiber stimuli (paired stimuli at 1 Hz) are shown for a slice from an adult male wild type mouse and a slice from an adult male mouse that overexpressed BDNF. Asterisks mark abnormal population spikes after the 10th pair of stimuli in the slice from the transgenic mouse (Croll et al., 1999).

Interestingly, entorhinal cortex also appeared hyperexcitable in slices of transgenic mice (Croll *et al.*, 1999), as it did in slices of normal male rats exposed to BDNF (Scharfman, 1997). The effects in entorhinal cortex may actually be more robust, in a way, because multiple field potentials could be evoked even by single stimuli (Scharfman, 1997; Croll *et al.,* 1999). This is intriguing given that these areas (CA3 and the entorhinal cortex) are those that exhibit vulnerability in animal models of epilepsy and human temporal lobe epilepsy (discussed further below). It is also intriguing because there are numerous studies which have demonstrated that seizures increase BDNF (Gall, 1993) and in patients with temporal lobe epilepsy there is an elevation in BDNF (Mathern *et al.*, 1997; Murray *et al.*, 2000). Thus, BDNF may initiate a positive feedback loop that leads to its upregulation and further ability to potentiate synaptic transmission (Binder *et al.*, 2001). This could be a substrate for epileptogenesis, and a novel one at that, given that prevailing hypotheses have only just begun to consider BDNF in epileptogenesis (Lahteinen *et al.*, 2002; He *et al.*, 2004; Tongiorgi *et al.*, 2004).

5. EFFECTS OF BDNF IN AN ANIMAL MODEL OF EPILSPY WITH MOSSY FIBER SPROUTING

As mentioned above, we have continued to pursue how BDNF acts in CA3 and one of the key issues is whether it acts on mossy fiber boutons. Also mentioned above, was

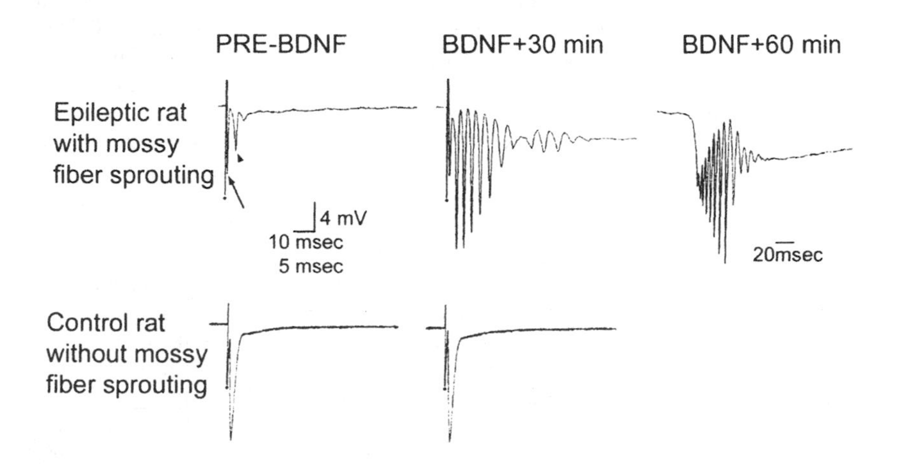

Figure 5. Effects of BDNF in an animal with chronic seizures (i.e., epileptic) and mossy fiber sprouting. Stimulation of the hilus to activate mossy fibers in an epileptic rat with mossy fiber sprouting (Top) evoked an antidromic population spike (arrow) followed by an orthodromic population spike (arrowhead). After BDNF was added to the buffer, the same stimulus evoked multiple population spikes (Top, center) and subsequently spontaneous population spike occur in bursts (Top, right). In contrast, only an antidromic spike was evoked by hilar stimulation in a slice from a control rat without sprouting (Bottom), even after supramaximal stimuli, and even after prolonged exposure to BDNF. No spontaneous activity occurred. Calibration for recordings from epileptic tissue, 10 msec; for control recordings, 5 msec (Scharfman *et al.*, 1999).

our interest in the possible relevance of mossy fiber BDNF to epileptiform activity. To address these "two birds" with one stone, so to speak, we chose to examine effects of BDNF after an experimental manipulation that induces mossy fibers to form synapses in a new location, the inner molecular layer. We reasoned that if there was an effect of BDNF that was specific to the mossy fiber boutons, it would be present in the new location. If it were not, we would not be able to conclude much, but if it were, we would have a compelling reason to associate BDNF actions with mossy fiber boutons. The reason why this approach harkened the adage about killing two birds with one stone was due to the fact that the experimental manipulation was one that led to an epileptic state. Thus, the same approach would allow us to potentially test the hypothesis that in the epileptic brain, BDNF may trigger seizures.

The experimental manipulation that we used is one with a long history: the muscarinic agonist pilocarpine is administered at a dose that leads to status epilepticus (continuous severe seizures; Turski *et al.*, 1983). Our laboratory uses the anticonvulsant diazepam after 1 hr of status to truncate status (Scharfman *et al.*, 2000, 2001, 2002b), because if left undisturbed, status may continue for many hours and animals have severe hippocampal damage. However, if truncated, there is much less damage in the dentate gyrus and CA3 region. In fact, many of the hilar mossy cells, considered to be one of the most vulnerable cell types to seizure-induced neuronal damage, survive in our animals (Scharfman *et al.*, 2001). Nevertheless, there is some hilar cell loss, and these animals do demonstrate two consequences of status epilepticus that are associated with seizure-induced neuronal damage: 1) chronic spontaneous seizures develop and last the lifetime of the rat (i.e., it develops epilepsy), and 2) the mossy fibers develop new collaterals that innervate an abnormal target region, the inner molecular layer (a phenomenon dubbed "mossy fiber sprouting"). The new mossy fibers in the inner molecular layer form synapses on both granule cells (Wenzel *et al.*, 2000; Buckmaster *et al.*, 2002) and interneurons (Sloviter, 1992; Kotti *et al.*, 1997). Synapses on granule cells are excitatory, but appear to have low probability of release (Scharfman *et al.*, 2003; Nadler, 2003). The net effect of mossy fiber sprouting on excitability is still a matter of controversy.

Using animals that had pilocarpine-induced status and subsequent repetitive spontaneous seizures, we made hippocampal slices and asked whether BDNF would 1) influence transmission from mossy fibers to granule cells, and 2) trigger spontaneous epileptiform events reflective of seizures (slices of course do not have seizures, but spontaneous epileptiform discharges recorded from slices, particularly long-lasting discharges, are often used to predict conditions that might relate to seizure activity *in vivo*). Indeed, this is what we found (Figure 5; Scharfman *et al.*, 1999). In animals with sprouting, hilar stimulation evoked an antidromic population spike in the granule cell layer that was followed by a second, orthodromic population spike (Figure 5). The second population spike was absent in animals without sprouting, or animals that had pilocarpine administered but failed to have status epilepticus (Figure 5). Presumably this second population spike reflects activation of recurrent collaterals on granule cells (Tauck and Nadler, 1985; Lynch and Sutula, 2000; Nadler, 2003). After BDNF exposure, the same stimulus that elicited one antidromic and one orthodromic population spike evoked an antidromic population spike followed by multiple, large amplitude, orthodromic population spikes (Figure 5). Indeed, after approximately 60 minutes, spontaneous bursts of population spikes were recorded from the granule cell layer. Intracellular recordings showed that these bursts reflected discharges of granule cells (Scharfman *et al.*, 1999), which was important to verify because volume transmission could theoretically have made it possible to record pyramidal cell discharges from the granule cell layer (Scharfman and Schwartzkroin, 1990). Other recordings indicated that the EPSPs underlying these bursts were generated from the inner molecular layer (Scharfman *et al.*, 1999). In extreme ventral slices, burst discharges were long lasting and particularly striking in the ventral (inferior) blade, where there appears to be greater sprouting and more BDNF (Scharfman *et al.*, 2002b).

6. EFFECTS OF BDNF INFUSION INTO THE DENTATE GYRUS: SEIZURES

These studies suggested that BDNF might be proconvulsant, particularly in the epileptic brain, but possibly in the normal brain as well. Indeed, sporadic spontaneous seizures had been witnessed in a few of the transgenic mice overexpressing BDNF. Perhaps more BDNF, applied directly to the mossy fiber area, would demonstrate more robust seizures. To test that hypothesis, we infused BDNF directly into the hilus using osmotic pumps. Interestingly, this did induce limbic seizures in approximately 1/3 of animals, witnessed by blinded investigators (Scharfman *et al.*, 2002a). They were not detected in animals infused with another large protein, albumin, at a similar concentration and infusion rate (Scharfman *et al.*, 2002a). We think that more BDNF-infused animals actually had seizures than the 1/3 that were witnessed, because animals were not examined 24 hr/day, so seizures could have been missed. In addition, we perfused the animals that had not been observed to have a seizure, and the hippocampi demonstrated pathology similar to a hippocampus with chronic spontaneous seizures (Scharfman *et al.*, 2002a). However, control animals did not exhibit abnormalities. One tool used to examine whether animals had a history of seizures was increased mossy fiber expression of neuropeptide Y, which develops in animal models of chronic limbic seizures (Sperk *et al.*, 1996). However, mossy fiber sprouting was minimal or not present, suggesting that severe seizures and numerous repetitive seizures had not necessarily occurred. This might be due to the fact that chronic infusion downregulated trkB receptors (as mentioned above), or simply that BDNF is not as potent a convulsant as other drugs that induce status and lead to an epileptic phenotype.

The studies described above demonstrated that BDNF had a robust effect on mossy fiber transmission, probably by an action of trkB receptors on mossy fiber boutons. In both the normal and epileptic brain, it appears that this can bias the network to a hyperexcitable state. And in fact, seizures were triggered in the normal rat *in vivo* by BDNF, and seizure-like events were triggered by exogenous BDNF in slices of epileptic rats. But a key issue was the use of exogenous BDNF to trigger the seizures. This led us to question whether endogenous BDNF would be sufficient to trigger seizures. If this was true, a much stronger argument could be made for a role of BDNF in epilepsy.

It should be noted that recently two laboratories have provided evidence using trkB transgenics or conditional knockouts that endogenous BDNF is quite likely to play a role in epileptogenesis. Thus, overexpression of the truncated form of trkB altered epileptogenesis and the seizures that ultimately resulted (Lahteinen *et al.*, 2002). In addition, conditional deletion of hippocampal trkB blocked epileptogenesis using the kindling model (He *et al.*, 2004). Moreover, microarray analyses of epileptic tissue have shown that the BDNF gene is commonly represented (Lahteinen *et al.*, 2004). However, microarrays have shown that many genes besides BDNF are induced during epileptogenesis (Lahteinen *et al.*, 2004). Furthermore, overexpression of the full-length form of trkB did not necessarily alter epileptogenesis (although it did modify the acute response to convulsants; Lahteinen *et al.*, 2003), and conditional deletion of BDNF did not modify kindling (He *et al.*, 2004). The latter result can be explained by the potential for alternate neurotrophins to bind to trkB if BDNF is not present (He *et al.*, 2004). In summary, it is clear that much still needs to be learned about the relationship between BDNF, epileptogenesis, and epilepsy.

7. THE INDUCTION OF BDNF BY ESTROGEN AND ITS CONSEQUENCES

Although there are many ways BDNF expression is regulated, one that came to our attention that was particularly intriguing was the reproductive hormone estrogen. It had been identified that estrogen can induce BDNF expression by an estrogen-sensitive response element on the BDNF gene (Sohrabji *et al.*, 1995). We found this interesting because many women are known to have seizures at times of their menstrual cycle when estrogen transiently rises (i.e., at ovulation; Herzog *et al.*, 1997). Other women with limbic epilepsy clearly have seizures that are influenced by their levels of estradiol (Bauer, 2001). Thus, we hypothesized that perhaps the normal fluctuations in estrogen in females provided an example of an endogenous mechanism that can induce BDNF and could trigger increased seizures susceptibility. This might occur only in a fraction of individuals, of course, because all women do not have epilepsy. Therefore, we hypothesized that all women might demonstrate BDNF induction after elevated estradiol, but only in those with an additional predisposition (such as abnormally high levels of BDNF induction by estradiol) would seizures occur. Other factors, that in themselves do not cause seizures but in addition to another factor such as high estradiol/BDNF would, could be genetic. Indeed, polymorphisms in the BDNF gene are linked to epilepsy (Kanemoto *et al.*, 2003), although not yet explored is the link to women with epilepsy. Structural abnormalities such as cortical dysplasias might also be a predisposing factor. In and of themselves they might not be sufficient to trigger seizures, but together with abnormally high BDNF they might lead to seizures.

What is the evidence that estradiol induces BDNF? After the identification of an estrogen response element on the BDNF gene, a number of studies examined the potential changes in BDNF expression after estradiol treatment. These studies demonstrated that in ovariectomized rats treated with estradiol, BDNF mRNA increases (Sohrabji *et al.*, 1995; Singh *et al.*, 1995). However, the functional effect of increased mRNA was not clear. Given that studies up to this time were conducted mostly in ovariectomized tissue, a setting that at best can be extrapolated only to the postmenopausal state, we chose to examine the intact female hippocampus. We focused on BDNF protein and potential functional implications.

To address potential changes due to elevated estradiol in the intact rat, we chose a comparison that is common to endocrinologists interested in dissecting the influence of estradiol from other reproductive steroids: a comparison of female brain on the mornings of proestrus, estrus and metestrus. For this approach, the standard 4-day estrous cycle was first established by daily vaginal cytologic exam (hormone levels were confirmed later by RIA; Scharfman *et al.*, 2003). The relative changes in hormones during the mornings of each day of the 4-day estrous cycle can be simplified as follows: 1) the morning of proestrus, when estradiol rises and is followed by a rise in progesterone in the afternoon, 2) the morning of estrus, when both estradiol and progesterone levels have returned to baseline over the course of the previous 12 -18 hrs, 3) the morning of metestrus (also referred to as diestrus 1), when both estradiol and progesterone have been low for the last 24 hrs, and 4) the morning of diestrus 2, when hormone levels are also low. The morning of proestrus as compared to the morning of metestrus is ideal to dissect the effects of estradiol, because the rise in progesterone, and other related hormones (LH, FSH) that occur at other times are not confounding.

We examined the expression of BDNF in animals at the cycle stages indicated above, as well as after ovariectomy, and in male rats for comparative purposes. We also examined mossy fiber transmission in these experimental groups. A summary of our findings are shown in Figure 6 (see also Scharfman *et al.*, 2003). First, BDNF protein expression does appear to rise on the morning of proestrus, consistent with a role of estradiol. Second, BDNF expression remains elevated through the morning of estrus, and appears to decline by the morning of metestrus. The persistent elevation is intriguing in light of the fact that mossy fiber transmission was abnormal on the morning of proestrus and estrus as well (Figure 6). This might mean that both estradiol and progesterone play roles, but it also may simply mean that after induction of mRNA, BDNF protein is maintained for up to 24 hours, regardless of the rapid decline in estradiol and changes in other reproductive steroids.

Remarkably, the slices from proestrous and estrous females behaved as if they were male rats with exogenous BDNF administered, or male mice with overexpression of BDNF. Thus, paired stimuli at 1 Hz to the mossy fibers, but no other CA3 input, evoked multiple population spikes. These were reversed by the trk inhibitor K252a, but not D-

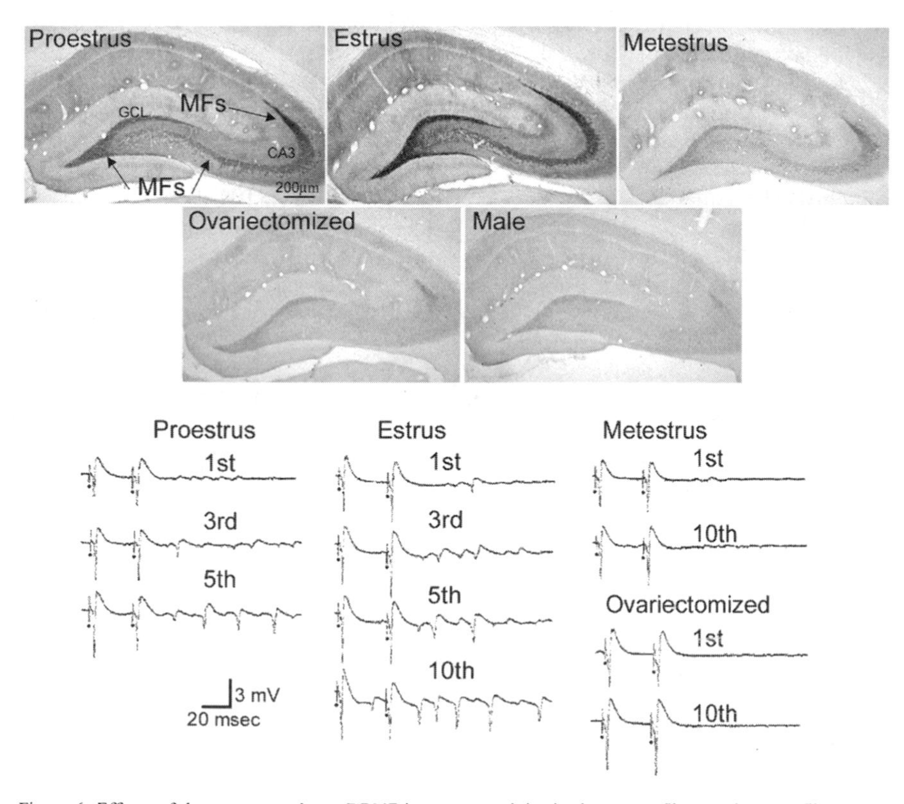

Figure 6. Effects of the estrous cycle on BDNF immunoreactivity in the mossy fibers and mossy fiber-evoked responses recorded extracellularly in the area CA3 cell layer. Top: Immunocytochemistry demonstrated an increase in mossy fiber BDNF on the mornings of proestrus and estrus relative to metestrus, ovariectomy, and BDNF expression in a male rat. Bottom: Mossy fiber-evoked responses recorded extracellularly in the area CA3 cell layer demonstrated that multiple population spikes could be evoked by pairs of mossy fiber stimuli at 1 Hz from slices prepared from rats on the morning of proestrus and estrus but not other cycle stages or after ovariectomy. Note that ovariectomized rats and male rats do express mossy fiber BDNF, but it is simply less than the proestrous or estrous rats; indeed, if the duration of incubation with reactants (DAB) is increased, mossy fiber BDNF in male rats is striking (Conner *et al.*, 1997). GCL= granule cell layer; MF = mossy fibers (Scharfman *et al.*, 2003).

APV (Scharfman *et al.*, 2003). Thus, BDNF protein, and physiological effects of BDNF, appear to fluctuate in concert across the estrous cycle of normal female rats, presumably stimulated by the proestrus surge in estradiol.

Although intriguing, the results raise several questions. One series of questions is how the results might be relevant to seizures in women with epilepsy. Although it will be a challenge to prove that studies in rats can be generalized to women, there are many possibilities. First, in women with hormone-sensitive seizures, is the hormone sensitivity due to induction of BDNF by natural fluctuations in estradiol? Second, at puberty, when epilepsy is often first diagnosed, might it be due to the induction of BDNF by estradiol? Third, does hormone replacement in postmenopausal women lead to increased BDNF in hippocampus and altered seizure susceptibility? Fourth, what is the role of BDNF in the male- might testosterone play a role analogous to estradiol?

Other questions relate to underlying mechanisms. As raised by Blurton-Jones et al. (2004), there is a puzzling mismatch between estrogen receptor localization and BDNF expressing cells. In their studies, which were notably in male rats, the point was made that classic estrogen receptors (ER), both ERα and ERβ, were not evident in granule cells (Blurton-Jones *et al.*, 2004), yet this is where mossy fiber BDNF is synthesized. Interestingly, ERα was present in granule cell endosomes in proestrous female rats (Milner *et al.*, 2001), providing a potential resolution to the paradox. However, it is usually assumed that the classic estrogen receptor that interacts with the genome is not a membrane receptor. A possibility raised by (Blurton-Jones *et al.*, 2004) is that estradiol acts indirectly on granule cells to regulate BDNF expression. Activity might be the mediator in this case, because estrogen receptors were found on a subset of GABAergic neurons (Blurton-Jones *et al.*, 2004), which innervate granule cells. Another possibility is that there is local regulation of BDNF synthesis by membrane estrogen receptors that bypass the gene. Indeed, there is increasing evidence for local control points in the cell for BDNF synthesis (Schratt *et al.*, 2004).

Other possibilities to consider are additional hormonal effects on BDNF expression, and cortisol is a prime candidate given that it rises during the morning of proestrus, and granule cells are well endowed with corticosteroid receptors. Moreover, stress clearly influences BDNF expression. The problem with this hypothesis is that stress appears to decrease BDNF in most studies to date (Smith *et al.*, 1995; Schaaf *et al.*, 1998). A few have shown that a transient elevation can occur before BDNF expression declines (Marmigere *et al.*, 2003), but the decline occurs within hours and hence would not explain the protracted elevation in BDNF that continues from proestrus throughout the morning of estrus.

8. CELLULAR MECHANISMS

Regarding cellular mechanisms, intracellular recordings have provided insight into potential reasons for underlying changes in mossy fiber transmission that lead to multiple population spikes. Thus, in slices which demonstrate multiple population spikes in response to a 1 Hz train of paired pulses to the mossy fibers, the intracellular correlate reveals that the initial responses, which are dominated by IPSPs or afterhyperpolarizations, are transformed as stimulation progresses into long lasting depolarizations with superimposed action potentials. An example of this is shown for a CA3 pyramidal cell from a slice of a proestrous rat vs. a metestrous rat in Figure 7. Nothing like this can be evoked by other inputs to the same CA3 pyramidal cells. It is unlikely that there are changes in intrinsic properties that explain the effect, because, for example, intrinsic properties of CA3 neurons do not appear to change after exposure to BDNF (MacLusky *et al.*, 2003).

Thus, the next step will be to examine what underlies the long-lasting depolarization. Is it a reflection of increased glutamate release, possibly so much that concomitant afterhyperpolarizations and IPSPs are unable to control neuronal discharge? Perhaps a decrement in glutamate transport works in conjunction with increased release to produce a long-lasting and large EPSP. Indeed, estradiol can influence glutamate

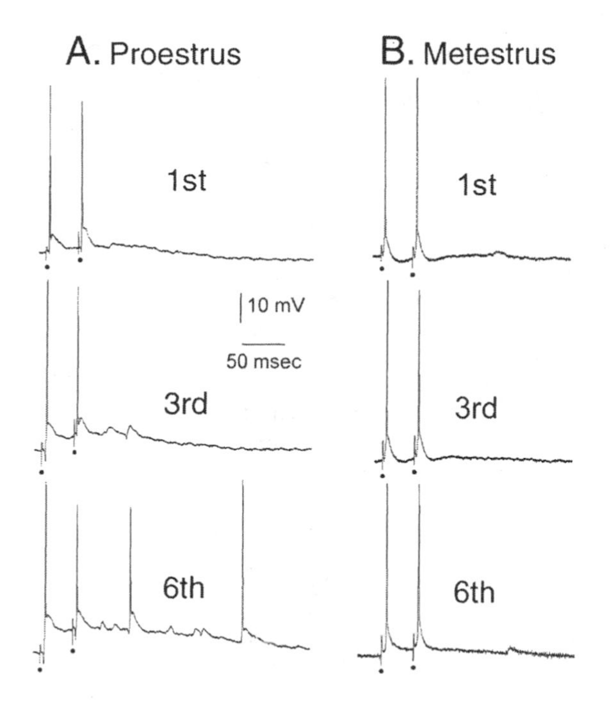

Figure 7. Recordings from CA3 pyramidal cells illustrate the intracellular correlate to extracellularly-recorded population spikes induced by repeated mossy fiber stimuli in the presence of increased BDNF. A. Responses to mossy fiber stimuli (paired pulses, 40 msec interstimulus interval, 1 Hz) are recorded from a CA3 pyramidal cell of a slice from a proestrous rat with elevated mossy fiber BDNF. Membrane potential, -72 mV. A long-lasting depolarization develops by the 3rd pair of stimuli and increases further by the 6th pair of stimuli. On this depolarization, action potentials are triggered that correspond to the multiple population spikes observed extracellularly, shown in Figure 6. B. Recordings from a CA3 pyramidal cell in a slice from a metestrous rat with relatively low mossy fiber BDNF. Membrane potential, -68 mV. No long-lasting depolarizations or extra action potentials are evoked. The cells from metestrous rats demonstrate similar responses as other CA3 pyramidal cells examined under other conditions when BDNF levels in mossy fibers are relatively low, such as male rats prior to exposure to exogenous BDNF.

transport, so this could be a mechanism that might be applicable to the actions of BDNF in female rats described above. But at least in cortical synaptosomes, high affinity transport of glutamate appears stronger at proestrus, not weaker (Mitrovic *et al.*, 1999), so this mechanism seems an unlikely explanation.

Another possibility is that IPSPs switch from hyperpolarizing to depolarizing, a possibility suggested by the fact that BDNF induces changes in KCC2 expression (Rivera *et al.*, 2004), and KCC2 is a key control of chloride entry during $GABA_A$ receptor-mediated IPSPs (Rivera *et al.*, 1999). In addition, there is a sex-related difference in KCC2 expression that has been documented outside the hippocampus (Galanopoulou *et al.*, 2003). However, this mechanism may not provide a good explanation, because IPSPs evoked by single stimuli to the mossy fiber across the estrous cycle had identical amplitude and reversal potentials (unpublished data). Nevertheless, a partial change in

KCC2 expression might make IPSPs in response to *single stimuli* relatively unaffected, but IPSPs evoked by *repetitive stimuli* much more labile. Indeed, the lability of IPSPs to repeated stimuli has long been known (Wong and Watkins, 1982; Deisz and Prince, 1989; Thompson and Gahwiler, 1989).

9. SUMMARY AND PERSPECTIVE

This chapter reviews studies of BDNF's effects on mossy fiber transmission that suggest a prominent role for this neurotrophin in regulation of this synapse. BDNF appears to increase the ability of mossy fibers to excite CA3 pyramidal neurons by an action on trkB receptors that leads to altered glutamate release. Under normal conditions, the result may be beneficial, because a slight increase in glutamate release may promote synaptogenesis during development, dendritic spine plasticity, as well as a long-lasting potentiation, analogous to what has been documented in CA1 (Figure 8).

However, too much BDNF may not be beneficial. Excess BDNF may arise following a variety of stimuli, insults, hormonal changes, or activity-induced upregulation. This may ultimately have negative consequences by increasing seizure susceptibility and leading to a chronic elevation in BDNF, similar to what has been detected in epileptic rats and human temporal lobe epilepsy (Figure 8). Under these conditions, any subsequent release of BDNF could trigger seizures theoretically. If correct, this hypothesis suggests that anticonvulsants that block BDNF upregulation or trkB may be a novel therapeutic strategy to treat epilepsy.

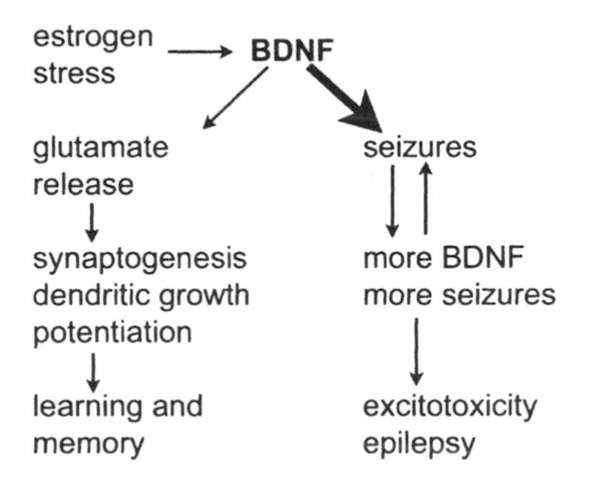

Figure 8. A working hypothesis for the role of BDNF in hippocampus. Several factors, such as estrogen and stress, can modulate BDNF levels in glutamatergic pathways such as the mossy fibers. Normal levels or small increases lead to enhanced glutamate release and potentiation, as well as normal development and plasticity of structure and function, such as learning. Excess levels lead to hyperexcitability and a positive feedback that ultimate raises BDNF levels further and promotes seizures and excitotoxicity. The two branchs of this schematic are not mutually exclusive, given that excess BDNF after an initial period of seizures may foster the changes in structure and function that underlie the epileptic state.

10. ACKNOWLEDGMENTS

I thank Regeneron Pharmaceuticals for providing materials critical to this work. I also thank Susan Croll, Neil MacLusky, Jeffrey Goodman, Thomas Mercurio and Marlene Wilson for their contributions to the studies discussed above. This work was supported by grant NS 37562 from the National Institutes of Health.

11. REFERENCES

Acsady, L., Kamondi, A., Sik, A., Freund, T., and Buzsaki, G., 1998, GABAergic cells are the major postsynaptic targets of mossy fibers in the rat hippocampus. *J Neurosci*, **18**, 3386.

Bauer, J., 2001, Interactions between hormones and epilepsy in female patients. *Epilepsia*, **42**, 20.

Bibel, M. and Barde, Y. A., 2000, Neurotrophins: key regulators of cell fate and cell shape in the vertebrate nervous system. *Genes Dev.*, **14**, 2919.

Binder, D. K., Croll, S. D., Gall, C. M., and Scharfman, H. E., 2001, BDNF and epilepsy: too much of a good thing? *Trends Neurosci*, **24**, 47.

Binder, D. K., Routbort, M. J., Ryan, T. E., Yancopoulos, G. D., and McNamara, J. O., 1999, Selective inhibition of kindling development by intraventricular administration of TrkB receptor body. *J Neurosci.*, **19**, 1424.

Black, I. B., 1999, Trophic regulation of synaptic plasticity. *J Neurobiol*, **41**, 108.

Blum, R., Kafitz, K. W., and Konnerth, A., 2002, Neurotrophin-evoked depolarization requires the sodium channel Na(V)1.9. *Nature*, **419**, 687.

Blurton-Jones, M., Kuan, P. N., and Tuszynski, M. H., 2004, Anatomical evidence for transsynaptic influences of estrogen on brain-derived neurotrophic factor expression. *J Comp Neurol.*, **468**, 347.

Brake, W. G., Alves, S. E., Dunlop, J. C., Lee, S. J., Bulloch, K., Allen, P. B., Greengard, P., and McEwen, B. S., 2001, Novel target sites for estrogen action in the dorsal hippocampus: an examination of synaptic proteins. *Endocrinology*, **142**, 1284.

Buckmaster, P. S., Zhang, G. F., and Yamawaki, R., 2002, Axon sprouting in a model of temporal lobe epilepsy creates a predominantly excitatory feedback circuit. *J Neurosci.*, **22**, 6650.

Chao, M. V., 2003, Neurotrophins and their receptors: a convergence point for many signalling pathways. *Nat Rev Neurosci.*, **4**, 299.

Cirulli, F., Berry, A., Chiarotti, F., and Alleva, E., 2004, Intrahippocampal administration of BDNF in adult rats affects short-term behavioral plasticity in the Morris water maze and performance in the elevated plus-maze. *Hippocampus*, **14**, 802.

Conner, J. M., Lauterborn, J. C., Yan, Q., Gall, C. M., and Varon, S., 1997, Distribution of brain-derived neurotrophic factor (BDNF) protein and mRNA in the normal adult rat CNS: evidence for anterograde axonal transport. *J. Neurosci.*, **17**, 2295.

Croll, S. D., Suri, C., Compton, D. L., Simmons, M. V., Yancopoulos, G. D., Lindsay, R. M., Wiegand, S. J., Rudge, J. S., and Scharfman, H. E., 1999, Brain-derived neurotrophic factor transgenic mice exhibit passive avoidance deficits, increased seizure severity and in vitro hyperexcitability in the hippocampus and entorhinal cortex. *Neuroscience*, **93**, 1491.

Danzer, S. C., Pan, E., Nef, S., Parada, L. F., and McNamara, J. O., 2004, Altered regulation of brain-derived neurotrophic factor protein in hippocampus following slice preparation. *Neuroscience*, **126**, 859.

Deisz, R. A. and Prince, D. A., 1989, Frequency-dependent depression of inhibition in guinea-pig neocortex in vitro by $GABA_B$ receptor feed-back on GABA release. *J Physiol.*, **412**, 513.

Drake, C. T., Milner, T. A., and Patterson, S. L., 1999, Ultrastructural localization of full-length trkB immunoreactivity in rat hippocampus suggests multiple roles in modulating activity-dependent synaptic plasticity. *J Neurosci*, **19**, 8009.

Frerking, M., Malenka, R. C., and Nicoll, R. A., 1998, Brain-derived neurotrophic factor (BDNF) modulates inhibitory, but not excitatory, transmission in the CA1 region of the hippocampus. *J Neurophysiol*, **80**, 3383.

Galanopoulou, A. S., Kyrozis, A., Claudio, O. I., Stanton, P. K., and Moshe, S. L., 2003, Sex-specific KCC2 expression and GABA(A) receptor function in rat substantia nigra. *Exp Neurol.*, **183**, 628.

Gall, C. M., 1993, Seizure-induced changes in neurotrophin expression: implications for epilepsy. *Exp Neurol.*, **124**, 150.

He, X. P., Kotloski, R., Nef, S., Luikart, B. W., Parada, L. F., and McNamara, J. O., 2004, Conditional deletion of TrkB but not BDNF prevents epileptogenesis in the kindling model. *Neuron.*, **43**, 31.

Herzog, A. G., Klein, P., and Ransil, B. J., 1997, Three patterns of catamenial epilepsy. *Epilepsia*, **38**, 1082.

Jovanic, J. N., Benfenati, F., Siow, Y. L., Sihra, T. S., Sanghera, J. S., Pelech, S. L., Greengard, P., and Czernik, A. J., 1996, Neurotrophins stimulate phosphorylation of synapsin I by MAP kinase and regulate synapsin I actin interactions. *Proc Natl Acad Sci USA*, **93**, 367.

Kafitz, K. W., Rose, C. R., Thoenen, H., and Konnerth, A., 1999, Neurotrophin-evoked rapid excitation through TrkB receptors. *Nature*, **401**, 918.

Kanemoto, K., Kawasaki, J., Tarao, Y., Kumaki, T., Oshima, T., Kaji, R., and Nishimura, M., 2003, Association of partial epilepsy with brain-derived neurotrophic factor (BDNF) gene polymorphisms. *Epilepsy Res.*, **53**, 255.

Kang, H. and Schuman, E. M., 1995, Long-lasting neurotrophin-induced enhancement of synaptic transmission in the adult hippocampus. *Science*, **267**, 1658.

Koponen, E., Voikar, V., Riekki, R., Saarelainen, T., Rauramaa, T., Rauvala, H., Taira, T., and Castren, E., 2004, Transgenic mice overexpressing the full-length neurotrophin receptor trkB exhibit increased activation of the trkB-PLCgamma pathway, reduced anxiety, and facilitated learning. *Mol Cell Neurosci.*, **26**, 166.

Korte, M., Staiger, V., Griesbeck, O., Thoenen, H., and Bonhoeffer, T., 1996, The involvement of brain-derived neurotrophic factor in hippocampal long-term potentiation revealed by gene targeting experiments. *J. Physiol.*, **90**, 157.

Kotti, T., RiekkinenSr., P. J., and Miettinen, R., 1997, Characterization of target cells for aberrant mossy fiber collaterals in the dentate gyrus of epileptic rat. *Exp Neurol.*, **146**, 323.

Kovalchuk, Y., Hanse, E., Kafitz, K. W., and Konnerth, A., 2002, Postsynaptic induction of BDNF-mediated long-term potentiation. *Science*, **295**, 1729.

Lahteinen, S., Pitkanen, A., Knuuttila, J., Toronen, P., and Castren, E., 2004, Brain-derived neurotrophic factor signaling modifies hippocampal gene expression during epileptogenesis in transgenic mice. *Eur J Neurosci.*, **19**, 3245.

Lahteinen, S., Pitkanen, A., Koponen, E., Saarelainen, T., and Castren, E., 2003, Exacerbated status epilepticus and acute cell loss, but no changes in epileptogenesis, in mice with increased brain-derived neurotrophic factor signaling. *Neuroscience*, **122**, 1081.

Lahteinen, S., Pitkanen, A., Saarelainen, T., Nissinen, J., Koponen, E., and Castren, E., 2002, Decreased BDNF signalling in transgenic mice reduces epileptogenesis. *Eur J Neurosci.*, **15**, 721.

Larmet, Y., Reibel, S., Carnahan, J., Nawa, H., Marescaux, C., and Depaulis, A., 1995, Protective effects of brain-derived neurotrophic factor on the development of hippocampal kindling in the rat. *NeuroReport*, **6**, 1937.

Lohof, A. M., Ip, N. Y., and Poo, M. M., 1993, Potentiation of developing neuromuscular synapses by the neurotrophins NT-3 and BDNF. *Nature*, **363**, 350.

Lynch M, Sutula T., 2000, Recurrent excitatory connectivity in the dentate gyrus of kindled and kainic acid-treated rats.*J Neurophysiol.* **83**, 693.

MacLuksy N. J., Mercurio T. C., Wilson M. A., Scharfman H. E., 2003, Changes in mossy fiber transmission across the female rat estrous cycle: role of BDNF and α7 nicotinic cholinergic receptors (α7nAChRS). *Abstract Viewer/Itinerary Planner.* Washington, DC: Society for Neuroscience, Online.

Marmigere, F., Givalois, L., Rage, F., Arancibia, S., and Tapia-Arancibia, L., 2003, Rapid induction of BDNF expression in the hippocampus during immobilization stress challenge in adult rats. *Hippocampus*, **13**, 646.

Mathern, G. W., Babb, T. L., Micevych, P. E., Blanco, C. E., and Pretorius, J. K., 1997, Granule cell mRNA levels for BDNF, NGF, and NT-3 correlate with neuron losses or supragranular mossy fiber sprouting in the chronically damaged and epileptic human hippocampus. *Mol Chem Neuropathol*, **30**, 53.

Messaoudi, E., Bardsen, K., Srebro, B., and Bramham, C. R., 1998, Acute intrahippocampal infusion of BDNF induces lasting potentiation of synaptic transmission in the rat dentate gyrus. *J Neurophysiol.*, **79**, 496.

Milner, T. A., McEwen, B. S., Hayashi, S., Li, C. J., Reagan, L. P., and Alves, S. E., 2001, Ultrastructural evidence that hippocampal alpha estrogen receptors are located at extranuclear sites. *J Comp Neurol*, **429**, 355.

Mitrovic, A. D., Maddison, J. E., and Johnston, G. A., 1999, Influence of the oestrous cycle on L-glutamate and L-aspartate transport in rat brain synaptosomes. *Neurochem Int*, **34**, 101.

Monteggia, L. M., Barrot, M., Powell, C. M., Berton, O., Galanis, V., Gemelli, T., Meuth, S., Nagy, A., Greene, R. W., and J.Nestler, E., 2004, Essential role of brain-derived neurotrophic factor in adult hippocampal function. *Proc Natl Acad Sci USA*, **101**, 10827.

Murray, K. D., Isackson, P. J., Eskin, T. A., King, M. A., and Montesinos, S. P., 2000, Altered mRNA expression for brain-derived neurotrophic factor and type II calcium/calmodulin-dependent protein kinase in the hippocampus of patients with intractable temporal lobe epilepsy. *J Comp Neurol*, **18**, 411.

Nadler, J. V., 2003, The recurrent mossy fiber pathway of the epileptic brain. *Neurochem Res.*, **28**, 1649.

Olofsdotter, K., Lindvall, O., and Asztely, F., 2000, Increased synaptic inhibition in dentate gyrus of mice with reduced levels of endogenous brain-derived neurotrophic factor. *Neuroscience*, **101**, 531.

Osehobo, P., Adams, B., Sazgar, M., Xu, Y., Racine, R. J., and Fahnestock, M., 1999, Brain-derived neurotrophic factor infusion delays amygdala and perforant path kindling without affecting paired-pulse measures of neuronal inhibition in adult rats. *Neurosci*, **92**, 1367.

Patterson, S. L., Abel, T., Deuel, T. A., Martin, K. C., Rose, J. C., and Kandel, E. R., 1996, Recombinant BDNF rescues deficits in basal synaptic transmission and hippocampal LTP in BDNF knockout mice. *Neuron*, **16**, 1137.

Reibel, S., Larmet, Y., Carnahan, J., Marescaux, C., and Depaulis, A., 2000, Endogenous control of hippocampal epileptogenesis: a molecular cascade involving brain-derived neurotrophic factor and neuropeptide Y. *Epilepsia*, **41**, S127.

Rivera, C., Voipio, J., Payne, J. A., Ruusuvuori, E., Lahtinen, H., Lamsa, K., Pirvola, U., Saarma, M., and Kaila, K., 1999, The K+/Cl-co-transporter KCC2 renders GABA hyperpolarizing during neuronal maturation. *Nature*, **397**, 251.

Rivera, C., Voipio, J., Thomas-Crusells, J., Li, H., Emri, Z., Sipila, S., Payne, J. A., Minichiello, L., Saarma, M., and Kaila, K., 2004, Mechanism of activity-dependent downregulation of the neuron-specific K-Cl cotransporter KCC2. *J Neurosci.*, **24**, 4683.

Rose, C. R., Blum, R., Pichler, B., Lepier, A., Kafitz, K. W., and Konnerth, A., 2003, Truncated TrkB-T1 mediates neurotrophin-evoked calcium signalling in glia cells. *Nature*, **426**, 74.

Rudge, J. S., Mather, P. E., Pasnikowski, E. M., Cai, N., Corcoran, T., Acheson, A., Anderson, K., Lindsay, R. M., and Wiegand, S. J., 1998, Endogenous BDNF protein is increased in adult rat hippocampas after a kainic acid induced excitotoxic insult but exogenous BDNF is not protective. *Expl Neurol.*, **149**, 398.

Schaaf, M. J., deJong, J., deKloet, E. R., and Vreugdenhil, E., 1998, Downregulation of BDNF mRNA and protein in the rat hippocampus by corticosterone. *Brain Res.*, **813**, 112.

Scharfman, H. E., 1997, Hyperexcitability in combined entorhinal/hippocampal slices of adult rat after exposure to brain-derived neurotrophic factor. *J Neurophysiol*, **78**, 1082.

Scharfman, H. E., Goodman, J. H., Solas, A. L., and Croll, S. D., 2002a, Spontaneous limbic seizures after intrahippocampal infusion of brain-derived neurotrophic factor. *Exp Neurol.*, **174**, 201.

Scharfman, H. E., Goodman, J. H., and Sollas, A. L., 1999, Actions of brain-derived neurotrophic factor in slices from rats with spontaneous seizures and mossy fiber sprouting in the dentate gyrus. *J Neurosci*, **19**, 5619.

Scharfman, H. E., Goodman, J. H., and Sollas, A. L., 2000, Granule-like neurons at the hilar/CA3 border after status epilepticus and their synchrony with area CA3 pyramidal cells: functional implications of seizure-induced neurogenesis. *J Neurosci*, **20**, 6144.

Scharfman, H. E., Kunkel, D. D., and Schwartzkroin, P. A., 1990, Synaptic connections of dentate granule cells and hilar neurons: results of paired intracellular recordings and intracellular horseradish peroxidase injections. *Neuroscience*, **37**, 693.

Scharfman, H. E., Mercurio, T. C., Goodman, J. H., Wilson, M. A., and MacLusky, N. J., 2003, Hippocampal excitability increases during the estrous cycle in the rat: a potential role for brain-derived neurotrophic factor. *J Neurosci.*, **23**, 11641.

Scharfman, H. E. and Schwartzkroin, P. A., 1990, Consequences of prolonged afferent stimulation of the rat fascia dentata: epileptiform activity in area CA3 of hippocampus. *Neuroscience*, **35**, 505.

Scharfman, H. E., Smith, K. L., Goodman, J. H., and Sollas, A. L., 2001, Survival of dentate hilar mossy cells after pilocarpine-induced seizures and their synchronized burst discharges with area CA3 pyramidal cells. *Neuroscience*, **104**, 741.

Scharfman, H. E., Solas, A. L., Smith, K. L., Jackson, M. B., and Goodman, J. H., 2002b, Structural and functional asymmetry in the normal and epileptic rat dentate gyrus. *J Comp Neurol.*, **454**, 424.

Schinder, A. F., Berninger, B., and Poo, M., 2000, Postsynaptic target specificity of neurotrophin-induced presynaptic potentiation. *Neuron.*, **25**, 151.

Schratt, G. M., Nigh, E. A., Chen, W. G., Hu, L., and Greenberg, M. E., 2004, BDNF regulates the translation of a select group of mRNAs by a mammalian target of rapamycin-phosphatidylinositol 3-kinase-dependent pathway during neuronal development. *J. Neurosci.*, **24**, 7366.

Singh, M., Meyer, E. M., and Simpkins, J. W., 1995, The effect of ovariectomy and estradiol replacement on brain-derived neurotrophic factor messenger ribonucleic acid expression in cortical and hippocampal brain regions of female Sprague-Dawley rats. *Endocrinology*, **136**, 2320.

Sloviter, R. S., 1992, Possible functional consequences of synaptic reorganization in the dentate gyrus of kainate-treated rats. *Neurosci Lett.*, **137**, 91.

Smith, M. A., Makino, S., Kvetnansky, R., and Post, R. M., 1995, Stress and glucocorticoids affect the expression of brain-derived neurotrophic factor and neurotrophin-3 mRNAs in the hippocampus. *J Neurosci.*, **15**, 1768.

Sohrabji, F., Miranda, R. C., and Toran-Allerand, C. D., 1995, Identification of a putative estrogen response element in the gene encoding brain-derived neurotrophic factor. *Proc Natl Acad Sci USA*, **92**, 11110.

Sperk, G., Bellmann, R., Gruber, B., Greber, S., Marksteiner, J., Roder, C., and Rupp, E., 1996, Neuropeptide Y expression in animal models of temporal lobe epilepsy. *Epilepsy Res Suppl.*, **12**, 197.

Tanaka, T., Saito, H., and Matsuki, N., 1997, Inhibition of $GABA_A$ synaptic responses by brain-derived neurotrophic factor (BDNF) in rat hippocampus. *J Neurosci*, **17**, 2959.

Tartaglia, N., Tyler, W. J., Neale, E., Pozzo-Miller, L., and Lu, B., 2001, Protein synthesis-dependent and -independent regulation of hippocampal synapses by brain-derived neurotrophic factor. *J Biol Chem.*, **276**, 37585.

Tauck, D. L. and Nadler, J. V., 1985, Evidence of functional mossy fiber sprouting in hippocampal formation of kainic acid-treated rats. *J Neurosci.*, **5**, 1016.

Thompson, S. M. and Gahwiler, B. H., 1989, Activity-dependent disinhibition. I. Repetivitve stimulation reduces IPSP driving force and conductance in the hippocampus in vitro. *J Neurophysiol.*, **61**, 501.

Tongiorgi, E., Armellin, M., Giulianini, P. G., Bregola, G., Zucchini, S., Paradiso, B., Steward, O., Cattaneo, A., and Simonato, M., 2004, Brain-derived neurotrophic factor mRNA and protein are targeted to discrete dendritic laminas by events that trigger epileptogenesis. *J Neurosci.*, **24**, 6824.

Turski, W. A., Cavalheiro, E. A., Schwarz, M., Czuczwar, S. J., Kleinrok, Z., and Turski, L., 1983, Limbic seizures produced by pilocarpine in rats: behavioural, electroencephalographic and neuropathological study. *Behav Brain Res.*, **9**, 315.

Tyler, W. J., Perrett, S. P., and Pozzo-Miller, L. D., 2002, The role of neurotrophins in neurotransmitter release. *The Neuroscientist*, **8**, 524.

Vicario-Abejon, C., Owens, D., McKay, R., and Segal, M., 2002, Role of neurotrophins in central synapse formation and stabilization. *Nature Reviews Neurosci*, **3**, 965.

von Kitzing E., Jonas P., Sakmann B., 1994, 1994, Quantal analysis of excitatory postsynaptic currents at the hippocampal mossy fiber-CA3 pyramidal cell synapse. *Adv Second Messenger Phosphoprotein Res.* **29**, 235.

Wenzel, H. J., Woolley, C. S., Robbins, C. A., and Schwartzkroin, P. A., 2000, Kainic acid-induces mossy fiber sprouting and synapse formation in the dentate gyrus of rats. *Hippocampus*, **10**, 244.

Wong, R. K. and Watkins, D. J., 1982, Cellular factors influencing GABA response in hippocampal pyramidal cells. *J Neurophysiol.*, **48**, 938.

Xu, B., Michalski, B., Racine, R. J., and Fahnestock, M., 2004, The effects of brain-derived neurotrophic factor (BDNF) administration on kindling induction, Trk expression and seizure-related morphological changes. *Neuroscience*, **126**, 521.

Yan, Q., Rosenfeld, R. D., Matheson, C. R., Hawkins, N., Lopez, O. T., Bennett, L., and Welcher, A. A., 1997, Expression of brain-derived neurotrophic factor (BDNF) protein in the adult rat central nervous system. *Neuroscience*, **78**, 431.

TRANSSYNAPTIC DIALOGUE BETWEEN EXCITATORY AND INHIBITORY HIPPOCAMPAL SYNAPSES VIA ENDOCANNABINOIDS

Pablo E. Castillo and Vivien Chevaleyre*

1. INTRODUCTION

The interaction between excitatory and inhibitory synapses is central in the control of neuronal excitability and synaptic plasticity. By regulating the level of depolarization and by shunting depolarizing currents, inhibitory synapses control the functional gain and modifiability of excitatory synapses (Freund and Buzsaki, 1996; Miles et al., 1996). Here, we discuss how the endocannabinoid retrograde signaling, by mediating long-term depression of inhibitory synaptic transmission (I-LTD) (Chevaleyre and Castillo, 2003), "translates" activity of excitatory synaptic inputs into a long-term disinhibition in the hippocampus.

Endocannabinoids are endogenous lipid molecules that act as retrograde messengers at many synapses in the central nervous system, by binding to presynaptic cannabinoid receptors (CB1Rs) and inhibiting neurotransmitter release (for recent reviews see Wilson and Nicoll, 2002; Freund et al., 2003; Piomelli, 2003). The CB1R is one of the most abundant G-protein-coupled receptor in the brain, and mediates most of the behavioral actions of Δ9-tetrahydrocannabinol (THC), the active principle of *Cannabis sativa*. Although the existence of an endogenous cannabinoid signaling system is well established, its physiological role is just beginning to be understood.

2. ENDOCANNABINOID-MEDIATED LONG-TERM DEPRESSION OF INHIBITORY SYNAPTIC TRANSMISSION (I-LTD)

It is becoming clear that endocannabinoids play an essential role in the induction of long-term plasticity in different brain structures (Gerdeman and Lovinger, 2003). For example, CB1Rs are necessary for the induction of LTD at excitatory synapses in

* P. E. Castillo and V. Chevaleyre, Department of Neuroscience, Albert Einstein College of Medicine, Bronx, New York 10461, USA. E-mail: pcastill@aecom.yu.edu

striatum (Gerdeman et al., 2002), nucleus accumbens (Robbe et al., 2002) and neocortex (Sjostrom et al., 2003), and at inhibitory synapses in the amygdala (Marsicano et al., 2002). In the hippocampus (Fig. 1A), repetitive activation of glutamatergic fibers that normally induces long-term potentiation (LTP) at excitatory synapses also triggers long-term depression at inhibitory synapses (I-LTD) via retrograde endocannabinoid signaling (Chevaleyre and Castillo, 2003). High frequency stimulation (HFS) and theta burst stimulation (TBS) in *stratum radiatum* of the CA1 area both induce a reduction of the amplitude of evoked inhibitory synaptic currents (IPSCs), or inhibitory synaptic potentials (IPSPs), monitored in pyramidal cells for more than an hour. Indeed, I-LTD is a robust phenomenon that is observed in rat and mouse hippocampus, and can occur at room temperature and at 35°C. The long-lasting depression of inhibitory synaptic transmission is associated with an enhancement in paired-pulse ratio (PPR) (Fig. 1B) and failure rate (Chevaleyre and Castillo, unpublished data), strongly suggesting that I-LTD is likely due to a persistent reduction of evoked GABA release.

2.1 Mechanism of induction of I-LTD

CB1R activation depresses inhibitory synaptic transmission by reducing GABA release (Katona et al., 1999; Hajos et al., 2000; Hoffman and Lupica, 2000). Interestingly, immunocytochemical studies in the hippocampus have shown that CB1Rs are mostly localized at presynaptic terminals of GABAergic interneurons that innervate pyramidal cells (Katona et al., 1999; Tsou et al., 1999; Hajos et al., 2000; Irving et al., 2000). In addition, virtually all hippocampal CB1R-immunoreactive neurons have so far proven to be cholecystokinin-containing GABAergic interneurons (Tsou et al., 1999). Three pieces of evidence indicate that the long-lasting reduction of evoked GABA release requires the activation of CB1Rs. First, I-LTD is completely abolished in presence of the CB1R selective antagonist AM251 (2 μM, Fig. 1C). Second, the depression of GABA release induced by CB1R agonists and I-LTD occlude each other, suggesting that both depressions share a common mechanism. Third, I-LTD is absent in CB1R knockout mice (Chevaleyre and Castillo, unpublished data). While several putative ligands for cannabinoid receptors, including anandamide (arachidonoyl ethanolamide), 2-AG (2-arachidonoyl glycerol), and noladin ether (2-arachidonyl glyceryl ether) have been identified (Mechoulam et al., 1998; Howlett et al., 2002; Pertwee and Ross, 2002), evidence is accumulating that 2-AG is the most efficacious endogenous natural ligand for cannabinoid receptors in the brain. It has been reported in hippocampal slices that HFS produces a selective enhancement of the endocannabinoid 2-AG, but not of anandamide (Stella et al., 1997). In agreement with this finding, I-LTD is abolished by the blockade of the enzyme DAG lipase that synthesizes 2-AG (Chevaleyre and Castillo, 2003). Thus, CB1R activation is an essential step in the induction of I-LTD, and 2-AG seems to be the endogenous ligand that mediates this form of plasticity.

How does repetitive stimulation in *s. radiatum* trigger 2-AG release and ultimately I-LTD? The usual triggering signal for long-term plasticity at inhibitory synapses is a rise of postsynaptic Ca^{2+} concentration most commonly through the activation of NMDARs (Stelzer et al., 1987; Wang and Stelzer, 1996; Caillard et al., 1999a; Lu et al., 2000) or voltage gated Ca^{2+} channels (Caillard et al., 1999b), although in some cases, the release of Ca^{2+} from internal stores may also be required (Komatsu, 1996; Caillard et al., 2000; Holmgren and Zilberter, 2001). In contrast, the induction of I-LTD seems to rely

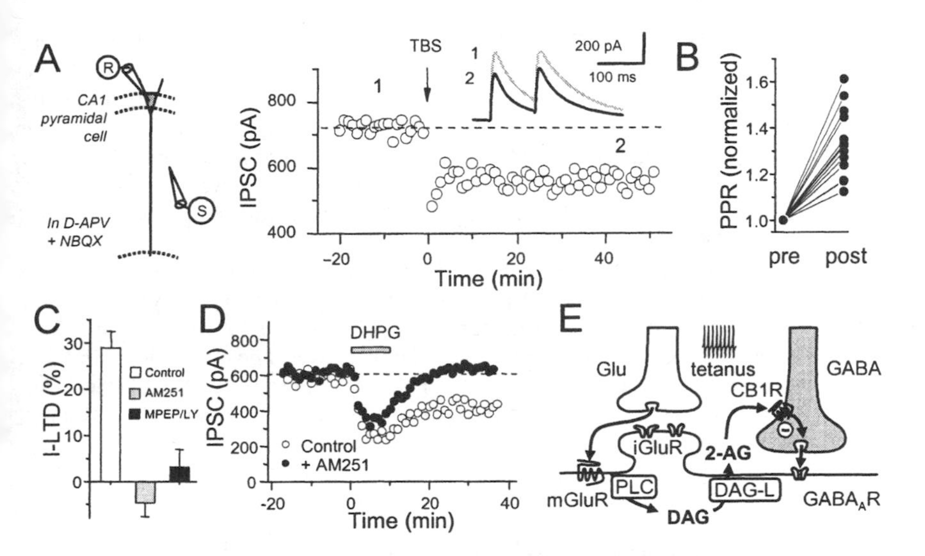

Figure 1. Heterosynaptic long-term depression of hippocampal GABAergic synapses (I-LTD). **A:** Theta-burst stimulation (TBS), consisting of a series of 10 bursts of 5 stimuli (100 Hz within the burst, 200 ms inter-burst interval), triggers long-term depression of IPSC (I-LTD) in CA1 pyramidal neurons in rat hippocampal slices. IPSCs were recorded in presence of 25 μM D-APV and 10 μM NBQX (+10 mV holding potential). Averaged sample traces taken during the experiment (indicated by numbers) are shown on top. **B:** Summary graph of the change in paired-pulse ratio (PPR) obtained after I-LTD induction in 11 cells recorded as in A. **C:** Summary graph of I-LTD magnitude in control conditions, in presence of 2 μM AM251 and 4 μM MPEP/100 μM LY367385, to block CB1Rs and group I mGluRs, respectively. **D:** Bath application of group I mGluR agonist DHPG for 10 minutes triggers long-term depression after washout in control condition but not when CB1Rs are blocked by AM251. **E:** Proposed mechanism of induction of I-LTD.

on a different and novel mechanism (Chevaleyre and Castillo, 2003). I-LTD does not depend on postsynaptic voltage-dependent processes as it can be triggered while voltage-clamping CA1 pyramidal cells at different holding potentials from -60 to +10 mV. Consistent with this observation, I-LTD does not require NMDAR activation as it is unaffected by D-APV (25 μM), or a rise in intracellular Ca^{2+} since it is unaffected after loading the postsynaptic cell with the calcium chelator BAPTA (20 mM). Rather, the activation of group I metabotropic glutamatergic receptors (mGluRs), which are coupled to PLC, is necessary and sufficient for the induction of I-LTD. As indicated in figure 1C, I-LTD is blocked by selective antagonists of mGluR subtypes 1 and 5 (LY367385 and MPEP, respectively) and it is also mimicked by exogenous activation of these receptors (Fig. 1D). Also, the mGluR-induced depression and I-LTD occlude one another, suggesting that both depressions share a common step. These findings are consistent with previous observations in hippocampus and cerebellum showing that the release of endocannabinoids can be driven via mGluRs independently of any postsynaptic Ca^{2+} rise (Maejima et al., 2001; Varma et al., 2001; Ohno-Shosaku et al., 2002). In theory, mGluRs that mediate I-LTD could localize presynaptically at GABAergic terminals or postsynaptically on the CA1 pyramidal cell. Previous observations suggested that mGluRs1/5 are expressed on the dendritic tree of these cells but have been difficult to find at GABAergic terminals (Shigemoto et al., 1997). Indeed, loading the postsynaptic cell with the G-protein blocker GDP-γS (2 mM), a manipulation that interferes with the signaling pathway downstream mGluR1/5 activation, strongly depresses I-LTD (Chevaleyre and Castillo, 2003). This observation not only indicates that I-LTD requires postsynaptic mGluR1/5 activation but also, most of the signaling that mediates I-LTD is generated from the postsynaptic cell with minimal, if any, contribution from neighboring cells. Finally, the long-term depression induced by the transient activation of mGluR1/5 is mediated by CB1Rs as it is blocked by AM-251, indicating that mGluRs1/5 act upstream from CB1R activation (Fig. 1D).

In conclusion, I-LTD is a heterosynaptic form of plasticity at GABAergic synapses which induction requires activation of postsynaptic mGluR1/5 on CA1 pyramidal cells and presynaptic CB1R on GABAergic terminals (Fig. 1E). Glutamate release from Schaffer collaterals as a result of repetitive stimulation activates postsynaptic mGluR1/5, leading to PLC activation and DAG formation. DAG is then converted by the DAG lipase (DAG-L) into 2-AG which inhibits GABA release by acting on presynaptic CB1Rs.

2.2 Mechanism of expression of I-LTD

The mechanism by which CB1R activation triggers a long-lasting reduction of GABA release is unknown. Diverse signal transduction pathways downstream from CB1R activation could be involved. For instance, the CB1R is coupled to Pertussis toxin sensitive G proteins ($G_{i/o}$) and has been shown to inhibit adenylyl cyclase, activate mitogen-activated protein kinases, reduce voltage dependent Ca^{2+} currents and modulate several K^{+} conductances (Schlicker and Kathmann, 2001; Howlett et al., 2002). There is also evidence that the inhibitory effect of CB1R agonists on transmitter release may involve sites downstream from Ca^{2+} entry; i.e. the release machinery (Vaughan et al., 1999; Takahashi and Linden, 2000; Diana and Marty, 2004). Future studies should identify which of these pathway(s) and target(s) underlie the persistent reduction of GABA release that occurs during I-LTD.

2.3 I-LTD is a local phenomenon

There is good evidence that somatic and dendritic inhibitory synapses onto hippocampal pyramidal cells may have different functional roles (Miles et al., 1996; McBain and Fisahn, 2001). Somatic inhibition is well suited for suppressing repetitive generation of sodium-dependent action potentials, whereas dendritic inhibition is more effective in suppressing dendritically generated calcium-dependent action potentials (Miles et al., 1996). In addition, dendritic inhibition may also play an important role in regulating the induction of synaptic plasticity at excitatory synapses (Yuste and Tank, 1996; Magee et al., 1998; Shepherd, 2003) (see below). Interestingly, I-LTD is only elicited by extracellular stimulation in *s. radiatum* but not in *s. pyramidale* (Fig. 2A). Because I-LTD is triggered by repetitive activation of glutamatergic inputs, this observation likely reflects the lack of such inputs to the somatic area of pyramidal cells (Megias et al., 2001). Thus, I-LTD is an endocannabinoid-mediated form of synaptic plasticity selectively associated with dendritic inhibitory inputs, where high levels of CB1R expression have been previously reported (Katona et al., 1999; Egertova and Elphick, 2000; Hajos et al., 2000).

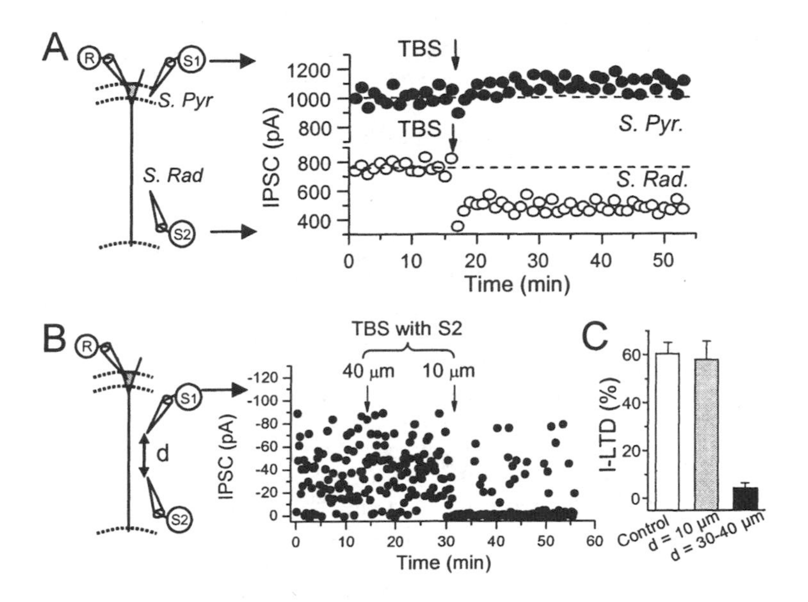

Figure 2. I-LTD is a highly localized phenomenon within the dendritic compartment of CA1 pyramidal cells. **A:** TBS in *stratum radiatum* (S. Rad) but not *stratum pyramidale* (S. Pyr), induces I-LTD. IPSC evoked by alternate stimulation in S. Rad. and S. Pyr. were recorded from the same postsynaptic cell. **B:** (Left) Arrangement of recording and stimulating pipettes used to study I-LTD spread. IPSCs were evoked by focal stimulation through S1 while S2 was only used to deliver TBS and positioned at a variable distance (d) from S1 along the apical dendrite. In these experiments, the presence of functional CB1Rs on the recruited GABAergic fibers was first verified by testing for DSI (depolarization-induced suppression of inhibition), a transient depression of inhibition (Llano et al., 1991; Pitler and Alger, 1992) that is also mediated by CB1Rs (Kreitzer and Regehr, 2001; Ohno-Shosaku et al., 2001; Wilson and Nicoll, 2001). Focal stimulation was obtained by reducing stimulus strength to elicit small amplitude (~30 pA) IPSCs. (Right) TBS applied 40 μm from S1 was unable to induce I-LTD. However, when applied 10 μm from S1, TBS induced robust I-LTD. **C:** Summary graph of I-LTD magnitude obtained when S1 and S2 where separated by 10 μm or 30-40 μm. The magnitude of control I-LTD obtained under these experimental conditions (i.e. using focal stimulation and mostly recruiting CB1R-positive inhibitory inputs) is shown for comparison purposes.

How local is I-LTD along the dendritic tree? Both the heterosynaptic nature of I-LTD and its dependence on retrograde signaling by diffusible messengers raise questions about the synapse-specificity of this form of plasticity. Whether I-LTD is a local or diffuse phenomenon has important implications for the dendritic integration of synaptic inputs and the potential control of plasticity at excitatory synapses. Theoretically, I-LTD could spread along the dendrite as a result of: 1, diffusion of 2-AG; 2, spread of postsynaptic signals downstream from mGluR activation; or 3, glutamate spillover (Kullmann and Asztely, 1998). However, preliminary observations using focal stimulation in *s. radiatum* in rat hippocampal slices strongly suggest that I-LTD is a highly localized phenomenon (Chevaleyre and Castillo, unpublished results). By stimulating a small number of synaptic inputs –excitatory and inhibitory– close to the dendritic tree, we found that endocannabinoid signaling that mediates I-LTD is remarkably restricted to a very small dendritic area of the pyramidal cell (Fig 2B,C). This property indicates that synaptically-driven release of endocannabinoids may exert a subtle point-to-point regulation of dendritic inhibitory inputs.

3. DSI AND I-LTD: SHORT vs. LONG-TERM ENDOCANNABINOID-MEDIATED FORMS OF SYNAPTIC PLASTICITY.

Endocannabinoids are also known to mediate *D*epolarization-induced *S*uppression of *I*nhibition (DSI), a transient depression of IPSCs induced by a brief depolarization of CA1 pyramidal cells (Ohno-Shosaku et al., 2001; Wilson and Nicoll, 2001). Although both DSI and I-LTD are mediated by endocannabinoid retrograde signaling, these two forms of synaptic plasticity differ in their spatial and temporal properties, induction mechanisms, and ultimately, in their functional impact.

1) While DSI typically lasts less than 30 seconds, I-LTD is a long-lasting (>1 hour) form of synaptic plasticity (Chevaleyre and Castillo, 2003). The different time course of these two endocannabinoid-mediated phenomena seems to be based on the duration of CB1R activation during induction. Because CB1R does not desensitize within one minute, it is likely that endocannabinoids are only released during a few seconds after postsynaptic depolarization. The longer duration of CB1R activation is presumably due to longer production of endocannabinoids that occurs as a result of mGluR activation. If I-LTD induction requires CB1R activation during certain period of time, pharmacological blockade of these receptors within this time window should reduce or block I-LTD. As shown in Fig. 3, application of the CB1R antagonist AM251 at different time intervals post-HFS reveals that in contrast to DSI, CB1Rs need to be activated for 5 to 10 minutes in order to trigger I-LTD. These findings are consistent with previous observations suggesting that 2-AG levels remain elevated for several minutes as a result of HFS of hippocampal slices (Stella et al., 1997). Interestingly, once the I-LTD is established (>10 min post-HFS), it becomes independent of CB1R activation.

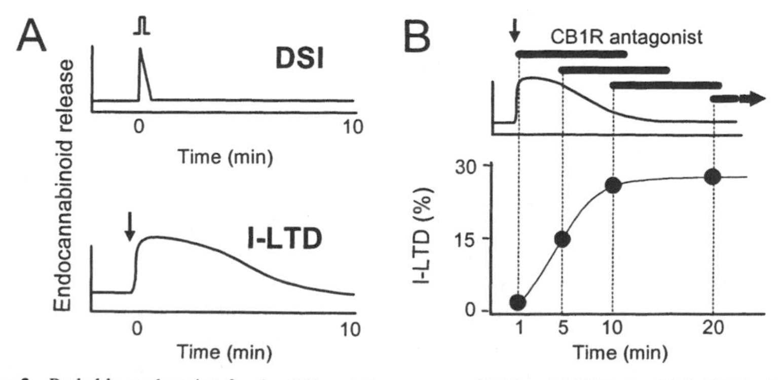

Figure 3. Probable explanation for the different time-course of DSI and I-LTD. **A:** While DSI is due to brief release of endocannabinoids (top), I-LTD seems to require the release of 2-AG for several minutes following induction stimulation (bottom). Thus, longer CB1R activation may be necessary for the induction of I-LTD. To test for this time requirement, the CB1R antagonist AM251 (4 μM, black horizontal bars) was applied at different time points after induction stimulation. **B:** Summary graph of the magnitude of I-LTD after AM251 application. The time points indicate the onset of AM251 application.

2) While DSI globally affect most CB1R-sensitive GABAergic inputs to the target neuron (Chevaleyre and Castillo, 2003), I-LTD is a more localized reduction of synaptic efficacy that only affects inhibitory inputs close to the Schaffer collateral to CA1 pyramidal cell synapses. This differential spatial distribution is likely due to the different mechanisms of induction of these two phenomena. DSI is triggered by large intracellular calcium rises that likely occur as a result of postsynaptic depolarization (i.e. firing of the postsynaptic neuron). Consequently, endocannabinoids produced during DSI target GABAergic inputs located throughout the somato-dendritic compartments of that neuron. In contrast, I-LTD is induced by local synaptic activity and mGluR1/5 activation and does not require a rise in intracellular Ca^{2+}. The resulting endocannabinoid 2-AG acts on GABAergic inputs located near the glutamatergic afferent synapses. Finally, while endocannabinoids produced during DSI may reach synapses on neighboring cells (Wilson et al., 2001), 2-AG produced as a result of repetitive activation of glutamatergic afferents only acts on GABAergic synapses on the target cell (Chevaleyre and Castillo, 2003).

3) I-LTD and DSI are likely mediated by different endocannabinoids. While most evidence indicates that 2-AG, produced via DAG lipase, is the natural endocannabinoid that mediates I-LTD, the nature of the endocannabinoid mediating DSI remains unclear. In contrast to I-LTD, DSI is unaffected by the blockade of PLC or DAG-lipase, suggesting that DSI would be mediated by another endocannabinoid than 2-AG. Alternatively, the biosynthetic pathways of 2-AG may differ, depending on the cells and types of stimuli (Sugiura et al., 2002). If 2-AG indeed mediates DSI (Kim and Alger, 2004), it is most likely produced through a different pathway.

DSI is a single-cell phenomenon that seems to be independent of presynaptic activity. Therefore, an attractive possibility is that the coincidence of two events may be required for induction of a longer-lasting synaptic depression such as I-LTD; that is, endocannabinoid retrograde signaling *and* the activation of presynaptic inhibitory interneurons. For example, presynaptic Ca^{2+} influx could be essential to induce I-LTD, presumably by modulating some signaling pathway downstream from CB1Rs.

Table 1. Different properties between DSI and I-LTD

	DSI	**I-LTD**
Duration	5-30 seconds	> 1 hour
Spatial distribution	Global (somatic & dendritic)	Local (only dendritic)
Induction mechanism	Postsynaptic depolarization	Afferent activation
Role of mGluRs	Regulatory	Mandatory
Dependency on postsynaptic Ca^{2+} rise	Yes	No
Endocannabinoid	Anandamide? 2-AG? Other?	2-AG

In conclusion, two independent endocannabinoid-mediated phenomena exist in parallel and may play different roles. Their most relevant differences are summarized in Table 1. The different temporal and spatial characteristics of I-LTD and DSI on synaptic inhibitory efficacy greatly enlarge the repertoire of effects that can be mediated by endocannabinoids.

4. PHYSIOLOGICAL RELEVANCE: ENDOCANNABINOIDS AND SYNAPTIC PLASTICITY IN THE HIPPOCAMPUS.

Changes in hippocampal synaptic plasticity may underlie disruptive effects of cannabinoids on learning and memory(Sullivan, 2000). Certainly, to explain these effects, a better understanding of the endogenous cannabinoid system is needed. Previous studies in the hippocampus have reported that cannabinoid agonists impede the induction of LTP (Collins et al., 1995; Terranova et al., 1995; Stella et al., 1997; Misner and Sullivan, 1999) and LTD at excitatory synapses (Misner and Sullivan, 1999). This action may result from a decrease in glutamate release (Misner and Sullivan, 1999). In contrast, endocannabinoids released during DSI may actually facilitate the induction of LTP (Carlson et al., 2002). While LTP seems to be enhanced in CB1R knockout mice (Bohme et al., 2000), the acute blockade of CB1Rs has no effect on the induction of LTP in vitro (Carlson et al., 2002; Chevaleyre and Castillo, 2003). Thus, the role of endocannabinoids in regulating long-term plasticity at hippocampal excitatory synapses is not entirely clear. We have recently provided evidence that the endocannabinoid-mediated I-LTD not only may underlie changes in CA1 pyramidal cell excitability but may also exert long-lasting modulatory actions on the induction of LTP at excitatory synapses.

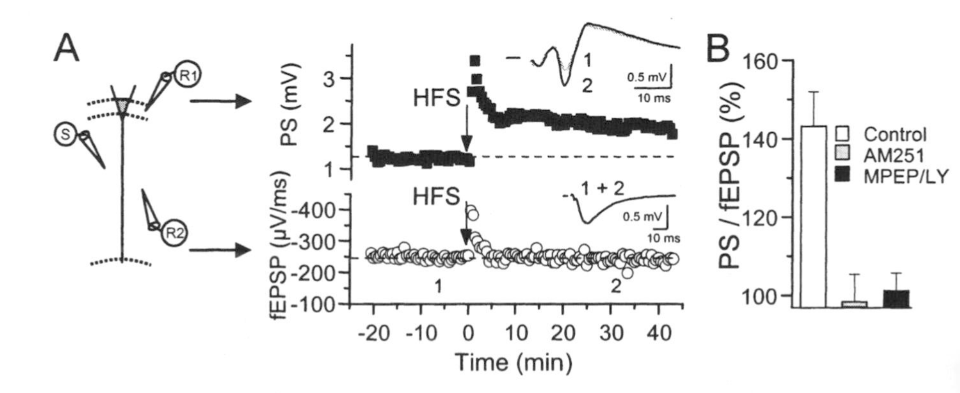

Figure 4. I-LTD is involved in E-S coupling potentiation. **A:** (Left) Arrangement of stimulating and recording pipettes used to monitor extracellular synaptic potentials (fEPSP) and population spike (PS). (Right) Representative experiment showing that, under recording conditions of intact inhibition, TBS induced robust LTP of the PS but not of the fEPSP. Averaged sample traces at the time indicated by numbers are shown on top. **B:** Summary graph of the E-S coupling potentiation (PS/EPSP) obtained after TBS in control conditions or in presence of CB1R or group I mGluR antagonists.

4.1 Role of I-LTD in E-S coupling potentiation.

GABAergic transmission plays a critical role in the regulation of neuronal excitability. The first description of LTP at excitatory synapses (Bliss and Lømo, 1973) already suggested that LTP has two components: (*1*) an increase in the excitatory postsynaptic potential (EPSP), and (*2*) an increase in the ability of an equal sized EPSP to fire an action potential. The latter is the so-called E-S coupling component of LTP that can be easily assessed by simultaneously recording synaptic (field EPSP) and population spike (PS) responses before and after induction stimulation. As shown in the experiment illustrated in figure 4, HFS may potentiate the PS component even in the absence of any potentiation of the EPSP. E-S potentiation has been usually linked to a persistent depression of inhibitory drive (Abraham et al., 1987; Chavez-Noriega et al., 1989; Lu et al., 2000), although changes in intrinsic excitability have been also suggested (Pugliese et al., 1994; Daoudal and Debanne, 2003; Sourdet et al., 2003). A previous report suggested that a calcineurin-mediated and NMDAR-dependent LTD of GABAergic inhibition, presumably due to postsynaptic changes in the sensitivity to GABA, explains E-S coupling potentiation (Lu et al., 2000). However, this observation is hard to reconcile with a previous report indicating that E-S coupling potentiation is NMDAR-independent (Bernard and Wheal, 1995). In our hands, E-S potentiation was abolished when HFS was delivered under experimental conditions known to block I-LTD such as CB1R or mGluR1/5 blockade (Fig. 4B). In addition, transient application of group I mGluR agonist DHPG, which is sufficient to trigger endocannabinoid-mediated long-term depression of inhibitory synaptic transmission (see figure 1), also induced a long-term increase in E-S coupling. As expected for a CB1R-mediated phenomenon, the DHPG-induced E-S potentiation was blocked in presence of AM251. Thus, the long-lasting disinhibition that occurs during the endocannabinoid-mediated and NMDAR-independent I-LTD can account for the enhancement in CA1 pyramidal cell excitability associated with LTP at excitatory synapses.

4.2 Endocannabinoids as mediators of metaplasticity in the hippocampus

It is commonly believed that the history of activity of a given neuron influences its future responses to synaptic inputs (Bienenstock et al., 1982). Previous studies have suggested that prior synaptic activity, which by itself does not trigger long-term changes in synaptic strength, can elicit a persistent modification ($\geq$ 1 hour) in the ability to induce subsequent synaptic plasticity, a phenomenon known as metaplasticity (Abraham and Bear, 1996). Given the well established effects of inhibition in facilitating LTP induction (Wigstrom and Gustafsson, 1983), enduring changes in GABAergic synaptic strength (Gaiarsa et al., 2002) may change subsequent synaptic modifiability, thereby forming the basis of some forms of metaplasticity (Abraham and Bear, 1996; Abraham and Tate, 1997). Most evidence that inhibition reduces the threshold for LTP induction is derived from analyzing the effects of GABA-A receptor antagonists. Therefore, whether or not specific synaptic inhibitory inputs regulate the induction of LTP under physiological conditions is less clear. Preliminary data from our laboratory indicates that by suppressing inhibition in a restricted dendritic area, endocannabinoids can locally and persistently facilitate the induction of LTP at excitatory synapses. Taking advantage on the fact that 10 Hz repetitive stimulation in CA1 *s. radiatum* selectively triggers I-LTD without inducing long-term plasticity at excitatory synapses (Fig 5A), we have recently found that priming synapses with a 10 Hz stimulation protocol facilitates subsequent induction of LTP (Fig. 5B). As expected if this facilitation is due to I-LTD, manipulations that block the induction of I-LTD, such as CB1R or mGluR1/5 pharmacological blockade, also abolished the priming effect (Fig. 5C). Also, no facilitation of LTP induction was observed in CB1R knockout mice. Finally, the facilitation induced by priming shares I-LTD's spatial and temporal features, that is, it is localized to the postsynaptic region exhibiting I-LTD, and shows a similar time course. Thus, the disinhibition caused by synaptically-driven release of endocannabinoids not only enhances excitability (Chevaleyre and Castillo, 2003) but also mediates metaplasticity.

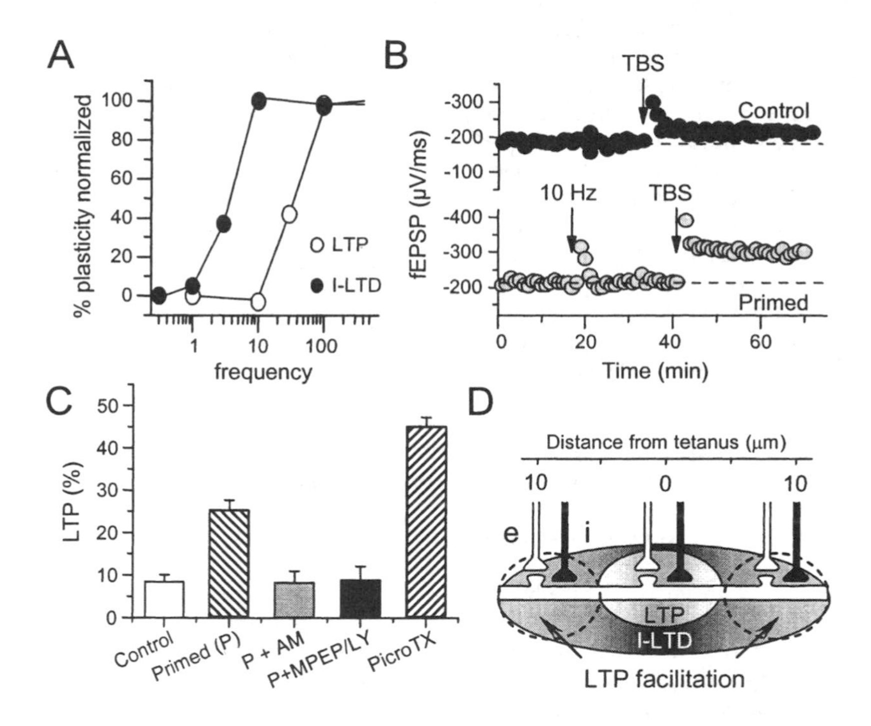

Figure 5. Endocannabinoid-mediated metaplasticity in the hippocampus. **A:** Frequency-response plot of the magnitude of LTP (white circles, excitatory synapses) and I-LTD (black circles, inhibitory synapses), both induced with two bursts of 100 stimuli. I-LTD can be triggered independently of LTP by a 10 Hz stimulating protocol. **B:** Representative experiment in which two independent pathways were monitored in the CA1 area using weak stimulation (fEPSP amplitude less than 0.3 mV), one was primed with a 10 Hz stimulation protocol whereas the naïve pathway served as control. Inhibition was intact in these experiments. TBS applied 20 minutes post-priming induced larger LTP than in the naïve pathway. **C:** Summary graph of the magnitude of LTP observed in control conditions, after priming (P), or after priming in presence of AM251 and MPEP/LY367385 to block CB1Rs and group I mGluRs respectively. For comparison, the magnitude of LTP in presence of the $GABA_AR$ antagonist pricrotoxin is also shown. **D:** Local facilitatory effects of I-LTD on LTP induction at Sch-CA1 synapses. The cartoon illustrates excitatory (*e*, white) and inhibitory inputs (*i*, black) impinging on the apical dendrite of a CA1 pyramidal cell. Local activation of excitatory inputs triggers LTP in a highly restricted area (<10 μm from the stimulating site) and at the same time, it triggers I-LTD in a slightly larger area, thereby facilitating LTP induction of neighboring excitatory inputs.

Changes in the induction threshold for synaptic plasticity are commonly specific to those synapses that were previously active (Abraham and Bear, 1996). Although the hippocampal endocannabinoid-mediated metaplasticity is a highly localized phenomenon, it may also target adjacent naïve synapses that are not necessarily activated during its induction. The close anatomical proximity between glutamatergic and CB1R-containing GABAergic synapses on the dendritic tree (Hajos et al., 2000), is entirely consistent with the interaction between excitatory and inhibitory inputs that is mediated by local endocannabinoid signaling. The fact that the activation of glutamatergic fibers triggers local release of endocannabinoids clearly suggests that this process is well-suited to regulate neighboring synapses within a narrow band of the dendrite. Interestingly, LTP seems to be more restricted than I-LTD. Because patterns of activity that induce LTP at excitatory synapses also induce I-LTD, we postulate that LTP is not only a change in synaptic strength but also a persistent facilitation of surrounding excitatory synapses mediated by endocannabinoids (Fig. 5D).

4.3 Effects of THC *in vivo* exposure on endocannabinoid-mediated forms of plasticity

Disruptive effects on cognitive functions such as learning and memory are one of the most widely described actions of marijuana consumption or cannabinoids injections (Sullivan, 2000; Lichtman et al., 2002). Yet, the molecular mechanisms underlying these effects are poorly understood. There is growing evidence that addictive drugs can alter synaptic plasticity in different brain structures (Ungless et al., 2001; Melis et al., 2002; Nestler, 2002; Hoffman et al., 2003; Fourgeaud et al., 2004; Kauer, 2004; Mato et al., 2004). Previous studies have shown that chronic treatment with cannabinoid decreases the number of CB1Rs in the hippocampus (Romero et al., 1997; Breivogel et al., 1999). In that way, cannabinoids could affect the ability of inhibitory synapses to express I-LTD. We have recently reported (Mato et al., 2004) that a single *in vivo* injection of a low dose of THC (3 mg/Kg), the principal psychoactive ingredient of marijuana, causes a profound depression of endocannabinoid-dependent forms of synaptic plasticity. In the hippocampus, I-LTD was abolished and DSI was highly reduced one day post-THC injection both in rats (Fig. 6) and mice (Mato et al., 2004).

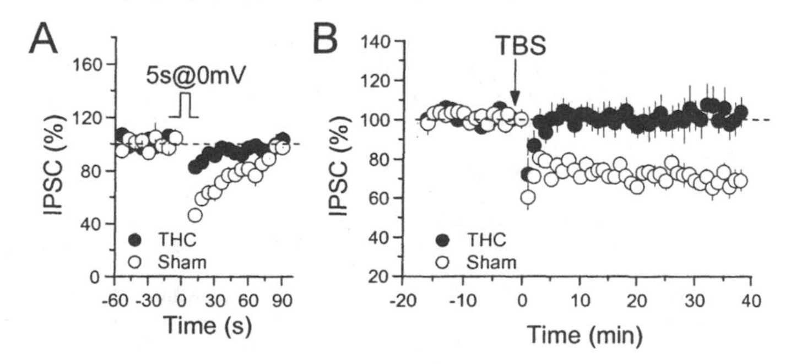

Figure 6. Single in vivo administration of THC abolishes endocannabinoid-mediated synaptic plasticity in the hippocampus. Summary graph of DSI (A) and I-LTD (B) obtained one day after THC (black circles) or vehicle (white circles) injection in rats.

Similar blockade was observed in the endocannabinoid-mediated LTD of glutamatergic synapses in the nucleus accumbens (Mato et al., 2004). These effects are reversible within three days post-injection and are associated with a transient decrease in the functional properties but not in the number of CB1Rs. Given the important role of the endocannabinoid retrograde system in hippocampal synaptic plasticity, it is likely that changes induced by THC exposure on I-LTD and DSI may explain the deleterious effects of cannabinoids on memory formation.

5. CONCLUDING REMARKS

Excitatory inputs in the hippocampus can feedback onto inhibitory synapses and induce long-term changes in GABAergic synaptic efficacy via retrograde endocannabinoid signaling (I-LTD). The disinhibition caused by synaptically-driven release of endocannabinoids not only enhances excitability (Chevaleyre and Castillo, 2003) but also mediates metaplasticity. These findings highlight the relevance of endocannabinoids in activity-dependent synaptic plasticity and reveal the variety of mechanisms by which these messengers may contribute to the storage of information in the brain. The profound repercussion of THC exposure on endocannabinoid-mediated synaptic plasticity in the hippocampus provides a mechanism by which cannabis derivatives may alter cognitive functions.

6. ACKNOWLEDGMENTS

This work was supported by NIH (National Institutes of Drug Abuse). P.E.C. is a Biomedical Pew Scholar. V.C. is an Epilepsy Foundation postdoctoral fellow.

7. REFERENCES

Abraham WC, Bear MF (1996) Metaplasticity: the plasticity of synaptic plasticity. Trends Neurosci 19:126-130.

Abraham WC, Tate WP (1997) Metaplasticity: a new vista across the field of synaptic plasticity. Prog Neurobiol 52:303-323.

Abraham WC, Gustafsson B, Wigstrom H (1987) Long-term potentiation involves enhanced synaptic excitation relative to synaptic inhibition in guinea-pig hippocampus. J Physiol 394:367-380.

Bernard C, Wheal HV (1995) Expression of EPSP/spike potentiation following low frequency and tetanic stimulation in the CA1 area of the rat hippocampus. J Neurosci 15:6542-6551.

Bienenstock EL, Cooper LN, Munro PW (1982) Theory for the development of neuron selectivity: orientation specificity and binocular interaction in visual cortex. J Neurosci 2:32-48.

Bliss TV, Lømo T (1973) Long-lasting potentiation of synaptic transmission in the dentate area of the anaesthetized rabbit following stimulation of the perforant path. J Physiol 232:331-356.

Bohme GA, Laville M, Ledent C, Parmentier M, Imperato A (2000) Enhanced long-term potentiation in mice lacking cannabinoid CB1 receptors. Neuroscience 95:5-7.

Breivogel CS, Childers SR, Deadwyler SA, Hampson RE, Vogt LJ, Sim-Selley LJ (1999) Chronic delta9-tetrahydrocannabinol treatment produces a time-dependent loss of cannabinoid receptors and cannabinoid receptor-activated G proteins in rat brain. J Neurochem 73:2447-2459.

Caillard O, Ben-Ari Y, Gaiarsa J-L (1999a) Mechanisms of Induction and Expression of Long-Term Depression at GABAergic Synapses in the Neonatal Rat Hippocampus. J Neurosci 19:7568-7577.

Caillard O, Ben-Ari Y, Gaiarsa JL (1999b) Long-term potentiation of GABAergic synaptic transmission in neonatal rat hippocampus. J Physiol 518 (Pt 1):109-119.

Caillard O, Ben-Ari Y, Gaiarsa JL (2000) Activation of presynaptic and postsynaptic ryanodine-sensitive calcium stores is required for the induction of long-term depression at GABAergic synapses in the neonatal rat hippocampus amphetamine. J Neurosci 20:RC94.

Carlson G, Wang Y, Alger BE (2002) Endocannabinoids facilitate the induction of LTP in the hippocampus. Nat Neurosci 5:723-724.

Chavez-Noriega LE, Bliss TV, Halliwell JV (1989) The EPSP-spike (E-S) component of long-term potentiation in the rat hippocampal slice is modulated by GABAergic but not cholinergic mechanisms. Neurosci Lett 104:58-64.

Chevaleyre V, Castillo PE (2003) Heterosynaptic LTD of hippocampal GABAergic synapses: a novel role of endocannabinoids in regulating excitability. Neuron 38:461-472.

Collins DR, Pertwee RG, Davies SN (1995) Prevention by the cannabinoid antagonist, SR141716A, of cannabinoid-mediated blockade of long-term potentiation in the rat hippocampal slice. Br J Pharmacol 115:869-870.

Daoudal G, Debanne D (2003) Long-term plasticity of intrinsic excitability: learning rules and mechanisms. Learn Mem 10:456-465.

Diana MA, Marty A (2004) Endocannabinoid-mediated short-term synaptic plasticity: depolarization-induced suppression of inhibition (DSI) and depolarization-induced suppression of excitation (DSE). Br J Pharmacol 142:9-19.

Egertova M, Elphick MR (2000) Localisation of cannabinoid receptors in the rat brain using antibodies to the intracellular C-terminal tail of CB. J Comp Neurol 422:159-171.

Fourgeaud L, Mato S, Bouchet D, Hemar A, Worley PF, Manzoni OJ (2004) A single in vivo exposure to cocaine abolishes endocannabinoid-mediated long-term depression in the nucleus accumbens. J Neurosci 24:6939-6945.

Freund TF, Buzsaki G (1996) Interneurons of the hippocampus. Hippocampus 6:347-470.

Freund TF, Katona I, Piomelli D (2003) Role of endogenous cannabinoids in synaptic signaling. Physiol Rev 83:1017-1066.

Gaiarsa JL, Caillard O, Ben-Ari Y (2002) Long-term plasticity at GABAergic and glycinergic synapses: mechanisms and functional significance. Trends Neurosci 25:564-570.

Gerdeman GL, Lovinger DM (2003) Emerging roles for endocannabinoids in long-term synaptic plasticity. Br J Pharmacol 140:781-789.

Gerdeman GL, Ronesi J, Lovinger DM (2002) Postsynaptic endocannabinoid release is critical to long-term depression in the striatum. Nat Neurosci 5:446-451.

Hajos N, Katona I, Naiem SS, MacKie K, Ledent C, Mody I, Freund TF (2000) Cannabinoids inhibit hippocampal GABAergic transmission and network oscillations. Eur J Neurosci 12:3239-3249.

Hoffman AF, Lupica CR (2000) Mechanisms of cannabinoid inhibition of GABA(A) synaptic transmission in the hippocampus. J Neurosci 20:2470-2479.

Hoffman AF, Oz M, Caulder T, Lupica CR (2003) Functional tolerance and blockade of long-term depression at synapses in the nucleus accumbens after chronic cannabinoid exposure. J Neurosci 23:4815-4820.

Holmgren CD, Zilberter Y (2001) Coincident spiking activity induces long-term changes in inhibition of neocortical pyramidal cells. J Neurosci 21:8270-8277.

Howlett AC, Barth F, Bonner TI, Cabral G, Casellas P, Devane WA, Felder CC, Herkenham M, Mackie K, Martin BR, Mechoulam R, Pertwee RG (2002) International Union of Pharmacology. XXVII. Classification of Cannabinoid Receptors. Pharmacol Rev 54:161-202.

Irving AJ, Coutts AA, Harvey J, Rae MG, Mackie K, Bewick GS, Pertwee RG (2000) Functional expression of cell surface cannabinoid CB(1) receptors on presynaptic inhibitory terminals in cultured rat hippocampal neurons. Neuroscience 98:253-262.

Katona I, Sperlagh B, Sik A, Kafalvi A, Vizi ES, Mackie K, Freund TF (1999) Presynaptically located CB1 cannabinoid receptors regulate GABA release from axon terminals of specific hippocampal interneurons. J Neurosci 19:4544-4558.

Kauer JA (2004) Learning mechanisms in addiction: synaptic plasticity in the ventral tegmental area as a result of exposure to drugs of abuse. Annu Rev Physiol 66:447-475.

Kim J, Alger BE (2004) Inhibition of cyclooxygenase-2 potentiates retrograde endocannabinoid effects in hippocampus. Nat Neurosci 7:697-698.

Komatsu Y (1996) GABAB receptors, monoamine receptors, and postsynaptic inositol trisphosphate-induced Ca2+ release are involved in the induction of long-term potentiation at visual cortical inhibitory synapses. J Neurosci 16:6342-6352.

Kreitzer AC, Regehr WG (2001) Retrograde inhibition of presynaptic calcium influx by endogenous cannabinoids at excitatory synapses onto Purkinje cells. Neuron 29:717-727.

Kullmann DM, Asztely F (1998) Extrasynaptic glutamate spillover in the hippocampus: evidence and implications. Trends Neurosci 21:8-14.

Lichtman AH, Varvel SA, Martin BR (2002) Endocannabinoids in cognition and dependence. Prostaglandins Leukot Essent Fatty Acids 66:269-285.

Llano I, Leresche N, Marty A (1991) Calcium entry increases the sensitivity of cerebellar Purkinje cells to applied GABA and decreases inhibitory synaptic currents. Neuron 6:565-574.

Lu YM, Mansuy IM, Kandel ER, Roder J (2000) Calcineurin-mediated LTD of GABAergic inhibition underlies the increased excitability of CA1 neurons associated with LTP. Neuron 26:197-205.

Maejima T, Hashimoto K, Yoshida T, Aiba A, Kano M (2001) Presynaptic inhibition caused by retrograde signal from metabotropic glutamate to cannabinoid receptors. Neuron 31:463-475.

Magee J, Hoffman D, Colbert C, Johnston D (1998) Electrical and calcium signaling in dendrites of hippocampal pyramidal neurons. Annu Rev Physiol 60:327-346.

Marsicano G, Wotjak CT, Azad SC, Bisogno T, Rammes G, Cascio MG, Hermann H, Tang J, Hofmann C, Zieglgansberger W, Di Marzo V, Lutz B (2002) The endogenous cannabinoid system controls extinction of aversive memories. Nature 418:530-534.

Mato S, Chevaleyre V, Robbe D, Pazos A, Castillo PE, Manzoni OJ (2004) A single in-vivo exposure to Delta9THC blocks endocannabinoid-mediated synaptic plasticity. Nat Neurosci.

McBain CJ, Fisahn A (2001) Interneurons unbound. Nat Rev Neurosci 2:11-23.

Mechoulam R, Fride E, Di Marzo V (1998) Endocannabinoids. Eur J Pharmacol 359:1-18.

Megias M, Emri Z, Freund TF, Gulyas AI (2001) Total number and distribution of inhibitory and excitatory synapses on hippocampal CA1 pyramidal cells. Neuroscience 102:527-540.

Melis M, Camarini R, Ungless MA, Bonci A (2002) Long-lasting potentiation of GABAergic synapses in dopamine neurons after a single in vivo ethanol exposure. J Neurosci 22:2074-2082.

Miles R, Toth K, Gulyas AI, Hajos N, Freund TF (1996) Differences between somatic and dendritic inhibition in the hippocampus. Neuron 16:815-823.

Misner DL, Sullivan JM (1999) Mechanism of cannabinoid effects on long-term potentiation and depression in hippocampal CA1 neurons. J Neurosci 19:6795-6805.

Nestler EJ (2002) Common molecular and cellular substrates of addiction and memory. Neurobiol Learn Mem 78:637-647.

Ohno-Shosaku T, Maejima T, Kano M (2001) Endogenous cannabinoids mediate retrograde signals from depolarized postsynaptic neurons to presynaptic terminals. Neuron 29:729-738.

Ohno-Shosaku T, Shosaku J, Tsubokawa H, Kano M (2002) Cooperative endocannabinoid production by neuronal depolarization and group I metabotropic glutamate receptor activation. Eur J Neurosci 15:953-961.

Pertwee RG, Ross RA (2002) Cannabinoid receptors and their ligands. Prostaglandins Leukot Essent Fatty Acids 66:101-121.

Piomelli D (2003) The molecular logic of endocannabinoid signalling. Nat Rev Neurosci 4:873-884.

Pitler TA, Alger BE (1992) Postsynaptic spike firing reduces synaptic GABAA responses in hippocampal pyramidal cells. J Neurosci 12:4122-4132.

Pugliese AM, Ballerini L, Passani MB, Corradetti R (1994) EPSP-spike potentiation during primed burst-induced long-term potentiation in the CA1 region of rat hippocampal slices. Neuroscience 62:1021-1032.

Robbe D, Kopf M, Remaury A, Bockaert J, Manzoni OJ (2002) Endogenous cannabinoids mediate long-term synaptic depression in the nucleus accumbens. Proc Natl Acad Sci U S A 99:8384-8388.

Romero J, Garcia-Palomero E, Castro JG, Garcia-Gil L, Ramos JA, Fernandez-Ruiz JJ (1997) Effects of chronic exposure to delta9-tetrahydrocannabinol on cannabinoid receptor binding and mRNA levels in several rat brain regions. Brain Res Mol Brain Res 46:100-108.

Schlicker E, Kathmann M (2001) Modulation of transmitter release via presynaptic cannabinoid receptors. Trends Pharmacol Sci 22:565-572.

Shepherd GM (2003) Information Processing in Complex Dendrites. In: Fundamental Neuroscience (Larry R. Squire FEB, Susan K. McConnell, James L. Roberts, Nicholas C. Spitzer, Michael J. Zigmond, ed). New York: Academic Press.

Shigemoto R, Kinoshita A, Wada E, Nomura S, Ohishi H, Takada M, Flor PJ, Neki A, Abe T, Nakanishi S, Mizuno N (1997) Differential Presynaptic Localization of Metabotropic Glutamate Receptor Subtypes in the Rat Hippocampus. J Neurosci 17:7503-7522.

Sjostrom PJ, Turrigiano GG, Nelson SB (2003) Neocortical LTD via coincident activation of presynaptic NMDA and cannabinoid receptors. Neuron 39:641-654.

Sourdet V, Russier M, Daoudal G, Ankri N, Debanne D (2003) Long-term enhancement of neuronal excitability and temporal fidelity mediated by metabotropic glutamate receptor subtype 5. J Neurosci 23:10238-10248.

Stella N, Schweitzer P, Piomelli D (1997) A second endogenous cannabinoid that modulates long-term potentiation. Nature 388:773-778.

Stelzer A, Slater NT, ten Bruggencate G (1987) Activation of NMDA receptors blocks GABAergic inhibition in an in vitro model of epilepsy. Nature 326:698-701.

Sugiura T, Kobayashi Y, Oka S, Waku K (2002) Biosynthesis and degradation of anandamide and 2-arachidonoylglycerol and their possible physiological significance. Prostaglandins Leukot Essent Fatty Acids 66:173-192.

Sullivan JM (2000) Cellular and molecular mechanisms underlying learning and memory impairments produced by cannabinoids. Learn Mem 7:132-139.

Takahashi KA, Linden DJ (2000) Cannabinoid receptor modulation of synapses received by cerebellar Purkinje cells. J Neurophysiol 83:1167-1180.

Terranova JP, Michaud JC, Le Fur G, Soubrie P (1995) Inhibition of long-term potentiation in rat hippocampal slices by anandamide and WIN55212-2: reversal by SR141716 A, a selective antagonist of CB1 cannabinoid receptors. Naunyn Schmiedebergs Arch Pharmacol 352:576-579.

Tsou K, Mackie K, Sanudo-Pena MC, Walker JM (1999) Cannabinoid CB1 receptors are localized primarily on cholecystokinin- containing GABAergic interneurons in the rat hippocampal formation. Neuroscience 93:969-975.

Ungless MA, Whistler JL, Malenka RC, Bonci A (2001) Single cocaine exposure in vivo induces long-term potentiation in dopamine neurons. Nature 411:583-587.

Varma N, Carlson GC, Ledent C, Alger BE (2001) Metabotropic glutamate receptors drive the endocannabinoid system in hippocampus. J Neurosci 21:RC188.

Vaughan CW, McGregor IS, Christie MJ (1999) Cannabinoid receptor activation inhibits GABAergic neurotransmission in rostral ventromedial medulla neurons in vitro. Br J Pharmacol 127:935-940.

Wang JH, Stelzer A (1996) Shared calcium signaling pathways in the induction of long-term potentiation and synaptic disinhibition in CA1 pyramidal cell dendrites. J Neurophysiol 75:1687-1702.

Wigstrom H, Gustafsson B (1983) Facilitated induction of hippocampal long-lasting potentiation during blockade of inhibition. Nature 301:603-604.

Wilson RI, Nicoll RA (2001) Endogenous cannabinoids mediate retrograde signalling at hippocampal synapses. Nature 410:588-592.

Wilson RI, Nicoll RA (2002) Endocannabinoid signaling in the brain. Science 296:678-682.

Wilson RI, Kunos G, Nicoll RA (2001) Presynaptic specificity of endocannabinoid signaling in the hippocampus. Neuron 31:453-462.

Yuste R, Tank DW (1996) Dendritic integration in mammalian neurons, a century after Cajal. Neuron 16:701-716.

TALKING BACK: ENDOCANNABINOID RETROGRADE SIGNALING ADJUSTS SYNAPTIC EFFICACY

David M. Lovinger*

1. INTRODUCTION

Synaptic transmission is the dominant form of cell-cell communication within the nervous system. The presynaptic terminal of one cell resides in close apposition to the postsynaptic elements of a second cell with a separation of only ~20-50 nm. The synaptic connection is especially physically stable, being cemented by proteins that span the gap between the cells. One has only to examine electron micrographs of brain synapses to appreciate the web of electron dense material connecting the two sides of the synapse (see Spacek and Harris, 1998; Tarrant and Routtenberg, 1997 for example). Much of this material is likely made up of extracellular portions of proteins. Indeed, pre- and postsynaptic elements in the mature brain are so tightly bound together that they are stable even following fractionation procedures that separate subcellular compartments (e.g. during the preparation of synaptosomes). The stability of synapses and the small separation between cells at these specializations creates a permissive environment for transcellular molecular communication.

The most widely known form of communication at synapses involves vesicular release of small neurotransmitter molecules from the presynaptic element that act on postsynaptic receptors to alter neuronal activity (Cooper, Bloom and Roth, 2003; Nicholls et al., 2001) synaptic communication in this "anterograde" direction can take the form of fast communication, wherein the neurotransmitter activates a ligand-gated

---

*D.M. Lovinger, Laboratory for Integrative Neuroscience, National Institute on Alcohol Abuse and Alcoholism/National Institutes of Health, Rockville, MD USA. E-mail: lovindav@willco.niaaa.nih.gov

ion channel that directly affects postsynaptic membrane conductance and membrane potential. Anterograde transmission also produces "neuromodulation" in which the transmitter activates a G-protein-coupled receptor that indirectly influences activity or intracellular signaling in the postsynaptic neuron.

In addition to anterograde transmission, other molecular interactions are involved in transsynaptic communication. Some of these interactions involve traditional small molecule secretion and activation of receptors. For example, presynaptic "autoreceptors" are activated by neurotransmitters released from the same synaptic terminal. Heterosynaptic depression often involves activation of presynaptic receptors by a neurotransmitter released from a nearby presynaptic terminal (Dunwiddie and Lovinger, 1993; Miller, 1998). Other transsynaptic interactions involve direct protein-protein binding. There are now many examples of protein pairs that are linked within the synaptic gap itself, with one protein being produced presynaptically and the interacting protein produced postsynaptically. These transsynaptic protein interactions serve not only to stabilize the synapse structurally and functionally (Ichtchenko et al., 1990; Fu et al., 2003; Scheiffele et al., 2000; Dearn et al., 2003), but some of the proteins, such as the Eph/Ephrin pairs, can serve a receptor-like role, stimulating intracellular signaling pathways (Contractor et al., 2002; Wilkinson, 2001; Palmer and Klein, 2003). In addition, it is quite possible that rearrangement of these transsynaptic protein pairs is important for changes in synaptic morphology. One possible mechanism for transsynaptic communication involving these proteins is "trans-endocytosis, in which membranes from two different cells are taken up into a single cell (see Spacek and Harris, 2004 for discussion). Thus, transsynaptic protein pairs serve a variety of important functions that regulate not only synapse formation, but synaptic efficacy as well.

While the focus of work on small molecule-mediated synaptic transmission has been on traditional pre- to postsynaptic communication, small molecules released from the postsynaptic neuron also act as neurotransmitters or neuromodulators, traveling in a "retrograde" direction across the synapse to alter presynaptic function. A retrograde transmission mechanism would be indicated if postsynaptic activation produced a change in presynaptic function. This type of signaling has now been demonstrated at a variety of synapses in the brain. For example, growth factors released from postsynaptic cells act on receptors located on presynaptic terminals to ensure cell survival and promote and stabilize synapse formation (see for example Heerssen and Segal, 2002). Growth factors may also participate in changing synaptic efficacy, although this function may involve both anterograde and retrograde transmission (Schuman, 1999; Lu, 2003; Zakharenko, 2003). Traditional neurotransmitters, molecules such as nitric oxide, and lipid metabolites also appear to participate in retrograde signaling that alters synaptic transmission (Dahl and Sarvey, 1989; Feinmark et al., 2003; Gage et al., 1997; Kreitzer and Regehr, 2001; Madison and Schuman, 1995, Ohno-Shosaku et al., 2001; Wilson and Nicoll, 2001). Retrograde signaling by these molecules produces neuromodulatory effects on the presynaptic terminal that bring about changes in synaptic transmission lasting for seconds (short-term) to minutes or longer (long-term). The remainder of this review will focus on retrograde signaling mediated by lipid metabolites known as endogenous cannabinoids, or endocannabinoids, that produces long-term changes in presynaptic function. Some mention will also be made of shorter-term presynaptic changes involving endocannabinoid signaling.

2. ENDOCANNABINOIDS: ARACHIDONYL AGONISTS AT CB RECEPTORS

The profound intoxicating effects of cannabinoid drugs have long been known to humans. However, the molecular mechanisms through which these widely-abused substances produce intoxication remained unclear until the 1980s. At that time evidence for a cannabinoid receptor began to accumulate based on radioligand binding studies and other neurochemical approaches (Devane et al., 1988; Herkenham et al., 1990; Howlett et al., 1988). With the advent of advanced cDNA cloning techniques, a variety of cell surface receptors were discovered that were capable of activating GTP-binding proteins (G-proteins). Among these was a receptor, originally discovered based on homology to other class I G-protein coupled receptors, that was discovered to interact with Δ9-THC and synthetic cannabinoid drugs (Matsuda et al., 1990). This receptor, now termed the cannabinoid 1 or CB1 receptor, has become well established as the major target molecule for cannabinoid actions in the brain.

The CB1 receptor most often activates the Gi/o class of G-protein, and this gives rise to a well known pattern of neuronal responses including calcium channel inhibition and inhibition of adenylate cyclase (Howlett et al., 1988; Mackie and Hille, 1992; Matsuda et al., 1990). Adenylate cyclase inhibition can, in turn, lead to downstream effects on other intracellular signaling pathways such as extracellular signal-regulated kinase (Davis et al., 2003; Derkinderen et al., 2003). CB1-mediated enhancement of activation of voltage-gated K+ channels has also been reported in hippocampal neurons (Deadwyler et al., 1993). Activation of GIRK-type potassium channels, which are normally activated by β/γ G-protein subunits subsequent to Gi/o activation, has been observed in heterologous systems (Henry and Chavkin, 1995), but oddly has not yet been seen in central neurons. Immunological and electrophysiological experiments indicate that the majority of CB1 receptors reside on presynaptic terminals in the CNS where they can inhibit synaptic transmission by depressing neurotransmitter release (Gerdeman and Lovinger, 2001; Huang et al., 2001; Hoffman and Lupica, 2001; Katona et al., 1999; Kreitzer and Regehr, 2001; Levenes et al., 1998; Ohno-Shosaku et al., 2001; Robbe et al., 2001; Rodriguez et al., 2000; Shen et al., 1996; Szabo et al., 1998; Wilson and Nicoll, 2001; Takahashi and Linden, 2000), a common action of receptors that activate Gi/o type G-proteins (Dunwiddie and Lovinger, 1993; Miller, 1998).

The psychoactive effects of cannabinoids and the identification of their receptor aroused suspicion that endogenous compounds capable of activating these receptors might exist, as was previously shown in pioneering studies of opiate receptors and their endogenous agonists (Pert and Snyder 1973). However, it has not always proven to be the case that binding sites for psychoactive compounds correspond to those for endogenous substances (for example, in the case of the phencyclidine binding site). In the case of the CB1 receptor, the analogy to the opiate receptor is apt in that the receptors are members of the same structural superfamily, and the cannabinoid site appeared to be the primary agonist binding site on the receptor. Within a few years of CB1 receptor identification evidence began to accumulate for the existence of lipid metabolites with agonist activity at this receptor. The first such compound to be identified was arachidonyl ethanolamide, or anandamide (figure 1). DeVane, Mechoulam and colleagues isolated anandamide during a screen for endogenous compounds with affinity for cannabinoid receptors (DeVane et al. 1992). Subsequent studies demonstrated that this compound has psychotropic effects mediated by the CB1 receptor (Fride and

Mechoulam, 1993; Crawley et al., 1993). In the following years work by Waku and colleagues as well as the Mechoulam group demonstrated that 2-arachidonyl glycerol (2-AG) (figure 1) also acts as an agonist at CB1 receptors (Mechoulam et al., 1995; Sugiura et al., 1995). Several other putative endogenous cannabinoids, or endocannabinoids, have since been proposed (Freund et al., 2003; Mechoulam, 2002; Porter et al., 2002), but the jury is still out as to whether these compounds are acceptable candidates for mediation of physiological endocannabinoid signaling.

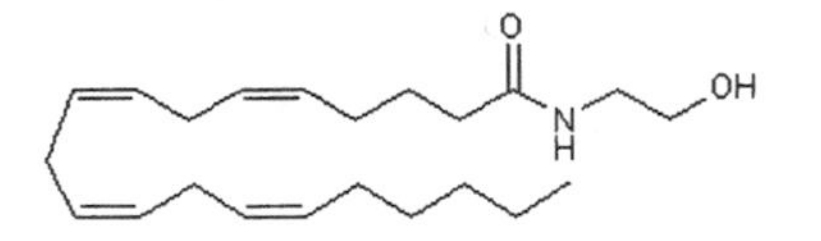

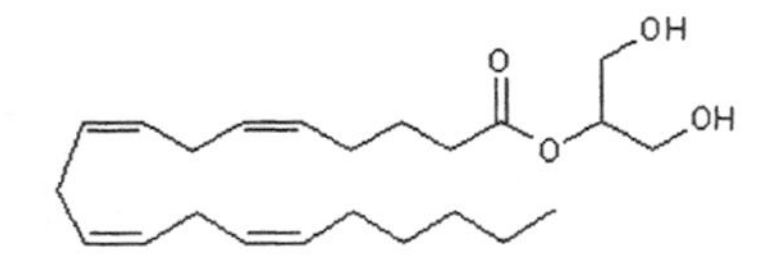

Figure 1. Endocannabinoids.

Evidence for regulated synthesis, release, cellular uptake and catabolism of anandamide and 2-AG has now been reported in a variety of cell and tissue types, including several brain regions (see Fride, 2002; Wilson and Nicoll, 2002 for review). The exact molecular identity of many of the molecules involved in these processes is not yet known, however there is much useful information already in the literature. In the case of anandamide, it appears that synthesis from arachidonic acid-containing lipid precursors can proceed via a two step process involving acylation of the phospholipid PE to form an n-arachidonyl-phosphatidylethanolamine (NAPE). This phospholipid can then be cleaved by phospholipase D to yield anandamide (Cadas et al., 1996; Sugiura et al., 1996; Di Marzo et al., 1996). Mechanisms involved in release are less well understood, but could involve simple transmembrane diffusion or facilitated diffusion-style transport (see discussion later in this review). The exact mechanism for endocannabinoid uptake is also unclear, but solid evidence has now accumulated that indicates the presence of a temperature-dependent and saturable uptake system, at least in certain cells (Di Marzo et al., 1994; Beltramo et al., 1997; Fegley et al., 2004; Hillard et al., 1997; but see also Glaser et al., 2003). Once inside the cell, anandamide is degraded by fatty acid amidohydrolase (FAAH) (Desarnaud et al., 1995; Ueda et al., 1995; Hillard et al., 1995; Cravatt et al., 1996). In the case of 2-AG, the release mechanism may be quite similar to that for anandamide, and there is evidence that cellular uptake of the two endocannabinoids occurs via the same mechanism (Piomelli et al., 1990). Synthesis of 2-AG can occur via either PLC or PLA1 mediated phospholipid hydrolysis (see Sugiura et al., 2002 for review). The PLC-catalyzed degradation yields diacylglycerol (DAG) that is then degraded by DAG lipase to 2-AG (Allen et al., 1992; Bisogno et al., 1997; Di Marzo et al., 1996; Sugiura et al., 2002). Enzymatic inactivation of 2-AG appears to involve a monoglycerol lipase that has 2-AG as one of its substrates (Dinh et al., 2002). One other fatty acid-metabolizing enzyme, namely cyclooxygenase 2 (COX-2) has also been shown to bind and break down both anandamide and 2-AG (Kozak et al. 2002) The cellular locations of the different enzymes involved in endocannabinoid metabolism varies among brain regions, giving rise to a variety of possible transcellular signaling scenarios involving both neurons and glia (Gulyas et al., 2004; Egertova et al., 1998; Egertova et al., 2003). The remainder of the present review will focus on one type of communication, namely retrograde synaptic signaling.

3. ENDOCANNABINOID-MEDIATED RETROGRADE SIGNALING IN SHORT- AND LONG-TERM PRESYNAPTIC PLASTICITY

There is now abundant evidence that endocannabinoids serve a retrograde signaling role in several areas of the nervous system. The first evidence has come from characterization of a short-lasting form of synaptic depression known as depolarization-induced suppression of inhibition (DSI). In the early 1990s several groups noted that depolarization of neurons produced decreased GABAergic inhibitory synaptic input (Llano et al., 1991; Pitler and Alger, 1992). This "suppression of inhibition" persisted for less than a minute after depolarization. Several pieces of evidence indicated that DSI expression involves decreased release of GABA from presynaptic terminals impinging the on the depolarized postsynaptic neuron. Evidence for this scenario includes increased paired-pulse facilitation ratio of synaptic currents, as well as decreases in the frequency of quantal GABAergic inhibitory postsynaptic currents in the absence of alterations in their amplitude, changes that signal decreases in quantal release (Alger et al., 1996; Pitler and Alger, 1994; Vincent et al., 1992). However, DSI was activated by postsynaptic depolarization even in the absence of explicit presynaptic activation. This raised the possibility that a signal arising from the postsynaptic neuron, a so-called retrograde signal, linked postsynaptic activation to depression of neurotransmitter release presynaptically. Subsequent to the discovery of DSI, depolarization-induced suppression of excitatory transmission (DSE) was also observed at parallel fiber and climbing fiber inputs to cerebellar Purkinje neurons (Kreitzer and Regehr, 2001; Maejima et al., 2001; Ohno-Shosaku et al., 2002). DSI and DSE have now been observed in the cerebral cortex and substantia nigra, respectively, in addition to the cerebellum and hippocampus (Melis et al., 2003; Trettel and Levine, 2003)

In the search for a retrograde signaling element in DSI several factors were considered. The signaling was rapid, suggesting a local effect. Postsynaptic depolarization and increased postsynaptic calcium concentration are required for DSI induction. These factors suggested to several investigators that an endocannabinoid might be the retrograde signaling molecule in this form of plasticity (Kreitzer and Regehr, 2001; Ohno-Shosaku et al., 2001; Wilson and Nicoll, 2001). As previously discussed, endocannabinoids are synthesized in a calcium-dependent manner (although calcium-independent mechanisms of synthesis also exist, Maejima et al., 2001; Kim et al., 2002). It was also known that the majority of CB1 receptors reside on presynaptic terminals, including terminals of GABAergic neurons (Egertova et al., 1998; Katona et al., 1999). These receptors inhibit synaptic transmission by reducing neurotransmitter release, making them good candidates for targeting by a retrograde signaling molecule.

Three research groups nearly simultaneously found evidence that endocannabinoids serve as retrograde signals mediating DSI. In the hippocampus, Wilson and Nicoll (2001) and Kano and colleagues (Ohno-Shosaku et al., 2001) generated a variety of evidence that endocannabinoids and CB1 receptors mediate DSI. Kreitzer and Regehr

(2001) demonstrated DSE at parallel and climbing fiber inputs onto Purkinje neurons in cerebellum, and provided similar evidence for mediation by an endocannabinoid and the CB1 receptor. The pattern of cellular and molecular events involved in DSI and DSE are similar at most synapses. Postsynaptic depolarization triggers an increase in postsynaptic calcium levels. This, in turn, stimulates endocannabinoid synthesis and perhaps release as well. The endocannabinoid traverses the synapse and activates presynaptic CB1 receptors. The CB1 receptors inhibit transmitter release. There is evidence in the Calyx of Held, hippocampus and cerebellum that inhibition of release involves reduction in the function of voltage-gated calcium channels. In cerebellum, Regehr and colleagues showed that CB1 activation reduced presynaptic calcium concentrations during afferent activation, and these investigators have recently provided clear evidence that this mechanism is involved in inhibition of transmission (Brown et al., 2004). In hippocampus, Nicoll and colleagues showed that DSI was occluded by blockade of N-type calcium channels, suggesting that the inhibition of transmission involved inhibition of these channels (Wilson et al., 2001). At the Calyx of Held in the brainstem, a CB1-mediated decrease in voltage-gated calcium channel function was recently demonstrated (Kushmerick et al., 2004). This is the most direct evidence to date for an effect of the CB1R on presynaptic calcium channels. CB1 receptors can potentiate the function of voltage-gated K^+ channels (Deadwyler et al. 1993), and this activation may play a role in presynaptic depression (Alger et al., 1996; Robbe et al., 2001; Diana and Marty, 2003). However, most experiments that suggest such a mechanism utilize potassium channel blockers, and the actions of these compounds are often difficult to interpret (see Brown et al., 2004 for discussion of this issue).

The duration of DSI is most likely controlled by the time course of endocannabinoid presence in the synapse. Evidence supporting this idea includes the findings that blockade of endocannabinoid metabolism prolongs DSI (Kim and Alger, 2004), while decreasing endocannabinoid reuptake produces DSI-like depression (Wilson and Nicoll, 2001). Inhibition of FAAH does not seem to affect DSI, but a recent report demonstrates that two inhibitors of COX-2 prolong DSI in hippocampus (Kim and Alger, 2004). The effect of inhibiting MGL remains to be determined. These studies beg the question of which endocannabinoid mediates DSI. At present this is not known, either for DSI or for other endocannabinoid-dependent forms of synaptic plasticity (see discussion below). However, it is highly probable that tools to assess the roles of different endocannabinoids in plasticity will become available within the next few years.

Retrograde signaling by endocannabinoids has also been implicated in several forms of long-term synaptic depression (LTD) in the brain (see Gerdeman and Lovinger, 2003 for discussion). Most forms of endocannabinoid-dependent LTD share some or all of the following common features including dependence on postsynaptic depolarization and increased postsynaptic intracellular calcium, activation of postsynaptic metabotropic glutamate receptors (mGluRs) and, of course, release and presynaptic actions of endocannabinoids (Gerdeman and Lovinger, 2001; Marsicano et al., 2001; Chevaleyre and Castillo, 2002; Robbe et al., 2001; Sjorstrom et al., 2002).

Many mGluR-dependent forms of long-term synaptic plasticity have been observed since the discovery of this type of receptor. One such form was striatal LTD, first characterized in the early 1990s (Calabresi et al., 1992a; Lovinger et al., 1993). Striatal LTD is induced by high frequency stimulation of cortical afferents that synapse onto striatal medium spiny neurons, which are the projection neurons of this subcortical structure. This form of LTD has been examined by several different laboratories, and consensus has formed about several aspects of its induction and expression. In addition to activation of mGluRs, induction of striatal LTD requires postsynaptic depolarization, a rise in postsynaptic calcium concentration, dopamine release and activation of dopamine receptors, including D2 receptors (see Calabresi et al., 2000; Gerdeman and Lovinger, 2003 for review).

In the late 1990s, evidence began to accumulate indicating that striatal LTD involved a presynaptic expression mechanism. Measurements of paired pulse facilitation ratios, the coefficient of variation of excitatory postsynaptic currents (EPSCs) and the frequency of spontaneous EPSCs all indicated a decrease in neurotransmitter release during LTD expression (Choi and Lovinger, 1997a,b). The increase in paired-pulse facilitation indicated that a decrease in release probability underlies LTD expression. This was supported by the fact that the increase in PPF is prevented by treatments that block LTD (Choi and Lovinger, 1997a).

Evidence that an endocannabinoid plays a crucial role in striatal LTD included experiments demonstrating that this form of plasticity is absent in CB1 knockout mice and in striatal slices treated with a CB1 antagonist (Gerdeman et al., 2002). These findings indicated that CB1 receptors were involved in LTD induction, and we had previously observed that CB1 agonists produced presynaptic inhibition at these synapses (Gerdeman and Lovinger, 2001), and this inhibition was lost in the CB1 knockout mouse (Gerdeman et al. 2002). Thus, it was reasonable to suspect that the presynaptic CB1 receptors present at this synapse might be involved in LTD induction. Evidence for a postsynaptic locus of endocannabinoid release in striatal LTD came from a number of experiments. First, postsynaptic manipulations that prevent depolarization or increases in postsynaptic calcium concentrations prevent LTD induction (Calabresi et al, 1992b; Choi and Lovinger, 1997a,b). Similar evidence has been collected in several other brain regions where endocannabinoid-dependent LTD has been observed (i.e. Nucleus accumbens, Visual Cortex and Hippocampus) (Castillo et al., 2003; Robbe et al., 2002; Sjostrom et al., 2003). Endocannabinoid synthesis can be driven by increased intracellular calcium, and thus it is attractive to think that the role of postsynaptic calcium rises is to stimulate endocannabinoid synthesis. Second, Gerdeman et al. (2002) demonstrated that extracellular application of endocannabinoid uptake inhibitors restores LTD even when the postsynaptic cell is filled with a calcium chelator (a condition that would normally block LTD induction). This finding indicates that the most likely effect of postsynaptic calcium chelation is to prevent production of a postsynaptic endocannabinoid, and that this can be overcome if the endocannabinoid is free to come from other cellular sources surrounding the neuron under study. The mGluRs involved in striatal LTD induction are the group I subtypes (mGluRs 1 and 5) (Gubellini et al., 2001, Sung et al., 2001). These receptors are known to reside postsynaptically on the striatal medium spiny neurons (Testa et al., 1995; Testa et al., 1998). Thus, mGluRs most likely participate in LTD induction by stimulating endocannabinoid synthesis. Our laboratory has also gathered evidence that blockade of endocannabinoid release from the postsynaptic neuron prevents LTD induction (Gerdeman et al., 2002; Ronesi et al., 2004). This will be discussed in the following section. Figure 2 presents a summary of the cellular and molecular events that are common to endocannabinoid-dependent LTD at excitatory synapses.

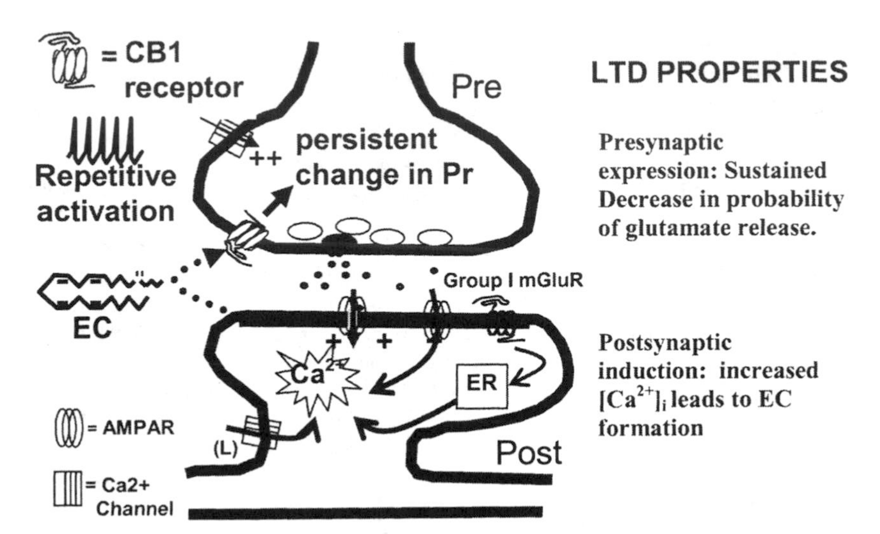

Figure 2. Cellular mechanisms involved in endocannabinoid-dependent LTD at an excitatory synapse. Repetitive presynaptic activation stimulates glutamatergic transmission that activates AMPA receptors and mGluRs. Postsynaptic depolarization activates L-type calcium channels and the resultant calcium influx sums with other signals to increase the postsynaptic intracellular calcium concentration. Increased calcium stimulates endocannabinoid synthesis. The endocannabinoid is released from the postsynaptic neuron and traverses the synapse to act on presynaptic CB1 receptors. Activation of the CB1 receptor, probably in conjunction with other presynaptic signals produced by repetitive activation, produces a long-lasting decrease in the probability of glutamate release. AMPAR = AMPA-type glutamate receptor, EC = endocannabinoid, ER = endoplasmic reticulum, mGluR = metabotropic glutamate receptor, Post = postsynaptic, Pr = probability of release, Pre = presynaptic.

4. POSSIBLE MECHANISMS OF ENDOCANNABINOID RELEASE AND TRANSSYNAPTIC EGRESS

Endocannabinoids are lipophilic molecules, and this raises several significant issues regarding their movement within and between cells. Vesicular concentration of endocannabinoids is unlikely given the fact that the molecules can diffuse across membranes, and given enough time would likely redistribute in and out of the vesicle (although a sufficiently avid transport system could, perhaps, concentrate the molecules). However, evidence from DSI studies in hippocampal neurons indicates that inhibitors of vesicular release do not alter endocannabinoid release (Wilson and Nicoll, 2001). Thus, it is worthwhile exploring other routes of release from the postsynaptic neuron during retrograde transmission.

As mentioned above, endocannabinoids can theoretically freely diffuse through the plasma membrane, and cellular uptake of anandamide has been demonstrated in a number of cell types (Deutsch and Chin, 1993; Di Marzo et al., 1994; Hillard and Jarrahian, 2000). Cellular anandamide uptake is saturable, temperature-dependent and inhibited by certain anandamide analogs (Di Marzo, 1994; Beltramo et al., 1997; Fegley et al, 2004; Hillard and Jarrahian, 2000). In some cells the transmembrane movement of endocannabinoids is less avid than in others. For example, maximal anandamide uptake into HeLa cells is almost an order of magnitude lower than that observed in cerebellar granule neurons and endothelial cells (Hillard and Jarrahian, 2000). These characteristics of endocannabinoid transport have generated speculation on the possible existence of an anandamide transport system (AMT). The AMT exhibits properties consistent with those of facilitated diffusion (Hillard et al., 1997). It must be noted that the molecular composition of the AMT has not been identified (and there is even some doubt about its existence, e.g. Glaser et al., 2003). However, the evidence for its existence is reasonably compelling (see Fegley et al, 2004; Hillard and Jarrahian, 2000; McFarland et al., 2004 for recent evidence supporting a specific transporter). The existence of an AMT-like transporter might allow endocannabinoids to move across the membrane via two different mechanisms, either direct diffusion or facilitated diffusion via the AMT. Regarding the question of endocannabinoid release from cells, there is evidence that an AMT-like transporter can mediate movement of these compounds from an intracellular to an extracellular site (Hillard et al., 1997; Maccarrone et al., 2002). Thus, it is tempting to speculate that an AMT-like system might be involved in endocannabinoid release under some conditions.

We have gathered evidence that an AMT-like transport mechanism mediates postsynaptic release of the endocannabinoid messenger molecule involved in induction of striatal LTD (Ronesi et al., 2004). If a transport inhibitor is loaded into the postsynaptic neuron, it blocks LTD induction. This effect has been observed using two different AMT blockers that differ in their nonspecific effects (Ronesi et al. 2004). In contrast to the effects of intracellular application of these inhibitors, applying the AMT inhibitors in the extracellular solution does not block LTD induction, and in fact can enhance induction by endocannabinoids coming from distal sights (Gerdeman et al. 2002). The reasons for the differential effect of externally and internally applied AMT inhibitors are not yet clear. However, at least two reasonable explanations exist. First, the molecular nature of the membrane transporters involved in endocannabinoid import versus export may differ. If two such transporters exist then one may be sensitive to internal blockers and the other to external blockers. This would make some sense given that transporters involved in cellular uptake would be more likely to have an external endocannabinoid binding site, while those involved in release from cells would likely be specialized for internal binding. The AMT inhibitors are all very similar to anandamide in molecular structure, and therefore probably interact with the endocannabinoid binding site on the transporter. This idea has some merit, but evidence from transport studies indicates that the AMT involved in cellular uptake works via facilitated diffusion. Thus, there is no need to postulate unidirectional transport, and it is probable that the same transporter is involved in both endocannabinoid uptake and release. Furthermore, the findings of Maccarone et al. (2002) in endothelial cells suggest that extracellular transport inhibitors can prevent anandamide release under certain conditions. This argues against the presence of a transporter with only intracellular substrate binding sites.

If only one transporter exists, then the ability of intracellular versus extracellular AMT inhibitors to act on release versus uptake, respectively, probably depends on the relative concentrations of the substrate and inhibitor, as well as the functional state of the transporter itself. If intracellular endocannabinoid levels are high, then high intracellular concentrations of blockers might be necessary to prevent release. In the case where endocannabinoids are high in the extracellular space, then high concentrations of antagonist may be needed to prevent uptake. This would likely occur if there is some competition between substrate and inhibitor for a single binding site on the transporter, as if transported substrate moving in one direction can "knock out" an inhibitor with an opposing concentration gradient if the concentration of substrate is considerably higher. One prediction of this model is that extracellular AMT inhibitor could prevent release if present at very high concentrations relative to intracellular substrate levels, as may have been the case in the Maccarrone et al. (2002) studies. The same would be true of intracellular inhibitor effects on uptake. However, this hypothesis cannot be practically tested in the brain slice/LTD model where concentrations of endogenous substrate cannot be controlled and there are limits on the concentrations of blocker that can be applied. In addition, the concentration gradient of endocannabinoids may influence functional states of the transporter, with high external concentrations promoting an outward-facing binding site, and high internal concentrations promoting an inward-facing binding site. In this case the affinity for inhibitor might change with the different substrate concentrations.

It is important to stress, however, that our understanding of endocannabinoid transmembrane transport is in its infancy. It is quite possible that the AMT differs in molecular nature from other small molecule transporters, and may not behave in a manner similar to those proteins. It is also possible that an AMT protein as such does not exist, and that endocannabinoid uptake and release are controlled by diffusion coupled with local concentration or degradation mechanisms (Glaser et al., 2003) or by lipid microdomains (McFarland et al., 2004). If these scenarios are correct then our findings with the AMT inhibitors might indicate interference with these concentration or diffusion mechanisms. This would not be too surprising given that the inhibitors share most of molecular features of anandamide, and thus could potentially inhibit endocannabinoid interactions with a variety of other molecules (although it should be noted that the inhibitors we have used do no alter endocannabinoid synthesis). It is also important to stress that we have not observed inhibition of DSI induction by AMT inhibitors (Ronesi et al. 2004). This observation suggests that not all endocannabinoid-dependent forms of synaptic plasticity depend on postsynaptic release involving an AMT-like transporter. It will be interesting to determine more about the release mechanisms involved in endocannabinoid-dependent DSI versus LTD.

Another challenging and unresolved issue related to retrograde transport is the mechanism by which the lipophilic endocannabinoid crosses the synapse to interact with a presynaptic receptor. The nagging question in this context is how a highly lipophilic fatty acid derivative moves through the extracellular environment. Estimated octanol-water partition coefficient (Log P) values for anandamide and 2-AG are 6.31 and 8.01, respectively [calculated using the KowWin 9LogKow Log P Calculation algorithm at http://www.syrres.com/esc/est_kowdemo.htm]. These endocannabinoids would likely be concentrated in cell membranes at concentrations several orders of magnitude higher than those present in aqueous solution in the extracellular space. Such a molecule would be unlikely to enter a hydrophilic solution in appreciable quantities unless the concentration gradient was huge. Even given these physico-chemical considerations, it is apparent that anandamide can partition into extracellular solutions, as picomole quantities can be recovered using intracerebral microdialysis (Giuffrida et al. 1999). Thus, it might be the case that generation of large amounts of endocannabinoid compounds results in a small proportion moving into a hydrophilic extracellular environment. On the other hand, electrophysiological experiments examining DSI and LTD suggest that endocannabinoids generated by a single cell do not produce actions at synapses more than 20 μm from the cell of origin (Brown et al., 2003; Wilson and Nicoll, 2001), indicating a steep drop in concentration with distance from the release site. It is not clear if this limited spread is due to preference for hydrophobic environments or limitations produced by cellular uptake. The observation that cannabinoid uptake inhibition can re-establish synaptic depression in which endocannabinoid production is blocked by calcium chelation (Gerdeman et al., 2002) suggests uptake may be one of the major factors limiting the distance of endocannabinoid diffusion.

However, there is not necessarily any need to invoke a scenario in which endocannabinoids must freely diffuse through a hydrophilic extracellular space given that the extracellular environment likely contains a variety of hydrophobic or partially hydrophobic molecules that could facilitate trans-cellular endocannabinoid movement. The simplest idea is that a fatty acid binding "carrier" protein might help remove the endocannabinoid from one membrane and facilitate its presentation to a receptor on the membrane of another cell. Obviously, there is abundant precedent for such a carrier system, as serum albumin has long been known to mediate intercellular fatty acid transport in blood. Indeed, serum albumin can serve as an artificial carrier to deliver endocannabinoids and cannabinoid drugs to neurons in brain slice preparations (Ronesi et al., 2004). No specific endocannabinoid-binding carrier proteins have yet been identified, but it is reasonable to speculate that such proteins might exist.

Another possibility is that endocannabinoids might bind to hydrophobic elements of proteins that span the synapse. It has long been appreciated from electron micrographic images that electron dense material is concentrated in the synaptic cleft itself, and that this material likely contains protein and glycoprotein molecules (see Spacek and Harris, 1998). Several pairs of proteins that can form transsynaptic linkages have now been identified, including the Eph-Ephrin pairs and the nuerexin-neuroligin pairs (Ichtchenko et al., 1990; Fu et al., 2003; Scheiffele et al., 2000; Dearn et al., 2003; Contractor et al., 2002; Wilkinson, 2001; Palmer and Klein, 2003). It is easy to conceive that endocannabinoids might traverse the syanapse by binding to hydrophobic regions formed by these transsynaptic proteins on their journey from the postsynaptic release site to the presynaptic receptor. Alternatively, endocannabinoids could reach the presynaptic terminal via a process of uptake similar to trans-endocytosis (Spacek and Harris, 2004). However, to date there is no specific evidence for such routes of endocannabinoid movement, and it remains to be determined if any transsynaptic protein pairs have significant extracellular regions that could interact with hydrophobic molecules. Nevertheless, it might be worth examining the role of transynaptic proteins in retrograde endocannabinoid signaling. Endocannabinoid signaling, transsynaptic protein function and trans-endocytosis may all work together as part of the rich dialogue of synaptic transmission and synaptic plasticity (Colley and Routtenberg, 1993).

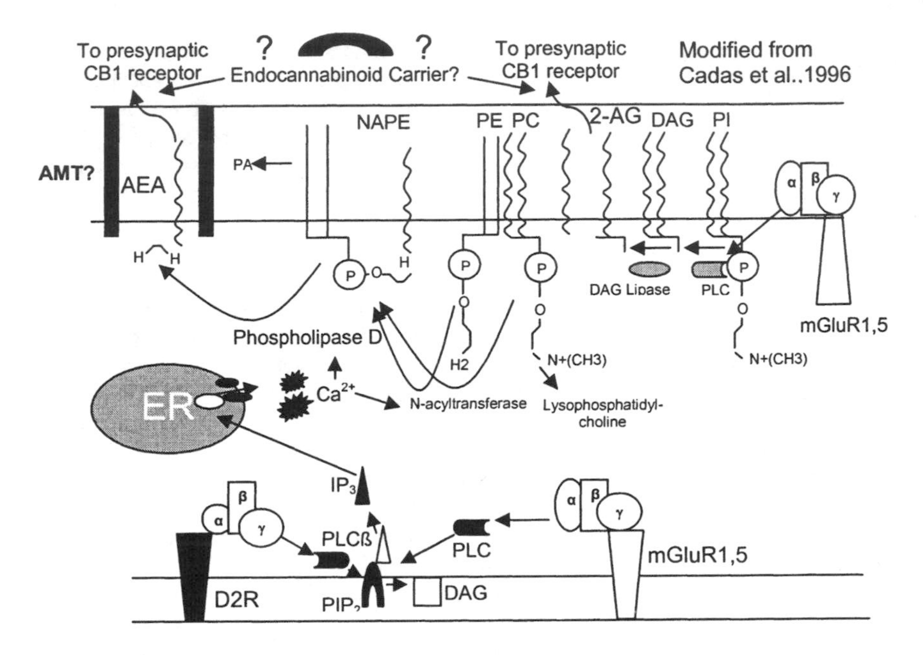

Figure 3. Schematic diagram of possible endocannabinoid production, release and transsynaptic movement mechanisms involved in striatal LTD. 2-AG = 2-arachidonyl glycerol, AEA = anandamide, AMT = anandamide membrane transporter, α = alpha subunit of heterotrimeric G-protein, β = beta G-protein subunit, γ = gamma G-protein subunit, DAG = diacylglycerol, D2R = D2 dopamine receptor, ER = endoplasmic reticulum, mGluR1,5 = metabotropic glutamate receptor 1 or 5, NAPE = n-acetylphosphatidylethanolamine, PC = phosphatidylcholine, PE = phosphatidyl ethanolamine, PLC = phospholipase C, PIP2 = phosphatidylinositol-bis-phosphate.

5. CONCLUSIONS

In addition to the well-known anterograde form of synaptic transmission, neurons participate in a variety of forms of neuromodulatory synaptic communication. Among these is retrograde synaptic communication in which the postsynaptic neuron alters presynaptic function via either chemical or direct mechanical signals. Endocannabinoids, fatty acid metabolites of arachidonyl lipds, are the chemical agents that mediate retrograde synaptic depression at a variety of synapses in the nervous system. Endocannabinoid-dependent synaptic depression involves postsynaptic synthesis of the compound via phospholipases that are activated by metabotropic receptors and/or calcium. The endocannabinoid is released from the postsynaptic neuron, perhaps via a plasma membrane transport system, and acts on presynaptic CB1 cannabinoid receptors to inhibit transmission. Retrograde endocannabinoid signaling can be terminated by uptake and/or enzymatic degradation. Both short- and long-term forms of endocannabinoid-dependent synaptic depression have been extensively documented. Short-lasting synaptic depression, including DSI and DSE is likely mediated by CB1 activation leading to modulation of presynaptic calcium channels, although modulation of potassium channels may also play a role. Longer-lasting cannabinoid-dependent LTD likely involves activation of presynaptic intracellular signaling pathways subsequent to CB1 activation. The endocannabinoid-activated signals may work in conjunction with other presynaptic signals to bring about LTD. Much remains to be determined about the identity of the endocannabiniods that mediate short- and long-term synaptic depression. In addition, the factors that regulate the function of enzymes involved in endocannabinoid synthesis and degradation are not well understood. Mechanisms of endocannabinoid uptake and trans-synaptic movement are also poorly characterized. The next few years should bring about exciting new information about all these aspects of endocannabinoid signaling and the role of this signaling in synaptic depression and cognitive function.

6. ACKNOWLEDGEMENTS

This work was supported by the NIAAA intramural research program.

7. REFERENCES

Alger BE, Pitler TA, Wagner JJ, Martin LA, Morishita W, Kirov SA, Lenz RA (1996) Retrograde signaling in depolarization-induced suppression of inhibition in rat hippocampal CA1 cells, J Physiol (Lond) 496[Pt 1]: 197-209.

Allen AC, Gammon CM, Ousley AH, McCarthy KD, Morrell P (1992) Bradykinin stimulates arachidonic acid release through the sequential actions of an sn-1 diacylglycerol lipase and a monoacylglycerol lipase, J Neurochem, 58(3):1130-1139.

Beltramo M, Stella N, Calignano A, Lin SY, Makriyannis A, Piomelli D. (1997) Functional role of high-affinity anandamide transport, as revealed by selective inhibition, Science 277:1094-1097.

Berke JD, Hyman SE (2000) Addiction, dopamine, and the molecular mechanisms of memory, Neuron 25:515-532.

Bisogno T, Sepep N, Melck D, Maurelli S, De Petrocellis L, Di Marzo V (1997) Biosynthesis, release and degradation of the novel endogenous cannabimimetic metabolite 2-arachidonoylglycerol in mouse neuroblastoma cells, Biochem J, 322 (Pt 2):671-677.

Brown SP, Safo PK, Regehr WG. (2004) Endocannabinoids inhibit transmission at granule cell to Purkinje cell synapses by modulating three types of presynaptic calcium channels, J Neurosci, 24(24):5623-5631.

Cadas H, Gaillet S, Beltramo M, Venance L, Piomelli D (1996) Biosynthesis of an endogenous cannabinoid precursor in neurons and its control by calcium and cAMP, J Neurosci 16: 3934-3942.

Calabresi P, Maj R, Mercuri NB, Bernardi G. (1992a) Coactivation of D1 and D2 dopamine receptors is required for long-term synaptic depression in the striatum. Neurosci Lett 142(1):95-99.

Calabresi P, Maj R, Pisani A, Mercuri NB, Bernardi G (1992b) Long-term synaptic depression in the striatum: physiological and pharmacological characterization, J Neurosci 12:4224-4233.

Calabresi Pl, Centonze D, Gubellini P, Marfia GA, Pisani A, Sancesario G, Bernardi G (2000) Synaptic transmission in the striatum: from plasticity to neurodegeneration, Prog Neurobiol, 61(3):231-265.

Chevaleyre V, Castillo PE (2003) Heterosynaptic LTD of hippocampal GABAergic synapses: a novel role of endocannabinoids in regulating excitability, Neuron, 38:461-472.

Choi S, Lovinger DM (1997a) Decreased probability of neurotransmitter release underlies striatal long-term depression and postnatal development of corticostriatal synapses, Proc. Natl. Acad. Sci. USA 94: 2665-2670.

Choi S, Lovinger DM (1997b) Decreased frequency but not amplitude of quantal synaptic responses associated with expression of corticostriatal long-term depression, J Neurosci 17:8613-8620.

Colley PA, Routtenberg A (1993) Long-term potentiation as synaptic dialogue, Brain Res Brain Res Rev, 18(1):115-122.

Contractor A, Rogers C, Maron C, Henkemeyer M, Swanson GT, Heinemann SF (2002) Trans-synaptic Eph receptor-ephrin signaling in hippocampal mossy fiber LTP, Science 296(5574):1864-1869.

Cooper JR, Bloom FE, Roth RH (2003) The Biochemical Basis of Neuropharmacology, Oxford University Press, Oxford, UK.

Cravatt BF, Giang DK, Mayfield SP, Boger DL, Lerner RA, Gilula NB (1996) Molecular characterization of an enzyme that degrades neuromodulatory fatty-acid amides, Nature (London) 384:83-87.

Crawley JN, Corwin RL, Robinson JK, Felder CC, Devane WA, Axelrod J (1993) Anandamide, an endogenous ligand of the cannabinoid receptor, induces hypomotility and hypothermia in vivo in rodents, Pharmacol Biochem Behav, 46(4):967-972.

Dahl D, Sarvey JM (1989) Norepinephrine induces pathway-specific long-lasting potentiation and depression in the hippocampal dentate gyrus, Proc Natl Acad Sci USA 86(12):4776-4780.

Davis MI, Ronesi J, Lovinger DM (2003) A predominant role for inhibition of the adenylate cyclase/protein kinase A pathway in ERK activation by cannabinoid receptor 1 in N1E-115 neuroblastoma cells, J Biol Chem 278(49):48973-48980.

Deadwyler SA, Hampson RE, Bennett BA, Edwards TA, Mu J, Pacheco MA, Ward SJ, Childers SR (1993) Cannabinoids modulate potassium current in cultured hippocampal neurons, Receptors Channels 1(2):121-134.

Dean C, Scholl FG, Choih J, DeMaria S, Berger J, Isacoff E, Scheiffele P (2003) Neurexin mediates the assembly of presynaptic terminals, Nat Neurosci 6:708–716.

Derkinderen P, Valjent E, Toutant M, Corvol JC, Enslen J, Ledent C, Trzaskos J, Caboche J, Girault JA (2003) Regulation of extracellular signal-regulated kinase by cannabinoids in hippocampus, J Neurosci 23(6):2371-2382.

Desarnaud F, Cadas H, Piomelli D (1995) Anandamide amidohydrolase activity in rat brain microsomes. Identification and partial characterization, J Biol Chem 270:6030-6035.

Devane WA, Dysarz FA 3rd, Johnson MR, Melvin LS, Howlett AC (1988) Determination and characterization of a cannabinoid receptor in rat brain. Mol Pharmacol 34(5):605-613.

Devane WA, Hanus L, Breuer A, Pertwee RG, Stevenson LA, Griffin G, Gibson D, Mandelbaum A, Etinger A, Mechoulam R (1992) Isolation and structure of a brain constituent that binds to the cannabinoid receptor, Science 258:1946-1949.

Diana MA, Marty A (2003) Characterization of depolarization-induced suppression of inhibition using paired interneuron-Purkinje cell recordings. J Neurosci 23:5906-5918.

Di Marzo V, Fontana A, Cadas H, Schinelli S, Cimino G, Schwartz JC, Piomelli D (1994) Formation and inactivation of endogenous cannabinoid anandamide in central neurons, Nature 372:686-691.

Di Marzo V, De Petrocellis L, Sugiura T, Waku K. (1996) Potential biosynthetic connections between the two cannabimimetic eicosanoids, anandamide and 2-arachidonoyl-glycerol, in mouse neuroblastoma cells, Biochem Biophys Res Commun 227(1):281-288.

Dinh TP, Carpenter D, Leslien FM, Freund TF, Katona I, Sensi SL, Kathuria S, Piomelli D (2002) Brain monoglyceride lipase participating in endocannabinoid inactivation, Proc Natl Acad Sci USA 99(16):10819-10824.

Dunwiddie TV, Lovinger DM, Eds. (1993) Presynaptic Receptors in the Mammalian Brain, 1993, Birkhäuser, Boston, USA.

Egertová M, Giang DK, Cravatt BF, Elphick MR (1998) A new perspective on cannabinoid signalling: complementary localization of fatty acid amide hydrolase and the CB_1 receptor in rat brain. Proc R Soc Lond B 265:2081–2085.

Egertova M, Cravatt BF, Elphick MR. (2003) Comparative analysis of fatty acid amide hydrolase and cb(1) cannabinoid receptor expression in the mouse brain: evidence of a widespread role for fatty acid amide hydrolase in regulation of endocannabinoid signaling, Neuroscience 119(2):481-496.

Fegley D, Kathuria S, Mercier R, Li C, Goutopoulos A, Makriyannis A, Piomelli D (2004) Anandamide transport is independent of fatty-acid amide hydrolase activity and is blocked by the hydrolysis-resistant inhibitor AM1172, Proc Natl Acad Sci USA 101(23):8756-8761.

Feinmark SJ, Begum R, Tsvetkov E, Goussakov I, Funk CD, Siegelbaum SA, Bolshakov VY (2003) 12-lipoxygenase metabolites of arachidonic acid mediate metabotropic glutamate receptor-dependent long-term depression at hippocampal CA3-CA1 synapses, J Neurosci 23(36): 11427-11435.

Freund TF, Katona I, Piomelli D (2003) Role of endogenous cannabinoids in synaptic signaling, Physiol Rev 83:1017-1066.

Fride E, Mechoulam R. (1993) Pharmacological activity of the cannabinoid receptor agonist, anandamide, a brain constituent, Eur J Pharmacol.231(2):313-4.

Fride E (2002) Endocannabinoids in the central nervous system--an overview, Prostaglandins Leukot Essent Fatty Acids 66(2-3):221-233.

Fu Z, Washbourne P, Ortinski P, Vicini S. (2003) Functional excitatory synapses in HEK293 cells expressing neuroligin and glutamate receptors, J Neurophysiol 90(6):3950-3957.

Gage AT, Reyes M, Stanton PK. (1997) Nitric-oxide-guanylyl-cyclase-dependent and –independent components of multiple forms of long-term synaptic depression, Hippocampus 7(3):286-295.

Gerdeman G, Lovinger DM (2001) CB1 cannabinoid receptor inhibits synaptic release of glutamate in rat dorsolateral striatum, J. Neurophysiol 85:468-471.

Gerdeman GL, Ronesi J, Lovinger DM (2002) Postsynaptic endocannabinoid release is critical to long-term depression in the striatum, Nat Neurosci 5:446-451.

Gerdeman GL, Lovinger DM (2003) Emerging roles for endocannabinoids in long-term synaptic plasticity, Br J Pharmacol 140(5):781-789.

Giuffrida A, Parsons LH, Kerr TM, Rodriguez de Fonseca F, Navarro M, Piomelli D (1999) Dopamine activation of endogenous cannabinoid signaling in dorsal striatum., Nat. Neurosci 2: 358-363.

Glaser ST, Abumrad NA, Fatade F, Kaczocha M, Studholme KM, Deutsch DG (2003) Evidence against the presence of an anandamide transporter, Proc Natl Acad Sci USA 100(7):4269-4274.

Gubellini P, Saulle E, Centonze D, Bonsi P, Pisani A, Bernardi G, Conquet F, Calabresi P (2001) Selective involvement of mGlu1 receptors in corticostriatal LTD, Neuropharmacology 40:839-846.

Gulyas AI, Cravatt BF, Bracey MH, Dinh TP, Piomelli D, Boscia F, Freund TF (2004) Segregation of two endocannabinoid-hydrolyzing enzymes into pre- and postsynaptic compartments in the rat hippocampus, cerebellum and amygdala, Eur J Neurosci 20(2):441-458.

Heerssen HM, Segal RA (2002) Location, location, location: a spatial view of neurotrophin signal transduction, Trends Neurosci 2002 25(3):160-165.

Henry DJ, Chavkin C. (1995) Activation of inwardly rectifying potassium channels (GIRK1) by co-expressed rat brain cannabinoid receptors in Xenopus oocytes, Neurosci Lett 86(2-3):91-94.

Herkenham M, Lynn AB, Little MD, Johnson MR, Melvin LS, de Costa BR, Rice KC (1990) Cannabinoid receptor localization in brain, Proc Natl Acad Sci USA 87(5):1932-1936.

Herkenham M, Lynn AB, de Costa BR, Richfield EK (1991) Neuronal localization of cannabinoid receptors in the basal ganglia of the rat, Brain Res 547:267-274.

Hillard CJ, Wilkinson DM, Edgemond WS, Campbell WB (1995) Characterization of the kinetics and distribution of N-arachidonylethanolamine (anandamide) hydrolysis by rat brain, Biochim Biophys Acta 1257:249-256.

Hillard CJ, Edgemond WS, Jarrahian A, Campbell WB (1997) Accumulation of N-arachidonoylethanolamine (anandamide) into cerebellar granule cells occurs via facilitated diffusion, J Neurochem 69:631–638.

Hillard CJ, Jarrahian A (2000) The movement of N-arachidonoylethanolamine (anandamide) across cellular membranes, Chem. Phys. Lipids 108:123-134.

Hoffman AF, Lupica CR (2001) Direct actions of cannabinoids on synaptic transmission in the nucleus accumbens: a comparison with opioids, J Neurophysiol 85:72-83.

Huang CC, Lo SW, Hsu KS (2001) Presynaptic mechanisms underlying cannabinoid inhibition of excitatory synaptic transmission in rat striatal neurons, J Physiol 532:731-748.

Howlett AC, Johnson MR, Melvin LS, Milne GM. (1988) Nonclassical cannabinoid analgesics inhibit adenylate cyclase: development of a cannabinoid receptor model. Mol Pharmacol 33(3):297-302.

Ichtchenko K, Hata Y, Nguyen T, Ullrich B, Moomaw C, Sudhof TC (1990) Neuroligin 1: a splice site-specific ligand for beta-neurexins, Cell 81:435–443.

Katona I, Sperlagh B, Sik A, Kafalvi A, Vizi ES, Mackie K, Freund TF (1999) Presynaptically located CB1 cannabinoid receptors regulate GABA release from axon terminals of specific hippocampal interneurons, J Neurosci 19:4544-4558.

Kim J, Isokawa M, Ledent C, Alger BE (2002) Activation of muscarinic acetylcholine receptors enhances the release of endogenous cannabinoids in the hippocampus, J Neurosci 22:10182-10191.

Kim J, Alger BE. (2004) Inhibition of cyclooxygenase-2 potentiates retrograde endocannabinoid effects in hippocampus, Nat Neurosci 7(7):697-698.

Kozak KR, Crews BC, Morrow JD, Wang LH, Ma YH, Weinander R, Jakobsson PJ, Marnett LJ (2002) Metabolism of the endocannabinoids, 2-arachidonylglycerol and anandamide, into prostaglandin, thromboxane, and prostacyclin glycerol esters and ethanolamides, J Biol Chem 277(47):44877-44885.

Kreitzer AC, Regehr WG (2001) Retrograde inhibition of presynaptic calcium influx by endogenous cannabinoids at excitatory synapses onto purkinje cells, Neuron 29:717-727.

Kushmerick C, Price GD, Taschenberger H, Puente N, Renden R, Wadiche JI, Duvoisin RM, Grandes P, von Gersdorff H (2004) Retroinhibition of presynaptic Ca^{2+} currents by endocannabinoids released via postsynaptic mGluR activation at a calyx synapse, J Neurosci. 24(26):5955-5965.

Levenes C et al. (1998) Cannabinoids decrease excitatory synaptic transmission and impair long- term depression in rat cerebellar Purkinje cells, J Physiol 510:867-79.

Llano I, Leresche N, Marty A (1991) Calcium entry increases the sensitivity of cerebellar Purkinje cells to applied GABA and decreases inhibitory synaptic currents, Neuron, 6:565–574.

Lovinger DM, Tyler EC, Merritt (1993) Short- and long-term synaptic depression in rat neostriatum, J Neurophysiol 70:1937-1949.

Lu B (2003) BDNF and activity-dependent synaptic modulation, Learn Mem 10(2):86-98.

Maccarrone M, Bari M, Battista N, Finazzi-Agrò A (2002) Estrogen stimulates arachidonoylethanolamide release from human endothelial cells and platelet activation, Blood, 100:4040-4048.

Mackie K, Hille B. (1992) Cannabinoids inhibit N-type calcium channels in neuroblastoma-glioma cells, Proc Natl Acad Sci USA 89(9):3825-3829.

Madison DV, Schuman EM. (1995) Diffusible messengers and intercellular signaling: locally distributed synaptic potentiation in the hippocampus, Curr Top Microbiol Immunol 196:5-6.

Maejima T, Hashimoto K, Yoshida T, Aiba A, Kano M (2001) Presynaptic inhibition caused by retrograde signal from metabotropic glutamate to cannabinoid receptors, Neuron, 31:463-475.

Matsuda LA, Lolait SJ, Brownstein MJ, Young AC, Bonner TI (1990) Structure of a cannabinoid receptor and functional expression of the cloned cDNA, Nature 346:561-564.

McFarland MJ, Porter AC, Rakhshan FR, Rawat DS, Gibbs RA, Barker EL (2004) A role for caveolae/lipid rafts in the uptake and recycling of the endogenous cannabinoid anandamide, J Biol Chem, in press.

Mechoulam R, Ben-Shabat S, Hanus L, Ligumsky M, Kaminski NE, Schatz AR, Gopher A, Almog S, Martin BR, Compton DR et al. (1995) Identification of an endogenous 2-monoglyceride, present in canine gut, that binds to cannabinoid receptors. Biochem Pharmacol., 50(1):83-90.

Mechoulam R (2002) Discovery of endocannabinoids and some random thoughts on their possible roles in neuroprotection and aggression, Prostaglandins Leukot Essent Fatty Acids, 66:93-99.

Melis M, Pistis M, Perra S, Muntoni AL, Pillolla G, Gessa GL (2004) Endocannabinoids mediate presynaptic inhibition of glutamatergic transmission in rat ventral tegmental area dopamine neurons through activation of CB1 receptors, J Neurosci., 24(1):53-62.

Miller RJ (1998) Presynaptic receptors. Annu Rev Pharmacol Toxicol 38:201-227.

Nicholls JG, Martin AR, Wallace BG, Fuchs PA (2001) From Neuron to Brain, 4th edition, Sinauer Associates, Sunderland, MA, USA.

Ohno-Shosaku T, Maejima T, Kano M (2001) Endogenous cannabinoids mediate retrograde signals from depolarized postsynaptic neurons to presynaptic terminals, Neuron 29:729-738.

Ohno-Shosaku T, Shosaku J, Tsubokawa H, Kano M (2002) Cooperative endocannabinoid production by neuronal depolarization and group I metabotropic glutamate receptor activation, Eur J Neurosci 15:953-961.

Ohno-Shosaku T, Tsubokawa H, Mizushima I, Yoneda N, Zimmer A, Kano M (2002) Presynaptic cannabinoid sensitivity is a major determinant of depolarization-induced retrograde suppression at hippocampal synapses, J Neurosci 22(10):3864-3872.

Palmer A, Klein R (2003) Multiple roles of ephrins in morphogenesis, neuronal networking, and brain function, Genes Dev 17:1429–1450.

Pert CB, Snyder SH. (1973) Opiate receptor: demonstration in nervous tissue, Science 179(77): 1011-1014.

Pitler TA, Alger BE (1992) Postsynaptic spike firing reduces synaptic $GABA_A$ responses in hippocampal pyramidal cells, J Neurosci 12:4122–4132.

Pitler TA, Alger BE. (1994) Depolarization-induced suppression of GABAergic inhibition in rat hippocampal pyramidal cells: G protein involvement in a presynaptic mechanism, Neuron 13(6):1447-1455.

Porter AC, Sauer JM, Knierman MD, Becker GW, Berna MJ, Bao J, Nomikos GG, Carter P, Bymaster FP, Leese AB, Felder CC (2002) Characterization of a novel endocannabinoid, virodhamine, with antagonist activity at the CB1 receptor, J Pharmacol Exp Ther 301(3):1020-1024.

Robbe D, Alonso G, Duchamp F, Bockaert J, Manzoni OJ (2001) Localization and mechanisms of action of cannabinoid receptors at the glutamatergic synapses of the mouse nucleus accumbens, J Neurosci 21:109-116.

Robbe D, Kopf M, Remaury A, Bockaert J, Manzoni OJ (2002) Endogenous cannabinoids mediate long-term synaptic depression in the nucleus accumbens, Proc Natl Acad Sci USA 99: 8384-8388.

Rodriguez JJ, Mackie K, Pickel VM (2000) Ultrastructural localization of the CB1 cannabinoid receptor in μ-opioid receptor patches of the rat caudate putamen nucleus, J Neurosci 21:823-833.

Ronesi J, Gerdeman GL, Lovinger DM. (2004) Disruption of endocannabinoid release and striatal long-term depression by postsynaptic blockade of endocannabinoid membrane transport, J Neurosci 24(7):1673-1679.

Scheiffele P, Fan J, Choih J, Fetter R, Serafini T (2000) Neuroligin expressed in nonneuronal cells triggers presynaptic development in contacting axons, Cell 101:657–669.

Schuman EM. (1999) Neurotrophin regulation of synaptic transmission, Curr Opin Neurobiol 9(1):105-109.

Shen M, Piser TM, Seybold VS, Thayer SA (1996) Cannabinoid receptor agonists inhibit glutamatergic synaptic transmission in rat hippocampal cultures, J Neurosci 16:4322-4334.

Sjostrom PJ, Turrigiano GG, Nelson SB. (2003) Neocortical LTD via coincident activation of presynaptic NMDA and cannabinoid receptors, Neuron 14;39(4):641-654.

Spacek J, Harris KM (1998) Three-dimensional organization of cell adhesion junctions at synapses and dendritic spines in area CA1 of the rat hippocampus, J Comp Neurol 393(1):58-68.

Spacek J, Harris KM (2004) Trans-endocytosis via spinules in adult rat hippocampus, J Neurosci 24(17):4233-4241.

Sugiura T, Kondo S, Sukagawa A, Nakane S, Shinoda A, Itoh K, Yamashita A, Waku K (1995) 2-Arachidonoylglycerol: a possible endogenous cannabinoid receptor ligand in brain. Biochem Biophys Res Commun 215(1):89-97.

Sugiura T, Kondo S, Sukagawa A, Tonegawa T, Nakane S, Yamashita A, Waku K (1996) Enzymatic synthesis of anandamide, an endogenous cannabinoid receptor ligand, through N-acylphosphatidylethanolamine pathway in testis: Involvement of a Ca^{2+}-dependent transacylase and phosphodiesterase activities, Biochem Biophys Res Commun 218:113–117.

Sugiura T, Kobayashi Y, Oka S, Waku K (2002) Biosynthesis and degradation of anandamide and 2-arachidonoylglycerol and their possible physiological significance, Prostaglandins Leukot Essent Fatty Acids 66(2-3):173-192.

Sung KW et al. (2001) Activation of group I mGluRs is necessary for induction of long-term depression at striatal synapses, J Neurophysiol 86:2405-2412.

Szabo B et al. (1998) Inhibition of GABAergic inhibitory postsynaptic currents by cannabinoids in rat corpus striatum, Neuroscience 85:395-403.

Tarrant SB, Routtenberg A (1977) The synaptic spinule in the dendritic spine: electron microscopic study of the hippocampal dentate gyrus, Tissue Cell 9(3):461-473.

Testa CM, Friberg IK, Weiss SW, Standaert DG. (1998) Immunohistochemical localization of metabotropic glutamate receptors mGluR1a and mGluR2/3 in the rat basal ganglia, J Comp Neurol 390(1):5-19.

Testa CM, Standaert DG, Landwehrmeyer GB, Penney JB Jr, Young AB (1995) Differential expression of mGluR5 metabotropic glutamate receptor mRNA by rat striatal neurons, J Comp Neurol 354(2):241-252.

Trettel J, Levine ES (2003) Endocannabinoids mediate rapid retrograde signaling at interneuron-pyramidal neuron synapses of the neocortex, J. Neurophysiol 89:2334–2338.

Ueda N, Kurahashi Y, Yamamoto S, Tokunaga T (1995) Partial purification and characterization of the porcine brain enzyme hydrolyzing and synthesizing anandamide, J Biol Chem 270:23823-23827.

Vincent P, Armstrong CM, Marty A. (1992) Inhibitory synaptic currents in rat cerebellar Purkinje cells: modulation by postsynaptic depolarization. J Physiol 456:453-471.

Wilkinson DG (2001) Multiple roles of EPH receptors and ephrins in neural development, Nat Rev Neurosci 2:155–164.

Wilson RI, Nicoll RA (2001) Endogenous cannabinoids mediate retrograde signalling at hippocampal synapses, Nature 410:588-592.

Wilson RI, Kunos G, Nicoll RA .(2001) Presynaptic specificity of endocannabinoid signaling in the hippocampus, Neuron 31(3):453-62.

Wilson RI, Nicoll RA. (2002) Endocannabinoid signaling in the brain, Science, 296(5568): 678-682.

Zakharenko SS, Patterson SL, Dragatsis I, Zeitlin SO, Siegelbaum SA, Kandel ER, Morozov A (2003) Presynaptic BDNF required for a presynaptic but not postsynaptic component of LTP at hippocampal CA1-CA3 synapses, Neuron 39(6):975-990.

SYNAPTIC VESICLE RECYCLING AS A SUBSTRATE FOR NEURAL PLASTICITY

Tuhin Virmani and Ege T. Kavalali*

1. INTRODUCTION

Chemical synapses are the principle nodes of communication between neurons and are critical for the processing and storage of information in the brain. They consist of two functionally and structurally distinct compartments: presynaptic terminals and postsynaptic specializations. Presynaptic terminals store and release neurotransmitter substances in membranous organelles named synaptic vesicles, whereas postsynaptic structures contain signaling molecules responsible for generation of neuronal responses to released neurotransmitters. This chapter is primarily devoted to discussion of the cellular mechanisms that can underlie forms of synaptic plasticity that arise from changes in the structure and function of presynaptic terminals. Here, we will discuss prevailing concepts regarding presynaptic forms of plasticity rather than providing an exhaustive review of existing literature. Several recent review papers give an excellent account of the topic and the literature, some of which will not be covered here in detail (Atwood and Karunanithi, 2002; Zucker and Regehr, 2002; Jahn et al., 2003; Murthy and De Camilli, 2003).

Presynaptic terminals are exceptionally dynamic vesicle trafficking machines. Most forms of short-term synaptic plasticity (milliseconds to minutes) as well as some forms of long-term plasticity arise from persistent alterations in the dynamics of vesicle trafficking in presynaptic terminals. What makes presynaptic forms of plasticity particularly interesting is that they do not only increase or decrease the amplitude of synaptic responses, but also cause frequency-dependent changes in chemical neurotransmission. In this manner, plasticity can alter the information coding in neural circuits beyond simple scaling of synaptic responses.

Over the past few years, several studies have demonstrated that neurotransmission is not sustained by mobilization of vesicles from a large reservoir but by constant recycling of a handful of vesicles. Therefore, the rate and the pathway of vesicle trafficking can critically determine synaptic efficacy during activity. However, very little is known about the processes that regulate vesicle trafficking in the synapse beyond vesicle fusion. Accordingly, the studies on presynaptic forms of plasticity have extensively focused on alterations at the fusion step. Therefore we currently have only a rudimentary understanding of the changes in other steps within the synaptic vesicle cycle associated with long-term synaptic plasticity. Nevertheless, we have increasing information on the complexity of presynaptic vesicle trafficking events that comprise the synaptic vesicle cycle. Therefore, before discussing recent experimental findings and arising concepts on

* Tuhin Virmani and Ege T. Kavalali, Center for Basic Neuroscience, U.T. Southwestern Medical Center, 6000 Harry Hines Boulevard, Dallas, TX 75390-9111, USA

presynaptic forms of plasticity, we will overview the basic properties of presynaptic structure and synaptic vesicle trafficking.

2. OVERVIEW OF PRESYNAPTIC ULTRASTRUCTURE

In the presynaptic terminals neurotransmitter substances are packaged into synaptic vesicles that are organelles with a diameter of 35 to 40 nm. Typical nerve endings in the central nervous system (CNS) contain up to 100-200 synaptic vesicles. Synaptic vesicles recycle locally within the terminal, independent of the cell body, or other membrane compartments in the secretory pathway. In the simplest case, vesicles in a CNS terminal can be divided into two pools (Figure 1). The first pool contains a relatively small fraction (5-10%) of vesicles close to release sites. These vesicles can be released by rapid uncaging of intrasynaptic Ca^{2+} (Schneggenburger et al., 1999), a 10-millisecond Ca^{2+}-current pulse (Wu and Borst, 1999), a brief high-frequency train of action potentials (Murthy and Stevens, 1999) or by hypertonic stimulation (Rosenmund and Stevens, 1996). This release ready pool of vesicles is referred to as the immediately releasable pool or the readily releasable pool (RRP). RRP vesicles are considered to be in a morphologically docked state, although not all morphologically docked vesicles are necessarily release competent at any given time (Schikorski and Stevens, 2001). A "priming" step in addition to the morphological docking is required to make vesicles fully release competent (Jahn et al., 2003). In contrast to this prevailing view on the functional architecture of the synapse, a recent study in the frog neuromuscular junction showed that the docked vesicle pool might not correspond to the readily releasable pool and some vesicles which are not docked and a distance away from the release site may have a priority to fuse over vesicles that appear to be docked (Rizzoli and Betz, 2004).

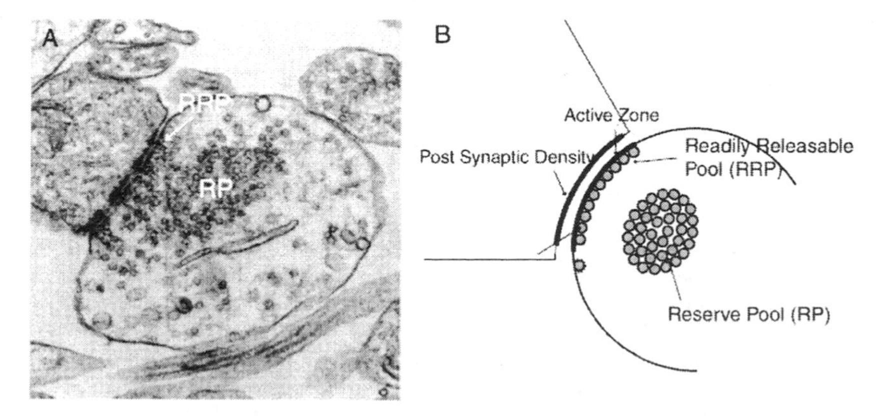

Figure 1. Synaptic ultrastructure and synaptic vesicle pool hierarchy. A. Electron micrograph of a hippocampal synapse in culture (courtesy of Dr. Xinran Liu). B. Putative organization of synaptic vesicle pools with respect to the active zone.

A secondary pool of vesicles, the reserve pool (RP), is spatially distant from the release sites and constantly replaces the vesicles in the RRP that have been exocytosed. The rate of replenishment of RRP vesicles from the reserve pool is a critical parameter that determines the response of synapses to repetitive stimulation. Recent evidence indicates that intrasynaptic Ca^{2+} can facilitate the rate of replenishment (Dittman and Regehr, 1998; Stevens and Wesseling, 1998; Wang and Kaczmarek, 1998). The number of vesicles contained in the RRP is a critical parameter that regulates the probability of release, which is defined as the probability that a presynaptic action potential can result in an exocytotic event. Therefore, the rate and pathways by which vesicles in the RRP are replenished is also a crucial determinant of presynaptic efficacy and of several forms of short- and long-term synaptic plasticity. In addition, several lines of evidence support the presence of the non-recycling pool of vesicles in the synapse. Mechanisms that can render this "resting" (or "dormant") pool functional remain to be determined (Sudhof, 2000; Harata et al., 2001).

Within the synapse, neurotransmitter release is restricted to active zones. The presynaptic active zone is precisely aligned with the postsynaptic neurotransmitter receptors and the post-synaptic density (PSD). The fusion of synaptic vesicles with these electron-dense regions of the presynaptic plasma membrane is spatially and temporally highly regulated. After docking at the active zone synaptic vesicles undergo a series of priming reactions to mature to a fusion-competent state. Recent evidence suggests that docking and priming reactions can occur fast within 300 ms (Zenisek et al., 2000). At this point, the influx of Ca^{2+} ions through voltage-gated Ca^{2+}-channels in response to action potentials triggers rapid exocytosis of fusion-competent vesicles. The initial release of vesicles from the RRP or docked-primed pool, and subsequent replenishment and release from the reserve pool results in biphasic release kinetics, with a fast release phase corresponding to release from the RRP and a slower release phase due to the mobilization and release of vesicles from the reserve pool. Such differences in mobilization rates from different pools have been observed in a number of secretory systems both containing active zones (e.g. central synapses, neuromuscular junction) and lacking active zones (e.g. adrenal chromaffin cells) (Neher, 1998).

3. PROPERTIES OF PRESYNAPTIC VESICLE RECYCLING IN RELATION TO PLASTICITY

3.1. Parallel Pathways For Vesicle Recycling In CNS Synapses

In the classical model of vesicle recycling at the synapse vesicles fuse with the plasma membrane at the active zone, release their neurotransmitter and proceed to completely collapse onto the membrane (Heuser and Reese, 1973; Sudhof, 1995; Cremona and De Camilli, 1997). This process is thought to occur on the time scale of milliseconds. Subsequently clathrin, through its adaptor proteins, is recruited to the membrane and forms coated vesicles that bud off the plasma membrane with the help of the GTPase dynamin. Endocytosis of clathrin-coated vesicles may also occur from membrane infoldings or endosomal cisternae (Koenig and Ikeda, 1996; Takei et al., 1996; Richards et al., 2000). A v-type ATPase then lowers the pH in these vesicles and subsequently the vesicles are refilled with neurotransmitter. The re-acidification and neurotransmitter filling steps are thought to be rapid, and the entire process for exocytosed vesicles to be re-available for release is thought to occur within 40 and 90 seconds (Betz and Bewick, 1993; Liu and Tsien, 1995; Ryan and Smith, 1995). Recent studies provide evidence of an additional fast pathway of vesicle recycling that prevents vesicle depletion and maintains synaptic responses in the face of repetitive stimulation (Delgado et al., 2000; Pyle et al., 2000; Sudhof, 2000; Sara et al., 2002; Aravanis et al., 2003; Gandhi and Stevens, 2003). This process may involve a pathway referred to as "kiss and run" in which the fusion pore opens transiently without complete collapse of the vesicle (Ceccarelli et al., 1973; Henkel and Almers, 1996; Valtorta et al., 2001). In central synapses, this hypothesis is supported by reports studying the effects of hypertonic challenge and strong electrical stimulation (Klingauf et al., 1998; Kavalali et al., 1999; Pyle et al., 2000; Stevens and Williams, 2000) but see (Sankaranarayanan and Ryan, 2000, 2001). Recent evidence suggests that these pathways may be modulated, and that the synaptic plasma membrane protein synaptotagmin 7 may play a role in this process (Virmani et al., 2003).

3.2. The Relationship Between Vesicle Recycling And Short-Term Synaptic Depression

Most synapses following sustained stimulation show a rapid depression during which postsynaptic responses decrease to a plateau. The underlying cause of this depression is thought to be limitations in the number of functional vesicles present in a synapse, although the potential role of an active process that limits the rate of synaptic vesicle fusion at high frequencies cannot be excluded. In support of the limitation of vesicle number hypothesis, several studies have shown that central synapses have a small number of functional synaptic vesicles especially during early stages of maturation after synaptogenesis (Vaughn, 1989; Fiala et al., 1998; Mozhayeva et al., 2002).

A current model of this form of short-term plasticity suggests that the early phase of synaptic depression results from depletion of vesicles in the RRP coupled with a decrease in release probability. According to this view, during the depressed plateau phase, synapses release vesicles that transiently populate the RRP from the reserve pool (Zucker and Regehr, 2002). This current linear model of short-term synaptic plasticity does not take into account the reuse of recycled vesicles during synaptic depression. This omission is supported by experiments conducted in neuromuscular junction and hippocampal synapses that estimated the time required for recycled synaptic vesicles to mix with the non-released population of vesicles and rejoin neurotransmission to be between 40 and 90 s (Betz and Bewick, 1992, 1993; Ryan et al., 1993; Liu and Tsien, 1995). However, recent studies have re-examined this question in hippocampal synapses and showed that RRP vesicles were reused within seconds (Pyle et al., 2000; Sara et al., 2002). Thus, restrictions imposed by limited synaptic vesicle supply during sustained stimulation may be partly compensated by recycling and reuse of synaptic vesicles that have undergone exocytosis (Sudhof, 2000; Harata et al., 2001).

3.3. Studies On Molecular Determinants Of Synaptic Vesicle Recycling

The most compelling evidence for the contribution of recycled vesicles to synaptic transmission comes from the analysis of *Drosophila* and mouse mutants of proteins involved in endocytosis. For instance, several studies using the *Drosophila* temperature sensitive dynamin mutant *shibire*, have provided strong support for a relationship between vesicle recycling and synaptic release during sustained stimulation. Dynamin is a GTPase that serves to pinch vesicles off the plasma membrane following recruitment of clathrin and its adaptor proteins onto fused vesicles. Using this system, Koenig and Ikeda showed that vesicle recycling can occur through two pathways, one at the active zone of the *Drosophila* neuromuscular junction, the other in its vicinity (Koenig and Ikeda, 1996, 1999). The active zone and non-active zone pathways were shown to populate different pools of vesicles at different time scales, with the active zone pathway repopulating the RRP within 1 minute of moving from the non-permissive (no dynamin function) to the permissive temperature (dynamin is functional), while reserve pool replenishment from the slower pathway occurred on the order of 25 minutes. Recent experiments in *Drosophila* neuromuscular junction compared the rate of synaptic depression between wild type and *shibire* mutant flies. In this study synapses from *shibire* mutant flies at non-permissive temperatures rapidly depressed without a plateau phase in response to high frequency stimulation. The kinetic difference between this rate of depression and the normal rate of depression when endocytosis is intact revealed a recycling rate of one to two vesicles per second per active zone (Delgado et al., 2000), a rate considerably faster than previous estimates and in line with estimations from hippocampal synapses (Sara et al., 2002).

Endophilin is another protein that plays a role in clathrin-mediated endocytosis. A recent study on the null mutant of the *Drosophila* homolog of endophilin (Verstreken et al., 2002) showed that at the *Drosophila* neuromuscular junction in spite of an increase in fast synaptic depression, neurotransmission was still maintained under prolonged stimulation. This would indicate that vesicle recycling was still maintained at these synapses in contrast to *shibire* mutant flies where complete depletion of the recycling pool was observed. This finding suggests that a clathrin independent endocytic mechanism can sustain neurotransmission. The size of the total recycling pool was also decreased at these neuromuscular junction synapses a common theme seen in all studies inhibiting clathrin-mediated endocytosis.

Studies using mouse mutants of molecules in the clathrin-dependent pathway also indicate a significant contribution of recycled vesicles to synaptic transmission. These studies also suggest the presence of a clathrin-independent pathway for vesicle recycling. For instance, mice deficient in Synaptojanin 1, a polyphosphoinositide phosphatase that facilitates the removal of the clathrin coat, had decreased survival and several neurological deficits (Cremona et al., 1999). These mice also had defects following prolonged stimulation in which the amplitude of the slow phase of synaptic depression was increased in both hippocampal slice preparations and inhibitory cortical neurons in culture (Cremona et al., 1999; Luthi et al., 2001). This data suggest that vesicles were unable to repopulate the reserve pool. As there is still neurotransmission occurring at these synapses, albeit at a lower steady state amplitude, this is consistent with a mode of recycling occurring which may not require the normal functioning of synaptojanin.

Amphiphysin is another adaptor protein at the synapse that can recruit coat proteins to the plasma membrane; in addition, it can also bring dynamin and synaptojanin to the forming clathrin coat. In mice deficient in this molecule, a form of vesicle recycling was still intact although the total size of the recycling pool decreased due to disruption of clathrin-mediated endocytosis (Di Paolo et al., 2002).

Taken together genetic interventions that disrupt endocytic proteins dynamin and clathrin argue for a strong relationship between vesicle recycling and synaptic output during sustained activity. Furthermore these studies suggest that a clathrin-independent pathway may operate in parallel and sustain neurotransmission and vesicle recycling. However several questions remain unanswered. For instance, little is known about the contribution of recycled vesicles to neurotransmission under changing stimulation patterns or following activation of second messenger pathways that strongly regulate neurotransmission. Moreover, although a consensus is emerging on the presence of a "kiss and run" like mechanism that can retrieve vesicles "fast" and preserve their molecular identity, nothing is known on the molecular basis of this pathway except for the implications that it may not involve clathrin. Can this pathway be simply due to reversibility of fusion? Or does it require retrieval by a molecular complex other than the clathrin coat?

3.4. The Role Of Synaptotagmins In Synaptic Vesicle Recycling

A critical hallmark of the synaptic vesicle cycle is the tight coupling of exo- and endocytosis. Recent studies have extensively focused on synaptotagmins as potential mediators of this coupling. Synaptotagmins are characterized by an N-terminal transmembrane domain, a central linker and two C-terminal C2 domains (Marqueze et al., 2000; Sudhof, 2002). These proteins have been extensively studied as Ca^{2+} sensors for vesicle exocytosis, primarily through the characterization of synaptotagmin 1 (Geppert et al., 1994; Fernandez-Chacon et al., 2001). While synaptotagmin 1 and 2 are located on the synaptic vesicle, synaptotagmins 3, 6 and 7 are present on the synaptic plasma membrane (Butz et al., 1999; Sugita et al., 2001). The C2 domain of synaptotagmin has a high affinity binding site for AP-2 and possibly stonin, two proteins believed to be important in clathrin-mediated endocytosis (Zhang et al., 1994; Li et al., 1995; Martina et al., 2001). In a recent study, rapid light induced inactivation of synaptotagmin 1 in the *Drosophila* neuromuscular junction impaired delayed endocytosis supporting the biochemical results discussed above (Poskanzer et al., 2003).

Among the plasma membrane synaptotagmins, synaptotagmin 7 is of particular interest since the truncated splice variant of synaptotagmin 7 (syt7B), produced due to a conserved stop codon in the second exon of the alternatively spliced region (Sugita et al., 2001), inhibits receptor mediated endocytosis in a number of non-neuronal cell lines (von Poser et al., 2000). Another splice variant of the same protein containing both C2 domains (syt7A) had no effect in the same system. The coincidence of two findings, namely the inhibition of receptor-mediated endocytosis by truncated synaptotagmin 7 variants in transfected fibroblasts (von Poser et al., 2000) and the discovery of the natural occurrence of such variants by alternative splicing in neurons (Sugita et al., 2001) raised the possibility that alternative splicing of synaptotagmin 7 may regulate synaptic vesicle recycling. Therefore, we have recently utilized these molecules to isolate and study a potential clathrin-independent fast vesicle-recycling pathway in hippocampal synapses.

We found that synapses containing the syt7B at high concentrations showed an increased rate of synaptic vesicle endocytosis after a brief stimulus pulse. These faster endocytosing vesicles also recycle faster as demonstrated by a 2-fold increase in the number of vesicles loaded with the styryl dye FM1-43 that become re-available for release within 30s compared to synapses with control levels of protein. In contrast, the presence of syt7A at high levels results in no change in basal endocytosis, yet vesicles are targeted towards slower recycling pathways, a phenomenon we observed both using FM dyes and electrophysiology. These two lines of evidence suggest that alternative splice forms of synaptotagmin 7 may target endocytosing vesicles to fast or slow recycling pathways (Figure 2), with the level of the various splice variants fine tuning this process dependent upon synapse type, function or activity experienced by the synapse.

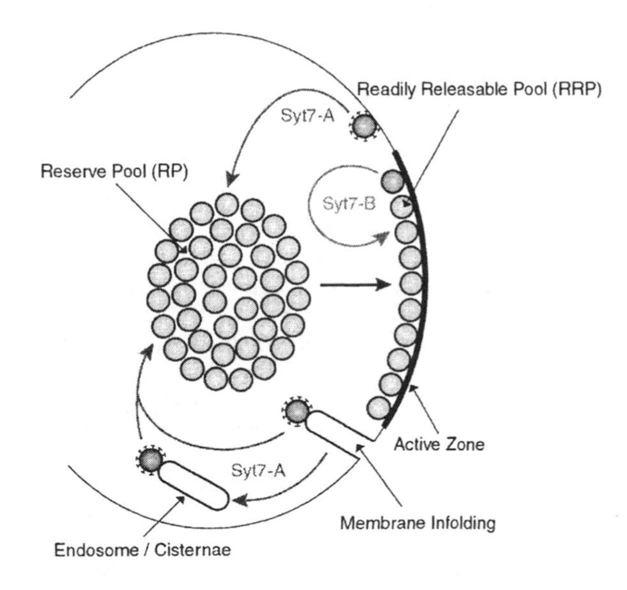

Figure 2. The role of synaptotagmin 7 splice variants in directing vesicles towards kinetically distinct recycling pathways. The short splice variant of synaptotagmin 7 (sy7-B) lacking the two C2 domains directs vesicles towards a fast recycling pathway. The full-length splice variant, synaptotagmin7-A, targets vesicles towards slower trafficking pathways. These include classical clathrin mediated endocytosis of individual vesicles, as well as slower pathways uncovered by synaptotagmin7-A overexpression where vesicles bud off membrane infoldings and/or endosomal cisternae (Virmani et al., 2003).

Interestingly, gene chip analysis of the mouse deficient in the Ca^{2+} channel α_{1A} subunit showed that synaptotagmin 7 was upregulated despite the fact that most synaptic proteins were unaltered. This suggests that synaptotagmin 7 contributes to the proper maintenance of synaptic vesicle recycling in these mice despite the absence of a key Ca^{2+} influx pathway (Piedras-Renteria et al., 2004). This finding supports the premise that synaptotagmin 7 levels and alternative splicing patterns can be altered by the activity history of a neuron thus rendering the vesicle recycling properties of its synaptic terminals plastic.

4. PRESYNAPTIC FORMS OF LONG-TERM SYNAPTIC PLASTICITY

Studies on presynaptic forms of long-term synaptic plasticity focus primarily on the modification of three parameters. First, the formation or unsilencing of new release sites or synapses. Second, the increase in the release probability by increasing the number of vesicles in the RRP or augmenting the fusion propensity of each vesicle. Lastly, the regulation of fusion pore kinetics thus modifying the rate of neurotransmitter diffusion out of synaptic vesicles. In this section we will discuss these three possibilities using their prevailing examples in the literature.

Formation of new release sites or new synapses is conceptually the simplest form of presynaptic modification that would result in long-term synaptic plasticity in the mammalian nervous system. Activity-dependent synapse formation has been elegantly visualized as the emergence of new dendritic spines using two-photon microscopy in the cortex and hippocampus (Trachtenberg et al., 2002; Zito and Svoboda, 2002). Electrophysiological studies have also suggested formation of new release sites or unsilencing of already existing synapses as an alternative possibility. For instance, Bolshakov and colleagues used the method of failure analysis to monitor synaptic activity between CA3-CA1 cell pairs before and after pharmacologically induced late phase LTP (L-LTP) by the membrane permeable analog of cAMP (Sp-cAMPs). In this way, they could show that in addition to an increase in release probability (a primarily presynaptic phenomenon that also occurs in early stages of LTP), there was an increase in the number of quanta released by a single presynaptic action potential (Bolshakov et al., 1997). This finding is consistent with an increase in the number of release sites between the cell pair during the LTP inducing stimulus, either by formation of new active zones within existing synapses or an increase in the total number of synapses, which release in a coordinated manner. These results were further supported by experiments performed in dissociated hippocampal cultures by using the styryl dye FM1-43 to directly monitor presynaptic terminals (Ma et al., 1999). In this study, authors showed that after Sp-cAMPs treatment, there was an increase in the number of FM1-43 labeled synaptic puncta without an increase in the total number of synaptic boutons in the CA1-CA3 cultures. This suggested that during L-LTP presynaptically silent synapses could be converted to actively recycling synapses. Interestingly, this phenomenon was blocked by the presence of the NMDA receptor blocker AP-5 during the course of the entire experiment, which indicated that a postsynaptic signal was required for the presynaptic modifications to occur. Retrograde information flow across the synapse as suggested by these findings is a recurring theme in synaptic function. Reciprocal interaction between pre- and postsynaptic elements is necessary for synapse formation. Retrograde flow of information across the synapse is also critical for normal synapse function and stability. The following sections will further discuss this concept.

4.1. Synaptic Cross-Talk

The most commonly studied form of long term synaptic plasticity is in the hippocampus between the Schaffer collateral axons that originate in the CA3 region and synapse onto CA1 pyramidal cell dendrites. This form of plasticity requires activation of postsynaptic NMDA receptors. In the long term, the enhanced plasticity can be sustained by increased AMPA receptors trafficking in the postsynaptic site (Malinow and Malenka, 2002). However, several studies have also reported concomitant changes in presynaptic vesicle recycling associated with this form of long-term potentiation. Here we will discuss these observations and their implications for potential plastic targets within the synaptic vesicle cycle.

4.1.1. Enhanced Synaptic Vesicle Recycling After LTP Inducing Stimulation

As discussed earlier, the vesicle-recycling pathway in central synapses plays an important role in maintaining the efficacy of neurotransmission during activity. Consequently, modulation of this pathway provides an attractive substrate for encoding plasticity in synapses. Most of the early work on LTP was performed using electrophysiology to monitor presynaptic output through activation of postsynaptic receptors, which resulted in a long debate on the exact locus of the changes seen during LTP expression. The advent of tools that directly report presynaptic changes provided evidence that at least in a set of central synapses, presynaptic potentiation may occur.

In an early study Malgaroli and colleagues utilized monoclonal and polyclonal antibodies raised against the lumenal domain of the synaptic vesicle protein, synaptotagmin 1, to study spontaneous vesicle recycling before and after LTP (Malgaroli et al., 1995). Monitoring spontaneous antibody uptake before, and 10 minutes after a 30s application of 50 μM glutamate to induce LTP in CA1-CA3 hippocampal cultures, the authors observed a two-fold increase in antibody uptake in induced synapses compared to mock stimulated synapses. Interestingly, weaker synapses potentiated more than stronger synapses suggesting that there may be a limitation on the resources available to larger synapses. While this study only looked at spontaneous vesicle cycling, it provided direct evidence that the synaptic vesicle cycle (coupling of exo- and endocytosis) in weak synapses could be potentiated during LTP inducing stimuli.

To address whether evoked vesicle recycling pathways were also enhanced during LTP, Ryan and colleagues (Ryan et al., 1996) used the styryl dye FM1-43 to monitor synapses before and after 30s long tetanus at 50Hz. Using a brief 20 action potential stimulation to label synapses with the FM1-43, they found an increase in dye uptake after the LTP inducing stimulus that lasted up to 60 minutes. This potentiation in dye uptake was blocked when AP-5 and the AMPA receptor antagonist CNQX were present in the bath, again suggesting the role of a retrograde molecule transmitting information back to the presynaptic terminal. The authors also observed an inverse correlation between initial release probability and potentiation of release probability after the tetanus similar to that observed in the study described above. However, a more prolonged 1200 action potential train that labels the total population of recycling synaptic vesicles did not reveal a change. This result suggests that in response to the tetanus these synapses increase their release probability without any change in the total vesicle pool size.

A later study in hippocampal slices showed that the induction of LTP by 200Hz tetanus or the potassium channel blocker tetraethylammonium in high Ca^{2+} (TEA-LTP) caused an increase in the rate of FM1-43 release compared to the baseline rate of dye loss (Zakharenko et al., 2001). This increase in release rate was partially blocked both by either AP-5 or the L-type Ca^{2+} channel blocker nitrendipine during the course of the tetanus. Interestingly, when the authors used 50Hz or 100Hz tetanus to induce LTP, despite the potentiation of the field EPSP, there was no change in the rate of dye release. This result indicated that Ca^{2+} influx through NMDA channels is required for the presynaptic modification, and the level for Ca^{2+} influx may affect the extent of the modification, presumably by producing varying levels of second messengers.

4.1.2. Nitric Oxide As A Retrograde Messenger

The dependence of LTP associated presynaptic modifications on postsynaptic NMDA receptor signaling raises the question on how the postsynaptic site signals back to the presynaptic terminal. One possibility comes from the fact that activation of NMDA receptors and resulting Ca^{2+} influx leads to Ca^{2+}-calmodulin dependent activation of neuronal nitric oxide synthase (nNOS). The subsequent production of nitric oxide (NO), which can freely diffuse across membranes and reach the presynaptic compartment, makes it an ideal candidate as a retrograde messenger. In support of this hypothesis, a series of studies have shown that LTP was susceptible to interference with NO synthesis as well as with activation of guanylate cyclase and cGMP generation which is a direct consequence of NO signaling (Schuman and Madison, 1991; Zhuo et al., 1994; Arancio et al., 1995 but see Selig et al., 1996).

How then does cGMP alter presynaptic neurotransmitter release? Arancio and colleagues (Arancio et al., 1995) studied the role of cGMP analogs in the induction of LTP in primary CA3-CA1 hippocampal cultures. They showed that cGMP functioned presynaptically, in an activity dependent manner, to induce long-term potentiation of evoked responses as well as an increase in spontaneous release events. Later work by Smith and colleagues, using hippocampal cultures, linked cGMP signaling to the generation of phosphatidyl inositol-3,4-diphosphate (PIP_2) which is a critical regulator of synaptic vesicle recycling (Micheva et al., 2001). By expressing the phospholipase C (PLC)-δ1 pleckstrin homology (PH) domain fused to GFP, the authors showed that under normal conditions the GFP was localized to the plasma membrane of cultured hippocampal neurons due to the PH domain PIP_2 interaction. However, upon stimulation they observed a fluorescence increase in synaptic boutons that were visualized with the styryl dye FM4-64. This activity-dependent redistribution of PH-GFP required new synthesis PIP_2 since it was blocked by the PI-4 kinase inhibitor phenularsine oxide (PAO). The NMDA antagonist AP-5 and the NO scavenger carboxy-PTIO could also block the PH-GFP redistribution. This finding provided a putative link between presynaptic PIP_2 metabolism and postsynaptic NMDA receptor activation via retrograde NO signaling.

Recently, the same group extended this study and linked the retrograde message of NO directly to changes in presynaptic vesicle recycling (Micheva et al., 2003). Using a fusion protein of the synaptic vesicle protein synaptobrevin/VAMP and a modified GFP with enhanced pH sensitivity (synapto-pHlourin) (Miesenbock et al., 1998) the authors showed that the application of drugs that inhibited the production and retrograde transmission of the NO signal (AP-5, TRIM and carboxy-PTIO) resulted in a slow down of vesicle endocytosis. In contrast, an increase in NO signaling via DEANO (NO donor) or 8-bromo-cGMP applications resulted in an increase in endocytosis. These data provided a putative mechanism for potentiation of presynaptic vesicle recycling via postsynaptic signaling.

Postsynaptic NMDA receptors → post-synaptic Ca^{2+} influx → calmodulin → nNOS → NO → presynaptic guanylyl cyclase → presynaptic cGMP → PIP2 synthesis → increase in vesicle endocytosis

These experiments further implied that this pathway may also play an important role in the maintenance of efficient vesicle recycling (through augmented synaptic vesicle retrieval) during intense stimulation in addition to its role in LTP via presynaptic cGMP increase.

4.1.3. Regulation Of Synaptic Vesicle Fusion Pore During Long Term Synaptic Plasticity

A number of studies have documented the presence of "silent synapses" in multiple brain regions. These synapses characteristically lack fast AMPA synaptic responses but still contain the slow NMDA responses (Malinow and Malenka, 2002). As a result, they fail to transmit under physiological conditions due to Mg^{++} block of the NMDA channel. However after LTP induction the silent synapses can be unsilenced by insertion of AMPA receptors. Recent studies have proposed an alternative presynaptic mechanism for conversion of these "silent synapses" to active ones (Choi et al., 2000). Using the rapidly dissociating competitive NMDA receptor antagonist L-AP5, the authors showed that in synapses lacking AMPA responses, this drug was able to effectively inhibit 99% of NMDA currents measured during postsynaptic depolarization to +40 mV. However, after LTP, once AMPA currents have emerged, the drug was only 50% effective in blocking NMDA responses. Assuming equal competition between the inhibitor and glutamate, the authors ascribed this decrease in inhibition to an increase in cleft glutamate concentration after the LTP inducing stimulus. They also showed that application of cyclothiazide (CTZ), a drug that blocks AMPA receptor desensitization, could unmask a slow AMPA current synapses in silent synapses that presumably lack AMPA channels (but see Montgomery et al., 2001). Taken together, this study suggested that "silent synapses" could be accounted for by a presynaptic mechanism in which neurotransmitter is released through a narrow non-expanding fusion pore that only allows limited glutamate into the cleft. Since NMDA receptors have a high affinity for glutamate, they can be activated, whereas low affinity AMPA receptors show no response. During LTP the release machinery is adapted such that fusion occurs through an expanding fusion pore resulting in complete collapse of vesicles and flooding of the cleft with neurotransmitter, resulting in the unmasking of AMPA responses.

This work was later supported by two other studies. The first study by Renger et al. showed that in synapses that lack AMPA responses to stimulation, or have slow responses, ionotophoretic application of a short glutamate pulse (100 nA, 1 ms) resulted in an AMPA type response, whereas application of a weaker but longer duration (10 nA, 10 ms) glutamate pulse resulted in an NMDA type response (Renger et al., 2001). These distinct glutamate profiles were used to mimic glutamate levels in the synaptic cleft after glutamate release from a completely collapsing vesicle versus a non-expanding fusion pore. The authors also showed that immature synapses have more AMPA-silent synapses than mature synapses in culture. Disruption of the synaptic vesicle protein synaptobrevin/VAMP by tetanus toxin application switched synapses to an AMPA silent variant while postsynaptic AMPA receptors still responded to exogenously applied glutamate.

In a recent study on the mechanisms underlying metabotropic glutamate receptor (mGluR) induced long-term depression (LTD), Zakharenko and colleagues used FM1-43 to monitor the presynaptic vesicle recycling before and after LTD induction (Zakharenko et al., 2002). They showed three major changes occurring after LTD induction, a reduction in the number of boutons that take up FM1-43, an increase in the number of dye loaded synapses that fail to destain (lose FM1-43) and a reduction in the rate of FM1-43 release. These effects were inhibited by blockers of mGluR dependent LTD such as MPEP (mGluR antagonist) and anisomysin (protein synthesis inhibitor) but not by AP-5. To further examine the mechanism of these changes, the authors performed two experiments. First, they loaded FM1-43 and FM4-64 (a red shifted analog of FM1-43) in two rounds of stimulation after LTD. In synapses where FM1-43 remained trapped after the first round of loading and destaining, subsequent FM4-64 loading resulted in quenching of the FM1-43 fluorescence via Fluorescent Resonance Energy Transfer (FRET) due to the overlap in the emission spectra of FM1-43 and the excitation spectra of FM4-64. Additionally, they used the faster departitioning dye FM2-10 in place of FM1-43, which showed less decrease in dye release kinetics after LTD induction compared to pre-LTD release. These two experiments led the authors to propose the hypothesis that after LTD, release of neurotransmitter shifts from a mode of complete collapse (allows FM dye to escape) to a kiss-and-run or non-expanding fusion pore mode (more FM1-43 is trapped). This is the opposite of what is observed during synapse maturation (Renger et al., 2001) and during LTP (Choi et al., 2000).

Alhtough these three studies provide evidence for a modulation of fusion pore dynamics during synapse maturation and long term plasticity, they do not rule out post-synaptic modifications also occurring during LTP/LTD. The molecular composition of the synaptic fusion pore is yet to be elucidated, in addition to the mechanisms and signaling pathways that may serve to modulate the different states of the fusion pore. The observation that synapses displaying evoked AMPA currents can be converted to AMPA silent synapses by cleaving the SNARE protein synaptobrevin/VAMP suggests that the basic synaptic machinery may play a role in either pore formation or modulation, but this finding is not consistent with the observation of normal AMPA responses in the mouse knockout of synaptobrevin (Schoch et al., 2001).

In light of the expanding literature supporting a postsynaptic silent synapse model for LTP, it seems likely that the synapses possess multiple mechanisms that allow them to express different forms of plasticity with different functional outcomes.

4.2. Purely Presynaptic Forms Of Long-Term Plasticity

While retrograde signaling may play a role in certain forms of postsynaptic NMDA receptor dependent LTP, there are synapses where the locus for LTP is primarily presynaptic. The hippocampal mossy fiber synapse is probably the most extensively studied example of these synapses. However a purely presynaptic locus for LTP has also been found to exist at cerebellar parallel fiber and corticothalamic synapses (Castillo et al., 1997; Castillo et al., 2002). This form of LTP is triggered by an influx of calcium into the presynaptic terminal followed by the downstream activation of protein kinase A (PKA), and ultimately the increased release of glutamate.

The major molecular players involved in this pathway are the synaptic vesicle associated protein Rab3A, and the active zone scaffold protein RIM. Rab3 proteins are small GTP binding proteins that reversibly attach to the membranes of synaptic vesicles. Along with synaptotagmin 1, they are thought to regulate the calcium dependent release of synaptic vesicles. While synaptotagmin 1 is important for triggering release, analysis of Rab3A knockout mice showed that this protein functions in a Ca^{2+} dependent manner to limit vesicle release. Additionally, in mice lacking expression of Rab3A, Castillo and colleagues showed that while synaptic transmission and short-term plasticity were normal at the mossy fiber synapse, LTP was essentially absent. This suggests that normally in the presence of Rab3A, LTP inducing stimuli partially inactivates Rab3A through a PKA dependent effector (i.e. RIM see below) resulting in the increased release of glutamate from the terminal.

The putative link between Rab3A and PKA is the protein RIM. RIM has been found to bind to the GTP-bound form of Rab3A. RIM also binds to several other active zone proteins including munc-13, RIM binding protein (RIM-BP) and α-liprins. Munc13 is essential for the priming step of synaptic vesicle exocytosis. α-liprins are proteins that bind to receptor tyrosine kinases and have been found to be essential for the normal functioning of active zones in *C. elegans*. The function of RIM-BP is unknown. In mice the knockout of the most abundant isoform, RIM1α, resulted in a 60% reduction of munc13-1 expression while certain postsynaptic proteins, GRIP, SynGAP, PSD95 and SHANK were increased. However, the synaptic structure appeared to be normal in these mice. The mossy fiber synapses lacking RIM1α, like those lacking Rab3A, were unable to express LTP. In addition, short-term plasticity, measured by paired-pulse facilitation and frequency facilitation were also unaffected similar to what was observed in Rab3A knockout mice. Increased calcium concentration or more robust induction protocols were unable to overcome this deficit in RIM1α mice.

At the parallel fiber synapse in the cerebellum the analysis of Rab3A and RIM1α deficient mice also showed similar results. Mice deficient in either of these proteins were unable to express LTP at this synapse. This finding strongly suggests that the mechanisms for presynaptic forms of LTP are similar in different synapses. Recent experiments in cultured cerebellar granule cell-purkinje cell synapses showed that phosphorylation of RIM at serine 413 by PKA was necessary for the expression of LTP thus pinpointing a simple molecular mechanism as the main cause of presynaptic LTP (Lonart et al., 2003).

Here it is interesting to note that in RIM1α and Rab3A deficient mice the phenotype of CA3-CA1 synapses, where LTP is primarily postsynaptic was entirely different. In this case, there was increased paired pulse facilitation (PPF) at excitatory synapses in mice deficient in either protein. However, in inhibitory synapses there was an increase in paired pulse depression (PPD) in the RIM1α knockout not observed in Rab3A knockout mice (short-term plasticity is generally unaffected in mossy fiber LTP). In addition, post tetanic potentiation, seen after intense stimulation due to Ca^{2+} build up in the presynaptic terminals, was enhanced in RIM1α knockout mice but was unchanged in Rab3A knockout mice. Release probability was also increased in RIM1α deficient mice but unchanged in Rab3A knockout mice. These divergent results from the same mouse lines in two different types of synapses suggest that the same proteins may be able to modulate synaptic plasticity in fundamentally different ways depending on the complement of other proteins and factors present in a given synapse.

5. REMAINING QUESTIONS: HETEROGENEITY AND REGULATION OF VESICLE RECYCLING

As the preceding discussion clearly demonstrates we are now only beginning to visualize the complexity of presynaptic signaling. Several mechanistic questions on the machinery that underlie synaptic vesicle endocytosis and recycling remain to be addressed. Currently, very little is known about the molecules that direct the synaptic vesicle cycle. Most of the information, albeit incomplete, is on the clathrin-mediated endocytosis. Research on this pathway focuses on the process of clathrin coat assembly and disassembly, the role of clathrin adaptor molecules, the exact function of dynamin in vesicle fission from the plasma membrane as well as the role of PIP_2 (and various components of the PIP_2 pathway) in tagging synaptic vesicles for retrieval and trafficking. However, there is currently no information on the molecular basis of fast endocytosis or the putative "kiss and run" pathway. A commonly referred to version of the "kiss and run" hypothesis proposes that there is extremely tight coupling between exocytosis and endocytosis and that synaptic vesicles are retrieved intact without loss of their molecular identity. From this perspective, it is tempting to speculate that the fusion machinery itself, such as SNARE proteins and their interaction partners, can be critical players in this pathway. In the case of clathrin-mediated endocytosis, analogous evidence suggests that synaptotagmin 1 could itself regulate the Ca^{2+} dependent exocytosis and subsequent endocytosis of vesicles by recruitment of clathrin adaptor proteins.

Better understanding of the underlying mechanism of "kiss and run" will allow us to address questions on the identity and regulation of synaptic fusion pore. It is highly likely that proteins that give rise to "kiss and run" pathway will also be involved in limiting neurotransmitter release during multiple forms of synaptic plasticity discussed above. Knowing the components of the "kiss and run" pathway will then help us to find out whether fusion pore regulation is a widespread means for plasticity and if all synapses are capable of regulating their fusion pore dynamics.

More molecular information on vesicle trafficking pathways in the synapse may also give clues on the regulation and heterogeneity of vesicle recycling amongst different synapses. Activity-dependent and synapse type specific variations in the levels and alternative splicing patterns of the key proteins involved in these pathways may underlie plasticity of vesicle recycling. For instance, recent findings on the role of synaptotagmin 7 suggest an appealing potential mechanism that can regulate synaptic vesicle recycling. In cultured hippocampal neurons, expression of the two splice variants of this protein, with or without the C2 domains that bind to calcium and clathrin adaptor AP2, regulated the rate of vesicle recycling in a bi-directional manner. Synaptotagmin 7 full-length protein targeted vesicles to slow recycling pathways, possibly involving endosomal intermediates, while splice variants lacking the C2 domains resulted in both fast vesicle endocytosis and recycling. Thus, the different levels of these splice variants could encode a molecular switch that can regulate the rate of vesicle recycling in a synapse. The baseline levels of the different variants in different brain regions or synapse populations could result in differing rates of recycling. Furthermore, an activity dependent process could potentially regulate the ratio of these variants and render the pathway of vesicle recycling adjustable. Upregulation of short C2 domain lacking forms, or down-regulation of full-length forms, in the presence of increased activity, could shift a slow recycling synapse to a fast recycling synapse. Permanent changes in the baseline levels of these molecules in a synapse could thus dramatically alter the input-output characteristics of a synapse.

Is synaptic vesicle recycling truly heterogeneous amongst synapses that have different structures and perform distinct functions? Synapses have diverse structural attributes so it is possible that their vesicle recycling pathways are not identical and may have evolved differently to serve different functional needs. For instance, two widely studied synapses the neuromuscular junction and the calyx of Held in the auditory brain stem both have a large number of vesicles (>10,000) and multiple active zones. In contrast, the small excitatory synapses in the CNS have in the order of 100 vesicles and usually a small number of active zones. Functionally, large synapses appear to be designed to sustain transmission in a failure proof manner, whereas small central synapses may purposefully have a more stochastic nature to increase their dynamic range for plasticity. However, the miniaturization of CNS synapses also makes them more prone to vesicle depletion during sustained activity. Despite these expectations, small central synapses are significantly resistant to depletion primarily due to rapid vesicle recycling and reuse that can maintain a sizable functional vesicle population even under intense stimulation (Sara et al., 2002). In contrast, some large synapses, such as retinal bipolar cell terminals are more susceptible to vesicle depletion under strong stimulation (Royle and Lagnado, 2003). As suggested previously, this structural divergence may be partly due to synapse specific regulation of vesicle recycling pathways (Harata et al., 2001). Recent evidence suggests that even in the large synapses only a handful of vesicles are actively recycling and the sustained neurotransmission heavily depends vesicle recycling (de Lange et al., 2003; Richards et al., 2003). Clearly, the question of functional and molecular diversity of synapses is an important one that will need to be addressed in a systematic manner in future studies.

The important role of synaptic vesicle recycling in the maintenance of efficient neurotransmission at the synapse makes it an ideal candidate for regulation during plasticity. Although much effort has been spent to dissect the kinetics of vesicle recycling in presynaptic terminals, activity dependent regulation of this essential process is largely unknown. Furthermore, the potential impact of the regulation of vesicle recycling on synaptic depression is also unclear. It is generally thought that fast recycling is up-regulated with increasing activity and Ca^{2+} influx. However, this hypothesis has been extrapolated from experiments measuring endocytosis, which is merely the first step in vesicle recycling. Therefore, the role of activity on the eventual reuse of vesicles remains to be studied in detail.

The functional and molecular diversity between structurally divergent synapses may also be applicable to synapses that share apparently similar structures. The question remains as to whether all small excitatory glutamatergic synapses in different regions in the brain have similar recycling mechanisms. The data described above for the differences in pre- vs. postsynaptic LTP suggest that there may indeed be some subtle differences. These differences would provide an ideal substrate for regulation of short-term plasticity. Small synapses in brain regions where baseline levels of activity are high would require higher rates of synaptic vesicle recycling to maintain efficient transmission, compared to regions where baseline activity are lower. As we have indicated in the beginning, fine-tuning of presynaptic vesicle recycling provides synapses with a powerful means to alter their time and frequency-dependent behavior thus profoundly influence information coding in the CNS.

6. REFERENCES

Arancio O, Kandel ER, Hawkins RD (1995) Activity-dependent long-term enhancement of transmitter release by presynaptic 3',5'-cyclic GMP in cultured hippocampal neurons. Nature 376:74-80.

Aravanis AM, Pyle JL, Tsien RW (2003) Single synaptic vesicles fusing transiently and successively without loss of identity. Nature 423:643-647.

Atwood HL, Karunanithi S (2002) Diversification of synaptic strength: presynaptic elements. Nat Rev Neurosci 3:497-516.

Betz WJ, Bewick GS (1992) Optical analysis of synaptic vesicle recycling at the frog neuromuscular junction. Science 255:200-203.

Betz WJ, Bewick GS (1993) Optical monitoring of transmitter release and synaptic vesicle recycling at the frog neuromuscular junction. J Physiol 460:287-309.

Bolshakov VY, Golan H, Kandel ER, Siegelbaum SA (1997) Recruitment of new sites of synaptic transmission during the cAMP-dependent late phase of LTP at CA3-CA1 synapses in the hippocampus. Neuron 19:635-651.

Butz S, Fernandez-Chacon R, Schmitz F, Jahn R, Sudhof TC (1999) The subcellular localizations of atypical synaptotagmins III and VI. Synaptotagmin III is enriched in synapses and synaptic plasma membranes but not in synaptic vesicles. J Biol Chem 274:18290-18296.

Castillo PE, Schoch S, Schmitz F, Sudhof TC, Malenka RC (2002) RIM1alpha is required for presynaptic long-term potentiation. Nature 415:327-330.
Castillo PE, Janz R, Sudhof TC, Tzounopoulos T, Malenka RC, Nicoll RA (1997) Rab3A is essential for mossy fibre long-term potentiation in the hippocampus. Nature 388:590-593.
Ceccarelli B, Hurlbut WP, Mauro A (1973) Turnover of transmitter and synaptic vesicles at the frog neuromuscular junction. J Cell Biol 57:499-524.
Choi S, Klingauf J, Tsien RW (2000) Postfusional regulation of cleft glutamate concentration during LTP at 'silent synapses'. Nat Neurosci 3:330-336.
Cremona O, De Camilli P (1997) Synaptic vesicle endocytosis. Curr Opin Neurobiol 7:323-330.
Cremona O, Di Paolo G, Wenk MR, Luthi A, Kim WT, Takei K, Daniell L, Nemoto Y, Shears SB, Flavell RA, McCormick DA, De Camilli P (1999) Essential role of phosphoinositide metabolism in synaptic vesicle recycling. Cell 99:179-188.
de Lange RP, de Roos AD, Borst JG (2003) Two modes of vesicle recycling in the rat calyx of Held. J Neurosci 23:10164-10173.
Delgado R, Maureira C, Oliva C, Kidokoro Y, Labarca P (2000) Size of vesicle pools, rates of mobilization, and recycling at neuromuscular synapses of a Drosophila mutant, shibire. Neuron 28:941-953.
Di Paolo G, Sankaranarayanan S, Wenk MR, Daniell L, Perucco E, Caldarone BJ, Flavell R, Picciotto MR, Ryan TA, Cremona O, De Camilli P (2002) Decreased synaptic vesicle recycling efficiency and cognitive deficits in amphiphysin 1 knockout mice. Neuron 33:789-804.
Dittman JS, Regehr WG (1998) Calcium dependence and recovery kinetics of presynaptic depression at the climbing fiber to Purkinje cell synapse. J Neurosci 18:6147-6162.
Fernandez-Chacon R, Konigstorfer A, Gerber SH, Garcia J, Matos MF, Stevens CF, Brose N, Rizo J, Rosenmund C, Sudhof TC (2001) Synaptotagmin I functions as a calcium regulator of release probability. Nature 410:41-49.
Fiala JC, Feinberg M, Popov V, Harris KM (1998) Synaptogenesis via dendritic filopodia in developing hippocampal area CA1. J Neurosci 18:8900-8911.
Gandhi SP, Stevens CF (2003) Three modes of synaptic vesicular recycling revealed by single-vesicle imaging. Nature 423:607-613.
Geppert M, Goda Y, Hammer RE, Li C, Rosahl TW, Stevens CF, Sudhof TC (1994) Synaptotagmin I: a major Ca2+ sensor for transmitter release at a central synapse. Cell 79:717-727.
Harata N, Pyle JL, Aravanis AM, Mozhayeva M, Kavalali ET, Tsien RW (2001) Limited numbers of recycling vesicles in small CNS nerve terminals: implications for neural signaling and vesicular cycling. Trends Neurosci 24:637-643.
Henkel AW, Almers W (1996) Fast steps in exocytosis and endocytosis studied by capacitance measurements in endocrine cells. Curr Opin Neurobiol 6:350-357.
Heuser JE, Reese TS (1973) Evidence for recycling of synaptic vesicle membrane during transmitter release at the frog neuromuscular junction. J Cell Biol 57:315-344.
Jahn R, Lang T, Sudhof TC (2003) Membrane fusion. Cell 112:519-533.
Kavalali ET, Klingauf J, Tsien RW (1999) Properties of fast endocytosis at hippocampal synapses. Philos Trans R Soc Lond B Biol Sci 354:337-346.
Klingauf J, Kavalali ET, Tsien RW (1998) Kinetics and regulation of fast endocytosis at hippocampal synapses. Nature 394:581-585.
Koenig JH, Ikeda K (1996) Synaptic vesicles have two distinct recycling pathways. J Cell Biol 135:797-808.
Koenig JH, Ikeda K (1999) Contribution of active zone subpopulation of vesicles to evoked and spontaneous release. J Neurophysiol 81:1495-1505.
Li C, Ullrich B, Zhang JZ, Anderson RG, Brose N, Sudhof TC (1995) Ca(2+)-dependent and -independent activities of neural and non-neural synaptotagmins. Nature 375:594-599.
Liu G, Tsien RW (1995) Properties of synaptic transmission at single hippocampal synaptic boutons. Nature 375:404-408.
Lonart G, Schoch S, Kaeser PS, Larkin CJ, Sudhof TC, Linden DJ (2003) Phosphorylation of RIM1alpha by PKA triggers presynaptic long-term potentiation at cerebellar parallel fiber synapses. Cell 115:49-60.
Luthi A, Di Paolo G, Cremona O, Daniell L, De Camilli P, McCormick DA (2001) Synaptojanin 1 contributes to maintaining the stability of GABAergic transmission in primary cultures of cortical neurons. J Neurosci 21:9101-9111.
Ma L, Zablow L, Kandel ER, Siegelbaum SA (1999) Cyclic AMP induces functional presynaptic boutons in hippocampal CA3-CA1 neuronal cultures. Nat Neurosci 2:24-30.
Malgaroli A, Ting AE, Wendland B, Bergamaschi A, Villa A, Tsien RW, Scheller RH (1995) Presynaptic component of long-term potentiation visualized at individual hippocampal synapses. Science 268:1624-1628.

Malinow R, Malenka RC (2002) AMPA receptor trafficking and synaptic plasticity. Annu Rev Neurosci 25:103-126.
Marqueze B, Berton F, Seagar M (2000) Synaptotagmins in membrane traffic: which vesicles do the tagmins tag? Biochimie 82:409-420.
Martina JA, Bonangelino CJ, Aguilar RC, Bonifacino JS (2001) Stonin 2: an adaptor-like protein that interacts with components of the endocytic machinery. J Cell Biol 153:1111-1120.
Micheva KD, Holz RW, Smith SJ (2001) Regulation of presynaptic phosphatidylinositol 4,5-biphosphate by neuronal activity. J Cell Biol 154:355-368.
Micheva KD, Buchanan J, Holz RW, Smith SJ (2003) Retrograde regulation of synaptic vesicle endocytosis and recycling. Nat Neurosci 6:925-932.
Miesenbock G, De Angelis DA, Rothman JE (1998) Visualizing secretion and synaptic transmission with pH-sensitive green fluorescent proteins. Nature 394:192-195.
Montgomery JM, Pavlidis P, Madison DV (2001) Pair recordings reveal all-silent synaptic connections and the postsynaptic expression of long-term potentiation. Neuron 29:691-701.
Mozhayeva MG, Sara Y, Liu X, Kavalali ET (2002) Development of vesicle pools during maturation of hippocampal synapses. J Neurosci 22:654-665.
Murthy VN, Stevens CF (1999) Reversal of synaptic vesicle docking at central synapses. Nat Neurosci 2:503-507.
Murthy VN, De Camilli P (2003) Cell biology of the presynaptic terminal. Annu Rev Neurosci 26:701-728.
Neher E (1998) Vesicle pools and Ca2+ microdomains: new tools for understanding their roles in neurotransmitter release. Neuron 20:389-399.
Piedras-Renteria ES, Pyle JL, Diehn M, Glickfeld LL, Harata NC, Cao Y, Kavalali ET, Brown PO, Tsien RW (2004) Presynaptic homeostasis at CNS nerve terminals compensates for lack of a key Ca2+ entry pathway. Proc Natl Acad Sci U S A 101:3609-3614.
Poskanzer KE, Marek KW, Sweeney ST, Davis GW (2003) Synaptotagmin I is necessary for compensatory synaptic vesicle endocytosis in vivo. Nature 426:559-563.
Pyle JL, Kavalali ET, Piedras-Renteria ES, Tsien RW (2000) Rapid reuse of readily releasable pool vesicles at hippocampal synapses. Neuron 28:221-231.
Renger JJ, Egles C, Liu G (2001) A developmental switch in neurotransmitter flux enhances synaptic efficacy by affecting AMPA receptor activation. Neuron 29:469-484.
Richards DA, Guatimosim C, Betz WJ (2000) Two endocytic recycling routes selectively fill two vesicle pools in frog motor nerve terminals. Neuron 27:551-559.
Richards DA, Guatimosim C, Rizzoli SO, Betz WJ (2003) Synaptic vesicle pools at the frog neuromuscular junction. Neuron 39:529-541.
Rizzoli SO, Betz WJ (2004) The structural organization of the readily releasable pool of synaptic vesicles. Science 303:2037-2039.
Rosenmund C, Stevens CF (1996) Definition of the readily releasable pool of vesicles at hippocampal synapses. Neuron 16:1197-1207.
Royle SJ, Lagnado L (2003) Endocytosis at the synaptic terminal. J Physiol 553:345-355.
Ryan TA, Smith SJ (1995) Vesicle pool mobilization during action potential firing at hippocampal synapses. Neuron 14:983-989.
Ryan TA, Ziv NE, Smith SJ (1996) Potentiation of evoked vesicle turnover at individually resolved synaptic boutons. Neuron 17:125-134.
Ryan TA, Reuter H, Wendland B, Schweizer FE, Tsien RW, Smith SJ (1993) The kinetics of synaptic vesicle recycling measured at single presynaptic boutons. Neuron 11:713-724.
Sankaranarayanan S, Ryan TA (2000) Real-time measurements of vesicle-SNARE recycling in synapses of the central nervous system. Nat Cell Biol 2:197-204.
Sankaranarayanan S, Ryan TA (2001) Calcium accelerates endocytosis of vSNAREs at hippocampal synapses. Nat Neurosci 4:129-136.
Sara Y, Mozhayeva MG, Liu X, Kavalali ET (2002) Fast vesicle recycling supports neurotransmission during sustained stimulation at hippocampal synapses. J Neurosci 22:1608-1617.
Schikorski T, Stevens CF (2001) Morphological correlates of functionally defined synaptic vesicle populations. Nat Neurosci 4:391-395.
Schneggenburger R, Meyer AC, Neher E (1999) Released fraction and total size of a pool of immediately available transmitter quanta at a calyx synapse. Neuron 23:399-409.
Schoch S, Deak F, Konigstorfer A, Mozhayeva M, Sara Y, Sudhof TC, Kavalali ET (2001) SNARE function analyzed in synaptobrevin/VAMP knockout mice. Science 294:1117-1122.
Schuman EM, Madison DV (1991) A requirement for the intercellular messenger nitric oxide in long-term potentiation. Science 254:1503-1506.

Selig DK, Segal MR, Liao D, Malenka RC, Malinow R, Nicoll RA, Lisman JE (1996) Examination of the role of cGMP in long-term potentiation in the CA1 region of the hippocampus. Learn Mem 3:42-48.
Stevens CF, Wesseling JF (1998) Activity-dependent modulation of the rate at which synaptic vesicles become available to undergo exocytosis. Neuron 21:415-424.
Stevens CF, Williams JH (2000) "Kiss and run" exocytosis at hippocampal synapses. Proc Natl Acad Sci U S A 97:12828-12833.
Sudhof TC (1995) The synaptic vesicle cycle: a cascade of protein-protein interactions. Nature 375:645-653.
Sudhof TC (2000) The synaptic vesicle cycle revisited. Neuron 28:317-320.
Sudhof TC (2002) Synaptotagmins: why so many? J Biol Chem 277:7629-7632.
Sugita S, Han W, Butz S, Liu X, Fernandez-Chacon R, Lao Y, Sudhof TC (2001) Synaptotagmin VII as a plasma membrane Ca(2+) sensor in exocytosis. Neuron 30:459-473.
Takei K, Mundigl O, Daniell L, De Camilli P (1996) The synaptic vesicle cycle: a single vesicle budding step involving clathrin and dynamin. J Cell Biol 133:1237-1250.
Trachtenberg JT, Chen BE, Knott GW, Feng G, Sanes JR, Welker E, Svoboda K (2002) Long-term in vivo imaging of experience-dependent synaptic plasticity in adult cortex. Nature 420:788-794.
Valtorta F, Meldolesi J, Fesce R (2001) Synaptic vesicles: is kissing a matter of competence? Trends Cell Biol 11:324-328.
Vaughn JE (1989) Fine structure of synaptogenesis in the vertebrate central nervous system. Synapse 3:255-285.
Verstreken P, Kjaerulff O, Lloyd TE, Atkinson R, Zhou Y, Meinertzhagen IA, Bellen HJ (2002) Endophilin mutations block clathrin-mediated endocytosis but not neurotransmitter release. Cell 109:101-112.
Virmani T, Han W, Liu X, Sudhof TC, Kavalali ET (2003) Synaptotagmin 7 splice variants differentially regulate synaptic vesicle recycling. Embo J 22:5347-5357.
von Poser C, Zhang JZ, Mineo C, Ding W, Ying Y, Sudhof TC, Anderson RG (2000) Synaptotagmin regulation of coated pit assembly. J Biol Chem 275:30916-30924.
Wang LY, Kaczmarek LK (1998) High-frequency firing helps replenish the readily releasable pool of synaptic vesicles. Nature 394:384-388.
Wu LG, Borst JG (1999) The reduced release probability of releasable vesicles during recovery from short-term synaptic depression. Neuron 23:821-832.
Zakharenko SS, Zablow L, Siegelbaum SA (2001) Visualization of changes in presynaptic function during long-term synaptic plasticity. Nat Neurosci 4:711-717.
Zakharenko SS, Zablow L, Siegelbaum SA (2002) Altered presynaptic vesicle release and cycling during mGluR-dependent LTD. Neuron 35:1099-1110.
Zenisek D, Steyer JA, Almers W (2000) Transport, capture and exocytosis of single synaptic vesicles at active zones. Nature 406:849-854.
Zhang JZ, Davletov BA, Sudhof TC, Anderson RG (1994) Synaptotagmin I is a high affinity receptor for clathrin AP-2: implications for membrane recycling. Cell 78:751-760.
Zhuo M, Hu Y, Schultz C, Kandel ER, Hawkins RD (1994) Role of guanylyl cyclase and cGMP-dependent protein kinase in long-term potentiation. Nature 368:635-639.
Zito K, Svoboda K (2002) Activity-dependent synaptogenesis in the adult Mammalian cortex. Neuron 35:1015-1017.
Zucker RS, Regehr WG (2002) Short-term synaptic plasticity. Annu Rev Physiol 64:355-405.

RETROGRADE MESSENGERS IN LONG-TERM PLASTICITY OF PRESYNAPTIC GLUTAMATE RELEASE IN HIPPOCAMPUS

Andreas Kyrozis, Karima Benameur, Xiao-lei Zhang, Jochen Winterer, Wolfgang Müller and Patric K. Stanton*

1. INTRODUCTION

Activity-dependent synaptic plasticity, defined as bidirectional reversible changes of synaptic strength in response to particular patterns of synaptic activity, has long been suggested as a cellular correlate of learning and memory. Since the first demonstration of long-term potentiation (LTP) in the mammalian central nervous system, specifically in the hippocampal dentate gyrus (Bliss and Lømo, 1973), considerable research effort has been directed to elucidating the complex biochemical and subcellular ultrastructural mechanisms of synaptic plasticity. The complexity derives partially from the multiplicity of signaling pathways that can operate in a single synapse, often under varying experimental conditions. These pathways may function independently (each contributing partially to sum in observed plasticity) or in coordination (both required). Implication of one pathway, therefore, cannot be necessarily be considered evidence against a role for another.

At some synapses, including the extensively studied hippocampal Schaffer collateral - CA1 synapse, long-term plasticity has been shown to require postsynaptic depolarization in the induction phase, but to involve presynaptic alterations in transmitter release in the subsequent maintenance phase. There are, in principle, two broad mechanisms that could account for these findings. First, both presynaptic and postsynaptic events may be required independently during induction and maintenance phases to produce the full sum of plastic changes observed. Alternatively, postsynaptic events during induction may generate chemical signals that travel in a retrograde

* A. Kyrozis, Dept Neurology, Albert Einstein Coll Med, Bronx, NY 10641; K. Benameur, Stroke Branch, NINDS, NIH, Bethesda, MD 20814; Jochen Winterer and Wolfgang Müller, Neuroscience Research Institute, Charité, Humboldt University, Berlin, D-10117, Germany; Xiao-lei Zhang and Patric K. Stanton, Dept Cell Biology & Anatomy, New York Med Coll, Valhalla, NY 10595. E-mail: patric_stanton@nymc.edu

direction to the presynaptic terminal and trigger alterations there in glutamate release that are expressed in the maintenance phase.

The concept of retrograde messengers in synaptic transmission is now a well accepted exception to the "neuron doctrine", which stipulates that neurotransmitters are packed within synaptic vesicles in specialized presynaptic terminals, released after action potential invasion and entry of Ca^{2+} into the terminal, and bind to postsynaptically located receptors resulting in the gating of receptor-associated channels or triggering of biochemical processses in the postsynaptic cell. Retrograde signaling generally does not appear to follow these rules. Since conventional synaptic vesicles do not appear to be present in spines and other postsynaptic sites, retrograde messengers seem less likely to be vesicle-packaged, water-soluble molecules, but more likely to be lipid-soluble and able to cross membranes from one cell to another. Such unconventional messengers include small gaseous molecules, notably nitric oxide (NO) and carbon monoxide (CO), as well as fatty acids and related molecules like arachidonic acid, platelet activating factor (PAF) and endocannabinoids (anandamide and 2-arachidonoyl glycerol).

In this chapter, we will focus on the role of NO as a putative retrograde messenger in long-term plasticity at the synapse between Schaffer collateral axons of CA3 pyramidal cells and apical dendrites of CA1 pyramidal cells in the mammalian hippocampus. The putative targets of NO in processes that lead to modulation of transmitter release will also be reviewed. Finally, possible roles for other gaseous and lipid retrograde messengers at the same synapse, namely CO, arachidonic acid and PAF, will be discussed.

Evidence for plasticity involving NO-mediated retrograde signaling has also been obtained in the medial vestibular nucleus (Grassi and Pettorossi, 2000; Pettorossi and Grassi, 2001) and at cerebellar mossy fiber - granule cell synapses (Maffei et al., 2003). Moreover, NO can act as an intercellular messenger in directions other than retrograde. Studies suggest that NO has postsynaptic targets in hippocampal dentate gyrus LTP (Wu et al., 1997, 1998), corticostriatal LTP and LTD (Calabresi et al., 1999; Centonze et al., 2003; Doreulee et al., 2003) and cerebellar parallel fiber-Purkinje cell LTD (Lev-Ram et al., 1997; Hartell, 2002; Lev-Ram et al., 2003), Interestingly, LTP at the latter synapse may also require NO acting on a presynaptic target (Jacoby et al., 2001). NO also modulates plasticity via intercellular communication where the direction is still unclear, in the neocortex (Nowicky and Bindman, 1993; Wakatsuki et al., 1998; Haul et al., 1999) and amygdala (Watanabe et al., 1995; Abe et al., 1996).

2. NO AS A RETROGRADE MESSENGER IN PRESYNAPTIC LTP AT SCHAFFER COLLATERAL-CA1 SYNAPSES

2.1 LTP and the need for a retrograde messenger

The Schaffer collateral-CA1 synapse is probably the most intensively examined mammalian synapse in studies of activity-dependent long-term plasticity. LTP is usually induced by high frequency (tetanic) stimulation of the Schaffer collaterals which leads to significant depolarization of the postsynaptic cell, conforming with the Hebbian rule that postulates synchronous presynaptic and postsynaptic activity is required to strengthen a synapse (Hebb, 1949; Bliss and Collingridge, 1993). Depolarization relieves the voltage-dependent Mg^{2+} block of the NMDA receptor with consequent Ca^{2+} entry and activation of Ca^{2+}-dependent enzymes in the postsynaptic neuron (Mayer et al., 1984; Bliss and Collingridge, 1993).

Although the computational role of the postsynaptic dendrite as the compartment that senses pairing of pre- and postsynaptic activity that induces LTP is well established, the locus (or loci) of expression of LTP has been much debated. Early studies addressing the question of presynaptic versus postsynaptic expression have relied on indirect electrophysiological measures of changes in transmitter release, such as quantal analysis and paired-pulse facilitation, that require assumptions that are not satisfied. Some suggested a presynaptic change (Bekkers and Stevens, 1990; Malinow and Tsien, 1990; Stevens and Wang, 1994; Clark and Collingridge, 1995), and others postsynaptic (Foster and McNaughton, 1991; Isaac et al., 1995) or mixed changes (Foster and McNaughton, 1991; Kullmann and Nicoll, 1992). More recent biochemical and structural data have solidified the view that LTP is accompanied by changes in postsynaptic AMPA receptors, such as single channel conductance alterations (Benke et al., 1998), phosphorylation and trafficking (Malenka and Nicoll, 1999; Malinow et al., 2000; Soderling and Derkach, 2000; Sheng and Lee, 2001) as well as remodeling of postsynaptic spines (Engert and Bonhoeffer, 1999; Shi et al., 1999; Toni et al., 1999).

Changes in transmitter release during LTP have been suggested in a number of different ways in hippocampal dissociated cell cultures and slices. LTP has been shown to be associated with increased glutamate concentrations in the synaptic cleft (Choi et al., 2000). Recently, transmitter release has been assessed directly using the fluorescent dye FM1-43, which labels synaptic vesicles (Betz and Bewick, 1992; Ryan et al., 1993), and methods have been developed to use this dye in brain slices (Pyle et al., 1999; Stanton et al., 2001, 2003; Zakharenko et al., 2001). LTP produced an increase in FM1-43 destaining from Schaffer collateral terminals loaded by synaptic stimulation in hippocampal slices (Zakharenko et al., 2001), suggesting enhanced transmitter release. The combination of a postsynaptic of induction and a presynaptic locus of expression is what prompted the search for candidate retrograde messenger molecules. NO has emerged as the most likely messenger in the induction of at least one form of LTP, and of LTD, at the Schaffer collateral-CA1 synapse, as well as at other synapses in the CNS.

Garthwaite and Garthwaite (1987) first proposed the hypothesis that NO functions as an intercellular messenger that generates cyclic GMP (cGMP) in response to stimulation of neuronal glutamate receptors. NO is generated by the oxidation of arginine to NO and citrulline, a reaction catalyzed by an enzyme long known as NADPH-diaphorase (Scherer-Singler et al., 1983), but now called the Ca^{2+}/calmodulin-dependent enzyme NO synthase (NOS) (Bredt and Snyder, 1992). In the nervous system, NOS immunoreactivity is detected in selected cell populations (Bredt et al., 1990; Bredt et al., 1991). The most prevalent NOS isoform is nNOS (or NOS-I), but eNOS (NOS-III) is also found.

There has been disagreement over localization of NOS in the Schaffer collateral-CA1 pyramidal cell synapse. Rat hippocampal nNOS was in some reports detected in interneurons, but not in CA1 pyramidal cells (Bredt et al., 1991; Valtschanoff et al., 1993; Lin and Totterdell, 1998; Lumme et al., 2000), but other studies describe expression in CA1 pyramidal neurons as well (Endoh et al., 1994; Wendland et al., 1994; Lopez-Figueroa et al., 1998). In a recent light and electron microscopic study in Sprague-Dawley rats, nNOS was found in about 8% of these synapses, ultrastructurally localized inside the postsynaptic plasma membrane of axospinous synapses (Burette et al., 2002).

Rat and mouse eNOS localization studies have also yielded apparently conflicting results, with some detecting it only in endothelial cells (Stanarius et al., 1997; Topel et al., 1998; Blackshaw et al., 2003), and others in CA1 pyramidal cells as well (Dinerman et al., 1994; O'Dell et al., 1994; Kantor et al., 1996; Doyle and Slater, 1997; Teichert et al., 2000). Some of the confusion may be due to different species and strains of experimental animals. In a recent study explicitly addressing this issue, nNOS was highly expressed in CA1 of humans and some mouse strains, but scantily in rats and different mouse strains (Blackshaw et al., 2003).

2.2 NO involvement in Schaffer collateral-CA1 LTP

Four types of evidence have been offered to suggest a role for NO in LTP at the Schaffer-CA1 synapse; effects of NOS inhibition, of NO scavengers and of NO donors, as well as demonstration of increases in extracellular [NO] during induction of LTP.

Bath application of NOS inhibitors has, in several studies, been shown to block or significantly reduce the amplitude of LTP in hippocampal slices (Bohme et al., 1991; O'Dell et al., 1991; Schuman and Madison, 1991; Bon et al., 1992; Haley et al., 1992; Boulton et al., 1995; Haley et al., 1996; Malen and Chapman, 1997) as well as *in vivo* (Doyle et al., 1996; Holscher, 1999), but there are other studies that have failed to show any effect of NOS inhibition on LTP (Kato and Zorumski, 1993; Cummings et al., 1994). Coapplication of the NOS substrate L-nitroarginine partially reversed this effect, suggesting that it was indeed a result of NOS inhibition (Bohme et al., 1991; O'Dell et al., 1991; Schuman and Madison, 1991; Haley et al., 1992). In some studies, a NOS inhibitor was also applied intracellularly into single CA1 pyramidal cells, where it impaired LTP at synapses on that neuron, suggesting a postsynaptic locus for required NOS (O'Dell et al., 1991; Schuman and Madison, 1991).

Demonstration of a role for NOS in the induction of LTP does not mean that a rise in [NO] is necessary; it could also be that constitutively generated NO may simply be permissive for the actions of other phasically activated biochemical pathways to trigger the changes in LTP. In a recent study (Bon and Garthwaite, 2003), the issue was extensively investigated by combinations of application of a NO donor and a NOS inhibitor. It was concluded that NO probably acts both as a phasically active signaling molecule during the induction phase of LTP, and as a constitutively active molecule before and for about 15 min after delivery of a stimulus train.

In some studies, bath application of NOS inhibitors failed to block induction of LTP (Cummings et al., 1994) and even enhanced it, possibly through enhancing NMDA receptor activation (Kato and Zorumski, 1993). This discrepancy could be due, at least in part, to a multiplicity of pathways that can elicit LTP and variations in experimental protocols that activate them to differing degrees.

Mice with targeted mutations of either eNOS, nNOS or both isoforms have been generated and tested for alterations in their magnitude of LTP. In nNOS-lacking mice, LTP was normal or only slightly reduced, compared to controls (O'Dell et al., 1994; Son et al., 1996). LTP was still reduced by a NOS inhibitor, suggesting that eNOS, rather than nNOS, might be the required isoform (O'Dell et al., 1994). In Son et al., (1996), however, LTP was essentially intact in eNOS mutants as well, whereas a significant reduction was only observed in double mutants lacking both isoforms. It was concluded that at least one of the isoforms is required for NO-dependent LTP, but they can compensate for each other. The expression of a residual LTP component in double mutants suggested that additional, NO-independent components of LTP coexist with NO-dependent forms. In a more recent study employing eNOS mutants, LTP induction by a weak tetanus was abolished, while a strong tetanus still elicited LTP, implying that eNOS can be necessary for a form of LTP selectively evoked by weaker stimulus trains (Wilson et al., 1999).

Experiments with stable mutations suffer from the disadvantage that compensatory mechanisms may be activated that obscure significant roles of the mutated protein. A more acute genetic manipulation has been used to examine the role of eNOS. Introduction of recombinant adenoviral vectors containing DNA coding for a dominant negative eNOS has been shown to inhibit LTP (Kantor et al., 1996), which could be rescued by introduction of eNOS fused to a transmembrane protein. It was concluded that eNOS, preferentially membrane-targeted, is required for LTP.

There are several reports of NO donors inducing LTP-like synaptic enhancement when paired with subthreshhold presynaptic stimulation in the CA1 region of hippocampal slices (Bohme et al., 1991; Bon et al., 1992; Zhuo et al., 1993; Zhuo et al., 1994a; Malen and Chapman, 1997). This enhancement was not blocked by NMDA receptor antagonists, indicating that NO is downstream of NMDA receptor activation (Zhuo et al., 1993; Malen and Chapman, 1997). The enhancement generally occluded tetanus-induced LTP, suggesting a biochemical convergence (Bohme et al., 1991; Bon et al., 1992). Similarly, the NO precursor L-arginine rescued LTP from block by a NOS inhibitor *in vitro* (Haley et al., 1992) and *in vivo* (Doyle et al., 1996). However, another study reported that pairing of photolytic release of NO with either weak tetanic stimulation or strong stimulation in the presence of an NMDA receptor antagonist did *not* lead to LTP-like enhancement, arguing against a role for NO (Murphy et al., 1994).

If a candidate retrograde messenger such as NO is generated postsynaptically and believed to cause a presynaptic effect, extracellular application of a membrane-impermeable scavenger should prevent the transsynaptic intercellular spread of NO and, hence, impair LTP. Hemoglobin (Hb), which binds NO, has been shown in several studies to significantly attenuate or block induction of LTP (O'Dell et al., 1991; Schuman and Madison, 1991; Haley et al., 1992). It should be noted, however, that Hb will bind other oxidative reactive species and has been reported to affect CA1 electrophysiology independent of its binding NO (Yip et al., 1996; Yip and Sastry, 2000). Nevertheless, more specific scavengers have also been able to block tetanically-induced LTP when applied either extracellularly or within single CA1 neurons (Ko and Kelly, 1999).

NO has recently been directly visualized by using the indicator DAQ, which increases its fluorescence when reacted with NO (von Bohlen et al., 2002). In this study, DAQ fluorescence was found to increase during induction of LTP, and the NOS inhibitor l-NAME reduced both DAQ fluorescence increases and LTP. These results are consistent with the hypothesis that a phasic increase in NO, rather than its constitutive presence alone, is required for LTP.

The question of LTP dependence on NO has been part of a broader question on the presynaptic versus postsynaptic expression of LTP. It has gradually become accepted that there are multiple induction mechanisms leading to multiple expression forms of LTP, interacting in complex ways, and their relative contribution to the sum of LTP observed may depend on experimental conditions that differ significantly between laboratories. Variables such as animal species and strain, age, recording temperature, stimulus protocol and stimulus history are known to be important determinants of the amount of NO-dependent and NO-independent LTP expressed. With the exception of stimulus protocol, few studies have methodically examined the roles of these parameters.

Species and strain differences could derive, in part, by differences in expression of nNOS and eNOS (see above). In a study examining the effect of rat strain, LTP was blocked by NOS inhibitors in Wistar, but not in Sprague-Dawley or Long-Evans, rats (Holscher, 2002). Animal age and recording temperature are also considered to be significant factors, with immature animals and lower temperatures favoring NO-dependent forms of LTP. Both of these parameters were assessed in one report (Williams et al., 1993), where in young rats either a NOS inhibitor or the NO scavenger hemoglobin partially inhibited LTP at room temperature, but were ineffective at 29-30°C, or in adult rats at room temperature. A number of studies have shown that inhibitors of the NO pathway blocks LTP induced by weaker, but not stronger, tetani (Haley et al., 1993; Lum-Ragan and Gribkoff, 1993; Wilson et al., 1999; Zhuo et al., 1999; but see also Gribkoff and Lum-Ragan, 1992)). It is not clear which mechanism is responsible for the NO-independent LTP induced by strong tetani, but it may involve CO as retrograde messenger (Zhuo et al., 1999; see below) or may be located postsynaptically.

2.3 NO effector pathways

(i) Soluble guanylyl cyclse ⇒ cyclic GMP ⇒ PKG

Soluble guanylyl cyclase (sGC) and its product cyclic GMP are principal mediators of signal transduction by NO in many systems (for review, see Denninger and Marletta, 1999; Wedel and Garbers, 2001). sGC is widely distibuted in the brain (Burgunder and Cheung, 1994; Ibarra et al., 2001). NOS and NO-stimulated cGMP accumulations tend to have complementary distributions. In some areas, NOS is located in postsynaptic structures and cGMP in presynaptic ones, whereas in others the loci are reversed (Southam and Garthwaite, 1993). This pattern is consistent with the hypothesis that NO can function as either a retrograde or anterograde intercellular messenger in different areas. A recent study in hippocampus shows a high degree of NOS and sGC colocalization, although both enzymes were present at a minority of synaptic sites (8-9% in CA1; (Burette et al., 2002). In the same study, electron microscopy demonstrated NOS in postsynaptic sites and sGC in presynaptic sites, consistent with the hypothesis that NO is generated postsynaptically and acts as a retrograde messenger.

If induction of LTP requires the activation of sGC, this event should, in principle, be biochemically detectable. There are two reports where LTP-inducing tetani elicited an increase in both sGC activity and cGMP production (Chetkovich et al., 1993; Monfort et al., 2002). The effect was blocked by either a NO inhibitor or the scavenger Hb (Chetkovich et al., 1993).

Cyclic GMP has several molecular targets, most notably cGMP-dependent protein kinases (PKG), phosphodiesterases (PDEs), and cyclic nucleotide gated channels (CNGCs) (for review, see Lucas et al., 2000), and the intracellular Ca^{2+} store mobilizer cyclic ADP-ribose (cADPR; Galione et al., 1993). Antagonists of cADPR binding to calcium stores failed to block the induction of LTP, even while it greatly reduced LTD (Reyes-Harde et al., 1999b), rendering unlikely a role for this pathway in LTP. Among the other cGMP effectors, only PKG has been given serious consideration and will be discussed below.

In testing whether NO promotes LTP via production of cGMP, the potent and selective sGC inhibitor ODQ (Garthwaite et al., 1995) has been variously reported to attenuate (Zhuo et al., 1994b; Boulton et al., 1995; Son et al., 1998; Monfort et al., 2002), and not to attenuate (Schuman et al., 1994; Gage et al., 1997; Wu et al., 1998), induction of LTP. cGMP analogs have in some studies facilitated LTP; in others not. Dibutyryl cGMP partially restored LTP in the presence of a NOS inhibitor (Haley et al., 1992). cGMP analogs coupled with weak tetanus produce long-term enhancement (Zhuo et al., 1994b; Son et al., 1998). In the latter study, the enhancement was reduced, but not blocked, by the NMDA receptor blocker AP5, suggesting that NMDAR activation has at least 2 actions required for LTP, one dependent and one independent of cGMP (Son et al., 1998). At the same time, other studies have failed to replicate this effect (Schuman et al., 1994; Selig et al., 1996), or even observed an enhancement of LTD (Gage et al., 1997; Reyes-Harde et al., 1999a; Wu et al., 1998).

In similarly variable results, blockade of LTP induction has been reported by structurally dissimilar PKG inhibitors Rp-8Br-cGMS (Zhuo et al., 1994b; Son et al., 1998; Monfort et al., 2002) and KT5823, while in other studies, PKG inhibition failed to block LTP (Schuman et al., 1994; Reyes-Harde et al., 1999a,b). In line with these latter findings, mice mutant for PKG have been reported to express normal NO-dependent LTP (Kleppisch et al., 1999).

Recently a drug with a novel type of action was tested for its effects on LTP. YC-1 acts by increasing the sensitivity of sGC to NO (as well as to CO; Ko et al., 1994; Friebe et al., 1996). YC-1 markedly enhanced potentiation induced by a weak tetanic stimulus (Chien et al., 2003), and this potentiation was significantly reduced by a NOS inhibitor, suggesting that NO, rather than CO, is required. It was also greatly reduced by a PKG inhibitor KT5823 (Chien et al., 2003).

Mice expressing mutant eNOS and deficiencies in LTP, a cGMP analog coupled with tetanus failed to induce LTP. Therefore, although LTP was apparently NO-dependent in these mice, it did not appear to be acting via the sGC-cGMP pathway (Wilson et al., 1999).

While the reasons for these variable results concerning the role of NO, cGMP and PKG in inducing LTP are unclear, some possibilities have been considered. The duration of elevations in [cGMP] may be critical, since brief, but not prolonged, perfusion with 8-Br-cGMP before weak tetanic stimulation produced long-lasting potentiation (Son et al., 1998). Smaller or slower increases in [cGMP] can, in fact, lead to long-term depression (Zhuo et al., 1994a; Gage et al., 1997), suggesting that, at intermediate levels, depression and potentiation may cancel out. The state of NMDA receptors may also be critical, since long-lasting potentiation induced by pairing application of cGMP analogs with synaptic activity is reduced when NMDA receptors are blocked (Son et al., 1998). This result suggests that NMDAR receptors can act partially through elevating [cGMP], and partially through additional mechanisms which are not well understood and may be less well controlled. In general, it seems clear that LTP consists of multiple components which can be differentially activated by different stimulus protocols.

(ii) ADP-ribosyltransferase

Besides sGC, another NO target, ADP-ribosyltransferase (ADPRT), has also been suggested to play a role in the induction of LTP. Mono-ADPRTs are cytosolic enzymes that catalyze attachment of a single ADP-ribose moiety to a loosely characterized population of proteins (Ueda and Hayaishi, 1985). It has been reported that ADP-ribosylation of particular proteins is induced by bath application of an NO donor, and is significantly reduced in hippocampal slices that have undergone LTP, compared to control slices, suggesting that LTP is accompanied by ADP-ribosylation (Duman et al., 1993). In one of the earliest studies implicating NO in the induction of LTP, bath application of two different ADPRT inhibitors blocked LTP, whereas postsynaptic injection of one of them was ineffective, suggesting that presynaptic ADPRT is required for LTP (Schuman et al., 1994). Perhaps the most convenient reconciliation of evidence implicating NO in the induction of *both* LTP and LTD (see below) would be if ADPRT was a target unique to induction of LTP, while sGC might either be selectively activated during induction of LTD, or be permissive for both forms of plasticity.

(iii) LTP in dissociated hippocampal cell cultures

NO-dependent LTP has also been observed at synapses in cultures of dissociated hippocampal cells. Compared to slices, this preparation permits easier access to the presynaptic neuron, which can be filled with appropriate drugs. First, it was observed that an NO donor enhanced spontaneous transmitter release (O'Dell et al., 1991). An elegant series of experiments was performed supporting a role for NO as a retrograde messenger (Arancio et al., 1996). In these studies, bath, presynaptic or postsynaptic application of a NO scavenger blocked LTP. In contrast, bath or postsynaptic, but *not* presynaptic, application of a NOS inhibitor blocked induction of LTP, suggesting that NOS is activated postsynaptically. NO donor application caused potentiation that was blocked by presynaptic, but not postsynaptic, application of a NO scavenger, indicating that NO acts presynaptically. Also, controlled localized either presynaptic or postsynaptic NO release (by UV-illumination of caged NO) coupled with a weak tetanus induced potentiation. Taken together, their data supports a model where NO is produced postsynaptically, diffuses to and acts on the presynaptic terminal to produce LTP (Arancio et al., 1996).

Evidence for presynaptic sGC and PKG as NO-activated effectors has been supplied by the same group. A presynaptically or bath-applied cGMP analog produced activity-dependent potentiation (Arancio et al., 1995), while presynaptic, but not postsynaptic, injection of a PKG inhibiting peptide blocked LTP (Arancio et al., 2001). It remains unclear whether, as part of reactive changes that can occur in culture situations, a presynaptic NO-cGMP-PKG pathway may be selectively up-regulated.

3. NO AS A RETROGRADE MESSENGER INDUCING PRESYNAPTIC LTD AT SCHAFFER COLLATERAL-CA1 SYNAPSES

3.1 LTD and the need for a retrograde messenger

LTD, like LTP before it, is beginning to generate interest as a form of plasticity with potential importance to memory processing and storage. LTD has clear potential for preventing saturation of synaptic strength that could ensue after repeated potentiation, leading to loss of information stored as *differences* in synaptic strength. Thus, LTD is considered necessary to maintain synapses in an optimal computational range

(Bienenstock et al., 1982; Wexler and Stanton, 1993; Stanton, 1996) and to maximize information extraction from covarying signal streams (Sejnowski, 1977). Unlike LTP, which has been studied for about 3 decades, LTD has only relatively recently attracted similar interest, triggered by the discovery of associative (Stanton and Sejnowski, 1989) and monosynaptic (Dudek and Bear, 1993) forms of LTD.

As is the case for LTP at the same synapse, the induction of LTD induction has been shown to require an increase in [Ca^{2+}] in the postsynaptic neuron (Mulkey and Malenka, 1992), but the magnitude of the increase that produces LTD is generally lower than that resulting in LTP (Neveu and Zucker, 1996). In most studies, the source of a significant fraction of this Ca^{2+} is influx is through NMDA channels (Dudek and Bear, 1993; Wexler and Stanton, 1993), but release from intracellular Ca^{2+} stores (both pre- and postsynaptic) have also been shown to make necessary contributions to the induction of LTD (O'Mara et al., 1995; Reyes and Stanton, 1996). A distinct form of LTD has been described at Schaffer collateral-CA1 synapses, independent of activation of NMDARs, but triggered by activation of group I metabotropic glutamate receptors (mGluRI) (Oliet et al., 1997). Like the more extensively studied NMDAR dependent form, it also requires a rise in postsynaptic [Ca^{2+}]. The two forms can both be expressed at these synapses, but experimental conditions used in most studies favor the NMDAR-dependent form, which is the form most studied for possible NO involvement. The following discussion will focus on the NMDAR-dependent form unless otherwise noted. It should be kept in mind, however, that even within the NMDAR-LTD form there is probably divergence of mechanisms after NMDAR activation, leading to expression of different alterations underlying LTD.

As with LTP, it is now clear that depolarization of the postsynaptic cell is usually required for induction of LTD, but the locus (or loci) of expression are still under debate. Physiological studies using quantal analysis or related approaches have supplied indirect evidence for presynaptic changes (Bolshakov and Siegelbaum, 1994; Oliet et al., 1996) and postsynaptic alterations, such as changes in AMPA receptor-gated channel conductances (Lee et al., 1998) and insertion and removal of AMPA receptors (in cultured cells, Carroll et al., 1999). On the presynaptic side, the first direct structural evidence of reduced transmitter release was recently obtained for a form of a cGMP-dependent form of chemical LTD (see below, Stanton et al., 2001), and for low frequency stimulus-induced LTD (Zakharenko et al., 2002; Stanton et al., 2003), by using the fluorescent dye FM1-43 to label synaptic vesicles directly (Betz and Bewick, 1992; Ryan et al., 1993; Pyle et al., 1999). Both these forms of LTD were found to be accompanied by a decrease in the rate of vesicular release of fluorescent dye, indicating decreased transmitter release.

3.2 NO involvement in Schaffer collateral-CA1 LTD

Several lines of evidence support a role for NO in induction of LTD at the Schaffer collateral-CA1 synapse by a number of stimulus protocols. Application of NOS blockers has been demonstrated to inhibit homosynaptic LTD induced by low frequency stimulation (Izumi and Zorumski, 1993; Santschi et al., 1999; Stanton et al., 2003), associative LTD induced by asynchronous stimulation of two separate inputs (Otani and Connor, 1995) and depotentiation after LTP induction (Izumi and Zorumski, 1993; Doyle et al., 1996). Additionally, bath application of hemoglobin, a membrane-impermeable

NO scavenger, also blocks induction of homosynaptic LTD (Izumi and Zorumski, 1993; Stanton et al., 2003).

Furthermore, in some studies, NO donors were able to induce long-lasting depression, mimicking low frequency stimulation (Izumi and Zorumski, 1993; Boulton et al., 1994). In other cases, such agents did not, by themselves, produce long-lasting alterations in synaptic transmission, but markedly enhanced LTD when paired with low frequency stimulation that would otherwise induce weak or no LTD (Zhuo et al., 1994a; Gage et al., 1997; Reyes-Harde et al., 1999a). Reminiscent of LTP, NOS inhibitors have been variously reported to block (Santschi et al., 1999) and fail to block (Cummings et al., 1994; Malen and Chapman, 1997) the induction of homosynaptic LTD or depotentiation after LTP (Bashir and Collingridge, 1994), and NO donors have also been reported not to facilitate LTD (Malen and Chapman, 1997). These variable findings, so similar to LTP, lead us again to invoke multiple forms of LTD, only some of which are NO-dependent, being expressed to varying degrees depending on differing experimental conditions. There have yet to be any detailed studies explicitly testing conditions that might determine the NO-dependence of LTD across stimulus protocols.

3.3 Involvement of the NO ⇒ cGMP ⇒ PKG pathway in LTD

As in LTP, the putative presynaptic NO effector that has attracted the most attention is sGC. In testing the importance of sGC to NO-mediated synaptic depression and LTD, experiments have measured [cGMP] after raising [NO], and examined effects of sGC inhibition and blockade of cGMP breakdown on synaptic transmission.

NO donors have been shown to elevate [cGMP] in hippocampal slices. The accumulation occurred mainly in a network of varicose fibers throughout CA1, consistent with a presynaptic location of sGC (Boulton et al., 1994).

sGC inhibition by ODQ has been shown to block the induction of LTD by low frequency stimulation (Gage et al., 1997). Depotentiation after previous LTP induction was reduced by about half, suggesting again the simultaneous expression of multiple forms of NO-dependent and independent LTD, and their sensitivity to the recent history of the synapse (Gage et al., 1997). ODQ also blocked LTD induced by pairing an NO donor with subthreshhold low frequency stimulation (Reyes-Harde et al., 1999a), thus demonstrating that NO can act through activating sGC to promote LTD. Inclusion of ODQ in postsynaptic cells through the recording electrode did not have an effect on LTD (Gage et al., 1997), suggesting a presynaptic locus for the sGC required for LTD.

Increasing cGMP availability through exogenous application of hydrolysis-resistant cell-permeant analogs, or decreasing its breakdown, can reversibly inhibit glutamate release and, when paired with some other stimulation, enhance LTD. Application of cGMP analogs paired with a weak low frequency stimulus train induced significantly larger LTD than stimulus alone (Zhuo et al., 1994a; Gage et al., 1997). Application of the type V phosphodiesterase (which selectively degrades cGMP) inhibitor zaprinast also produces a reversible depression of synaptic responses (Boulton et al., 1994; Santschi et al., 1999). Interestingly, this depression became irreversible if the cAMP-dependent protein kinase (PKA) inhibitor H89 was co-applied with zaprinast ("chemical LTD", Santschi et al., 1999). This type of LTD occluded low frequency stimulation - induced LTD, further supporting the hypothesis that a significant portion of the latter is mediated through activation of a cGMP-PKG pathway (Santschi et al., 1999; Reyes-Harde et al., 1999a).

Biochemical approaches have also shown decreases in transmitter release produced by raising [cGMP]. Zaprinast-induced increases in [cGMP] leads to decrease in transmitter release as assessed by FM1-43 fluorescence, and this decrease is long-lasting only when paired with concomitant inhibition of PKA (Stanton et al., 2001). Similarly, zaprinast paired with H89 has been shown to produce LTD of glutamate release from isolated rat hippocampal presynaptic terminals, suggesting that synaptosomes possess all the machinery necessary for cGMP (provided PKA activity is inhibited) to persistently depress vesicular transmitter release (Bailey et al., 2003).

There is evidence for two distinct targets of cGMP that are likely to play important roles in the induction of LTD; PKG and cyclic ADP-ribose (cADPR). PKG is the best characterized target of cGMP and, as in NO-dependent LTP, has been investigated as a possible effector in NO-dependent LTD. Studies suggest that induction of LTD by low frequency stimulation can be blocked by bath application of either of two structurally dissimilar PKG antagonists, KT5823 or Rp-8pCPT-cGMP (Reyes-Harde et al., 1999a). KT5823 also blocked LTD induced by pairing an NO donor with submaximal low frequency stimulation (Reyes-Harde et al., 1999a), suggesting that NO promotes LTD through activation of PKG. Furthermore, chemical LTD induced by coapplication of the PDEV inhibitor zaprinast with PKA inhibition was also blocked by bath-applied, but not by postsynaptically injected, KT5823, suggesting that the site of action of cGMP and PKG is action presynaptic (Santschi et al., 1999).

Biochemical evidence also support a role for PKG in induction of LTD. KT5823 has been found to significantly reduce LTD of presynaptic vesicular release imaged using FM1-43 (Stanton et al., 2003), and chemical LTD of glutamate release from isolated synaptosomes (Bailey et al., 2003).

A second cGMP-stimulated effector that has been implicated in the induction of LTD is the second messenger cADPR, which stimulates or facilitates release of Ca^{2+} from intracellular stores. Besides Ca^{2+} entry through NMDA receptor-gated channels, Ca^{2+} release from both presynaptic and postsynaptic stores have also been shown to be required for induction of Schaffer collateral LTD (Reyes and Stanton, 1996). There are two distinct, functionally defined types of intracellular Ca^{2+} stores; one released by the PLC metabolic product IP3, the second by either Ca^{2+}-induced Ca^{2+} release or the exogenous plant product ryanodine. cADPR appears to be a principal endogenous activator of release from this ryanodine receptor-gated store (Galione and Churchill, 2002). Bath application of ryanodine (which depletes the ryanodine calcium pool) before low frequency stimulation, blocked induction of LTD, while its injection into postsynaptic CA1 pyramidal neuron did not (Reyes and Stanton, 1996; Reyes-Harde et al., 1999a), indicating a requirement for Ca^{2+} release from *presynaptic* ryanodine stores. Ryanodine also blocked LTD induced by pairing an NO donor with submaximal low frequency stimulation (Reyes-Harde et al., 1999a), suggesting that the actions of NO also require release from ryanodine stores.

Interest in a possible role for cADPR in LTD was fostered by the discoveries (in sea urchin eggs and mammalian cells) that NO, by raising [cGMP], leads to activation of ADP-ribosyl cyclase/hydrolase, a difunctional enzyme that catalyzes the production of cADPR and release of Ca^{2+} from ryanodine stores (Galione et al., 1993; Willmott et al.,

1996; Galione and Churchill, 2002). Recently, it has been shown that elevating [cGMP] in hippocampal slices does generate cADPR, and that bath application of a cell permeant cADPR analog that blocks binding to ryanodine stores does impair induction of LTD (Reyes-Harde et al., 1999a). Similar to the data summarized previously, postsynaptic injection of a cADPR binding inhibitor failed to block induction of LTD (Reyes-Harde et al., 1999b), consistent with a presynaptic site of action.

3.4 Proposed mechanism

Taking all the available data together, our working hypothesis for the biochemical cascade underlying NO-dependent LTD is summarized in **figure 1**.

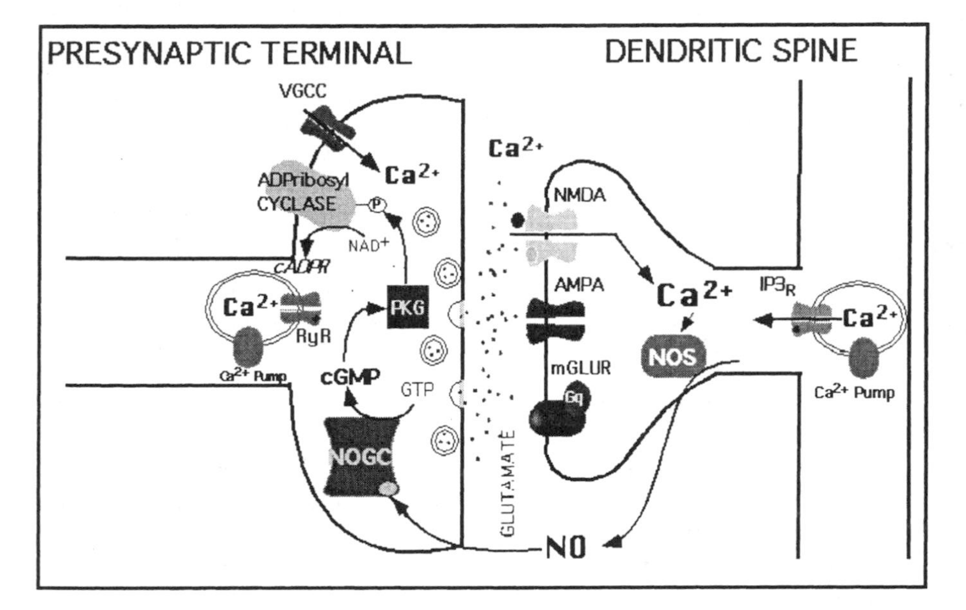

Figure 1. A proposed scheme in which induction of LTD begins with Ca^{2+} entry via NMDA receptors and mGluR-triggered, IP_3-mediated release of stored Ca^{2+}, which activates a Ca^{2+}/calmodulin-dependent nitric oxide synthase (NOS). NOS produces the retrograde messenger nitric oxide (NO), which diffuses to the presynaptic terminal. NO activates soluble guanylate cyclase (NOGC) to increase [cGMP]. One target of cGMP is PKG which, when activated, can phosphorylate many serine/threonine residues. A potential target of PKG is a cyclase (CD38) which produces cADP-ribose from β-NAD^{+}. Finally, cADP-ribose can modulate release of Ca^{2+} from intracellular stores by binding to the ryanodine receptors on these stores, making this Ca^{2+} available to activate CaMKII, presumable leading, by further unknown steps, to reduced transmitter release. (reprinted with permission from Reyes-Harde et al., 1999)

Abbreviations: AMPA, a-amino-3-hydroxy-5-methylisoxazole-4-propionic acid receptor; cADPR, cyclic adenosine 5'-diphosphate ribose; CaMKII, Ca^{2+}/calmodulin-dependent protein kinase II; cGMP, guanosine 3',5' cyclic monophosphate; GTP, guanosine triphosphate; Gq, stimulatory G-protein; IP_3R, inositol trisphosphate receptor; mGluR, metabotropic glutamate receptors; NMDA, N-methyl-D-aspartate receptor; NOS, nitric oxide synthase; NO, nitric oxide; NOGC, nitric oxide activated guanylate cyclase; PKG, cyclic GMP-dependent protein kinase; RyR, ryanodine receptors; VGCC, voltage-gated calcium channels.

Glutamate activates NMDA receptors, gating Ca^{2+} entry which sums with additional Ca^{2+} released from mGluR-activated postsynaptic IP3 stores, leading to NOS activation and postsynaptic generation of NO, which diffuses to the presynaptic terminal (and perhaps other sites as well). There, it activates sGC and synthesis of cGMP, which in turn activates PKG which activates ADPribosyl cyclase/hydrolase. The latter generates cADPR, which releases Ca^{2+} from the ryanodine-sensitive presynaptic store. Additional presynaptic molecular targets of PKG and Ca^{2+} that produce long-term reductions in glutamate release remain to be elucidated, though there is evidence that one such Ca^{2+} target is Ca^{2+}/calmodulin-dependent protein kinase II (Stanton and Gage, 1996).

3.5 Could NO be necessary for inducing both LTD & LTP at the same synapse?

The involvement of NO in both LTP and LTD at the same synapse raises the question of what determines the direction of change in synaptic strength. The simplest answer could be that NO is a "plasticity enabler", with the direction of presynaptic and/or postsynaptic alterations being determined by additional biochemical pathways converging with the NO pathway on the pyramidal cell dendritic spine and/or presynaptic Schaffer collateral terminal. At the same time, baseline experimental conditions often differ significantly between laboratories, and seemingly subtle differences could alter biochemical pathways and their interactions with the NO pathway, thereby changing the direction of plastic changes. Such variables include animal species, age, slice origin along the dorsoventral axis, slice placement (submersion or interface chamber), temperature and ionic concentrations.

One parameter that might account, at least in part, for the direction of plasticity is stimulation frequency/intensity. It is plausible that, while both high and low frequency stimulation will generate NO, stimulus frequency and intensity will determine which additional pathways are activated along with the production of NO. Haley et al. (1993) found that LTP induced by a weak tetanus could be blocked by NOS inhibitors, while that induced by a strong one could not, and Williams et al. (1993) showed that the ability of NOS inhibitors to impair induction of LTP was temperature and age dependent. The role of stimulation frequency in determining the direction of NO-dependent LTP plasticity was explicitly tested by (Zhuo et al., 1994a). LTP was induced by pairing an NO donor with a weak tetanus (50Hz), which was by itself unable to induce LTP. LTD was induced if the NO donor or a cGMP analog was paired with very low frequency stimulation (0.25Hz). In *in vivo* experiments, a NOS inhibitor blocked both tetanus-induced LTP and the subsequent LFS-induced depotentiation (Doyle et al., 1996).

More recently, the complex effects of exogenous NO in both potentiation and depression in CA1 has been studied in more detail. Paired with stimulation at 0.2Hz, NO donors produced a depression of synaptic transmission. On washout of NO, the depression gave way to a persistent potentiation, that at least partially occluded subsequent tetanus-induced LTP. At a lower stimulation frequency of 0.033 Hz, depression was unaltered, but no rebound potentiation took place and subsequent tetanus-induced LTP was normal (Bon and Garthwaite, 2001a, b). The biochemical mechanisms that determine whether NO-cGMP pathway activation promotes or results in induction of LTP or LTD are unknown. Evidence exists suggesting three second messengers that can interact with NO-cGMP signaling; cAMP, Ca^{2+} and ADPRT, discussed below.

(i) cAMP

The NO⇒cGMP⇒PKG and cAMP⇒PKA pathways have antagonistic regulatory mechanisms and opposing actions in many excitable cell types (Sperelakis et al., 1994; Wexler et al., 1998; Buttner and Siegelbaum, 2003). As already described, at the Schaffer collateral-CA1 synapse, increasing [cGMP] leads to decreased transmitter release, but this LTD will only persist if accompanied by inhibition of the cAMP⇒PKA pathway (Reyes-Harde et al., 1999b; Santschi et al., 1999a; Stanton et al., 2001). Preliminary data from our laboratory suggest that group II metabotropic glutamate receptors and adenosine A1 receptors, both of which are coupled to Gi, may play physiologic roles in reducing [cAMP], a factor that appears necessary for inducing LTD. On the LTP side, combined increases in [cAMP] and [cGMP] have been implicated in LTP. However, studies have focused more on *post*synaptic [cAMP], which seems to be most relevant to the late, protein synthesis-dependent phase of LTP (Lu et al., 1999), whereas the interaction of presynaptic cAMP with presynaptic cGMP in modulating plasticity of glutamate release is less well understood. There are studies where raising [cAMP] did lead to synaptic response enhancement of apparently presynaptic origin (Chavez-Noriega and Stevens, 1994). In conclusion, data are consistent with the hypothesis that one determinant of the long-term effects of elevating [cGMP] either pre- or postsynaptically may well be concomitant elevation or inhibition of cAMP production and PKA activity.

(ii) [Ca^{2+}] in the presynaptic terminal

Unlike the postsynaptic dendritic spine, where high [Ca^{2+}] is thought to be necessary for the induction of LTP, while intermediate [Ca^{2+}] leads to LTD (and low [Ca^{2+}] to no change), there is much less known about how [Ca^{2+}] in the presynaptic terminal affects messenger systems that regulate long-term plasticity of transmitter release.

Interestingly, Ca^{2+} is known to interact with the NO pathway and at least some of its downstream effectors. It is noteworthy, for example, that sGC, a prime NO target candidate in both LTP and LTD, is inhibited by high [Ca^{2+}] (Garbers and Lowe, 1994) and that in several other biological processes Ca^{2+} & cGMP have antagonistic roles (review by Lucas et al., 2000). Additionally, Ca^{2+} biphasically regulates ryanodine receptors, with moderate levels having an activating effect while high levels inhibit further release (Lee et al., 1994). These effects lead to speculation that, within a window of moderate presynaptic [Ca^{2+}] elicited by low frequency stimulation, NO is able to effectively activate sGC while ryanodyne receptors can still be stimulated synergistically by cADPR and Ca^{2+}. On the other hand, when [Ca^{2+}] is high, such as during tetanic stimulation, sGC and ryanodyne receptors may both be more inhibited, blocking the presumed LTD pathway. NO may then activate a target other than sGC (possibly ADPRT), to induce LTP. There is still much work to do to understand the complex [Ca^{2+}] modulation of presynaptic sGC and other components critical to LTP and LTD. We hypothesize that **a)** dynamic switching of NO from a messenger that elevates [cGMP] to one that activates ADPRT, coupled with **b)** pairing of increases in [cGMP] with

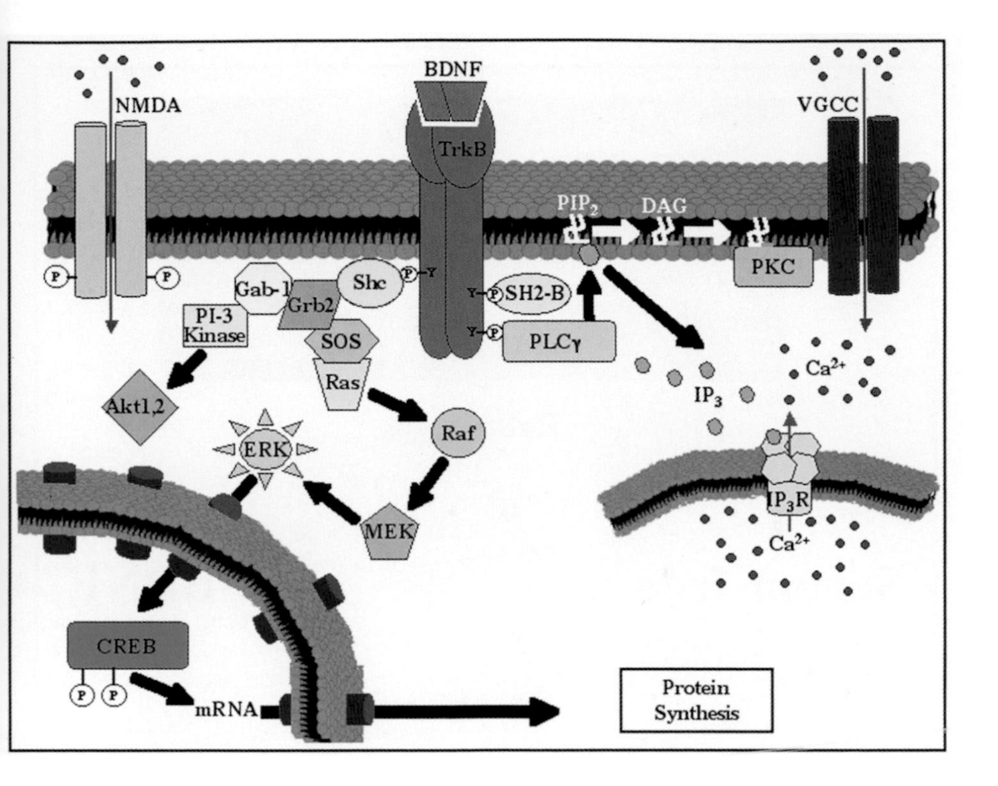

igure 1 (Amaral and Pozzo-Miller). Intracellular signaling cascades activated by BDNF-stimulated utophosphorylation of TrkB receptors may affect prominent pathways responsible for elevations of tracellular Ca^{2+} levels in neurons. Activation of the PLC-γ/IP_3 signaling pathway leads to Ca^{2+} mobilization om IP_3-sensitive intracellular stores. In addition, NMDA receptors and voltage-gated Ca^{2+} channels could be irectly phosphorylated by activated TrkB receptors or indirectly through the Ras-MAPK or PI3-K pathways. ombined, these covalent modifications potentially enhance Ca^{2+} signals within spines and dendrites. (A black nd white version of this figure appears on p. 186.)

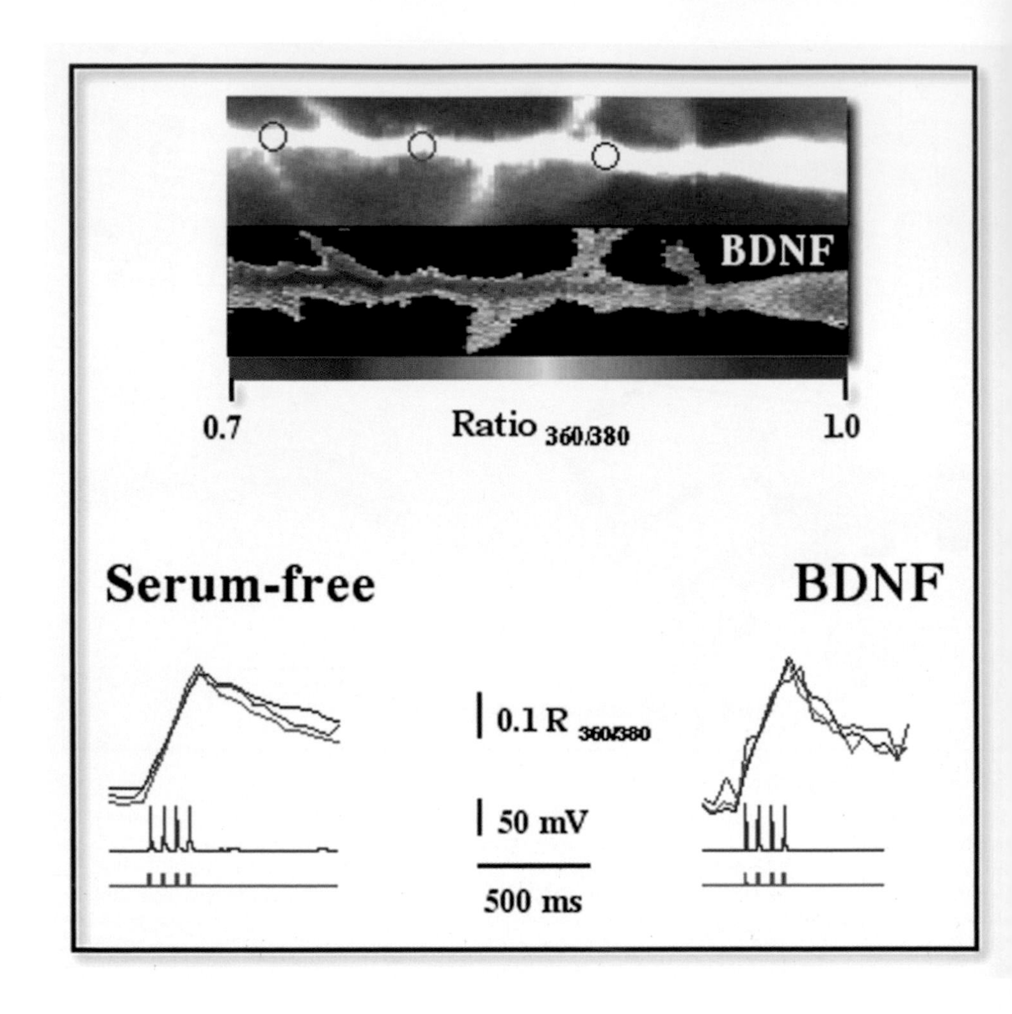

Figure 2 (Amaral and Pozzo-Miller). Dendritic Ca^{2+} elevations evoked by trains of bAPs. The fluorescence image (bis-fura-2, 380nm excitation) shows an example of an apical spiny dendrite, in this case from a BDNF-treated CA1 pyramidal neuron. The colors of the individual ROIs correspond to the Ca^{2+} traces shown below. The lower image is a pixel-by-pixel ratio (360/380nm) image displayed using the pseudocolor scale shown below. Top traces show Ca^{2+} levels expressed as bis-fura-2 ratios within the color-coded ROIs shown above. The traces below are from simultaneous whole-cell recordings of membrane voltage (red) and current (blue) in the current-clamp mode. The whole-cell pipette contained a K-gluconate solution and 200μM bis-fura-2. Digital imaging was performed with a cooled CCD camera at 20-33 frames-per-second. Short trains of brief depolarizing current pulses (5ms) delivered by the whole-cell recording electrode elicited action potentials. (A black and white version of this figure appears on p. 190.)

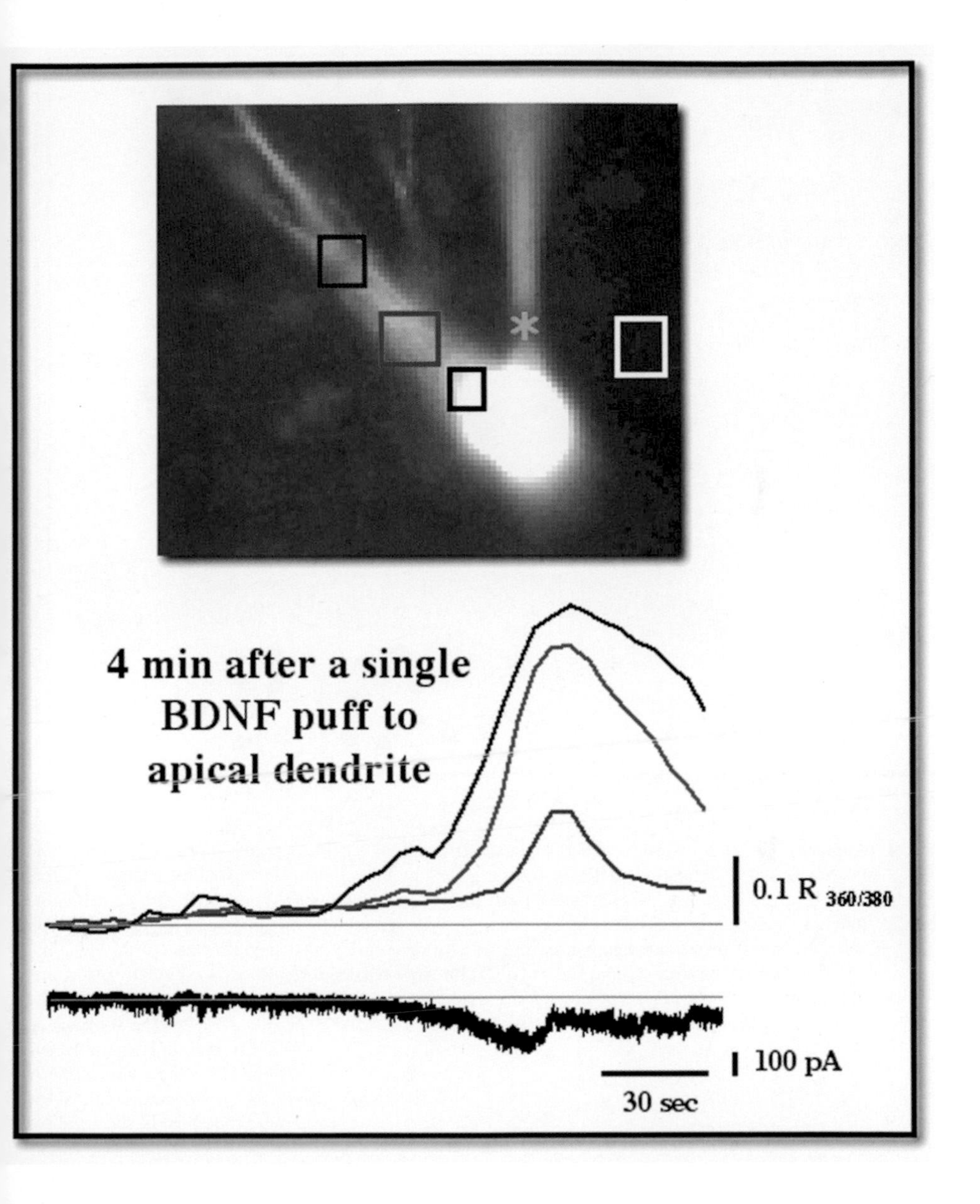

Figure 3 (Amaral and Pozzo-Miller). BDNF induces slowly developing intracellular Ca^{2+} elevations, which are associated with inward currents that have similar kinetics. The fluorescence image (380nm excitation) shows a CA1 pyramidal neuron filled with the Ca^{2+} indicator bis-fura-2 through the whole-cell recording electrode (asterisk). The traces represent intracellular Ca^{2+} levels within the color-coded ROIs and the simultaneously recorded membrane current. The whole-cell electrode contained a Cs-gluconate solution and 100μM bis-fura-2; voltage-gated Na^{+} channels were blocked by TTX (500nM); the holding voltage was -65mV. BDNF was applied by pressure ejection (25sec, 30psi) onto the apical dendrite of the neuron under recording. (A black and white version of this figure appears on p. 192.)

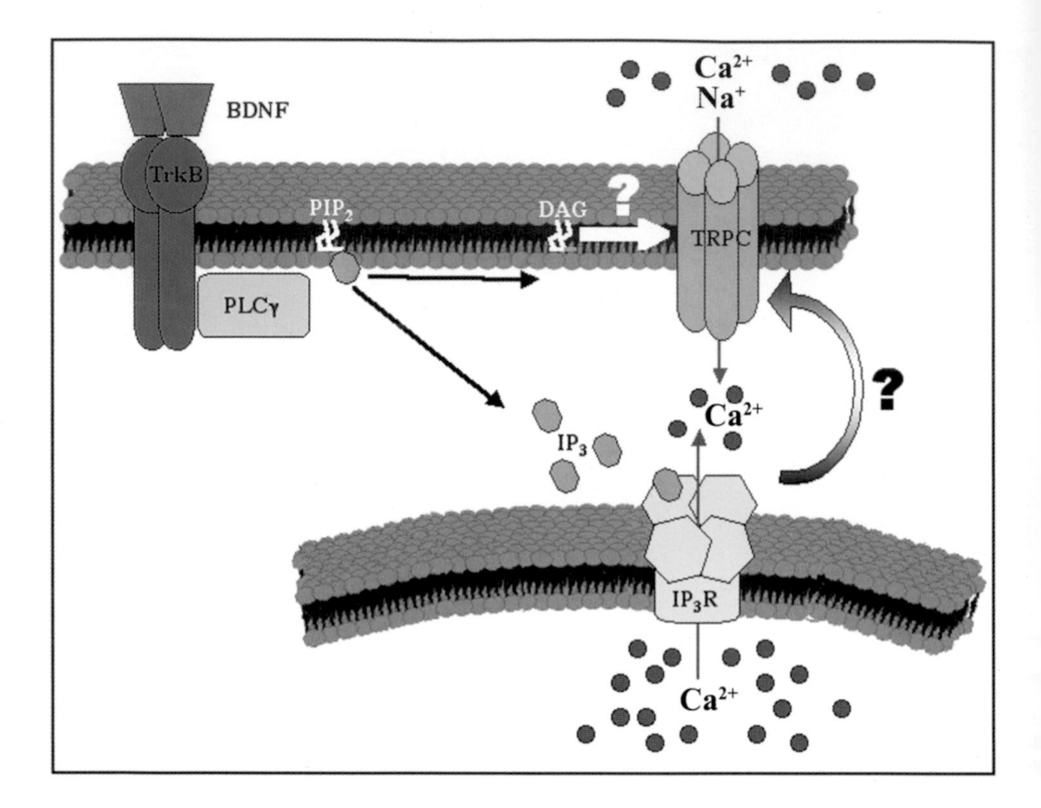

Figure 4 (Amaral and Pozzo-Miller). TrkB-initiated stimulation of PLCγ causes PIP_2 hydrolysis and formation of DAG and IP_3. Activation of IP_3Rs leads to Ca^{2+} mobilization from intracellular stores. Native TRPC channels expressed in hippocampal neurons, most likely heteromultimeric, may be gated by a diffusible factor, Ca^{2+} ions released from IP_3-sensitive Ca^{2+} stores, a physical interaction with activated IP_3Rs, or by DAG itself. TRPC channels are known to mediate a non-selective cationic current that requires intact IP_3R signaling, full intracellular Ca^{2+} stores, and extracellular Ca^{2+} ions. Preliminary results exemplified in Figure 3 suggest that BDNF-induced dendritic Ca^{2+} elevations include a Ca^{2+} mobilization component as well as Ca^{2+} entry from the extracellular space. (A black and white version of this figure appears on p. 194.)

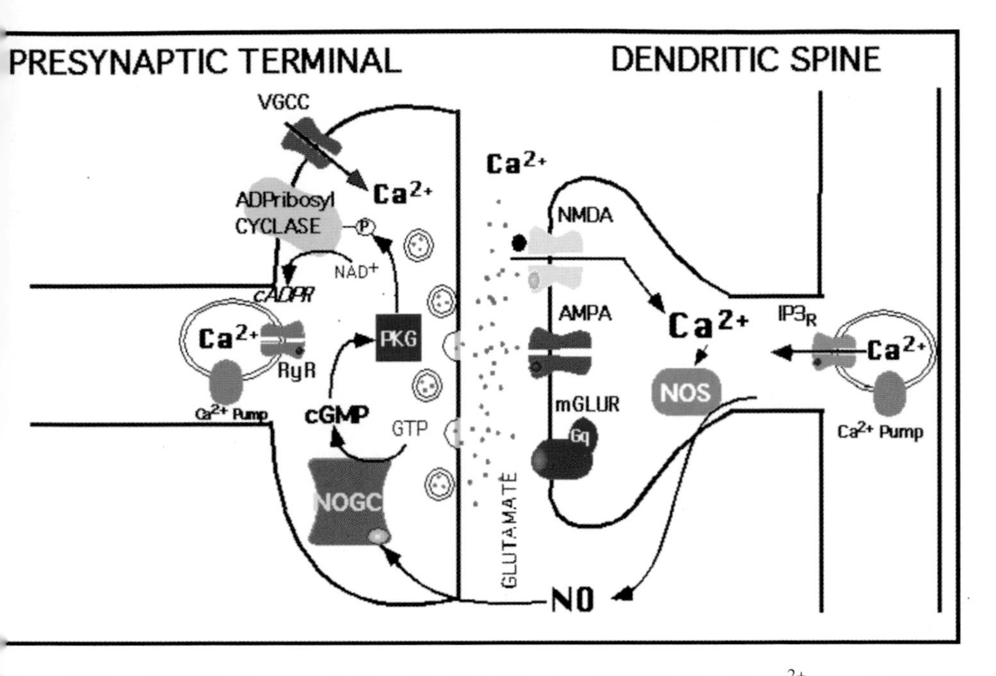

igure 1 (Kyrozis et al.). A proposed scheme in which induction of LTD begins with Ca^{2+} entry via NMDA eceptors and mGluR-triggered, IP_3-mediated release of stored Ca^{2+}, which activates a Ca^{2+}/calmodulin-lependent nitric oxide synthase (NOS). NOS produces the retrograde messenger nitric oxide (NO), which liffuses to the presynaptic terminal. NO activates soluble guanylate cyclase (NOGC) to increase [cGMP]. One arget of cGMP is PKG which, when activated, can phosphorylate many serine/threonine residues. A potential arget of PKG is a cyclase (CD38) which produces cADP-ribose from β-NAD^+. Finally, cADP-ribose can nodulate release of Ca^{2+} from intracellular stores by binding to the ryanodine receptors on these stores, making his Ca^{2+} available to activate CaMKII, presumable leading, by further unknown steps, to reduced transmitter elease. (reprinted with permission from Reyes-Harde et al., 1999). (A black and white version of this figure ippears on p. 284.)

Abbreviations: AMPA, a-amino-3-hydroxy-5-methylisoxazole-4-propionic acid receptor; cADPR, cyclic idenosine 5'-diphosphate ribose; CaMKII, Ca^{2+}/calmodulin-dependent protein kinase II; cGMP, guanosine 3',5' cyclic nonophosphate; GTP, guanosine triphosphate; Gq, stimulatory G-protein; IP_3R, inositol trisphosphate receptor; mGluR, netabotropic glutamate receptors; NMDA, N-methyl-D-aspartate receptor; NOS, nitric oxide synthase; NO, nitric oxide; NOGC, nitric oxide activated guanylate cyclase; PKG, cyclic GMP-dependent protein kinase; RyR, ryanodine receptors; VGCC, voltage-gated calcium channels.

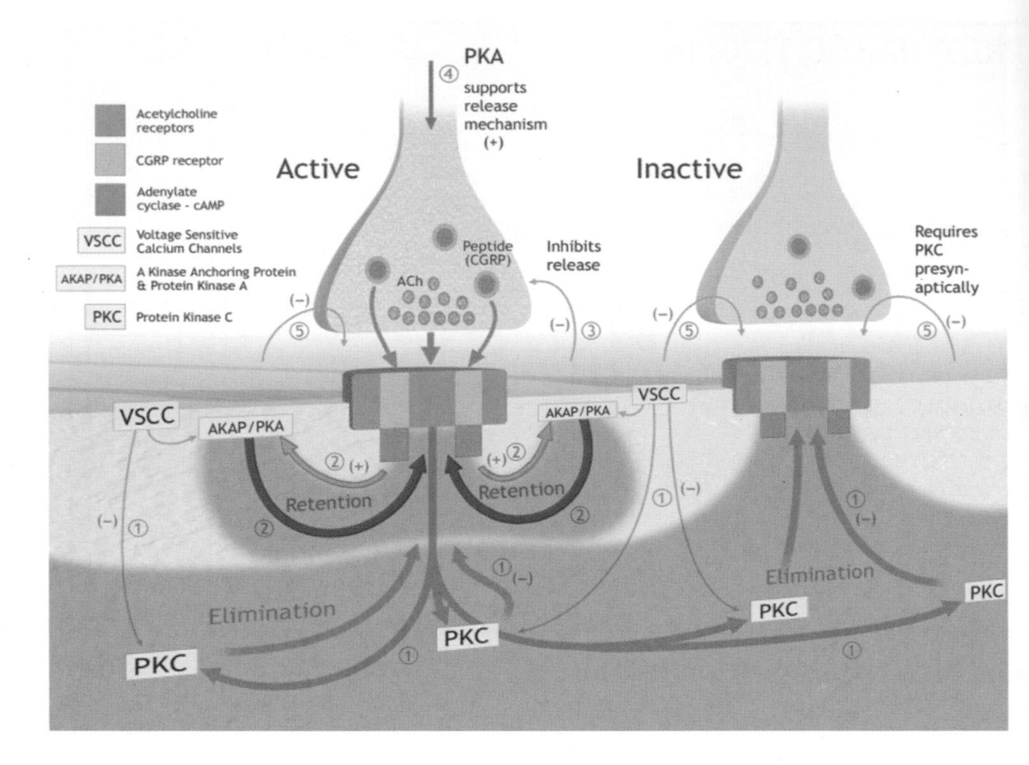

Figure 9 (Nelson *et al.*). Schematic diagram illustrating the hypothesized opposing kinase actions that generate the selective activity-dependent retention and loss of active and inactive neural inputs, respectively, on a common postsynaptic cell. From Li et al., 2002. (A black and white version of this figure appears on p. 451.)

varying levels of glutamatergic and adenosinergic inhibition of adenylate cyclase, probably explains a good portion of the complexity of NO-dependent synaptic plasticity.

4. OTHER PUTATIVE RETROGRADE MESSENGERS IN CA1

4.1 CO

CO is another small, membrane-permeant, endogenously synthesized gaseous molecule chemically similar to NO. The mRNA for its synthetic enzyme, heme oxygenase, is abundant in brain (isoform HO-2), especially in the hippocampus, cerebellum and olfactory cells, and largely overlaps with mRNA for sGC (Verma et al., 1993). These properties render it another attractive candidate for interneuronal retrograde messenger status. At the Schaffer collateral-CA1 synapse, application of CO paired with a weak tetanic stimulation produced long-lasting potentiation, similar to NO (Zhuo et al., 1993). Inhibitors of HO-2 have also been reported to block the induction of LTP in field CA1 (Stevens and Wang, 1993; Zhuo et al., 1993) and the dentate gyrus (Ikegaya et al., 1994). However, the potential role of CO as retrograde messenger mediating LTP induction has been more questionable for several reasons. HO inhibitors used have doubtful specificity and, most importantly, can also inhibit NOS (Meffert et al., 1994; Otani and Connor, 1995) and sGC (Luo and Vincent, 1994), raising the possibility that their actions may be entirely NO-mediated. Furthermore, HO-2 inhibitors have been found to not only block induction of LTP, but to also reverse established LTP (Stevens and Wang, 1993), suggesting that CO may acts constitutively (in a tonic role) rather than as an inductor. Supporting this hypothesis, Zhuo et al., (1999) found that HO-2 is constitutively active, but not stimulated by LTP-inducing stimuli. These authors concluded that LTP requires NO as a phasic mediator (consistent with studies described above), plus tonic HO activity (Zhuo et al., 1999).

4.2 Arachidonic Acid

Another membrane-permeant molecule considered quite early as a possible retrograde messenger playing an important role in inducing LTP is arachidonic acid (AA). Arachidonic acid (AA) is liberated from membrane phospholipids by the action of the Ca^{2+}-dependent enzyme phospholipase A_2 (PLA_2), which is highly expressed in hippocampal neurons (Kishimoto et al., 1999). AA is further metabolized to a variety of bioactive substances via the lipoxygenase and cycloxygenase pathways. AA and some of its 12-lipoxygenase metabolites are membrane-permeant lipid molecules that are known to play important roles in presynaptic regulation in invertebrates (Piomelli et al., 1987).

AA application was reported to facilitate induction of LTP (Williams et al., 1989). Such facilitation, however, was blocked by NMDA receptor antagonists (O'Dell et al., 1991), suggesting that AA acts upstream of NMDA receptor activation, perhaps by enhancing glutamate release during tetanic stimulation. LTP was inhibited by PLA_2 & lipoxygenase inhibitors, but the relative lack of specificity of the drugs used raises doubts whether AA was the messenger involved (O'Dell et al., 1991). In part because of the emergence of NO as a more likely retrograde messenger in LTP, the role of AA has not been further investigated. Recently, mice deficient in 12-lipoxygenase were shown to express apparently normal LTP (Feinmark et al., 2003), raising further doubts about a necessary role for AA in LTP.

In contrast to LTP, there is new evidence supporting a role for AA and/or one of its lipoxygenase metabolites as retrograde messengers in an mGluR-dependent form of LTD. This form requires postsynaptic depolarization and an increase in postsynaptic [Ca^{2+}] (similar to NMDA receptor-dependent LTD) and is thought, on the basis of electrophysiological observations, to be expressed by a decrease in transmitter release, suggesting the presence of a retrograde messenger (Bolshakov and Siegelbaum, 1994; Oliet et al., 1997; Feinmark et al., 2003). AA application has been reported to facilitate induction of mGluR-dependent LTD (Bolshakov and Siegelbaum, 1995). In one study, the non-selective alkylating agent bromophenacylbromide, and non-selective lipoxygenase inhibitor nordihydroguaiaretic acid were reported to block induction of LTD (Normandin et al.,1996), but use of a much more selective inhibitor 3-(4-octadecyl)-benzoylacrylic acid (OBAA) failed to produce any such block (Stanton, 1995). These discrepancies could be due to differences in amount of NMDAR and mGluR-dependent forms of LTD, and/or to a lack of knowledge of which AA metabolites are involved.

Recently, new pharmacologic and genetic tools have begun to allow a more detailed evaluation of the role of AA pathways in mGluR-LTD (Feinmark et al., 2003). Bath application of the 12-lipoxygenase metabolite 12(S)-hydroperoxyeicosa-5Z, 8Z, 10E, 14Z-tetraenoic acid (12(S)-HPETE) induced a long-lasting depression that occluded low frequency stimulus-induced LTD. In this study, LTD was blocked by a 12-lipoxygenase inhibitor, and was impaired in 12-lipoxygenase knockout mice. Finally, LTD-inducing protocols led to accumulation of a metabolite of 12(S)-HPETE. These data make 12(S)-HPETE a more attractive new candidate as a retrograde messenger acting selectively in the induction of mGluR-dependent LTD (Feinmark et al., 2003).

4.3 Platelet-Activating Factor

As is the case with arachidonic acid and its metabolites, the membrane-permeant phospholipid messenger platelet-activating factor (PAF) is also synthesized in neurons after Ca^{2+}-mediated activation of PLA_2. These properties make PAF another legitimate candidate retrograde/intercellular messenger (for review, see Bazan, 2003). At the Schaffer collateral-CA1 synapse, bath application of PAF analogs has been reported to increas transmitter release in some studies (Clark et al., 1992; Wieraszko et al., 1993) but not others (Kobayashi et al., 1999). In one report, PAF injected into the postsynaptic CA1 pyramidal neuron also increased transmitter release, prima facae evidence that PAF can act as a retrograde messenger (Kato et al., 1994). LTP-inducing high frequency stimulation induced a significant rise in PAF (Kornecki et al., 1996). Tetanus-induced LTP and PAF-induced potentiation occlude each other, suggesting a common pathway (Wieraszko et al., 1993). PAF receptor antagonists block the induction of LTP in some studies (Arai and Lynch, 1992; Kondratskaya et al., 2004), but not others (Wieraszko et al., 1993; Kobayashi et al., 1999). Transgenic mice deficient in PAF receptor reportedly exhibited normal LTP (Kobayashi et al., 1999); a distressingly common observation that again leads to invoking of compensatory mechanisms or redundant plasticity mechanisms that make PAF non-essential for plasticity expression.

5. SELECTIVE PLASTICITY OF RELEASE FROM THE READILY-RELEASABLE VESICLE POOL IN SCHAFFER COLLATERAL TERMINALS

For some time, substantial indirect evidence has suggested that, in addition to a clear collection of alterations in postsynaptic function, persistent changes in presynaptic glutamate release are also associated with long-term plasticity of synaptic transmission. Some 20 years ago, Bliss and colleagues began producing a stream of data showing long-lasting increases in release of radiolabelled glutamate during expression of LTP (Lynch et al., 1985; Feasey et al., 1986; Bliss et al., 1986,1987; Errington et al., 1987) and classical conditioning (Laroche et al., 1987; Lynch et al., 1990). Quantal analysis studies have supported presynaptic (Bekkers and Stevens, 1990; Malinow and Tsien, 1990; Foster and McNaughton, 1991) and postsynaptic (Kullmann and Nicoll, 1992; Kuhnt et al., 1992; Voronin et al., 1992a,b) sites of alterations underlying LTP. Later, indirect evidence appeared suggesting that LTD may also be expressed, at least in part, by presynaptic alterations in glutamate release (Gage and Stanton, 1996; Reyes and Stanton, 1996; Reyes-Harde et al., 1999; Goda and Stevens, 1998). The FM series of styryl dyes, which have high affinities for presynaptic transmitter vesicles, have been successfully used in isolated neuronal systems to directly visualize transmitter release from vesicles (Betz and Bewick, 1992; Ryan et al., 1993), but high background fluorescence has limited their use in brain slices with more intact circuitry. Recent advances using two-photon microscopy to selectively excite dyes in structures as small as presynaptic terminals have enabled, for the first time, direct imaging of vesicular release from Schaffer collateral terminals in hippocampal slices (Stanton et al., 2001, 2003; Zakharenko et al., 2001, 2002).

To utilize FM dyes to study plasticity of presynaptic vesicular release, it is necessary to resolve the conundrum of having to depolarize terminals to cause vesicle fusion and allow the dye access to vesicle interiors for loading. This means that stimulus-evoked long-term plasticity has to be induced first, followed by a brief depolarization to load the dye, and then a second stimulus train to measure the rate of unloading. It was by no means clear that plastic changes would persist through the loading stimulus, but was worth essaying. There are a number of commonly-used methods for stimulating vesicular release from distinct pools of vesicles to selectively load these pools. The use of high [K^+] (i.e. 45mM for 10-15 min) will cause near-total vesicle release and load all vesicles. A more modest hypertonic shock (i.e. 800 mOsm for 2 min) will selectively release only those vesicles in the so-called "readily-releasable pool" (RRP; Rosenmund and Stevens, 1996; Stanton et al., 2003), which has been shown electrophysiologically, optically and by electron microscopy to load a subpopulation of vesicles (~25-30% of the total) that are either docked or within 200nm of the active release zone. A comparison of the rates of release of an FM dye from terminals loaded with each of these methods allowed us to determine whether the induction of LTP or LTD is associated with a change in vesicular release probability, and if the RRP is selectively targeted. We found inducing LTP and LTD produces marked, highly selective, changes in rates of release from the RRP, with little or no effect on reserve vesicle pools (Stanton et al., 2003).

Figure 2 shows the expression of LTD and LTP of presynaptic release from the RRP, and some significant properties of this presynaptic plasticity. In these experiments, we first induced either LTD with a low-frequency train of Schaffer collateral stimuli (2Hz/10min), or LTP with high-frequency trains (4x100Hz/500msec), and then (15 min later) loaded the RRP with FM1-43 and blocked all further synaptic potentials with

CNQX. **Figure 2A** illustrates typical two-photon laser scanning images taken from *stratum radiatum* of field CA1 during a series of destaining stimulus bursts (10Hz/5sec each 30 sec) in a control slices (top row) versus a slice where LTD had been induced prior to loading of the RRP with FM1-43 (bottom row). Brightly fluorescent spherical clusters (mean diameter 1.14 ± 0.03 μm, n=640) that become dimmer upon Schaffer collateral stimulation (~90% of these puncta) are release sites. The rate of vesicular release was markedly slower in the slice where LTD had been previously induced.

We have conducted a series of studies examining the messenger cascades associated with the induction of LTD and LTP of vesicular release at Schaffer collateral-CA1 synapses in hippocampal slices. Comparison of the rate of destaining of Schaffer collateral terminals loaded with FM1-43 by high K^+ depolarization (total vesicle pool) to rates of release from the sucrose-loaded RRP shows that both LTD and LTP produce a selective change in rate of release from the RRP (Stanton et al., 2003). **Figures 2B** and **C** illustrate mean release rate profiles from the RRP before and after the induction of LTD (**2B**) and LTP (**2C**). In the case of LTD, a 2Hz/10 min stimulus train evoked LTD that was associated with a marked reduction in the rate of release from the RRP. The NMDA receptor antagonist AP5 almost completely blocked the induction of LTD of synaptic transmission by this stimulus protocol, and also completely blocked the effect of LTD on presynaptic release. While these data strongly suggest that presynaptic LTD requires NMDA receptor activation, they do not dismiss the possibility that postsynaptic forms of LTD might also require NMDA receptor activation. Indeed, there is experimental evidence indicating that there are distinct NMDA receptor-dependent and mGluRI receptor-dependent forms of LTD (Oliet et al., 1997); our data indicates that the NMDA receptor-dependent form combines expression of both synapse-specific presynaptic and postsynaptic changes that reduce synaptic strength.

Figure 2C illustrates the effect of two LTP induction protocols on RRP release. A "weak" induction protocol consisted of four 50Hz/500msec trains of stimuli at a stimulus intensity producing 50% of a maximal EPSP, while a "strong" protocol consisted of six 100Hz/500msec trains at 80% of maximal EPSP intensity. There is a clear-cut difference in the effect of these two LTP induction protocols on presynaptic vesicular release from the RRP. The "weak" LTP protocol produced significant LTP (149 ± 8.7% of pre-tetanus baseline) that was not associated with any change in release probability. In sharp contrast, the "strong" LTP induction protocol produced larger LTP (195 ± 4.9% of pre-tetanus baseline) that was associated with a marked increase in the rate of FM1-43 release from Schaffer collateral presynaptic terminals. Induction of both weak and strong LTP of evoked EPSPs was blocked by the NMDA receptor antagonist AP5, and presynaptic LTP of RRP release associated with strong LTP was also NMDA receptor-dependent (**Fig. 2C**). These data lead us to conclude that, while weak LTP appears to be wholly postsynaptic in locus of expression, strong LTP can consist of a mixture of presynaptic and postsynaptic long-term changes, both requiring NMDA receptor activation.

The development of techniques for imaging presynaptic vesicular release from the RRP in brain slices allowed us direct confirmation of the indirect data described previously that outlined a NO-guanylyl cyclase-cGMP-PKG pathway underlying a presynaptic form of LTD, and to test the vesicle pool(s) targeted by this pathway. In this scheme, postsynaptic Ca^{2+} influx into CA1 pyramidal neuron apical dendrites activates Ca^{2+}/calmodulin activated NOS. NO diffuses freely across short distances, its range of action limited to distances on the order of ~50μM by its volatility (Philippides et al.,

2000; see also Engert and Bonhoeffer, 1997). Within this distance, the presynaptic terminal is one, but not the only, potential target of NO action. To test the importance of this pathway for long-term plasticity of RRP release, we tested the effect of the competitive NOS inhibitor L-Nitroarginine (L-NA; 100 μM) on the induction of LTD. As shown in **figure 2D**, L-NA completely blocked LTD of presynaptic release from the RRP. In this study, we found that L-NA produced a partial (~50%) block of LTD of Schaffer collateral-CA1 synaptic transmission (Stanton et al., 2003), consistent with the conclusion that there are both NO-dependent and NO-independent components to LTD. When we examined the ability of the extracellular NO scavenger hemoglobin (Hb; 100 μm) to impair induction of LTD, it produced a partial blockade of both LTD (not shown; Stanton et al., 2003) and of the effects of LTD on RRP release mechanisms (**Fig. 2E**). While this data shows a 50% residual presynaptic LTD component, this could be due either to involvement of other retrograde messengers, or to incomplete scavenging of NO. Overall, the body of our data clearly support the cascade illustrated in **figure 1** as importantly involved in LTD of presynaptic glutamate release from the RRP pool of vesicles at Schaffer collateral terminals in field CA1.

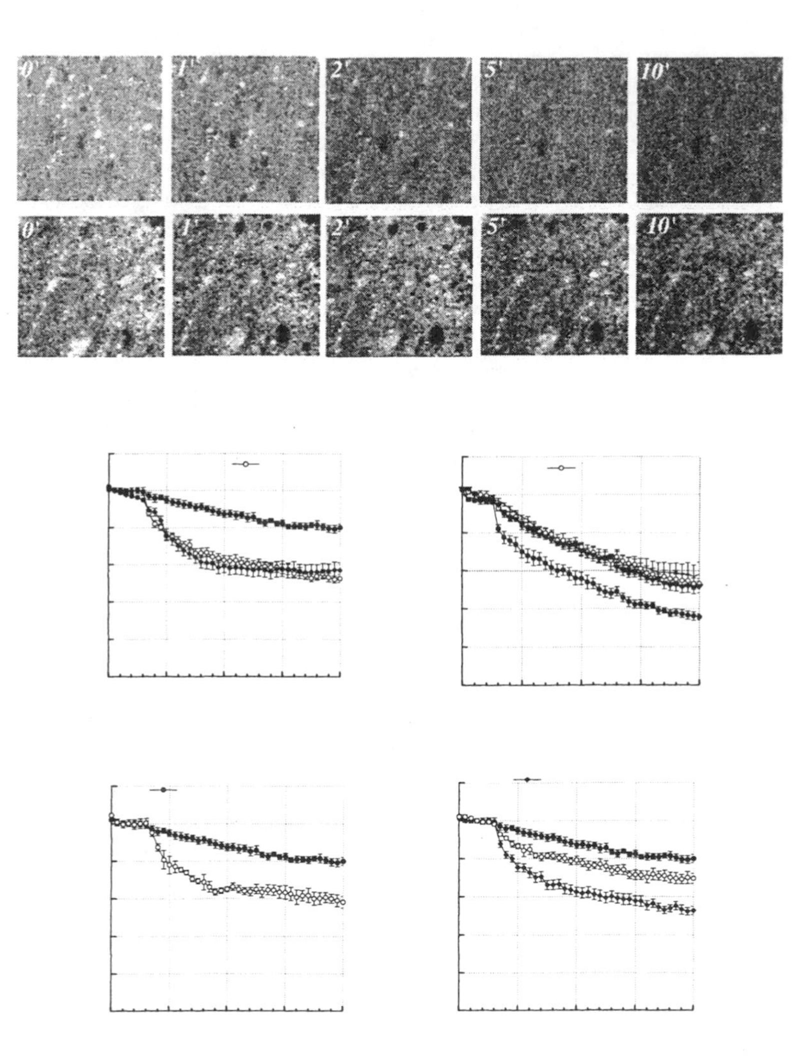

Figure 2. Presynaptic LTD and LTP of release from the readily-releasable vesicle pool of Schafffer collateral terminals. **A:** Two-photon excitation fluorescent images of readily-releasable vesicle pool puncta in stratum radiatum of field CA1 in a control slice (Upper Row), versus a slice where LTD was induced (Lower Row). Numbers represent time in minutes following start of unloading stimulation. **B:** Time courses of Schaffer collateral stimulus-evoked (solid bar; 10 Hz/5 s bursts each 30 s) FM1-43 destaining from the RRP in control slices (○; n=7) , LTD slices (●; n=6), and slices where LTD was blocked with D-AP5 (10μM). **C:** **D:** Time course of RRP release (solid bar; 10 Hz/5 s bursts each 30s) in slices treated with the NOS inhibitor L-nitroarginine (L-NA;10 μM;○; n=4) compared to controls (●; n=5). **E:** The NO scavenger hemoglobin (Hb; 10 μM;○; n=4) partially blocked the reduction in RRP release seen in LTD slices compared to control LTD (●; n=5) and unstimulated slices (◆; n=6). (modified from Stanton et al. (2003))

6. CONCLUSIONS

As the healthy debate over the sites and mechanisms expressing long-term changes in synaptic strength continues, some points are coming into clearer focus. It seems evident that there are multiple types of alterations that lead to persistent increases (LTP) and decreases (LTD) in synaptic strength, and that they can be co-activated to varying degrees by different stimulus patterns. They focus, not surprisingly, on the synapse, since this is the clearest way to ensure the synapse-specificity of changes, and retention of the information that those synapses contain. However, it is clear that both postsynaptic and presynaptic long-term alterations in function can occur, and that they run the gamut from rapid, local events such as phosphorylation and insertion/removal of glutamate receptors from existing dendritic spine receptor pools, to nuclear transport of signals that alter gene expression, to physical remodeling of both presynaptic and postsynaptic structure. This volume contains a number of excellent and complementary views of such changes, but is by no means exhaustive of the wonderous complexity of them.

The question of the utility, and even necessity, for diffusible messenger molecules localized in the synapse is driven by some important functional constraints. First, if a synapse needs to calculate the *covariance* (Sejnowski, 1977; Stanton and Sejnowski, 1989) between a synaptic input and its postsynaptic target, then only the postsynaptic dendritic spine has access at any given moment to both sets of information; how depolarized (hence how activated) the postsynaptic neuron is, and how much glutamate was released (hence how activated) its presynaptic input is. Second, there are several reasons why long-term plasticity of presynaptic transmitter release may be functionally useful. Plasticity of release may have a different time course than plasticity of postsynaptic spine responsiveness. The presynaptic terminal may continue to supply glutamate that supports long-term signaling to the postsynaptic dendritic spine, and delayed consolidation of changes in transmission and/or structure. Also, we should not exclude the possibility that retrograde signaling does not stop at the presynaptic terminal, but continues back to the nucleus of the presynaptic neuron. Again, this volume makes clear that there is vast potential complexity to the number and properties of mechanisms activated by retrograde messengers.

It is typically assumed that retrograde messenger molecules will be membrane-permeant (whether gaseous or lipid), so that they can be released from postsynaptic dendrites that lack rapid vesicular release machinery, but compounds such as NO and endocannabinoids indicate that such molecules can act either within the presynaptic terminal, or at receptors on its surface.

While we endeavored in this chapter to give a global overview of evidence implicating a number of candidate retrograde messenger molecules in the induction of multiple forms of both LTP and LTD, a focus of our work has been in testing and characterizing the importance of the NO⇒sGC⇒cGMP⇒PKG biochemical cascade as a retrograde messenger system producing long-term changes, in particular LTD, of glutamate release. While we have mostly examined the Schaffer collateral-CA1 synapse, we also have evidence that this pathway has similar functions at mossy fiber-CA3 synapses. This data in no way contradicts data from other brain areas, such as the striatum or cerebellum, that suggest additional postsynaptic roles for this cascade in plasticity, or even the coexistence of presynaptic and postsynaptic effects of NO.

It is clear from our work that the dynamic co-regulation of [cGMP] and [cAMP] are both key factors in determining the magnitude and direction of presynaptic (and, quite probably, postsynaptic) long-term plasticity. Under most conditions, elevating [cGMP] alone appears to produce only transient depression of presynaptic release, while adding active physiological or pharmacological inhibition of adenylate cyclase or PKA is able to convert this depression into LTD. This is consistent with observations that PKA has a much higher tonic activity level than PKG, and that cAMP activation of PKA causes long-lasting potentiation (Stanton and Sarvey, 1985; Slack and Pockett, 1991; Matthies and Reymann, 1993), and even appearance of new presynaptic release sites (Ma et al., 1999).

Our most recent studies utilizing two-photon laser-scanning microscopy to directly image Schaffer collateral vesicular release rates have led to some interesting new conclusions. First, the readily-releasable vesicle pool appears to be a selective target of long-term plastic changes that can lead to both increases and decreases in glutamate release. EM and fluorescence data (Stanton et al., 2003) supplies evidence for alterations in release probability that are probably due to both changes in numbers of RRP vesicles in the pool (Goussakov et al., 2000; Schikorski and Srevens, 2001; Murthy et al., 2001), and in functional state of vesicle fusion machinery proteins. In addition, we also observed that LTD is associated with an increase in the percentage of labeled boutons that did not release *at all* (Stanton et al., 2003), suggesting that terminals can turn completely off, and silent ones may be available for recruitment.

With respect to modulation of existing active zone function, there are a number of rates that could be modified (**Fig. 3**). These include the rate of transfer from the reserve pool to the RRP (*1*), priming and release from the RRP (*2*), the preferential recycling of vesicles back into the RRP, perhaps following "kiss and run" release (*3*) (Pyle et al., 2000), and return of vesicles to the reserve pool for later conversion back into the RRP (*4*). Of these, any of the first three could result in a *selective* decrease in release kinetics from the RRP, though the presence of an effect on release during the first stimulus burst favors an action on priming and/or p_r (**2**). On the other hand, an effect on rate ***(1)*** should produce an additional component of LTD of release from the K^+-loaded vesicle pools not accounted for by sucrose loading the RRP. Comparison of stimulus-evoked release time courses of the RRP and reserve vesicle pools suggests a late divergence after 10 min of discontinuous evoked release (Stanton et al., 2003). We interpreted this as suggesting an effect on transfer from the reserve pool to the RRP, but that this rate would be too slow to account for initial LTD of RRP release. The rate of refilling of the RRP has been shown to be dependent on firing frequency (Wang and Kaczmarek, 1998), while RRP size can be altered at mossy fiber-CA3 synapses by seizure activity (Goussakov et al., 2000), suggesting that RRP recycling rates are a likely target for long-term modification underlying activity-dependent long-term plasticity.

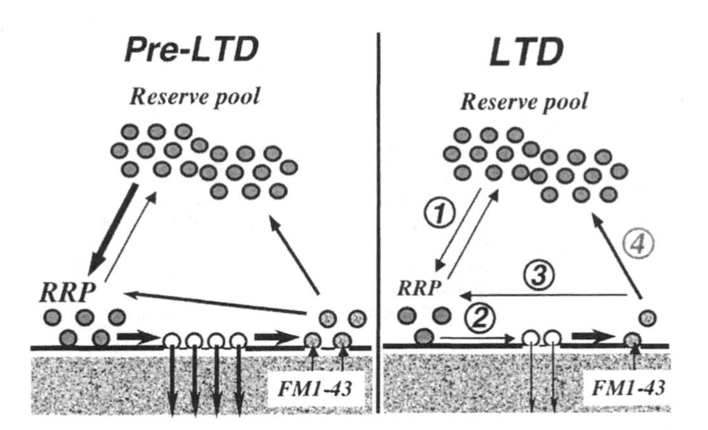

Figure 3. Potential sites of modifications underlying presynaptic LTD of release selectively targeting the RRP. Vesicular transmitter release targets before (Pre-LTD) and during (LTD) expression of LTD: 1) Transfer from the reserve vesicle pool to the readily-releasable pool (RRP); 2) Priming and release of docked vesicles; 3) "Kiss and run" recycling of vesicles preferentially into the RRP; and 4) Recycling of vesicles into the reserve pool. While a reduction in the rates of any of these steps could produce presynaptic LTD, our observation that LTD preferentially reduces release from the RRP, without altering reserve pool size or early release kinetics suggests that the rates of RRP priming and release probability (p_r, 2), and/or re-entry (3), are reduced during LTD. (reprinted with permission from Stanton et al. (2003))

7. ACKNOWLEDGMENTS

The authors would like to thank the many laboratory members and collaborators who contributed to studies described here, including Drs. Christopher Bailey, Ruth Empson, Allison Gage, Anthony Galione, Gregor Laube, Dave Hall, Can Nguyen, Barry Potter, Ivan Raginov, Magali Reyes, Linda Santschi, Rüdiger Veh, and Eric Wexler. This work was supported by NIH grant NS44421 (PKS), the Alexander von Humboldt Foundation (PKS), Grass and Onassis Foundations (AK) and the Deutsche Forschungsgemeinschaft (WM). Most importantly, we gratefully acknowledge Dr. John Sarvey, whose scientific influence is evident throughout our work, and whose extraordinary humanity was a wonderful and sadly-missed part of our lives. Thanks, John.

8. REFERENCES

Abe K, Watanabe Y, Saito H (1996) Differential role of nitric oxide in long-term potentiation in the medial and lateral amygdala. Eur J Pharmacol 297:43-46.
Arai A, Lynch G (1992) Antagonists of the Platelet-activating Factor Receptor Block Long-term Potentiation in Hippocampal Slices. Eur J Neurosci 4:411-419.
Arancio O, Kandel ER, Hawkins RD (1995) Activity-dependent long-term enhancement of transmitter release by presynaptic 3',5'-cyclic GMP in cultured hippocampal neurons. Nature 376:74-80.
Arancio O, Kiebler M, Lee CJ, Lev-Ram V, Tsien RY, Kandel ER, Hawkins RD (1996) Nitric oxide acts directly in the presynaptic neuron to produce long-term potentiation in cultured hippocampal neurons. Cell 87:1025-1035.
Arancio O, Antonova I, Gambaryan S, Lohmann SM, Wood JS, Lawrence DS, Hawkins RD (2001) Presynaptic role of cGMP-dependent protein kinase during long-lasting potentiation. J Neurosci 21:143-149.
Bailey CP, Trejos JA, Schanne FA, Stanton PK (2003) Pairing elevation of [cyclic GMP] with inhibition of PKA produces long-term depression of glutamate release from isolated rat hippocampal presynaptic terminals. Eur J Neurosci 17:903-908.
Bashir ZI, Collingridge GL (1994) An investigation of depotentiation of long-term potentiation in the CA1 region of the hippocampus. Exp Brain Res 100:437-443.
Bazan NG (2003) Synaptic lipid signaling: significance of polyunsaturated fatty acids and platelet-activating factor. J Lipid Res 44:2221-2233.
Bekkers JM, Stevens CF (1990) Presynaptic mechanism for long-term potentiation in the hippocampus. Nature 346:724-729.
Benke TA, Luthi A, Isaac JT, Collingridge GL (1998) Modulation of AMPA receptor unitary conductance by synaptic activity. Nature 393:793-797.
Betz WJ, Bewick GS (1992) Optical analysis of synaptic vesicle recycling at the frog neuromuscular junction. Science 255:200-203.
Bienenstock EL, Cooper LN, Munro PW (1982) Theory for the development of neuron selectivity: orientation specificity and binocular interaction in visual cortex. J Neurosci 2:32-48.
Blackshaw S, Eliasson MJ, Sawa A, Watkins CC, Krug D, Gupta A, Arai T, Ferrante RJ, Snyder SH (2003) Species, strain and developmental variations in hippocampal neuronal and endothelial nitric oxide synthase clarify discrepancies in nitric oxide-dependent synaptic plasticity. Neurosci 119:979-990.
Bliss TV, Collingridge GL (1993) A synaptic model of memory: long-term potentiation in the hippocampus. Nature 361:31-39.
Bliss TV, Douglas RM, Errington ML, Lynch MA (1986) Correlation between long-term potentiation and release of endogenous amino acids from dentate gyrus of anaesthetized rats. J Physiol 377:391-408.
Bliss TV, Errington ML, Laroche S, Lynch MA (1987) Increase in K^+-stimulated, Ca^{2+}-dependent release of $[^3H]$glutamate from rat dentate gyrus three days after induction of long-term potentiation. Neurosci Lett 83:107-112.
Bliss TV, Lømo T (1973) Long-lasting potentiation of synaptic transmission in the dentate area of the anaesthetized rabbit following stimulation of the perforant path. J Physiol 232:331-356.
Bohme GA, Bon C, Stutzmann JM, Doble A, Blanchard JC (1991) Possible involvement of nitric oxide in long-term potentiation. Eur J Pharmacol 199:379-381.
Bolshakov VY, Siegelbaum SA (1994) Postsynaptic induction and presynaptic expression of hippocampal long-term depression. Science 264:1148-1152.
Bolshakov VY, Siegelbaum SA (1995) Hippocampal long-term depression: arachidonic acid as a potential retrograde messenger. Neuropharmacology 34:1581-1587.
Bon C, Bohme GA, Doble A, Stutzmann JM, Blanchard JC (1992) A Role for Nitric Oxide in Long-term Potentiation. Eur J Neurosci 4:420-424.
Bon CL, Garthwaite J (2001a) Nitric oxide-induced potentiation of CA1 hippocampal synaptic transmission during baseline stimulation is strictly frequency-dependent. Neuropharmacology 40:501-507.
Bon CL, Garthwaite J (2001b) Exogenous nitric oxide causes potentiation of hippocampal synaptic transmission during low-frequency stimulation via the endogenous nitric oxide-cGMP pathway. Eur J Neurosci 14:585-594.
Bon CL, Garthwaite J (2003) On the role of nitric oxide in hippocampal long-term potentiation. J Neurosci 23:1941-1948.
Boulton CL, Southam E, Garthwaite J (1995) Nitric oxide-dependent long-term potentiation is blocked by a specific inhibitor of soluble guanylyl cyclase. Neuroscience 69:699-703.
Boulton CL, Irving AJ, Southam E, Potier B, Garthwaite J, Collingridge GL (1994) The nitric oxide--cyclic GMP pathway and synaptic depression in rat hippocampal slices. Eur J Neurosci 6:1528-1535.

Bredt DS, Snyder SH (1992) Nitric oxide, a novel neuronal messenger. Neuron 8:3-11.
Bredt DS, Hwang PM, Snyder SH (1990) Localization of nitric oxide synthase indicating a neural role for nitric oxide. Nature 347:768-770.
Bredt DS, Glatt CE, Hwang PM, Fotuhi M, Dawson TM, Snyder SH (1991) Nitric oxide synthase protein and mRNA are discretely localized in neuronal populations of the mammalian CNS together with NADPH diaphorase. Neuron 7:615-624.
Burette A, Zabel U, Weinberg RJ, Schmidt HH, Valtschanoff JG (2002) Synaptic localization of nitric oxide synthase and soluble guanylyl cyclase in the hippocampus. J Neurosci 22:8961-8970.
Burgunder JM, Cheung PT (1994) Expression of soluble guanylyl cyclase gene in adult rat brain. Eur J Neurosci 6:211-217.
Buttner N, Siegelbaum SA (2003) Antagonistic modulation of a hyperpolarization-activated Cl^- current in Aplysia sensory neurons by SCP(B) and FMRFamide. J Neurophysiol 90:586-598.
Calabresi P, Gubellini P, Centonze D, Sancesario G, Morello M, Giorgi M, Pisani A, Bernardi G (1999) A critical role of the nitric oxide/cGMP pathway in corticostriatal long-term depression. J Neurosci 19:2489-2499.
Carroll RC, Lissin DV, von Zastrow M, Nicoll RA, Malenka RC (1999) Rapid redistribution of glutamate receptors contributes to long-term depression in hippocampal cultures. Nat Neurosci 2:454-460.
Centonze D, Gubellini P, Pisani A, Bernardi G, Calabresi P (2003) Dopamine, acetylcholine and nitric oxide systems interact to induce corticostriatal synaptic plasticity. Rev Neurosci 14:207-216.
Chavez-Noriega LE, Stevens CF (1994) Increased transmitter release at excitatory synapses produced by direct activation of adenylate cyclase in rat hippocampal slices. J Neurosci 14:310-317.
Chetkovich DM, Klann E, Sweatt JD (1993) Nitric oxide synthase-independent long-term potentiation in area CA1 of hippocampus. Neuroreport 4:919-922.
Chien WL, Liang KC, Teng CM, Kuo SC, Lee FY, Fu WM (2003) Enhancement of long-term potentiation by a potent nitric oxide-guanylyl cyclase activator, 3-(5-hydroxymethyl-2-furyl)-1-benzyl-indazole. Mol Pharmacol 63:1322-1328.
Choi S, Klingauf J, Tsien RW (2000) Postfusional regulation of cleft glutamate concentration during LTP at 'silent synapses'. Nat Neurosci 3:330-336.
Clark GD, Happel LT, Zorumski CF, Bazan NG (1992) Enhancement of hippocampal excitatory synaptic transmission by platelet-activating factor. Neuron 9:1211-1216.
Clark KA, Collingridge GL (1995) Synaptic potentiation of dual-component excitatory postsynaptic currents in the rat hippocampus. J Physiol 482 (Pt 1):39-52.
Cummings JA, Nicola SM, Malenka RC (1994) Induction in the rat hippocampus of long-term potentiation (LTP) and long-term depression (LTD) in the presence of a nitric oxide synthase inhibitor. Neurosci Lett 176:110-114.
Denninger JW, Marletta MA (1999) Guanylate cyclase and the NO/cGMP signaling pathway. Biochim Biophys Acta 1411:334-350.
Dinerman JL, Dawson TM, Schell MJ, Snowman A, Snyder SH (1994) Endothelial nitric oxide synthase localized to hippocampal pyramidal cells: implications for synaptic plasticity. Proc Natl Acad Sci U S A 91:4214-4218.
Doreulee N, Sergeeva OA, Yanovsky Y, Chepkova AN, Selbach O, Godecke A, Schrader J, Haas HL (2003) Cortico-striatal synaptic plasticity in endothelial nitric oxide synthase deficient mice. Brain Res 964:159-163.
Doyle C, Holscher C, Rowan MJ, Anwyl R (1996) The selective neuronal NO synthase inhibitor 7-nitro-indazole blocks both long-term potentiation and depotentiation of field EPSPs in rat hippocampal CA1 in vivo. J Neurosci 16:418-424.
Doyle CA, Slater P (1997) Localization of neuronal and endothelial nitric oxide synthase isoforms in human hippocampus. Neuroscience 76:387-395.
Dudek SM, Bear MF (1993) Bidirectional long-term modification of synaptic effectiveness in the adult and immature hippocampus. J Neurosci 13:2910-2918.
Duman RS, Terwilliger RZ, Nestler EJ (1993) Alterations in nitric oxide-stimulated endogenous ADP-ribosylation associated with long-term potentiation in rat hippocampus. J Neurochem 61:1542-1545.
Endoh M, Maiese K, Wagner JA (1994) Expression of the neural form of nitric oxide synthase by CA1 hippocampal neurons and other central nervous system neurons. Neuroscience 63:679-689.
Engert F, Bonhoeffer T (1997) Synapse specificity of long-term potentiation breaks down at short distances. Nature 388:279-284.
Engert F, Bonhoeffer T (1999) Dendritic spine changes associated with hippocampal long-term synaptic plasticity. Nature 399:66-70.

Errington ML, Lynch MA, Bliss TV (1987) Long-term potentiation in the dentate gyrus: induction and increased glutamate release are blocked by D(-)aminophosphonovalerate. 20:279-284.

Feasey KJ, Lynch MA, Bliss TV (1986) Long-term potentiation is associated with an increase in calcium-dependent, potassium-stimulated release of [^{14}C]glutamate from hippocampal slices: an ex vivo study in the rat. Brain Res. 364:39-44.

Feinmark SJ, Begum R, Tsvetkov E, Goussakov I, Funk CD, Siegelbaum SA, Bolshakov VY (2003) 12-lipoxygenase metabolites of arachidonic acid mediate metabotropic glutamate receptor-dependent long-term depression at hippocampal CA3-CA1 synapses. J Neurosci 23:11427-11435.

Foster TC, McNaughton BL (1991) Long-term enhancement of CA1 synaptic transmission is due to increased quantal size, not quantal content. Hippocampus 1:79-91.

Friebe A, Schultz G, Koesling D (1996) Sensitizing soluble guanylyl cyclase to become a highly CO-sensitive enzyme. Embo J 15:6863-6868.

Gage AT, Reyes M, Stanton PK (1997) Nitric-oxide-guanylyl-cyclase-dependent and -independent components of multiple forms of long-term synaptic depression. Hippocampus 7:286-295.

Galione A, Churchill GC (2002) Interactions between calcium release pathways: multiple messengers and multiple stores. Cell Calcium 32:343-354.

Galione A, White A, Willmott N, Turner M, Potter BV, Watson SP (1993) cGMP mobilizes intracellular Ca^{2+} in sea urchin eggs by stimulating cyclic ADP-ribose synthesis. Nature 365:456-459.

Garbers DL, Lowe DG (1994) Guanylyl cyclase receptors. J Biol Chem 269:30741-30744.

Garthwaite J, Garthwaite G (1987) Cellular origins of cyclic GMP responses to excitatory amino acid receptor agonists in rat cerebellum in vitro. J Neurochem 48:29-39.

Garthwaite J, Southam E, Boulton CL, Nielsen EB, Schmidt K, Mayer B (1995) Potent and selective inhibition of nitric oxide-sensitive guanylyl cyclase by 1H-[1,2,4]oxadiazolo[4,3-a]quinoxalin-1-one. Mol Pharmacol 48:184-188.

Goussakov IV, Fink K, Elger CE, Beck H (2000) Metaplasticity of mossy fiber synaptic transmission involves altered release probability. J Neurosci 20:3434-3441.

Grassi S, Pettorossi VE (2000) Role of nitric oxide in long-term potentiation of the rat medial vestibular nuclei. Neuroscience 101:157-164.

Gribkoff VK, Lum-Ragan JT (1992) Evidence for nitric oxide synthase inhibitor-sensitive and insensitive hippocampal synaptic potentiation. J Neurophysiol 68:639-642.

Haley JE, Wilcox GL, Chapman PF (1992) The role of nitric oxide in hippocampal long-term potentiation. Neuron 8:211-216.

Haley JE, Malen PL, Chapman PF (1993) Nitric oxide synthase inhibitors block long-term potentiation induced by weak but not strong tetanic stimulation at physiological brain temperatures in rat hippocampal slices. Neurosci Lett 160:85-88.

Haley JE, Schaible E, Pavlidis P, Murdock A, Madison DV (1996) Basal and apical synapses of CA1 pyramidal cells employ different LTP induction mechanisms. Learn Mem 3:289-295.

Hartell NA (2002) Parallel fiber plasticity. Cerebellum 1:3-18.

Haul S, Godecke A, Schrader J, Haas HL, Luhmann HJ (1999) Impairment of neocortical long-term potentiation in mice deficient of endothelial nitric oxide synthase. J Neurophysiol 81:494-497.

Hebb DO (1949) *The Organization of Behavior,* Wiley, New York.

Holscher C (1999) Nitric oxide is required for expression of LTP that is induced by stimulation phase-locked with theta rhythm. Eur J Neurosci 11:335-343.

Holscher C (2002) Different strains of rats show different sensitivity to block of long-term potentiation by nitric oxide synthase inhibitors. Eur J Pharmacol 457:99-106.

Ibarra C, Nedvetsky PI, Gerlach M, Riederer P, Schmidt HH (2001) Regional and age-dependent expression of the nitric oxide receptor, soluble guanylyl cyclase, in the human brain. Brain Res 907:54-60.

Ikegaya Y, Saito H, Matsuki N (1994) Involvement of carbon monoxide in long-term potentiation in the dentate gyrus of anesthetized rats. Jpn J Pharmacol 64:225-227.

Isaac JT, Nicoll RA, Malenka RC (1995) Evidence for silent synapses: implications for the expression of LTP. Neuron 15:427-434.

Izumi Y, Zorumski CF (1993) Nitric oxide and long-term synaptic depression in the rat hippocampus. Neuroreport 4:1131-1134.

Jacoby S, Sims RE, Hartell NA (2001) Nitric oxide is required for the induction and heterosynaptic spread of long-term potentiation in rat cerebellar slices. J Physiol 535:825-839.

Kantor DB, Lanzrein M, Stary SJ, Sandoval GM, Smith WB, Sullivan BM, Davidson N, Schuman EM (1996) A role for endothelial NO synthase in LTP revealed by adenovirus-mediated inhibition and rescue. Science 274:1744-1748.

Kato K, Zorumski CF (1993) Nitric oxide inhibitors facilitate the induction of hippocampal long-term potentiation by modulating NMDA responses. J Neurophysiol 70:1260-1263.
Kato K, Clark GD, Bazan NG, Zorumski CF (1994) Platelet-activating factor as a potential retrograde messenger in CA1 hippocampal long-term potentiation. Nature 367:175-179.
Kishimoto K, Matsumura K, Kataoka Y, Morii H, Watanabe Y (1999) Localization of cytosolic phospholipase A2 messenger RNA mainly in neurons in the rat brain. Neuroscience 92:1061-1077.
Kleppisch T, Pfeifer A, Klatt P, Ruth P, Montkowski A, Fassler R, Hofmann F (1999) Long-term potentiation in the hippocampal CA1 region of mice lacking cGMP-dependent kinases is normal and susceptible to inhibition of nitric oxide synthase. J Neurosci 19:48-55.
Ko FN, Wu CC, Kuo SC, Lee FY, Teng CM (1994) YC-1, a novel activator of platelet guanylate cyclase. Blood 84:4226-4233.
Ko GY, Kelly PT (1999) Nitric oxide acts as a postsynaptic signaling molecule in calcium/calmodulin-induced synaptic potentiation in hippocampal CA1 pyramidal neurons. J Neurosci 19:6784-6794.
Kobayashi K, Ishii S, Kume K, Takahashi T, Shimizu T, Manabe T (1999) Platelet-activating factor receptor is not required for long-term potentiation in the hippocampal CA1 region. Eur J Neurosci 11:1313-16.
Kondratskaya EL, Pankratov YV, Lalo UV, Chatterjee SS, Krishtal OA (2004) Inhibition of hippocampal LTP by ginkgolide B is mediated by its blocking action on PAF rather than glycine receptors. Neurochem Int 44:171-177.
Kornecki E, Wieraszko A, Chan J, Ehrlich YH (1996) Platelet activating factor (PAF) in memory formation: role as a retrograde messenger in long-term potentiation. J Lipid Mediat Cell Signal 14:115-126.
Kuhnt U, Hess G, Voronin LL (1992) Statistical analysis of large excitatory postsynaptic potentials recorded in guinea pig hippocampal slices: binomial model. Exp Brain Res 89:265-274.
Kullmann DM, Nicoll RA (1992) Long-term potentiation is associated with increases in quantal content and quantal amplitude. Nature 357:240-244.
Laroche S, Errington ML, Lynch MA, Bliss TV (1987) Increase in [3H]glutamate release from slices of dentate gyrus and hippocampus following classical conditioning in the rat. Behav Brain Res 25:23-29.
Lee HC, Aarhus R, Graeff R, Gurnack ME, Walseth TF (1994) Cyclic ADP ribose activation of the ryanodine receptor is mediated by calmodulin. Nature 370:307-309.
Lee HK, Kameyama K, Huganir RL, Bear MF (1998) NMDA induces long-term synaptic depression and dephosphorylation of the GluR1 subunit of AMPA receptors in hippocampus. Neuron 21:1151-1162.
Lev-Ram V, Mehta SB, Kleinfeld D, Tsien RY (2003) Reversing cerebellar long-term depression. Proc Natl Acad Sci U S A 100:15989-15993.
Lev-Ram V, Jiang T, Wood J, Lawrence DS, Tsien RY (1997) Synergies and coincidence requirements between NO, cGMP, and Ca2+ in the induction of cerebellar long-term depression. Neuron 18:1025-1038.
Lin H, Totterdell S (1998) Light and electron microscopic study of neuronal nitric oxide synthase-immunoreactive neurons in the rat subiculum. J Comp Neurol 395:195-208.
Lopez-Figueroa MO, Itoi K, Watson SJ (1998) Regulation of nitric oxide synthase messenger RNA expression in the rat hippocampus by glucocorticoids. Neuroscience 87:439-446.
Lu YF, Kandel ER, Hawkins RD (1999) Nitric oxide signaling contributes to late-phase LTP and CREB phosphorylation in the hippocampus. J Neurosci 19:10250-10261.
Lucas KA, Pitari GM, Kazerounian S, Ruiz-Stewart I, Park J, Schulz S, Chepenik KP, Waldman SA (2000) Guanylyl cyclases and signaling by cyclic GMP. Pharmacol Rev 52:375-414.
Lum-Ragan JT, Gribkoff VK (1993) The sensitivity of hippocampal long-term potentiation to nitric oxide synthase inhibitors is dependent upon the pattern of conditioning stimulation. Neurosci 57:973-983.
Lumme A, Soinila S, Sadeniemi M, Halonen T, Vanhatalo S (2000) Nitric oxide synthase immunoreactivity in the rat hippocampus after status epilepticus induced by perforant pathway stimulation. Brain Res 871:303-310.
Luo D, Vincent SR (1994) Metalloporphyrins inhibit nitric oxide-dependent cGMP formation in vivo. Eur J Pharmacol 267:263-267.
Lynch MA, Errington ML, Bliss TV (1985) Long-term potentiation of synaptic transmission in the dentate gyrus: increased release of [^{14}C]glutamate without increase in receptor binding. Neurosci Lett 62:123-129.
Lynch MA, Errington ML, Clements MP, Bliss TV, Redini-Del Negro C, Laroche S (1990) Increases in glutamate release and phosphoinositide metabolism associated with long-term potentiation and classical conditioning. Prog Brain Res 83:251-256.
Ma L, Zablow L, Kandel ER, Siegelbaum SA (1999) Cyclic AMP induces functional presynaptic boutons in hippocampal CA3-CA1 neuronal cultures. Nat Neurosci 2:24-30.

Maffei A, Prestori F, Shibuki K, Rossi P, Taglietti V, D'Angelo E (2003) NO enhances presynaptic currents during cerebellar mossy fiber-granule cell LTP. J Neurophysiol 90:2478-2483.

Malen PL, Chapman PF (1997) Nitric oxide facilitates long-term potentiation, but not long-term depression. J Neurosci 17:2645-2651.

Malenka RC, Nicoll RA (1999) Long-term potentiation--a decade of progress? Science 285:1870-1874.

Malinow R, Tsien RW (1990) Presynaptic enhancement shown by whole-cell recordings of long-term potentiation in hippocampal slices. Nature 346:177-180.

Malinow R, Mainen ZF, Hayashi Y (2000) LTP mechanisms: from silence to four-lane traffic. Curr Opin Neurobiol 10:352-357.

Matthies H, Reymann KG (1993) Protein kinase A inhibitors prevent the maintenance of hippocampal long-term potentiation. Neuroreport 4:712-714.

Mayer ML, Westbrook GL, Guthrie PB (1984) Voltage-dependent block by Mg^{2+} of NMDA responses in spinal cord neurones. Nature 309:261-263.

Meffert MK, Haley JE, Schuman EM, Schulman H, Madison DV (1994) Inhibition of hippocampal heme oxygenase, nitric oxide synthase, and long-term potentiation by metalloporphyrins. Neuron 13:1225-1233.

Monfort P, Munoz MD, Kosenko E, Felipo V (2002) Long-term potentiation in hippocampus involves sequential activation of soluble guanylate cyclase, cGMP-dependent protein kinase, and cGMP-degrading phosphodiesterase. J Neurosci 22:10116-10122.

Mulkey RM, Malenka RC (1992) Mechanisms underlying induction of homosynaptic long-term depression in area CA1 of the hippocampus. Neuron 9:967-975.

Murphy KP, Williams JH, Bettache N, Bliss TV (1994) Photolytic release of nitric oxide modulates NMDA receptor-mediated transmission but does not induce long-term potentiation at hippocampal synapses. Neuropharmacology 33:1375-1385.

Murthy VN, Sckikorski T, Stevens CF, Zhu Y (2001) Inactivity produces increases in neurotransmitter release and synapse size. Neuron 32:673-682.

Neveu D, Zucker RS (1996) Postsynaptic levels of $[Ca^{2+}]_i$ needed to trigger LTD and LTP. Neuron 16:619-629.

Normandin M, Gagne J, Bernard J, Elie R, Miceli D, Baudry M, Massicotte G (1996) Involvement of the 12-lipoxygenase pathway of arachidonic acid metabolism in homosynaptic long-term depression of the rat hippocampus. Brain Res 730:40-46.

Nowicky AV, Bindman LJ (1993) The nitric oxide synthase inhibitor, N-monomethyl-L-arginine blocks induction of a long-term potentiation-like phenomenon in rat medial frontal cortical neurons in vitro. J Neurophysiol 70:1255-1259.

O'Dell TJ, Hawkins RD, Kandel ER, Arancio O (1991) Tests of the roles of two diffusible substances in long-term potentiation: evidence for nitric oxide as a possible early retrograde messenger. Proc Natl Acad Sci U S A 88:11285-11289.

O'Dell TJ, Huang PL, Dawson TM, Dinerman JL, Snyder SH, Kandel ER, Fishman MC (1994) Endothelial NOS and the blockade of LTP by NOS inhibitors in mice lacking neuronal NOS. Science 265:542-6.

O'Mara SM, Rowan MJ, Anwyl R (1995) Metabotropic glutamate receptor-induced homosynaptic long-term depression and depotentiation in the dentate gyrus of the rat hippocampus in vitro. Neuropharmacology 34:983-989.

Oliet SH, Malenka RC, Nicoll RA (1996) Bidirectional control of quantal size by synaptic activity in the hippocampus. Science 271:1294-1297.

Oliet SH, Malenka RC, Nicoll RA (1997) Two distinct forms of long-term depression coexist in CA1 hippocampal pyramidal cells. Neuron 18:969-982.

Otani S, Connor JA (1995) Long-term depression of naive synapses in adult hippocampus induced by asynchronous synaptic activity. J Neurophysiol 73:2596-2601.

Pettorossi VE, Grassi S (2001) Different contributions of platelet-activating factor and nitric oxide in long-term potentiation of the rat medial vestibular nuclei. Acta Otolaryngol Suppl 545:160-165.

Philippides A, Husbands P, O'Shea M (2000) Four-dimensional neuronal signaling by nitric oxide: a computational analysis. J Neurosci 20:1199-1207.

Piomelli D, Volterra A, Dale N, Siegelbaum SA, Kandel ER, Schwartz JH, Belardetti F (1987) Lipoxygenase metabolites of arachidonic acid as second messengers for presynaptic inhibition of Aplysia sensory cells. Nature 328:38-43.

Pyle JL, Kavalali ET, Choi S, Tsien RW (1999) Visualization of synaptic activity in hippocampal slices with FM1-43 enabled by fluorescence quenching. Neuron 24:803-808.

Pyle JL, Kavalali ET, Piedras-Renteria ES, Tsien RW (2000) Rapid reuse of readily releasable pool vesicles at hippocampal synapses. Neuron 28:221-231.

Reyes M, Stanton PK (1996) Induction of hippocampal long-term depression requires release of Ca^{2+} from separate presynaptic and postsynaptic intracellular stores. J Neurosci 16:5951-5960.

Reyes-Harde M, Potter BV, Galione A, Stanton PK (1999a) Induction of hippocampal LTD requires nitric-oxide-stimulated PKG activity and Ca^{2+} release from cyclic ADP-ribose-sensitive stores. J Neurophysiol 82:1569-1576.

Reyes-Harde M, Empson R, Potter BV, Galione A, Stanton PK (1999b) Evidence of a role for cyclic ADP-ribose in long-term synaptic depression in hippocampus. Proc Natl Acad Sci USA 96:4061-4066.

Rosenmund C, Stevens CF (1996) Definition of the readily-releasable pool of vesicles at hippocampal synapses. Neuron 16:1197-1207.

Ryan TA, Reuter H, Wendland B, Schweizer FE, Tsien RW, Smith SJ (1993) The kinetics of synaptic vesicle recycling measured at single presynaptic boutons. Neuron 11:713-724.

Santschi L, Reyes-Harde M, Stanton PK (1999) Chemically induced, activity-independent LTD elicited by simultaneous activation of PKG and inhibition of PKA. J Neurophysiol 82:1577-1589.

Scherer-Singler U, Vincent SR, Kimura H, McGeer EG (1983) Demonstration of a unique population of neurons with NADPH-diaphorase histochemistry. J Neurosci Methods 9:229-234.

Schikorski T, Stevens CF (2001) Morphological correlates of functionally defined synaptic vesicle populations. Nat Neurosci 4:391-395.

Schuman EM, Madison DV (1991) A requirement for the intercellular messenger nitric oxide in long-term potentiation. Science 254:1503-1506.

Schuman EM, Meffert MK, Schulman H, Madison DV (1994) An ADP-ribosyltransferase as a potential target for nitric oxide action in hippocampal long-term potentiation. Proc Natl Acad Sci USA 91:11958-11962.

Sejnowski TJ (1977) Storing covariance with nonlinearly interacting neurons. J Math Biol 4:303-321.

Selig DK, Segal MR, Liao D, Malenka RC, Malinow R, Nicoll RA, Lisman JE (1996) Examination of the role of cGMP in long-term potentiation in the CA1 region of the hippocampus. Learn Mem 3:42-48.

Sheng M, Lee SH (2001) AMPA receptor trafficking and the control of synaptic transmission. Cell 105:825-828.

Shi SH, Hayashi Y, Petralia RS, Zaman SH, Wenthold RJ, Svoboda K, Malinow R (1999) Rapid spine delivery and redistribution of AMPA receptors after synaptic NMDA receptor activation. Science 284:1811-1816.

Slack JR, Pockett S (1991) Cyclic AMP induces long-term increase in synaptic efficacy in CA1 region of rat hippocampus. Neurosci Lett 130:69-72.

Soderling TR, Derkach VA (2000) Postsynaptic protein phosphorylation and LTP. Trends Neurosci 23:75-80.

Son H, Lu YF, Zhuo M, Arancio O, Kandel ER, Hawkins RD (1998) The specific role of cGMP in hippocampal LTP. Learn Mem 5:231-245.

Son H, Hawkins RD, Martin K, Kiebler M, Huang PL, Fishman MC, Kandel ER (1996) Long-term potentiation is reduced in mice that are doubly mutant in endothelial and neuronal nitric oxide synthase. Cell 87:1015-1023.

Southam E, Garthwaite J (1993) The nitric oxide-cyclic GMP signalling pathway in rat brain. Neuropharmacology 32:1267-1277.

Sperelakis N, Xiong Z, Haddad G, Masuda H (1994) Regulation of slow calcium channels of myocardial cells and vascular smooth muscle cells by cyclic nucleotides and phosphorylation. Mol Cell Biochem 140:103-117.

Stanarius A, Topel I, Schulz S, Noack H, Wolf G (1997) Immunocytochemistry of endothelial nitric oxide synthase in the rat brain: a light and electron microscopical study using the tyramide signal amplification technique. Acta Histochem 99:411-429.

Stanton PK (1995) Phospholipase A2 activation is not required for long-term synaptic depression. Eur J Pharmacol 273:R7-9.

Stanton PK (1996) LTD, LTP, and the sliding threshold for long-term synaptic plasticity. Hippocampus 6:35-42.

Stanton PK, Gage AT (1996) Distinct synaptic loci of Ca2+/calmodulin-dependent protein kinase II necessary for long-term potentiation and depression. J Neurophysiol 76:2097-2101.

Stanton PK, Heinemann U, Muller W (2001) FM1-43 imaging reveals cGMP-dependent long-term depression of presynaptic transmitter release. J Neurosci 21:RC167.

Stanton PK, Sarvey JM (1985a) Blockade of norepinephrine-induced long-lasting potentiation in the hippocampal dentate gyrus by an inhibitor of protein synthesis. Brain Res 361:276-283.

Stanton PK, Sarvey JM (1985b) The effect of high-frequency electrical stimulation and norepinephrine on cyclic AMP levels in normal versus norepinephrine-depleted rat hippocampal slices. Brain Res 358:343-348.

Stanton PK, Sejnowski TJ (1989) Associative long-term depression in the hippocampus induced by hebbian covariance. Nature 339:215-218.

Stanton PK, Winterer J, Bailey CP, Kyrozis A, Raginov I, Laube G, Veh RW, Nguyen CQ, Muller W (2003) Long-term depression of presynaptic release from the readily releasable vesicle pool induced by NMDA receptor-dependent retrograde nitric oxide. J Neurosci 23:5936-5944.

Stevens CF, Wang Y (1993) Reversal of long-term potentiation by inhibitors of haem oxygenase. Nature 364:147-149.

Stevens CF, Wang Y (1994) Changes in reliability of synaptic function as a mechanism for plasticity. Nature 371:704-707.

Teichert AM, Miller TL, Tai SC, Wang Y, Bei X, Robb GB, Phillips MJ, Marsden PA (2000) In vivo expression profile of an endothelial nitric oxide synthase promoter-reporter transgene. Am J Physiol Heart Circ Physiol 278:H1352-1361.

Toni N, Buchs PA, Nikonenko I, Bron CR, Muller D (1999) LTP promotes formation of multiple spine synapses between a single axon terminal and a dendrite. Nature 402:421-425.

Topel I, Stanarius A, Wolf G (1998) Distribution of the endothelial constitutive nitric oxide synthase in the developing rat brain: an immunohistochemical study. Brain Res 788:43-48.

Ueda K, Hayaishi O (1985) ADP-ribosylation. Annu Rev Biochem 54:73-100.

Valtschanoff JG, Weinberg RJ, Kharazia VN, Nakane M, Schmidt HH (1993) Neurons in rat hippocampus that synthesize nitric oxide. J Comp Neurol 331:111-121.

Verma A, Hirsch DJ, Glatt CE, Ronnett GV, Snyder SH (1993) Carbon monoxide: a putative neural messenger. Science 259:381-384.

von Bohlen und Halbach O, Albrecht D, Heinemann U, Schuchmann S (2002) Spatial nitric oxide imaging using 1,2-diaminoanthraquinone to investigate the involvement of nitric oxide in long-term potentiation in rat brain slices. Neuroimage 15:633-639.

Voronin LL, Kuhnt U, Gusev AG (1992a) Analysis of fluctuations of "minimal" excitatory postsynaptic potentials during long-term potentiation in guinea pig hippocampal slices. Exp Brain Res 89:288-299.

Voronin LL, Kuhnt U, Gusev AG, Hess G (1992b) Quantal analysis of long-term potentiation of "minimal" excitatory postsynaptic potentials in guinea pig hippocampal slices: binomial approach. Exp Brain Res 89:275-287.

Wakatsuki H, Gomi H, Kudoh M, Kimura S, Takahashi K, Takeda M, Shibuki K (1998) Layer-specific NO dependence of long-term potentiation and biased NO release in layer V in the rat auditory cortex. J Physiol 513 (Pt 1):71-81.

Wang LY, Kaczmarek LK (1998) High-frequency firing helps replenish the readily releasable pool of synaptic vesicles. Nature 394:384-388.

Watanabe Y, Saito H, Abe K (1995) Nitric oxide is involved in long-term potentiation in the medial but not lateral amygdala neuron synapses in vitro. Brain Res 688:233-236.

Wedel B, Garbers D (2001) The guanylyl cyclase family at Y2K. Annu Rev Physiol 63:215-233.

Wendland B, Schweizer FE, Ryan TA, Nakane M, Murad F, Scheller RH, Tsien RW (1994) Existence of nitric oxide synthase in rat hippocampal pyramidal cells. Proc Natl Acad Sci U S A 91:2151-2155.

Wexler EM, Stanton PK (1993) Priming of homosynaptic long-term depression in hippocampus by previous synaptic activity. Neuroreport 4:591-594.

Wexler EM, Stanton PK, Nawy S (1998) Nitric oxide depresses GABAA receptor function via coactivation of cGMP-dependent kinase and phosphodiesterase. J Neurosci 18:2342-2349.

Wieraszko A, Li G, Kornecki E, Hogan MV, Ehrlich YH (1993) Long-term potentiation in the hippocampus induced by platelet-activating factor. Neuron 10:553-557.

Williams JH, Errington ML, Lynch MA, Bliss TV (1989) Arachidonic acid induces a long-term activity-dependent enhancement of synaptic transmission in the hippocampus. Nature 341:739-742.

Williams JH, Li YG, Nayak A, Errington ML, Murphy KP, Bliss TV (1993) The suppression of long-term potentiation in rat hippocampus by inhibitors of nitric oxide synthase is temperature and age dependent. Neuron 11:877-884.

Willmott N, Sethi JK, Walseth TF, Lee HC, White AM, Galione A (1996) Nitric oxide-induced mobilization of intracellular calcium via the cyclic ADP-ribose signaling pathway. J Biol Chem 271:3699-3705.

Wilson RI, Godecke A, Brown RE, Schrader J, Haas HL (1999) Mice deficient in endothelial nitric oxide synthase exhibit a selective deficit in hippocampal long-term potentiation. Neurosci 90:1157-1165.

Wu J, Wang Y, Rowan MJ, Anwyl R (1997) Evidence for involvement of the neuronal isoform of nitric oxide synthase during induction of long-term potentiation and long-term depression in the rat dentate gyrus in vitro. Neuroscience 78:393-398.

Wu J, Wang Y, Rowan MJ, Anwyl R (1998) Evidence for involvement of the cGMP-protein kinase G signaling system in the induction of long-term depression, but not long-term potentiation, in the dentate gyrus in vitro. J Neurosci 18:3589-3596.

Yip S, Sastry BR (2000) Effects of hemoglobin and its breakdown products on synaptic transmission in rat hippocampal CA1 neurons. Brain Res 864:1-12.

Yip S, Ip JK, Sastry BR (1996) Electrophysiological actions of hemoglobin on rat hippocampal CA1 pyramidal neurons. Brain Res 713:134-142.

Zakharenko SS, Zablow L, Siegelbaum SA (2001) Visualization of changes in presynaptic function during long-term synaptic plasticity. Nat Neurosci 4:711-717.

Zakharenko SS, Zablow L, Siegelbaum SA (2002) mGluR-dependent LTD alters mode of presynaptic exocytosis. Neuron 35:1099-1110.

Zhuo M, Kandel ER, Hawkins RD (1994a) Nitric oxide and cGMP can produce either synaptic depression or potentiation depending on the frequency of presynaptic stimulation in the hippocampus. Neuroreport 5:1033-1036.

Zhuo M, Small SA, Kandel ER, Hawkins RD (1993) Nitric oxide and carbon monoxide produce activity-dependent long-term synaptic enhancement in hippocampus. Science 260:1946-1950.

Zhuo M, Laitinen JT, Li XC, Hawkins RD (1999) On the respective roles of nitric oxide and carbon monoxide in long-term potentiation in the hippocampus. Learn Mem 6:63-76.

Zhuo M, Hu Y, Schultz C, Kandel ER, Hawkins RD (1994b) Role of guanylyl cyclase and cGMP-dependent protein kinase in long-term potentiation. Nature 368:635-639.

HIPPOCAMPAL LONG-TERM DEPRESSION AS A DECLARATIVE MEMORY MECHANISM

Denise Manahan-Vaughan*

1. SYNOPSIS

Hippocampal long-term depression (LTD) consists of a robust activity-dependent reduction in synaptic strength which is typically induced by prolonged low frequency afferent stimulation. Although much emphasis has been placed on long-term potentiation (LTP) as the putative cellular mechanism for spatial memory formation in the hippocampus, recent evidence points to LTD as an additional hippocampal memory mechanism. In contrast to the initial hypothesis that it serves to prevent synapse saturation, by means, for example, of enabling reversal of LTP, considerable evidence points to the fact that LTD is not mechanistically the reverse of LTP. Although several common processes are shared by LTD and LTP, such as changes in intracellular calcium and induction of signaling cascades, molecular, biochemical, electrophysiological and pharmacological studies all indicate that several quite distinct induction and maintenance mechanisms are required for LTD. Most intriguing is the new evidence that whereas LTP may serve to encode space, LTD encodes the features of space. Indeed it has even been demonstrated that LTD correlates exactly with the spatial learning ability of rats. LTD may thus contribute a very specific component to the formation of spatial memory. It is therefore quite likely that LTP and LTD work together in the formation of a spatial memory trace.

---

* Denise Manahan-Vaughan[1,2], PhD, [1]Learning and Memory Research, Faculty of Medicine; [2]International Graduate School of Neuroscience, Ruhr University Bochum,Universitaetsstrasse 150, FNO 1/116, 44780 Bochum. E-mail: denise.manahan-vaughan@charite.de

2. INTRODUCTION

A pivotal issue in brain research deals with identification and characterisation of the mechanisms of learning and memory formation and storage. It is probable that modifications of synaptic connectivity, in the form of synaptic plasticity, underlie the acquisition of memory traces and their permanent storage (Bear, 1996). Declarative memory, in the form of spatial memory, is encoded by the hippocampus (Eichenbaum, 1997; Tulving and Markowitsch, 1998), and this structure is of particular interest due to its expression of multiple forms of synaptic plasticity. The existence of multiple hippocampal plasticity forms, including the two primary forms of activity-dependent synaptic plasticity, long-term potentiation (LTP) (Bliss and Lømo, 1973) and long-term depression (LTD) (Dunwiddie and Lynch, 1978; Dudek and Bear, 1992), suggest that declarative memory is encoded by more than one form of plasticity, or that distinct forms of plasticity encode distinct forms of memory. It has been postulated that the cellular mechanisms which underlie LTD expression in the cerebral cortex, together with those underlying long-term potentiation (LTP), are responsible for information storage by the hippocampus (Bear, 1996; Bear and Abraham, 1996). On the other hand, LTP is widely believed to comprise *the* cellular mechanism underlying spatial memory. Often overlooked, is the fact that LTD, as is the case with LTP, fulfills three crucial criteria which recommend its consideration as a memory mechanism: it is input-specific (Dudek and Bear, 1992), it is associative (Stanton and Sejnowski, 1989; Stanton, 1996) and it persists for prolonged periods (Manahan-Vaughan, 1997; Manahan-Vaughan and Braunewell, 1999). Furthermore, LTD is protein synthesis dependent (Manahan-Vaughan et al, 2001), as is the case for LTP (Stanton and Sarvey, 1984; Frey et al, 1988) and long-term memory (Flood et al, 1981; Fride et al, 1989) and it occurs in close association with declarative memory events (Manahan-Vaughan and Braunewell, 1999; Kemp and Manahan-Vaughan, 2004). For a number of years, LTD was postulated to serve to reverse LTP (Christie et al, 1994; Stanton, 1996; Wagner and Alger, 1996), or to comprise the cellular mechanism for forgetting (Tsumoto, 1993). Indeed, one cannot exclude that certain forms of „negative“ plasticity such as depotentiation comprise these mechanisms. However, evidence is emerging that not only do LTD and depotentiation integrate distinctly different mechanisms for their expression (Fitzjohn et al, 1998; Zhuo et al, 1999; Lee et al, 2000), hippocampal LTD may in fact encode very specific aspects of spatial memory. This article will focus on hippocampal LTD, as a putative cognitive mechanism, with particular emphasis on LTD in the hippocampal CA1 region.

3. INDUCTION OF LTD IN THE HIPPOCAMPUS OF FREELY MOVING RATS

In the hippocampus, *in vivo*, LTD in the hippocampal CA1 region is most effectively induced by afferent stimulation at a frequency of 1Hz (900 pulses) (Heynen et al, 1996; Manahan-Vaughan, 1997, 2000). In both the CA1 region and dentate gyrus a reduction in basal synaptic transmission to roughly 70% of pre-induction values is induced which, *in vivo*, has been followed for 7-10 days in Wistar rats (Manahan-Vaughan and Braunewell, 1999; Naie and Manahan-Vaughan, 2004) (Fig. 1). Weaker, albeit persistent, LTD results from 2 Hz stimulation (900 pulses) in vivo, whereas stimulation using a similar pulse number but higher stimulation frequency is less successful (Manahan-Vaughan, 2000). In addition, varying the pulse number given at 1 or 2 Hz does not improve responses.

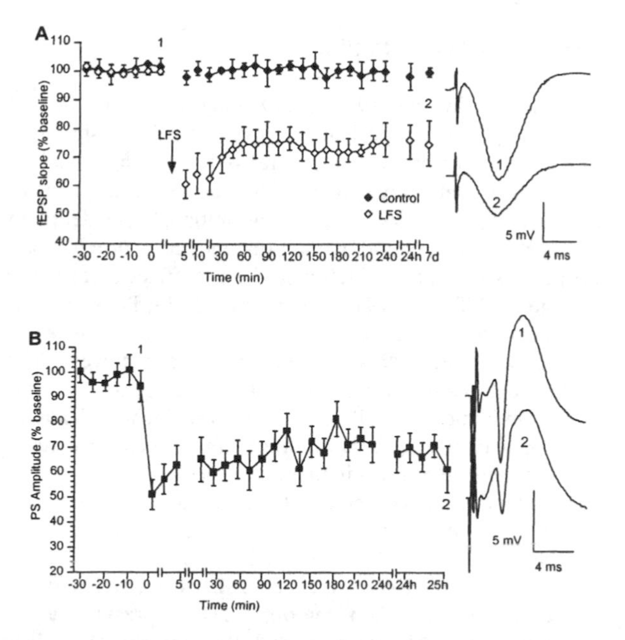

Figure 1. Robust LTD occurs in both hippocampal CA1 region and dentate gyrus of freely moving rats. Application of low frequency stimulation (1 Hz, 900 pulses) to the schaffer collaterals or medial perforant path induces robust LTD at stratum radiatum (A) or dentate gyrus granule cell synapses (B), respectively.

Compared to the depression obtained at 1Hz (900 pulses) (Manahan-Vaughan, 2000). Reliable CA1 LTD has been reported to follow paired-pulse stimulation in vivo (Thiels et al, 1994, 1996; Doyere et al, 1996). Here, for example, a train of 200 pairs of pulses delivered at 0.5 Hz elicits robust LTD in anesthetised rats if the interstimulus interval of the paired pulses is short (25 ms). Longer stimulus intervals are less effective (Thiels et al, 1994).

In contrast to the data described above, it has been reported that Wistar (Doyle et al, 1997) and Sprague Dawley rats (Errington et al, 1992) do not express LTD in certain laboratories. Failure to induce LTD in the CA1 region of Long Evans rats has also been reported (Staubli and Scafidi, 1997). Strain-dependent variations in the expression of LTD occur, which might explain these conflicting reports (Manahan-Vaughan and Braunewell, 1999; Manahan-Vaughan, 2000). In a study where three rat strains were subjected to identical experimental conditions it was found that whereas Wistar and Sprague Dawley rats express robust LTD (albeit of differing magnitude), Hooded Lister rats fail to express LTD following the same induction protocol (1Hz, 900 pulses) (Manahan-Vaughan, 2000). Wistar rats from differing breeding sources also express differing degrees of synaptic plasticity (Manahan-Vaughan, unpublished data). Thus, differences within rat strains accross laboratories may explain inconsistencies in LTD induction. This in turn may relate to the very different level of cognitive performance demonstrated by rats of differing strains (Andrews, 1996).

4. LTD- A BRIEF MECHANISTIC OVERVIEW

LTD was first identified in the 1970's, not long after the first detailed descriptions of LTP. Lynch et al (1977) , described heterosynaptic depression in the CA1 region *in vitro*, which appeared to result from a generalised depression of the neuron (Dunwiddie et al, 1978). Evidence that homosynaptic LTD occurs, came later (Stanton and Sejnowski, 1989; Dudek and Bear, 1992). LTD was initially considered to be of lesser importance for memory processes due to the fact that induction *in vitro*, or *in vivo*, led to rather brief (< 2h) periods of LTD –which would not suffice for formation of a long-term memory trace (Alger and Teyler, 1976; Alger et al, 1978; Pockett et al, 1990; Christie and Abraham, 1992) and due to failures of attempts to induce persistent LTD *in vivo* (Errington et al, 1992; Doyle et al, 1997). Contrasting reports however described LTD which could endure for hours (Abraham and Goddard, 1983; Colbert et al, 1992; Heynen et al, 1996; Manahan-Vaughan, 1997) and days (Manahan-Vaughan and Braunewell, 1999) in the hippocampus. These findings contributed to a revision of thinking with regard to the possible role of LTD in memory processing.

In the quarter of a century since its identification, substantial progress has been made in our understanding of the mechanisms underlying LTD. It occurs in all three hippocampal regions *in vitro* (Dudek and Bear, 1992; Thiels et al, 1994; Kobayashi, 1996) and *in vivo* (Christie and Abraham, 1992; Thiels et al, 1994; Derrick and Martinez, 1996; Manahan-Vaughan, 2000). In the CA3 region, LTD is presynaptic, NMDA receptor-independent and requires the activation of mGluRs (Fitzjohn et al, 1998), for successful induction to occur. In the CA1 region and dentate gyrus, activation of postsynaptic N-methyl-D-aspartate (NMDA) receptors as well as metabotropic glutamate receptors (mGluR) is typically required (Dudek and Bear, 1992; Mulkey and Malenka, 1992; Thiels et al, 1996; Manahan-Vaughan, 1997; Otani and Connor, 1998). In addition, in the CA1 region of juvenile rats *in vitro*, an exclusively NMDA receptor-dependent form and an exclusively mGluR-dependent form of LTD have been described (Oliet et al, 1997; Nicoll et al, 1998), whereas an NMDA receptor independent LTD also exists in the dentate gyrus (Trommer et al, 1996; Wang et al, 1997).

For qualification as a putative memory mechanism, synaptic plasticity is expected to fulfill a number of criteria which are considered prerequisite. These criteria are based on observations about mechanisms involved in learning and memory. Thus, it is expected, for example, that neurotransmitter systems which are involved in learning will also regulate plasticity, that synaptic plasticity exhibits phases equivalent to phases of memory, and that synaptic plasticity be protein synthesis-dependent.

As is the case with LTP, it has also been found that LTD fulfills all of these criteria thus qualifying LTD as a candidate mechanism for memory (Braunewell and Manahan-Vaughan, 2001) (Table 1). Intriguingly, LTP and LTD, in the CA1 region, share common molecular mechanisms, at least for the induction phase of these forms of synaptic plasticity: namely activation of NMDA-receptors and increase in postsynaptic calcium levels (Morris et al, 1989; Dudek and Bear, 1992).

Criteria	Declarative Memory	LTP	LTD
Input specific	+ (Hebb)	+	+
Occurs in distinct phases[1]	+	+	+
NMDA receptor dependent[2]	+	+	+
mGlu receptor dependent[3]	+	+	+
Requirement, intracellular calcium elevation[4]	+	+	+
Kinase dependent[5]	+	+	+
Phosphatase dependent[6]	+	+	+
Immediate early gene dependent[7]	+	+	+
Protein synthesis dependent[8]	+	+	+

Table 1. LTP and LTD as candidate memory mechanisms.
The criteria which have been used to qualify LTP as a memory mechanism are all fulfilled by LTD. In both cases a strong correlation with declarative memory is evident.
([1]Morris et al, 1986; Dudek and Bear, 1992; Frey et al, 1993; Manahan-Vaughan, 1997; Manahan-Vaughan et al, 2002; [2]Wigstrom & Gustafsson 1986; Dudek and Bear, 1992; Manahan-Vaughan, 1997; [3]Bortolotto et al, 1995; Manahan-Vaughan, 1997; Otani and Connor, 1998; Lu et al, 1997; Naie & Manahan-Vaughan, 2004; [4]Cummings et al, 1996; Mizumori et al, 1987a,b; [5]Bank et al, 1989; Silva et al, 1992; Soderling. 1993; Brandon et al, 1995; Nguyen & Wu, 2003; [6]Mulkey and Malenka, 1992; Mulkey et al, 1993, 1994; Wang & Stelzer, 1994; Mansuy et al, 1998; [7]Dragunow, 1996; Abraham et al, 1994; Bozon et al, 2003; Fleischman et al, 2003; [8]Frey et al, 1988; Manahan-Vaughan et al, 2000; Stanton & Sarvey, 1984; Mizumori et al, 1985)

For LTP, a high Ca^{2+} influx preferentially leads to activation of protein kinases such as PKC and Ca^{2+}-calmodulin kinase II (CaMKII). For LTD, a moderate Ca^{2+} influx preferentially leads to activation of calcium-dependent protein phosphatases (Mulkey and Malenka, 1992; Mulkey et al, 1993, 1994) such as protein phosphatase 1 (PP1) und protein phosphatase 2A (PP2A) and to inactivation of kinases such as CaMKII and PKM (proteolytically activated protein kinase C). Despite the common core mechanisms for induction of LTP and LTD, additional intracellular mechanisms are involved which strongly distinguish these forms of plasticity from each other. For example, in contrast to early speculation that LTP and LTD are the functional inverse of one another, it has become apparent that they are associated with phosphorylation and dephosphorylation, respectively, of distinct GluR1 phosphorylation sites (Lee et al, 2000).

LTD in the CA1 region can be sub-divided in at least four different phases: the initial phase, induction phase, the maintenance phase and the late protein synthesis-dependent phase (Dudek and Bear, 1992; Manahan-Vaughan, 1997; Manahan-Vaughan et al, 2002). Aside from the well documented role of glutamate, CA1 LTD is regulated by multiple neurotransmitter systems which are essential for learning and memory, such as dopamine (Chen et al, 1996; Chen and Tonegawa, 1997), serotonin (Bockaert et al, 1998; Kemp and Manahan-Vaughan, 2004), acetylcholine (Segal and Auerbach, 1997; Radcliffe and Dani, 1998; Fujii and Sumikawa , 2001) and noradrenaline (Katsuki et al, 1997; Izumi and Zorumski, 1999).

4.1 Neurotransmitter Receptors That Are Critically Required For Hippocampal LTD

All significant memory-relevant neurotransmitter receptors come into question as candidates for the mediation of learning-facilitated hippocampal LTD, and much work remains to be done with regard to this issue. A multitude of studies have demonstrated how application of agonists of various neurotransmitter receptors can influence the induction or expression of either LTD or LTP. A significant effect obtained with a neurotransmitter antagonist offers a better argument however, that the neurotransmitter receptor in question is essential for the plasticity form concerned. Bearing this in mind, some particular candidates hold promise as being exclusively important for LTD, all of which couple via a G-protein to adenylyl cyclase: the group II and group III metabotropic glutamate receptors-given their selective role in LTD and not LTP (Manahan-Vaughan, 1997; Klausnitzer et al, 2004) and the serotonin, 5HT4 receptor, which despite a well documented role in hippocampus-based memory (Buhot, 1997; Terry et al, 1998; Bockaert et al, 1998), does not appear to be critically involved in LTP (Kulla and Manahan-Vaughan, 2002; Kemp and Manahan-Vaughan, 2004). Intriguingly, although application of a 5HT4 receptor antagonist has no effect on CA1 LTP, the antagonist facilitates the intermediate phases of LTD, whereas agonist activation (in a concentration which does not affect basal synaptic transmission) inhibits both LTD and learning (Kemp and Manahan-Vaughan, unpublished data).

Neurotransmitter receptor	LTP	LTD
NMDA[1]	-	-
mGluR1/5[2]	-	-
mGluR2/3[3]	o	-
AP4-sensitive mGluR[4]	o	-
Dopamine D1-like[5]	-	-
Dopamine D2-like[6]	-	+
5HT1A[7]	+	n.k.
5HT2[8]	+	n.k
5HT3[9]	+	n.k.
5HT4[10]	o	+
α-adrenoceptors[11]	-	n.k
β-adrenoceptors[12]	o/-	-
Nicotinic receptors[13]	-/+	n.k.
Muscarinic receptors[14]	-	n.k.
Adenosine A1[15]	+	+
Adenosine A2[16]	-	n.k.

Table 2. Summary of the neurotransmitter receptor antagonists which have been used to identify a role for specific receptors in homosynaptic NMDA receptor-dependent LTP and LTD in the hippocampus (-, inhibits; +, enhances; o, no effect; n.k., not known). ([1]Collingridge et al, 1988; Dudek & Bear, 1992; [2,3]Manahan-Vaughan, 1997; [4]Manahan-Vaughan, 2000; Klausnitzer et al, 2004; [5]Chen et al, 1995; Huang & Kandel, 1995; [6]Chen et al 1996; Frey et al, 1989; [7]Sakai & Tanaka, 1993; [8]Wang & Arvanov, 1988; [9]Staubli & Xu, 1995; [10]Kemp & Manahan-Vaughan, 2004; [11]Katsuki et al, 1997; [12]Bramham et al, 1997; Straube & Frey, 2003; [13]Fuji et al, 2000; [14]Kobayashi et al, 1997; [15,16]Fuji et al, 2000; Kemp & Bashir, 1997).

From agonist and antagonist studies it would appear on occasions to be the case that when modulation of a neurotransmitter receptor prevents LTP, it will enhance LTD or vise versa (Chen et al 1996; Frey et al, 1989; Manahan-Vaughan,1998; Kulla et al, 1999) in agreement with earlier theories that LTD comprises the flipside of LTP (Christie et al, 1994). Interestingly, however, it is also often the case that antagonists that block LTP also block LTD, suggesting that even downstream of the NMDA receptor many common signaling pathways are shared (Chen et al, 1995; Huang & Kandel, 1995;

Manahan-Vaughan, 1997). More recent data, indicate however, that induction of LTP or LTD lead to different changes at distinct phosphorylation sites on AMPA receptors (Lee et al, 2000), supporting in turn that LTD is not simply a reversal mechanism for LTP.

5. LTD AND LEARNING

The hippocampus is critically required for the formation of spatial memory. However, a role in non-spatial memory has also been postulated (Eichenbaum, 1997). Strong evidence exists that LTP is involved in spatial learning but it is unclear whether LTP may also contribute to hippocampus-dependent forms of non-spatial learning. Indications that LTP may be correlated with spatial forms of declarative memory come mainly from pharmacological and knockout studies. Thus, transgenic animals that lack LTP often show deficits in spatial memory (Lu et al, 1997; Jia et al, 1998; Rampon et al, 2000), and spatial memory and contextual fear memory are prevented by the NMDA receptor antagonist aminophosphonopentanoic acid (AP5) in concentrations which also block LTP (Kim et al, 1991; Morris et al, 1989). On the other hand, similar concentrations of AP5 also block LTD in vivo (Manahan-Vaughan, 1997). The question thus arises as to whether LTP and LTD encode different types of declarative memory, or whether they function together to create memory engrams.

Although the basic expression mechanisms of LTD and LTP appear similar, their physiological context seems to differ dramatically. Whereas high levels of postsynaptic activation favour induction of LTP, low levels of activation favour induction of LTD (Bear, 1996). Neurotransmitter receptors (e.g. group II mGlu receptors) or signaling mechanisms (e.g. the catalytic and regulatory PKA subunit) which are essential for induction of LTD are not required for LTP (Brandon et al, 1997; Manahan-Vaughan, 1997). Behavioral states which favour LTD (mild stress, exploration) tend to impede LTP (Xu et al, 1997; Manahan-Vaughan and Braunewell, 1999) . Thus, it is not surprising that LTD, in contrast to LTP, is not required for contextual fear memory, and it is also very likely that LTD plays an important role in alternative memory forms. Intriguingly however, a role for LTD in spatial learning is emerging (Manahan-Vaughan and Braunewell, 1999; Nakao et al, 2002; Kemp and Manahan-Vaughan, 2004).

5.1 Hippocampal LTD and Novelty Acquisition

The first indication that LTD may be involved in spatial learning derived from in vivo studies in the rat CA1 region where behavioral learning in a novel environment was shown to result in enhanced LTD induction in vivo (Manahan-Vaughan and Braunewell, 1999). In this study, two rat strains were compared: Wistar rats which readily express LTD following LFS given at 1Hz (900 pulses), and Hooded Lister rats that express short-term depression (STD) following this stimulation protocol. When the animals were exposed to a novel-"stimulus- rich" environment during LFS, LTD was induced in the rat strain (Hooded Lister) which was apparently LTD-resistant, whereas LTD was enhanced in the rat strain (Wistar) which normally expressed LTD following LFS (Figure 2). Thus, behavioral learning, in the form of novelty acquisition, facilitated expression of LTD. This could suggest that hippocampal LTD encodes recognition memory. However, distinct cortical areas are believed to process very distinct aspects of recognition memory

(Brown and Aggleton, 2001), such as the perirhinal cortex for object recognition (Wan et al, 1999) and the piriform cortex for odor recognition (Wilson and Stevenson, 2003). Thus, although it may not process recognition of the distinct features characterising a stimulus, the hippocampus may act as a novelty detector: conducting mismatch predictions by comparing stored information with new incoming cues (Lisman and Otmakhova, 2001). This hippocampal role may be fulfilled by LTD.

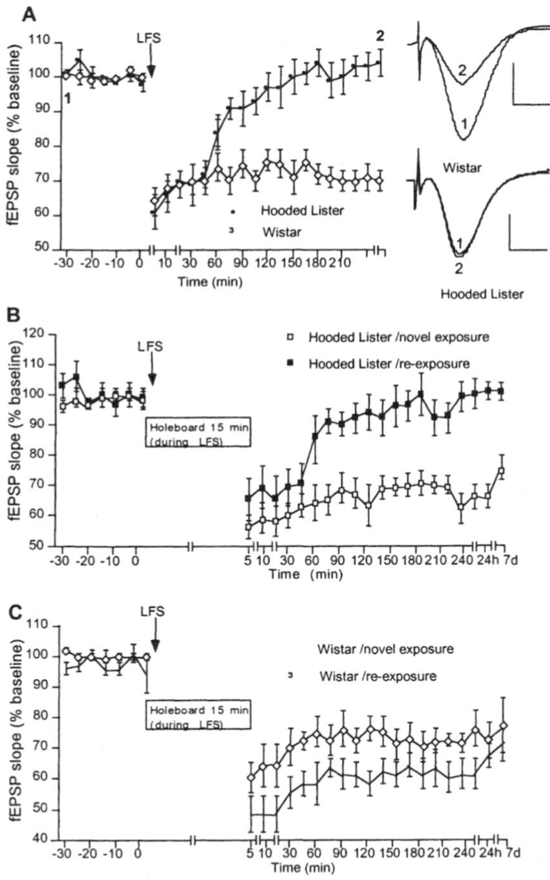

Figure 2. LTD is facilitated by learning in rats.
Wistar rats express robust LTD in the CA1 region following low frequency stimulation at 1Hz (A). Hooded Lister rats express short-term depression following the same stimulation protocol (A). Exposure to a novel environment containing unfamilar objects strongly facilitates LTD in both Hooded Lister (B) and Wistar rat strains (C). Rexposure to the same environment 7 days after the first exposure does not reinforce LTD (Manahan-Vaughan and Braunewell, 1999).

5.2 HIPPOCAMPAL LTD AS A SPATIAL MEMORY MECHANISM

Another possibility is that the hippocampus processes spatial features of object recognition. In a recent study it was demonstrated that although rexposure to a familiar constellation of objects failed to facilitate LTD, reexposure to the objects in a novel constellation repeatedly enhanced LTD (Kemp and Manahan-Vaughan, 2004). Furthermore, whereas an empty novel environment inhibited LTD induction, LTP was enhanced (Figure 3). In addition, LTP was inhibited by a novel environment containing novel objects (Kemp and Manahan-Vaughan, 2004). An association between object recognition, novelty and LTD is also supported by knockout studies of the NR1 subunit of the NMDA receptor. Here, object recognition and LTD expression were impaired, however, exposure to a novel environment facilitated a recovery in the memory deficits (Rampon et al, 2000). Other studies have shown that the degree of expression of hippocampal LTD can predict competence in spatial learning in a water maze: the better the LTD, the better the performance in the water maze (Nakao et al, 2002).

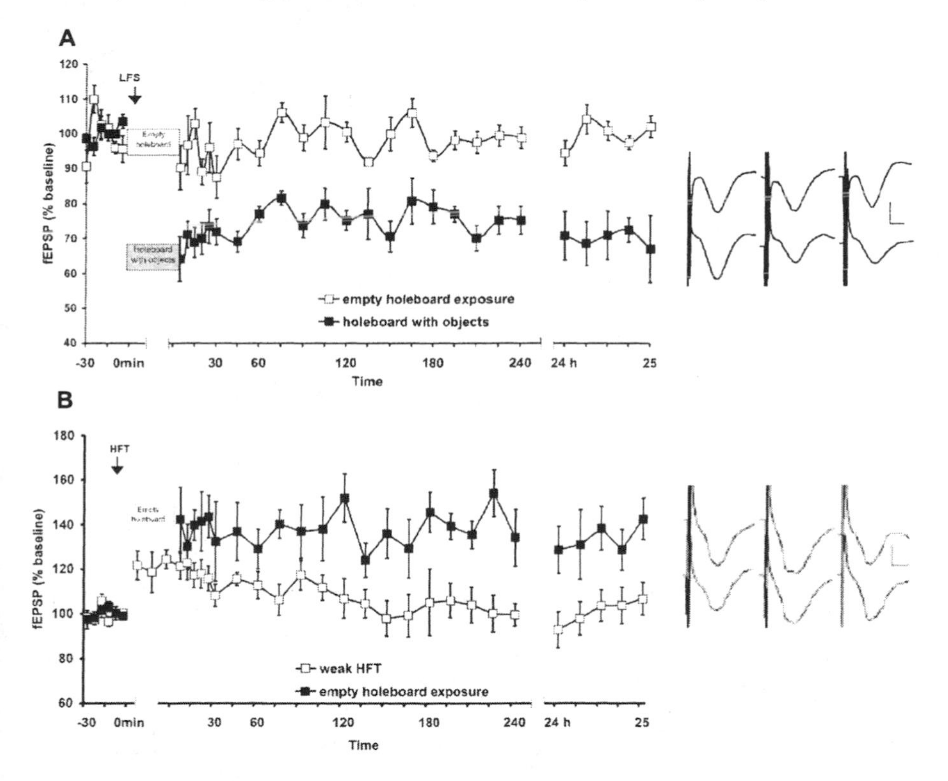

Fig 3. LTD is inhibited and LTP facilitated by exploration of empty space.
LTD is induced by low frequency stimualltion (1 Hz) in Wistar rats. LTP is induced by tetanisation (100 Hz). Exposure of the animals to an empty novel environment inhibits LTD (A) but facilitates LTP (B) (Kemp and Manahan-Vaughan, 2004).

All of these data point towards a very defined role for hippocampal LTD in spatial memory. It would appear that whereas LTP encodes space itself, LTD may encode features of space. This is an extremely intriguing possibility, as it suggests that in keeping with the predictions made in the 1990's (Bear, 1996; Bear and Abraham, 1006), LTP and LTD do indeed work together to form a spatial memory trace.

6. CONCLUSIONS

LTD has not yet been exposed to the intense scrutiny undergone by LTP as a putative learning mechanism. Thus, much work remains to be done. A marked absence in the evaluation of LTD in knockout and transgenic studies leaves many questions unanswered (see Braunewell and Manahan-Vaughan, 2001 for review). However, current data point to the likelihood that LTD is as worthy a candidate mechanisms for declarative memory as its counterpart, LTP. In fact it would appear that LTD and LTP cooperate together to form specific forms of memory traces, with LTD serving to bring context to spatial information encoded by LTP. Further extensive studies will be required to clarify this postulate. Given the diametrically opposed responses of LTP and LTD to behavioral states such as stress, and the apparent requirement of LTP but not LTD for contextual fear memory, it is also possible that LTP and LTD also encode quite distinct forms of learning. The association of LTD and LTP with both quite distinct, and cooperative forms, of information storage would be extremely useful from a physiological and cognitive perspective. Such a functional flexibility would serve to greatly enhance the range of synaptic modification available to a population of synapses, and in turn, substantially increase the scope and capacity of information storage within a neural network.

7. ACKNOWLEDGEMENTS

Work in the author's laboratory is supported by grants from the German research Foundation (Deutsche Forschungsgemeinschaft) (Ma 1843, SFB 515/B8, SFB509/B13, SFB 6006/C4, Graduiertenkolleg 736).

8. REFERENCES

Abraham WC, Christie BR, Logan B, Lawlor P, Dragunow M (1994) Immediate early gene expression associated with the persistence of heterosynaptic long-term depression in the hippocampus. Proc Natl Acad Sci USA. 91:10049-10053.

Abraham WC, Goddard GV (1983) Asymmetric relationships between homosynaptic long-term potentiation and heterosynaptic long-term depression. Nature. 305:717-719.

Alger BE, Megela AL, Teyler TJ (1978) Transient heterosynaptic depression in the hippocampal slice. Brain Res Bull. 3:181-184.

Alger BE, Teyler TJ (1976) Long-term and short-term plasticity in the CA1, CA3, and dentate regions of the rat hippocampal slice. Brain Res. 110:463-480.

Andrews JS (1996) Possible confounding influence of strain, age and gender on cognitive performance in rats. Brain. Res. Cogn. Brain. Res. 3: 251-267.

Bank B, LoTurco JJ, Alkon DL (1989) Learning-induced activation of protein kinase C: A molecular memory trace.Mol Neurobiol. 3:55-70.

Bear MF (1996) A synaptic basis for memory storage in the cerebral cortex. Proc. Natl. Acad. Sci. USA 93: 13453-13459.

Bear MF, Abraham WC (1996) Long-term depression in hippocampus. Annu Rev. Neurosci. 19:437-462.

Bliss TVP, Lømo T (1973) Long-lasting potentiation of synaptic transmission in the dentate area of anaesthetized rabbit following stimulation of the perforant path. J. Physiol. 232: 331-356.

Bockaert J, Claeysen S, Sebben M, Dumuis A (1998) 5-HT4 receptors: gene, transduction and effects on olfactory memory. Ann. N. Y. Acad. Sci. 861:1-15.

Bortolotto ZA, Bashir ZI, Davies CH, Taira T, Kaila K, Collingridge GL (1995) Studies on the role of metabotropic glutamate receptors in long-term potentiation: some methodological considerations. J Neurosci Methods. 59:19-24.

Bozon B, Kelly A, Josselyn SA, Silva AJ, Davis S, Laroche S (2003) MAPK, CREB and zif268 are all required for the consolidation of recognition memory.Philos Trans R Soc Lond B Biol Sci. 358:805-814.

Bramham C.R, Bacher-Svendsen K, Sarvey JM (1997) LTP in the lateral perforant path is beta-adrenergic receptor-dependent. Neuroreport. 8:719-724.

Brandon EP, Idzerda RL, McKnight GS (1997) PKA isoforms, neural pathways, and behaviour: making the connection. Curr. Opin. Neurobiol. 3:397-403.

Brandon EP, Zhuo M, Huang YY, Qi M, Gerhold KA, Burton KA, Kandel ER, McKnight GS, Idzerda RL (1995) Hippocampal long-term depression and depotentiation are defective in mice carrying a targeted disruption of the gene encoding the RI beta subunit of cAMP-dependent protein kinase. Proc Natl Acad Sci USA 92:8851-8855.

Braunewell KH, and Manahan-Vaughan D (2001) Long-term depression: a cellular basis for learning? Rev. Neurosci. 12:121-40.

Brown MW, Aggleton JP (2001) Recognition memory: what are the roles of the perirhinal cortex and hippocampus? Nat. Rev. Neurosci. 2:51-61.

Buhot MC (1997) Serotonin receptors in cognitive behaviors. Curr .Opin. Neurobiol. 7:243-54.

Chen Z, Ito K, Fujii S, Miura M, Furuse H, Sasaki H, Kaneko K, Kato H, and Miyakawa H (1996) Roles of dopamine receptors in long-term depression: enhancement via D1 receptors and inhibition via D2 receptors. Receptors Channels. 4:1-8.

Chen C, Tonegawa S (1997) Molecular genetic analysis of synaptic plasticity, activity-dependent neural development, learning, and memory in the mammalian brain. Annu. Rev. Neurosci. 20:157-184.

Chen Z, Fujii S, Ito K, Kato H, Kaneko K, Miyakawa H (1995) Activation of dopamine D1 receptors enhances long-term depression of synaptic transmission induced by low frequency stimulation in rat hippocampal CA1 neurons. Neurosci Lett. 188:195-198.

Chen Z, Ito K, Fujii S, Miura M, Furuse H, Sasaki H, Kaneko K, Kato H, Miyakawa H (1996) Roles of dopamine receptors in long-term depression: enhancement via D1 receptors and inhibition via D2 receptors. Receptors Channels. 4:1-8.

Christie BR, Abraham WC (1992) Priming of associative long-term depression in the dentate gyrus by theta frequency synaptic activity. Neuron 9:79-84.

Christie BR, Kerr DS, Abraham WC (1994) Flip side of synaptic plasticity: long-term depression mechanisms in the hippocampus. Hippocampus. 4:127-135.

Colbert CM, Burger BS, Levy WB (1992) Longevity of synaptic depression in the hippocampal dentate gyrus. Brain Res. 571:159-161.

Collingridge GL, Herron CE, Lester RA (1988) Frequency-dependent N-methyl-D-aspartate receptor-mediated synaptic transmission in rat hippocampus. J Physiol. 399:301-312.

Conquet F, Bashir ZI, Davies CH, Daniel H, Ferraguti F, Bordi F, Franz-Bacon K, Reggiani A, Matarese V, Conde F, et al. (1994) Motor deficit and impairment of synaptic plasticity in mice lacking mGluR1. Nature 372:237-243.

Cummings JA, Mulkey RM, Nicoll RA, Malenka RC (1996) Ca^{2+} signaling requirements for long-term depression in the hippocampus. Neuron. 16:825-833.

Derrick BE, Martinez JL Jr (1996) Associative, bidirectional modifications at the hippocampal mossy fibre-CA3 synapse. Nature 381:429-434.

Doyere V, Errington ML, Laroche S, Bliss TV, (1996) Low-frequency trains of paired stimuli induce long-term depression in area CA1 but not in dentate gyrus of the intact rat. Hippocampus 6:52-57.

Doyle CA, Cullen WK, Rowan MJ, Anwyl RA (1997) Low-frequency stimulation induces homosynaptic depotentiation but not long-term depression of synaptic transmission in the adult anaesthetized and awake rat hippocampus in vivo. Neuroscience 77: 75-85.

Dragunow M (1996) A role for immediate-early transcription factors in learning and memory. Behav Genet. 26:293-299.

Dunwiddie T, Lynch G (1987) Long-term potentiation and depression of synaptic responses in the rat hippocampus: localization and frequency dependency. J Physiol. 276:353-367.

Dudek SM, Bear MF (1992) Homosynaptic long-term depression in area CA1 of hippocampus and the effects of NMDA receptor blockade. Proc. Natl. Acad. Sci. USA. 89: 4363-4367.

Eichenbaum H (1997) Declarative memory:insights from cognitive neurobiology.Annu.Rev.Psychol. 48:547-72.

Errington ML, Bliss TVP, Richter-Levin G, Yenk K, Doyere V, Laroche S (1991) Stimulation at 1-5 Hz does not produce long-term depression or depotentiation in the hippocampus of the adult rat in vivo. J. Neurophysiol. 74: 1793-1799.

Fitzjohn SM, Bortolotto ZA, Palmer MJ, Doherty AJ, Ornstein PL, Schoepp DD, Kingston AE, Lodge D, Collingridge GL (1998) The potent mGlu receptor antagonist LY341495 identifies roles for both cloned and novel mGlu receptors in hippocampal synaptic plasticity. Neuropharmacology. 37:1445-1458.

Fleischmann A, Hvalby O, Jensen V, Strekalova T, Zacher C, Layer LE, Kvello A, Reschke M, Spanagel R, Sprengel R, Wagner EF, Gass P (2003) Impaired long-term memory and NR2A-type NMDA receptor-dependent synaptic plasticity in mice lacking c-Fos in the CNS. J Neurosci. 23:9116-9122.

Flood JF, Landry DW, Bennett EL, Jarvik ME (1981) Long-term memory: disruption by inhibitors of protein synthesis and cytoplasmic flow. Pharmacol .Biochem. Behav. 15:289-296.

Frey U, Hartmann S, Matthies H (1989) Domperidone, an inhibitor of the D2-receptor, blocks a late phase of an electrically induced long-term potentiation in the CA1-region in rats. Biomed Biochim Acta.48:473-476.

Frey U, Huang YY, Kandel ER (1993) Effects of cAMP simulate a late stage of LTP in hippocampal CA1 neurons. Science. 260:1661-1664.

Frey U, Krug M, Reymann KG, Matthies H (1988) Related Articles, Links Abstract Anisomycin, an inhibitor of protein synthesis, blocks late phases of LTP phenomena in the hippocampal CA1 region in vitro. Brain Res. 452:57-65.

Fride E, Ben-Or S, Allweis C (1989) Mitochondrial protein synthesis may be involved in long-term memory formation. Pharmacol. Biochem. Behav. 32:873-878.

Fujii S, Kato H, Ito K, Itoh S, Yamazaki Y, Sasaki H, Kuroda Y (2000) Effects of A1 and A2 adenosine receptor antagonists on the induction and reversal of long-term potentiation in guinea pig hippocampal slices of CA1 neurons. Cell Mol Neurobiol. 20:331-50.

Fujii S, Ji Z, Sumikawa K (2000) Inactivation of alpha7 ACh receptors and activation of non-alpha7 ACh receptors both contribute to long term potentiation induction in the hippocampal CA1 region. Neurosci Lett. 286:134-138.

Fujii S, Sumikawa K (2001) Nicotine accelerates reversal of long-term potentiation and enhances long-term depression in the rat hippocampal CA1 region. Brain Res. 2001; 894:340-346.

Heynen AJ, Abraham WC, Bear MF (1996) Bidirectional modification of CA1 synapses in the adult hippocampus in vivo. Nature 381: 163-166.

Huang YY, Kandel ER (1995) D1/D5 receptor agonists induce a protein synthesis-dependent late potentiation in the CA1 region of the hippocampus. Proc Natl Acad Sci U S A. 92:2446-2450.

Izumi Y, Zorumski CF (1999) Norepinephrine promotes long-term potentiation in the adult rat hippocampus in vitro.Synapse. 31:196-202.

Jia Z, Lu Y, Henderson J, Taverna F, Romano C, Abramow-Newerly W, Wojtowicz JM, Roder J (1998) Selective abolition of the NMDA component of long-term potentiation in mice lacking mGluR5. Learn. Mem. 5:331-343.

Katsuki H, Izumi Y, Zorumski CF (1997) Noradrenergic regulation of synaptic plasticity in the hippocampal CA1 region. J. Neurophysiol. 77:3013-3020.

Kemp N, Bashir ZI (1997) A role for adenosine in the regulation of long-term depression in the adult rat hippocampus in vitro. Neurosci Lett. 225:189-192.

Kemp A, Manahan-Vaughan D (2004) Hippocampal long-term depression and long-term potentiation encode different aspects of novelty acquisition. Proc. Natl. Acad. Sci. U S A. 101:8192-8197.

Klausnitzer J, Kulla A, Manahan-Vaughan D (2004) Role of the group III metabotropic glutamate receptor in LTP, depotentiation and LTD in dentate gyrus of freely moving rats. Neuropharmacology. 46:160-170.

Kim JJ, DeCola JP, Landeira-Fernandez J, Fanselow MS (1991) N-methyl-D-aspartate receptor antagonist APV blocks acquisition but not expression of fear conditioning. Behav. Neurosci. 105:126-133.

Kobayashi K, Manabe T, Takahashi T (1996) Presynaptic long-term depression at the hippocampal mossy fiber-CA3 synapse. Science. 273:648-650.

Kobayashi M, Ohno M, Shibata S, Yamamoto T, Watanabe S (1997) Concurrent blockade of beta-adrenergic and muscarinic receptors suppresses synergistically long-term potentiation of population spikes in the rat hippocampal CA1 region. Brain Res. 777:242-246.

Kulla A, Manahan-Vaughan D (2002) Modulation by serotonin 5-HT(4) receptors of long-term potentiation and depotentiation in the dentate gyrus of freely moving rats. Cereb. Cortex. 12:150-62.

Kulla A, Reymann KG, Manahan-Vaughan D (1999) Time-dependent induction of depotentiation in the dentate gyrus of freely moving rats: involvement of group 2 metabotropic glutamate receptors. Eur J Neurosci. 11:3864-3872.

Lee HK, Barbarosie M, Kameyama K, Bear MF, Huganir RL (2000) Regulation of distinct AMPA receptor phosphorylation sites during bidirectional synaptic plasticity. Nature. 2000; 405:955-959.

Lisman JE, Otmakhova NA (2001) Storage, recall, and novelty detection of sequences by the hippocampus: elaborating on the SOCRATIC model to account for normal and aberrant effects of dopamine. Hippocampus. 11:551-568.

Lu YM, Jia Z, Janus C, Henderson JT, Gerlai R, Wojtowicz JM, Roder JC (1997) Mice lacking metabotropic glutamate receptor 5 show impaired learning and reduced CA1 long-term potentiation (LTP) but normal CA3 LTP. J. Neurosci. 17: 5196-5205.

Lynch GS, Dunwiddie T, and Gribkoff V (1977) Heterosynaptic depression: a postsynaptic correltate of long-term potentiation. Nature 266: 737-739.

Manahan-Vaughan D (1997) Group 1 and 2 metabotropic glutamate receptors play differential roles in hippocampal long-term depression and long-term potentiation in freely moving rats. J. Neurosci. 17: 3303-3311.

Manahan-Vaughan D (1998) Priming of group 2 metabotropic glutamate receptors facilitates induction of long-term depression in the dentate gyrus of freely moving rats. Neuropharmacology. 37:1459-1464.

Manahan-Vaughan D (2000) Long-term depression in freely moving rats is dependent upon strain variation, induction protocol and behavioral state. Cereb. Cortex 10:482-487.

Manahan-Vaughan D, Braunewell KH (1999) Novelty acquisition is associated with induction of hippocampal long-term depression. Proc. Natl. Acad. Sci. USA 96:8739-8744.

Manahan-Vaughan D, Kulla A, Frey U (2000) Requirement of translation but not transcription for the maintenance of long-term depression in the CA1 region of freely moving rats. J. Neurosci. 20:8572-8576.

Mansuy IM, Mayford M, Jacob B, Kandel ER, Bach ME (1998) Restricted and regulated overexpression reveals calcineurin as a key component in the transition from short-term to long-term memory. Cell 92:39-49.

Mizumori SJ, Channon V, Rosenzweig MR, Bennett EL (1987) Short- and long-term components of working memory in the rat. Behav. Neurosci. 101:782-789.

Mizumori SJ, Rosenzweig MR, Bennett EL (1985) Long-term working memory in the rat: effects of hippocampally applied anisomycin. Behav. Neurosci. 99:220-232.

Mizumori SJ, Sakai DH, Rosenzweig MR, Bennett EL, Wittreich P (1987) Investigations into the neuropharmacological basis of temporal stages of memory formation in mice trained in an active avoidance task. Behav. Brain Res. 23:239-250.

Morris RG, Anderson E, Lynch GS, Baudry M (1989) Synaptic plasticity and learning: selective impairment of learning rats and blockade of long-term potentiation in vivo by the N-methyl-D-aspartate receptor antagonist AP5. J. Neurosci. 9:3040-3057.

Mulkey RM, Malenka RC (1992) Mechanisms underlying induction of homosynaptic long-term depression in area CA1 of the hippocampus. Neuron 9: 967-975.

Mulkey RM, Endo S, Shenolikar S, Malenka RC (1994) Involvement of a calcineurin/inhibitor-1 phosphatase cascade in hippocampal long-term depression. Nature 369:486-488.

Mulkey RM, Herron CE, Malenka RC (1993) An essential role for protein phosphatases in hippocampal long-term depression. Science 261:1051-1055

Nakao K, Ikegaya Y, Yamada MK, Nishiyama N, Matsuki N (2002) Hippocampal long-term depression as an index of spatial working memory. Eur. J. Neurosci. 16:970-974.

Nguyen PV, Woo NH (2003) Regulation of hippocampal synaptic plasticity by cyclic AMP-dependent protein kinases. Prog Neurobiol. 71:401-437.

Nicoll RA, Oliet SH, Malenka RC (1998) NMDA receptor-dependent and metabotropic glutamate receptor-dependent forms of long-term depression coexist in CA1hippocampal pyramidal cells. Neurobiol. Learn. Mem. 70:62-72.

Oliet SH, Malenka RC, Nicoll RA (1997) Two distinct forms of long-term depression coexist in CA1 hippocampal pyramidal cells. Neuron 18:969-982

Otani S, Connor JA (1998) Requirement of rapid Ca^{2+} entry and synaptic activation of metabotropic glutamate receptors for the induction of long-term depression in adult rat hippocampus. J. Physiol. 511:761-770.
Pockett S, Brookes NH, Bindman LJ (1990) Long-term depression at synapses in slices of rat hippocampus can be induced by bursts of postsynaptic activity. Exp. Brain Res. 80:196-200.
Radcliffe KA, Dani JA (1998) Nicotinic stimulation produces multiple forms of increased glutamatergic synaptic transmission. J. Neurosci. 18:7075-7083.
Rampon C, Tang YP, Goodhouse J, Shimizu E, Kyin M, Tsien JZ (2000) Enrichment induces structural changes and recovery from nonspatial memory deficits in CA1 NMDAR1-knockout mice. Nat. Neurosci. 3:238-244.
Sakai N, Tanaka C (1993) Inhibitory modulation of long-term potentiation via the 5-HT1A receptor in slices of the rat hippocampal dentate gyrus. Brain Res. 613:326-30.
Segal M, Auerbach, JM (1997) Muscarinic receptors involved in hippocampal plasticity. Life Sci. 60:1085-1091.
Silva AJ, Paylor R, Wehner JM, Tonegawa S (1992) Impaired spatial learning in alpha-calcium-calmodulin kinase II mutant mice. Science. 257:206-211.
Soderling TR (1993) Calcium/calmodulin-dependent protein kinase II: role in learning and memory. Mol Cell Biochem. 127-128:93-101.
Stanton PK (1996) LTD, LTP, and the sliding threshold for long-term synaptic plasticity.Hippocampus 6:35-42.
Stanton PK, Sarvey JM (1984) Blockade of long-term potentiation in rat hippocampal CA1 region by inhibitors of protein synthesis. J. Neurosci. 4:3080-3088.
Stanton PK, Sejnowski TJ (1989) Associative long-term depression in the hippocampus induced by hebbian covariance. Nature. 339:215-218.
Staubli U, Scafidi J (1997) Studies on long-term depression in area CA1 of the anesthetized and freely moving rat. J. Neurosci. 17:4820-4928.
Staubli U, Xu FB (1995) Effects of 5-HT3 receptor antagonism on hippocampal theta rhythm, memory, and LTP induction in the freely moving rat. J Neurosci. 15:445-52.
Straube T, Frey JU (2003) Involvement of beta-adrenergic receptors in protein synthesis-dependent late long-term potentiation (LTP) in the dentate gyrus of freely moving rats: the critical role of the LTP induction strength. J. Neurosci. 119:473-479.
Sweatt JD (1999) Toward a molecular explanation for long-term potentiation. Learn. Mem. 6:399-416.
Terry AV Jr, Buccafusco JJ, Jackson WJ, Prendergast MA, Fontana DJ, Wong EH, Bonhaus DW, Weller P, Eglen RM (1998) Enhanced delayed matching performance in younger and older macaques administered the 5-HT4 receptor agonist, RS 17017. Psychopharmacology (Berlin) 135:407-15
Thiels E, Barrionuevo G, Berger TW (1994) Excitatory stimulation during postsynaptic inhibition induces long-term depression in hippocampus in vivo. J. Neurophysiol. 72:3009-3016.
Thiels E, Xie X, Zeckel MF, Barrionuevo G, Berger T (1996) NMDA receptor-dependent LTD in different subfields of hippocampus in vivo and in vitro. Hippocampus 6: 43-51.
Thiels E, Norman ED, Barrionuevo G, Klann E (1998) Transient and persistent increases in protein phosphatase activity during long-term depression in the adult hippocampus in vivo. Neuroscience 86:1023-1029.
Trommer BL, Liu YB, Pasternak JF (1996) Long-term depression at the medial perforant path-granule cell synapse in developing rat dentate gyrus. Brain Res. Dev. Brain Res. 96:97-108.
Tsumoto T (1993) Long-term depression in cerebral cortex: a possible substrate of "forgetting" that should not be forgotten. Neurosci. Res. 16: 263-270.
Tulving E, Markowitsch HJ (1998) Episodic and declarative memory: role of the hippocampus. Hippocampus. 8:198-204.
Wagner JJ, Alger BE (1996) Homosynaptic LTD and depotentiation: do they differ in name only? Hippocampus 6: 24-29.
Wan H, Aggleton JP, Brown MW (1999) Different contributions of the hippocampus and perirhinal cortex to recognition memory. J. Neurosci. 19:1142-1148.
Wang RY, Arvanov VL (1998) M100907, a highly selective 5-HT2A receptor antagonist and a potential atypical antipsychotic drug, facilitates induction of long-term potentiation in area CA1 of the rat hippocampal slice. Brain Res. 779:309-313.
Wang Y, Rowan MJ, Anwyl R (1997) Induction of LTD in the dentate gyrus in vitro is NMDA receptor independent, but dependent on Ca^{2+} influx via low-voltage-activated Ca^{2+} channels and release of Ca^{2+} from intracellular stores. J. Neurophysiol. 77: 812-825.
Wang JH, Stelzer A (1994) Inhibition of phosphatase 2B prevents expression of hippocampal long-term potentiation. Neuroreport. 5:2377-2380.

Wigstrom H, Gustafsson B (1986) Postsynaptic control of hippocampal long-term potentiation. J Physiol (Paris). 81:228-236.

Wilson DA, Stevenson RJ (2003) The fundamental role of memory in olfactory perception. Trends Neurosci. 26:243-247.

Xu L, Anwyl R, Rowan MJ (1997) Behavioural stress facilitates the induction of long-term depression in the hippocampus. Nature 387:497-500.

Zhuo M, Zhang W, Son H, Mansuy I, Sobel RA, Seidman J, Kandel ER (1999) A selective role of calcineurin aalpha in synaptic depotentiation in hippocampus. Proc. Natl. Acad. Sci . USA 96:4650-4655.

NMDA RECEPTORS: FROM PROTEIN-PROTEIN INTERACTIONS TO TRANSACTIVATION

John F. MacDonald*, Suhas A. Kotecha, Wei-Yang Lu and Michael F. Jackson

1. TRANSMITTERS AND CO-TRANSMITTERS

In the nervous systems of many invertebrates there is little or no anatomical redundancy in the number of neurons participating in an individual functional pathway. For example, a single neuron may serve to form an entire functional pathway. This contrasts with mammalian central nervous systems where a single pathway may be composed of numerous apparently redundant neurons, their axon and synapses. Likewise the effector is often duplicated in multiple postsynaptic neurons or muscle cells. The synaptic transmitter released at these synapses is also redundant in that the majority of the cells within a given pathway will release the same transmitter. Integration of signals between functional pathways or between neurons in the same functional pathway can be achieved by a variety of mechanisms but the interface provided by chemical synapses and released transmitter is undoubtedly the major one. This review will consider a range of potential mechanisms whereby excitatory synaptic receptors become targets of transmitters, growth factors and modulators.

Each neuron, in line with Dale's Principle, should release the same transmitter at each of its synaptic connections; albeit, nothing is implicit in this statement that more than a single transmitter is released from a neuron and it is recognized that more than one transmitter or "co-transmitter" can be released from an individual neuron (Salter and De Koninck, 1999). The semantics of whether a transmitter is a co-transmitter or a modulator is somewhat arbitrary; although, for the purposes of this discussion a co-transmitter will be considered to be one released from the same neuron as the primary transmitter whereas a modulator can be considered a transmitter which modulates the activity of receptors that recognize a second independent transmitter. The release of more than one transmitter from a single synaptic bouton implies that receptors for each of the transmitters will be found either in the immediate proximity of pre- or postsynaptic

*J.F. MacDonald, S.A. Kotecha Wei-Yang Lu and M.F. Jackson, Departments of Physiology and Anesthesia, University of Toronto and Sunnybrook Hospital, 1 King's College Circle, Toronto, Ontario, M5S 1A8, Canada. E-mail: j.macdonald@utoronto.ca

structures or at least within perisynaptic or periannular locations. Depending on the ultimate mobility of the released transmitter it is possible that a transmitter may "spill over" into extrasynaptic regions or even to nearby synapses (Kullmann et al., 1999; Rusakov and Kullmann, 1998; Clark and Cull-Candy, 2002) and thus the transmitter becomes a "modulator" under the definition given previously.

2. GLUTAMATE RECEPTORS AND EXCITATORY TRANSMISSION

The vast majority of excitatory synapses within the mammalian central nervous system utilize L-glutamate as the transmitter regardless of whether or not the neurons participate within the same functional pathway. This ubiquity of glutamatergic postsynaptic excitatory synaptic potentials (EPSPs) implies that glutamate receptors are likely to be important targets of various co-transmitters and modulators. Furthermore, regulation of postsynaptic glutamate receptor function and expression is the first step in the induction of various forms of synaptic plasticity in cortical pathways. Excitatory postsynaptic currents (EPSCs), the voltage-clamped events underlying epsps (**Fig. 1**), have a large and rapid α-amino-3-hydroxy-5-methyl-4-isoxazole propionate receptor (AMPAR) component ($EPSC_{AMPA}$) and a much smaller and slower N-methyl-D-asparate receptor (NMDAR) component ($EPSC_{NMDA}$) (Clements et al., 1992; Gasic and Hollmann, 1992; Seeburg, 1993) (**Fig. 2**). Activation of NMDARs uniquely require the binding of the co-agonist, glycine. Glutamate receptors are anchored at dendritic spine synapses by scaffolding proteins of the postsynaptic density (PSD) and this complex consists of scaffolding, adaptor and signal transduction proteins (Sheng, 2001; Scannevin and Huganir, 2000). NMDARs in the PSD form a large multi-protein complex together with a variety of signal proteins (Husi et al., 2000).

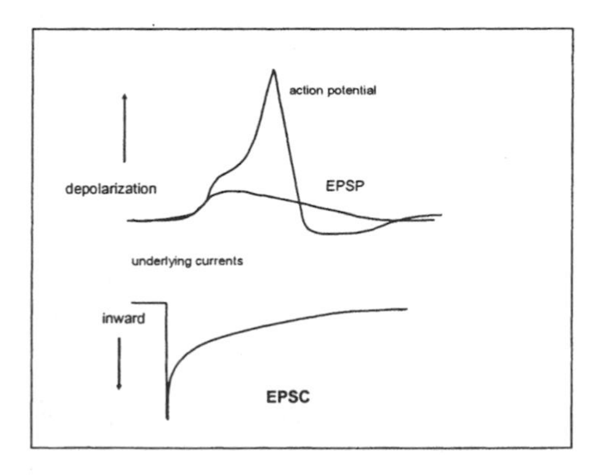

Figure 1. Most excitatory synaptic potentials (EPSPs) are generated by postsynaptic glutamate receptors. Summation of epsps can reach threshold and evoke and action potential. Using voltage-clamp techniques the currents flowing through glutamate receptors can be quantified as inward currents near resting membrane potential.

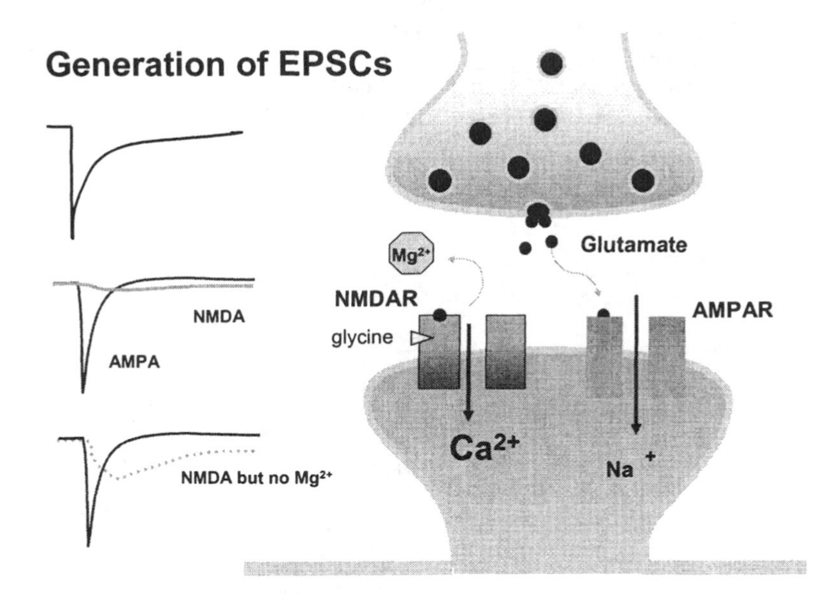

Figure 2. AMPARs and NMDARs are structurally related tetrameric non-selective cation channels (permeable to Na^+ and K^+) but near resting membrane potential about 10% of NMDAR-mediated current is carried by the inward movement of Ca^{2+}. Due to the voltage-dependent block of NMDARs by extracellular Mg^{2+} the contribution by NMDARs is greatly reduced and the majority of the excitatory postsynaptic current results from activation of AMPARs. Depolarization or removal of Mg^{2+} relieves this block and enhances the currents carried by NMDARs. Gating to the open state of the channels is contingent upon the binding of the co-agonist glycine.

3. RECEPTORS AND SYNAPTIC PLASTICITY IN THE HIPPOCAMPUS

Long-term potentiation (LTP) of excitatory synapses of the CA1 region of the hippocampus serves as the leading cellular and molecular model of learning and memory (Malenka and Nicoll, 1999; Lynch, 2004). At CA1 synapses (i.e. Schaffer collateral-CA1 (Gustafsson and Wigstrom, 1988; Collingridge and Bliss, 1995; Nicoll and Malenka, 1999) the induction of LTP, or its inhibitory counterpart long-term depression (LTD), requires entry of Ca^{2+} via NMDARs (Collingridge and Bliss, 1995) and activation of Ca^{2+}-calmodulin dependent kinase II (CamKII) (**Fig. 3**). Changes in post-translational properties (i.e. phosphorylation) (Soderling and Derkach, 2000) and/or altered numbers of AMPARs likely contribute to the induction of LTP and LTD (Luscher et al., 2000). LTP is maintained by stimulation of the mitogen-activated protein kinase (MAPK) cascade, altered gene expression (Sweatt, 1999; Albright et al., 2000; Thomas and Huganir, 2004) and changes in synaptic morphology (Hering and Sheng, 2001). CamKII and protein kinase C (PKC) are Ca^{2+}-dependent enzymes which regulate glutamate receptors (Wang et al., 1994; Carvalho et al., 1999; Swope et al., 1999; Soderling and Derkach, 2000) and play central roles in the induction of LTP (Lynch, 2004). They also participate in various pathways that target the MAPK pathway with respect to cellular growth (Gutkind, 1998) and LTP (Thomas and Huganir, 2004).

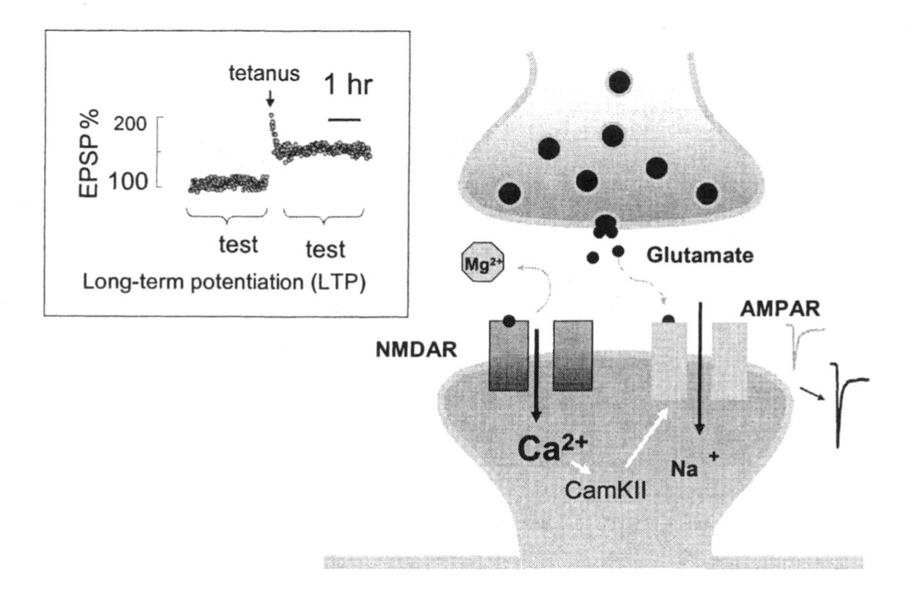

Figure 3. The strong depolarization induced by the tetanic stimulation leads of a relief of the Mg^{2+} block, an influx of Ca^{2+} into the spine and the activation of CamKII, leading to enhanced AMPAR phosphorylation and increased delivery of AMPARs to the synapse. CamKII may be upstream of MAPK with respect to LTP (Thomas and Huganir, 2004). Following the induction of LTP the EPSCs mediated by AMPA are potentiated. Test refers to a low frequency stimulation that evokes a stable amplitude EPSP prior to the tetanus. After the tetanus this same stimulation evokes a larger EPSP (LTP).

NMDARs are Ca^{2+} permeable non-selective cation channels and their Ca^{2+} signal induces LTP at CA1 synapses (Soderling and Derkach, 2000) through stimulation of various enzyme cascades including CamKII, PKC and the Ca^{2+}-dependent tyrosine kinase, CAKβ also called Pyk2 (Soderling and Derkach, 2000; Salter and Kalia, 2004). Even a single EPSC increases the Ca^{2+} in individual postsynaptic spines of CA1 neurons (Kovalchuk et al., 2000) although early in development many synapses appear to lack a functional AMPA component and serve as "silent synapses" (Voronin and Cherubini, 2003; Kullmann, 2003; Isaac et al., 1999). Conversion of silent synapses to active synapses is associated with NMDAR stimulation and may provide a mechanism whereby the degree of synaptic activity determines the strength and endurance of particular inputs to the neuron (van Zundert et al., 2004; Zhou and Poo, 2004). The localized release of glutamate at single small spines acting via NMDARs is sufficient to induce an increase in spine volume that parallels LTP. This structural plasticity is restricted to relatively small spines and requires dynamic actin filament interactions (Matsuzaki et al., 2004).

4. NMDARS AND INTERACTIONS VIA SIGNAL TRANSDUCTION CASCADES

In most hippocampal pyramidal neurons combinations of dimeric NR1-NR2A or NR1-NR2B or trimeric combinations of NR1-NR2A-NR2B subunits form the NMDAR (Dingledine et al., 1999; Sheng and Kim, 2002). Recent evidence indicates that NMDARs containing the NR2B subunit may interact preferentially with SAP103 to form predominantly extrasynaptic receptors whilst a PSD-95 protein interaction with NR2A containing receptors may target this combination to adult synapses (van Zundert et al., 2004). Early in neonatal life there is a predominance of NR2B containing NMDARs in both extrasynaptic and synaptic compartments. However, this gradually changes with the enhanced expression of NR2A containing receptors targeted predominantly to synaptic sites. In adult hippocampal neurons NMDARs are likely compartmentalized with NR2A receptors forming the bulk of synaptic receptors and NR2B containing receptors predominating in non-synaptic or extrasynaptic regions. Acute movements of NMDARs between the synaptic and extrasynaptic compartments (Tovar and Westbrook, 1999) is likely related to their degree of mobility in the membrane which can be influenced by receptor activity and kinases such as PKC (Groc et al., 2004) as well as the potential specific trafficking of NMDARs through endocytosis and vesicular-mediated insertion (Nong et al., 2004). The definition of extrasynaptic receptors is vague as it is likely that synaptically released glutamate may reach at least a subset of these receptors as a consequence of "spill over" (van Zundert et al., 2004).

Raymond et al. (Li et al., 2002) have shown that functional differences between these populations of receptors can be detected in cultured neurons. In particular, the extrasynaptic NR2B population demonstrates a calcium-independent rundown which is prevented by enhanced tyrosine phosphorylation (Li et al., 2002). This rundown or calcium-independent desensitization likely result from the internalization of NMDARs which is also tyrosine kinase dependent (Vissel et al., 2001). It may also be related to the internalization primed by the binding of the co-transmitter glycine to the NMDAR (Nong et al., 2003). Trafficking of NR2A and NR2B subunits also differs in these neurons and presumably in synaptic versus extrasynaptic compartments (Hawkins et al., 2004; Prybylowski and Wenthold, 2004; Wenthold et al., 2003; Nong et al., 2004).

The physiological role of NR2B versus NR2A subunits is clearly more complex than a simple compartmentalization of receptors implies because activation of NR2A containing receptors is required for the induction of LTP whilst NR2B receptors and not NR2A receptors are responsible for LTD in the CA1 region (Liu et al., 2004). This is, however, consistent with observations that in a culture model activation of synaptic NMDARs is associated with LTP and LTD results from activation of extrasynaptic receptors by exogenously applied NMDA (Lu et al., 2001).

Metabotropic glutamate receptors (mGluRs) are also found at CA1 synapses (**Fig. 4**). These receptors potentially couple to Gq and Gi and perhaps Gs families of G-proteins (Conn and Pin, 1997). There are at least 8 mGluRs (plus various mRNA splice variants) and they have been classified into three groups (I,II,III) based on sequence and functional homologies. G-proteins are activated through interactions with heptahelical proteins or G-protein coupled receptors (GPCRs). Four major gene families (Downes and Gautam, 1999) constitute this family of proteins: 1) $G\alpha_s$, stimulate adenylyl cyclase 2) $G\alpha_{i/o/t}$, inhibit adenylyl cyclase 3) $G\alpha q/_{11}$, activate phospholipase C, 1,2,-diacylglycerol (DAG)-dependent PKC and inositol 1,4,5-trisphosphate (IP3)-dependent release of Ca^{2+} 4) $G\alpha_{12}$, regulate low-molecular weight G-proteins of the Rho-family. Some GPCRs (i.e.$G\alpha q/_{11}$) function as growth-inducing receptors (Gudermann et al., 2000). GPCRs regulate gene expression through stimulation of various signal pathways including the MAPK cascade (Gutkind, 1998; Hall et al., 1999; Della et al., 1999; Luttrell et al., 1999). $G\alpha s$ and $G\alpha i$ can also bind to and directly activate Src, the proto-oncogene product (Ma and Huang, 2002; Ma et al., 2000). *"It now seems likely that each heptahelical receptor may activate its own relatively specific set of intracellular signaling pathways, including both G-protein-dependent and G-protein-independent mechanisms" (Hall et al., 1999).* For example the metabotropic glutamate receptor, mGluR1a, acts via a G-protein ***independent*** mechanism to activate Src (Heuss et al., 1999). GPCRs also regulate gene expression through the activation of non-receptor tyrosine kinases (i.e. Src) and MAPK cascades (Gutkind, 1998; Hall et al., 1999; Luttrell et al., 1999).

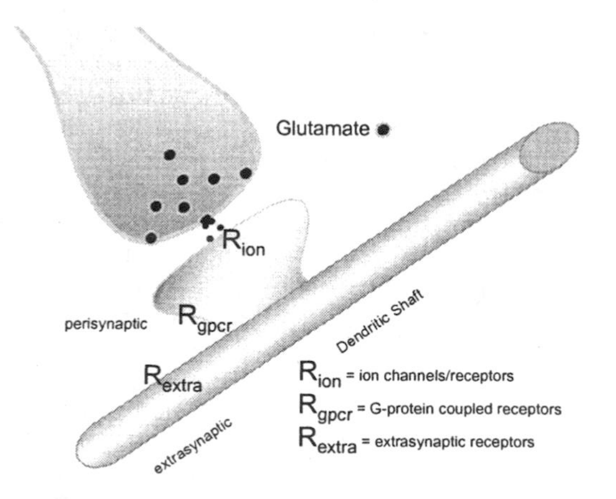

Figure 4. Postsynaptic receptors serve as both ion channels (R_{ion}) and GPCR (R_{gpcr}) at conventional synapses. Some of these receptors are remote from the postsynaptic structure and act as extrasynaptic receptors (R_{extra}).

Group I mGluRs (i.e. mGluR1a, mGluR5) stimulate PLC, inositol phosphate hydrolysis, release of Ca^{2+} and presumably activation of PKC. PKC regulates postsynaptic GluRs (Soderling and Derkach, 2000; MacDonald et al., 2001; Kotecha and MacDonald, 2003) and hippocampal LTP (MacDonald et al., 2001). Group II and III mGluRs generally inhibit adenylate cyclase (AC) activity. In some systems, mGluRs have also been reported to stimulate AC (Conn and Pin, 1997). Metabotropic GluRs have a wide variety of actions on central neurons including effects on transmitter release as well as actions on various voltage and ligand-gated ion channels(Conn and Pin, 1997; Conn, 2003). These receptors are located within the periannular (perisynaptic) region of excitatory synapses in close proximity to and surrounding NMDARs and AMPARs (Lujan et al., 1996) and are likely activated by synaptically released glutamate (Conn and Pin, 1997). Group I mGluRs are physically anchored in the PSD by the protein Homer, which also binds to Shank which in turn is linked via cortactin binding to filamentous actin (Garner et al., 2000). NMDARs also link to Shank via GKAP providing evidence that NMDARs and mGluRs are closely associated with each other in the PSD (Garner et al., 2000; Li and Sheng, 2003). This structural evidence suggests potential functional interactions between mGluRs and NMDARs. There are reports that mGluRs (presumably group I) enhance NMDAR responses expressed in *Xenopus* oocytes (Kelso et al., 1992) and central neurons via activation of PLC (Aniksztejn et al., 1991; Ben-Ari et al., 1992; Aronica et al., 1993; Fitzjohn et al., 1996). In some cases, the effect has been attributed to a PKC-dependent mechanism (Kelso et al., 1992; Aniksztejn et al., 1991; Aniksztejn et al., 1992) and in others to a PKC-independent, hydrolysis of inositol phosphate and release of intracellular Ca^{2+} (Rahman and Neuman, 1996). In direct contradiction, stimulation of mGluRs has also been reported to inhibit NMDAR-evoked responses in cerebellar (Courtney and Nicholls, 1992) and striatal neurons (Colwell and Levine, 1994). Applications of phorbol esters have also been reported to inhibit NMDAR-evoked responses in CA1 neurons (Markram and Segal, 1992). The role of mGluRs in LTP remains ambiguous (Bortolotto et al., 1999; Bortolotto and Collingridge, 1999) but mGluRs are responsible for a distinct form of LTD and they may play an important role in Fragile X syndrome (Bear et al., 2004). The LTD associated with activation of group I mGluRs requires protein synthesis and operates via activation of a subset of MAPKs, specifically p38MAPK, whilst NMDA-induced LTD does not (Gallagher et al., 2004). In contrast, the CA3 region is enriched in mGluR1a. In CA3 neurons Heuss et al. (Heuss et al., 1999) report that mGluR1a acts via an entirely G-protein independent mechanism to activate Src which then stimulates cation channels underlying a slow epsps.

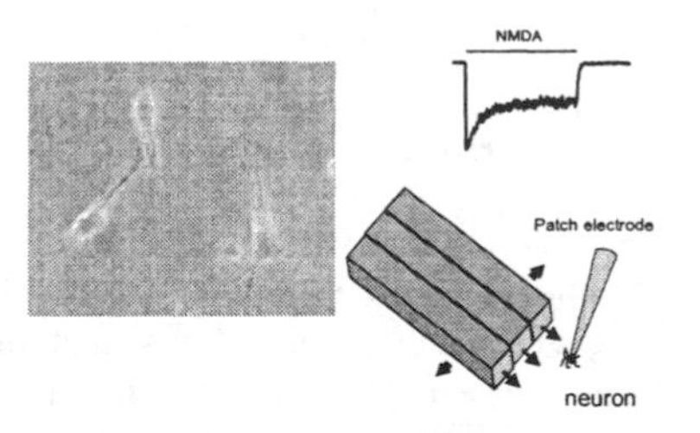

Figure 5. Shown are several acutely isolated CA1 pyramidal neurons (phase contrast) with major dendritic processes attached. A single neuron is patch clamped and lifted into the stream of solution flowing from a complex of 3 glass barrels. The barrels are computer controlled and rapid changes in solution can be achieved using this technique. Applications of NMDA evoke inward currents at -60 mV which peak and then reach a steady state after 2 second.

CA1 pyramidal cells express group I mGluRs (Conn and Pin, 1997; Bortolotto and Collingridge, 1999; Minakami et al., 1993; Romano et al., 1995) and it is reported that they are rich in mGluR5 (Shigemoto et al., 1993). Metabotropic GluR5s are located in the PSD and dendritic spines of CA1 neurons (Romano et al., 1995). In hippocampal slice recordings, the broad-spectrum group I mGluR agonist trans-ACPD (aminocyclopentane-trans-1,3-dicarboxylic acid) has reported to potentiate NMDAR-mediated responses (Ben-Ari et al., 1992) and the mGluR5 agonist CHPG ((RS)-2-chloro-5-hydroxyphenylglycine) is reported to induce a slowly-developing and long-lasting potentiation of NMDAR-responses and excitatory synaptic transmission via activation of PKC (Doherty et al., 1997; Valenti et al., 2002). In mGluR5 knockout mice LTP of NMDAR-mediated EPSCs is absent but can be restored through pharmacological activation of PKC (Jia et al., 1998). Kotecha et al. (Kotecha et al., 2003), using acutely isolated CA1 pyramidal neurons (**Fig. 5**) from young rats, demonstrated that the current evoked by a rapid application of NMDA was enhanced by activation of mGluR5 receptors (**Fig. 6**).

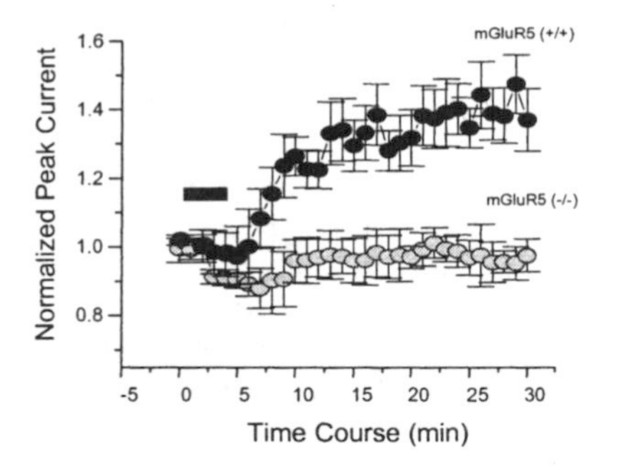

Figure 6. A plot of normalized peak NMDAR currents against the time course of the recordings. Results are averaged for a population of cells taken from wild type mice, mGluR5(+/+) and transgenic mice lacking mGluR5 (-/-). The mGluR5 agonist CHPG was applied during the period indicated by the black bar. Note that the potentiation begin after the application and continues well afterwards in mGluR5 (+/+) neurons.

This potentiation was dependent upon release of intracellular Ca^{2+} via IP_3Rs, activation of a G_q-mediated signal transduction pathway that includes the sequential stimulation PKC and the tyrosine kinases CAKβ and Src (see below). It was absent in transgenic mice lacking the mGluR5 receptor (**Fig. 6**). Interestingly, a co-incident influx of Ca^{2+} via NMDARs was required for stimulation of this signal transduction pathway. As the time between mGluR5 activation and the application of NMDA was increased the potentiation was lost and NMDAR mediated currents were depressed. Strong intracellular buffering of Ca^{2+} also eliminated the potentiation supporting the interpretation that a critical concentration Ca^{2+} or a specific microdomain of Ca^{2+} is required to trigger this signal transduction pathway (**Fig. 7**). These results were confirmed by Grishin et al. (Grishin et al., 2004) in a study that showed that the degree of intracellular Ca^{2+} buffering determines the balance between potentiation and depression of NMDAR currents mediated by mGluR5 receptors. These authors also demonstrated that different neurons (e.g. CA1 versus CA3 neurons) displayed different sensitivities of this signal transduction pathway. Whether or not such signaling has been described as PKC-dependent or PKC-independent the ultimate regulation of NMDARs appears to be mediated via Src. Indeed, Src appears to play the role of "hub" for such signaling (Ali and Salter, 2001). The relationship of this interaction between NMDARs and mGluR5 has not yet been fully explored in terms of the mechanisms of synaptic plasticity at CA1 synapses although the role for the CAKβ and Src in the induction of CA1 LTP is substantial (Huang et al., 2001). A tyrosine kinase dependent enhancement of NMDARs receptor insertion has also been associated with the expression of LTP at adult CA1 synapses with little change in AMPARs suggesting that alterations of NMDARs (Grosshans and Browning, 2001) play a much more fundamental role in the long-term enhancement of synaptic efficacy than was previously believed.

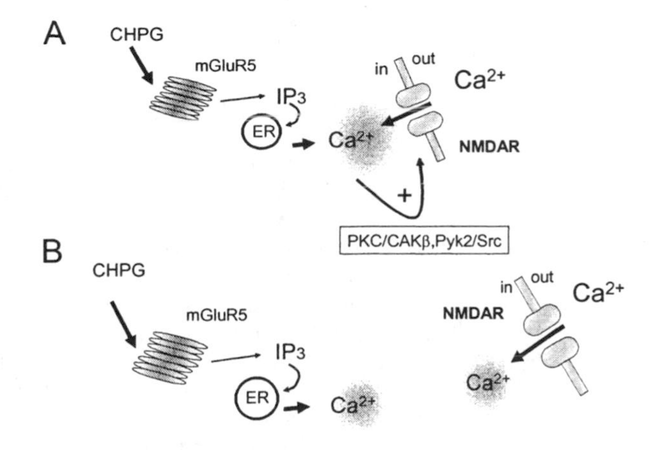

Figure 7. A illustrates a schema where the increase in intracellular Ca^{2+} evoked by mGluR5 activation must overlap temporally and/or spatially with the influx of Ca^{2+} through NMDARs in order for potentiation of NMDAR activity to occur. If this overlap does not occur then NMDAR activity is depressed by a Ca^{2+}-calmodulin dependent process.

5. GPCRS AND REGULATION OF HIPPOCAMPAL NMDARS.

Cholinergic neurons of the forebrain innervate the hippocampus (Levey, 1996; Bartus et al., 1982; Coyle et al., 1983) and contribute to learning and memory (Levey, 1996). The loss of these cholinergic neurons may also underlie memory loss in

Alzheimer's disease (Levey, 1996; Bartus et al., 1982; Coyle et al., 1983). Drugs that enhance this cholinergic transmission help to preserve memory in these patients (Levey, 1996). CA1 pyramidal neurons possess postsynaptic cholinergic receptors (muscarine receptors), which are thought to enhance NMDARs and excitatory synaptic transmission at CA1 synapses (Levey, 1996).

M1 receptors are by far the most abundant subtype of muscarinic receptor found in CA1 (Marino et al., 1998) and they are detected in dendritic spines and putative glutamatergic synapses (Rouse et al., 1998). M1 has been reported to co-localize with NR1a subunits at CA1 synapses (Rouse et al., 1998). CA1 cells also express M3, M4 and M5 (Levey et al., 1995) receptors.

There are reports that carbachol (a mixed nicotinic and muscarinic agonist) potentiates NMDAR responses in hippocampal slice neurons by a PKC-**independent** mechanism involving instead a IP_3-dependent release of intracellular Ca^{2+} (Harvey et al., 1993; Markram and Segal, 1992) an effect likely mediated by M1 receptors (Marino et al., 1998). In contrast, we have shown that muscarine enhances NMDAR-induced currents in isolated CA1 pyramidal neurons through a PKC-**dependent** activation of the tyrosine kinase, Src (M1-PKC-Pyk2 or CAKβ-Src) (Lu et al., 1999). We have shown that two different GPCRs, i) muscarinic M1 receptors and ii) lysophosphatidic acid receptors (LPA, a growth inducing Edg receptor (Chun et al., 2000)) enhance peak NMDAR-currents in isolated CA1 neurons of the rat by sequentially activating PKC and the non-receptor tyrosine kinase, Src (Lu et al., 1999; Huang et al., 2001) (**Fig. 8**). The effects of muscarine were selectively blocked by an M1 directed toxin (unpublished). Miniature $EPSCs_{NMDA}$ in cultured neurons (Lu et al., 1999) and evoked $EPSCs_{NMDA}$ in hippocampal slices were also enhanced by this muscarine-activated kinase cascade (Lu et al., 1998). In recordings of organotypic slices muscarine also potentiates NMDAR responses by a mechanism that is regulated by intracellular Ca^{2+} concentrations (Grishin et al., 2004).

There is extensive evidence that Src upregulates NMDAR channel activity (Salter and Kalia, 2004) and Src is required for induction of LTP at CA1 synapses (Lu et al., 1998). Phorbol esters acting via PKC increase tyrosine phosphorylation of NR2A (76% increase) and 2B (41% increase) subunits in CA1 hippocampal mini slices (Grosshans and Browning, 2001). In rat hippocampal mini slices (Parfitt et al., 1992), little of the detected NR2A, unlike NR2B, was associated with NR1 subunits (functional receptors must contain both NR1 and NR2 subunits) and in cultured hippocampal neurons only 40 to 50% of NR2A is expressed on the cell surface (Hall and Soderling, 1997). A similar lack of NR2A surface expression was observed in mini slices (Grosshans and Browning, 2001). LTP is associated with an increased surface expression NMDARs as a consequence of enhanced NR1 translocation to the surface(Grosshans et al., 2002). Therefore, GPCRs such as M1 can target Src to regulate NMDAR activity and ultimately control the induction of LTP at CA1 synapses.

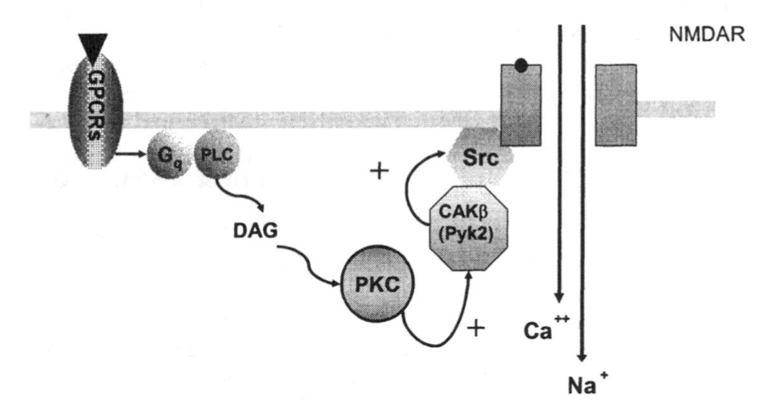

Figure 8. M1 receptors and other Gq coupled receptors found in CA1 pyramidal neurons stimulate phospholipase Cβ which produces diacyl glycerol (DAG) leading to PKC activation. PKC and Ca^{2+} then lead to phosphorylation and association of Pyk2 and Src and cause enhancement of NMDAR activity.

Another GPCR mediated mechanism of enhancing NMDAR responses and transmission at CA1 synapses has also been reported. Pituitary adenylate cyclase-activating polypeptide (PACAP) receptor activation employing the ligand PACAP(1-38), can enhance or depress responses to NMDA (Liu and Madsen, 1997) and potentiate (or depress) excitatory field potentials of CA1 synapses (Roberto et al., 2001; Yaka et al., 2003) depending upon ligand concentration. In a cell expression system PACAP receptors also inhibit Src (Chang et al., 2002). Some, but by no means all PACAP receptors, are thought to act via Gs and stimulation of adenylate cyclase (Spengler et al., 1993). PKA activation can in turn disrupt the protein-protein interactions of RACK1 (receptor for activated kinase 1) with tyrosine kinases such as fyn leading to an enhanced tyrosine phosphorylation of NR2B (Yaka et al., 2003; Yaka et al., 2002). One possibility is that the sequences of fyn that recognize and bind to RACK1 are mimicked by sequences in the C-terminus of NR2B (Yaka et al., 2003; Yaka et al., 2002). Therefore, disinhibition of RACK1 binding to fyn should then allow NR2B containing receptors to replace RACK1 and permit a fyn-induced phosphorylation and potentiation of NMDAR responses. Evidence supporting this hypothesis includes the demonstration that the PACAP potentiation was blocked by introduction of peptides corresponding to the RACK1 binding sequence that binds to fyn and NR2B (Yaka et al., 2003). However, the RACK1 binding sequence observed in fyn is also entirely duplicated in Src (Salter and

Kalia, 2004). In this regard, we have found that PACAP(1-38) enhances NMDA currents and this potentiation is block by selective inhibitors of PKC or CAKβ or by the use of the selective Src inhibitor peptide, Src(40-58) in acutely isolated CA1 neurons (Macdonald, Weerapura & MacDonald, unpublished results). Furthermore, this potentiation was attenuated in neurons taken from the CA1 region of transgenic mice lacking phospholipase Cβ1. These results provide strong support for a role for Gq proteins in the signaling cascade rather than Gs. Therefore the functional PACAP receptor enhancement of NMDARs is mediated in large part by the Gq, PKC, Cakβ, and Src signaling pathway which it shares in common with M1, LPA and mGluR5. This is not to say that additional signal cascades may also regulate NMDARs via tyrosine kinases and phosphatases.

6. RECEPTOR-RECEPTOR INTERACTIONS: INDIRECT AND DIRECT

Most postsynaptic glutamate receptors are almost certainly part of a protein complex. Demonstrations of direct protein-protein interactions between the receptors and scaffolding proteins, adaptors and signaling enzymes has laid to rest the concept of isolated receptor proteins. Many of these protein interactions are related to the insertion, localization and internalization and degradation or recycling of receptors (Malenka, 2003; Prybylowski and Wenthold, 2004; Perez-Otano and Ehlers, 2004). Specific protein-protein interactions have also been described for NMDARs (Sheng and Kim, 2002; Nong et al., 2004). For example, within the PSD mGluRs are bound to Homer which binds to IP_3Rs or to Shank which binds to GKAP which in turn binds to NMDARs (and not AMPARs). Such indirect binding relationships between two different types of glutamate receptors (ion channel and GPCR) within the same postsynaptic structure imply possible functional interactions between the receptors. The most parsimonious interpretation of this relationship is that the receptors are held in sufficient micro-proximity to permit the localized release of Ca^{2+} and modulation of the activity of Ca^{2+}-dependent kinases and phosphatases to cause either co-regulation of both receptors or to allow one receptor to regulate the other's function. For example, NMDAR activation regulates phosphorylation and desensitization of mGluRs (Alagarsamy et al., 1999b; Alagarsamy et al., 1999a). Such findings emphasize that the proximity of microdomains of ions and enzymatic activity engender specificity of actions. It is possible that different populations of proteins and microdomains of activity constitute unique signaling complexes for extrasynaptic versus synaptic NMDARs. This concept may also extend to pathological conditions where NR2B containing (a presumably extrasynaptic) receptors have been found to be ideal targets for preventing the intracellular influx associated with excitotoxicity (Aarts et al., 2002; Hardingham et al., 2002; Ning et al., 2004). Surprisingly, blocking NR2A containing receptors maybe entirely counterproductive with respect to saving neurons (Hardingham and Bading, 2003; Vanhoutte and Bading, 2003).

Even more intriguing has been the demonstration of direct interactions between receptor proteins of two different transmitters, one an ion channel and the other a GPCR. Liu and colleagues have provided convincing evidence of direct binding and functional interactions between dopamine and $GABA_A$ receptors (D5- $GABA_A$) (Liu et al., 2000) and between dopamine and NMDARs (D1-NMDAR) (Lee et al., 2002). Specifically the D1 protein binds, via two separate sites, to the NR1 and NR2A subunits, respectively. In an expression system NR1-NR2A mediated currents were inhibited by applications of a D1 agonist and these effects were blocked by intracellular perfusion of a peptide (t-3) which mimics the binding site of the D1 receptor with NR2A subunits. In contrast, a t-2 peptide mimicking the addition of D1 agonist the NR1 interaction prevented NMDA-induced cell death in cultured neurons (Lee et al., 2002). These effects were independent of G-protein coupled signaling. With these effects it is unclear if glutamate can be classified as the primary transmitter and dopamine as the modulator or vice versa or both. If direct receptor-receptor interactions are to occur in central neurons then the receptor proteins must be co-localized (**Fig. 9**), at least during part of their trafficking cycle. These interactions may occur in specific cellular compartments such as extrasynaptic versus synaptic compartments. NR2A-D1 interactions could be synaptic and NR1-D1 could be either extrasynaptic and/or synaptic. At this juncture the evidence that the NR2A interaction can actually occur within the postsynaptic density at functional synapses is still lacking. Co-internalization or insertion of receptors is yet another possibility (Ferguson, 2001) reflecting a rather different process of signaling.

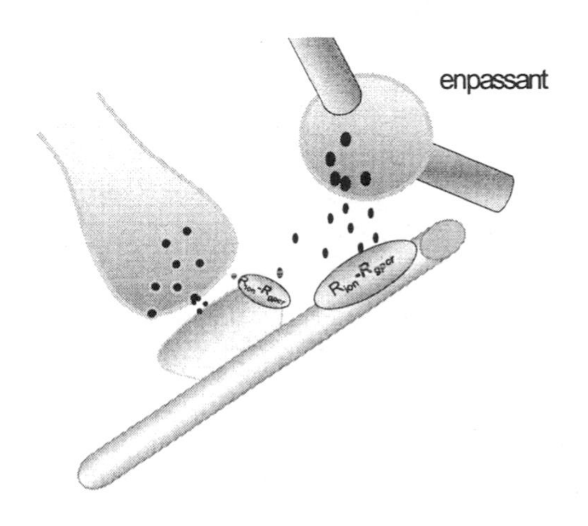

Figure 9. If direct binding between ion channel receptors and GPCR are to mediate functional interactions then their complexes must either be located synaptically or extrasynaptically. Of course these complexes may also form intracellularly and influence delivery of receptors as well as potentially altering endocytosis.

7. REMOTE SIGNALING BETWEEN GPCRS AND RECEPTOR CHANNELS

The physical relationship between many GPCRs (e.g. dopaminergic, muscarinic, PACAPergic etc.) and excitatory synapses is not well understood. These GPCRS may or may not be located in the subsynaptic region of excitatory synapses. If direct receptor-receptor binding serve to modulate synaptic transmission then the GPCRs would have to be located within or at least very close to the PSD and synaptic NMDARs. Some GPCRs, such as mGluRs, will be located at the periphery of the PSD and in such a position as to promote the previously described Gq signal transduction cascades. However, there is no reason to assume that other GPCRs are located within or even close to the PSD. It is likely that many GPCRs would be found in extrasynaptic compartments (with respect to NMDARs) and/or perhaps in the loosely defined postsynaptic regions of enpassant type synapses (i.e.. dopaminergic). This raises the question of if or how do remotely located GPCRs signal to the excitatory spine synapses and the receptors of the PSD? (**Fig. 10**).

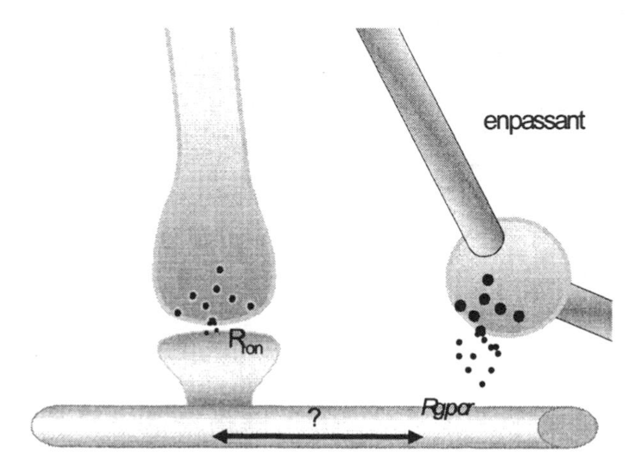

Figure 10. Ion channel receptors for glutamate (R_{ion}) are found in both the synaptic and extrasynaptic compartments. Assuming that the modulating transmitter is released at a enpassant synapse and that its receptors ($\boldsymbol{R_{gpcr}}$) are located a relatively long distance from synaptic glutamate channels and the PSD how can effective functional signal be established between transmitter systems?

8. REMOTE SIGNALING BETWEEN DOPAMINE AND NMDARS VIA TRANSACTIVATION

The CA1 region is enriched in D4 receptors (a D2 class receptor subtype) and dopamine depresses $EPSPs_{NMDA}$ evoked by stimulation of the Schaffer-collateral or the Perforant Path (Otmakhova and Lisman, 1998). Very little is understood about signaling via the D2 class in neurons although inhibition of adenylyl cyclase is the conventional signal pathway attributed to these receptors. Atypical and typical neuroleptic drugs appear to act by blocking dopamine receptors of the D2 class (i.e. D2 or D4) and upregulation of the D4 subtype of receptor has been implicated in Schizophrenia and attention deficit hyperactivity disorder (Wong and Van Tol, 2003b; Wong and Van Tol, 2003a). A reduced NMDAR function has also been linked to Schizophrenia (Javitt and Zukin, 1991; Zukin and Javitt, 1993; Zylberman et al., 1995).

In several early studies we showed that activation of platelet-derived growth factor receptors (PDGFRβ), a receptor tyrosine kinase, acutely inhibits NMDA-evoked currents in isolated and cultured hippocampal neurons (Lei et al., 1999; Valenzuela et al., 1996). These rather poorly named receptors and their ligands are wide spread through the central nervous system, including the hippocampus. PDGFRs provide potent mitogenic signals and are known to do this through activation of MAPKs (ERK1/2, extracellular receptor kinases 1 and 2). Furthermore, a number of GPCRs are also known to be mitogens by virtue of their ability to stimulate PDGFRs in the absence of a PDGR ligand by a process called "transactivation" (**Fig. 11**). Transactivation occurs by mechanisms that are not well understood (Shah and Catt, 2004). However, GPCRs promote auto tyrosine phosphorylation of growth factor receptors which is the step required to form the docking sites for various signaling molecules (i.e. phospholipase Cγ, $PtdIns_3$-K, Src etc.) (Shah and Catt, 2004). Activated PDGFRs then signal to the genome by several different and complex cascades (Heldin et al., 1998). GPCRs are also know to transactivate Trk neurotrophin receptors such as NGF and BDNF that potentially regulate neuronal growth and proliferation (Lee and Chao, 2001; Rajagopal et al., 2004).

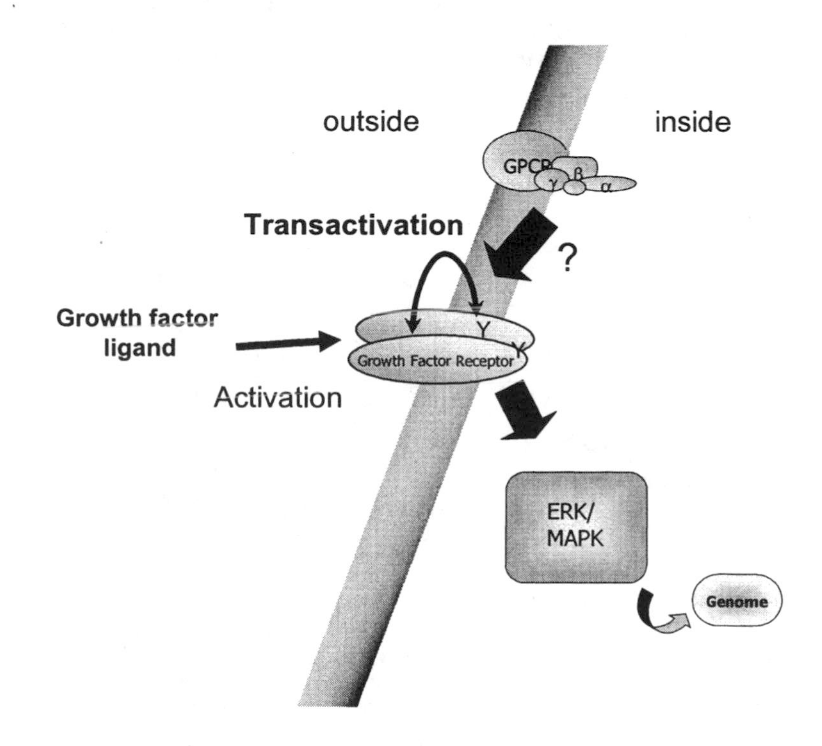

Figure 11. Release of a growth factor ligand causes dimerization and autophosphorylation of its receptors. Phosphorylated receptors then signal to the MAPK (ERKs) cascade and modify transcription of proteins. By mechanisms that are not well understood some GPCRs can stimulate growth factor autophosphorylation and MAPK signaling in the absence of the growth factor ligand (transactivation).

Recently we demonstrated that the selective D2 class agonist, quinpirole, inhibits NMDAR activity to about the same degree and with similar kinetics to what we observed with PDGFβ ligand (Kotecha et al., 2002). Furthermore, the inhibition of NMDARs by quinpirole was blocked by selective inhibitors of PDGFR tyrosine kinase activity, as was the quinpirole induced tyrosine phosphorylation of PDGFRs in the CA1 region of the hippocampus. D4 and D2 receptor transactivation of recombinant PDGFRs was also shown in cell lines (Oak et al., 2001). This D4 receptor induced transactivation of NMDARs was dependent upon phospholipase Cγ and stimulation of calcium-calmodulin (**Fig. 12**). Our results raised the question as to whether or not transactivation is a more general feature of cell signaling by D2 receptors in other regions of the cortex such as the prefrontal cortex. This region is of considerable interest given its potential role in the positive symptoms of Schizophrenia. In preliminary experiments we have found PDGFR transactivation occurs in the prefrontal cortex but unlike the hippocampus it requires activation of the D2 subtype of receptor rather than the D4.

Transactivation provides a unique mechanism for GPCR modulation of excitatory synaptic transmission. Essentially, a GPCR can "highjack" the signaling cascade of a growth factor receptor in order to mediate its own modulation of another transmitter system. This allows signals to be initiated from GPCRs located remotely from the synaptic spines as they are from the transcriptional apparatus of the neuron.

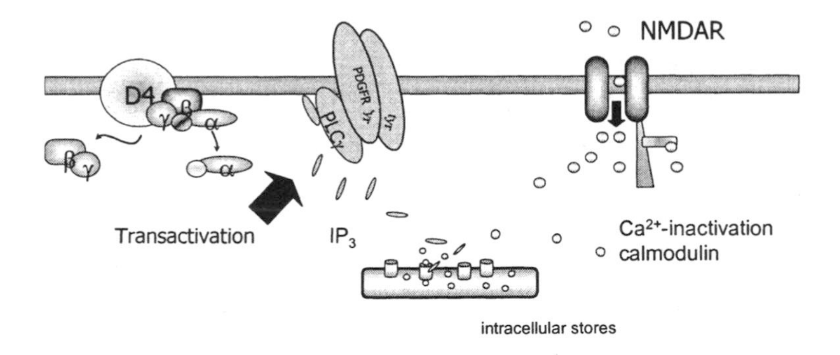

Figure 12. D4 receptors via trimeric G-proteins cause transactivation of PDGFRβ receptors in CA1 hippocampal neurons. Autophosphorylated PDGFRs (although dimerization may or may not occur) bind to and stimulate phospholipase Cγ leading to the release of intracellular calcium and a Ca^{2+}-calmodulin-dependent inactivation of NMDARs. By this mechanism a GPCR located remotely from excitatory synaptic spines can still have a modulatory effect on transmission.

9. REFERENCES

Aarts M, Liu Y, Liu L, Besshoh S, Arundine M, Gurd JW, Wang YT, Salter MW, Tymianski M (2002) Treatment of ischemic brain damage by perturbing NMDA receptor- PSD-95 protein interactions. Science 298: 846-850.

Alagarsamy S, Marino MJ, Rouse ST, Gereau RW, Heinemann SF, Conn PJ (1999a) Activation of NMDA receptors reverses desensitization of mGluR5 in native and recombinant systems. Nat Neurosci 2: 234-240.

Alagarsamy S, Rouse ST, Gereau RW, Heinemann SF, Smith Y, Conn PJ (1999b) Activation of N-methyl-D-aspartate receptors reverses desensitization of metabotropic glutamate receptor, mGluR5, in native and recombinant systems. Ann N Y Acad Sci 868: 526-530.

Albright TD, Jessell TM, Kandel ER, Posner MI (2000) Neural science: a century of progress and the mysteries that remain. Neuron 25 Suppl: S1-55.

Ali DW, Salter MW (2001) NMDA receptor regulation by Src kinase signalling in excitatory synaptic transmission and plasticity. Curr Opin Neurobiol 11: 336-342.

Aniksztejn L, Bregestovski P, Ben-Ari Y (1991) Selective activation of quisqualate metabotropic receptor potentiates NMDA but not AMPA responses. Eur J Pharmacol 205: 327-328.

Aniksztejn L, Otani S, Ben-Ari Y (1992) Quisqualate metabotropic receptors modulate NMDA currents and facilitate induction of long term potentiation through protein kinase C. Eur J Neurosci 4: 500-505.

Aronica E, Dell'Albani P, Condorelli DF, Nicoletti F, Hack N, Balazs R (1993) Mechanisms underlying developmental changes in the expression of metabotropic glutamate receptors in cultured cerebellar granule cells: homologous desensitization and interactive effects involving N-methyl-D- aspartate receptors. Mol Pharmacol 44: 981-989.

Bartus RT, Dean RL, III, Beer B, Lippa AS (1982) The cholinergic hypothesis of geriatric memory dysfunction. Science 217: 408-414.

Bear MF, Huber KM, Warren ST (2004) The mGluR theory of fragile X mental retardation. Trends Neurosci 27: 370-377.

Ben-Ari Y, Aniksztejn L, Bregestovski P (1992) Protein kinase C modulation of NMDA currents: an important link for LTP induction. Trends Neurosci 15: 333-339.

Bortolotto ZA, Collingridge GL (1999) Evidence that a novel metabotropic glutamate receptor mediates the induction of long-term potentiation at CA1 synapses in the hippocampus. Biochem Soc Trans 27: 170-174.

Bortolotto ZA, Fitzjohn SM, Collingridge GL (1999) Roles of metabotropic glutamate receptors in LTP and LTD in the hippocampus. Curr Opin Neurobiol 9: 299-304.

Carvalho AL, Kameyama K, Huganir RL (1999) Characterization of phosphorylation sites on the glutamate receptor 4 subunit of the AMPA receptors. J Neurosci 19: 4748-4754.

Chang BY, Harte RA, Cartwright CA (2002) RACK1: a novel substrate for the Src protein-tyrosine kinase. Oncogene 21: 7619-7629.

Chun J, Weiner JA, Fukushima N, Contos JJ, Zhang G, Kimura Y, Dubin A, Ishii I, Hecht JH, Akita C, Kaushal D (2000) Neurobiology of receptor-mediated lysophospholipid signaling. From the first lysophospholipid receptor to roles in nervous system function and development. Ann N Y Acad Sci 905: 110-117.

Clark BA, Cull-Candy SG (2002) Activity-dependent recruitment of extrasynaptic NMDA receptor activation at an AMPA receptor-only synapse. J Neurosci 22: 4428-4436.

Clements JD, Lester RAJ, Tong G, Jahr CE, Westbrook GL (1992) The time course of glutamate in the synaptic cleft. Science 258: 1498-1501.

Collingridge GL, Bliss TV (1995) Memories of NMDA receptors and LTP. Trends Neurosci 18: 54-56.

Colwell CS, Levine MS (1994) Metabotropic glutamate receptors modulate *N*-methyl-D-aspartate receptor function in neostriatal neurons. Neuroscience 61: 497-507.

Conn PJ (2003) Physiological roles and therapeutic potential of metabotropic glutamate receptors. Ann N Y Acad Sci 1003: 12-21.

Conn PJ, Pin JP (1997) Pharmacology and functions of metabotropic glutamate receptors. Annu Rev Pharmacol Toxicol 37: 205-237.

Courtney MJ, Nicholls DG (1992) Interactions between phospholipase C-coupled and *N*-methyl- D-aspartate receptors in cultured cerebellar granule cells: Protein kinase C mediated inhibition of *N*-methyl-D-aspartate responses. J Neurochem 59: 983-992.

Coyle JT, Price DL, DeLong MR (1983) Alzheimer's disease: a disorder of cortical cholinergic innervation. Science 219: 1184-1190.

Della RG, Maudsley S, Daaka Y, Lefkowitz RJ, Luttrell LM (1999) Pleiotropic coupling of G protein-coupled receptors to the mitogen- activated protein kinase cascade. Role of focal adhesions and receptor tyrosine kinases. J Biol Chem 274: 13978-13984.

Dingledine R, Borges K, Bowie D, Traynelis SF (1999) The glutamate receptor ion channels. Pharmacol Rev 51: 7-61.

Doherty AJ, Palmer MJ, Henley JM, Collingridge GL, Jane DE (1997) (RS)-2-chloro-5-hydroxyphenylglycine (CHPG) activates mGlu5, but no mGlu1, receptors expressed in CHO cells and potentiates NMDA responses in the hippocampus [In Process Citation]. Neuropharmacology 36: 265-267.

Downes GB, Gautam N (1999) The G protein subunit gene families. Genomics 62: 544-552.

Ferguson SS (2001) Evolving concepts in G protein-coupled receptor endocytosis: the role in receptor desensitization and signaling. Pharmacol Rev 53: 1-24.

Fitzjohn SM, Irving AJ, Palmer MJ, Harvey J, Lodge D, Collingridge GL (1996) Activation of group I mGluRs potentiates NMDA responses in rat hippocampal slices. Neurosci Lett 203: 211-213.

Gallagher SM, Daly CA, Bear MF, Huber KM (2004) Extracellular signal-regulated protein kinase activation is required for metabotropic glutamate receptor-dependent long-term depression in hippocampal area CA1. J Neurosci 24: 4859-4864.

Garner CC, Nash J, Huganir RL (2000) PDZ domains in synapse assembly and signalling. Trends Cell Biol 10: 274-280.

Gasic GP, Hollmann M (1992) Molecular neurobiology of glutamate receptors. Annu Rev Physiol 54: 507-536.

Grishin AA, Gee CE, Gerber U, Benquet P (2004) Differential calcium-dependent modulation of NMDA currents in CA1 and CA3 hippocampal pyramidal cells. J Neurosci 24: 350-355.

Groc L, Heine M, Cognet L, Brickley K, Stephenson FA, Lounis B, Choquet D (2004) Differential activity-dependent regulation of the lateral mobilities of AMPA and NMDA receptors. Nat Neurosci 7: 695-696.

Grosshans DR, Browning MD (2001) Protein kinase C activation induces tyrosine phosphorylation of the NR2A and NR2B subunits of the NMDA receptor. J Neurochem 76: 737-744.

Grosshans DR, Clayton DA, Coultrap SJ, Browning MD (2002) LTP leads to rapid surface expression of NMDA but not AMPA receptors in adult rat CA1. Nat Neurosci 5: 27-33.

Gudermann T, Grosse R, Schultz G (2000) Contribution of receptor/G protein signaling to cell growth and transformation. Naunyn Schmiedebergs Arch Pharmacol 361: 345-362.

Gustafsson B, Wigstrom H (1988) Physiological mechanisms underlying long-term potentiation. Trends Neurosci 11: 156-162.

Gutkind JS (1998) The pathways connecting G protein-coupled receptors to the nucleus through divergent mitogen-activated protein kinase cascades. J Biol Chem 273: 1839-1842.

Hall RA, Premont RT, Lefkowitz RJ (1999) Heptahelical receptor signaling: beyond the G protein paradigm. J Cell Biol 145: 927-932.

Hall RA, Soderling TR (1997) Differential surface expression and phosphorylation of the N-methyl-D-aspartate receptor subunits NR1 and NR2 in cultured hippocampal neurons. J Biol Chem 272: 4135-4140.

Hardingham GE, Bading H (2003) The yin and yang of NMDA receptor signalling. Trends Neurosci 26: 81-89.

Hardingham GE, Fukunaga Y, Bading H (2002) Extrasynaptic NMDARs oppose synaptic NMDARs by triggering CREB shut-off and cell death pathways. Nat Neurosci 5: 405-414.

Harvey J, Balasubramaniam R, Collingridge GL (1993) Carbachol can potentiate *N*-methyl-D-aspartate responses in the rat hippocampus by a staurosporine and thapsigargin- insensitive mechanism. Neurosci Lett 162: 165-168.

Hawkins LM, Prybylowski K, Chang K, Moussan C, Stephenson FA, Wenthold RJ (2004) Export from the endoplasmic reticulum of assembled N-methyl-d-aspartic acid receptors is controlled by a motif in the c terminus of the NR2 subunit. J Biol Chem 279: 28903-28910.

Heldin CH, Ostman A, Ronnstrand L (1998) Signal transduction via platelet-derived growth factor receptors. Biochim Biophys Acta 1378: F79-113.

Hering H, Sheng M (2001) Dendritic spines: structure, dynamics and regulation. Nat Rev Neurosci 2: 880-888.

Heuss C, Scanziani M, Gahwiler BH, Gerber U (1999) G-protein-independent signaling mediated by metabotropic glutamate receptors. Nat Neurosci 2: 1070-1077.

Huang Y, Lu W, Ali DW, Pelkey KA, Pitcher GM, Lu YM, Aoto H, Roder JC, Sasaki T, Salter MW, MacDonald JF (2001) CAKbeta/Pyk2 kinase is a signaling link for induction of long-term potentiation in CA1 hippocampus. Neuron 29: 485-496.

Husi H, Ward MA, Choudhary JS, Blackstock WP, Grant SG (2000) Proteomic analysis of NMDA receptor-adhesion protein signaling complexes [In Process Citation]. Nat Neurosci 3: 661-669.

Isaac JT, Nicoll RA, Malenka RC (1999) Silent glutamatergic synapses in the mammalian brain. Can J Physiol Pharmacol 77: 735-737.

Javitt DC, Zukin SR (1991) Recent advances in the phencyclidine model of schizophrenia. Am J Psychiatry 148: 1301-1308.

Jia Z, Lu Y, Henderson J, Taverna F, Romano C, Abramow-Newerly W, Wojtowicz JM, Roder J (1998) Selective abolition of the NMDA component of long-term potentiation in mice lacking mGluR5 [In Process Citation]. Learn Mem 5: 331-343.

Kelso SR, Nelson TE, Leonard JP (1992) Protein kinase C-mediated enhancement of NMDA currents by metabotropic glutamate receptors in Xenopus oocytes. J Physiol (Lond) 449:705-18: 705-718.

Kotecha SA, Jackson MF, Al Mahrouki A, Roder JC, Orser BA, MacDonald JF (2003) Co-stimulation of mGluR5 and N-Methyl-D-aspartate Receptors Is Required for Potentiation of Excitatory Synaptic Transmission in Hippocampal Neurons. J Biol Chem 278: 27742-27749.

Kotecha SA, MacDonald JF (2003) Signaling molecules and receptor transduction cascades that regulate NMDA receptor-mediated synaptic transmission. Int Rev Neurobiol 54: 51-106.

Kotecha SA, Oak JN, Jackson MF, Perez Y, Orser BA, Van Tol HHM, MacDonald JF (2002) A D2 class dopamine receptor transactivates a receptor tyrosine kinase to inhibit NMDA receptor transmission. Neuron 35: 1111-1122.

Kovalchuk Y, Eilers J, Lisman J, Konnerth A (2000) NMDA receptor-mediated subthreshold Ca(2+) signals in spines of hippocampal neurons. J Neurosci 20: 1791-1799.

Kullmann DM (2003) Silent synapses: what are they telling us about long-term potentiation? Philos Trans R Soc Lond B Biol Sci 358: 727-733.

Kullmann DM, Min MY, Asztely F, Rusakov DA (1999) Extracellular glutamate diffusion determines the occupancy of glutamate receptors at CA1 synapses in the hippocampus. Philos Trans R Soc Lond B Biol Sci 354: 395-402.

Lee FJ, Xue S, Pei L, Vukusic B, Chery N, Wang Y, Wang YT, Niznik HB, Yu XM, Liu F (2002) Dual regulation of NMDA receptor functions by direct protein-protein interactions with the dopamine Dl receptor. Cell 111: 219-230.

Lee FS, Chao MV (2001) Activation of Trk neurotrophin receptors in the absence of neurotrophins. Proc Natl Acad Sci U S A 98: 3555-3560.

Lei S, Lu WY, Xiong ZG, Orser BA, Valenzuela CF, MacDonald JF (1999) Platelet-derived growth factor receptor-induced feed-forward inhibition of excitatory transmission between hippocampal pyramidal neurons. J Biol Chem 274: 30617-30623.

Levey AI (1996) Muscarinic acetylcholine receptor expression in memory circuits: implications for treatment of Alzheimer disease. Proc Natl Acad Sci U S A 93: 13541-13546.

Levey AI, Edmunds SM, Hersch SM, Wiley RG, Heilman CJ (1995) Light and electron microscopic study of m2 muscarinic acetylcholine receptor in the basal forebrain of the rat. J Comp Neurol 351: 339-356.

Li B, Chen N, Luo T, Otsu Y, Murphy TH, Raymond LA (2002) Differential regulation of synaptic and extra-synaptic NMDA receptors. Nat Neurosci 5: 833-834.

Li Z, Sheng M (2003) Some assembly required: the development of neuronal synapses. Nat Rev Mol Cell Biol 4: 833-841.

Liu F, Wan Q, Pristupa ZB, Yu XM, Wang YT, Niznik HB (2000) Direct protein-protein coupling enables cross-talk between dopamine D5 and gamma-aminobutyric acid A receptors. Nature 403: 274-280.

Liu GJ, Madsen BW (1997) PACAP38 Modulates Activity of NMDA Receptors in Cultured Chick Cortical Neurons. J Neurophysiol 78: 2231-2234.

Liu L, Wong TP, Pozza MF, Lingenhoehl K, Wang Y, Sheng M, Auberson YP, Wang YT (2004) Role of NMDA receptor subtypes in governing the direction of hippocampal synaptic plasticity. Science 304: 1021-1024.

Lu W, Man H, Ju W, Trimble WS, MacDonald JF, Wang YT (2001) Activation of synaptic NMDA receptors induces membrane insertion of new AMPA receptors and LTP in cultured hippocampal neurons. Neuron 29: 243-254.

Lu WY, Xiong ZG, lei S, Orser BA, Dudek E, Browning MD, MacDonald JF (1999) G-protein-coupled receptors act via protein kinase C and Src to regulate NMDA receptors. Nat Neurosci 2: 331-338.

Lu YM, Roder JC, Davidow J, Salter MW (1998) Src Activation in the Induction of Long-Term Potentiation in CA1 Hippocampal Neurons. Science 279: 1363-1367.

Lujan R, Nusser Z, Roberts JD, Shigemoto R, Somogyi P (1996) Perisynaptic location of metabotropic glutamate receptors mGluR1 and mGluR5 on dendrites and dendritic spines in the rat hippocampus. Eur J Neurosci 8: 1488-1500.

Luscher C, Nicoll RA, Malenka RC, Muller D (2000) Synaptic plasticity and dynamic modulation of the postsynaptic membrane. Nat Neurosci 3: 545-550.

Luttrell LM, Daaka Y, Lefkowitz RJ (1999) Regulation of tyrosine kinase cascades by G-protein-coupled receptors. Curr Opin Cell Biol 11: 177-183.

Lynch MA (2004) Long-term potentiation and memory. Physiol Rev 84: 87-136.

Ma YC, Huang J, Ali S, Lowry W, Huang XY (2000) Src tyrosine kinase is a novel direct effector of G proteins. Cell 102: 635-646.

Ma YC, Huang XY (2002) Novel regulation and function of Src tyrosine kinase. Cell Mol Life Sci 59: 456-462.

MacDonald JF, Kotecha SA, Lu WY, Jackson MF (2001) Convergence of PKC-dependent kinase signal cascades on NMDA receptors. Curr Drug Targets 2: 299-312.

Malenka RC (2003) Synaptic plasticity and AMPA receptor trafficking. Ann N Y Acad Sci 1003: 1-11.

Malenka RC, Nicoll RA (1999) Long-term potentiation--a decade of progress? Science 285: 1870-1874.

Marino MJ, Rouse ST, Levey AI, Potter LT, Conn PJ (1998) Activation of the genetically defined m1 muscarinic receptor potentiates N-methyl-D-aspartate (NMDA) receptor currents in hippocampal pyramidal cells [In Process Citation]. Proc Natl Acad Sci U S A 95: 11465-11470.

Markram H, Segal M (1992) Activation of protein kinase C suppresses responses to NMDA in rat CA1 hippocampal neurones. J Physiol (Lond) 457:491-501: 491-501.

Matsuzaki M, Honkura N, Ellis-Davies GC, Kasai H (2004) Structural basis of long-term potentiation in single dendritic spines. Nature 429: 761-766.

Minakami R, Katsuki F, Sugiyama H (1993) A variant of metabotropic glutamate receptor subtype 5: an evolutionally conserved insertion with no termination codon. Biochem Biophys Res Commun 194: 622-627.

Nicoll RA, Malenka RC (1999) Expression mechanisms underlying NMDA receptor-dependent long-term potentiation. Ann N Y Acad Sci 868:515-25: 515-525.

Ning K, Pei L, Liao M, Liu B, Zhang Y, Jiang W, Mielke JG, Li L, Chen Y, El Hayek YH, Fehlings MG, Zhang X, Liu F, Eubanks J, Wan Q (2004) Dual neuroprotective signaling mediated by downregulating two distinct phosphatase activities of PTEN. J Neurosci 24: 4052-4060.

Nong Y, Huang YQ, Ju W, Kalia LV, Ahmadian G, Wang YT, Salter MW (2003) Glycine binding primes NMDA receptor internalization. Nature 422: 302-307.

Nong Y, Huang YQ, Salter MW (2004) NMDA receptors are movin' in. Current Opinion in Neurobiology 14: 353-361.

Oak, J. N., Lavine, N., and Van Tol, H. H. Dopamine D4 and D2L receptor stimulation of the mitogen-activated protein kinase pathway is dependent on transactivation of the PDGF receptor. Molecular Pharmacology . 2001.

Otmakhova NA, Lisman JE (1998) Dopamine selectively inhibits the direct cortical pathway to the CA1 hippocampal region. J Neurosci 19: 1437-1445.

Parfitt KD, Doze VA, Madison DV, Browning MD (1992) Isoproterenol increases the phosphorylation of the synapsins and increases synaptic transmission in dentate gyrus, but not in area CA1, of the hippocampus. Hippocampus 2: 59-64.

Perez-Otano I, Ehlers MD (2004) Learning from NMDA receptor trafficking: clues to the development and maturation of glutamatergic synapses. Neurosignals 13: 175-189.

Prybylowski K, Wenthold RJ (2004) N-Methyl-D-aspartate receptors: subunit assembly and trafficking to the synapse. J Biol Chem 279: 9673-9676.

Rahman S, Neuman RS (1996) Characterization of metabotropic glutamate receptor-mediated facilitation of N-methyl-D-aspartate depolarization of neocortical neurones. Br J Pharmacol 117: 675-683.

Rajagopal R, Chen ZY, Lee FS, Chao MV (2004) Transactivation of Trk neurotrophin receptors by g-protein-coupled receptor ligands occurs on intracellular membranes. J Neurosci 24: 6650-6658.

Roberto M, Scuri R, Brunelli M (2001) Differential effects of PACAP-38 on synaptic responses in rat hippocampal CA1 region. Learn Mem 8: 265-271.

Romano C, Sesma MA, McDonald CT, O'Malley K, Van den Pol AN, Olney JW (1995) Distribution of metabotropic glutamate receptor mGluR5 immunoreactivity in rat brain. J Comp Neurol 355: 455-469.

Rouse ST, Gilmor ML, Levey AI (1998) Differential presynaptic and postsynaptic expression of m1-m4 muscarinic acetylcholine receptors at the perforant pathway/granule cell synapse. Neuroscience 86: 221-232.

Rusakov DA, Kullmann DM (1998) Extrasynaptic glutamate diffusion in the hippocampus: ultrastructural constraints, uptake, and receptor activation. J Neurosci 18: 3158-3170.

Salter MW, De Koninck Y (1999) An ambiguous fast synapse: a new twist in the tale of two transmitters. Nat Neurosci 2: 199-200.

Salter MW, Kalia LV (2004) SRC kinases: A hub for NMDA receptor regulation. Nature Reviews Neuroscience 5: 317-328.

Scannevin RH, Huganir RL (2000) Postsynaptic organization and regulation of excitatory synapses. Nat Rev Neurosci 1: 133-141.
Seeburg PH (1993) The TiPS/TINS lecture: the molecular biology of mammalian glutamate receptor channels. Trends Pharmacol Sci 14: 297-303.
Shah BH, Catt KJ (2004) GPCR-mediated transactivation of RTKs in the CNS: mechanisms and consequences. Trends Neurosci 27: 48-53.
Sheng M (2001) Molecular organization of the postsynaptic specialization. Proc Natl Acad Sci U S A 98: 7058-7061.
Sheng M, Kim MJ (2002) Postsynaptic signaling and plasticity mechanisms. Science 298: 776-780.
Shigemoto R, Nomura S, Ohishi H, Sugihara H, Nakanishi S, Mizuno N (1993) Immunohistochemical localization of a metabotropic glutamate receptor, mGluR5, in the rat brain. Neurosci Lett 163: 53-57.
Soderling TR, Derkach VA (2000) Postsynaptic protein phosphorylation and LTP. Trends Neurosci 2000 Feb 23 (2):75 -80 23: 75-80.
Spengler D, Waeber C, Pantaloni C, Holsboer F, Bockaert J, Seeburg PH, Journot L (1993) Differential signal transduction by five splice variants of the PACAP receptor. Nature 365: 170-175.
Sweatt JD (1999) Toward a molecular explanation for long-term potentiation. Learn Mem 6: 399-416.
Swope SL, Moss SI, Raymond LA, Huganir RL (1999) Regulation of ligand-gated ion channels by protein phosphorylation. Adv Second Messenger Phosphoprotein Res 33:49-78: 49-78.
Thomas GM, Huganir RL (2004) MAPK cascade signalling and synaptic plasticity. Nat Rev Neurosci 5: 173-183.
Tovar KR, Westbrook GL (1999) The incorporation of NMDA receptors with a distinct subunit composition at nascent hippocampal synapses in vitro. J Neurosci 19: 4180-4188.
Valenti O, Conn PJ, Marino MJ (2002) Distinct physiological roles of the Gq-coupled metabotropic glutamate receptors Co-expressed in the same neuronal populations. J Cell Physiol 191: 125-137.
Valenzuela CF, Xiong Z, MacDonald JF, Weiner JL, Frazier CJ, Dunwiddie TV, Kazlauskas A, Whiting PJ, Harris RA (1996) Platelet-derived growth factor induces a long-term inhibition of N- methyl-D-aspartate receptor function. J Biol Chem 271: 16151-16159.
van Zundert B, Yoshii A, Constantine-Paton M (2004) Receptor compartmentalization and trafficking at glutamate synapses: a developmental proposal. Trends Neurosci 27: 428-437.
Vanhoutte P, Bading H (2003) Opposing roles of synaptic and extrasynaptic NMDA receptors in neuronal calcium signalling and BDNF gene regulation. Curr Opin Neurobiol 13: 366-371.
Vissel B, Krupp JJ, Heinemann SF, Westbrook GL (2001) A use-dependent tyrosine dephosphorylation of NMDA receptors is independent of ion flux. Nat Neurosci 4: 587-596.
Voronin LL, Cherubini E (2003) "Presynaptic silence" may be golden. Neuropharmacology 45: 439-449.
Wang LY, Dudek EM, Browning MD, MacDonald JF (1994) Modulation of AMPA/kainate receptors in cultured murine hippocampal neurones by protein kinase C. J Physiol (Lond) 475: 431-437.
Wenthold RJ, Prybylowski K, Standley S, Sans N, Petralia RS (2003) Trafficking of NMDA receptors. Annu Rev Pharmacol Toxicol 43: 335-358.
Wong AH, Van Tol HH (2003a) Schizophrenia: from phenomenology to neurobiology. Neurosci Biobehav Rev 27: 269-306.
Wong AH, Van Tol HH (2003b) The dopamine D4 receptors and mechanisms of antipsychotic atypicality. Prog Neuropsychopharmacol Biol Psychiatry 27: 1091-1099.
Yaka R, He DY, Phamluong K, Ron D (2003) Pituitary adenylate cyclase-activating polypeptide (PACAP(1-38)) enhances N-methyl-D-aspartate receptor function and brain-derived neurotrophic factor expression via RACK1. J Biol Chem 278: 9630-9638.
Yaka R, Thornton C, Vagts AJ, Phamluong K, Bonci A, Ron D (2002) NMDA receptor function is regulated by the inhibitory scaffolding protein, RACK1. Proc Natl Acad Sci U S A 99: 5710-5715.
Zhou Q, Poo MM (2004) Reversal and consolidation of activity-induced synaptic modifications. Trends Neurosci 27: 378-383.
Zukin SR, Javitt DC (1993) Phencyclidine receptor binding as a probe of NMDA receptor functioning: implications for drug abuse research. NIDA Res Monogr 133:1-12: 1-12.
Zylberman I, Javitt DC, Zukin SR (1995) Pharmacological augmentation of NMDA receptor function for treatment of schizophrenia. Ann N Y Acad Sci 757:487-91: 487-491.

THE PHASES OF LTP: THE NEW COMPLEXITIES

J. E. Lisman, S. Raghavachari, N. Otmakhov and N. A. Otmakhova *

1. INTRODUCTION

The discovery of LTP and LTD has led to an enormous effort to understand their underlying molecular mechanisms. We now know that hundreds of proteins and many second messenger systems are involved. Despite this progress, a coherent explanation of synaptic plasticity is lacking. One of the difficulties is that LTP itself is not a unitary process. Indeed, some work suggests that there may be up to three different phases of LTP, each involving different biochemical processes[26,71]. In this paper the evidence regarding the phases of LTP will be reviewed. We take special note of the process of "erasure" in which, under some conditions, a tetanus can reduce the LTP induced by a previous tetanus. This cannot be accounted for by standard models, according to which the contributions of different phases simply add. A possible explanation of "erasure" is proposed.

A second goal of this paper is take a long-range view of how this complexity will come to be "understood". It is relevant to consider in this regard the vertebrate rod phototransduction cascade because this is a process where a high degree of understanding has been achieved. All the proteins in this cascade have been identified14. Importantly, when the measured properties of the proteins are incorporated into computer simulations, the detailed kinetic and amplification properties of excitation can be accounted for[35]. It is therefore appropriate to say that the problem has been "solved". Success in the field of synaptic plasticity will similarly be gauged by the ability of simulations to provide a coherent account of the properties of LTP/LTD in molecular terms. Indeed the need for such models is particularly great in this field because of its complexity; whereas the rod cascade involves about 10 key proteins, synaptic plasticity appears to involve hundreds[61]. This is such a large number that it becomes impossible to form a mental picture of the biochemical network; computer simulations are required for keeping track of the

* All authors are from Department of Biology and Volen Center for Complex Systems, Brandeis University, Waltham, MA 02254. E-mail: Lisman@Binah.cc.Brandeis.edu

reactions. We therefore consider at the end of this article whether the necessary computational tools are available.

Many of the important initial experiments indicating the complexity of LTP, specifically the role of cAMP and the protein synthesis were done in the laboratory of John Sarvey. This chapter is dedicated in his honor.

2. THE MULTIPLICITY OF BIOCHEMICAL SIGNALS

The two most important second messengers involved in LTP and LTD appear to be Ca^{2+} and cAMP[10]. The evidence for the role of Ca^{2+} is based on the finding that Ca^{2+} buffers can block LTP/LTD[10] and that elevation of Ca^{2+} (by uncaging methods) is sufficient to induce LTP/LTD[46,74]. The factors that determine whether LTP or LTD occurs remain somewhat unclear, but the level of Ca^{2+} that is reached is clearly important; high Ca^{2+} elevation favors LTP whereas moderate elevation favors LTD and the related process, depotentiation[36,74]. Perhaps the most important Ca^{2+} sensor for strengthening the synapse is the kinase, CaMKII[37]. The important targets of CaMKII include GluR1 and CaMKII itself[37], however other critical targets that initiate the cascade leading to AMPA receptor insertion remain to be identified[39]. Another important Ca^{2+} sensor is Ca-dependent adenylyl cyclase, which produces cAMP[55]. The Ca^{2+} sensor for weakening is the Ca-activated phosphatase, calcineurin[38].

The first evidence that cAMP has a role in synaptic plasticity came from John Sarvey's laboratory. It was demonstrated that norepinephrine, acting through cAMP, could produce potentiation of the EPSP that persisted long after norephinephrine was removed[66]. Corespondingly, norepinephrine depletion suppressed tetanus-induced LTP[65]. Subsequent work indicated that cAMP affected potentiation through PKA[21,41]. We now know that the important targets of PKA include GluR1, Inhibitor-1 and CREB[47].

The first indication that protein synthesis might be important in LTP was the finding that strong synaptic stimulation evoked the synthesis of new proteins in the hippocampal slice[13,18]. It remained unclear, however, whether this synthesis had a functional significance for LTP. To address this question, Stanton and Sarvey[64] studied the effect of several different inhibitors of protein synthesis and found that LTP was greatly reduced. This was the first demonstration that LTP requires synthesis of new proteins and suggested that in addition to modulation, construction or expansion of synapses was likely to occur. Very recent work has provided the first real-time picture of the persistent enlargement of synaptic spines by LTP[40].

3. THE PHASES OF LTP

The idea that LTP might have different phases with different mechanisms originated in Germany (Krug et al., 1984; Frey et al., 1988; Frey et al., 1991; Matthies and Reymann, 1993) and received strong support from the laboratory of Kandel (Frey et al., 1993; Huang and Kandel, 1994; Nguyen et al., 1994; Huang et al., 1996b). These phases were separated on the basis of their different dependence on PKA and protein synthesis, different time courses and different pattern of stimulation required for their induction. In initial work, two phases were distinguished. It was found that several tetani produced larger and longer LTP compared to a single tetanus and that this larger LTP decreased 2-3 hours after induction if protein synthesis inhibitors were applied during the induction (Krug et al., 1984). This late stage of LTP was found to also depend on D1 dopamine receptors, cAMP, PKA, and CREB phosphorylation (Frey et al., 1991; Matthies and Reymann, 1993; Nguyen et al., 1994; Huang and Kandel, 1995; Huang et al., 1996a; Impey et al., 1996; Abel et al., 1997; Kandel, 2001).

More recently an intermediate PKA-dependent but protein synthesis independent phase was described (Winder et al., 1998). These experiments involved pharmacological inhibition of protein kinase A (PKA) or genetic upregulation of the Ca-activated phosphatase, calcineurin (both manipultations have a common effect on a target protein, I1; see below). Neither perturbation significantly reduced the LTP evoked by one tetanus. However, if two or more tetani were given either perturbation reduced LTP relative to that evoked by the same number of tetani without perturbation. Importantly, however, 2 tetani LTP was not affected by protein synthesis inhibitors whereas the LTP by more than 2 tetani was affected (but see discussion of this issue below). It thus seemed that 2tet LTP revealed a phase that was sensitive to PKA but which was not protein synthesis dependent. Based on these findings, the authors proposed that LTP is the summation of three phases (Fig.1). The early phase is evoked by a single tetanus and does not depend on calcineurin or PKA. The intermediate phase is evoked if two or more tetani are given and is controlled by calcineurin and PKA. If more than two tetani are given, the late phase is evoked. This phase is the only one that requires protein synthesis. The late phase develops slowly and maintains LTP after the other phases have faded.

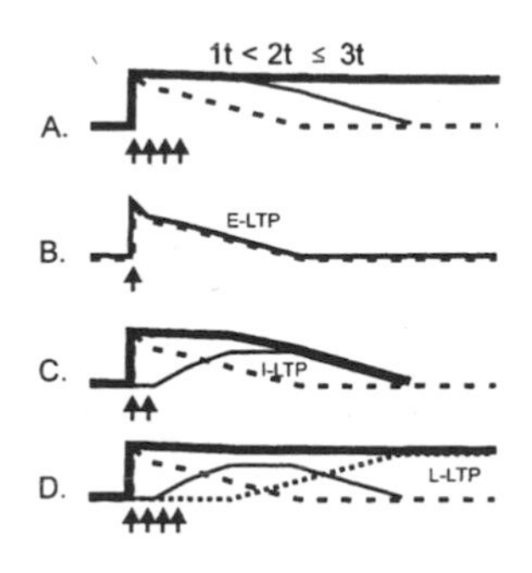

Fig. 1 The 3-phase model of LTP (adapted from (Winder et al., 1998)). A) The duration of LTP depends on the number of tetani (1-3). B.-D. shows how early (dashed line), intermediate (thin line) and late (dotted line) phases sum to form the observed LTP (thick line). The number of arrows indicates the number of tetani.

4. SOME PHENOMENA CANNOT BE EXPLAINED BY THE CONCEPT OF PURELY ADDITIVE PHASES

At the heart of the three-phase model (Winder et al., 1998) is the idea that increasing the number of tetani **adds** to what has been produced by the previous tetanus (Fig. 1D). Although this appears to be true in many studies, it is not always the case. In particular, under conditions of PKA inhibition, the tetani that follow the first may actually decrease the LTP produced by the previous tetani. Fig. 2 shows four experiments illustrating this phenomenon. In each case, the LTP produced by multiple tetani is **less** than that evoked by fewer tetani. This has the important implication that subsequent tetani can **erase** some of the LTP induced by previous ones. Erasure is especially dramatic in the data of Blitzer (Blitzer et al., 1995) (Fig. 2D). In this study a single tetanus produced LTP (~ 80% increase) that was PKA-independent (was not affected by PKA inhibitor). Three tetani produced LTP of a similar size (~80%). However, when PKA blocker was present LTP magnitude after three tetani was only ~ 40%. This suggests that the first tetanus induced PKA-independent LTP that was erased by the next two tetani in the presence pf the PKA inhibitor. This erasure phenomenon has not been previously noted and requires explanation. Another example of erasure occurs when many repetitive tetani are given; the first several repetitions result in saturating LTP however, with many further repetitions, the LTP level slowly diminishes (Abraham and Huggett, 1997).

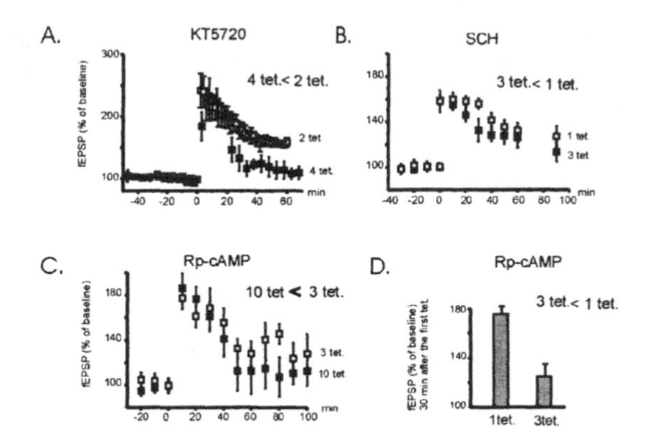

Fig. 2 Evidence that the LTP evoked by large number of tetani can be smaller than that evoked by fewer tetani. Experiments were done under conditions expected to reduce PKA activity. A. PKA inhibitor, KT5720; data replotted from figures 4A, D (Winder et al., 1998). B. SCH reduces activation of D1 dopamine receptor and should thereby reduce the concentration of cAMP. Data replotted from figures 4A, B (Huang and Kandel, 1995) C. PKA inhibitor, Rp-cAMP. Data replotted from figures 5A1, A2 (Nguyen and Kandel, 1996). D. Rp-cAMP applied intracellularly. Plotted from data of (Blitzer et al., 1995).

We will argue that these erasure phenomena can be explained in terms of what is already known about kinase/phosphatase interactions in LTP/LTD and the process of "metaplasticity" (Abraham and Bear, 1996). Metaplasticity occurs when previous neuronal activation affects the results of subsequent stimulation. In the following sections we will briefly review the role of kinase/phosphatase system in bidirectional plasticity and the role of metaplasticity. We will then argue how these ideas can be merged to explain erasure.

5. THE ANTAGONISTIC ROLE OF PHOSPHATASES AND KINASES IN BIDIRECTIONAL PLASTICITY

The scheme in Fig. 3 summarizes our current understanding of the bidirectional modification of synaptic strength (Lisman, 1994; Roberson et al., 1996). During intensive synaptic stimulation, Ca^{2+} -activated kinase (probably CaMKII) promotes induction of

LTP, but this is opposed by a Ca^{2+}-dependent phosphatase cascade (PP1 activity initiated by calcineurin; Fig. 3). Whether the synaptic strength increases or decreases will depend on the quantitative ratio of phosphatase and kinase activities during the induction process. During LTP induction, the Ca^{2+} elevation is high and this causes the kinase process to dominate, leading to synaptic strengthening. During LTD induction, Ca^{2+} elevation is only moderate. Because calcineurin is particularly sensitive to Ca^{2+}, the phosphatase pathway is selectively activated. As a result PP1 activity gets stronger than the kinase activity, thus causing synaptic weakening.

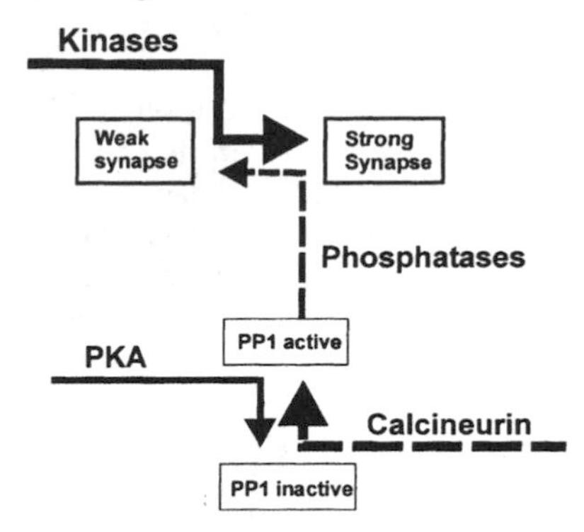

Fig. 3 Role of kinases and phosphatases in the bidirectional control of synaptic strength. Adapted from (Lisman, 1994).

In this scheme the major role of cAMP-dependent processes is the control of PP1 activity. This is important because PP1 is the only phosphatase able to dephosphorylate the activated CaMKII of the PSD. PP1 is inhibited by phosphorylated inhibitor 1 (I1P). The enzyme that phosphorylates I1 is cAMP-dependent PKA. On the other hand, calcineurin dephosphorylates I1P into I1, which has low affinity to PP1. This results in the activation of PP1 and allows it to counteract CaMKII phosphorylation. This model of bidirectional plasticity is consistent with substantial experimental data on the involvement of Ca, CaMKII and phosphatases in LTP (Lisman et al., 2002).

6. METAPLASTICITY: A PLASTIC STATE AFTER SINGLE TETANUS

The properties of synaptic plasticity are not fixed, but can be changed by prior synaptic activity or neuromodulation. These changes have been termed "metaplasticity" (Abraham and Bear, 1996). One of the first demonstrations of metaplasticity was that a brief tetanus that is too weak to induce LTP can prevent a subsequent strong tetanus from inducing LTP (Huang et al., 1992). There is now abundant evidence that this type of metaplasticity makes it subsequently easier to induce weakening (depotentiation) in addition to making it harder to induce LTP (Wexler and Stanton, 1993; Wagner and Alger, 1995; Staubli and Chun, 1996; Holland and Wagner, 1998; Kang-Park et al., 2003). Therefore, metaplasticity unlevels the playing field; initial tetani create a short period during which it is easy to induce depotentiation by either low-frequency stimulation or (under some conditions) by additional tetani. It appears that this type of metaplasticity is due to changes in the balance between kinase and phosphatase in favor of activation of phosphatase (O'Dell and Kandel, 1994; Staubli and Chun, 1996; Otmakhova and Lisman, 1998; Moody et al., 1999; Kang-Park et al., 2003). The kinase can still win during periods of subsequent LTP induction, but the metaplasticity that develops gives a boost to the weakening process: it makes the induction of additional LTP more difficult with each following tetanus and it makes it easier to produce synaptic weakening. We will develop the idea that this type of metaplasticity might be involved in the erasure phenomenon (see below).

In order to reconcile the existing data, it must be assumed that metaplasticity is regulated by at least two processes. The protein, I1, in its phosphorylated state (I1p) inhibits PP1. If the enhanced phosphatase involved in metaplasticity were due solely to effects on I1, it would be expected that I1p concentration would fall after multiple tetani. To the contrary, however, I1p concentration is raised after three tetani (Blitzer et al., 1998). This indicates that metaplasticity must involve *additional factors* that can powerfully regulate the phosphatase/kinase balance. Thus, the phosphatase/kinase balance is affected by at least two processes----one process involves the control of I1 by calcineurin and PKA and tends to weaken the phosphatase by increasing the amount of I1p (see previous chapter); the second operates by an unknown mechanism. This mechanism is activated by initial tetani and tends to increase the phosphatase/kinase ratio (by increasing the phosphatase or decreasing the kinase) during the following tetani or LFS. The net effect of these two factors is normally to enhance the phosphatase relative to the kinase, making it easier to produce weakening. Therefore, even though PP1 activity is decreased by initial tetani (Blitzer et al., 1998), the net effect of metaplasticity processes is to create conditions that lead to a rise in the phosphatase/kinase ratio during the following tetani.

7. WHY MULTIPLE TETANI PRODUCE ERASURE IN THE PRESENCE OF PKA INHIBITOR

Given these ideas about bidirectional control and metaplasticity, one can suggest why erasure occurs. Under normal circumstances, during one tetanus LTP metaplasticity is not yet a factor and the kinase wins. Similarly, when subsequent tetani are given under normal conditions the kinase/phosphatase ratio, though reduced by metaplasticity, still is high enough to make the kinase the winner. However, when the I1-dependent process is altered experimentally to favor the phosphatase (by inhibiting PKA or activating calcineurin) this finally tips the balance so that phosphatase activity is **stronger** than the kinase activity and produces the synaptic weakening (erasure) during the multiple tetani seen in Fig. 2.

8. WHY IS PKA NOT STRONGLY INVOLVED IN A SINGLE-TET LTP?

We now turn to a second difficult issue in understanding bidirectional synaptic modification. If low Ca^{2+} activates the phosphatase cascade, then why doesn't high Ca^{2+} also do so and prevent the kinase cascade from winning during single-tet LTP? One reason the phosphatase might NOT be activated at high Ca^{2+} is that its dependence on Ca^{2+} is bell-shaped. Indeed, it was proposed that the fall of phosphatase activity at high

Ca^{2+} is due to Ca^{2+}-dependent activation of adenylate cyclase; the resulting activation of PKA would oppose the action of calcineurin in I1 and thereby prevent PP1 activation (Lisman, 1989). Consistent with this, direct measurements show that cAMP and PKA activity increase as a result of single tetanus (Blitzer et al., 1995; Roberson et al., 1996). Nevertheless, the fact remains that PKA inhibition (Blitzer et al., 1995; Winder et al., 1998), inhibition of adenylyl cyclase (Otmakhova and Lisman, 1998) or inhibition of both adenylyl cyclase and PKA do not produce very dramatic inhibition of 1-tet LTP (Otmakhova et al., 2000). Thus, some other reason must be sought for why PP1 is not strongly activated during a single tetanus.

A second possible reason for lack of PP1 activation (and weak PKA involvement) in early LTP is that phosphatase activation may be slow because it integrates over many stimuli. During a single 1sec tetanus, the build up PP1 activity may be too small to reach the threshold for beating the kinase. The slow build up may be a reason why induction of LTD normally requires a prolonged low-frequency stimulation (Mockett et al., 2002) or long period of directly raising intracellular Ca^{2+} by photolysis (Yang et al., 1999).

This raises the question of whether there is any plausible reason for why activation of PP1 may occur with a delay? One scenario arises from consideration of a spatial aspect of cAMP production. During a standard LTP induction paradigm the stimulus pulse is generally adjusted to be below threshold for inducing a postsynaptic spike. Since it takes less than a hundred synapses to induce a spike, it is safe to conclude that less than 100 synapses are active. This is less than 1% of the total synapses on a CA1 pyramidal cell, a calculation that emphasizes the spatial sparseness of the active synapses that activate the biochemical reactions. In many cases, sparseness is not an issue because the relevant reactions occur within groups of enzymes that are immobilized at the synapse. However, in other cases involving diffusing species, sparseness could become a major issue and this would be the case for cAMP. Thus, even though Ca^{2+} elevation is local near active synapse, the cAMP produced by locally activated Ca-activated adenylate cyclase would rapidly (within seconds) diffuse into the nearby dendrite. Similarly, although calcineurin appears to be tethered to synapses, its target, I1, is diffusible. For such diffusible species, building up a substantial change in local concentration will take many stimuli. To see why this is case, consider what would happen if a brief pulse of calcineurin activation dephoshorylated all of the I1 in a spine. After the fall of Ca, a small molecules like I1p diffuse into spines from the parent dendrite within about 100 msec, thus preventing PP1 activation for a significant duration. With repeated stimuli, the basal pool of I1p in nearby dendritic regions would be slowly depleted and this would lead to prolonged PP1 activation, perhaps sufficient to induce LTD. We are intrigued by the fact that a simple counterintuitive prediction of the above model appears to be true; if one simply raises stimulus strength substantially (thereby activating many more synapses), a stimulus that can induce LTP now induces LTD (Barr et al., 1995; Kang-Park et al., 2003). This would follow from the fact that now the spacing between active synapses is so small that the dendritic concentration of I1p is lowered substantially by a single tetanus and that the phosphatase now can therefore beat the kinase.

These considerations illustrate the importance of thinking of active synapses as sources and sinks of diffusible molecule that regulate plasticity. It will therefore be critical to take into account the distance between active synapses and the diffusion of key molecule to understand plasticity.

9. MULTIPLE WAYS OF THE INVOLVEMENT OF cAMP IN LTP

Recent discoveries have revealed more complexities of cAMP action and it is useful to enumerate these here. As discussed above, brief inhibition of PKA does not dramatically reduce early LTP. However, longer applications of PKA inhibitors (several hours) produces more pronounced inhibition of 1 tet LTP (Otmakhova et al., 2000). Moreover, monoamine depletion of animals also may result in complete inhibition of 1 tet LTP (Yang et al., 2002). A possible explanation for the stronger effect of the prolonged suppression of the cAMP pathway might come from the results of Malinow (Esteban et al., 2003) indicating that basal PKA-dependent phosphorylation of GluR1 is required but not sufficient for LTP. Therefore, prolonged inhibition of the cAMP pathway may reduce this basal PKA-dependent phosphorylation and more dramatically affect early LTP. In addition, at younger ages (less than P10), LTP is primarily due to activation of cAMP-dependent processes, rather than CaMKII-dependent processes (Yasuda et al., 2003). In adult animals, PKA-dependent phosphorylation of GluR1 receptors is also required during the increase in synaptic strength induced by dedepression protocols (Lee et al., 2000). Finally, since cAMP/PKA can downregulate some K channels, this can have a secondary facilitatory effect on LTP induction (Johnston et al., 2003). This is not a complete list, but is sufficient to demonstrate the complex network of cAMP-dependent processes involved in early LTP.

Interestingly, new data indicate that cAMP-PKA/PP1-dependent "gating" mechanism is also involved in controlling the protein synthesis-dependent late phase of LTP (Woo et al., 2002). That is consistent with the finding that CREB phosphorylation is under control of the balance between multiple kinases and phosphatases ((Bito et al., 1996) see also discussion in (Woo et al., 2002)).

One important model for the role of cAMP in late phase of LTP is that it leads to PKA-dependent activation of transcription factor CREB and the following increase in protein synthesis of specific set of proteins (the total protein syntheses may be even decreased after induction of late LTP (Chotiner et al., 2003)). Importantly, it was suggested that this late phase can be induced in isolation of the previous phases by raising intracellular level of cAMP that directly activate transcription/translation machinery leading to LTP induction. A recent study, however, indicate that the mechanisms of induction of LTP by cAMP might be more complicated and require simultaneous activation of NMDA-dependent processes (Otmakhov et al., 2004). It is interesting to note, that these effects in CA1 may closely resemble the potentiation induced in the dentate gyrus by noradrenaline application, a form of LTP that depends on cAMP elevation. This potentiation is also blocked by NMDA antagonists (Sarvey et al., 1989).

10. HOW EARLY IS LATE LTP: THE ROLE OF PROTEIN SYNTHESIS

According to the multiple-phase model, only the late stage is dependent on protein synthesis. The activation of protein synthesis occurs as follows: multiple tetani raise the concentration of cAMP, activate PKA and thereby enhance the phosphorylation of the nuclear transcription factor CREB. These proteins are necessary for maintaining synaptic strength after the early phase fades and is replaced by the late phase, probably through the structural enhancement of synaptic connections (Huang et al., 1996). This model gives a general explanation of why LTP declines after several hours if protein synthesis is blocked. There are many studies, however, that demonstrate that inhibition of protein synthesis before and during LTP induction may significantly decrease LTP **as early as 30 min** after (Winder et al., 1998) or, under some conditions, even **immediately after** the induction (Stanton and Sarvey, 1984) instead of 2 hrs. Interference with the CREB pathway (and therefore protein synthesis) also dramatically decreases early LTP (Bourtchuladze et al., 1994). In addition, some experiments suggest that late phase LTP can be induced immediately after the tetanus by prior activation of a protein synthesis (Frey and Morris, 1997). Other experiments show that activation or protein synthesis and late LTP can occur after a single tetanus (Osten et al., 1996) or a single 3 sec theta burst stimulation (Abel et al., 1997).

New data also indicate that the process of activation of protein synthesis is much more complex that previously thought. In addition to cAMP/PKA several other protein has been implicated such as mGluR (Raymond et al., 2000), PKC, MAPK (Roberson et al., 1999), CaMKII (Miller et al., 2002; Atkins et al., 2003), BDNF (Patterson et al., 2001) and mTOR/p70S6K pathway (Cammalleri et al., 2003). Some of these enzymes are also clearly involved in early LTP, but it now appears that they are also involved in activation of transcription and regulation of local dendritic translation, which is also critical for the development of late LTP (Cammalleri et al., 2003).

Taken together, these new findings point to the complexity of the processes underlying different LTP phases. These phases does not seem to be clearly separable processes but instead significantly overlap in time and in the underlying biochemical systems. The understanding of such processes will require careful quantitative analysis.

11. TOWARDS A UNIFIED PICTURE OF MULTIPLE LTP PROCESSES

The goal of the field of synaptic plasticity would be to describe the biochemical steps necessary for induction of the various phases of LTP and the process that underlie metaplasticity. A successful model should describe all of the core problems: 1) How do interactions of various signals during induction determine the sign and persistence of LTP? 2) What are the maintenance mechanisms that confer stability to synaptic changes; these are the targets of cascades activated during induction 3) How do the maintenance mechanisms affect the expression mechanisms; i.e. the presynaptic and postsynaptic processes that control the release and detection of glutamate? Included in the last category are questions about receptor modulation, trafficking and anchoring, as well as the growth processes that can enlarge the synaptic area.

The number of reactions involved is likely to be very large; indeed the literature on LTP is already so formidable that there are separate subfields with little interaction with the other. In order to unify this material and to understanding the interactions of a large number of reactions, computer simulations will be a useful and probably necessary tool. Generating the computer simulation capability necessary to describe an LTP and LTD experiment is itself a formidable undertaking. We now turn to a discussion of this issue.

12. WHAT KIND OF COMPUTER SIMULATIONS WILL BE REQUIRED TO UNDERSTAND SYNAPTIC LTP/LTD?

We first review the small number of previous attempts to explore how known biochemical reactions produce synaptic plasticity. These have variously addressed the two core problems in plasticity: that of persistence and bidirectionality. In initial attempts at understanding the biochemistry of plasticity, it was assumed for simplicity that reactions were responding to steady-state concentrations of Ca^{2+} (Lisman, 1989). It was shown that calcineurin activation and adenylate cyclase action could yield a bell shaped dependence of PP1 on Ca. This made it possible to show how low Ca^{2+} could promote synaptic weakening whereas high Ca^{2+} promoted LTP and that CaMKII could act as a molecular switch capable of long-term information storage. More recently, models of CaMKII have used dynamic inputs to demonstrate switching of this molecule (Okamoto and Ichikawa, 2000; Zhabotinsky, 2000). Other work has attempted to explain aspects of spike-timing and tetanus-induced plasticity based on simple analytical equations for Ca^{2+} entry which were then used to show how different levels of Ca^{2+} differentially produced LTP or LTD depending on the phosphorylation and dephosphorylation of the CaMKII and PKA sites on GluR1 (Castellani et al., 2001; Shouval et al., 2002). In these models, no attempt was made to understand how modifications of the receptor could be maintained persistently (i.e. there is no switch). The work of Bhalla and Iyengar (Bhalla and Iyengar, 1999) is noteworthy because it is the most ambitious for including a very large number of biochemical reactions that appear involved. This model had temporal dynamics and provided a different interpretation of how a bistable switch might occur. It did not, however, address the issue of bidirectional plasticity. It should be emphasized that LTD is generally induced by very long periods of repetitive stimulation. None of the models so far developed give any insight into why such long periods of stimulation are required. The existence of these long time scales will, however, pose especially difficult computational problems. An additional complexity that must be modeled is membrane voltage since voltage can control Ca^{2+} entry, which in turn affects the biochemistry of plasticity (see above). Moreover, recent work in cortex suggests that while the occurrence of postsynaptic spikes during an EPSP produces LTP, the burst of spikes after EPSP causes LTD (Birtoli and Ulrich, 2004). Bhalla has recently extended the previous model of plasticity (Bhalla, 2002) to explicitly include Ca^{2+} diffusion and the dynamics of membrane voltage for a compartmental model of a CA1 pyramidal neuron. This model was able to produce distinct spatial-temporal patterns of reactant concentrations in response to temporally distinct stimuli. However, this model did not explore how these patterns could be translated into persistent bidirectional changes in synaptic strengths.

What specifications should a state-of-the-art simulation have to allow the development of models of LTP and LTD? What follows is our current thinking on this matter.

1. The model should be able to describe both the spatial and temporal aspects of the biochemistry. In this way, we can consider in detail the diffusion of multiple species (both intracellular and extracellular) of signaling molecules and second messengers along a dendritic segment and within the special chemical compartment created by spines.
2. Because of the voltage-dependence of NMDA channels and calcium channels, Ca^{2+} entry into the spine depends critically on the membrane voltage, which is driven by a combination of synaptic and voltage-dependent conductance. Furthermore, the spine

compartment may be at a different voltage than the dendrite (Raghavachari and Lisman, 2004b). This requires that the membrane voltage be explicitly modeled in the dendrite and spine compartments. The synaptic input necessary to induce LTP/LTD is driven by hundreds of synapses.

3. The presynaptic dynamical process of facilitation and feedback inhibition of GABA onto excitatory terminals are thought to be important in LTP induction
4. Previous work has shown that the autophosphorylating action of CaMKII and the antagonistic action of the phosphatase, PP1 form the basis of a bistable system that can be switched by changes in Ca^{2+} level. This switch is thought to be responsible for the maintenance of LTP. All previous models have modeled the switch using ordinary differential equations that assume continuous chemical concentration and deterministic dynamics. These assumptions are violated in the cell because the postsynaptic density of a single synapse typically contains only tens of molecules of the kinase and phosphatase. However, we have recently developed a mean-field model of the switch that accounts for stochastic fluctuations due to small numbers of reactants and protein turnover (Miller et al., 2004). As result, it may not be necessary to directly model the stochastic fluctuations of the switch. This would greatly simplify computations.
5. A number of biochemical reactions involved in synaptic plasticity take place between reactants in the cytoplasm and (virtually) immobilized reactants in the membrane or the PSD. Examples include the phosphorylation of AMPA receptors and the dephosphorylating actions of PP1. Recent work has shown that the spatial aspects of such reactions require modification of the assumption all the reactants are in the bulk phase and uniformly dispersed (Chotiner et al., 2003). Importantly, concentrations of enzymes within the PSD may be misleading since the spatial proximity of enzymes within the PSD may also determine their interactions.
6. Finally, the model must be able to account for processes that have long time-scales (~15 min for the induction of LTD)

Given these constraints, the mathematical model is a system of reaction-diffusion equations, representing the second messengers and other diffusible reactants coupled to a probabilistic master equation for the switch and the cable-equation for the membrane voltage. A number of methods exist to simplify these models, which essentially exploit differences in time and spatial scales, when these scales are separable.

An important issue is whether stochastic fluctuations of must be dealt with in models that describe LTP/LTD. Monte Carlo models have been developed (MCell) by Tom Bartol, Joel Stiles and co-workers (Bartol et al., 1991; Stiles et al., 1996; Franks et al., 2002)) which provide a spatially detailed simulation of neurotransmitter-mediated synaptic transmission. For *space-independent* unimolecular transitions such as unbinding and conformation changes, MCell computes time-evolved probabilities and makes decisions in a fashion similar to the Gillespie method. For *space-dependent* associations, on the other hand, MCell's algorithm is a unique and highly optimized integration of probabilistic diffusion methods and testing for binding transitions. Molecules move by random walk and there is a test for all possible transitions (binding, unbinding, conformation changes) at each step using Monte Carlo probabilities derived from bulk solution rate constants. We have recently used similar Monte Carlo techniques to construct a highly detailed model of AMPA transmission in the CA1 region of the hippocampus (Raghavachari and Lisman, 2004a). Monte Carlo methods are satisfactory at simulating interactions between membrane-bound and freely diffusing reactants. Another advantage is the ease with which conformational transitions, such as receptor opening, and covalent modifications, such as phosphorylation, can be handled. These models can be extended to include multiple diffusing species and complex biochemical reaction networks superimposed on increasingly complex spatial arrangements of participating molecules.

13. CONCLUSION

The work begun by John Sarvey on the biochemical processes involved in synaptic plasticity has progressed rapidly. The LTP field is now blessed with a great deal of data showing how the different phases of LTP and the synaptic weakening processes are affected by pharmacological and genetic perturbations. There is the feeling in the memory field that the cast of characters involved in LTP has been largely identified. These include second messenger cascades, kinases, phosphatases, channels, structural proteins, a vesicle delivery system, regulators of gene transcription and translation, and mediators of synaptic growth. The challenge ahead is to understand how this cast works together to make the play memorable.

14. ACKNOWLEDGEMENTS

This work was supported by the W.M. Keck Foundation and by National Institutes of Health Grants 2 R01 NS-27337 and P50 MH60450.

15. REFERENCES

Abel T, Nguyen PV, Barad M, Deuel TA, Kandel ER, Bourtchouladze R (1997) Genetic demonstration of a role for PKA in the late phase of LTP and in hippocampus-based long-term memory. Cell 88:615-626.

Abraham WC, Huggett A (1997) Induction and reversal of long-term potentiation by repeated high-frequency stimulation in rat hippocampal slices. Hippocampus 7:137-145.

Atkins CM, Nozaki N, Shigeri Y, Soderling TR (2003) CaMKII regulates cytoplasmic protein synthesis mediated by CPEB. In: Neuroscience Meeting, p 583.519. New Orleans.

Barr DS, Lambert NA, Hoyt KL, Moore SD, Wilson WA (1995) Induction and reversal of long-term potentiation by low- and high- intensity theta pattern stimulation. J Neurosci 15:5402-5410.

Bartol TM, Jr., Land BR, Salpeter EE, Salpeter MM (1991) Monte Carlo simulation of miniature endplate current generation in the vertebrate neuromuscular junction. Biophys J 59:1290-1307.

Bhalla US (2002) Biochemical signaling networks decode temporal patterns of synaptic input. J Comput Neurosci 13:49-62.

Bhalla US, Iyengar R (1999) Emergent properties of networks of biological signaling pathways. Science 283:381-387.

Birtoli B, Ulrich D (2004) Firing mode-dependent synaptic plasticity in rat neocortical pyramidal neurons. J Neurosci 24:4935-4940.

Bito H, Deisseroth K, Tsien RW (1996) CREB phosphorylation and dephosphorylation: a Ca(2+)- and stimulus duration-dependent switch for hippocampal gene expression. Cell 87:1203-1214.

Blitzer RD, Wong T, Nouranifar R, Iyengar R, Landau EM (1995) Postsynaptic cAMP pathway gates early LTP in hippocampal CA1 region. Neuron 15:1403-1414.

Blitzer RD, Connor JH, Brown GP, Wong T, Shenolikar S, Iyengar R, Landau EM (1998) Gating of CaMKII by cAMP-regulated protein phosphatase activity during LTP. Science 280:1940-1942.

Cammalleri M, Lutjens R, Berton F, King AR, Simpson C, Francesconi W, Sanna PP (2003) Time-restricted role for dendritic activation of the mTOR-p70S6K pathway in the induction of late-phase long-term potentiation in the CA1. Proc Natl Acad Sci U S A 100:14368-14373.

Castellani GC, Quinlan EM, Cooper LN, Shouval HZ (2001) A biophysical model of bidirectional synaptic plasticity: dependence on AMPA and NMDA receptors. Proc Natl Acad Sci U S A 98:12772-12777.

Chotiner JK, Khorasani H, Nairn AC, O'Dell TJ, Watson JB (2003) Adenylyl cyclase-dependent form of chemical long-term potentiation triggers translational regulation at the elongation step. Neuroscience 116:743-752.

Esteban JA, Shi SH, Wilson C, Nuriya M, Huganir RL, Malinow R (2003) PKA phosphorylation of AMPA receptor subunits controls synaptic trafficking underlying plasticity. Nat Neurosci.

Franks KM, Bartol TM, Jr., Sejnowski TJ (2002) A Monte Carlo model reveals independent signaling at central glutamatergic synapses. Biophys J 83:2333-2348.

Frey U, Huang YY, Kandel ER (1993) Effects of cAMP simulate a late stage of LTP in hippocampal CA1 neurons. Science 260:1661-1664.

Frey U, Krug M, Reymann KG, Matthies H (1988) Anisomycin, an inhibitor of protein synthesis, blocks late phases of LTP phenomena in the hippocampal CA1 region in vitro. Brain Res 452:57-65.

Frey U, Matthies H, Reymann KG, Matthies H (1991) The effect of dopaminergic D1 receptor blockade during tetanization on the expression of long-term potentiation in the rat CA1 region in vitro. Neurosci Lett 129:111-114.

Holland LL, Wagner JJ (1998) Primed facilitation of homosynaptic long-term depression and depotentiation in rat hippocampus. J Neurosci 18:887-894.

Huang YY, Kandel ER (1995) D1/D5 receptor agonists induce a protein synthesis-dependent late potentiation in the CA1 region of the hippocampus. Proc Natl Acad Sci U S A 92:2446-2450.

Huang YY, Nguyen PV, Abel T, Kandel ER (1996a) Long-lasting forms of synaptic potentiation in the mammalian hippocampus. Learn Mem 3:74-85.

Huang Y-Y, Kandel ER (1994) Recruitment of Long-lasting and Protein Kinase A-dependent Long-term Potentiation in the CA1 Region of Hippocampus Requires Repeated Tetanization. Learning & Memory 1:74-82.

Huang Y-Y, Nguyen PV, Abel T, Kandel ER (1996b) Long lasting forms of synaptic potentiation in the mamalian hippocampus. Learning & Memory 3:74-85.

Impey S, Mark M, Villacres EC, Poser S, Chavkin C, Storm DR (1996) Induction of CRE-mediated gene expression by stimuli that generate long- lasting LTP in area CA1 of the hippocampus. Neuron 16:973-982.

Johnston D, Christie BR, Frick A, Gray R, Hoffman DA, Schexnayder LK, Watanabe S, Yuan LL (2003) Active dendrites, potassium channels and synaptic plasticity. Philos Trans R Soc Lond B Biol Sci 358:667-674.

Kandel ER (2001) The molecular biology of memory storage: a dialogue between genes and synapses. Science 294:1030-1038.

Kang-Park MH, Sarda MA, Jones KH, Moore SD, Shenolikar S, Clark S, Wilson WA (2003) Protein phosphatases mediate depotentiation induced by high-intensity theta-burst stimulation. J Neurophysiol 89:684-690.

Krug M, Lossner B, Ott T (1984) Anisomycin blocks the late phase of long-term potentiation in the dentate gyrus of freely moving rats. Brain Res Bull 13:39-42.

Lee HK, Barbarosie M, Kameyama K, Bear MF, Huganir RL (2000) Regulation of distinct AMPA receptor phosphorylation sites during bidirectional synaptic plasticity. Nature 405:955-959.

Lisman J (1989) A mechanism for the Hebb and the anti-Hebb processes underlying learning and memory. Proc Natl Acad Sci U S A 86:9574-9578.

Lisman J, Schulman H, H. C (2002) The molecular basis of CaMKII function in synaptic and behavioral memory. In: Nat Rev Neurosci, pp 175-190.

Matthies H, Reymann KG (1993) Protein kinase A inhibitors prevent the maintenance of hippocampal long-term potentiation. Neuroreport 4:712-714.

Miller P, Lisman JE, Zhabotinsky A, Wang X-J (2004) in preparation.

Miller S, Yasuda M, Coats JK, Jones Y, Martone ME, Mayford M (2002) Disruption of dendritic translation of CaMKIIalpha impairs stabilization of synaptic plasticity and memory consolidation. Neuron 36:507-519.

Mockett B, Coussens C, Abraham WC (2002) NMDA receptor-mediated metaplasticity during the induction of long-term depression by low-frequency stimulation. Eur J Neurosci 15:1819-1826.

Moody TD, Carlisle HJ, O'Dell TJ (1999) A nitric oxide-independent and beta-adrenergic receptor-sensitive form of metaplasticity limits theta-frequency stimulation-induced LTP in the hippocampal CA1 region. Learn Mem 6:619-633.

Nguyen PV, Abel T, Kandel ER (1994) Requirement of a critical period of transcription for induction of a late phase of LTP. Science 265:1104-1107.

O'Dell T, Kandel E (1994) Low-frequency stimulation erases LTP through an NMDA Receptor-mediated Activation on Protein Phosphatases. Learning and Memory 1:129-139.

Okamoto H, Ichikawa K (2000) Switching characteristics of a model for biochemical-reaction networks describing autophosphorylation versus dephosphorylation of Ca2+/calmodulin-dependent protein kinase II. Biol Cybern 82:35-47.

Otmakhov N, Khibnik L, Otmakhova N, Carpenter S, Riahi S, Asrican B, Lisman J (2004) Forskolin-induced LTP in the CA1 hippocampal region is NMDA receptor dependent. J Neurophysiol 91:1955-1962.

Otmakhova NA, Lisman JE (1998) D1/D5 Dopamine Receptors Inhibit Depotentiation at CA1 Synapses via cAMP-Dependent Mechanism. J Neurosci 18:1270-1279.

Otmakhova NA, Otmakhov N, Mortenson LH, Lisman JE (2000) Inhibition of the cAMP pathway decreases early long-term potentiation at CA1 hippocampal synapses. J Neurosci 20:4446-4451.

Patterson SL, Pittenger C, Morozov A, Martin KC, Scanlin H, Drake C, Kandel ER (2001) Some Forms of cAMP-Mediated Long-Lasting Potentiation Are Associated with Release of BDNF and Nuclear Translocation of Phospho-MAP Kinase. Neuron 32:123-140.

Raghavachari S, Lisman JE (2004a) Properties of Quantal Transmission at CA1 Synapses. J Neurophysiol.

Raghavachari S, Lisman JE (2004b) Properties of Quantal Transmission at CA1 Synapses. J Neurophysiol April 28.

Raymond CR, Thompson VL, Tate WP, Abraham WC (2000) Metabotropic glutamate receptors trigger homosynaptic protein synthesis to prolong long-term potentiation. J Neurosci 20:969-976.

Roberson ED, English JD, Sweatt JD (1996) A biochemist's view of long-term potentiation. Learn Mem 3:1-24.

Roberson ED, English JD, Adams JP, Selcher JC, Kondratick C, Sweatt JD (1999) The mitogen-activated protein kinase cascade couples PKA and PKC to cAMP response element binding protein phosphorylation in area CA1 of hippocampus. J Neurosci 19:4337-4348.

Sarvey JM, Burgard EC, Decker G (1989) Long-term potentiation: studies in the hippocampal slice. J Neurosci Methods 28:109-124.

Shouval HZ, Bear MF, Cooper LN (2002) A unified model of NMDA receptor-dependent bidirectional synaptic plasticity. Proc Natl Acad Sci U S A 99:10831-10836.

Stanton PK, Sarvey JM (1984) Blockade of long-term potentiation in rat hippocampal CA1 region by inhibitors of protein synthesis. J Neurosci 4:3080-3088.

Staubli U, Chun D (1996) Proactive and retroactive effects on LTP produced by theta pulse stimulation: mechanisms and characteristics of LTP reversal in vitro. Learning and Memory 3:96-105.

Stiles JR, Van Helden D, Bartol TM, Jr., Salpeter EE, Salpeter MM (1996) Miniature endplate current rise times less than 100 microseconds from improved dual recordings can be modeled with passive acetylcholine diffusion from a synaptic vesicle. Proc Natl Acad Sci U S A 93:5747-5752.

Wagner JJ, Alger BE (1995) GABAergic and developmental influences on homosynaptic LTD and depotentiation in rat hippocampus. J Neurosci 15:1577-1586.

Wexler EM, Stanton PK (1993) Priming of homosynaptic long-term depression in hippocampus by previous synaptic activity. Neuroreport 4:591-594.

Winder DG, Mansuy IM, Osman M, Moallem TM, Kandel ER (1998) Genetic and pharmacological evidence for a novel, intermediate phase of long-term potentiation suppressed by calcineurin. Cell 92:25-37.

Woo NH, Abel T, Nguyen PV (2002) Genetic and pharmacological demonstration of a role for cyclic AMP-dependent protein kinase-mediated suppression of protein phosphatases in gating the expression of late LTP. Eur J Neurosci 16:1871-1876.

Yang HW, Lin YW, Yen CD, Min MY (2002) Change in bi-directional plasticity at CA1 synapses in hippocampal slices taken from 6-hydroxydopamine-treated rats: the role of endogenous norepinephrine. Eur J Neurosci 16:1117-1128.

Yang SN, Tang YG, Zucker RS (1999) Selective induction of LTP and LTD by postsynaptic [Ca2+]i elevation. J Neurophysiol 81:781-787.

Yasuda H, Barth AL, Stellwagen D, Malenka RC (2003) A developmental switch in the signaling cascades for LTP induction. Nat Neurosci 6:15-16.

Zhabotinsky AM (2000) Bistability in the Ca(2+)/calmodulin-dependent protein kinase-phosphatase system. Biophys J 79:2211-2221.

CREB: A CORNERSTONE OF MEMORY CONSOLIDATION?

Sheena A. Josselyn, Mahta Mortezavi, Alcino J. Silva*

1. INTRODUCTION

It has long been appreciated that memories can persist for dramatically different lengths of times, from seconds and minutes to a lifetime. Not only do these forms of memory differ in their persistence but they also have distinct molecular requirements. Short-term memory (STM) persists on a time scale of minutes to hours and is thought to be mediated by covalent modifications of existing synaptic molecules, such as the phosphorylation or dephosphorylation of enzymes, receptors or ion channels (Stork and Welzl, 1999). In contrast, long-term memory (LTM) persists for days or longer, and is thought to be mediated by the growth of new synapses and the restructuring of existing synapses (Bailey and Chen, 1989). There is extensive evidence from a wide variety of species that, unlike STM, LTM requires the transcription and translation of new proteins (Davis and Squire, 1984; Matthies, 1989). This raises the question of which transcription factors may mediate this process.

One attractive candidate for coupling the neuronal activation that occurs during learning with the gene expression required for LTM is cAMP (cyclic adenosine 3',5'-monophosphate) responsive element binding protein (CREB). CREB is a family of transcription factors that modulates the transcription of genes with cAMP responsive elements (CRE) located in their promoter regions. Although a wealth of convergent evidence from studies using invertebrate and vertebrate species shows that the CREB family of genes is important for LTM, recent studies have questioned whether the mammalian CREB gene itself, a single gene member of the CREB family, is truly critical to LTM formation (Balschun et al., 2003). This chapter provides a detailed examination of the evidence both for and against the involvement of CREB in LTM, particularly

* S.A. Josselyn and M. Mortezavi, Program in Integrative Biology and Brain & Behavior, Hospital for Sick Children Res Inst, 555 University Ave, Toronto, ON Canada M5G 1X8; A.J. Silva, Depts Neurobiology, Psychology and Psychiatry and Brain Res Institute, 695 Young Drive South, Gonda Building, UCLA, Los Angeles, CA 90095-1761. E-mail: Silvaa@mednet.ucla.edu

focusing on memory in mammals. Before this, however, we provide a brief overview of the structure and function of CREB.

2. STRUCTURE AND FUNCTION OF CREB

CREB is a member of a family of structurally related transcription factors. In mammals, at least three genes encode the CREB-like proteins, *CREB*, *CREM* (cAMP Response Element Modulator) and *ATF-1* (Activating Transcription Factor 1) (Foulkes and Sassone-Corsi, 1992; Hoeffler et al., 1988; Rehfuss et al., 1991). The mouse and human *CREB* gene is comprised of 11 exons (Cole et al., 1992; Hoeffler, 1992; Waeber et al., 1991), and alternative splicing generates the three major activator isoforms of CREB: α, d, and β (Blendy et al., 1996; Gonzalez and Montminy, 1989; Yamamoto et al., 1990). In addition to these transcriptional activators, the CREB family also includes transcriptional repressors. For example, the CREM gene codes several isoforms that repress CRE-dependent transcription: the CREM a, b and g isoforms as well as the inducible cyclic AMP early repressor (ICER) (Foulkes et al., 1991; Molina et al., 1993).

CREB, CREM and ATF1 proteins share a conserved basic leucine zipper (bZip) domain that is responsible for dimerization between CREB family members and DNA binding (Busch and Sassone-Corsi, 1990). Two glutamine-rich constitutive activation domains (Q1 and Q2) flank a kinase-inducible transactivation domain (KID) that contains a key phosphorylation site, initially characterized as being regulated by protein kinase A (PKA).

Although CREB may be phosphorylated at a number of residues [such as Ser129 or Ser142, see ((Sun et al., 1994; Giebler et al., 2000; Kornhauser et al., 2002)], the activation of transcriptional activity requires phosphorylation at Ser133 in the KID domain [for review see (Fimia et al., 1998)]. Mutation at this site (to a non-phosphorylatable Alanine) abolishes the stimulus-induced transcriptional activation of CREB (Gonzalez and Montminy, 1989). The phosphorylation state of Ser133 is regulated by a balance between the actions of protein phosphatases and kinases, which remove or add phosphate groups, respectively. A diverse array of kinases (such as cAMP/PKA, MAP kinases, CaM kinases and others) propagate signals from the plasma membrane to the nucleus where they phosphorylate CREB at Ser133 (Bito et al., 1996; Bacskai et al., 1993; Dash et al., 1991; Hagiwara et al., 1993; Chen et al., 1992; Finkbeiner et al., 1997; Xing et al., 1996; Wu et al., 2001; Hardingham et al., 2001). On the other hand, protein phosphatase 1 (PP1) and PP2A dephosphorylate this residue (Hagiwara et al., 1992; Bito et al., 1996; Wadzinski et al., 1993). Phosphorylation of CREB at Ser133 supports binding with a protein complex, that includes CREB-binding protein (CBP) and p300, and subsequent recruitment of the basal transcription machinery required to promote transcription (Chrivia et al., 1993).

The first studies to implicate CREB in memory were performed in the invertebrate *Aplysia*.

3. CREB AND LONG-TERM PLASTICITY IN SNAILS

Mechanical stimulation of the siphon or mantle shelf of the marine snail *Aplysia californica* produces a gill withdrawal response. This reflex exhibits several forms of plasticity that persist for different lengths of time, thus paralleling memory in mammals. For instance, a single shock to the tail produces transient sensitization that persists for several minutes (similar to STM). On the other hand, 5 or more shocks produces sensitization lasting days or longer (similar to LTM) (Bailey et al., 1992; Carew et al., 1971; Cleary et al., 1998; Frost et al., 1988; Goelet et al., 1986; Kandel, 1981; Pinsker et al., 1970).

The role of CREB in this memory-like phenomenon was evaluated using a cellular analog of sensitization in cultured neurons referred to as short- and long-term facilitation. Injection of a reporter gene driven by a CRE-containing promoter shows that stimuli that produce long-term facilitation trigger CRE-mediated transcription, while stimuli that are not sufficient to produce long-term facilitation do not (Kaang et al., 1993). Similarly, training that induces LTM in the pond snail *Lymnaea stagnalis* (using an appetitive classical conditioning protocol) also phosphorylates CREB (Ribeiro et al., 2003). Therefore, stimuli that induce memory-like phenomenon in snails engage CRE-mediated transcription.

The *Aplysia* CREB1 gene encodes three proteins (ApCREB1a, ApCREB1b and ApCREB1c) (Bartsch et al., 1995). ApCREB1a shares structural and functional homology with CREB transactivators in mammals, while ApCREB1b resembles mammalian ICER, a repressor of CREB transcription. Blocking ApCREB1a or enhancing ApCREB1b function in cultured sensory neurons blocks long-term facilitation (but not short-term facilitation) (Bartsch et al., 1995; Dash et al., 1990). Consistent with this, disrupting CREB function impairs LTM in the pond snail (Sadamoto et al., 2004). On the other hand, enhancing ApCREB1a or inhibiting ApCREB1b function lowers the threshold for producing long-term facilitation (Bartsch et al., 1995; Martin et al., 1997). Consistent with the earlier electrophysiological results in *Aplysia*, behavioral results in fruit flies also suggest that CREB manipulations can both impair and enhance LTM, findings that demonstrates a critical role for CREB-like proteins in memory consolidation.

4. CREB AND MEMORY IN FLIES

Memory in fruit flies (*Drosophila melanogaster*) may be assessed following associative training. For instance, in an olfactory conditioning paradigm, flies learn to avoid a previously neutral odor that was paired with shock in favor of another odor that was not paired with shock (Tully, 1991). Multiple training trials in this paradigm produces LTM that lasts several hours and requires protein synthesis (Tully et al., 1994).

In the mid 1970s, behavioral screens of mutagenized flies using this paradigm led to the isolation and characterization of several genes involved in learning and memory (Dudai et al., 1976; Quinn et al., 1974; Quinn et al., 1979; Livingstone et al., 1984). Mutants identified by this screen were subsequently determined to have disruptions in components of the cAMP/PKA signaling cascade that is directly upstream of CREB (Tully, 1991; Byers et al., 1981; Levin et al., 1992).

CREB itself was shown to be involved in fruit fly memory by Yin, Tully and colleagues using a reverse genetics approach (Yin et al., 1994). Acutely disrupting CRE-mediated transcription by transgenic overexpression of a CREB transcriptional repressor (dCREB2b) blocked the LTM, but not STM, produced by multiple training trials. This suggests that the protein synthesis required for LTM in this task is mediated, at least in part, by CREB-dependent transcription. However, it is important to note that the approaches used in both the *Aplysia* and *Drosophila* experiments could have affected a number of CREB-like transcription factors. For example, over-expression of the CREB repressor in flies could have interfered with a range of CREB-like transcription factors that recognize promoter elements similar to those bound by CREB itself.

Spaced training (in which training trials are separated by long intervals) is required to produce maximal LTM in species ranging from *Aplysia* (Carew and Kandel, 1973; Pinsker et al., 1973; Frost and Kandel, 1995; Cleary et al., 1998; Mauelshagen et al., 1998) to human (Ebbinghaus, 1885). The same holds true in flies: multiple spaced training is necessary for maximal LTM, whereas the same number of trials presented in a massed fashion (with short or no time between successive trials) produces strong STM but weak LTM. However, massed training is sufficient to produce strong LTM if a CREB activator (dCREB2a) is overexpressed. Indeed, a single training trial produces LTM in these CREB-overexpressing flies (Yin et al., 1995), perhaps the fly equivalent of a 'photographic' or 'flashbulb' memory (Yin and Tully, 1996). Transgenic flies overexpressing a mutant activator, where Ser231 (similar to Ser133 of the mammalian CREB gene) was replaced by an Alanine, do not show LTM after one training trial, indicating that phosphorylation of CREB at this residue (and CRE-mediated transcription) is required for the enhancement of LTM (Yin et al., 1995). More recently, dCREB2 was found to be differentially expressed following spaced, rather than massed, training in this task (Dubnau et al., 2003). Together, these results show that CREB manipulations can both impair and enhance LTM formation in *Drosophila*. Furthermore, these findings indicate that the temporal characteristics of training may be a critical variable in determining the type of memory induced and whether that memory critically depends on intact CREB function.

5. CREB AND MEMORY IN MAMMALS

Studies using invertebrates have made fundamental strides towards resolving the role of CREB in LTM. Parallel studies in mammals have also shown that CREB is important for memory. Mammalian memory has been assessed using several memory tasks [including fear conditioning, recognition memory and spatial memory] that may differentially rely on distinct brain regions. In addition to different memory tests, different molecular approaches have also been used to study CREB function in mammals, including mutant mice (knockouts and transgenics), antisense and viral vectors. Each method alters CREB function in a distinct way and, in total, they provide a considerable body of evidence for the role of CREB-dependent transcription in LTM.

5.1 Deletions of CREB by homologous recombination

The homozygous deletion of the three major CREB isoforms (α, β and d; $CREB^{null}$) produced by a deletion in the bZIP domain (exon 10) of the *CREB* gene results in fully formed mice that die at birth due to atelectasis of the lung (Rudolph et al., 1998). However, inserting a neomycin-resistance cassette (neo) into exon 2 mice produces viable mice that mature into adulthood but have a deletion of the two main isoforms of CREB (a and d) (Hummler et al., 1994). The insertion of the neo gene into exon 2 does not disrupt the translation of CREBβ (which begins at exon 4). In fact, these $CREB^{ad}$

mice have upregulated levels of CREBβ as well as CREM activator (τ) and repressor (α and β) isoforms (Blendy et al. 1996), perhaps explaining why these mice develop normally into adulthood. Despite these upregulations, however, CRE-DNA binding is virtually abolished (by > 90%) (Walters and Blendy, 2001; Pandey et al., 2000) and the levels of CREB protein are dramatically reduced (roughly 85-90% reduction compared to controls) in the brains of $CREB^{ad}$ mice (Walters and Blendy, 2001Walters, 2003 #242).

Crossing a $CREB^{ad}$ mouse with a mouse heterozygous for the $CREB^{null}$ mutation results in mice that carry a single allele for the CREBβ isoform referred to as $CREB^{comp}$ mice (Gass et al., 1998). Similar to the $CREB^{ad}$ mice, these mice mature normally into adulthood and have upregulated levels of CREM.

Conditional CREB knockout mice have also been generated by inserting loxP recognition sequences around exon 10 of the *CREB* gene (Balschun et al., 2003; Mantamadiotis et al., 2002). Recombination of the floxed CREB allele (leading to a null allele that encodes an unstable truncated CREB protein devoid of DNA-binding and dimerization domains) occurs in cells with sufficient levels of Cre recombinase. Mice homozygous for $CREB^{loxP}$ were crossed with transgenic mice expressing Cre recombinase under two different promoters, aCaMKII and nestin. The aCaMKII promoter directs expression of Cre recombinase in postnatal excitatory forebrain neurons whereas the nestin promoter directs expression in all brain cells earlier in development (roughly embryonic day 12.5) (Mantamadiotis et al., 2002). Crossing the $CREB^{loxP}$ mice with a line of transgenic mice expressing Cre recombinase fused to a C-terminal fragment of a progesterone receptor under the aCaMKII promoter produced mice ($CREB^{CaMKCre7}$) with a loss of CREB protein in roughly 70-80% of CA1 (and other forebrain) neurons (as determined by CREB immunoreactivity in brain slices) (Balschun et al., 2003; Mantamadiotis et al., 2002). On the other hand, $CREB^{NesCre}$ mice show a loss of CREB in all brain regions and have a dwarf phenotype (Balschun et al., 2003; Mantamadiotis et al., 2002). Similar to $CREB^{\alpha d}$ mice, these conditional CREB knockout mice have upregulated levels of CREM (Mantamadiotis et al., 2002).

5.2 Decreasing CREB function by transgenic expression of a dominant interfering form of CREB

Another method of disrupting CREB function involves the transgenic expression of a dominant negative allele that blocks the function of all CREB isoforms as well as CREM and ATF1. For instance, Rammes et al. (2000) developed transgenic mice that express a phosphorylation-defective CREB transgene (M-CREB) in which the primary phosphorylation site for transcriptional activation, Ser133, is mutated to an Ala (S133A). This mutated form of CREB cannot be activated by phosphorylation at Ser133, but still binds and occupies CRE sites, thus repressing endogenous CREB function (Gonzalez and Montminy, 1989). Expression of this CREB repressor was constitutively driven by the aCaMKII promoter (Rammes et al., 2000). Three lines of transgenic mice were generated and showed transgene expression in the hippocampus, amygdala and cortex. Whether this manipulation caused a decrease of CRE-mediated transcription was not reported.

5.3 Inducible and region specific manipulations of CREB function

Temporal control over the S133A CREB repressor was achieved by fusing the CREB repressor with a ligand-binding domain (LBD) of a human estrogen receptor with a G521R mutation (LBD^{G521R}) in transgenic mice (Kida et al., 2002). The activity of this mutated LBD is regulated not by estrogen but by the synthetic ligand, tamoxifen (Danielian et al., 1993; Logie and Stewart, 1995; Feil et al., 1996). In the absence of tamoxifen, the LBD^{G521R}-$CREB^{S133A}$ fusion protein is bound to heat shock proteins and is therefore inactive (Feil et al., 1996). However, administration of tamoxifen activates this inducible CREB repressor ($CREB^{IR}$) fusion protein, allowing it to compete with endogenous CREB and disrupt CRE-mediated transcription. Region specific expression was achieved by using a promoter (aCaMKII promoter) that is only active in a subset of brain regions, including the hippocampus, amygdala and cortex (Kida et al., 2002) Disruption of CREB function in these mice is shown by a decrease in mRNA for 14-3-3eta, a downstream target of CREB. Importantly, unlike the CREB "knock-out" mice ($CREB^{ad}$, $CREB^{comp}$, $CREB^{NesCre}$ mice), $CREB^{IR}$ mice do not show upregulated levels of CREM, perhaps due to the acute nature of the CREB disruption.

Pittenger and colleagues (Pittenger et al., 2002) used another dominant negative form of CREB to inducibly repress CREB function in forebrain neurons. K-CREB, a mutant form of CREB that dimerizes with other CREB family members but does not bind DNA (Xie et al., 1997; Walton et al., 1992), was expressed under the aCaMKII promoter in combination with the tetracycline transactivator (tTA/*tetO*) system, in which expression of the transgene is turned off in the presence of doxycycline (a tetracycline analog). One line of mice (referred to as dCA1-KCREB transgenic mice) showed expression of the transgene in dorsal CA1 regions of the hippocampus (Pittenger et al., 2002). The downregulation of enkephalin mRNA levels confirms the disruption of CRE-mediated transcription in these mice.

Mice have also been engineered that have enhanced CREB function. Fusing CREB with the transactivation domain of HSV VP16 produced a form of CREB that is 25-fold more active that wildtype (WT) CREB *in vitro* (Barco et al., 2002). Transgenic mice expressing this constitutively active form of CREB under the control of the aCaMKII promoter in combination with the tetracycline transactivator system were developed. CRE-mediated transcription in this mouse is increased as evidenced by enhanced expression of several CRE-regulated genes, including BDNF, dynorphin and c-fos (Barco et al., 2002).

5.4 Antisense knockdown of CREB function

Antisense oligonucleotides have also been used to disrupt CREB function in a temporally and spatially restricted manner in otherwise normal rats and mice (Wahlestedt, 1994). Synthetic oligonucleotides that are the genetic complement (antisense) of the mRNA encoding a specific protein are rapidly taken up by neurons (Ogawa and Pfaff, 1996), base-pair to their cognate mRNAs, thereby blocking the synthesis of specific proteins (Ghosh and Cohen, 1992). Typically, the effects of CREB antisense oligonucleotides are compared to two control treatments; oligonucleotides of the same base composition but in a randomized order (scrambled oligonucleotides) and

PBS. Injection of CREB antisense acutely and significantly decreased endogenous CREB protein (from 39-97%) in the brain region of interest (Lamprecht et al., 1997; Guzowski and McGaugh, 1997).

5.5 Viral vectors to manipulate CREB function

Spatial and temporal control of the CREB function may also be achieved through the use of viral vectors. This method exploits the natural ability of viruses, such as herpes simplex virus (HSV), to insert DNA specifically into neurons (Simonato et al., 2000). Replication-defective HSV vectors expressing wild-type (HSV-CREB) or mutant forms of CREB (HSV-M-CREB) have been engineered. Infusing HSV-CREB into the brains of rats and mice not only increases the level of CREB protein but also CRE-mediated transcription [as shown by increases in CRE-regulated transcription of an endogenous (dynorphin), and reporter gene (in CRE-reporter mice, see below)] (Barrot et al., 2002; Carlezon et al., 1998). On the other hand, HSV-M-CREB decreases CRE-mediated transcription (Barrot et al., 2002) and the level of dynorphin mRNA (Carlezon et al., 1998). In addition to studies of the role of CREB in memory, this approach has also been used to show that CREB is a key transcription factor in the neuronal adaptation to drugs of abuse (Carlezon et al., 1998) and ocular dominance plasticity (Mower et al., 2002).

5.6 Detecting CREB activation during learning

The role of CREB in memory may be studied by manipulating CREB function (above) but also by using imaging techniques to examine CREB activation and CRE-mediated transcription following various types of stimulation. These studies are invaluable since they provide a powerful synergy between systems and molecular approaches. They can identify *where* and *when* CREB activation occurs following stimulation that induces memory.

For instance, transgenic mice with a β-galactosidase reporter construct under the regulation of a CRE-promoter (CRE-LacZ) have been used to identify the types of behavioural training and electrophysiological stimulation that induce CRE-mediated transcription (Impey et al., 1996; Impey et al., 1998). In addition, CREB activation may be monitored by examining the levels of CREB phosphorylated at Ser133 or downstream products of CRE-mediated transcription such as C/EBP (Taubenfeld et al., 2001a; Desmedt et al., 2003).

6. CONDITIONED FEAR MEMORY

Fear conditioning is a common method of assessing memory in rodents. Typically, mice or rats are placed in a conditioning chamber and a footshock is delivered. In the discrete cue version of this task, a tone that co-terminates with the footshock is added. Memory is assessed as the percentage of time rodents spend freezing (defined as the cessation of all movements except respiration) when placed back in the conditioning chamber (refereed to as contextual fear conditioning) or when the tone is played in a novel chamber (referred to as discrete cued fear conditioning). The neural systems critical for mediating fear conditioning have been well characterized. The amygdala is crucial for both cued and contextual fear conditioning as lesioning or inhibiting protein synthesis in this structure disrupts LTM (LeDoux, 2000; Davis, 1992; Fanselow and Gale, 2003; Schafe and LeDoux, 2000). However, the role for the hippocampus in fear conditioning is not as straightforward (see below).

The first study to examine the role of CREB in mammalian memory used the $CREB^{\alpha d}$ mice (Bourtchuladze et al., 1994). $CREB^{ad}$ mice on a mixed 129 x C57Bl\6 background showed normal conditioned freezing to both a tone and context previously paired with footshock when tested shortly (<1 hour) after training. However, both contextual and cued fear conditioning were impaired in another group of mice tested 24 hours after training (Bourtchuladze et al., 1994). The deficit in LTM for fear conditioning has been replicated using both freezing (Kogan et al., 1997; Graves et al., 2002; Frankland et al., 2004) and fear-potentiated startle (Falls et al., 2000) to measure conditioned fear. One study using F1 hybrids (129 x C57Bl\6) also noted a deficit in STM (Graves et al., 2002). It is also important to note that the LTM deficit may be overcome by additional spaced training trials, highlighting the importance of the temporal dynamics of the training protocol to the CREB phenotype (Kogan et al., 1997).

Using $CREB^{ad}$ mice from a different genetic background (FVB/N x C57Bl\6), Gass and colleagues (1998) noted normal STM and mild LTM deficits in both cued and contextual fear conditioning (Gass et al., 1998). However, as pointed out by Graves and colleagues (2002), mice from this genetic background are generally poorer learners, potentially masking a greater effect of the $CREB^{ad}$ mutation. Nevertheless, $CREB^{comp}$ mice (with only one CREBb allele) showed a severe LTM deficit in both context and cued fear conditioning (Gass et al., 1998). This finding indicates that gene dosage is another important variable. That is, mice with a greater disruption in CREB function ($CREB^{comp}$ versus $CREB^{ad}$ mice) show a greater impairment in LTM.

Transgenic mice overexpressing a dominant negative form of CREB have also been tested using fear conditioning. One of 3 lines of mice that constitutively express the CREB repressor in the amygdala and hippocampus showed impaired LTM for cued fear conditioning (Rammes et al., 2000). However, as the "time not moving" and "distance traveled" during the test session were used as the index of memory, rather than the more sensitive measure of time spent freezing, in these hybrid mice (FVB/N x C57Bl\6), it is possible that a subtle phenotype in the other lines could have been obscured. Furthermore, in these transgenic mice the CREB repressor is expressed constitutively (driven by the aCaMKII promoter) perhaps leading to a compensatory upregulation by other CREB family members.

This potential limitation, however, was addressed using a transgenic mouse that allows precise temporal control over the expression of the CREB repressor (Kida et al., 2002). Importantly, these $CREB^{IR}$ transgenic mice develop with normal CREB levels. CREB function is acutely and reversibly disrupted a short time (6 hours) prior to training. Mice are tested 24 hours later, with presumably normal CREB function. LTM, but not STM, for both cued and contextual fear conditioning was impaired in these transgenic mice (Kida et al., 2002). Consistent with this, acutely disrupting CREB function in the CA1 region of the hippocampus through the use of antisense oligonucleotides disrupted LTM for contextual fear (Athos et al., 2002). Together, these findings show that acutely disrupting CREB function impairs LTM for conditioned fear. Moreover, as the disruptions of CREB are temporary, these data also provide compelling evidence that the

LTM phenotype observed in mutant mice with a chronic disruption of CREB (CREB^{ad} or CREB^{comp} mice) cannot be solely attributed to developmental abnormalities.

Therefore, the finding of normal LTM for contextual fear conditioning reported in 3 other types of CREB mutant mice may be surprising. Normal LTM for contextual fear conditioning was observed in dCA1-KCREB mice [expressing a CREB repressor in the dorsal CA1 region of the hippocampus (Pittenger et al., 2002)] and two lines of conditional CREB knockout mice [CREB^{CaMKCre7}, with a deletion of CREB in 70-80% of CA1 neurons of the hippocampus; and CREB^{NesCre}, with a deletion of CREB in the nervous system (Balschun et al., 2003)]. These results, however, may not be unexpected in light of results showing that the hippocampus is not strictly required for contextual fear conditioning. While post-training lesions of dorsal hippocampus dramatically impair contextual fear conditioning, pre-training lesions do not (Frankland et al., 1998; Maren et al., 1997). Normal LTM for contextual fear conditioning task in these hippocampal-lesioned rodents may result from the use of elemental (rather than contextual) strategies (Frankland et al., 1998; Maren et al., 1997). Importantly, the three lines of CREB mutant mice that have normal LTM for contextual fear have chronic disruptions in CREB that could be analogous to "pre-training lesions" of CREB. Thus, unlike antisense that acutely disrupts CREB function, CREB function is chronically impaired in both the CREB^{CaMKCre7} and CREB^{NesCre} mice. Even in the inducible dCA1-KCREB mice, the transgene is turned "on" for a significant period of time before training (e.g., the mice are not fed doxycycline). Taken together, these findings suggest that this method of contextual fear conditioning (in which mice may acquire the task bys using non-hippocampal elemental-based strategies) may not be sufficiently sensitive to show potential LTM deficits in mice with more chronic pre-training perturbations in hippocampal CREB function.

To address the potential limitation of contextual fear conditioning, a variation of this task that critically relies on the hippocampus was used to evaluate the role of CREB in contextual LTM. Unlike standard contextual conditioning, this context pre-exposure task temporally separates the acquisition of context information from its association with the footshock (Wiltgen et al., 2001; Matus-Amat et al., 2004; Frankland et al., 2004). Animals that were pre-exposed to a training context the day (Day 0) before receiving a shock (Day 1) immediately following replacement in that context show high levels of freezing, whereas animals not pre-exposed to the context show low levels of freezing following this immediate shock (Day 1) (Wiltgen et al., 2001; Matus-Amat et al., 2004; Frankland et al., 2004). This phenomenon is sometimes referred to as the immediate shock deficit. Infusion of anisomycin into the dorsal hippocampus following context pre-exposure (on Day 0) results in low freezing levels following the immediate shock (on Day 1) indicating that protein synthesis in the dorsal hippocampus is vital for the context pre-exposure rescue of the immediate shock deficit (Barrientos et al., 2002). Perhaps similarly, CREB^{ad} mice show low levels of freezing following an immediate shock despite being pre-exposed to the training context the day before. Freezing levels in the CREB^{ad} mice were similar to those observed in WT mice that were not pre-exposed to the training context (Frankland et al., 2004). Therefore, using this task that critically depends on protein synthesis in the hippocampus, CREB was shown to be important for LTM for contextual memory.

The results of imaging studies show that behavioural training that produces LTM for conditioned fear also increases the levels of phosphorylated CREB and CRE-mediated transcription. Thus, CRE-LacZ reporter mouse showed an increase in CRE-mediated transcription in the CA1 region of the dorsal hippocampus and amygdala following fear training (Impey et al., 1998; Athos et al., 2002). Similar training also produced an increase the levels of pCREB and CRE-regulated genes (such as C/EBP) (Impey et al., 1998; Taubenfeld et al., 2001b; Taubenfeld et al., 1999; Viola et al., 2000; Stanciu et al., 2001; Cammarota et al., 2000; Bevilaqua et al., 1999). Although these data linking CREB activation to behavioral stimuli that induce LTM are correlative, they converge with findings from many studies that showing that disrupting CREB function (through a variety of different approaches) produces deficits in LTM for conditioned fear. But, what are the effects of increasing CREB function on LTM for conditioned fear?

To answer this question, Josselyn and colleagues (2001) used HSV-CREB to increase CREB levels and function in a subpopulation of amygdala neurons (roughly 20-30% of neurons in the lateral nucleus of the amygdala) (Josselyn et al., 2001). Rats were trained for cued fear conditioning using a massed training protocol that typically produces weak STM but no or weak LTM. However, rats infused with HSV-CREB in the amygdala two days prior to training show robust LTM, similar to levels produced by spaced training. The finding of enhanced cued fear memory following infusion of HSV-CREB into the amygdala was recently replicated (Wallace et al., 2004). Thus, acutely increasing CREB levels and function can enhance memory, a finding consistent with results in *Drosophila* (Yin et al., 1995) and *Aplysia* (Bartsch et al., 1995).

It is interesting to note that transgenic mice with chronic overexpression of CREB in cerebellar Purkinje neurons, however, show decreased CRE-mediated transcription and decreased performance in motor learning tasks (Brodie et al., 2004). This result is similar to the findings that chronic expression of the CREB repressor produces no change in LTM while acute expression produces a deficit in LTM. Together, these results highlight the tightly regulated nature of the CREB system and how this affects the impact of CREB manipulations in LTM.

7. RECOGNITION MEMORY (OBJECT AND SOCIAL RECOGNITION)

Novel object recognition is a task that relies on the natural exploratory behavior of mice. Typically, training consists of exposing mice to two identical objects for a short period of time (e.g., 15 min). Memory for this task is shown by mice spending more time exploring a novel object rather than the familiar object used in training. Two groups of researchers showed that decreasing CREB function specifically disrupt LTM, but not STM, in this task. Thus, dCA1-KCREB mice (Pittenger et al., 2002) and $CREB^{IR}$ mice (Bozon et al., 2003) have disrupted LTM for object recognition.

Consistent with previous data using a variety of learning tasks and species, spaced, rather than massed, training produces maximal LTM for object recognition (Genoux et al., 2002). Furthermore, spaced but not massed object recognition training is associated with increased levels of phosphorylated CREB and CRE-mediated transcription in the hippocampus and cortex (Genoux et al., 2002). However, in transgenic mice with decreased PP1 activity (a phosphatase that dephosphorylates CREB at Ser133) massed training alone is sufficient to produce both robust LTM and an increase in CRE-mediated transcription (Genoux et al., 2002). A similar pattern of results was also observed in transgenic mice that overexpress type-1 adenylyl cylase

(Wang et al., 2004). These mice show elevated levels of pCREB and LTM for object recognition following a single training trial (whereas the WT littermate controls do not).

Similar results have also been found using a social recognition task. Training in this memory test consists of exposing a mouse to a conspecific. Memory is inferred if the amount of time a mouse spends exploring this familiar conspecifics is less than the time spent exploring a novel conspecific. LTM, but not STM, for social recognition is impaired in both the $CREB^{\alpha d}$ and $CREB^{IR}$ mice (Kogan et al. 2000(Falls et al., 2000) Therefore, using two types of recognition tasks, several groups of researchers have shown that CREB is activated following spaced training whereas disrupting CREB impaired LTM and increasing CREB function enhanced LTM.

8. FOOD PREFERENCE MEMORY

Rodents develop a preference for foods recently smelled on the breath of other rodents (Bunsey and Eichenbaum, 1995; Galef et al., 1988; Galef et al., 1983). Training that produces LTM for social transmission of food preference increases the level of pCREB in the hippocampus (Countryman et al., 2004). Consistent with this, $CREB^{\alpha d}$ mice show intact STM, but impaired LTM, in this task (Kogan et al., 1997). Using a longer "training time" (10 minutes) with which to interact with the breath of a demonstrator mouse, Gass and colleagues (1998) showed normal LTM in both $CREB^{ad}$ and $CREB^{comp}$ mice in a C57Bl\6 x FVB/N background. These results show that the deficits in CREB mutant mice are dependent on the training conditions of the tasks, and perhaps, even on the genetic background of the animals tested. This and other results demonstrate that CREB is not the only transcription factor mediating LTM formation, and that, under certain circumstances, loss of CREB can be compensated by other genes.

9. OLFACTORY MEMORY

Memory may also be assessed using an olfactory conditioning task in neonatal rats. In this task an odor is paired with an appetitive (such as a stroke of the back) or an aversive (footshock) stimulus. Both appetitive and aversive olfactory conditioning are associated with an increase in pCREB levels in the olfactory bulbs (McLean et al., 1999; Zhang et al., 2003). Moreover, disrupting CREB function (via infusion of CREB antisense oligonucleotides or HSV-M-CREB) produces a specific LTM deficit for olfactory conditioning (Zhang et al., 2003; Yuan et al., 2003).

10. CONDITIONED TASTE AVERSION MEMORY

In the conditioned taste aversion paradigm, ingestion of a novel taste is paired with transient sickness (produced by injection of lithium chloride, LiCl). Memory for this association is demonstrated by the animal avoiding that taste on subsequent presentations (Garcia et al., 1955). Previous studies show that LTM for conditioned taste aversion depends on protein synthesis in the amygdala (Yamamoto and Fujimoto, 1991; Frankland and Josselyn, 2004) and insular cortex (Rosenblum et al., 1993).

The results from many studies that use different techniques to disrupt CREB function (mutant mice and antisense) converge to show that CREB is important in LTM for conditioned taste aversion. Thus, $CREB^{ad}$ (Frankland and Josselyn, 2004), $CREB^{IR}$

(Frankland and Josselyn, 2004) and CREB^{NesCre} (Balschun et al., 2003) mice show disrupted LTM for conditioned taste aversion. CREB^{CaMKCre7} mice and CREB^{comp} mice have not been tested in this paradigm. Acutely disrupting CREB function in the amygdala through the use of antisense similarly disrupts LTM but not STM for conditioned taste aversion (Lamprecht et al., 1997). Furthermore, training that produces conditioned taste aversion (pairing the novel taste with LiCl) also induced robust CREB activation (phosphorylation) in the lateral nucleus of the amygdala (Swank, 2000). Similar increases are not observed if rats are exposed to the taste or LiCl alone, indicating that activation of CREB is related to associative learning. Unlike some other memory tasks, conditioned taste aversion places few performance demands on the subject. Therefore conclusions regarding the role of CREB in memory may be drawn independent of potential effects on other behaviors, such as motor behaviour. Virtually all experimental data from many different labs examining the role of CREB in conditioned taste aversion memory point to the same conclusion; that CREB is critical for LTM.

11. SPATIAL MEMORY

In the hidden platform version of the Morris water maze rodents learn to find a platform submerged in a pool of opaque water by using spatial cues in the experimental room (Morris et al., 1982). Typically, spatial memory is assessed during a probe trial in which the platform is removed and the percentage of time the animals spends searching in the spatial location where the platform was previously positioned (target quadrant) is measured. This form of spatial memory is sensitive to hippocampal lesions (Sutherland et al., 1982; Morris et al., 1982).

In the first study to use the Morris water maze to study the role of CREB in spatial memory, CREB^{ad} mice were trained with 1 trial a day for 15 days (Bourtchuladze et al., 1994). At the beginning of training, the latencies to reach the platform for both WT and CREB^{ad} mice were the similarly long. As training progressed, the time to reach the platform grew shorter and shorter in WT mice (indicating learning) while the latencies of CREB^{ad} mice remained long. As predicted from these training latencies, WT mice searched selectively in the target quadrant during the probe test while CREB^{ad} mice searched randomly (Bourtchuladze et al., 1994). Moreover, WT mice crossed the exact previous location of platform more often than the CREB^{ad} mice. The spatial memory deficit in CREB^{ad} mice was replicated using similar, as well as more intense, training schedules (Bourtchuladze et al., 1994; Kogan et al., 1997). However, similar to other memory tests discussed above, the role of CREB in spatial memory is sensitive to the training parameters. Thus, increasing the time between trials to 10 minutes or 1 hour (rather than 1 minute) overcomes the spatial memory deficit in CREB^{ad} mice (Kogan et al., 1997; Gass et al., 1998).

Spatial memory deficits have also been observed using two different techniques to significantly disrupt CREB function in the hippocampus. Mice that overexpress a dominant negative form of CREB in the dorsal CA1 region of the hippocampus (dCA1-KCREB), also showed impaired spatial memory as measured by several variables (percentage time spent in the target quadrant, proximity to the training location of the platform and platform crossing) (Pittenger et al., 2002). A critical control experiment showed that other dCA1-KCREB transgenic mice fed doxycycline (thereby turning the transgene off) for 2 weeks before beginning of an identical Morris maze experiment had normal spatial memory. Thus, the phenotype in these mutant mice cannot be attributed to development deficits. In addition, disrupted spatial memory was found following infusion of CREB antisense into the dorsal hippocampus of otherwise normal rats (Guzowski and McGaugh, 1997). Consistent with this, a treatment that significantly decreased the level of phosphorylated CREB (active) in the hippocampus of adult mice (prenatal administration of morphine) also produced a deficit in spatial memory (Yang et al., 2003). Therefore, many labs have shown that disrupting CREB function in the hippocampus disrupts spatial memory.

Balschun and colleagues (2003) evaluated the impact of various CREB mutations (using all CREB mutants studied by this group including those of Gass et al. 1988) in a large analysis of spatial memory. The percentage of time spent in the target quadrant during the probe trial for WT mice (n=38) was 35.78%, $CREB^{ad}$ (n=10) 28.5%, $CREB^{comp}$ (n=13) 26.39%, $CREB^{NesCre}$ (n=14) 31.11% and $CREB^{CaMKCre7}$ (n=12) 40.87%. Although these authors interpret this pattern of results as indicating that CREB is not necessary for spatial memory, an alternative interpretation is that the spatial memory phenotype is determined by the degree of CREB disruption. In support of this interpretation, the $CREB^{comp}$ mice (with a loss of both CREBa and d alleles as well as one CREBb allele) show poorer spatial memory than the $CREB^{ad}$ mice (with a loss of both CREBa and d alleles but two intact CREBb alleles) (see also Gass et al. 1988). Similarly, the $CREB^{NesCre}$ mice (with a virtually complete deletion in CREB) showed poorer spatial memory than the $CREB^{CaMKCre7}$ mice (with a loss of CREB in 70-80% of CA1 neurons). This latter finding of normal spatial memory in the $CREB^{CaMKCre7}$ mice suggests, therefore, that the remaining 20-30% of CA1 neurons with normal levels of CREB is sufficient to support normal spatial memory. Together, these findings suggest that CREB function may be disrupted in a larger portion of hippocampal neurons by transgenic overexpression of a dominant negative form of CREB (as in dCA1-KCREB transgenic mice) or CREB antisense than the "mosaic" pattern of CREB deletion reported in the hippocampus of $CREB^{CaMKCre7}$ mice.

12. SYNAPTIC PLASTICITY: LONG-TERM POTENTIATION (LTP) AND LONG-TERM DEPRESSION (LTD)

Repetitive electrical activity can induce persistent increases in the strength of synaptic transmission, referred to as long-term potentiation (LTP). LTP exhibits several characteristics that make it an attractive model of memory (long-lasting, co-operativity, associativity, input specificity, reversibility, etc.) (Bliss and Collingridge, 1993; Maren and Baudry, 1995). Although numerous reviews have drawn attention to the difficulties strictly linking LTP to memory (Shors and Matzel, 1997; Lynch, 1998), LTP remains the strongest cellular model of learning-related plasticity in the mammalian nervous system (Martin et al., 2000; Moser et al., 1998; Goosens and Maren, 2002).

The best studied form of LTP is between CA3 and CA1 pyramidal neurons of the hippocampus. An extensive literature now demonstrates that a large number of genetic and pharmacologic manipulations that disrupt LTP also impair learning, while manipulations that result in modest increases in CA3/CA1 LTP often also result in striking hippocampal memory enhancements (Matynia et al., 2002; Kandel, 2001; Kaplan and Abel, 2003; Silva, 2003). In an interesting parallel to memory, this form of LTP exhibits at least two mechanistically and functionally distinct phases (Bliss and

Collingridge, 1993; Chapman, 2001): E-LTP is typically induced by a single high-frequency tetanus train of 100 Hz stimulation, lasts for approximately 1-3 hours and does not require protein synthesis. On the other hand, L-LTP is typically induced by 3 or 4 trains of 100 Hz stimulation, persists for several hours and requires activation of transcription and protein synthesis (Stanton and Sarvey, 1984; Frey et al., 1988; Nguyen et al., 1994; Frey et al., 1993; Kandel, 2001). Therefore both the early forms of LTP and memory do not require transcription while the later forms do.

In another interesting parallel to memory, stimulation that produces L-LTP, but not E-LTP, increases the levels of b-galactosidase in a CRE-LacZ reporter transgenic mouse (Impey et al., 1996). The finding that L-LTP-producing stimulation engages CRE-mediated transcription has also been shown in hippocampal slices (Matthies et al., 1997; Lu et al., 1999; Ahmed et al., 2004), amygdala slices (Huang et al., 2000; Lin et al., 2001), dissociated hippocampal neurons (Bito et al., 1996; Deisseroth et al., 1996) and the hippocampus *in vivo* (Schulz et al., 1999; Davis et al., 2000).

Bourtchuladze et al. (1994) first studied the role of CREB in LTP by stimulating Schaffer collaterals with a train of 100 pulses at 100 Hz. Although the post-tetanic potential (measured 15-20 seconds after the tetanus) did not differ in $CREB^{ad}$ mice, the resulting LTP was unstable (beginning10 min after tetanic conditioning and continuing throughout the 2-hour test session). However, similar to the results from memory experiments, increasing the interval between tetani can overcome the LTP deficits observed in the $CREB^{ad}$ mutants (A.J.S and J.H.K., unpublished observations). The temporal spacing between successive tetanic trains also alters the PKA-dependence of LTP (Scharf et al., 2002; Woo et al., 2003).

Therefore, it is perhaps not surprising that LTP produced by a spaced training protocol (3 stimulus trains of 100 Hz with a 10-min intertrain interval) was normal in $CREB^{ad}$, $CREB^{comp}$, dCA1-KCREB mice and two conditional CREB knockout mice ($CREB^{CaMKCre7}$ and $CREB^{NesCre}$) (Gass et al., 1998; Balschun et al., 2003; Pittenger et al., 2002). However, the dCA1-KCREB mice showed deficient forskolin-induced, DA-regulated, and heterosynaptically-induced forms of LTP (Pittenger et al., 2002; Huang et al., 2004). These findings highlight the importance of using several LTP-inducing paradigms to thoroughly investigate the role of CREB in synaptic plasticity.

Therefore, decreasing CREB function disrupts some forms of LTP. But what are the effects of increasing CREB function? This question was answered by using transgenic mice overexpressing a dominant active form of CREB (CREBVP16). In these transgenic mice, stimulation that normally induces E-LTP produced strong L-LTP (Barco et al., 2002). Moreover, crossing these mice with increased CREB function to mice with a mutation in CBP rescues the deficit in L-LTP observed in the CBP mutant mice (Alarcon et al., 2004). Similarly, transgenic mice overexpressing type-1 adenylyl cyclase (with elevated levels of pCREB) also showed L-LTP following a single train of high-frequency stimulation that produces only E-LTP in WT littermate controls (Wang et al., 2004). These findings add to the conclusions based on experiments performed in flies, slugs and rats; stimulation that normally induces short-lived changes in memory or LTP produce long-lived changes if CREB function is increased (Yin et al., 1995; Bartsch et al., 1995; Josselyn et al., 2001). Together, these data show that CREB can enhance both synaptic plasticity and memory.

In combination with LTP, long-term depression (LTD) is thought to regulate the fine-tuning of synaptic weights that may be critical for learning (Bear and Malenka, 1994; Linden and Connor, 1995). Bidirectional control of synaptic strength maintains synaptic efficacy within a dynamic operating range so that synapses can remain modifiable in response to electrical activity. LTD is typically induced by low-frequency stimulation (Barrionuevo et al., 1980; Staubli and Lynch, 1990; Dudek and Bear, 1992). Similar to LTP, LTD is blocked by protein synthesis inhibitors (Ahn et al., 1999) and stimulation that induces LTD is associated with increased levels of CREB phosphorylation in the hippocampus (Deisseroth et al., 1996). Furthermore, a dominant negative form of CREB inhibited LTD in dissociated neurons (Ahn et al., 1999) and differences in LTD were also noted between CREB mutant mice ($CREB^{comp}$, $CREB^{CaMKCre7}$, $CREB^{NesCre}$) and WT controls (Balschun et al., 2003). Mutants were reported to be more susceptible to low-frequency stimulation (Balschun et al., 2003), suggesting important differences from WT mice in the fine-tuning of synaptic plasticity.

13. SUMMARY AND CONCLUSIONS

The critical importance of CREB in memory has been convincingly shown in several invertebrate species. However, the results from mammalian studies are more complex. Reasons for this complexity may include the larger numbers of genes in the mammalian CREB family and the differences in methods used to alter CREB function.

The preponderance of evidence from the many studies that used a variety of memory tasks and different techniques to disrupt memory demonstrates unequivocally that CREB has an important role in LTM in mammals. The ultimate products of CREB-mediated transcription may support the structural and functional processes necessary for LTM in species ranging from snails to flies to rodents and, perhaps, humans.

However, this conclusion is accompanied by two important caveats. The first is that the CREB memory phenotype is sensitive to the training parameters (specifically the spacing between successive trials). This sensitivity was shown in several memory paradigms (both behavioural and electrophysiological) and may be related to the fundamental differences between the types of memory produced by massed and spaced training. Massed training generally induces weak LTM. However, massed training alone induces robust LTM in flies, rats and mice if CREB function is increased (Yin et al., 1995; Josselyn et al., 2001; Genoux et al., 2002). On the other hand, multiple spaced trials can overcome the LTM deficit produced by impairing CREB function in many memory paradigms. This suggests that spaced training protocols may recruit additional molecular mechanisms, including other transcription factors.

The second stipulation is that the extent of CREB disruption determines the presence and degree of LTM impairment. In general, greater disruptions in CREB produce greater disruptions in LTM. For instance, the $CREB^{comp}$ (with only 1 CREB b allele) mice show a larger impairment than the $CREB^{ad}$ mice (with 2 CREB b alleles) in many memory tasks (Gass et al., 1998). Furthermore, $CREB^{NesCre}$ mice (with a brain wide deficit in CREB) show larger spatial memory impairment than $CREB^{CaMKCre7}$ mice (with a deletion of CREB in only 70-80% of CA1 neurons) (Balschun et al., 2003). Indeed, these results show that the remaining 20-30% of CA1 neurons with normal CREB levels can support normal spatial memory. Consistent with this is the observation that increasing CREB function in a small proportion of neurons in the lateral amygdala of a rat is sufficient to induce LTM for conditioning fear following suboptimal training

(Josselyn et al., 2001). Together, these data suggest that in terms of CREB and LTM, all that is needed is a few good neurons.

14. REFERENCES

Ahmed T, Frey S, Frey JU (2004) Regulation of the phosphodiesterase PDE4B3-isotype during long-term potentiation in the area dentata in vivo. Neuroscience 124:857-867.

Ahn S, Ginty DD, Linden DJ (1999) A late phase of cerebellar long-term depression requires activation of CaMKIV and CREB. Neuron 23:559-568.

Alarcon JM, Malleret G, Touzani K, Vronskaya S, Ishii S, Kandel ER, Barco A (2004) Chromatin, acetylation, memory, and LTP are impaired in CBP+/- mice: a model for the cognitive deficit in Rubinstein-Taybi syndrome and its amelioration. Neuron 42:947-959.

Athos J, Impey S, Pineda VV, Chen X, Storm DR (2002) Hippocampal CRE-mediated gene expression is required for contextual memory formation. Nat. Neurosci. 5:1119-1120.

Bacskai BJ, Hochner B, Mahaut-Smith M, Adams SR, Kaang BK, Kandel ER, Tsien RY (1993) Spatially resolved dynamics of cAMP and protein kinase A subunits in Aplysia sensory neurons. Science 260:222-226.

Bailey CH, Chen M(1989) Time course of structural changes at identified sensory neuron synapses during long-term sensitization in Aplysia. J. Neurosci. 9:1774-1780.

Bailey CH, Montarolo P, Chen M, Kandel ER, Schacher S, (1992) Inhibitors of protein and RNA synthesis block structural changes that accompany long-term heterosynaptic plasticity in Aplysia. Neuron 9:749-758.

Balschun D, Wolfer DP, Gass P, Mantamadiotis T, Welzl H, Schutz G, Frey JU, Lipp HP (2003) Does cAMP response element-binding protein have a pivotal role in hippocampal synaptic plasticity and hippocampus-dependent memory? J. Neurosci. 23:6304-6314.

Barco A, Alarcon JM, Kandel ER (2002) Expression of constitutively active CREB protein facilitates the late phase of long-term potentiation by enhancing synaptic capture. Cell 108:689-703.

Barrientos RM, O'Reilly RC, Rudy JW (2002) Memory for context is impaired by injecting anisomycin into dorsal hippocampus following context exploration. Behav. Brain Res. 134:299-306.

Barrionuevo G, Schottler F, Lynch G (1980) The effects of repetitive low frequency stimulation on control and "potentiated" synaptic responses in the hippocampus. Life Sci. 27:2385-2391.

Barrot M, Olivier JD, Perrotti LI, DiLeone RJ, Berton O, Eisch AJ, Impey S, Storm DR, Neve RL, Yin JC, Zachariou V, Nestler EJ (2002) CREB activity in the nucleus accumbens shell controls gating of behavioral responses to emotional stimuli. Proc. Natl. Acad. Sci. USA 99:11435-11440.

Bartsch D, Ghirardi M, Skehel PA, Karl KA, Herder SP, Chen M, Bailey CH, Kandel ER (1995) Aplysia CREB2 represses long-term facilitation: relief of repression converts transient facilitation into long-term functional and structural change. Cell 83:979-992.

Bear MF, Malenka RC (1994) Synaptic plasticity: LTP and LTD. Curr. Opin. Neurobiol. 4:389-399.

Bevilaqua LR, Cammarota M, Paratcha G, de Stein ML, Izquierdo I, Medina JH (1999) Experience-dependent increase in cAMP-responsive element binding protein in synaptic and nonsynaptic mitochondria of the rat hippocampus. Eur. J. Neurosci. 11:3753-3756.

Bito H, Deisseroth K, Tsien RW (1996) CREB phosphorylation and dephosphorylation: a Ca(2+)- and stimulus duration-dependent switch for hippocampal gene expression. Cell 87:1203-1214.

Blendy JA, Kaestner KH, Schmid W, Gass P, Schutz G (1996) Targeting of the CREB gene leads to up-regulation of a novel CREB mRNA isoform. Embo. J. 15:1098-1106.

Bliss TV, Collingridge GL (1993) A synaptic model of memory: long-term potentiation in the hippocampus. Nature 361:31-39.

Bourtchuladze R, Frenguelli B, Blendy J, Cioffi D, Schutz G, Silva AJ (1994) Deficient long-term memory in mice with a targeted mutation of the cAMP-responsive element-binding protein. Cell 79:59-68.

Bozon B, Kelly A, Josselyn SA, Silva AJ, Davis S, Laroche S (2003) MAPK, CREB and zif268 are all required for the consolidation of recognition memory. Philos. Trans. R. Soc. Lond. B. Biol. Sci. 358:805-814.

Brodie CR, Khaliq M, Yin JC, Brent Clark H, Orr HT, Boland LM (2004) Overexpression of CREB reduces CRE-mediated transcription: behavioral and cellular analyses in transgenic mice. Mol. Cell. Neurosci. 25:602-611.

Bunsey M, Eichenbaum H (1995) Selective damage to the hippocampal region blocks long-term retention of a natural and nonspatial stimulus-stimulus association. Hippocampus 5:546-556.

Busch SJ, Sassone-Corsi P (1990) Dimers, leucine zippers and DNA-binding domains. Trends Genet. 6:36-40.

Byers D, Davis RL, Kiger JA Jr (1981) Defect in cyclic AMP phosphodiesterase due to the dunce mutation of learning in Drosophila melanogaster. Nature 289:79-81.

Cammarota M, Bevilaqua LR, Ardenghi P, Paratcha G, Levi de Stein M, Izquierdo I, Medina JH (2000) Learning-associated activation of nuclear MAPK, CREB and Elk-1, along with Fos production, in the rat hippocampus after a one-trial avoidance learning: abolition by NMDA receptor blockade. Brain Res. Mol. Brain Res. 76:36-46.

Carew TJ, Castellucci VF, Kandel ER (1971) An analysis of dishabituation and sensitization of the gill-withdrawal reflex in Aplysia. Int. J. Neurosci. 2:79-98.

Carew TJ, Kandel ER (1973) Acquisition and retention of long-term habituation in Aplysia: correlation of behavioral and cellular processes. Science 182:1158-1160.

Carlezon WA Jr, Thome J, Olson VG, Lane-Ladd SB, Brodkin ES, Hiroi N, Duman RS, Neve RL, Nestler EJ (1998) Regulation of cocaine reward by CREB. Science 282:2272-2275.

Chapman PF (2001) The diversity of synaptic plasticity. Nat. Neurosci. 4:556-558.

Chen RH, Sarnecki C, Blenis J (1992) Nuclear localization and regulation of erk- and rsk-encoded protein kinases. Mol. Cell. Biol. 12:915-927.

Chrivia JC, Kwok RP, Lamb N, Hagiwara M, Montminy MR, Goodman RH (1993) Phosphorylated CREB binds specifically to the nuclear protein CBP. Nature 365:855-859.

Cleary LJ, Lee WL, Byrne JH (1998) Cellular correlates of long-term sensitization in Aplysia. J. Neurosci. 18:5988-5998.

Cole TJ, Copeland NG, Gilbert DJ, Jenkins NA, Schutz G, Ruppert S (1992) The mouse CREB (cAMP responsive element binding protein) gene: structure, promoter analysis, and chromosomal localization. Genomics 13:974-982.

Countryman RA, Orlowski JD, Brightwell JJ, Oskowitz AZ, Colombo PJ (2004) CREB phsophorylation and c-Fos expression in the hippocampus of rats during acquisition and recall of a socially transmitted food preference. Hippocampus, In press.

Danielian PS, White R, Hoare SA, Fawell SE, Parker MG (1993) Identification of residues in the estrogen receptor that confer differential sensitivity to estrogen and hydroxytamoxifen. Mol. Endocrinol. 7:232-240.

Dash PK, Hochner B, Kandel ER (1990) Injection of the cAMP-responsive element into the nucleus of Aplysia sensory neurons blocks long-term facilitation. Nature 345:718-721.

Dash PK, Karl KA, Colicos MA, Prywes R, Kandel ER (1991) cAMP response element-binding protein is activated by Ca2+/calmodulin- as well as cAMP-dependent protein kinase. Proc. Natl. Acad. Sci. USA 88:5061-5065.

Davis HP, Squire LR (1984) Protein synthesis and memory: a review. Psychol. Bull. 96:518-559.

Davis M (1992) The role of the amygdala in fear and anxiety. Annu. Rev. Neurosci. 15:353-375.

Davis S, Vanhoutte P, Pages C, Caboche J, Laroche S (2000) The MAPK/ERK cascade targets both Elk-1 and cAMP response element-binding protein to control long-term potentiation-dependent gene expression in the dentate gyrus in vivo. J. Neurosci. 20:4563-4572.

Deisseroth K, Bito H, Tsien RW (1996) Signaling from synapse to nucleus: postsynaptic CREB phosphorylation during multiple forms of hippocampal synaptic plasticity. Neuron 16:89-101.

Desmedt A, Hazvi S, Dudai Y (2003) Differential pattern of cAMP response element-binding protein activation in the rat brain after conditioned aversion as a function of the associative process engaged: taste versus context association. J. Neurosci. 23:6102-6110.

Dubnau J, Chiang AS, Grady L, Barditch J, Gossweiler S, McNeil J, Smith P, Buldoc F, Scott R, Certa U, Broger C, Tully T (2003) The staufen/pumilio pathway is involved in Drosophila long-term memory. Curr. Biol. 13:286-296.

Dudai Y, Jan YN, Byers D, Quinn WG, Benzer S (1976) dunce, a mutant of Drosophila deficient in learning. Pro.c Natl. Acad. Sci. USA 73:1684-1688.

Dudek SM, Bear MF (1992) Homosynaptic long-term depression in area CA1 of hippocampus and effects of N-methyl-D-aspartate receptor blockade. Proc. Natl. Acad. Sci. USA 89:4363-4367.

Ebbinghaus H (1885) Uber das Gedachtnis. New York: Dover.

Falls WA, Kogan JH, Silva AJ, Willott JF, Carlson S, Turner JG (2000) Fear-potentiated startle, but not prepulse inhibition of startle, is impaired in CREB alphadelta-/- mutant mice. Behav. Neurosci. 114:998-1004.

Fanselow MS, Gale, GD (2003) The amygdala, fear, and memory. Ann. N. Y. Acad. Sci. 985:125-134.

Feil R, Brocard J, Mascrez B, LeMeur M, Metzger D, Chambon P (1996) Ligand-activated site-specific recombination in mice. Proc. Natl. Acad. Sci. USA 93:10887-10890.

Fimia GM, De Cesare D, Sassone-Corsi P (1998) Mechanisms of activation by CREB and CREM: phosphorylation, CBP, and a novel coactivator, ACT. Cold Spring Harb. Symp. Quant. Biol. 63:631-642.

Finkbeiner S, Tavazoie SF, Maloratsky A, Jacobs KM, Harris KM, Greenberg ME (1997) CREB: a major mediator of neuronal neurotrophin responses. Neuron 19:1031-1047.

Foulkes NS, Borrelli E, Sassone-Corsi P (1991) CREM gene: use of alternative DNA-binding domains generates multiple antagonists of cAMP-induced transcription. Cell 64:739-749.

Foulkes NS, Sassone-Corsi P (1992) More is better: activators and repressors from the same gene. Cell 68:411-414.

Frankland PW, Cestari V, Filipkowski RK, McDonald RJ, Silva AJ (1998) The dorsal hippocampus is essential for context discrimination but not for contextual conditioning. Behav. Neurosci. 112:863-874.

Frankland PW, Josselyn SA (2004) CREB and long-term memory. In: Memories are made of these: from messengers to molecules (Ed. by Riedel, G. & Platt, B.). Georgetown, Tx: Landes Bioscience.

Frankland PW, Josselyn SA, Anagnostaras SG, Kogan JH, Takahashi E, Silva AJ (2004) Consolidation of CS and US representations in associative fear conditioning. Hippocampus 14:557-559.

Frey U, Huang YY, Kandel ER (1993) Effects of cAMP simulate a late stage of LTP in hippocampal CA1 neurons. Science 260:1661-1664.

Frey U, Krug M, Reymann KG, Matthies H (1988) Anisomycin, an inhibitor of protein synthesis, blocks late phases of LTP phenomena in the hippocampal CA1 region in vitro. Brain Res. 452:57-65.

Frost WN, Clark GA, Kandel ER (1988) Parallel processing of short-term memory for sensitization in Aplysia. J. Neurobiol. 19:297-334.

Frost WN, Kandel ER (1995) Structure of the network mediating siphon-elicited siphon withdrawal in Aplysia. J. Neurophysiol. 73:2413-2427.

Galef BG Jr, Mason JR, Preti G, Bean NJ (1988) Carbon disulfide: a semiochemical mediating socially-induced diet choice in rats. Physiol. Behav. 42:119-124.

Galef BG Jr, Wigmore SW, Kennett DJ (1983) A failure to find socially mediated taste aversion learning in Norway rats (R. norvegicus). J. Comp. Psychol. 97:358-363.

Garcia J, Kimeldorf DJ, Koelling RA (1955) Conditioned aversion to saccharin resulting from exposure to gamma radiation. Science 122:157-158.

Gass P, Wolfer DP, Balschun D, Rudolph D, Frey U, Lipp HP, Schutz G (1998) Deficits in memory tasks of mice with CREB mutations depend on gene dosage. Learn. Mem. 5:274-288.

Genoux D, Haditsch U, Knobloch M, Michalon A, Storm D, Mansuy IM (2002) Protein phosphatase 1 is a molecular constraint on learning and memory. Nature 418:970-975.

Ghosh MK, Cohen JS (1992) Oligodeoxynucleotides as antisense inhibitors of gene expression. Prog. Nucleic Acid Res. Mol. Biol. 42:79-126.

Giebler HA, Lemasson I, Nyborg JK (2000) p53 recruitment of CREB binding protein mediated through phosphorylated CREB: a novel pathway of tumor suppressor regulation. Mol. Cell. Biol. 20:4849-4858.

Goelet P, Castellucci VF, Schacher S, Kandel ER (1986) The long and the short of long-term memory--a molecular framework. Nature 322:419-422.

Gonzalez GA, Montminy MR (1989) Cyclic AMP stimulates somatostatin gene transcription by phosphorylation of CREB at serine 133. Cell 59:675-680.

Goosens KA, Maren S (2002) Long-term potentiation as a substrate for memory: evidence from studies of amygdaloid plasticity and Pavlovian fear conditioning. Hippocampus 12:592-599.

Graves L, Dalvi A, Lucki I, Blendy JA, Abel T (2002) Behavioral analysis of CREB alphadelta mutation on a B6/129 F1 hybrid background. Hippocampus 12:18-26.

Guzowski JF, McGaugh JL (1997) Antisense oligodeoxynucleotide-mediated disruption of hippocampal cAMP response element binding protein levels impairs consolidation of memory for water maze training. Proc. Natl. Acad. Sci. USA 94:2693-2698.

Hagiwara M, Alberts A, Brindle P, Meinkoth J, Feramisco J, Deng T, Karin M, Shenolikar S, Montminy M (1992) Transcriptional attenuation following cAMP induction requires PP-1-mediated dephosphorylation of CREB. Cell 70:105-113.

Hagiwara M, Brindle P, Harootunian A, Armstrong R, Rivier J, Vale W, Tsien R, Montminy MR (1993) Coupling of hormonal stimulation and transcription via the cyclic AMP-responsive factor CREB is rate limited by nuclear entry of protein kinase A. Mol. Cell. Biol. 13:4852-4859.

Hardingham GE, Arnold FJ, Bading H (2001) Nuclear calcium signaling controls CREB-mediated gene expression triggered by synaptic activity. Nat. Neurosci. 4:261-267.

Hoeffler JP (1992) Structure/function relationships of CREB/ATF proteins. J. Invest. Dermatol. 98:21S-28S.

Hoeffler JP, Meyer TE, Yun Y, Jameson JL, Habener JF (1988) Cyclic AMP-responsive DNA-binding protein: structure based on a cloned placental cDNA. Science 242:1430-1433.

Huang YY, Martin KC, Kandel ER (2000) Both protein kinase A and mitogen-activated protein kinase are required in the amygdala for the macromolecular synthesis-dependent late phase of long-term potentiation. J. Neurosci. 20:6317-6325.

Huang YY, Pittenger C, Kandel ER (2004) A form of long-lasting, learning-related synaptic plasticity in the hippocampus induced by heterosynaptic low-frequency pairing. Proc. Natl. Acad. Sci. USA 101:859-864.

Hummler E, Cole TJ, Blendy JA, Ganss R, Aguzzi A, Schmid W, Beermann F, Schutz G (1994) Targeted mutation of the CREB gene: compensation within the CREB/ATF family of transcription factors. Proc. Natl. Acad. Sci. USA 91:5647-5651.

Impey S, Mark M, Villacres EC, Poser S, Chavkin C, Storm DR (1996) Induction of CRE-mediated gene expression by stimuli that generate long-lasting LTP in area CA1 of the hippocampus. Neuron 16:973-982.

Impey S, Smith DM, Obrietan K, Donahue R, Wade C, Storm DR (1998) Stimulation of cAMP response element (CRE)-mediated transcription during contextual learning. Nat. Neurosci. 1:595-601.

Josselyn SA, Shi C, Carlezon WA Jr, Neve RL, Nestler EJ, Davis M (2001) Long-term memory is facilitated by cAMP response element-binding protein overexpression in the amygdala. J. Neurosci. 21:2404-2412.

Kaang BK, Kandel ER, Grant SG (1993) Activation of cAMP-responsive genes by stimuli that produce long-term facilitation in Aplysia sensory neurons. Neuron 10:427-435.

Kandel ER (1981) Calcium and the control of synaptic strength by learning. Nature 293:697-700.

Kandel ER (2001) The molecular biology of memory storage: a dialogue between genes and synapses. Science 294:1030-1038.

Kaplan MP, Abel T (2003) Genetic approaches to the study of synaptic plasticity and memory storage. CNS Spectr. 8:597-610.

Kida S, Josselyn SA, de Ortiz SP, Kogan JH, Chevere I, Masushige S, Silva AJ (2002) CREB required for the stability of new and reactivated fear memories. Nat. Neurosci. 5:348-355.

Kogan JH, Frankland PW, Blendy JA, Coblentz J, Marowitz Z, Schutz G, Silva AJ (1997) Spaced training induces normal long-term memory in CREB mutant mice. Curr. Biol. 7:1-11.

Kornhauser JM, Cowan CW, Shaywitz AJ, Dolmetsch RE, Griffith EC, Hu LS, Haddad C, Xia Z, Greenberg ME (2002) CREB transcriptional activity in neurons is regulated by multiple, calcium-specific phosphorylation events. Neuron 34:221-233.

Lamprecht R, Hazvi S, Dudai Y (1997) cAMP response element-binding protein in the amygdala is required for long- but not short-term conditioned taste aversion memory. J. Neurosci. 17:8443-8450.

LeDoux JE (2000) Emotion circuits in the brain. Annu. Rev. Neurosci. 23:155-184.

Levin LR, Han PL, Hwang PM, Feinstein PG, Davis RL, Reed RR (1992) The Drosophila learning and memory gene rutabaga encodes a Ca2+/Calmodulin-responsive adenylyl cyclase. Cell 68:479-489.

Lin CH, Yeh SH, Lu KT, Leu TH, Chang WC, Gean PW (2001) A role for the PI-3 kinase signaling pathway in fear conditioning and synaptic plasticity in the amygdala. Neuron 31:841-851.

Linden DJ, Connor JA (1995) Long-term synaptic depression. Annu. Rev. Neurosci. 18:319-357.

Livingstone MS, Sziber PP, Quinn WG(1984) Loss of calcium/calmodulin responsiveness in adenylate cyclase of rutabaga, a Drosophila learning mutant. Cell 37:205-215.

Logie C, Stewart AF (1995) Ligand-regulated site-specific recombination. Proc. Natl. Acad. Sci. USA 92:5940-5944.

Lu YF, Kandel ER, Hawkins RD (1999) Nitric oxide signaling contributes to late-phase LTP and CREB phosphorylation in the hippocampus. J. Neurosci. 19:10250-10261.

Lynch G (1998) Memory and the brain: unexpected chemistries and a new pharmacology. Neurobiol. Learn. Mem. 70:82-100.

Mantamadiotis T, Lemberger T, Bleckmann SC, Kern H, Kretz O, Martin-Villalba A, Tronche F, Kellendonk C, Gau D, Kapfhammer J, Otto C, Schmid W, Schutz G (2002) Disruption of CREB function in brain leads to neurodegeneration. Nat. Genet. 31:47-54.

Maren S, Aharonov G, Fanselow MS (1997) Neurotoxic lesions of the dorsal hippocampus and Pavlovian fear conditioning in rats. Behav. Brain Res. 88:261-274.

Maren S, Baudry M (1995) Properties and mechanisms of long-term synaptic plasticity in the mammalian brain: relationships to learning and memory. Neurobiol. Learn. Mem. 63:1-18.

Martin KC, Casadio A, Zhu H, Yaping E, Rose JC, Chen M, Bailey CH, Kandel ER (1997) Synapse-specific, long-term facilitation of aplysia sensory to motor synapses: a function for local protein synthesis in memory storage. Cell 91:927-938.

Martin SJ, Grimwood PD, Morris RG (2000) Synaptic plasticity and memory: an evaluation of the hypothesis. Annu. Rev. Neurosci. 23:649-711.

Matthies H (1989) In search of cellular mechanisms of memory. Prog. Neurobiol. 32:277-349.

Matthies H, Schulz S, Thiemann W, Siemer H, Schmidt H, Krug M, Hollt V (1997) Design of a multiple slice interface chamber and application for resolving the temporal pattern of CREB phosphorylation in hippocampal long-term potentiation. J. Neurosci. Methods 78:173-179.

Matus-Amat P, Higgins EA, Barrientos RM, Rudy JW (2004) The role of the dorsal hippocampus in the acquisition and retrieval of context memory representations. J. Neurosci. 24:2431-2439.

Matynia A, Kushner SA, Silva AJ (2002) Genetic approaches to molecular and cellular cognition: a focus on LTP and learning and memory. Annu. Rev. Genet. 36:687-720.

Mauelshagen J, Sherff CM, Carew TJ (1998) Differential induction of long-term synaptic facilitation by spaced and massed applications of serotonin at sensory neuron synapses of Aplysia californica. Learn. Mem. 5:246-256.

McLean JH, Harley CW, Darby-King A, Yuan Q (1999) pCREB in the neonate rat olfactory bulb is selectively and transiently increased by odor preference-conditioned training. Learn. Mem. 6:608-618.

Molina CA, Foulkes NS, Lalli E, Sassone-Corsi P (1993) Inducibility and negative autoregulation of CREM: an alternative promoter directs the expression of ICER, an early response repressor. Cell 75:875-886.

Morris RG, Garrud P, Rawlins JN, O'Keefe J (1982) Place navigation impaired in rats with hippocampal lesions. Nature 297:681-683.

Moser EI, Krobert KA, Moser MB, Morris RG (1998) Impaired spatial learning after saturation of long-term potentiation. Science 281:2038-2042.

Mower AF, Liao DS, Nestler EJ, Neve RL, Ramoa AS (2002) cAMP/Ca2+ response element-binding protein function is essential for ocular dominance plasticity. J. Neurosci. 22:2237-2245.

Nguyen PV, Abel T, Kandel ER (1994) Requirement of a critical period of transcription for induction of a late phase of LTP. Science 265:1104-1107.

Ogawa S, Pfaff DW (1996) Application of antisense DNA method for the study of molecular bases of brain function and behavior. Behav. Genet. 26:279-292.

Pandey SC, Mittal N, Silva AJ (2000) Blockade of cyclic AMP-responsive element DNA binding in the brain of CREB delta/alpha mutant mice. Neuroreport 11:2577-2580.

Pinsker H, Kupfermann I, Castellucci V, Kandel E (1970) Habituation and dishabituation of the gill-withdrawal reflex in Aplysia. Science 167:1740-1742.

Pinsker HM, Hening WA, Carew TJ, Kandel ER (1973) Long-term sensitization of a defensive withdrawal reflex in Aplysia. Science 182:1039-1042.

Pittenger C, Huang YY, Paletzki RF, Bourtchouladze R, Scanlin H, Vronskaya S, Kandel ER (2002) Reversible inhibition of CREB/ATF transcription factors in region CA1 of the dorsal hippocampus disrupts hippocampus-dependent spatial memory. Neuron 34:447-462.

Quinn WG, Harris WA, Benzer S (1974) Conditioned behavior in Drosophila melanogaster. Proc. Natl. Acad. Sci. USA 71:708-712.

Quinn WG, Sziber PP, Booker R (1979) The Drosophila memory mutant amnesiac. Nature 277:212-214.

Rammes G, Steckler T, Kresse A, Schutz G, Zieglgansberger W, Lutz B (2000) Synaptic plasticity in the basolateral amygdala in transgenic mice expressing dominant-negative cAMP response element-binding protein (CREB) in forebrain. Eur. J. Neurosci. 12:2534-2546.

Rehfuss RP, Walton KM, Loriaux MM, Goodman RH (1991) The cAMP-regulated enhancer-binding protein ATF-1 activates transcription in response to cAMP-dependent protein kinase A. J. Biol. Chem. 266:18431-18434.

Ribeiro MJ, Serfozo Z, Papp A, Kemenes I, O'Shea M, Yin JC, Benjamin PR, Kemenes G (2003) Cyclic AMP response element-binding (CREB)-like proteins in a molluscan brain: cellular localization and learning-induced phosphorylation. Eur. J. Neurosci. 18:1223-1234.

Rosenblum K, Meiri N, Dudai Y (1993) Taste memory: the role of protein synthesis in gustatory cortex. Behav. Neural Biol. 59:49-56.

Rudolph D, Tafuri A, Gass P, Hammerling GJ, Arnold B, Schutz G (1998) Impaired fetal T cell development and perinatal lethality in mice lacking the cAMP response element binding protein. Proc. Natl. Acad. Sci. USA 95:4481-4486.

Sadamoto H, Sato H, Kobayashi S, Murakami J, Aonuma H, Ando H, Fujito Y, Hamano K, Awaji M, Lukowiak K, Urano A, Ito E (2004) CREB in the pond snail Lymnaea stagnalis: cloning, gene expression, and function in identifiable neurons of the central nervous system. J. Neurobiol. 58:455-466.

Schafe GE, LeDoux JE (2000) Memory consolidation of auditory pavlovian fear conditioning requires protein synthesis and protein kinase A in the amygdala. J. Neurosci. 20:RC96.

Scharf MT, Woo NH, Lattal KM, Young JZ, Nguyen PV, Abel T (2002) Protein synthesis is required for the enhancement of long-term potentiation and long-term memory by spaced training. J. Neurophysiol. 87:2770-2777.

Schulz S, Siemer H, Krug M, Hollt V (1999) Direct evidence for biphasic cAMP responsive element-binding protein phosphorylation during long-term potentiation in the rat dentate gyrus in vivo. J. Neurosci. 19:5683-5692.

Shors TJ, Matzel LD (1997) Long-term potentiation: what's learning got to do with it? Behav. Brain Sci. 20:597-614.

Silva AJ (2003) Molecular and cellular cognitive studies of the role of synaptic plasticity in memory. J. Neurobiol. 54:224-237.

Simonato M, Manservigi R, Marconi P, Glorioso J (2000) Gene transfer into neurones for the molecular analysis of behaviour: focus on herpes simplex vectors. Trends Neurosci. 23:183-190.

Stanciu M, Radulovic J, Spiess J (2001) Phosphorylated cAMP response element binding protein in the mouse brain after fear conditioning: relationship to Fos production. Brain Res. Mol. Brain Res. 94:15-24.

Stanton PK, Sarvey JM (1984) Blockade of long-term potentiation in rat hippocampal CA1 region by inhibitors of protein synthesis. J. Neurosci. 4:3080-3088.

Staubli U, Lynch G (1990) Stable depression of potentiated synaptic responses in the hippocampus with 1-5 Hz stimulation. Brain Res. 513:113-118.

Stork O, Welzl H (1999) Memory formation and the regulation of gene expression. Cell. Mol. Life Sci. 55:575-592.

Sun P, Enslen H, Myung PS, Maurer RA (1994) Differential activation of CREB by Ca2+/calmodulin-dependent protein kinases type II and type IV involves phosphorylation of a site that negatively regulates activity. Genes Dev. 8:2527-2539.

Sutherland RJ, Kolb B, Whishaw IQ (1982) Spatial mapping: definitive disruption by hippocampal or medial frontal cortical damage in the rat. Neurosci. Lett. 31:271-276.

Swank MW (2000) Phosphorylation of MAP kinase and CREB in mouse cortex and amygdala during taste aversion learning. Neuroreport 11:1625-1630.

Taubenfeld SM, Milekic MH, Monti B, Alberini CM (2001a) The consolidation of new but not reactivated memory requires hippocampal C/EBPbeta. Nat. Neurosci. 4:813-818.

Taubenfeld SM, Wiig KA, Bear MF, Alberini CM (1999) A molecular correlate of memory and amnesia in the hippocampus. Nat. Neurosci. 2:309-310.

Taubenfeld SM, Wiig KA, Monti B, Dolan B, Pollonini G, Alberini CM (2001b) Fornix-dependent induction of hippocampal CCAAT enhancer-binding protein [beta] and [delta] Co-localizes with phosphorylated cAMP response element-binding protein and accompanies long-term memory consolidation. J. Neurosci. 21:84-91.

Tully T (1991) Physiology of mutations affecting learning and memory in Drosophila--the missing link between gene product and behavior. Trends Neurosci. 14:163-164.

Tully T, Preat T, Boynton SC, Del Vecchio M (1994) Genetic dissection of consolidated memory in Drosophila. Cell 79:35-47.

Viola H, Furman M, Izquierdo LA, Alonso M, Barros DM, de Souza MM, Izquierdo I, Medina JH (2000) Phosphorylated cAMP response element-binding protein as a molecular marker of memory processing in rat hippocampus: effect of novelty. J. Neurosci. 20:RC112.

Wadzinski BE, Wheat WH, Jaspers S, Peruski LF Jr, Lickteig RL, Johnson GL, Klemm DJ (1993) Nuclear protein phosphatase 2A dephosphorylates protein kinase A-phosphorylated CREB and regulates CREB transcriptional stimulation. Mol. Cell. Biol. 13:2822-2834.

Waeber G, Meyer TE, LeSieur M, Hermann HL, Gerard N, Habener JF (1991) Developmental stage-specific expression of cyclic adenosine 3',5'-monophosphate response element-binding protein CREB during spermatogenesis involves alternative exon splicing. Mol. Endocrinol. 5:1418-1430.

Wahlestedt C (1994) Antisense oligonucleotide strategies in neuropharmacology. Trends Pharmacol. Sci. 15:42-46.

Wallace TL, Stellitano KE, Neve RL, Duman RS (2004) Effects of cyclic adenosine monophosphate response element binding protein overexpression in the basolateral amygdala on behavioral models of depression and anxiety. Biol. Psychiatry 56:151-160.

Walters CL, Blendy JA (2001) Different requirements for cAMP response element binding protein in positive and negative reinforcing properties of drugs of abuse. J. Neurosci. 21:9438-9444.

Walton KM, Rehfuss RP, Chrivia JC, Lochner JE, Goodman RH (1992) A dominant repressor of cyclic adenosine 3',5'-monophosphate (cAMP)-regulated enhancer-binding protein activity inhibits the cAMP-mediated induction of the somatostatin promoter in vivo. Mol. Endocrinol. 6:647-655.

Wang H, Ferguson GD, Pineda VV, Cundiff PE, Storm DR (2004) Overexpression of type-1 adenylyl cyclase in mouse forebrain enhances recognition memory and LTP. Nat. Neurosci. 7:635-642.

Wiltgen BJ, Sanders MJ, Behne NS, Fanselow MS (2001) Sex differences, context preexposure, and the immediate shock deficit in Pavlovian context conditioning with mice. Behav. Neurosci. 115:26-32.

Woo NH, Duffy SN, Abel T, Nguyen PV (2003) Temporal spacing of synaptic stimulation critically modulates the dependence of LTP on cyclic AMP-dependent protein kinase. Hippocampus 13:293-300.

Wu GY, Deisseroth K, Tsien RW (2001) Activity-dependent CREB phosphorylation: convergence of a fast, sensitive calmodulin kinase pathway and a slow, less sensitive mitogen-activated protein kinase pathway. Proc. Natl. Acad. Sci. USA 98:2808-2813.

Xie S, Price JE, Luca M, Jean D, Ronai Z, Bar-Eli M (1997) Dominant-negative CREB inhibits tumor growth and metastasis of human melanoma cells. Oncogene 15:2069-2075.

Xing J, Ginty DD, Greenberg ME (1996) Coupling of the RAS-MAPK pathway to gene activation by RSK2, a growth factor-regulated CREB kinase. Science 273:959-963.

Yamamoto KK, Gonzalez GA, Menzel P, Rivier J, Montminy MR (1990) Characterization of a bipartite activator domain in transcription factor CREB. Cell 60:611-617.

Yamamoto T, Fujimoto Y (1991) Brain mechanisms of taste aversion learning in the rat. Brain Res. Bull. 27:403-406.

Yang SN, Huang LT, Wang CL, Chen WF, Yang CH, Lin SZ, Lai MC, Chen SJ, Tao PL (2003) Prenatal administration of morphine decreases CREBSerine-133 phosphorylation and synaptic plasticity range mediated by glutamatergic transmission in the hippocampal CA1 area of cognitive-deficient rat offspring. Hippocampus 13:915-921.

Yin JC, Del Vecchio M, Zhou H, Tully T (1995) CREB as a memory modulator: induced expression of a dCREB2 activator isoform enhances long-term memory in Drosophila. Cell 81:107-115.

Yin JC, Tully T (1996) CREB and the formation of long-term memory. Curr. Opin. Neurobiol. 6:264-268.

Yin JC, Wallach JS, Del Vecchio M, Wilder EL, Zhou H, Quinn WG, Tully T (1994) Induction of a dominant negative CREB transgene specifically blocks long-term memory in Drosophila. Cell 79:49-58.

Yuan Q, Harley CW, Darby-King A, Neve RL, McLean JH (2003) Early odor preference learning in the rat: bidirectional effects of cAMP response element-binding protein (CREB) and mutant CREB support a causal role for phosphorylated CREB. J. Neurosci. 23:4760-4765.

Zhang JJ, Okutani F, Inoue S, Kaba H (2003) Activation of the cyclic AMP response element-binding protein signaling pathway in the olfactory bulb is required for the acquisition of olfactory aversive learning in young rats. Neuroscience 117:707-713.

SYNAPTIC PLASTICITY IN THE CENTRAL NERVOUS SYSTEM: A ROLE FOR CALCIUM-REGULATED ADENYLYL CYCLASES

Gregory D. Ferguson, Josephine M. Atienza, and Daniel R. Storm*

1. INTRODUCTION

The mammalian brain contains approximately 100 billion neurons and up to 100 trillion synapses, providing it with an impressive number of potential sites for information processing and storage. However, it is not only the number of synapses and complexity of neural networks, but also the functionality of synapses that are a substrate for cognition. Neurons have the remarkable capacity to modulate the efficacy of neurotransmission in response to prior electrical activity, so-called synaptic plasticity. The strength and pattern of a stimulus dictates whether these changes last for milliseconds or persist for a lifetime. It is now generally accepted that activity-dependent modification of synaptic strength brought about by experiential learning is the basis of memory formation. In addition to its role in cognition, synaptic plasticity also regulates cortical and visual system development (Katz and Shatz, 1996). Abnormal synaptic plasticity may lead to cognitive deficits, drug addiction, and psychiatric disease (Winder, et al., 2002). The fundamental nature and importance of synaptic plasticity has generated much interest in molecular and cellular mechanisms that underlie this neuronal property.

Adenylyl cyclases (ACs) are a family of enzymes that catalyze the conversion of adenosine triphosphate (ATP) to adenosine 3'-5' monophosphate (cAMP). In brain, cAMP has numerous functions including transducing signals, modulating synaptic output, regulating gene expression, and controlling neuronal survival and repair. In this review, we will discuss the role of calcium-regulated ACs in central nervous system plasticity. Although most other ACs are also expressed in brain, calcium-regulated ACs have a predominant role in brain function because they directly couple calcium signals derived from neuronal activity to the production of cAMP. Two classes of calcium-regulated ACs have been identified in brain: calcium/calmodulin (CaM)-stimulated ACs, AC1 and AC8, and the calcium-inhibited AC3. Experimental evidence from our lab and others indicate that these enzymes contribute to neuroplasticity throughout the brain. For example, AC1 and AC8 modulate hippocampal long-term potentiation (LTP) and are

* Gregory D. Ferguson and Daniel R. Storm, Univ Washington, Dept Pharmacology, Seattle, WA 98195, Josephine M. Atienza, Sequenom Inc, San Diego, CA, 92037, USA. E-mail: dstorm@u.washington.edu

required for long-term memory. AC3 functions primarly in olfactory neurons, where it provides the cAMP signal required for odorant detection. In contrast to AC1 and AC8, AC3 is inhibited by calcium, a property that may allow the rapid reduction in cAMP important in desensitization.

2. EXPERIMENTAL

$AC1^{-/-}$ (Wu, et al., 1995), $AC8^{-/-}$ (Schaefer, et al., 2000), and $AC3^{-/-}$ (Wong, et al., 2000) mice were generated as described previously using standard gene targeting techniques. $AC1^{-/-}$ x $AC8^{-/-}$ double knockout mice were obtained as described previously (Wong, et al., 1999) by interbreeding AC1 and AC8 single mutant mice. AC1 overexpressing transgenic mice were obtained as described in using standard techniques involving pronuclear injection of an alpha calcium/calmodulin-dependent protein kinase II (CaMKII) promoter/AC1 cDNA construct.. (Wang, et al., 2004) Electrophysiological measurements were made in freshly cut transverse hippocampal slices as described in (Wang, et al., 2003). Electrolfactograms were generated through measurements in the main olfactory epithelium of $AC3^{-/-}$ mice as decribed in (Wong, et al., 2000). Animal behavior was performed as outlined in (Wang, et al., 2004).

3. RESULTS

3.1 Isolation and Structure of Adenylyl Cyclases

A milestone in the adenylyl cyclase field was the isolation and subsequent publication of the bovine AC1 cDNA sequence in 1989 (Krupinski, et al., 1989). The purification of an adenylyl cyclase protein from bovine brain allowed isolation of this cDNA. Brain calmodulin (CaM)-sensitive adenylyl cyclases were first separated in bovine brain using CaM-Sepharose affinity chromatography and were distinguishable from CaM-insensitive forms with a monoclonal antibody raised against CaM-sensitive AC (Rosenberg and Storm, 1987). Independently, two groups purified Ca^{2+}/CaM-stimulated adenylyl cyclase activity to near homogeneity using CaM-sepharose affinity chromatography or forskolin-agarose affinity chromatography (Pfeuffer, et al., 1985, Yeager, et al., 1985) . Additional members of the AC family, including AC8 and AC3, were isolated using sequence information and probes derived from AC1.

All adenylyl cyclases have a similar predicted membrane topology, consisting of a short cytoplasmic amino-terminus (N1), a six-pass transmembrane domain (M1), a 40 kD cytoplasmic loop (C1), another six-pass transmembrane domain (M2), and a cytoplasmic carboxy-terminus of approximately 35 kD (C2). The catalytic domain of adenylyl cyclases is bipartite; consisting of active domains in the C1 and C2 regions. The C1 and C2 domains are inactive individually but, when mixed together, AC enzyme activity and regulation is restored. Although the predicted molecular weight of mammalian adenylyl cyclases ranges from 110 to 130 kD, ACs have an apparent molecular weight on SDS-PAGE gels of ~150 kD due to heavy glycosylation in the extracellular loops of M1 and M2. Because of the number of transmembrane domains, attempts to crystallize and solve the structure of ACs have been unsuccessful. However, crystallization of the isolated C1 and C2 intracellular loops, which form heterodimers, provided a mechanism by which ACs catalyze the ATP to cAMP reaction and how G-proteins couple to and activate ACs

(Tesmer, et al., 1997). Thus, the membrane topology and tertiary structure of ACs has been inferred from hydropathy and domain analysis of the cDNA sequence. Overall, the predicted structure of the adenylyl cyclases resemble transporters and ion channels, but there is no evidence that these enzymes function as either (Reddy, et al., 1995). AC family members generally share 30-50 % amino acid identity throughout the entire protein, with higher levels of identity observed in the conserved regions of the protein including the catalytic regions and some portions of the transmembrane domain. Consistent with similarity in function, AC8 is more closely related to AC1 than is AC3. AC1 and AC8 share 67% identity in a conserved portion of the intracellular loop, while AC3 shares only 37% identity with AC1 across this same region.

3.2 Distribution and Regulatory Properties of the Calcium-Sensitive ACs

Adenylyl cyclases are a large family of enzymes that are expressed throughout the body and are regulated under varying conditions by a wide range of extracellular stimuli (for review see Xia and Storm, 1996). The diversity of this family is reflected in the calcium-regulated ACs, which are discussed in detail below (Table 1).

3.2.1 Adenylyl Cyclase Type 1

AC1 is expressed exclusively in brain, making it unique among the AC family. Most other ACs are expressed in brain, but are also detected outside the nervous system. Within brain, AC1 is found in hippocampus, neocortex, entorhinal cortex, cerebellum, olfactory bulb, and the pineal gland (Xia, et al., 1993). Of particular interest is the high levels of expression in the hippocampus; a region of the brain involved in learning and memory. AC1 levels in the rodent hippocampus increase dramatically during postnatal days 1-16, a period during development when synaptogenesis is reaching its maximum (Villacres, et al., 1995). The tissue specificity and developmentally regulated expression of the AC1 gene may be controlled by a 280 bp region just 5' to the transcriptional start site that contains a binary E box-like transcription factor binding site (Chan, et al., 2001). AC1 protein is detectable in the mossy fibers in the dentate gyrus of the macaque monkey, *Macaca nemestrin*, suggesting that AC1 is localized to axons in neurons of the hippocampus (Kumar, et al., 2001).

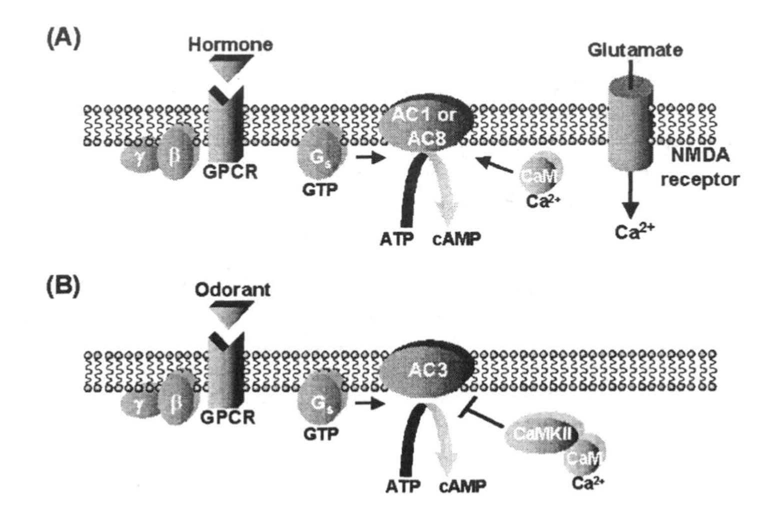

Figure 1. (A) General model for the activation of calcium-regulated adenylyl cyclases. In the presence of glutamate and post-synaptic depolarization, NMDA-type receptors allow calcium to enter the neuron. Calcium bound to calmodulin stimulates the membranous adenylyl cyclases to produce cAMP. Both AC1 and AC8 are stimulated by calcium and calmodulin. When already activated by calcium, AC1 but not AC8 can be synergistically stimulated through G-protein coupled receptors (GPCR) like beta-adrenergic receptors. Beta- and gamma-subunits of the G-protein do not participate in adenylyl cyclase activation. (B) AC3, when functioning in the olfactory epithelium, can be activated by G-protein coupled odorant receptors. In contrast to AC1 and AC8, AC3 is inhibited by calcium through calcium/CaM-sensitive kinase II via phosphorylation of serine 1076.

Calcium and CaM cooperate to stimulate AC1 enzymatic activity with an EC50 of 150 nM free Ca^{2+}, slightly above resting concentrations of calcium in neurons (Wu, et al., 1993). CaM interacts with and regulates AC1 from within the C1 loop region. Using peptide competitors and site-specific mutagenesis, the CaM binding site within this region was identified (Minocherhomjee, et al., 1987, Wu, et al., 1993). Mutation of AC1 residues F503 and K504 in the C1 domain decrease calcium sensitivity and stimulation *in vivo*, suggesting interaction with CaM within this region is required for maximal AC1 activation. In addition, a peptide corresponding to amino acids 495-522 of AC1 blocked Ca^{2+} /CaM stimulation of AC1 by CaM (Minocherhomjee, et al., 1987).

Interestingly, G_s-coupled receptor stimulation alone does not active AC1 (Wayman, et al., 1994). However, when G_s-coupled receptor activation is paired with an increase in intracellular calcium, a synergistic level of activation is observed in AC1. Similarly, co-activation of AC1 with calcium and beta-adrenergic agonists leads to synergistic stimulation of a cAMP response element (CRE)-mediated transcription (Impey, et al., 1994). Therefore, AC1 is synergistically activated by calcium and G_s-coupled receptor activation and functions as a coincidence detector to integrate these signals (Fig. 1A).

The levels of cAMP in the brain are tightly regulated by the opposing actions of ACs and phosphodiesterases, enzymes that catabolize cAMP. An optimal level of cAMP is required by neurons to support normal synaptic plasticity and memory formation. Alterations in the levels of cAMP can lead to abnormal plasticity, memory, or both. For example, genetic disruption of the inhibitory G-protein $G_{i\alpha 1}$ gene in mice leads to increases in AC activity and enhancements in hippocampal LTP, while reducing memory formation (Pineda, et al., 2004). Normally, AC1 enzymatic activity is reduced by inhibitory G-proteins and by calcium signals under certain conditions. AC1 activity is blocked by G_i-coupled somatostatin and dopamine D_2 receptor stimulation (Nielsen, et al., 1996). AC1 can also be inhibited also by calcium/CaM-dependent kinase IV (CaMKIV) *in vivo*. AC1 has two CaMKIV consensus phosphorylation sites, Ser-545 and Ser-552, near its CaM binding domain. Mutagenesis of either of these serine residues to alanine abolishes CaMKIV inhibition of AC1 (Wayman, et al., 1996). This sensitivity to CaMKIV inhibitory phosphorylation may have evolved as a mechanism by which neurons can reduce AC activity when calcium levels become excessive.

3.2.2 Adenylyl Cyclase Type 8

AC8 is expressed in brain, lung and parotid gland (Watson, et al., 2000). Within brain, AC8 is found in hippocampus, olfactory bulb, thalamus, habenula, cerebral cortex, and hypothalamus. This expression pattern can be recapitulated using a beta-galactosidase reporter gene under the transcriptional control of a 10 kilobase fragment of AC8 promoter (Muglia, et al., 1999). Interestingly, the AC8 promoter contains a CRE element and is regulated by the cAMP response element binding protein (CREB), suggesting that AC8 may be part of a positive feedback loop (Chao, et al., 2002). Like AC1, AC8 is stimulated by Ca^{2+} /CaM (Fig. 1A). However, AC8 is approximately 5 times less sensitive to calcium than AC1, with half-maximal stimulation at 800 nM free calcium (Nielsen, et al., 1996). The CaM binding domain of AC8 has been localized to the carboxy-terminal portion of the C2 region. Peptide competitors corresponding to this domain inhibit Ca^{2+}/CaM stimulation of AC8 enzymatic activity (Gu and Cooper, 1999). AC8 is not stimulated by G_s-coupled receptors nor is it synergistically stimulated by G_s-coupled receptors and calcium (Nielsen, et al., 1996). Serotonin stimulates AC8 activity *in vivo,* but this stimulation is mediated by increased intracellular calcium, not by G_s-coupled stimulation (Baker, et al., 1998). AC8 is not inhibited by G_i-coupled receptors nor by CaM kinase IV *in vivo*. Thus, AC8 functions purely as a low-affinity, stimulatory cAMP producing Ca^{2+} detector.

3.2.3 Adenylyl Cyclase Type 3

AC3 is expressed in heart, vascular smooth muscle, germ cells, brain, lung, and perhaps most importantly, the olfactory epithelium. AC3 is activated by G_s-coupled receptor stimulation (Defer, et al., 1998). However, sub-micromolar concentrations of free calcium inhibit G_s-coupled AC3 activation, suggesting that AC3 may be inhibited by a calcium-stimulated kinase (Wayman, et al., 1995b) (Fig. 1B). Indeed, CaM-kinase inhibitors antagonize the calcium-dependent inhibition of AC3. In addition, constitutively active CaMKII completely inhibits AC3 activation by the G_s-coupled beta-adrenergic agonist isoproterenol. Calcium increases in 293 cells (Wayman, et al., 1995a) or brief exposure of olfactory sensory neurons (OSNs) to odorants promote the phosphorylation of AC3 on a CaMKII consensus site at Ser 1076 (Wei, et al., 1998). This phosphorylation is prevented by CaM-kinase inhibitors and site-directed mutagenesis of the CaMKII consensus site (S1076A) blocks inhibition of AC3. Interestingly, cells expressing AC3 exhibit hormone-stimulated calcium and cAMP oscillations (Wayman, et al., 1995a). The mechanism of this oscillation appears to involve AC3 and cAMP effectors. Hormone stimulated, G-protein coupled increases in cAMP promote the activation of the cAMP-dependent protein kinase, protein kinase A (PKA). PKA phosphorylates and promotes the opening of the inositol trisphosphate (IP_3) receptor, which regulates calcium release from the endoplasmic reticulum (ER) stores. Increased intracellular calcium leads to the activation of CaMKII and the inhibition of AC3. Without AC3 acitivity, phosphodiesterases quickly reduce the levels of cAMP. PKA is then unable to keep IP_3 receptors phosphorylated, and calcium moves back into the ER. cAMP and calcium will continue to oscillate in these cells, with a periodicity of 3-5 minutes, provided an activator of AC3 is present. This calcium and cAMP feedback mechanism involving AC3 may provide the molecular basis for olfactory detection and desensitization.

Table 1. Distribution and Regulatory Properties of Calcium-Regulated Adenylyl Cyclases

	Calcium	G_s	Phosphorylation	Distribution	KO Phenotype
AC1	Stimulated 150 nM	Stimulated[1]	Inhibited/ CaMKIV Ser 545, Ser 552	brain, adrenal gland, retina	spatial learning deficit, mossy fiber LTP
AC3	Inhibited 200 nM	Stimulated	Inhibited/CaMKII Ser1076	olfactory epithelium, brain, retina, heart, smooth muscle, adrenal, spinal cord	anosmia, weak homozygote
AC8	Stimulated 800 nM	Insensitive	not determined	brain, lung, parotid	mossy fiber LTP deficit

[1]in the presence of calcium

3.3 Studies in Adenylyl Cyclase Knockout Mice.

Small molecule inhibitors of ACs, such as SQ22536 or SCH23390, are routinely used to study AC function. However, these compounds are generally effective only in the μM range and do not pharmacologically distinguish between members of the AC family. The lack of potent and specific inhibitors of ACs has necessitated the development of homozygous mutant "knockout" mouse strains to study the physiological roles of these genes *in vivo*.

3.3.1 Calcium-Regulated ACs are not Essential for Survival in Mice

$AC1^{-/-}$ (Wu, et al., 1995), $AC8^{-/-}$ (Schaefer, et al., 2000), and $AC1^{-/-}$ x $AC8^{-/-}$ double knockout (DKO) (Wong, et al., 1999) mutant mice are viable and have a normal lifespan, suggesting that Ca^{2+}/CaM-stimulated adenylyl cyclase activity is not essential for survival. In both $AC1^{-/-}$ and $AC8^{-/-}$ single knockout strains the Ca^{2+}/CaM-stimulated cyclase activity in brain is reduced by approximately 50%, indicating there is no compensation by either enzyme for the other (Wong, et al., 1999). Ca^{2+}/CaM-stimulated adenylyl cyclase activity is not observed in the brain of $AC1^{-/-}$ x $AC8^{-/-}$ double knockout mice, suggesting no other calcium-sensitive cyclases exist in brain.

AC1 plays a role in the developing retina and somatosensory cortex. The naturally occurring *barrelless* strain of mice harbors a mutation that deletes the AC1 gene (Abdel-Majid, et al., 1998). As implied by its name, *barrelless* mice do not develop normal barrel structures, even though the size of their whisker representation is normal (Welker, et al., 1996). Barrel structures are clusters of thalamocortical neurons in the somatosensory cortex, each cluster representing the receptive field of a single whisker. Interestingly, $AC1^{-/-}$ mice also fail to develop normal barrel structures (Abdel-Majid, et al., 1998). cAMP and PKA are required for the proper trafficking of the GluR1 subunit of the AMPA receptor (Esteban, et al., 2003). Phosphorylation of GluR1 by PKA is required for its transport from endosomes to the cell surface, where it participates in post synaptic excitatory responses. Therefore, the loss of AC1 activity and concomitant reduction in cAMP levels may explain why AMPA currents are greatly reduced in thalamocortical slices from *barrelless* mice (Lu, et al., 2003). *Barrelless* mice also have deficits in topographical maps of the retina that result from abnormal projection of retinal ganglion cells (Ravary, et al., 2003). Thus, AC1, through the production of cAMP, is important in the development of topographic maps in brain and in the trafficking of ion channels.

$AC3^{-/-}$ mice are significantly weakened at birth when compared to wild-type or heterozygous littermates (Wong, et al., 2000). Although $AC3^{-/-}$ mice are represented in near normal Mendelian ratios in litters from heterozygous parents (+/+, 24%; +/-, 56%; -/-, 20%), 80% of the homozygous mutant mice die within 48 hours. This high mortality rate can be ameliorated by reducing litter sizes with foster mothers. If an $AC3^{-/-}$ mouse survives through this critical period, however, they achieve normal size by 3 months of age.

3.3.2 Calcium-Stimulated Adenylyl Cyclases are Required for Some Forms of Synaptic Plasticity

Long-term potentiation (LTP) is a form of synaptic plasticity found in brain that may be a cellular substrate for memory formation. LTP is known to require cAMP and PKA, implying that AC activity is required. Brain slices from $AC1^{-/-}$ mice examined *in vitro* exhibit deficits in LTP at a variety of synapses. For example, in the Schaffer collateral pathway of the hippocampus, $AC1^{-/-}$ brain display reduced magnitude and rate of increase of excitatory post-synaptic potentials (EPSPs) in the first 20 minutes following tetanus (Wu, et al., 1995). By 3 hours post-tetanus, EPSP potentiation was not different from wild-type slices, suggesting the AC1 mutation has its strongest effect during the early stages of LTP. The Schaffer collateral pathway is a well-studied electrophysiological pathway in which axonal projections from CA3 pyramidal neurons project to and synapse on the dendritic fields of the pyramidal neurons in area CA1. Slices from $AC8^{-/-}$ mice also display a modest electrophysiological phenotype in the Schaffer collateral pathway. The fact that neither the $AC1^{-/-}$ or $AC8^{-/-}$ single mutant mice exhibit a substantial LTP phenotype indicates that these enzymes can compensate for each other. $AC1^{-/-}$ x $AC8^{-/-}$ double knockout slices display a complete loss of long-lasting LTP at the Schaffer collateral synapses in area CA1 (Wong, et al., 1999). This deficit is restricted to long-term synaptic plasticity; a short-term form of synaptic plasticity, paired-pulse facilitation, and the basal properties of neurotransmission are normal in $AC1^{-/-}$ x $AC8^{-/-}$ double knockout mice. Therefore, calcium-stimulated AC activity is required for CA1 LTP.

In contrast to Schaffer collateral LTP, which is dependent on NMDA-receptor activation and on post-synaptic calcium, mossy fiber LTP is dependent on presynaptic calcium, cAMP, and PKA activity. Mossy fibers are the axonal projections of the dentate granule cells that synapse on CA3 dendrites. It has been hypothesized that mossy fiber LTP requires a calcium-sensitive AC activity (Weisskopf, et al., 1994). Indeed, slices from $AC1^{-/-}$ display deficits in mossy fiber LTP (Villacres, et al., 1998). This deficit in $AC1^{-/-}$ mossy fiber LTP can be rescued by administration of the general adenylyl cyclase activator forskolin. Slices from $AC8^{-/-}$ mice also display deficits in mossy fiber LTP (Wang, et al., 2003). AC8 is targeted to synapses more readily than AC1 in cultured hippocampal neurons, perhaps allowing it to function at mossy fiber synapses despite its lower Ca^{2+} affinity.

Like mossy fiber synapses, parallel fiber/Purkinje cell synapses of the cerebellum display a presynaptic form of LTP. AC1 is expressed a high levels in the cerebellum and is targeted to synapses in cultured cerebellar neurons (Wang, et al., 2002, Xia, et al., 1991). Cerebellar preparations from $AC1^{-/-}$ mice display a complete loss of parallel fiber/Purkinje cell LTP induced by 4-8 Hz stimulation (Lev-Ram, et al., 2002); an observation that may correlate with the poor performance of $AC1^{-/-}$ mice on the accelerating rotarod (Storm, et al., 1998). Together these data indicate that the calcium-stimulated ACs play an important role in long lasting forms of plasticity at a variety of synapses throughout the brain.

3.3.3 Calcium-stimulated Adenylyl Cyclases are Necessary for Long-term Memory Formation

The cAMP/ PKA signal transduction pathway has an evolutionarily conserved role in long-term memory. For example, in the seasnail *Aplysia* increased cAMP in presynaptic sensory neurons via 5-HT receptor activation is sufficient to support facilitation of the gill withdrawl reflex (Brunelli, et al., 1976). Drosophila memory mutants have been mapped to genes that regulate or respond to the levels of intracellular cAMP: *rutabaga*, a calcium-sensitive adenylyl cyclase (AC), *dunce*, a cAMP phosphodiesterase (PDE), *amnesiac*, a AC-linked neuropeptide, and *PKA-RI*, a PKA regulatory subunit (for review see Davis, 1996). In mammals, transgenic mice overexpressing a dominant negative form of the PKA regulatory subunit (Abel, et al., 1997), R(AB), exhibit deficits in long term contextual memory. Moreover, elevating cAMP levels in brain artificially with rolipram, a PDE IV inhibitor, leads to enhanced contextual memory performance (Barad, et al., 1998). To examine the role of calcium-stimulated adenylyl cyclases in memory, $AC1^{-/-}$, $AC8^{-/-}$, and $AC1^{-/-}$ x $AC8^{-/-}$ double knockout mice were tested in a variety of memory tasks. $AC1^{-/-}$ and $AC8^{-/-}$ single knockout mice exhibit normal passive avoidance and contextual fear conditioning long-term memory (Wong, et al., 1999), suggesting these enzymes are functionally redundant in fear-associated memory formation. However, $AC1^{-/-}$ mice have a deficit in the Morris water maze (Wu, et al., 1995), a paradigm in which spatial memory is tested. Although these mice acquire the hidden platform task normally, they do not show a training quadrant preference during a 24-hour memory test. These data indicate AC1 has a role in the recall of spatial memory.

In contrast to the single adenylyl cyclase knockout mice, the $AC1^{-/-}$ x $AC8^{-/-}$ double knockout mice display a robust long-term memory deficit in the passive avoidance memory paradigm. The $AC1^{-/-}$ x $AC8^{-/-}$ double knockout mice are significantly impaired when tested for memory at 30 minutes, 24 hours, or 8 days after training. (Wong, et al., 1999) These mice perform normally during a 5-minute test, indicating these mice can learn the task normally and that short-term memory is intact. The passive avoidance memory impairment in the $AC1^{-/-}$ x $AC8^{-/-}$ double knockout mice can be rescued by unilateral administration of forskolin directly into CA1 region of the hippocampus, indicating the loss of Ca^{2+}-stimulated adenylyl cyclase activity can fully account for this memory phenotype. Taken together, these behavioral data indicate that Ca^{2+}/CaM-stimulated adenylyl cyclases are required for long-term memory, and that either AC1 or AC8 can produce the cAMP signal required for fear-conditioned memory.

AC1 is a key mediator of long-lasting synaptic plasticity and memory formation. This role is conserved in flies: the memory mutant *rutabaga* contains mutations in an AC1-like gene. In mice, AC1 is required for normal performance in the Morris water maze. Moreover, AC1 levels are increased in the hippocampus following Morris water maze training (Mons, et al., 2004), suggesting AC1 has a role in the encoding of memories. To determine if AC1 is sufficient to enhance cAMP-dependent plasticity and memory in mice, we developed a transgenic strain of mice harboring AC1 under the transcriptional control of the alpha-CaMKII promoter, which drives postnatal expression in the forebrain (Wang, et al., 2004). No overt phenotype was visible in the AC1 transgenic mice (AC1-TG). However, elevated levels of cAMP were measured in the hippocampus but not the cerebellum of AC1-TG mice, indicating the transgene is expressed in the proper brain regions and is enzymatically active (Fig. 2A). The CaMKII promoter is not active in cerebellar neurons. Also, cAMP production and AC1 activity were calcium-stimulated, confirming the identity of the transgene. PKA activity was also measured and is elevated in the hippocampus of the AC1-TG mice (data not shown).

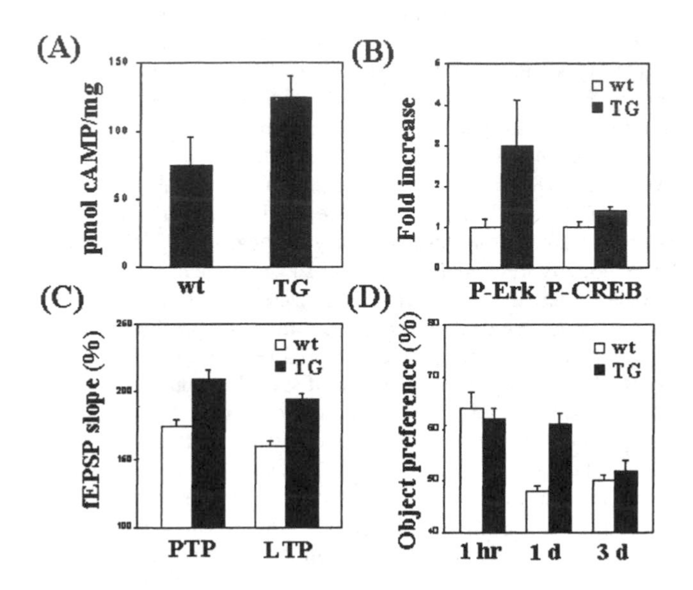

Figure 2. Transgenic mice overexpressing AC1 in forebrain display increased cAMP-mediated signal transduction, synaptic plasticity, and memory formation. (A) The levels of cAMP are elevated in the hippocampus of AC1 overexpressors (TG) relative to wild-type (wt) littermate control mice. (B) Enhanced Erk and CREB phosphorylation in untrained TG mice. (C) Elevated post-tetanic potentiation (PTP, measured 1 minute post-tetanus) and LTP (measured 60 minutes post-tetanus) in TG mice reflected in larger fEPSP slopes. (D) Object recognition memory in improved in TG mice. Mice were trained in a single 5-minute session and tested for recognition memory at 1 hour, 1 day, or 3 days. (Adapted from Wang, et al., 2004).

Increased cAMP and PKA activity promote the stimulation of the mitogen-activated protein kinase (MAPK) pathway, resulting in a 3-fold increase in Erk phosphorylation in the hippocampus of AC1-TG mice (Wang, et al., 2004) (Fig. 2B). The MAPK pathway can be stimulated through a cAMP-activated guanine nucleotide exchange factor, or through PKA which phosphorylates and activates B-raf in neurons (Grewal, et al., 2000). In addition, the levels of phosphorylated CREB were also increased in the hippocampus of AC1-TG . The MAPK pathway converges on the transcription factor CREB and the enhancer element to which in binds, the cAMP response-element (CRE). CREB is phosphorylated in hippocampus following training for contextual fear conditioning (Impey, et al., 1998b) and CREB function is required for LTM in flies (Yin, et al., 1994) and in mice (Kida, et al., 2002).

The AC1-TG mice also display enhanced plasticity in the form of post-tetanic potentiation (PTP) and LTP at Schaffer collateral synapses (Fig. 2C). PTP is a presynaptically regulated form of plasticity that lasts for minutes after stimulation. cAMP has a well established role in presynaptic forms of plasticity and neurotransmitter release mechanisms. The enhanced PTP is maintained in the AC1-TG, resulting in elevated LTP at 60 minutes post-tetanus. This LTP enhancement, because it is dependent on *de novo* gene expression, may be due to the stimulation of CRE-mediated transcription.

AC1-TG mice also exhibit enhanced object recognition memory when tested 1 day after a single 5-minute training session (Fig. 2D). Object recognition is a hippocampus-dependent form of memory in which mice are trained to discriminate between a novel and familiar object. Thus, elevated AC1 activity promotes enhanced synaptic plasticity and LTM, indicating AC1 is sufficient for these processes. The AC1-TG strain, along with a handful of other strains, is one of the few that display a well-correlated gain-of-function phenotype: that is, an increase in synaptic plasticity *and* enhanced LTM.

3.3.4 AC3 is Required for Detection of Odorants in the Main Olfactory Epithelium

cAMP is a critical second messenger in the olfactory system, suggesting ACs may have a role in the molecular mechanisms of odorant detection. Many odorants stimulate cAMP production in the cilia of the main olfactory epithelium (MOE) (Pace, et al., 1985). Several ACs, including AC2, AC3, and AC4, are expressed in the MOE that may account for this cAMP production. To evaluate the role of AC3 in olfactory responses, the AC3 gene was disrupted in mice (Wong, et al., 2000). Although AC2 and AC4 are expressed in MOE, electro-olfactogram responses to volatile odorants are completely absent in the main olfactory bulb of $AC3^{-/-}$ mice (Fig. 3A), indicating that AC3 and not AC2 or AC4 provide the cAMP signal required for odorant detection. In addition, $AC3^{-/-}$ mice perform

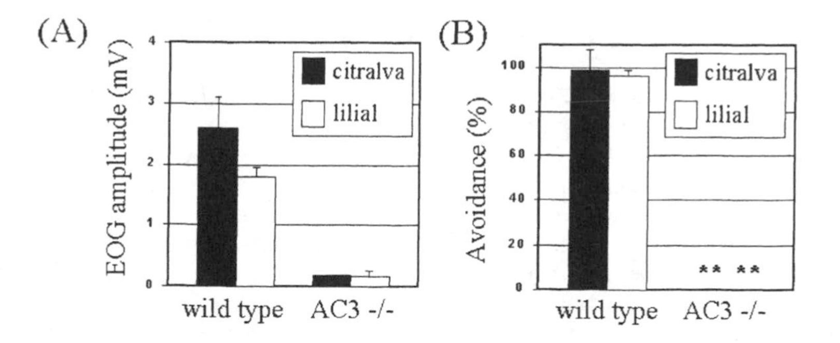

Figure 3. (A) Odorant-evoked electrophysiological responses are significantly reduced in the main olfactory bulb of AC3 (-/-) mice. Plotted are the citralva and lilial electro-olfactrogram (EOG) amplitudes. (B) Olfactory dependent learning is impaired in AC3 (-/-) mice. Mice were trained to associate citralva and lilial odors with a mild footshock and then tested for avoidance behavior. Percent avoidance values are shown. ** No avoidance behavior exhibited (Adapted from Wong et al, 2000).

poorly in several odorant-based behavioral tests (Wong, et al., 2000) (Fig. 3B). These experiments provided the first direct evidence that cAMP signaling and AC3 is required for olfaction. $AC3^{-/-}$ mice were also used to demonstrate that the vomeronasal organ (VNO), traditionally thought to detect only pheromones, can detect certain volatile odorants in the absence of MOE signaling (Trinh and Storm, 2003). Because AC3 is not expressed in VNO, this type of signaling is AC3-independent.

A rapid rise in cAMP in response to an olfactory cue is crucial because it alerts the animal to the presence of a specific odorant in its environment. However, odorant-stimulated cAMP peaks are transient in nature. Rapid desensitization of cAMP signaling is a critical property of the olfactory system that allows animals to respond to odorants. This property may be dependent upon AC3 and its unique regulatory properties such as inhibitory phosphorylation by CaMKII. This general model is supported by data showing that treatment of olfactory sensory neurons with CaM kinase II inhibitors impairs odor adaptation (Leinders-Zufall, et al., 1999). Multiple odorant-stimulated cAMP peaks may be required to detect gradients of odorants. An animal cannot sense an olfactory gradient unless it can terminate olfactory signaling and re-sample its environment during movement through an odorant gradient. If an animal can sample and compare an odorant signal in the time it takes to move through a gradient, it may modify its behavior by moving towards or away from the source of odorant. An animal may determine, at the molecular level, whether it is moving up or down a gradient by comparing the amplitude of the first cAMP transient to the second, second to the third, etc. Collectively, these data indicate that AC3 is ideally suited to couple odorant detection to the production of cAMP and to provide a molecular mechanism that contributes to the rapid decline in intracellular cAMP.

4. DISCUSSION

cAMP and AC1 have multiple short- and long-term effects in neurons. These changes are mediated primarily through the two main cAMP effectors in neurons: protein kinase A (PKA) and the cAMP-activated guanine nucleotide exchange factor (cAMP-GEF). cAMP and PKA exert short-term effects in neurons through phosphorylation of ion channels. For example, phosphorylation of the AMPA subunit GluR1 on serine 845 by PKA increases its channel open probability (Banke, et al., 2000), allowing increased AMPA-mediated current to flow into the neuron. Moreover, activity-induced PKA phosphorylation of GluR1 is required for intracellular transport and synaptic incorporation of latent the AMPA receptors that participate in "silent synapses" (Esteban, et al., 2003). These phosphorylation events and the physiological effects they mediate are typically transient and may therefore have a greater role in short-term forms of synaptic plasticity.

The necessity for AC1 and AC8 in long-lasting LTP (L-TLP) and long-term memory is likely due, at least in part, to the activation of cAMP-dependent signal transduction and transcriptional pathways, the basic elements of which are shown below (Fig. 4). The mitogen-activated protein kinase (MAPK) and CREB/CRE transcriptional pathways play central and important roles in plasticity and memory formation (Poseur and Storm, 2001). NMDA receptor activation by glutamate is required for Schaffer collateral L-LTP and for some forms of hippocampus-dependent memory (Morris, et al., 1986). Ca^{2+} entry via these receptors can stimulate adenylyl cyclase activity and cAMP production (Chetkovich, et al., 1991). Both PKA and cAMP GEF stimulate the MAPK pathway through Rap-1 (Dugan, et al., 1999). PKA is also required for the translocation of MAPK into the nucleus of neurons (Impey, et al., 1998a). Within the nucleus, MAPK phosphorylates and activates rsk2, the major CREB kinase in neurons. Phosphorylation of CREB by rsk2 and recruitment of the CREB binding protein transactivator, CBP, promote transcription from CRE-containing genes. Behavioral training in the contextual fear conditioning paradigm leads to the activation of a CRE-reporter *in vivo*. (Impey, et al., 1998b) In addition, CRE oligonucleotide decoys block LTM in mice. (Athos, et al., 2002) These data indicate that cAMP/PKA activity, CREB function, and CRE-mediated transcription are required for LTM. However, it is likely that genes regulated by other signaling pathways and through other enhancer elements will be important in plasticity and memory. It will be interesting and informative to analyze signal transduction, CRE-reporter activation, and gene expression profiles in AC1 gain- or loss-of-function mice. We have already shown that the MAPK pathway is potentiated in the AC1-TG strain.

As demonstrated by experiments using inhibitors of transcription, L-LTP and long-term memory require *de novo* gene transcription. Numerous studies have identified genes that are regulated by behavioral learning. Hypothesis-driven, candidate gene approaches have tended to be more reliable and informative. Contextual fear conditioning is an well-established hippocampus-dependent form of learning and memory. In this paradigm, a training context (the conditional stimulus, CS) is temporally paired with a mild aversive footshock (the unconditional stimulus, US), which leads to robust long-term memory after a single training session. Memory in this paradigm is expressed as freezing, a stereotypical fear response in which movement, except that required to breathe, cease (Fanselow, et al., 1989). Contextual training in this manner leads to the induction of a CRE-reporter gene in vivo and to transcriptional upregulation of many CRE containing genes including c-fos (Radulovic, et al., 1998), BDNF (Hall, et al., 2000), 14-3-3 eta (Kida, et al., 2002), and C/EBP-β (Taubenfeld, et al., 2001). Many of the same genes are similarly induced following training in the Morris water maze. A number of gene expression profiling experiments following behavioral training utilizing PCR- or DNA microarray techniques have been described. Although interesting, these studies are far more technically challenging, suffer from reliability issues, and can be best used to identify candidates that may have been previously unassociated with memory formation.

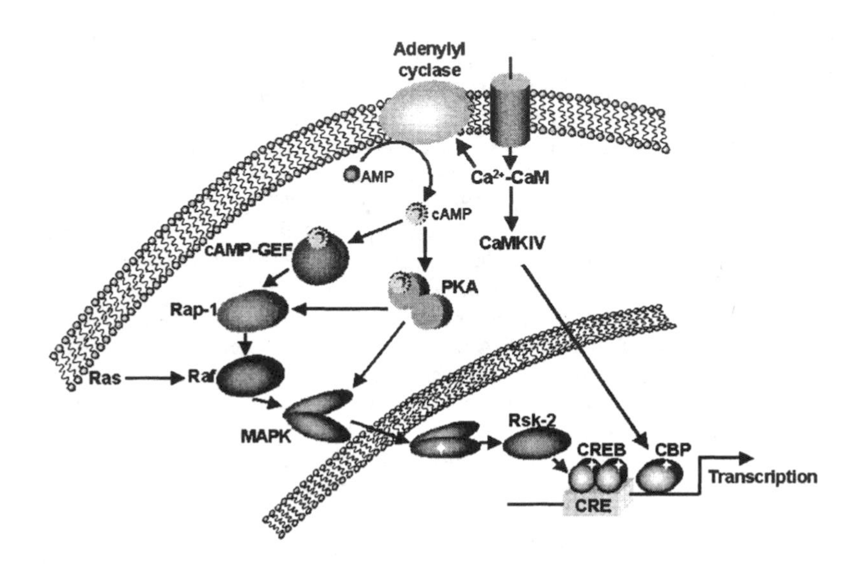

Figure 4. Calcium/calmodulin-stimulated adenylyl cyclases provide a critical link between neuronal activity and cAMP production. Post-synaptic calcium increases generated through NMDA receptors stimulate adenylyl cyclases and activate PKA and MAPK signaling transduction pathways, which converge on the CREB/CRE transcriptional pathway. Phosphorylation of the transcription factor CREB and the transactivator protein CBP leads to the expression of genes required for L-LTP and LTM. It is proposed that either AC1 or AC8 can provide the cAMP signal necessary for L-LTP and LTM.

The most convincing study of this type profiled Morris water maze-induced changes in gene expression over multiple time points, which allowed the authors to provide kinetics for the genes identified in the screen (Cavallaro, et al., 2002). For example, fibroblast growth factor (FGF)-18 is upregulated at multiple time points, within the windows of transcriptional activity required for memory formation. This study is one of the few to functionally validate an identified target. Interestingly, when administered exogenously into brain, FGF-18 enhanced performance in Morris water maze.

4.1 Calcium-Regulated Adenylyl Cyclases as Targets of Memory Enhancing Drugs

Individuals in the U.S. and the Western world are living longer, more productive lives. However, ageing is associated with normal cognitive decline. Age-associated memory impairment (AAMI) or the more severe mild cognitive impairment (MCI) are conditions now routinely diagnosed by physicians. In addition, the occurrence of neurodegenerative disorders such as Alzheimer's disease or senile dementia that are accompanied by memory loss are also increasing dramatically. Therefore, drugs that enhance memory are of great interest to pharmaceutical companies. Many of the companies that are specifically pursuing memory enhancing drugs, notably Helicon Therapetics and Memory Pharmaceuticals, are focusing on cAMP because of the wealth of data implicating this pathway in memory and plasticity. Indeed, PDE4 has been identified as a target in screens for modulators of the cAMP pathway (Tully, et al., 2003). Rolipram is an older PDE4 inhibitor that promotes increased cAMP in brain and can enhance memory in aged rats (Barad, et al., 1998). However, rolipram is of limited clinical value because of severe inflammatory side effects. The development of rolipram-like drugs with greater specificity and potency are being pursued. Are ACs good potential drug targets? AC8 is distributed throughout the body and is therefore not an ideal candidate as a drug target. In contrast, AC1 is an excellent candidate because of its neurospecific expression pattern and because it is decreased in the hippocampus of aged mice (Mons, et al., 2004). Moreover, we have shown that AC1 is sufficient for cAMP-mediated signal transduction, plasticity, and memory. It may be feasible to develop an activator of AC1. Ideally, one would want to develop a small-molecule that activates AC1 *only* in the presence of calcium, i.e. when accompanied by neuronal activity. This would ensure that cAMP production is occurring only at the appropriate time. Compounds of this type may have therapeutic value as memory enhancing drugs.

5. ACKNOWLEDGEMENTS

We thank members of the Storm lab for helpful discussions and contribution of data to this chapter.

6. REFERENCES

Abdel-Majid RM, Leong WL, Schalkwyk LC, Smallman DS, Wong ST, Storm DR, Fine A, Dobson MJ, Guernsey DL, Neumann PE (1998) Loss of adenylyl cyclase I activity disrupts patterning of mouse somatosensory cortex. Nat Genet 19:289-91,

Abel T, Nguyen PV, Barad M, Deuel TA, Kandel ER, Bourtchouladze R (1997) Genetic demonstration of a role for PKA in the late phase of LTP and in hippocampus-based long-term memory. Cell, 88:615-26,

Athos J, Impey S, Pineda VV, Chen X, Storm DR (2002) Hippocampal CRE-mediated gene expression is required for contextual memory formation. Nat Neurosci 5:1119-20,

Baker LP, Nielsen MD, Impey S, Metcalf MA, Poser SW, Chan G, Obrietan K, Hamblin MW, Storm DR (1998) Stimulation of type 1 and type 8 Ca2+/calmodulin-sensitive adenylyl cyclases by the Gs-coupled 5-hydroxytryptamine subtype 5-HT7A receptor. J Biol Chem 273:17469-76,

Banke TG, Bowie D, Lee H, Huganir RL, Schousboe A, Traynelis SF (2000) Control of GluR1 AMPA receptor function by cAMP-dependent protein kinase. J Neurosci 20:89-102.

Barad M, Bourtchouladze R, Winder DG, Golan H, Kandel ER 1998) Rolipram, a type IV-specific phosphodiesterase inhibitor, facilitates the establishment of long-lasting long-term potentiation and improves memory, Proc Natl Acad Sci U S A, 95, 15020-5.

Brunelli M, Castellucci V, Kandel ER (1976) Synaptic facilitation and behavioral sensitization in Aplysia: possible role of serotonin and cyclic AMP. Science 194:1178-81.

Cavallaro S, D'Agata V, Manickam P, Dufour F, Alkon DL (2002) Memory-specific temporal profiles of gene expression in the hippocampus. Proc Natl Acad Sci USA 99:16279-84.

Chan GC, Lernmark U, Xia Z, Storm DR (2001) DNA elements of the type 1 adenylyl cyclase gene locus enhance reporter gene expression in neurons and pinealocytes. Eur J Neurosci 13:2054-66.

Chao JR, Ni YG, Bolanos CA, Rahman Z, DiLeone RJ, Nestler EJ (2002) Characterization of the mouse adenylyl cyclase type VIII gene promoter: regulation by cAMP and CREB. Eur J Neurosci, 16:1284-94.

Chetkovich DM, Gray R, Johnston D, Sweatt JD (1991) N-methyl-D-aspartate receptor activation increases cAMP levels and voltage-gated Ca^{2+} channel activity in area CA1 of hippocampus. Proc Natl Acad Sci USA, 88:6467-71.

Davis RL)1996) Physiology and biochemistry of Drosophila learning mutants. Physiol Rev, 76:299-317.

Defer N, Marinx O, Poyard M, Lienard MO, Jegou B, Hanoune J (1998) The olfactory adenylyl cyclase type 3 is expressed in male germ cells, FEBS Lett, 424:216-20.

Dugan LL, Kim JS, Zhang Y, Bart RD, Sun Y, Holtzman DM, Gutmann DH (1999) Differential effects of cAMP in neurons and astrocytes. Role of B-raf. J Biol Chem, 274:25842-8.

Esteban JA, Shi SH, Wilson C, Nuriya M, Huganir RL, Malinow R (2003) PKA phosphorylation of AMPA receptor subunits controls synaptic trafficking underlying plasticity. Nat Neurosci, 6:136-43.

Fanselow MS, Calcagnetti DJ, Helmstetter FJ (1989) Modulation of appetitively and aversively motivated behavior by the kappa opioid antagonist MR2266. Behav Neurosci, 103:663-72.

Grewal SS, Horgan AM, York RD, Withers GS, Banker GA, Stork PJ (2000) Neuronal calcium activates a Rap1 and B-Raf signaling pathway via the cyclic adenosine monophosphate-dependent protein kinase. J Biol Chem, 275:3722-8.

Gu C, Cooper DM (1999) Calmodulin-binding sites on adenylyl cyclase type VIII. J Biol Chem, 274:8012-21.

Hall J, Thomas KL, Everitt BJ (2000) Rapid and selective induction of BDNF expression in the hippocampus during contextual learning. Nat Neurosci, 3:533-5.

Impey S, Wayman G, Wu Z, Storm DR (1994) Type I adenylyl cyclase functions as a coincidence detector for control of cyclic AMP response element-mediated transcription: synergistic regulation of transcription by Ca^{2+} and isoproterenol. Mol Cell Biol, 14:8272-81.

Impey S, Obrietan K, Wong ST, Poser S, Yano S, Wayman G, Deloulme JC, Chan G, Storm DR (1998a) Cross talk between ERK and PKA is required for Ca^{2+} stimulation of CREB-dependent transcription and ERK nuclear translocation. Neuron, 21:869-83.

Impey S, Smith DM, Obrietan K, Donahue R, Wade C, Storm DR (1998b) Stimulation of cAMP response element (CRE)-mediated transcription during contextual learning. Nat Neurosci, 1:595-601.

Katz LC, Shatz CJ (1996) Synaptic activity and the construction of cortical circuits. Science, 274:1133-8.

Kida S, Josselyn SA, de Ortiz SP, Kogan JH, Chevere I, Masushige S, Silva AJ (2002) CREB required for the stability of new and reactivated fear memories. Nat Neurosci, 5:348-55.

Krupinski J, Coussen F, Bakalyar HA, Tang WJ, Feinstein PG, Orth K, Slaughter C, Reed RR, Gilman AG (1989) Adenylyl cyclase amino acid sequence: possible channel- or transporter-like structure. Science, 244:1558-64.

Kumar PA, Baker LP, Storm DR, Bowden DM (2001) Expression of type I adenylyl cyclase in intrinsic pathways of the hippocampal formation of the macaque (Macaca nemestrina). Neurosci Lett 299:181-4.

Leinders-Zufall T, Ma M, Zufall F (1999) Impaired odor adaptation in olfactory receptor neurons after inhibition of $Ca^{2+/}$calmodulin kinase II. J Neurosci, 19:RC19.

Lev-Ram V, Wong ST, Storm DR, Tsien RY (2002) A new form of cerebellar long-term potentiation is postsynaptic and depends on nitric oxide but not cAMP. Proc Natl Acad Sci USA, 99:8389-93.

Lu HC, She WC, Plas DT, Neumann PE, Janz R, Crair MC (2003) Adenylyl cyclase I regulates AMPA receptor trafficking during mouse cortical 'barrel' map development. Nat Neurosci, 6:939-47.

Minocherhomjee M, Selfe S, Flowers NJ, Storm DR (1987) Direct interaction between the catalytic subunit of the calmodulin-sensitive adenylate cyclase from bovine brain with 125I-labeled wheat germ agglutinin and 125I-labeled calmodulin. Biochemistry, 26:4444-8.

Mons N, Segu L, Nogues X, Buhot MC (2004) Effects of age and spatial learning on adenylyl cyclase mRNA expression in the mouse hippocampus. Neurobiol Aging, 25:1095-106.

Morris RG, Anderson E, Lynch GS, Baudry M (1986) Selective impairment of learning and blockade of long-term potentiation by an N-methyl-D-aspartate receptor antagonist, AP5. Nature, 319:774-6.

Muglia LM, Schaefer ML, Vogt SK, Gurtner G, Imamura A, Muglia LJ (1999) The 5'-flanking region of the mouse adenylyl cyclase type VIII gene imparts tissue-specific expression in transgenic mice. J Neurosci, 19:2051-8.

Nielsen MD, Chan GC, Poser SW, Storm DR (1996) Differential regulation of type I and type VIII Ca^{2+}-stimulated adenylyl cyclases by Gi-coupled receptors in vivo. J Biol Chem, 271:33308-16.

Pace U, Hanski E, Salomon Y, Lancet D (1985) Odorant-sensitive adenylate cyclase may mediate olfactory reception. Nature, 316:255-8.

Pfeuffer E, Mollner, S, Pfeuffer T (1985) Adenylate cyclase from bovine brain cortex: purification and characterization of the catalytic unit. Embo J, 4:3675-9.

Pineda VV, Athos JI, Wang H, Celver J, Ippolito D, Boulay G, Birnbaumer, L, Storm, DR (2004) Removal of G(ialpha1) Constraints on Adenylyl Cyclase in the Hippocampus Enhances LTP and Impairs Memory Formation. Neuron, 41:153-63.

Poser S, Storm DR (2001) Role of Ca2+-stimulated adenylyl cyclases in LTP and memory formation. Int J Dev Neurosci, 19:387-94.

Radulovic J, Kammermeier J, Spiess J (1998) Relationship between fos production and classical fear conditioning: effects of novelty, latent inhibition, and unconditioned stimulus preexposure. J Neurosci, 18:7452-61.

Ravary A, Muzerelle A, Herve D, Pascoli V, Ba-Charvet KN, Girault JA, Welker E, Gaspar P, (2003) Adenylate cyclase 1 as a key actor in the refinement of retinal projection maps. J Neurosci, 23:2228-38.

Reddy R, Smith D, Wayman G, Wu Z, Villacres EC, Storm DR (1995) Voltage-sensitive adenylyl cyclase activity in cultured neurons. A calcium-independent phenomenon. J Biol Chem, 270:14340-6.

Rosenberg GB, Storm DR (1987) Immunological distinction between calmodulin-sensitive and calmodulin-insensitive adenylate cyclases. J Biol Chem, 262:7623-8.

Schaefer ML, Wong ST, Wozniak DF, Muglia LM, Liauw JA, Zhuo M, Nardi A, Hartman RE, Vogt SK, Luedke CE, Storm DR, Muglia LJ (2000) Altered stress-induced anxiety in adenylyl cyclase type VIII-deficient mice. J Neurosci, 20:4809-20.

Storm DR, Hansel C, Hacker B, Parent A, Linden DJ (1998) Impaired cerebellar long-term potentiation in type I adenylyl cyclase mutant mice. Neuron, 20:1199-210.

Taubenfeld SM, Wiig KA, Monti B, Dolan B, Pollonini G, Alberini CM (2001) Fornix-dependent induction of hippocampal CCAAT enhancer-binding protein [beta] and [delta] Co-localizes with phosphorylated cAMP response element-binding protein and accompanies long-term memory consolidation. J Neurosci, 21:84-91.

Tesmer JJ, Sunahara RK, Gilman AG, Sprang SR (1997) Crystal structure of the catalytic domains of adenylyl cyclase in a complex with Gsalpha.GTPgammaS. Science, 278:1907-16.

Trinh K, Storm DR (2003) Vomeronasal organ detects odorants in absence of signaling through main olfactory epithelium. Nat Neurosci, 6:519-25.

Tully T, Bourtchouladze R, Scott R, Tallman J (2003) Targeting the CREB pathway for memory enhancers. Nat Rev Drug Discov, 2:267-77.

Villacres EC, Wu Z, Hua W, Nielsen MD, Watters JJ, Yan C, Beavo J, Storm DR (1995) Developmentally expressed Ca(2+)-sensitive adenylyl cyclase activity is disrupted in the brains of type I adenylyl cyclase mutant mice. J Biol Chem, 270:14352-7.

Villacres EC, Wong ST, Chavkin C, Storm DR (1998) Type I adenylyl cyclase mutant mice have impaired mossy fiber long-term potentiation J Neurosci, 18:3186-94.

Wang H, Chan GC, Athos J, Storm DR (2002) Synaptic concentration of type-I adenylyl cyclase in cerebellar neurons. J Neurochem, 83:946-54.

Wang H, Pineda VV, Chan GC, Wong ST, Muglia LJ, Storm DR (2003) Type 8 adenylyl cyclase is targeted to excitatory synapses and required for mossy fiber long-term potentiation. J Neurosci, 23:9710-8.

Wang H, Ferguson GD, Pineda VV, Cundiff PE, Storm DR (2004) Overexpression of type-1 adenylyl cyclase in mouse forebrain enhances recognition memory and LTP. Nat Neurosci, 7:635-42.

Watson EL, Jacobson KL, Singh JC, Idzerda R, Ott SM, DiJulio DH, Wong ST, Storm DR (2000) The type 8 adenylyl cyclase is critical for Ca^{2+} stimulation of cAMP accumulation in mouse parotid acini. J Biol Chem, 275:14691-9.

Wayman GA, Impey S, Wu Z, Kindsvogel W, Prichard L, Storm DR (1994) Synergistic activation of the type I adenylyl cyclase by Ca2+ and Gs-coupled receptors in vivo. J Biol Chem, 269:25400-5.

Wayman GA, Hinds TR, Storm DR (1995a) Hormone stimulation of type III adenylyl cyclase induces Ca2+ oscillations in HEK-293 cells. J Biol Chem, 270:24108-15.

Wayman GA, Impey S, Storm DR (1995b) Ca^{2+} inhibition of type III adenylyl cyclase in vivo. J Biol Chem, 270:21480-6.

Wayman GA, Wei J, Wong S, Storm DR (1996) Regulation of type I adenylyl cyclase by calmodulin kinase IV in vivo. Mol Cell Biol, 16:6075-82.

Wei J, Zhao AZ, Chan GC, Baker LP, Impey S, Beavo JA, Storm DR (1998) Phosphorylation and inhibition of olfactory adenylyl cyclase by CaM kinase II in Neurons: a mechanism for attenuation of olfactory signals. Neuron, 21:495-504.

Weisskopf MG, Castillo PE, Zalutsky RA, Nicoll RA (1994) Mediation of hippocampal mossy fiber long-term potentiation by cyclic AMP. Science, 265:1878-82.

Welker E, Armstrong-James M, Bronchti G, Ourednik W, Gheorghita-Baechler F, Dubois R, Guernsey DL, Van der Loos H, Neumann PE (1996) Altered sensory processing in the somatosensory cortex of the mouse mutant barrelless. Science, 271:1864-7.

Winder DG, Egli RE, Schramm NL, Matthews RT (2002) Synaptic plasticity in drug reward circuitry. Curr Mol Med, 2:667-76.

Wong ST, Athos J, Figueroa XA, Pineda VV, Schaefer ML, Chavkin CC, Muglia LJ, Storm DR (1999) Calcium-stimulated adenylyl cyclase activity is critical for hippocampus-dependent long-term memory and late phase LTP. Neuron, 23:787-98.

Wong ST, Trinh K, Hacker B, Chan GC, Lowe G, Gaggar A, Xia Z, Gold GH, Storm DR (2000) Disruption of the type III adenylyl cyclase gene leads to peripheral and behavioral anosmia in transgenic mice. Neuron, 27:487-97.

Wu Z, Wong ST, Storms DR (1993) Modification of the calcium and calmodulin sensitivity of the type I adenylyl cyclase by mutagenesis of its calmodulin binding domain J Biol Chem, 268:23766-8.

Wu ZL, Thomas SA, Villacres EC, Xia Z, Simmons ML, Chavkin C, Palmiter RD, Storm DR (1995) Altered behavior and long-term potentiation in type I adenylyl cyclase mutant mice. Proc Natl Acad Sci USA, 92:220-4.

Xia Z, Choi EJ, Wang F, Blazynski C, Storm DR (1993) Type I calmodulin-sensitive adenylyl cyclase is neural specific. J Neurochem, 60:305-11.

Xia ZG, Refsdal CD, Merchant KM, Dorsa DM, Storm DR (1991) Distribution of mRNA for the calmodulin-sensitive adenylate cyclase in rat brain: expression in areas associated with learning and memory. Neuron, 6:431-43.

Yeager RE, Heideman W, Rosenberg GB, Storm DR (1985) Purification of the calmodulin-sensitive adenylate cyclase from bovine cerebral cortex. Biochemistry, 24:3776-83.

Yin JC, Wallach JS, Del Vecchio M, Wilder EL, Zhou H, Quinn WG, Tully T (1994) Induction of a dominant negative CREB transgene specifically blocks long-term memory in Drosophila. Cell, 79:49-58.

RAPID NUCLEAR RESPONSES TO ACTION POTENTIALS

J. Paige Adams, Eric Hudgins, Joseph J. Lundquist, Meilan Zhao, and Serena M. Dudek*

1. WHAT DOES THE NUCLEUS KNOW ABOUT MEMORIES?

Long-term potentiation (LTP) is a lasting increase in the strength of synaptic transmission that has often been used as a model of memory at the cellular level. Like some forms of memory, the later phases of LTP have been shown to be dependent on new RNA (Nguyen et al., 1994; Frey et al., 1996) and protein (Frey et al., 1988) synthesis. Yet since LTP is also thought to be synapse-specific, the question arises how changes taking place at one of 10,000 synapses in a pyramidal cell could be coded in a single nucleus. In 1995, John Lisman articulated the paradox by asking, "what does the nucleus know about memories (Lisman, 1995)?" Is information stored as a self-sustaining switch in the nucleus? Or is the nucleus responsible for up-regulating key components for a self-sustaining switch located at the synapse? Perhaps the former is true, but evidence has not yet emerged to substantiate or refute the idea in the context of hippocampal LTP or memory. In contrast, the latter scenario is beginning to enjoy an accumulation of experimental support for the notion of synaptic tagging with capture (Frey and Morris, 1997; Barco et al., 2002; Dudek and Fields, 2002) to consolidate the late phases of LTP. The story that is emerging is one in which the nucleus could rapidly respond to LTP-inducing stimulation and the tag at the modified synapse would be responsible for capturing the product. In this chapter, we will explore two aspects of the process: 1) how is the signal transmitted to the nucleus to initiate transcription in the context of late-phase LTP? and 2) what signals are fast enough to fit time constraints imposed by experimental evidence? Though Lisman didn't address how changes might occur in the nucleus, we believe that action potentials during LTP induction could play a central role in informing the nucleus that LTP had likely occurred. We hope to make the

* National Institute of Environmental Health Sciences, National Institutes of Health, Research Triangle Park, NC 27709. E-mail: dudek@niehs.nih.gov

case that without attending to signals from individual potentiated synapses, the nucleus can detect action potentials relating to LTP induction and respond by up-regulating the relevant genes accordingly.

2. INDUCTION OF TRANSCRIPTION CAN BE RAPID

The late-phase of LTP appears to be sensitive to transcriptional inhibitors only during a very brief period (on the order of a few minutes) following the induction events (Nguyen et al., 1994; Frey et al., 1996). After induction, LTP is completely insensitive to RNA synthesis inhibitors, making it unlikely that lengthy cascades of immediate early genes are at work in late-phase LTP. Accordingly, the nuclear response to LTP-inducing stimulation would be required to take place on an extremely rapid time scale. This is not an unreasonable idea, since examples of such rapid induction of immediate early genes in neurons have been observed within minutes of the stimulation. *Arc* and *zif-268*, for example, can be induced within 5 minutes of LTP-inducing stimulation and behavioral paradigms including learning and exploration (Guzowski et al., 1999). The gene for Homer, a protein that associates with metabotropic glutamate receptors, can be transcribed in as early as 16 minutes (Vazdarjanova et al., 2002), and other genes, such as that for Protein Kinase Mζ can be transcribed and translated within 40 minutes after LTP-inducing stimulation (Hernandez et al., 2003). Later phases of gene expression, between one and six hours, do occur with learning (Levenson et al., 2004b), but since the effects of RNA synthesis inhibitors are less clear in behaving animals, distinguishing which phase of gene expression is required for long term memory (LTM) has been difficult. One study, though, has shown that inhibition of Arc protein expression impairs the maintenance of LTP and memory, indicating that a rapidly transcribed gene can play a role in both (Guzowski et al., 2000). Based on these, and the studies performed on late-phase LTP, we propose that a molecule that is up-regulated within minutes of neuronal activity, such as *arc*, is most likely to be involved in consolidation events linked to the late-phase of LTP and perhaps long-term memory. While much has been learned about neuronal gene regulation several hours after activity, such as that for Brain Derived Neurotrophic Factor (BDNF) (Martinowich et al., 2003) (Levenson et al., 2004b), research on the mechanisms of rapid induction of gene transcription within minutes, like that required for late-phase LTP, has lagged behind.

3. IS THERE TIME TO GET A SIGNAL FROM THE SYNAPSE TO THE NUCLEUS?

Can we get a molecular signal from the synapse to the nucleus with this 15 minute time window? Among many proposed candidates, NF-κB has thus far been the only synapse-to-nucleus signal demonstrated to be perhaps up to the task. In non-neuronal cells, Rel/NF-κB transcription factors are important for diverse functions as cell proliferation and apoptosis (Gilmore, 1999), and in neurons, they have been implicated in neuroprotection (Mattson, 1997). Translocation to the nucleus stands as the primary means by which these transcription factors are regulated. It is thought that NF-κB subunits are bound to the inhibitory protein IκB, which is then degraded by the 26S proteosome upon activating stimuli. By labeling p65 (RelA), a subunit of NF-κB that resides at synapses, with green fluorescent protein (GFP), the movement of NF-κB throughout neurons can be followed after stimulation with NMDA, glutamate, kainate or potassium (Wellmann et al., 2001; Meffert et al., 2003). After such stimulation, GFPp65 was shown to translocate to the nucleus and was found to require CaMKII activity and submembrane calcium increases for the translocation (i.e. it was blocked by BAPTA, but not EGTA) (Meffert et al., 2003). Though an increase in fluorescence in the nucleus was observed as early as 30 minutes (Meffert et al., 2003), earlier time-points were not reported, making an assessment of p65's suitability for rapid regulation of transcription difficult. Even so, Meffert et al. (2003) showed that by bleaching distal portions of dendrites, they could measure movement of GFPp65 toward the cell body and estimate the rate: up to 35 nm per second (1.5-2 μm/minute) following stimulation with NMDA. Assuming this rate, p65 coming from a point midway in a pyramidal cell dendrite, 50 μm away from the nucleus, would take between 25 and 35 minutes to reach the nucleus. This is just outside of the limit given by studies using RNA synthesis inhibitors showing that inhibitors applied before, but not immediately after, high frequency stimulation can block late-phase LTP (Nguyen et al., 1994; Frey et al., 1996). Even if one assumes that the inhibitor may take up to 15-30 minutes to diffuse into the cell, the rate of p65 translocation is not likely to be fast enough to initiate transcription within this critical time window. Considering that protein synthesis relating to late-phase LTP also displays a critical time window and must be accomplished within 15 minutes (assuming a longer time-lag to get into the cell, 30 minutes) (Frey et al., 1988; Otani, 1989), current evidence does not favor the synapse-to-nucleus hypothesis for LTP. Hence, we believe that a role for a putative synapse-to-nucleus signal in late-phase LTP should not be presumed to be the case but instead should require evidence supporting the claim. What signal could be fast enough? Our alternative model posits that neuronal activity, either in the form of bursts or single pulses, is sufficient for such nuclear signaling and is quite definitely fast enough (Figure 1). Action potential-dependent signals could be in the form of the calcium-dependent activation of nuclear enzymes, such as CaMKIV (Hardingham et al., 2001), but translocations to the nucleus from the somatic cytoplasm or membrane are certainly not ruled out under such a model (Deisseroth et al., 1998; Dolmetsch et al., 2001).

4. ACTION POTENTIALS CAN RAPIDLY ACTIVATE SIGNALING

It is expected that the frequency of action potentials will turn out to be important for the encoding of learning- and/or LTP-related information at the nucleus. The role of frequency in encoding gene transcription has been addressed in dorsal root ganglion neurons (Eshete and Fields, 2001). In the context of hippocampal LTP, high frequencies are very effective at rapidly activating enzymes including CaM kinase (Fukunaga et al.,

1993), PKC (Klann et al., 1993), cAMP dependent protein kinase (Roberson and Sweatt, 1996), and the mitogen-activated protein kinase (MAPK) extracellular signal-regulated kinase (ERK) (English and Sweatt, 1996), while low frequencies are effective at activating others including ERK (> 3Hz and paired-pulse LTD stimulation (Dudek and Fields, 2001; Thiels et al., 2002)) and p38 MAP kinase (Bolshakov et al., 2000). Thus a reasonable hypothesis for encoding frequency in the nucleus involves either a recruitment of enzymes at high frequencies (and presumably high calcium) and/or the recruitment of phosphatases and other kinases at lower frequencies (and calcium levels), such as is likely to be the case for LTD and LTP at the synapse. What is not clear yet, however, is whether high frequencies, such as those typically used to induce LTP, are even necessary for induction of late-phase LTP-related transcription; recent ideas on the role of spike timing in plasticity replace the necessity of high frequency afferent activity with a critical timing of pre-synaptic activity and post-synaptic spiking (Abbott and Nelson, 2000). Further work will be required, though, to test the idea that lower frequencies of firing are sufficient for induction of late-phase LTP-related genes.

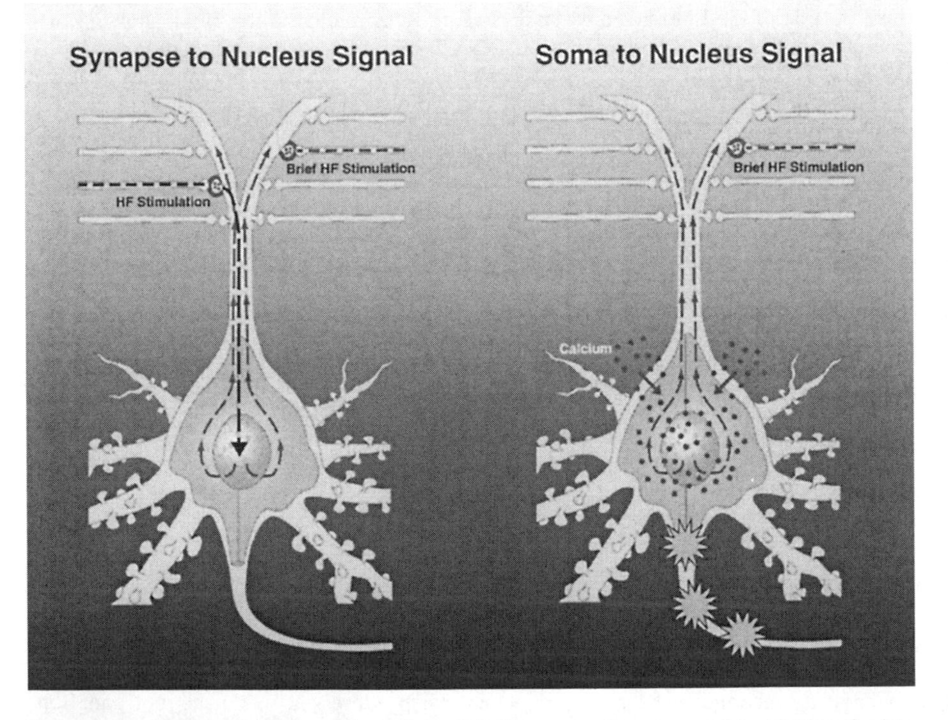

Figure 1. Two models of signaling to the nucleus. Left: The synapse-to-nucleus model posits that after late-LTP (High Frequency stimulation) a signal is made that travels to the nucleus, where a plasticity-related RNA is transcribed and the gene product (RNA or protein) is transported throughout the neuron. Only tagged synapses, including those where early-LTP was induced (brief HF stimulation), can take advantage of the gene product. Right: The soma-to-nucleus model proposes that when the cell fires action potentials at sufficient frequency and number, calcium enters, probably through L-type voltage sensitive calcium channels, to activate somatic and/or nuclear signals. The gene product goes out to the cell as in the first model to make late-phase LTP at all tagged synapses.

Two important assumptions are being made when one takes the view that a molecular synapse-to-nucleus signal is required for late-phase LTP: 1) if LTP is induced at a single synapse, enough of a signal is generated there to induce gene transcription. This would require that the few molecules made at a synapse (estimated 1/500,000th of the cell's volume) can find their way down to the nucleus and find the appropriate target there. If assumption (1) seems unlikely, one must then assume 2) that hundreds more synapses need to be potentiated simultaneously (or nearly so) in order to generate sufficient numbers of molecules to accumulate in the nucleus in concentrations that stoichiometrically favor transcription. If one maintains that LTP can occur at single synapses, action potentials present a signaling opportunity not to be overlooked. Assuming action potentials are sufficient for inducing the relevant transcription, one need only make two assumptions, also shared by the first group: 1) late-phase LTP requires new RNA synthesis, and 2) a synaptic tag needs to be generated at the potentiated synapse. To address the first assumption, evidence has accumulated to support the assumption that RNA synthesis is required for late-phase LTP in addition to the pharmacological studies using RNA synthesis inhibitors; experiments on dendrites severed from their cell bodies show that LTP induced in this preparation decays to baseline, indicating that something is necessary from the soma for late-phase LTP (Frey et al., 1989). The second assumption is necessary if one assumes the first.

One issue that has been raised with respect to activation of transcription factors is whether excitatory post-synaptic potentials (EPSPs), over action potentials, preferentially activate L-type calcium channels (Mermelstein et al., 2000). CyclicAMP response element binding protein (CREB) phosphorylation, it seems, relies on the presence of synaptic activity since action potentials induced by field stimulation were ineffective at inducing the response (Deisseroth et al., 1996), but see Dudek and Fields (2002). The difference may turn out to be that Dudek and Fields (2002) used antidromically delivered (theta) bursts of action potentials to induce CREB phosphorylation, which may be more effective at activating L-channels than constant rate stimulation. Other forms of signaling were also induced with the antidromic stimulation including phosphorylation of ERK and induction of Zif268 protein. Further evidence in support of the idea of action potential-dependent transcription events is that synapses need not be activated for the rescue of early-LTP in a Frey and Morris type experiment (Frey and Morris, 1997); action potentials delivered antidromically from the alveus were sufficient to protect early-LTP from decay (Dudek and Fields, 2002). While there is certainly some value to the argument of a synapse-to-nucleus signal vs. action potential-dependent signals, the subject of synaptic vs. action potential depolarization is not likely to be a problem in living animals, since action potentials are necessarily induced by (and be accompanied by) synaptic activity.

Whether action potentials themselves are important for such signaling to the nucleus is another issue. In at least one case (ERK phosphorylation), stimulation intensities sufficient to recruit action potentials are necessary for enzyme activation, and staining for the activated enzyme occurs in an all-or-none fashion throughout the neuron (Dudek and Fields, 2001). Moreover, drugs that block action potentials such as muscimol (to activate inhibitory synapses) and CNQX (to block excitatory synapses) completely blocked activation of ERK. Coupled with the result that antidromic activation of CA1 pyramidal cells can also activate ERK, one can conclude that action potentials are necessary and

sufficient for many types of signaling. The role of NMDA receptors should not be downplayed, however, since full blockade of theta-burst stimulation (TBS)-induced ERK activation required inhibitors of both L-channel- and NMDA receptor-dependent sources of calcium (Dudek and Fields, 2001). Since the action potentials also play a role in the full activation of the NMDA receptors, synaptic- and action potential-dependent sources of calcium are indeed likely to be exceedingly intertwined.

5. WHAT IF NMDA RECEPTORS PLAY A ROLE IN FIRING ACTION POTENTIALS?

Consider the problem of determining whether NMDA receptor activation and subsequent formation of a signal that travels to the nucleus is involved in a transcriptional event, either *in vivo* or in cell cultures. What if the transcriptional event is tied to action potential firing, instead? The absence of an effect of NMDA receptor blockers would support the hypothesis that action potentials were involved, but doesn't address whether or not the event was related to LTP, unless one were *assuming* a signal from the synapse was necessary for LTP. But what if NMDA receptor blockers somehow interfered with action potential generation? Action potentials would be reduced in number, as would the transcriptional read-out. As a result, since NMDA receptor blockers interfered with action potentials, the conclusion that NMDA receptors produce biochemical signals initiating the nuclear events would not be warranted until action potential number was controlled.

In fact the literature is full of examples of just this result. As early as 1988, studies have shown that 2-amino-5-phosphonovaleric acid (APV), 3-(+/-)-2-carboxypiperazin-4-yl]-propyl-1-phosphonic acid (CPP), and (+)-5-methyl-10,11-dihydro-5H-dibenzo[a,d]cyclo-hepten-5, 10-imine maleate (MK801) reduced population spikes recorded in the dentate gyrus (Abraham and Mason, 1988; Dahl et al., 1990). None of the drugs had any effect on the field EPSPs, indicating that the drug concentrations were in a selective range, and they did not have any effect on the antidromic spike, indicating that the drugs had no effect on intrinsic firing or membrane properties of the neurons. In the visual system *in vivo*, iontophoretic application of APV or CPP abolished or greatly reduced the visual responses of both X and Y cells in the dorsal lateral geniculate nucleus (dLGN) (Sillito et al., 1990), and minipump infusion of APV into the visual cortex suppressed neuronal responses to visual stimulation (Miller et al., 1989), but see Bear, *et al.* (1990). More recently, in slices of the LGN, NDMA receptors were found to regulate spiking in response to retinal input stimulation, particularly during repetitive stimulation (Blitz and Regehr, 2003). These data provide strong support for the view that NMDA receptors play an important role in the neuronal response to synaptic stimulation in the form of action potentials. The current bias toward the view of a synapse-to-nucleus signal for LTP-related transcription probably arose from fact that transcriptional events can be blocked with NMDA receptor antagonists. The conclusion was a good one using available data, but since of the role of NMDA receptors in action potential generation is becoming more apparent, it should be reevaluated. Only when the number of action potentials are controlled during NMDA receptor blockade can the conclusion be made that signals (other than electrical) are specifically generated at the synapse that later reach the nucleus to support late-phase LTP.

6. TRANSCRIPTION FACTORS REGULATED BY ACTIVITY

6.1. Regulation of *Zif268* during plasticity

One of the most interesting genes that is rapidly induced with neuronal activity is *zif268* (also known as *Krox24*, *NGFI-A*, *Egr-1*). *Zif268*, in contrast to many effector immediate early genes (IEGs) such as *Arc* and *BDNF*, is a regulatory IEG in that it regulates the transcription of other genes. *Zif268* is a member of the Egr family of zinc finger transcription factors, along with *Egr-2* (*Krox20*), which itself has been shown to be rapidly up-regulated with LTP-inducing stimulation in the hippocampus (Williams et al., 1995). Not only is *zif268* rapidly expressed after LTP-inducing stimulation (Richardson et al., 1992; Worley et al., 1993) and exploration of a novel environment (Wallace et al., 1995; Guzowski et al., 2001), but a role for *zif268* in the long-term maintenance of both LTP and memory has also been shown in mice lacking the gene (Jones et al., 2001). *Zif268* has a number of interesting regulatory elements including 2 cAMP response elements (CREs) and 6 serum response elements (SREs) (Tsai-Morris et al., 1988). This is particularly attractive because ERK, which is thought to play a role in both the activation of CREB, and SRE through Elk-1 (Davis et al., 2000), is also known to be rapidly activated with memory and LTP-inducing stimulation (reviewed in Adams and Sweatt, (Adams and Sweatt, 2002)). Curiously, the promoter for *zif268* also contains an AP-1 site, which requires fos-jun dimers for activation, but activity is required for the expression of *fos* and *jun*. Clearly, then, basal levels of Fos and Jun protein are sufficient for transcription of *zif268*, since it can be so rapidly induced after activity. Whether any specific signals from the synapse are required for *zif268* transcription has not been addressed.

Which genes depend on the Zif268 protein for their transcription? Though there are several possible candidates (Myers et al., 1998; Nagavarapu et al., 2001), one consideration is that while *zif268* is rapidly activated within our restricted time window for RNA synthesis in late-phase LTP, whether there is sufficient time for *zif268* translation and the subsequent transcription of downstream genes is an issue. Part of the difficulty in assessing a role for *zif268*-dependent cascades in late LTP is that detailed kinetic studies on the timing of LTP's sensitivity to RNA synthesis inhibitors have not been performed. Given the correlation between late-phase LTP, long-term memory and *zif268* transcription, this possibility should probably not be ruled out. One recent study, though, using antisense oligodeoxynuclotides, found that *zif268* was more important during the reconsolidation phase of memory than initial consolidation of memory; *BDNF* was found to have the opposite profile (Lee et al., 2004). BDNF protein, however, has not been detected after LTP and was slightly down-regulated following exploration of a novel environment (Walton et al., 1999). Whatever the relative roles of *BDNF* and *zif268*, the regulatory IEGs such as *zif268* represent a way that a large number of genes can be rapidly regulated through the activity-dependent transcription of a few, and could be responsible for synaptic effects not measurable in the slice preparation.

6.2. The cAMP Response Element Binding protein and activity

The importance of the Ca^{+2} and cAMP response element binding protein (CREB) in neuronal signaling has been demonstrated by several studies ranging from memory consolidation in *Drosophila*, long-term facilitation in *Aplysia*, and long-term potentiation in mice (Alberini et al., 1994; Bourtchuladze et al., 1994; Yin et al., 1994; Yin et al., 1995). In addition, the transcriptional activity of CREB, a member of the basic leucine zipper (bZIP) family of transcription factors, is necessary for proper neuorological function (Lonze and Ginty, 2002). In humans, mutations in the CREB binding protein (CBP) have been implicated in Rubinstein-Taybi syndrome, which is a rare form of mental retardation (Lacombe et al., 2001; Coupry et al., 2002). The severity of these disruptions is not surprising when one considers that CREs are upstream of several immediate early genes and a variety of other genes including those for synapsin I and PKM (Hoesche et al., 1995; Hernandez et al., 2003).

Activation of CREB is achieved by phosphorylation of Ser-133 following a variety of stimuli including electrical activity (Ginty et al., 1994; Deisseroth et al., 1996; Liu and Graybiel, 1996; Impey et al., 1998b). Other residues have also been demonstrated to be important for CREB functionality; both Ser-142 and Ser-143 can be phosphorylated *in vivo* (Deisseroth and Tsien, 2002). CaMK II appears to be important for Ser-142 phosphorylation, which has shown both transcriptional activation and inactivation properties, while casein kinase II acts upon Ser-143 (Gau et al., 2002; Kornhauser et al., 2002). While the 142/143 phosphorylated species may occur following K^+ depolarization, the exact role either of the individual sites plays in activity-dependent transcription is largely unknown. The path from an initial stimulus to a transcriptionally competent phosphorylated CREB species has been the subject of some debate, particularly regarding whether a signal from the synapse is required for CREB phosphorylation. In one model, calmodulin serves as the synapse-to-nucleus or cytosol-to-nucleus signal (Deisseroth et al., 1998; Mermelstein et al., 2001), but see Hardingham, *et al.* (2001). In either case, a current model of CREB activation in neurons begins with the influx of Ca^{+2} through NMDA and L-type voltage gated calcium channels (Sheng et al., 1988; Lerea et al., 1992; Bading et al., 1995; Shaywitz and Greenberg, 1999). The corresponding rise in intracellular Ca^{+2} can then lead to the activation of several signaling cascades that converge on CREB. The primary routes appear to include both a fast pathway through CaM kinase IV and a slower route through ERK (Bito et al., 1996; Impey et al., 1998b; Wu et al., 2001). In addition, activation of protein kinase A has also resulted in CREB phosphorylation, although there has been some question over the extent of its participation in activity-dependent CREB activation (Gonzalez and Montminy, 1989; Pokorska et al., 2003). Following phosphate incorporation on Ser-133, CREB can associate with a number of other proteins, most notably CBP (Vo and Goodman, 2001), which itself needs to be phosphorylated at S301 for NMDA receptor dependent transcription to occur (Impey et al., 2002). This protein interacts with a number of transcription factors and has histone acetyltransferase activity (Chan and La Thangue, 2001) (see below, section 7). Following complex assembly, RNA transcription can commence.

A variety of factors appear to be important for the deactivation of CREB in neurons. Both calcineurin and protein phosphatase-1 (PP1) impinge upon CREB to dephosphorylate Ser-133, thus reducing its activity (Bito et al., 1996). In addition, Bading and co-workers (2002) have suggested a mechanism for downregulating CREB activity through the action of extrasynaptic NMDA receptors (Hardingham et al., 2002). While the mechanism of this deactivation is unknown, it has been suggested that a class I histone deacetylase could stabilize the CREB/PP1 interaction (Vanhoutte and Bading, 2003).

Though one study that specifically tested whether CREB has a pivotal role in plasticity and hippocampal dependent memory has found it not to be the case (Balschun et al., 2003), studies have continued to come out in favor of the molecule having a role in both. CRE-mediated gene expression as assessed with a lac-z reporter does correlate with LTP maintenance (Impey et al., 1996) and learning (Impey et al., 1998a). And a gene with many CREs in its promotor, *c-fos*, is induced with odor discrimination (Hess et al., 1995a) and exploration (Hess et al., 1995b). In addition, although *arc* has no CREs in its promoter, *zif268* has two. Interestingly, the expression of a constitutively active CREB protein enhances the late-phase of LTP (Barco et al., 2002). When the critical genes of interest that get transported to the synapse and incorporated there are identified, the role of CREB will become evident.

6.3. Role of Rel/NF-κB in Long-Term Memory

When considering candidates for synapse-to-nucleus signals, the Rel/NF-κB family members come up frequently and from unexpected places. NF-κB and c-Rel are part of a family of Rel/ NF-κB proteins that contain related subunits. NF-κB contains the subunits p50 and p65, with p65 also known as RelA. Rel/NF-κB transcription factors are known to dimerize and are restrained to the cytoplasm in their inactive forms bound to IκB (Karin, 1999). Upon activation, IκB is degraded and the Rel/NF-κB dimer can translocate to the nucleus, where it can bind the consensus sequence 5'-GGGRN-YYYCC-3' (Chen and Ghosh, 1999; Pahl, 1999)(R is an unspecified purine; Y is an unspecified pyrimidine; and N is any nucleotide).

In a study looking at long-term memory in the crab *Chasmagnathus*, DNA binding activity for Rel/NF-κB was found to increase after training protocols that induced LTM (spaced training), but not short-term memory (massed training) (Freudenthal and Romano, 2000). This is particularly interesting because in an unbiased search using high-density oligonucleotide arrays, RNA extracted from the hippocampi of contextual fear-conditioned mice showed that a variety of genes were up-regulated (Levenson et al., 2004b). When the promoters were sequenced, c-Rel was found in the highest frequency among regulatory elements. In order to confirm findings from microarray data suggested that *c-Rel* is important in regulating memory-related genes, a series of transgenic *c-Rel* knockout mice were developed to study their behavior. Using cued conditioning, amygdala-dependent LTM was assessed and no difference in freezing behavior was observed, suggesting *c-Rel* is not involved in amygdala-dependent LTM formation. In contrast, *c-Rel* knockout mice were found to have delayed freezing behavior in a contextual fear conditioning test, indicating that *c-Rel* is an important element for hippocampus-dependent LTM. Also shown in that study, p65 knockout mice showed deficits in performance on a radial arm maze testing spatial learning. The tissue for the microarray study was taken at time points between 1 and 6 hours, which would have been plenty of time to get a signal from the synapse to the nucleus as was shown with p65 (Meffert et al., 2003). This raises the question as to the relevance of the genes regulated immediately after the training. What genes are turned on immediately and expressed in less than 1/2 hour after learning, and their role, is certainly of interest and could help determine the link between late-phase LTP and actual learning.

7. CROMATIN REMODELING AND ACTIVITY

Recently, the idea that synaptic or neuronal activity could regulate chromatin structure in the nucleus of neurons has had growing acceptance. Investigations have been performed on the activity-dependent regulation of chromatin structure by examining activity-dependent regulation of histone acetyl transferases (HATs) and histone deacytlases (HDACs) and changes in the post-translational modifications at the promoters of various genes known to be involved in learning and memory. HATs and HDACs, by regulating the acetylation states of histone proteins, are thought to make DNA more, or less, accessible to transcriptional regulators and polymerases. Two studies have recently been published showing evidence that HATs and HDACs can be regulated by activity. In the first study (Chawla et al., 2003), the authors show that spontaneous electrical activity in cultured hippocampal neurons causes the export of HDAC4 out of the nucleus and into the cytoplasm, thus effectively shutting the HDAC off and potentially increasing acetylation rates in the nucleus. In addition, activation of either NMDA receptors or calcium channels led to the export of another HDAC, HDAC5. Interestingly, both cases were dependent on the activity of CaMKII. In the second paper (Martinowich et al., 2003), the authors use a different method of inducing activity in neuronal cultures and examine the regulation of a specific gene, the *BDNF* gene. The data presented indicated that KCl-induced depolarization leads to a dissociation of an HDAC complex from the *BDNF* promoter and further, induces an increase in BDNF synthesis. The data from these two studies thus provide evidence that activity can regulate the acetylation/deacetylation rates of chromatin by regulating the availability of HATs and HDACs to the chromatin. As a result, activity can regulate the dynamic state of chromatin by modulating the ability of transcription, replication, and repair factors to access DNA.

A more daunting task has recently been attempted by several groups: to identify the histone code of post-translational modification patterns on promoters of genes implicated in learning and memory. In a first step toward this goal, two research groups have shown that seizure activity can lead to post-translational modifications on the promoters of several genes. The first group used epileptic seizures induced by pilocarpine in rats to examine acetylation events on the promoters of the *Glutamate Receptor 2 (GluR2)* and *BDNF* genes in hippocampal area CA3 (Huang et al., 2002). They found that seizure activity induced an increase in acetylation on H4 associated with the *BDNF* promoter, but a decrease in acetylation on H4 at the *GluR2* promoter. These acetylation changes occurred within 3 hours of onset of seizure activity. The second group used electoconvulsive seizures to examine acetylation events on the promoters of the *c-fos*, *BDNF*, and *CREB* genes in the hippocampus (Tsankova et al., 2004). They found that acetylation of H4 and acetylation in combination with phosphorylation of H3 increased on the *c-Fos* and the *BDNF* promoter, while the *CREB* promoter showed increased H4 acetylation and H3 acetylation alone; the mRNA for all three genes increased. The data from these two groups indicate that strong central nervous system activity can regulate the levels of histone acetylation and phosphorylation on specific promoters, and thus lead to increased expression of those gene products.

Another line of evidence showing that activity can regulate chromatin structure comes from investigations on the role of CBP. CBP has two functions: first, to act as a platform for transcription machinery at the CREB promoter, and second, to act as a histone acetyltransferase enzyme. CBP was first found to be involved in the activity-dependent regulation of chromatin structure in the invertebrate model organism *Aplysia californica* (Guan et al., 2002). The authors found that paradigms that induce long-term facilitation (LTF) at sensory-motor neuron synapses also induce the recruitment of CBP to the CAAT box enhancer binding protein (C/EBP) promoter. This in turn stimulates the acetylation of H3 and H4 on the promoter and increases the expression of C/EBP, an IEG known to be involved in the expression of LTF at these synapses. More recently, two groups have reported on the effects of the removal of CBP protein, using knock-out technology in the mouse. Both groups found that hippocampal long-term plasticity at the cellular level and long-term memory (LTM) at the behavioral level was affected by the removal of CBP from the system (Alarcon et al., 2004; Korzus et al., 2004). Though only one group found that levels of H2B acetylation, but not H3 or H4 acetylation, were decreased in the hippocampi of mutant mice (Alarcon et al., 2004), the data from both of these studies suggests that the HAT activity of CBP is important for learning and memory in both invertebrates and vertebrate systems.

More recently, researchers have begun to specifically examine how chromatin structure is regulated by activity. The first group to make the connection between ERK and chromatin was Swank and Sweatt (2001) showing that a robust insular cortex-dependent learning paradigm led to both ERK activation and an ERK-dependent increase in the acetylation of lysines on H2A and H4 (Swank and Sweatt, 2001). While this group did not observe any changes in the acetylation of lysines on H3 using this insular context-dependent learning paradigm, they and other groups have found changes in both acetylation and phosphorylation of H3 using other paradigms. Crosio *et al* (Crosio et al., 2003) found that activation of dopamine, muscarinic acetylcholine, and kainate receptors all led to a rapid but transient increase in acetylation and phosphorylation (within 15 minutes) of H3 in the hippocampus that correlated with an increase in ERK activity. In addition, these treatments induced an increase in *c-Fos* and *MAP kinase phosphatase (MKP)* transcription. Levenson *et al.* (Levenson et al., 2004a) have recently found that a hippocampal-dependent learning paradigm, contextual fear conditioning, induces both an increase in ERK activation as well as H3 acetylation within one hour of training. The increase in H3 acetylation is dependent on NMDA-receptor activation and ERK activation. Interestingly, increasing H3 acetylation biochemically by applying an HDAC inhibitor leads to an increase in both LTP and LTM (Alarcon et al., 2004; Levenson et al., 2004a). These data indicate that learning and memory are associated with changes in post-translational modifications on chromatin, and that these changes are dependent on the MAPK ERK. Though it is certainly possible that the IEGs do not require histone modification for transcription, what is missing at this point are studies showing that the changes can occur with LTP-inducing stimulation on a time scale rapid enough to support transcription of IEGs and fall within the time constraints of LTP's sensitivity to RNA synthesis inhibitors.

8. MAPPING ACTIVITY-DEPENDENT GENE REGULATION ONTO MEMORY

8.1 Distinguishing relevant activity

How can one distinguish between LTP- and/or learning-related neuronal activity and so-called "just activity"? Thresholds for frequency, which may include particular patterns of activity, and numbers of action potentials could work to distinguish the two. An example of how thresholds regulate signaling in the nucleus is found with the apparent threshold for activation of ERK (Dudek and Fields, 2001). Amazingly, the threshold number of pulses for activation of ERK is about 80 (presynaptic) pulses (in a theta burst stimulation, (TBS): 2 sets of 10 bursts of 4 pulses at 100Hz, delivered at 5 Hz, with an unknown number of postsynaptic action potentials). What is amazing is not that there is a threshold for ERK activation, but that the threshold is identical to that for the induction of late-phase LTP (Dudek, unpublished). The frequency threshold for ERK activation is about 3 Hz, but it can be lower if paired pulses are used (Thiels et al., 2002). This brings us back to the possibility that high frequencies may, or may not, be necessary for the relevant gene transcription since high frequencies are not required for spike-timing dependent plasticity (STDP) or LTP (if pairing is used). It may be that bursts are essential, however; while TBS, 100 Hz, and 5 Hz all are very good at activating ERK, only TBS appears to recruit voltage-dependent calcium channels to the degree that supports ERK phosphorylation in the presence of NMDA receptor blockers (Dudek and Fields, 2001). Bursts may also play a critical role in LTP induction (at the synapse), as well (Pike et al., 1999). Unfortunately, until we really know how LTP is induced *in vivo*, we cannot know if our ideas are sufficient –or necessary– for the gene expression related to late-phase LTP.

Given the conclusion that action potentials can rapidly regulate gene transcription, how does this mechanism support a role in memory? What if the nucleus doesn't need to know whether learning has occurred but just that an appropriate firing pattern had taken place? What if cells up-regulate relevant genes during behavior that is most likely to result in LTP somewhere on the cell's 10,000 synapses? This could be accomplished during hippocampal theta, for example, which takes place during a number of preparatory and locomotor behaviors, for example (Buzsaki, 2002). Interestingly, cells will fire at theta during voluntary wheel running, and at least one IEG can be up-regulated as well (*BDNF*) (Oliff et al., 1998). Though the case of a treadmill is highly atypical of an normal activity for an animal in the wild, any situation where an animal would be locomoting fast, such as during escape from a predator, may be a good time for any synaptic modifications to be consolidated, which could occur if the relevant active neurons had boosted transcription. Another role for activity may come during REM sleep, when *zif-268* is up-regulated in an experience-dependent fashion (Ribeiro et al., 1999). What is necessary in all these cases, though, is that the gene product gets back to the modified synapse, which has to have been tagged.

8.2. Synaptic tagging and capture

While it is true that a long-lasting switch in the nucleus can occur in situations such as maternal imprinting (Weaver et al., 2004), the more commonly accepted view on memory storage does not involve self-sustaining changes in the nucleus. The idea of a local synaptic modification that allows for the tagging of the potentiated (or depressed) synapse is truly a necessity if one is to assume synapse specificity together with a role for the nucleus. Whether genes need to be up-regulated for life, or just transcribed for a limited time was addressed in the first study to show that tagging exists in the hippocampus (Frey and Morris, 1997). In Frey and Morris's 1997 study, the synaptic tag, which could be induced with either early- or late-phase LTP-inducing stimulation, had only a limited lifespan: 1-2 hours. As a result of the limited nature of the tag, there may be no point in having the gene product around for much longer than the tag. The exception being is if one were to consider the induction of a gene to later associate a weaker, early-LTP stimulus with a stronger one earlier in time. Again, however, the gene being up-regulated for a lifetime would not be necessary. Alas, this brings us back to Lisman's argument that we still require a self-sustaining alteration at synapse for life. An interesting consideration is whether one needs a transcriptional event, with tagging and capture, to do it.

While the nature of the tagging (and self-sustaining switch) is unknown (but see Lisman, *et al.*, (2002)), several studies have approached the issue of tagging the modified synapse to capture the relevant gene product. In *Aplysia*, a tagging-like phenomenon has been described (Martin et al., 1997), and the transcription can be accompanied by a local protein synthesis (Sherff and Carew, 1999). There is no evidence contrary to the idea of transcription and local protein synthesis in hippocampus. In fact, evidence suggests that *arc* message is rapidly transported to the dendrites where it can be locally translated in regions of activated dendrites (Steward et al., 1998). Synthesis of the LTP-related gene appears to require CREB (Barco et al., 2002) and action potentials (Dudek and Fields, 2002), since both a constitutively active CREB and antidromic stimulation are sufficient to rescue early-LTP (at a "tagged" synapse) from decay. It is not clear yet to what degree the gene products differ between LTP and LTD (Kauderer and Kandel, 2000), since the late-phases of both seem to require transcription and may substitute for one another in a form of "cross-tagging" (Sajikumar and Frey, 2004). Elucidation of the specific gene product(s) will tell what role the nucleus plays in consolidation of LTP and possibly memory.

8.3. Predictions

Based on the idea that action potentials alone can regulate LTP-related gene expression, several predictions can be made that will support the model if true. First, action potentials, while not critical for LTP induction, should be critical for the late phase of LTP (and/or learning); they are sufficient (Dudek and Fields, 2002). Regrettably, this one is difficult to test. Second, if number and frequency of action potentials are controlled, synaptic stimulation without NMDA receptor activation should be able to substitute for LTP-inducing stimulation in a tagging experiment. This one is testable, and should be able to refute the idea outright if not true. Third and finally, similar experiments, while controlling for action potentials, should be able to be designed to separate nuclear events from synaptic events in late-phase LTP and LTM.

In the context of LTP, the nucleus may hear the synapses, but only together as a unit and not as individual voices.

9. ACKNOWLEDGEMENT

We would like to acknowledge Dr. John Sarvey, whose influence has been felt in the preparation of this manuscript (Dahl et al., 1990) as it will surely be felt in the writing of many future ones.

10. REFERENCES

Abbott L, Nelson S (2000) Synaptic plasticity: taming the beast. Nat Neurosci 3 Suppl:1178-1183.

Abraham WC, Mason SE (1988) Effects of the NMDA receptor/channel antagonists CPP and MK801 on hippocampal field potentials and long-term potentiation in anesthetized rats. Brain Res 462:40-46.

Adams JP, Sweatt JD (2002) Molecular psychology: roles for the ERK MAP kinase cascade in memory. Annual Review of Pharmacology and Toxicology 42:135-163.

Alarcon JM, Malleret G, Touzani K, Vronskaya S, Ishii S, Kandel ER, Barco A (2004) Chromatin acetylation, memory, and LTP are impaired in CBP+/- mice: a model for the cognitive deficit in Rubinstein-Taybi syndrome and its amelioration. Neuron 42:947-959.

Alberini CM, Ghirardi M, Metz R, Kandel ER (1994) C/EBP Is an Immediate-Early Gene Required for the Consolidation of Long-Term Facilitation in Aplysia. Cell 76:1099-1114.

Bading H, Segal MM, Sucher NJ, Dudek H, Lipton SA, Greenberg ME (1995) N-Methyl-D-Aspartate Receptors Are Critical for Mediating the Effects of Glutamate on Intracellular Calcium-Concentration and Immediate-Early Gene-Expression in Cultured Hippocampal-Neurons. Neuroscience 64:653-664.

Balschun D, Wolfer DP, Gass P, Mantamadiotis T, Welzl H, Schutz G, Frey JU, Lipp H-P (2003) Does the cAMP Response Element Binding Protein have a pivitol role in hippocampal synaptic plasticity and hippocampal-dependent memory? Journal of Neuroscience 23:6304-6314.

Barco A, Alarcon JM, Kandel ER (2002) Expression of constitutively active CREB protein facilitates the late phase of long-term potentiation by enhancing synaptic capture. Cell 108:689-703.

Bear M, Kleinschmidt A, Gu Q, Singer W (1990) Disruption of experience-dependent synaptic modifications in striate cortex by infusion of an NMDA receptor antagonist. J Neurosci 10:909-925.

Bito H, Deisseroth K, Tsien RW (1996) CREB phosphorylation and dephosphorylation: A Ca2(+)- and stimulus duration-dependent switch for hippocampal gene expression. Cell 87:1203-1214.

Blitz DM, Regehr WG (2003) Retinogeniculate synaptic properties controlling spike number and timing in relay neurons. J Neurophysiol 90:2438-2450.

Bolshakov VY, Carboni L, Cobb MH, Siegelbaum SA, Belardetti F (2000) Dual MAP kinase pathways mediate opposing forms of long-term plasticity at CA3-CA1 synapses. Nat Neurosci 3:1107-1112.

Bourtchuladze R, Frenguelli B, Blendy J, Cioffi D, Schutz G, Silva AJ (1994) Deficient Long-Term-Memory in Mice with a Targeted Mutation of the cAMP-Responsive Element-Binding Protein. Cell 79:59-68.

Buzsaki G (2002) Theta Oscillations in the hippocampus. Neuron 33:325-340.

Chan HM, La Thangue NB (2001) p300/CBP proteins: HATs for transcriptional bridges and scaffolds. Journal of Cell Science 114:2363-2373.

Chawla S, Vanhoutte P, Arnold FJ, Huang CL, Bading H (2003) Neuronal activity-dependent nucleocytoplasmic shuttling of HDAC4 and HDAC5. J Neurochem 85:151-159.

Chen FE, Ghosh G (1999) Regulation of DNA binding by Rel/NF-kappaB transcription factors: structural views. Oncogene 18:6845-6852.

Coupry I, Roudaut C, Stef M, Delrue MA, Marche M, Burgelin I, Taine L, Cruaud C, Lacombe D, Arveiler B (2002) Molecular analysis of the CBP gene in 60 patients with Rubinstein-Taybi syndrome. Journal of Medical Genetics 39:415-421.

Crosio C, Heitz E, Allis CD, Borrelli E, Sassone-Corsi P (2003) Chromatin remodeling and neuronal response: multiple signaling pathways induce specific histone H3 modifications and early gene expression in hippocampal neurons. j Cell Science 116:4905-4914.

Dahl D, Burgard EC, Sarvey JM (1990) NMDA receptor antagonists reduce medial, but not lateral, perforant path-evoked EPSPs in dentate gyrus of rat hippocampal slice. Experimental Brain Research 83:172-177.

Davis S, Vanhoutte P, Pages C, Caboche J, Laroche S (2000) The MAPK/ERK cascade targets both Elk-1 and cAMP response element-binding protein to control long-term potentiation-dependent gene expression in the dentate gyrus in vivo. J Neurosci 20:4563-4572.

Deisseroth K, Tsien RW (2002) Dynamic multiphosphorylation passwords for activity-dependent gene expression. Neuron 34:179-182.

Deisseroth K, Bito H, Tsien RW (1996) Signaling from synapse to nucleus: Postsynaptic CREB phosphorylation during multiple forms of hippocampal synaptic plasticity. Neuron 16:89-101.

Deisseroth K, Heist EK, Tsien RW (1998) Translocation of calmodulin to the nucleus supports CREB phosphorylation in hippocampal neurons. Nature 392:198-202.

Dolmetsch RE, Pajvani U, Fife K, Spotts JM, Greenberg ME (2001) Signaling to the nucleus by an L-type calcium channel-calmodulin complex through the MAP kinase pathway. Science 294:318-319.

Dudek SM, Fields RD (2001) Mitogen-activated protein kinase/extracellular signal-regulated kinase activation in somatodendritic compartments: roles of action potentials, frequency, and mode of calcium entry. J Neurosci 21:RC122.

Dudek SM, Fields RD (2002) Somatic action potentials are sufficient for late-phase LTP-related cell signaling. Proc Natl Acad Sci U S A 99:3962-3967.

English JD, Sweatt JD (1996) Activation of p42 mitogen-activated protein kinase in hippocampal long term potentiation. J Biol Chem 271:24329-24332.

Eshete F, Fields RD (2001) Spike frequency decoding and autonomous activation of Ca2+-calmodulin-dependent protein kinase II in dorsal root ganglion neurons. Journal of Neuroscience 21:6694-6705.

Freudenthal R, Romano A (2000) Participation of Rel/NF-KB transcription factors in long-term memory in the crab *Chasmagnathus*. Brain Res 855:274-281.

Frey JU, Morris RG (1997) Synaptic tagging and long-term potentiation. Nature 385:533-536.

Frey U, Krug M, Reymann KG, Matthies H (1988) Anisomycin, an inhibitor of protein synthesis, blocks late phases of LTP phenomena in the hippocampal CA1 region in vitro. Brain Res 452:57-65.

Frey U, Frey S, Schollmeier F, Krug M (1996) Influence of actinomycin D, a RNA synthesis inhibitor, on long-term potentiation in rat hippocampal neurons in vivo and in vitro. J Physiol 490:703-711.

Frey U, Krug M, Brodemann R, Reymann K, Matthies H (1989) Long-term potentiation induced in dendrites separated from rat's CA1 pyramidal somata does not establish a late phase. Neuroscience Letters 97:135-139.

Fukunaga K, Stoppini L, Miyamoto E, Muller D (1993) Long-term potentiation is associated with an increased activity of Ca2+/calmodulin-dependent protein kinase II. J Biol Chem 268:7863-7867.

Gau D, Lemberger T, von Gall C, Kretz O, Minh NL, Gass P, Schmid W, Schibler U, Korf HW, Schutz G (2002) Phosphorylation of CREB Ser142 regulates light-induced phase shifts of the circadian clock. Neuron 34:245-252.

Gilmore T (1999) The Rel/NF-kappaB signal transduction pathway: introduction. Oncogene 18:6842-6844.

Ginty DD, Bonni A, Greenberg ME (1994) Nerve Growth-Factor Activates a Ras-Dependent Protein-Kinase That Stimulates C-Fos Transcription Phosphorylation of CREB. Cell 77:713-725.

Gonzalez GA, Montminy MR (1989) Cyclic-Amp Stimulates Somatostatin Gene-Transcription by Phosphorylation of Creb at Serine-133. Cell 59:675-680.

Guan Z, Giustetto M, Lomvardas S, Kim J-H, Mimiaci MC, Schwartz JH, Thanos D, Kandel ER (2002) Integration of long-term-memory-related synaptic plasticity involves bidirectional regulation of gene expression and chromatin structure. Cell 111:483-493.

Guzowski JF, McNaughton BL, Barnes CA, Worley PF (1999) Environment-specific expression of the immediate-early gene Arc in hippocampal neuronal ensembles. Nat Neurosci 2:1120-1124.

Guzowski JF, Setlow B, Wagner EK, McGaugh JL (2001) Experience-dependent gene expression in the rat hippocampus after spatial learning: a comparison of the immediate-early genes Arc, c-fos, and zif268. J Neurosci 21:5089-5098.

Guzowski JF, Lyford GL, Stevenson GD, Houston FP, McGaugh JL, Worley PF, Barnes CA (2000) Inhibition of activity-dependent arc protein expression in the rat hippocampus impairs the maintenance of long-term potentiation and the consolidation of long-term memory. J Neurosci 20:3993-4001.

Hardingham GE, Arnold FJL, Bading H (2001) Nuclear calcium signaling controls CREB-mediated gene expression triggered by synaptic activity. Nature Neuroscience 4:261-267.

Hardingham GE, Fukunaga Y, Bading H (2002) Extrasynaptic NMDARs oppose synaptic NMDARs by triggering CREB shut-off and cell death pathways. Nature Neuroscience 5:405-414.

Hernandez AI, Blace N, Crary JF, Serrano PA, Leitges M, Libien JM, Weinstein G, Tcherapanov A, Sacktor TC (2003) Protein Kinase Mzeta synthesis from a brain mRNA encoding an independent protein kinase Czeta catalytic domain. J Biol Chem 278:40305-40316.

Hess US, Lynch G, Gall CM (1995a) Changes in c-fos mRNA expression in rat brain during odor discrimination learning: differential involvement of hippocampal subfields CA1 and CA3. J Neurosci 15:4786-4795.

Hess US, Lynch G, Gall CM (1995b) Regional patterns of c-fos mRNA expression in rat hippocampus following exploration of a novel environment versus performance of a well-learned discrimination. J Neurosci 15:7796-7809.

Hoesche C, Bartsch P, Kilimann MK (1995) The CRE Consensus Sequence in the Synapsin-I Gene Promoter Region Confers Constitutive Activation but No Regulation by cAMP in Neuroblastoma-Cells. Biochimica Et Biophysica Acta-Gene Structure and Expression 1261:249-256.

Huang Y, Doherty JJ, Dingledine R (2002) Altered histone acetylation at *Glutamate Receptor* 2 and *Brain-Derived Neurotrophic Factor* Genes is an early event triggered by status epilepticus. Journal of Neuroscience 22:8422-8428.

Impey S, Mark M, Villacres EC, Poser S, Chavkin C, Storm DR (1996) Induction of CRE-mediated gene expression by stimuli that generate long-lasting LTP in area CA1 of the hippocampus. Neuron 16:973-982.

Impey S, Smith DM, Obrietan K, Donahue R, Wade C, Storm DR (1998a) Stimulation of cAMP response element (CRE)-mediated transcription during contextual learning. Nat Neurosci 1:595-601.

Impey S, Obrietan K, Wong ST, Poser S, Yano S, Wayman G, Deloulme JC, Chan G, Storm DR (1998b) Cross talk between ERK and PKA is required for Ca2+ stimulation of CREB-dependent transcription and ERK nuclear translocation. Neuron 21:869-883.

Impey S, Fong AL, Wang YH, Cardinaux JR, Fass DM, Obrietan K, Wayman GA, Storm DR, Soderling TR, Goodman RH (2002) Phosphorylation of CBP mediates transcriptional activation by neural activity and CaM kinase IV. Neuron 34:235-244.

Jones MW, Errington ML, French PJ, Fine A, Bliss TV, Garel S, Charnay P, Bozon B, Laroche S, Davis S (2001) A requirement for the immediate early gene Zif268 in the expression of late LTP and long-term memories. Nat Neurosci 4:289-296.

Karin M (1999) How NF-kappaB is activated: the role of the IkappaB kinase (IKK) complex. Oncogene 18:6867-6874.

Kauderer BS, Kandel ER (2000) Capture of a protein synthesis-dependent component of long-term depression. Proc Natl Acad Sci U S A 97:13342-13347.

Klann E, Chen SJ, Sweatt JD (1993) Mechanism of protein kinase C activation during the induction and maintenance of long-term potentiation probed using a selective peptide substrate. Proc Natl Acad Sci U S A 90:8337-8341.

Kornhauser JM, Cowan CW, Shaywitz AJ, Dolmetsch RE, Griffith EC, Hu LS, Haddad C, Xia ZG, Greenberg ME (2002) CREB transcriptional activity in neurons is regulated by multiple, calcium-specific phosphorylation events. Neuron 34:221-233.

Korzus E, Rosenfeld MG, Mayford M (2004) CBP histone acetyltransferase activity is a critical component of memory consolidation. Neuron 42:961-972.

Lacombe D, Coupry I, Roudaut C, Stef M, Delrue MA, Taine L, Arveller B (2001) Molecular analysis of the CBP gene in 65 patients with Rubinstein-Taybi syndrome. American Journal of Human Genetics 69:614-614.

Lee JLC, Everitt BJ, Thomas KL (2004) Independent cellular processes for hippocampal memory consolidation and reconsolidation. Science 304:839-843.

Lerea LS, Butler LS, McNamara JO (1992) NMDA and Non-NMDA Receptor Mediated Increase of C-Fos Messenger-RNA in Dentate Gyrus Neurons Involves Calcium Influx Via Different Routes. Journal of Neuroscience 12:2973-2981.

Levenson JM, O'Riordan KJ, Brown KD, Trinh MA, Molfese DL, Sweatt JD (2004a) Regulation of histone acetylation during memory formation in the hippocampus. Journal of Biological Chemistry.

Levenson JM, Choi S, Lee SY, Cao YA, Ahn HJ, Worley KC, Pizzi M, Liou HC, Sweatt JD (2004b) A bioinformatics analysis of memory consolidation reveals involvement of the transcription factor c-rel. J Neurosci 24:3933-3943.

Lisman J (1995) What does the nucleus know about memories? Journal of NIH Research 7:43-46.

Lisman J, Schulman H, Cline H (2002) The molecular basis of CaMKII function in synaptic and behavioural memory. Nat Rev Neurosci 3:175-190.

Liu FC, Graybiel AM (1996) Spatiotemporal dynamics of CREB phosphorylation: Transient versus sustained phosphorylation in the developing striatum. Neuron 17:1133-1144.

Lonze BE, Ginty DD (2002) Function and regulation of CREB family transcription factors in the nervous system. Neuron 35:605-623.

Martin KC, Casadio A, Zhu H, Yaping E, Rose JC, Chen M, Bailey CH, Kandel ER (1997) Synapse-specific, long-term facilitation of aplysia sensory to motor synapses: a function for local protein synthesis in memory storage. Cell 91:927-938.

Martinowich K, Hattori D, Wu H, Fouse S, He F, Hu Y, Fan G, Sun YE (2003) DNA methylation-related chromatin remodeling in activity-dependent *Bdnf* gene regulation. Science 302:890-893.

Mattson M (1997) Neuroprotective signal transduction: relevance to stroke. Neurosci Biobehav Rev 21:193-206.

Meffert M, Chang JM, Wiltgen BJ, Fanselow MS, Baltimore D (2003) NF-KB functions in synaptic signaling and behavior. Nat Neurosci 6:1072-1078.

Mermelstein PG, Bito H, Deisseroth K, Tsien RW (2000) Critical dependence of cAMP response element-binding protein phosphorylation on L-type calcium channels supports a selective response to EPSPs in preference to action potentials. Journal of Neuroscience 20:266-273.

Mermelstein PG, Deisseroth K, Dasgupta N, Isaksen AL, Tsien RW (2001) Calmodulin priming: nuclear translocation of a calmodulin complex and the memory of prior neuronal activity. Proc Natl Acad Sci U S A 98:15342-15347.

Miller KD, Chapman B, Stryker MP (1989) Visual responses in adult cat visual cortex depend on N-methyl-D-aspartate receptors. Proc Natl Acad Sci U S A 86:5183-5187.

Myers SJ, Peters J, Huang Y, Comer MB, Barthel F, Dingledine R (1998) Transcriptional regulation of the GluR2 gene: neural-specific expression, multiple promoters, and regulatory elements. Journal of Neuroscience 18:6723-6739.

Nagavarapu U, Danthi S, Boyd RT (2001) Characterization of a rat neuronal nicotinic acetylcholine receptor alpha7 promoter. J Biol Chem 276:16749-16457.

Nguyen PV, Abel T, Kandel ER (1994) Requirement of a critical period of transcription for induction of a late phase of LTP. Science 265:1104-1107.

Oliff HS, Berchtold NC, Isackson P, Cotman CW (1998) Exercise-induced regulation of brain-derived neurotrophic factor (BDNF) transcripts in the rat hippocampus. Brain Res Mol Brain Res 61:147-153.

Otani S, Marshall, C.J, Tate, W.P, Goddard, G.V, Abraham, W.C (1989) Maintenance of long-term potentiation in rat dentate gyrus requires protein synthesis but not messenger RNA synthesis immediately post-tetanization. Neuroscience 28:519-526.

Pahl HL (1999) Activators and target genes of Rel/NF-kappaB transcription factors. Oncogene 18:6853-6866.

Pike FG, Meredith RM, Oldingand AWA, Paulsen OO (1999) Postsynaptic bursting is essential for 'Hebbian' induction of associative long-term potentiation at excitatory synapses in rat hippocampus. J Physiol 518:571-576.

Pokorska A, Vanhoutte P, Arnold FJL, Silvagno F, Hardingham GE, Bading H (2003) Synaptic activity induces signalling to CREB without increasing global levels of cAMP in hippocampal neurons. Journal of Neurochemistry 84:447-452.

Ribeiro S, Goyal V, Mello CV, Pavlides C (1999) Brain gene expression during REM sleep depends on prior waking experience. Learn Mem 6:500-508.

Richardson CL, Tate WP, Mason SE, Lawlor PA, Dragunow M, Abraham WC (1992) Correlation between the induction of an immediate early gene, zif/268, and long-term potentiation in the dentate gyrus. Brain Res 580:147-154.

Roberson ED, Sweatt JD (1996) Transient activation of cyclic AMP-dependent protein kinase during hippocampal long-term potentiation. J Biol Chem 271:30436-30441.

Sajikumar S, Frey JU (2004) Late-associativity, synaptic tagging, and the role of dopamine during LTP and LTD. Neurobiology of Learning and Memory 82:12-25.

Shaywitz AJ, Greenberg ME (1999) CREB: A stimulus-induced transcription factor activated by a diverse array of extracellular signals. Annual Review of Biochemistry 68:821-861.

Sheng M, Dougan ST, McFadden G, Greenberg ME (1988) Calcium and Growth-Factor Pathways of c-Fos Transcriptional Activation Require Distinct Upstream Regulatory Sequences. Molecular and Cellular Biology 8:2787-2796.

Sherff CM, Carew TJ (1999) Coincident induction of long-term facilitation in Aplysia: cooperativity between cell bodies and remote synapses. Science 285:1911-1914.

Sillito AM, Murphy PC, Salt TE, Moody CI (1990) Dependence of retinogeniculate transmission in cat on NMDA receptors. J Neurophysiol 63:347-355.

Steward O, Wallace CS, Lyford GL, Worley PF (1998) Synaptic activation causes the mRNA for the IEG Arc to localize selectively near activated postsynaptic sites on dendrites. Neuron 21:741-751.

Swank MW, Sweatt JD (2001) Increased histone acetyltransferase and lysine acetyltransferase activity and biphasic activation of the ERK/RSK cascade in insular cortex during novel taste learning. J Neurosci 21:3383-3391.

Thiels E, Kanterewicz BI, Norman ED, Trzaskos JM, Klann E (2002) Long-term depression in the adult hippocampus in vivo involves activation of extracellular signal-regulated kinase and phosphorylation of Elk-1. J Neurosci 22:2054-2062.

Tsai-Morris CH, Cao XM, Sukhatme VP (1988) 5' flanking sequence and genomic structure of Egr-1, a murine mitogen inducible zinc finger encoding gene. Nucleic Acids Res 16:8835-8846.

Tsankova NM, Kumar A, Nestler EJ (2004) Histone modifications at gene promoter regions in rat hippocampus after acute and chronic electroconvulsive seizures. J Neurosci 24:5603-5610.

Vanhoutte P, Bading H (2003) Opposing roles of synaptic and extrasynaptic NMDA receptors in neuronal calcium signalling and BDNF gene regulation. Current Opinion in Neurobiology 13:366-371.

Vazdarjanova A, McNaughton BL, Barnes CA, Worley PF, Guzowski JF (2002) Experience-dependent coincident expression of the effector immediate-early genes arc and Homer 1a in hippocampal and neocortical neuronal networks. J Neurosci 22:10067-10071.

Vo N, Goodman RH (2001) CREB-binding protein and p300 in transcriptional regulation. Journal of Biological Chemistry 276:13505-13508.

Wallace CS, Withers GS, Weiler IJ, George JM, Clayton DF, Greenough WT (1995) Correspondence between sites of NGFI-A induction and sites of morphological plasticity following exposure to environmental complexity. Brain Res Mol Brain Res 32:211-220.

Walton M, Henderson C, Mason-Parker S, Lawlor P, Abraham WC, Bilkey D, Dragunow M (1999) Immediate early gene transcription and synaptic modulation. J Neurosci Res 58:96106.

Weaver IC, Cervoni N, Champagne FA, D'Alessio AC, Sharma S, Seckl JR, Dymov S, Szyf M, Meaney MJ (2004) Epigenetic programming by maternal behavior. Nat Neurosci 7:847-854.

Wellmann H, Kaltschmidt B, Kaltschmidt C (2001) Retrograde transport of transcription factor NF-kappa B in living neurons. J Biol Chem 276:11821-11829.

Williams J, Dragunow M, Lawlor P, Mason S, Abraham WC, Leah J, Bravo R, Demmer J, Tate W (1995) Krox20 may play a key role in the stabilization of long-term potentiation. Brain Res Mol Brain Res 28:87-93.

Worley PF, Bhat RV, Baraban JM, Erickson CA, McNaughton BL, Barnes CA (1993) Thresholds for synaptic activation of transcription factors in hippocampus: correlation with long-term enhancement. J Neurosci 13:4776-4786.

Wu GY, Deisseroth K, Tsien RW (2001) Activity-dependent CREB phosphorylation: Convergence of a fast, sensitive calmodulin kinase pathway and a slow, less sensitive mitogen-activated protein kinase pathway. Proceedings of the National Academy of Sciences of the United States of America 98:2808-2813.

Yin JCP, Delvecchio M, Zhou H, Tully T (1995) CREB as a Memory Modulator - Induced Expression of a Dcreb2 Activator Isoform Enhances Long-Term-Memory in Drosophila. Cell 81:107-115.

Yin JCP, Wallach JS, Delvecchio M, Wilder EL, Zhou H, Quinn WG, Tully T (1994) Induction of a Dominant-Negative Creb Transgene Specifically Blocks Long-Term-Memory in Drosophila. Cell 79:49-58.

SYNAPTIC DIALOGUE: SUBSTRATE FOR PROTEIN-SYNTHESIS-INDEPENDENT LONG-TERM MEMORY

Matthew R. Holahan and Aryeh Routtenberg*

1. BACKGROUND

The present article makes three points concerning the mechanism of long-lasting memory. First, such a mechanism requires 'synaptic dialogue', a term coined by Colley and Routtenberg (1993) to describe a model of ongoing, self-regenerating, autocatalytic, bi-directional communication between the two sides of the synapse (see Figure 1). Inherent in this model is an intrinsic positive feedback system providing the capacity to maintain the newly altered 'state of the synapse' for long periods of time. Because it is a positive feedback system, it is essential to incorporate well-known regulatory mechanisms, such as protein phosphatases and proteases intercellularly and synaptic inhibition intracellularly to prevent runaway activity.

Second, the 'state of the synapse' is the sum of the regulated post-translational modifications (PTMs) of proteins already present at the synapse. PTMs vary considerably at the mechanistic level as they may involve, but are not limited to, protein phosphorylation, proteolysis, protein translocation or a polymerization reaction. Here we will focus on the protein phosphorylation mechanism with a further focus on protein kinase C (PKC) phosphorylation of pre- or postsynaptic substrates. The concatenation of these PTMs in each of the synaptic compartments represents the signaling system that is orchestrated to maintain the altered synaptic state.

Third, the maintenance of long-lasting synaptic change does not require *de novo* protein synthesis as an instructive mechanism. The main event that prolongs synaptic change is the PTM of proteins already present at the pre- and post-synaptic sides. Protein synthesis serves to replenish these modified proteins but it is the PTM of synaptic proteins that serves the main instructive event for long-term memory storage. A critical analysis of the *de novo* protein synthesis model for long-term memory storage has recently been completed (Routtenberg and Rekart, 2005) and so will not be discussed further here.

---

* Cresap Neuroscience Laboratory, Northwestern University, Evanston, IL 60208.
E-mail: aryeh@northwestern.edu

One driving force behind the development of the synaptic dialogue concept was the controversy concerning the pre- or postsynaptic location where persistence of synaptic plasticity underlying information storage took place (Kauer et al., 1988; Bekkers and Stevens, 1990; Bliss, 1990; Malinow and Tsien, 1990; Nicoll and Malenka, 1999; Isaac, 2003; Nicoll, 2003). On the presynaptic side, theories focused on increased transmitter release in vertebrates (Bliss and Collingridge, 1993 for review) and invertebrates (Byrne and Kandel, 1996), growth of pre-synaptic terminals (Hebb, 1949) or both release and growth (Benowitz and Routtenberg, 1997). Postsynaptic theories focused on glutamate receptors that were initially cryptic (e.g., Lynch and Baudry, 1984) then silent (Liao et al., 1995; Isaac et al., 1995; Voronin et al., 1996), tagged (Frey and Morris, 1997; 1998) or inserted (Shi et al., 1999). Spine outgrowth has also been postulated to be a component of the persistent, postsynaptic changes underlying synaptic plasticity (Cline, 2001; Wong and Ghosh, 2002; van Aelst and Cline, 2004). While there is a growing appreciation of the dialogue model since its publication in 1993 (cf. Voronin and Cherubini, 2003; Lisman, 2003; Roberts and Glanzman, 2003), there remain those who persist in the view of an exclusive postsynaptic mechanism as highlighted in a recent review by Nicoll (2003) that states, "...the preponderance of evidence favours a postsynaptic expression mechanism..."(pg. 721). Nicoll and Malenka (1999) have also stated that "...numerous experiments fail to support a presynaptic expression mechanism, many experiments do point to a postsynaptic expression mechanism." (pg. 515). Therefore, because postsynaptic theories are still alive and well, it remains important to emphasize the key concept of the dialogue hypothesis that plasticity emerges from the co-participation of both sides of the synapse.

2. SYNAPTIC DIALOGUE MODEL

A review of Pubmed indicates that when the terms 'synapse' and 'dialogue' are searched, the earliest reference is to the Colley and Routtenberg (1993) review. When the phrase "synaptic dialogue" is searched, 17 items in Pubmed are found, and the articles retrieved indicate bi-directional communication, not only between the pre- and postsynaptic sides but also between the synapse and nucleus (Kandel, 2001), evidence for which had been gathered several years earlier (Hendry et al., 1974; DiStefano et al., 1992; Kinney et al., 1996). The term dialogue is used in the present chapter to reflect an *ongoing* communication between pre- and postsynaptic sides. This gives rise to the major point of this article: persistence of a plastic change coordinated in the two synaptic compartments long after the initiating event has occurred.

3. OVERVIEW OF DIALOGUE STEPS

Figure 1 shows a modified version of the synaptic dialogue model as originally proposed by Colley and Routtenberg (1993). As can be seen in this diagram after initiating steps **1** (calcium influx) and **2** (calcium binding to endocytotic-exocytotic machinery), steps **3-10** constitute one cycle of the dialogue model followed by a recurrent cycle shown in Figure 1 as steps **11-18**. The implication is that this is an ongoing (as represented by the circle with green arrows encompassing the finer details; the "circle of persistence"), yet regulated (see later for description of R1 – R4) process of recurrent events.

Step 1 of the model is calcium (Ca^{++}) influx into the presynaptic terminal produced by an invading action potential and the consequent activation of voltage-gated calcium channels (VGCC). This influx in Ca^{++} activates Ca^{++}-sensor proteins (Fernández-Chacón et al., 2001), such as synaptophysin, that make up part of the exocytotic-endocytotic protein machinery (EPM in **Figure 1**; Step 2). Calcium binding to particular proteins of the EPM serves to induce the transport and docking of vesicles resulting in neurotransmitter (e.g., glutamate) release (Step 3).

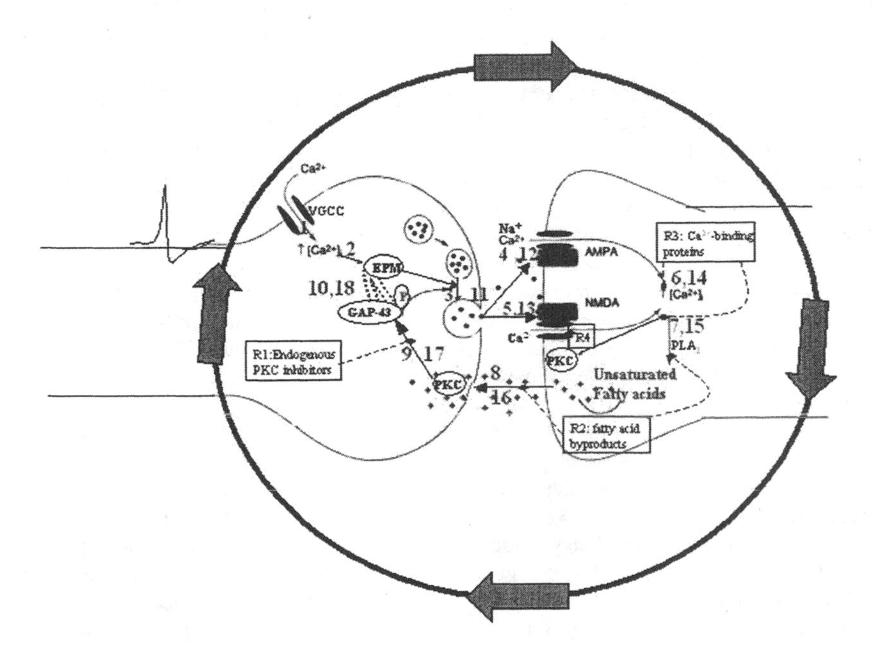

Figure 1. Representation of the synaptic dialogue model. Numbers in red reflect the different steps with 3 – 10 showing one cycle and steps 11 – 18 reflecting the recurrent nature of the dialogue model as suggested by the encompassing circle and green arrows (the "circle of persistence"). R1 – R4 in the boxes represent the regulatory mechanisms imposed on the dialogue model and acting on the pathways indicated that contain runaway activity. Abbreviations are as follows, VGCC: voltage-gated calcium channel; EPM: endocytotic-expcytotic protein machinery; PLA_2: phospholipase A2; PKC: protein kinase C; GAP-43: 43 kDa growth-associated protein. Note that this model is a highly schematic, simplified attempt to abstract certain basic features of the synapse as it adjusts to input.

Activation of postsynaptic glutamate receptors (Steps 4 and 5; AMPA and NMDA) by the released transmitter leads to a rise in intracellular postsynaptic Ca^{++} (Step 6) and the activation of multiple intracellular cascades including the activation of phospholipase A_2 (PLA_2; Step 7) resulting in elevated levels of unsaturated fatty acids (Step 7).

Unsaturated fatty acids are hypothesized to act as retrograde factors (Step 8) that influence activity in the presynaptic element. These membrane-permeant fatty acids (cf., Linden and Routtenberg, 1989) enter the presynaptic terminal and activate presynaptic

PKC, resulting in the phosphorylation of presynaptic PKC substrates (Step 9) including GAP-43. Phosphorylated GAP-43 interacts with the EPM (Step 10) as suggested by Haruta et al. (1997) and results in the release of more neurotransmitter (Step 11) and presynaptic axonal growth (not shown in the figure but discussed below).

The release of more neurotransmitter (Step 11) initiates the recurrent cycle by resulting in postsynaptic AMPA and NMDA receptor activation (Steps 12 and 13) and the rise in additional Ca^{++} (Step 14). Additional Ca^{++} is proposed to activate postsynaptic kinases including calcium/calmodulin-dependent protein kinase II (CaMKII; not shown in the figure but discussed below), protein kinase C (PKC) and additional unsaturated fatty acids via the PLA_2 pathway (Step 15). Activation of these kinases are hypothesized to modulate receptor dynamics and cytoskeletal rearrangements in the postsynaptic element (not shown in figure but discussed in text). Unsaturated fatty acids act as retrograde factors (Step 16) to activate presynaptic kinases leading to the phosphorylation of presynaptic substrates (Step 17) and interactions with the EPM (Step 18). While only two cycles of the dialogue model are presented in Figure 1, the "circle of persistence" (shown as the black ring with green arrows) is meant to reflect the ongoing nature of the outlined steps. This ongoing dialogue process underscores the point emphasized by Routtenberg and Rekart (in press) that maintenance of synaptic plasticity and the perseverance of information storage can be achieved in the absence of protein synthesis.

The dialogue model emphasizes persistent and coordinated PTMs on both the pre- and postsynaptic sides. Presynaptic mechanisms include, but are not limited to, alterations in transmitter release and changes in the stability of the cytoskeleton. Postsynaptic mechanisms include, but are not limited to, changes in receptor sensitivity, receptor numbers and cytoskeletal rearrangements. Phosphorylation of presynaptic GAP-43 serves a pivotal role in presynaptic alterations while *cis*-unsaturated fatty acids (e.g., arachidonic acid, oleate) serve a key role as retrograde factors to facilitate a continuous dialogue between pre- and postsynaptic elements. To insure control of this positive feedback process, there are a set of regulatory mechanisms (R1 – R4) that attenuate the recurrency of the dialogue so that it is always under control. These include the actions of endogenous kinase inhibitors to regulate PTMs of kinase substrates (R1), the control of fatty acid levels by lipases (R2), modulation of postsynaptic Ca^{++} levels by Ca^{++}-binding proteins (R3) and regulation of postsynaptic receptors by PKC (R4). While the components of the dialogue process are presented sequentially, it is hypothesized that once initiated, the pre- and postsynaptic processes likely occur in parallel (see Linden and Routtenberg, 1987; Colley and Routtenberg, 1993) with orthograde and retrograde factors keeping the parallel components 'informed' of the state of activity of the other side.

4. A STEP-BY-STEP APPROACH TO MAINTENANCE OF PLASTICITY THROUGH ONGOING DIALOGUE

4.1. Steps 1 and 2: activation of VGCC and EPM

A depolarizing event produced by stimulation of the presynaptic component leads to an increase in the probability that voltage-gated calcium channels (VGCCs) will open and allow the influx of Ca^{++} (Step 1 in **Fig. 1**; for recent review see Faber and Sah, 2003). One role for Ca^{++} influx through presynaptic VGCCs is the initiation of secretory processes for neurotransmitter release from presynaptic terminals. This occurs because this divalent cation binds to particular EPM proteins that regulate the release of neurotransmitter (Step 2 in **Fig. 1**). Synaptic vesicles are targeted to presynaptic release sites, dock to the plasma membrane, are primed for arrival of release signals followed by vesicle fusion and exocytosis of transmitter. Synaptophysin appears to be a critical part of the EPM (for review see Valtorta, et al., 2004; Südhof, 2004) as it is a presynaptic phosphoprotein that interacts with a variety of nerve terminal proteins (e.g., vesicle-associated membrane protein 2/synaptobrevin II). Injection of antisense oligonucleotides or antibodies raised against synaptophysin into *Xenopus* oocytes decreased the level of synaptophysin and resulted in a decrement in Ca^{++}-induced neurotransmitter release for up to 2 hours postinjection (Alder, et al., 1992). Overexpression of synaptophysin in neurons produced by injection of synthetic synaptophysin mRNA leads to elevations in the frequency of spontaneous synaptic currents (Alder, et al., 1995). Finally, electrophysiological recordings in hippocampal CA1 region revealed that synaptic plasticity (short- and long-term) was compromised in synaptophysin/synaptogyrin knockout mice (Janz, et al., 1999). These data indicate that synaptophysin may mediate fusion and exocytosis of synaptic vesicles for transmitter release in a Ca^{++}-dependent manner underscoring synaptic plasticity.

4.2. Step 3: neurotransmitter release

The synaptic dialogue process requires neurotransmitter release both during the initial stages (Step 3 in **Fig. 1**) and during the maintenance of persistent changes (Step 11 in **Fig. 1**). Measurements of glutamate release in vivo (Dolphin et al., 1982; Bliss et al., 1986) and in vitro (Lynch et al., 1985; Lynch and Bliss, 1986) have shown that after LTP induction, during the expression phase, there is an increase in evoked transmitter release from the presynaptic element. Quantal analysis of synaptic current fluctuations failure rates before and after LTP induction in hippocampal preparations suggests that enhanced synaptic efficacy recruits a presynaptic mechanism that increases the probability of transmitter release (Bekkers and Stevens, 1990; Malinow and Tsien, 1990; Stevens and Wang, 1994).

That enhanced release of transmitter underlies long-term memory storage processes is suggested by the fact that LTP attenuates paired-pulse facilitation (PPF). It is generally thought that PPF occurs because of increased transmitter mobilization and release probability (Schulz et al., 1994). After LTP, which occurs in part because of enhancement of these functions, PPF is attenuated by occlusion (Kleschevnikov et al., 1997; Kleschevnikov and Routtenberg, 2001;Voronin et al., 1992; Kuhnt and Voronin, 1994) in a variety of brain regions, including DG (Christie and Abraham, 1994), CA1 region of the hippocampus (Kleschevnikov et al., 1997; Wang and Kelly, 1997), subiculum (Commins et al., 1998), amygdala (Huang and Kandel, 1998), thalamus (Castro-Alamancos and Calcagnotto, 1999) and neocortex (Volgushev et al., 1997). Thus, neurotransmitter release should be considered an integral component of the dialogue process (Step 3).

Induction of LTP in dentate gyrus in vivo leading to a time-dependent reduction in PPF, is reversed by application of APV resulting in no change of PPF (Kleschevnikov and Routtenberg, 2001). Application of phorbol ester (a PKC activator), which did not by itself influence the level of PPF, increased and prolonged the post-tetanic occlusion of PPF when LTP was restored in the presence of APV. The most straightforward explanation is that this is a presynaptic phenomena, and that an increase of vesicle release probability occurred after the tetanus in the presence of the PKC activator (phorbol ester).

Given that PKC regulates transmitter release probability (Thomson, 2000) and that GAP-43 is a presynaptic target for PKC, it is suggested that PKC phosphorylation of presynaptic GAP-43 (Step 9 described below) leads to enhanced neurotransmitter release and provides a mechanism for the occluded PPF after LTP induction.

4.3. Steps 4 and 5: AMPA and NMDA receptor activation

According to the dialogue hypothesis there should be a parallel, commensurate, coordinated change in the postsynaptic element that is also responsible, along with presynaptic modifications, for maintaining persistent synaptic change (Zalutsky and Nicoll, 1990; Foster and McNaughton, 1994). Induction of LTP increased the size of spontaneous miniature potentials (Manabe et al., 1992) while quantal analysis of responses elicited by glutamate release suggested an increase in the quantal size (Liao et al., 1992). These findings suggested that LTP induction enhanced the postsynaptic response to glutamate arguing for a critical involvement for the AMPA and NMDA receptors underlying synaptic plasticity (Steps 4, 5 in **Fig. 1**).

One prevailing model for the above observations suggests the existence of "silent" AMPA receptors that are inserted into the postsynaptic membrane (Voronin et al., 1996; Malinow et al., 2000; Isaac, 2003). The silent synapse model suggests that LTP results in the insertion of AMPA receptors containing GluR1/2 subunits into the postsynaptic membrane (Shi et al., 2001). These receptors are then replaced with AMPA receptors composed of GluR 2/3 subunits (Malinow and Malenka, 2002) and the maintenance of the potentiated response ensues. In support of this contention, visualization of AMPA GluR1 receptor subunits showed that inducing LTP or increasing the activity of calcium/calmodulin-dependent protein kinase II (CaMKII) resulted in an increase in the number of labeled GluR1 AMPA subunits into the postsynaptic membrane (Hayashi et al., 2000). Experiments using hippocampal cultures found that induction of long-term *depression* (LTD) by bath application of low concentrations of glutamate resulted in a decrease in the frequency of miniature excitatory postsynaptic potentials and a rapid and selective redistribution of GluR1 AMPA subunits *away* from the synapse (Carroll et al., 1999; Lissin et al., 1999). These data suggest a bi-directional regulation of potentiated responses occurring with both AMPA receptor insertion and by the removal of AMPA receptors (R4 in **Fig. 1** described below).

4.4. Step 6: rise in postsynaptic Ca^{++}

The NMDA receptor is known to be a critical component for initiating plasticity at central synapses (Nicoll, 2003). NMDA receptors allow a rise in intracellular Ca^{++} levels (Step 6 in **Fig. 1**) when there is sufficient postsynaptic depolarization brought about by ligand binding to AMPA receptors (Step 4) to remove the Mg^{++} block from the NMDA channel. Activation of NMDA receptors (Step 5) leads to a potentiation that decays in less than 1 hr, suggesting the need for additional, downstream factors for the establishment of long-term synaptic alterations. An additional factor may be the rise in postsynaptic Ca^{++} concentrations (Step 6) that triggers a cascade of intracellular processes involving activation of PKC, Ca^{++}/ calmodulin-dependent kinase II, phospholiapse A_2 and other calcium-dependent processes. Glutamate receptor activation (both NMDA and the group I metabotropic glutamate receptor, mGluR1) also elevates fatty acid levels (for review see Murakami and Routtenberg, 2004). These intracellular cascades are thought to be critical components of the dialogue process for the maintenance of long-term synaptic change.

4.5. Step 7: liberation of fatty acids

Postsynaptic Ca^{++} influx through NMDA channels and mGluR1 receptors increases levels of unsaturated fatty acids (Pellerin and Wolfe, 1991;Collins et al., 1995) by activation of phospholipase A_2 (Dumuis et al., 1988; Step 7 in **Fig. 1**). Inhibitors of phospholipase A_2 have been found to block LTP in the dentate gyrus and CA1 hippocampal regions (R2 in **Fig. 1**; Linden, et al., 1987; Williams and Bliss, 1988;O'Dell et al., 1991) and induction of LTP stimulates the release of fatty acids in the hippocampus (Lynch et al., 1989). Exogenous application of unsaturated fatty acids coupled with weak stimulation (Linden, et al., 1987; O'Dell et al., 1991) or possibly alone (Williams et al., 1989) produces a long-lasting potentiation in the hippocampus. Inhibitors of fatty acid liberation *in vitro* abolish the regulation of AMPA-receptor binding by Ca^{++} (Menard et al., 2004) indicating that postsynaptic fatty acid liberation may modulate postsynaptic receptor kinetics as a regulatory mechanism. Another pathway to potentiation produced by elevated levels of fatty acids is the activation of PKC (Linden et al., 1987; Chalimoniuk et al., 2004). These results indicate that a fatty acid-sensitive cascade would have multiple actions both as a second messenger in the postsynaptic intracellular pathway regulating the dialogue process (R2) and as a retrograde factor to influence the presynaptic element (Step 8 described below).

4.5.1. PTMs and receptor dynamics

An important downstream factor in the maintenance of plasticity is the activation of CaMKII (Lisman, 1994; 2003). Ca^{++} influx through NMDA channels leads to autocatalytic phosphorylation of CaMKII and this phosphorylation state, and maintained kinase activity, can continue even in the presence of decreased Ca^{++} concentrations (Hanson and Schulman, 1992). This Ca^{++}-independent activity sustaining the phosphorylated state of the kinase suggests a mechanism for long-term PTM of postsynaptic substrates such as AMPA receptors (Shearman, et al., 1989b; not shown in figure).

PKC may also serve a critical role in the persistence of plasticity due to its authophosphorylation capabilities, similar to CaMKII. Experiments have shown that the late phase of LTP is sensitive to PKC inhibitors (Frey et al., 1988). Polymyxin B, a PKC inhibitor, blocked the maintenance of LTP without blocking the early, induction phase (Reymann et al., 1988) as did staurosporine (Matthies et al., 1991). These data suggest that PKC is required for the maintenance, rather than the induction, of plasticity (Colley et al., 1990). PKC activation has been shown to directly modulate NMDA channel kinetics (R4 in Figure 1) and insertion of new NMDA receptors into the plasma membrane (Lan et al., 2001). This PKC-dependent NMDA insertion has been found to follow LTP induction in CA1 hippocampal neurons (Grosshans et al., 2002). These PTMs of NMDA receptors would lead to the persistence of synaptic change thereby prolonging the dialogue process.

4.5.2. PTMs and dendritic outgrowth

The maintenance of synaptic plasticity at the postsynaptic element has also been associated with dendritic outgrowth. Anatomical analysis of the numbers and fine structures of spines suggested that a number of input-dependent procedures produce an increase in the size of the postsynaptic density (reviewed by Edwards, 1995). Early studies (Volkmar and Greenough, 1972; Greenough et al., 1973; 1979 Greenough and Volkmar, 1973; Globus et al., 1973;) found that exposing animals to enriched environments was associated with an increase in the numbers of dendritic arborizations (spines).

Outgrowth of spines may underlie the persistent changes brought about by synaptic potentiation (Chen et al., 2004) while others may include splitting of all ready present spines (Carlin and Siekevitz, 1983) or changes in the fine structure of the spine head (Boothe et al., 1979). Geinisman (1993), following earlier work on the synaptic spinule of Tarrant and Routtenberg (1977), suggested that a selective increase in the number of perforated spine apparatuses would produce synapses with higher efficacy for information storage. In hippocampal slices, LTP induction in CA1 synapses leads to the appearance of new postsynaptic dendritic spines (Engert and Bonhoeffer, 1999) while no spine outgrowth occurred in areas where potentiation was blocked. Visualization of hippocampal CA1 neurons expressing a green fluorescent protein showed that high frequency synaptic stimulation enhanced postsynaptic spine growth that was both long lasting and localized to the stimulated dendritic region (Maletic-Savatic et al., 1999). Disruption of actin polymerization (Dunaevsky et al., 1999) impaired the appearance of dendritic protrusions suggesting that actin may play an important role in dendritic outgrowth. These persistent changes in postsynaptic structure would facilitate the dialogue process by enhancing some aspect of the communication between the pre- and postsynaptic elements.

The postsynaptic PTM pathway for dendritic outgrowth appears to involve NMDA-induced activation of CaMKII and subsequent changes in actin dynamics. Treatment of tectal neurons with the NMDA receptor antagonist APV over a 24-hour period had no apparent affect on axon morphology but did decrease dendritic arborizations (Rajan et al., 1999). The development of this dendritic arbor was not blocked by either AMPA receptor blockade or treatment with the sodium channel blocker, TTX (Rajan and Cline, 1998; Richards et al., 2004). In an experiment that examined the role of CaMKII on dendritic growth in cultured hippocampal neurons (Maletic-Savatic et al., 1998), it was found that calcium-evoked dendritic growth was coincident with elevated levels of phosphorylated CaMKII and blockade of CaMKII activity blocked dendritic outgrowth. Expression of a constitutively active CaMKII mutant produced dendritic growth that was dependent on microtubule polymerization (Maletic-Savatic et al., 1998). These data indicate that phosphorylation of CaMKII plays a pivotal role in dendritic outgrowth via posttranslational modifications proteins all ready present at the postsynaptic side.

One role for PKC on the postsynaptic side will likely involve an interaction with CaMKII (Fong et al., 2002) or other kinases (Hering and Sheng, 2003) that regulates actin dynamics and spine outgrowth (Pilpel and Segal, 2004). One PKC substrate in the postsynaptic element is neurogranin (Represa et al., 1990), which contains a binding site for calmodulin that overlaps the PKC phosphorylation site. It is possible that PKC phosphorylation of neurogranin liberates calmodulin that goes on to affect actin polymerization and dendritic growth. In addition to its proposed interactions with CaMKII, PKC activation has been reported to induce synaptogenesis in dissociated culture neurons that is dependent on levels of free fatty acids (Hama et al., 2004). Therefore, PKC may be a critical kinase in the regulation of those kinases that modify postsynaptic receptor and actin dynamics to stabilize and regulate long-term synaptic changes.

4.6. Step 8: retrograde factors

Retrograde factors, along with conventional neurotransmitters, are here suggested to play a key role in coordinating the changes that occur on both sides of the synapse (Step 8 in **Fig. 1**). LTP-induced elevation of presynaptic GAP-43 phosphorylation was attenuated by postsynaptic NMDA receptor antagonists suggesting the existence of a retrograde signaling mechanism (Linden et al., 1988). Consistent with this proposal, Bliss and colleagues had found that LTP-induced enhanced neurotransmitter release was prevented by NMDA receptor blockade, providing additional support for the existence of a retrograde factor (Lynch and Bliss, 1986). It was later argued that unsaturated fatty acids were likely candidates (Williams et al., 1989). In support of this, it was shown that both oleate and arachidonate were potent activators of PKC, suggesting a target for the putative fatty acid retrograde factor (Linden et al., 1988; Linden and Routtenberg, 1989; Naor et al., 1988; Shearman et al., 1989a; 1991a).

One hypothesized signaling pathway for the dialogue model that links the pre- and postsynaptic elements is postsynaptic NMDA activation leads to presynaptic PKC activation via retrogradely transported unsaturated fatty acids (Step 8 in **Fig. 1**; for reviews see, Lynch et al., 1991; Medina and Izquierdo, 1995; Fitzsimonds and Poo, 1998; Schmidt, 2004). Training chicks on a passive avoidance task elevated concentrations of fatty acids 30, 50 and 75 min posttraining (Clements and Rose, 1996) while training rats on the Morris water maze (McGahon et al., 1996) or LTP induction in the dentate gyrus (McGahon and Lynch, 1996) increased the release of glutamate and resulted in activation of PKC in synaptosome preparations. In both cases (McGahon et al., 1996; McGahon and Lynch, 1996), the effect of unsaturated fatty acids on glutamate release and PKC activity was occluded in the trained rats and in the stimulated pathway. In rats that were not trained and in the control, non-stimulated pathway, fatty acid application stimulated both PKC activity and glutamate release. External application of *cis*-fatty acids that permeate the membrane and activate PKC modulated Na^+ channel currents while compounds that permeate the membrane but did not activate PKC (methyloleate) had no effect on Na^+ channel kinetics (Linden and Routtenberg, 1989). These data indicate that unsaturated fatty acids might trigger an intracellular PKC-dependent presynaptic change in glutamate release (Williams et al., 1989; Lynch and Voss, 1990) produced by training or LTP induction.

In aged animals fed a diet enriched in fatty acids, learning and memory were enhanced (Wong et al., 1989), a a long-lasting form of LTP was observed (McGahon et al., 1997) while depolarization-stimulated glutamate release was elevated compared to aged controls (see also Kotani et al., 2003 for similar findings). Additional reports indicated that unsaturated fatty acids could stimulate glutamate release (Breukel et al., 1997; McGahon and Lynch, 1998) as well as affect vesicular uptake of glutamate (Breukel et al., 1997; Roseth et al., 1998). Bath application of unsaturated fatty acids at concentrations that produce potentiation in hippocampal slices, produced PKC translocation and elevated phosphorylated GAP-43 levels (Luo and Vallano, 1995). This did not occur with bath application of a nitric acid analog. This unsaturated fatty acid-induced GAP-43 phosphorylation was shown to be dependent on the synergistic effects of Ca^{++} and diacylglycerol (Schaechter and Benowitz, 1993). These results indicate that unsaturated fatty acids, but not nitric oxide (Kamisaki et al., 1994), may exert their retrograde presynaptic effects on the phosphorylation of GAP-43 leading to enhanced transmitter release. This would serve as a critical element in the maintenance of synaptic change and long-term memory.

4.7. Step 9: activation of PKC

The PKC shown in Step 9 of Figure 1 is not meant to represent a particular member of the family of PKC isozymes that are known to exist. Indeed, the role of isozyme-specificity in plasticity is an important topic not covered here. Conventional PKCs have binding sites for and are regulated by diacylglycerol, phosphatidylserine and Ca^{++} (Lovinger and Routtenberg, 1988b; Huang, 1990; Shinomura et al., 1991; Shearman et al., 1991a). Activation of G-protein coupled receptors, tyrosine kinase receptors as well as Ca^{++} influx can lead to PKC activation and/ or modulation via stimulation of either phospholipase Cs to yield diacylglycerol or phospholipase D to yield phosphatidic acid then diacylglycerol (reviewed in Newton, 1995; 1997). In addition, fatty acids can modulate PKC activity (Linden and Routtenberg, 1989; Naor et al., 1988; Shearman et al., 1989a; 1991a; Sasaki et al., 1993). In the present model, PKC activation is crucial for the phosphorylation (i.e., PTM) of both pre- (Step 9) and postsynaptic substrates.

The discovery of the role of oleic acid arose when Murakami explored the moieties within diacylglycerol that would regulate PKC. Contrary to prevailing views concerning what effectors regulated PKC, it was reported that oleic acid by itself, without calcium, diacylglycerol, or phospholipid phosphotidylserine, was able to activate PKC (Murakami and Routtenberg, 1985; Murakami, et al., 1986). This activation was shown to phosphorylate GAP-43 in purified enzyme-substrate preparations (Wong, et al., 1989). It was also found that oleic acid and arachidonic acid could convert short-term potentiation (STP) to LTP in the intact preparation (Linden et al., 1987). Finally, behavioral studies demonstrated both improved learning and elevated levels of phosphorylated GAP-43 after a 10-week period of ingestion of a diet enriched in unsaturated fatty acids (corn but not olive oil), but not in saturated fatty acids (Wong et al., 1989). These studies suggested that the extent of PKC activation might be determined by the extent of conjoint elevation of several different elements (Routtenberg, 1997; Huang, 1990; Shinomura et al., 1991). This would allow for a regulatory mechanism between each of these elements so that no single change would be effective in driving the dialogue process (R1 in **Fig. 1**).

4.8. Step 10: enhanced neurotransmitter release

It appears that one major substrate for PKC is the presynaptic protein, GAP-43 (Chan et al., 1986; Rosenthal et al., 1987). GAP-43: **a)** is neuron-specific (Chan et al., 1986; Basi et al., 1987; Alexander et al., 1988; Liu and Storm, 1989; Skene and Virag, 1989; Zuber et al., 1989; Nielander et al., 1990; for reviews see Benowitz and Routtenberg, 1987;Benowitz and Routtenberg, 1997), **b)** is found in high concentrations in growth cones (Nelson et al., 1989), **c)** closely associated with regenerative and developmental growth (Skene, 1989; Strittmatter et al., 1995), **d)** has both a calmodulin-binding domain and PKC phosphorylation site (Alexander et al., 1988; de Graan et al., 1990), **e)** demonstrates a direct relation between its phosphorylation state and enhancement of LTP (Lovinger et al., 1985; Lovinger and Routtenberg, 1988a; Gianotti et al., 1992) and **f)** is linked to behavioral learning (Ehrlich and Routtenberg, 1974; 1975; Ehrlich et al., 1977; Routtenberg et al., 2000; Rekart, et al., 2004). These characteristics make GAP-43 an attractive substrate for the PKC-mediated altered state of the synapse.

PKC regulates exocytotic processes in the presynaptic terminal. In a recent review, Thomson (2000) noted that PKC activation lead to "an increase in the availability of vesicles for fusion and a decrease in Ca^{++} ...(and) increased Ca^{++} affinity for the release machinery (p.306)." Thomson also noted that "activation of PKC did increase (release probability) P and consequently reduced expression of facilitation without affecting the presynaptic action potential, the associated Ca^{++} influx or steady state Ca^{++} concentration $[Ca^{++}]_i$ ...p.306." Pavlidis, et al. (2000) demonstrated in pairs of CA3 pyramidal neurons in organotypic hippocampal slices that blockade of presynaptic protein kinases (including PKC) prevented the expression of LTP. Likewise, Tao, et al., (2000) showed in dissociated culture cells that the excitability of presynaptic terminals is regulated by PKC.

Earlier work (Routtenberg and Ehrlich, 1975; Ehrlich et al., 1977) and more recent work (Cammarota et al., 1997) using the step-down inhibitory avoidance task has indicated that there is an increase in phosphorylation of the PKC-substrate GAP-43 after training. More recent work has shown that PKC phosphorylation of substrate is important for information storage as indicated by enhanced learning in mice overexpressing GAP-43 that is phosphorylatable but not in mice overexpressing GAP-43 that cannot be phosphorylated (Routtenberg et al., 2000). Evidence using a fear conditioning paradigm has also shown that after training, GAP-43 phosphorylation is increased (Young et al., 2000; 2002). Izquierdo, et al., (2000) found a time-dependent blockade of 24 hr retention when PKC inhibitors were given up to 30 min after learning indicating that PKC was a key element in the learning process. Weeber, et al. (2000) showed that a reduction in amygdaloid PKCβ, which preferentially phosphorylates GAP-43 (Sheu et al., 1990), impaired fear conditioning, while Paratcha, et al. (2000) found that such conditioning selectively activated the PKCβI isoform. These data suggest that one posttranslational modification that may underlie persistent changes in synaptic plasticity is the phosphorylation of presynaptic GAP-43 resulting in enhanced neurotransmitter release.

4.8.1. axonal elongation

An additional presynaptic alteration underlying long-term synaptic enhancement is likely to be the remodeling of presynaptic terminals (Cantallops et al., 2000). In a demonstration of presumed presynaptic synaptogenesis (Ramírez-Amaya et al., 1999; 2001; Rekart, et al., 2003) it was found that water maze overtrained to find a hidden platform increased the area that stained positively for hippocampal mossy fiber axon terminals in the CA3 stratum oriens of adult Wistar rats, a location not normally invaded by mossy fibers. This result has been interpreted to suggest that presynaptic mossy fiber axonal terminal redistribution, or growth, may mediate persistent change in long-term memory storage (Ramírez-Amaya et al., 2001). It was recently found (Tashiro et al.,

2003) that presynaptic motility of mossy fiber axons in slice culture was differentially regulated by kainate receptor activation in young but not mature slices.

Posttranslational modifications of actin may be involved in axonal elongation (for reviews see Caroni, 1997; 2001a). Polymerization of monomeric actin in presynaptic terminals leads to plasma membrane extensions that are proposed to be a crucial step in axonal elongation (Meyer and Feldman, 2002). Experimental results have shown that inhibition of actin impairs synaptic plasticity (Yamazaki et al., 2001) and memory (Cox et al., 2003; Soderling et al., 2003; Fischer et al., 2004) while potentiation of synaptic potentials elevates polymerized actin levels (Fukazawa et al., 2003). The relationship between GAP-43 and memory is suggested from studies showing that overexpression of GAP-43 enhances axonal elongation (Aigner et al., 1995) and memory (Routtenberg et al., 2000). Phosphorylated GAP-43 has been found to stabilize actin filaments (He et al., 1997) and it is suggested that GAP-43 may interact with specific phospholipases (PIP_2) that promote actin dynamics and axonal elongation (Laux et al., 2000; Frey et al., 2000).

A model depicting the relationship between GAP-43 and actin has recently been proposed (Hatch, Chadderon, Rekart and Routtenberg, unpublished). The model suggests that activation of the presynaptic element by means of Ca^{++} influx results in the upregulation of GAP-43, which binds to and sequesters PIP_2 (Caroni, 2001b). Bound PIP_2 promotes the cleavage of long actin filaments and promotes actin cytoskeletal dynamics. Repeated activation of the presynaptic terminal by either repeated stimulation or feedback from postsynaptic retrograde signals causes additional Ca^{++} influx removing calmodulin from GAP-43 and allowing PKC to phosphorylate GAP-43. Phosphorylated GAP-43 is thought to have two end points: **1)** the release of more PIP_2 which sequesters actin binding proteins and **2)** the binding of actin filaments that keeps them apposed to the presynaptic membrane and stabilizes them for elongation (Kabir et al., 2001) resulting in axon terminal outgrowth (Ling et al., 2004). This process would require an ongoing synaptic dialogue (retrograde signals from postsynaptic element and presynaptic GAP-43 phosphorylation leading to neurotransmitter release and postsynaptic receptor activation) to maintain the actin polymerization and structural integrity for synaptic enhancement.

5. REGULATED CONTROL OF THE DIALOGUE PROCESS

The synaptic dialogue process must be subject to regulation to prevent the neural circuit from becoming uncontrolled. Homeostatic regulation of synaptic networks has been proposed (Turrigiano and Nelson, 2004) as a necessary mechanism to stabilize synaptic plasticity or, in cases of compensatory events, reduce it. Here we discuss a set of regulatory events, labeled R1- R4 on Figure 1, recognizing that this is a subset of the potential mechanisms that would dampen any attempts of the dialogue mechanism to achieve runaway positive feedback.

5.1. R-1: PKC-substrate regulation

One key step in the dialogue signaling mechanism that is a target for regulation involves endogenous inhibition of PKC. Sheu, et al., (1994) reported that the glial-derived S100b protein inhbitied in vitro phosphorylation of GAP-43 by purified PKC. This suggests that neuro-glial interactions exist that can regulate the activity of dialogue model by inhibiting the PKC phosphorylation of GAP-43. In addition to this PKC-substrate inhibition, Chan et al., (1986) reported that calmodulin inhibited PKC phosphorylation of GAP-43 suggesting another endogenous control mechanism mediated by the inhibitory control of PKC. In both cases, the regulation can occur on both sides of the synapse. A full discussion of this mechanism and its impact on regulating memory is found elsewhere (Routtenberg, 1993).

5.2. R-2: regulation by fatty acid metabolites and their breakdown products

Another regulatory mechanism includes the action of fatty acid metabolites working either postsynaptically or diffusing across the synapse and working presynaptically (R2 in **Fig. 1**). Stimulation protocols that lead to LTD result in measurable changes in the levels of 12-(S)-HETE (an arachidonic acid metabolite) in hippocampal slices (Feinmark et al., 2003). These same authors reported that inhibition of a second arachidonic metabolite (12-lipoxygenase) blocked induction of LTD and direct application of 12-(S)-HPETE to hippocampal slices induced LTD. Further work (Cunha et al., 2004) indicated that this regulatory mechanism might be mediated by glutamate release from the presynaptic element. This would raise the intriguing possibility that different unsaturated fatty acid metabolites exert a bi-directional control of synaptic plasticity at both the pre- and postsynaptic elements. This control mechanism may be related to the need for a synergistic effect of fatty acids, Ca^{++} and diacylglycerol on the activation of presynaptic PKC and the phosphorylation of GAP-43. Any reduction in the levels of one may cause reductions in the phosphorylation of GAP-43 protein thereby dampening axonal growth, transmitter release or both.

Cells normally break down fatty acids through an oxidation process, which may provide an additional, essential regulatory mechanism in the dialogue model. Each molecule of fatty acid is broken down completely by a cycle of reactions that trims two carbons from the carboxyl end. These cycles produce one molecule of acetyl CoA as a byproduct. Acetyl CoA then provides the majority of useful energy for the cell. Therefore, as the neurons are maintaining the dialogue process using fatty acids as a retrograde factor, the neurons also require energy. As the energy requirement increases, the need for acetyl CoA most likely also increases engaging oxidation cycles that will break down the free fatty acids (R2 in **Fig. 1**). This process not only serves to nourish the neuron but may also, as a consequence, lead to regulation of the dialogue process by quelling the retrograde signaling mechanisms.

5.3. R-3: Ca^{++}-binding proteins

The positive feedback system inherent in the dialogue model is also highly regulated by calcium-dependent phosphatases and binding proteins (R3 in **Fig. 1**). Calcium-binding proteins such as calbindin, parvalbumin, and calretinin serve an important regulatory mechanism that would dampen runaway activity in the positive feedback dialogue model. Different levels of Ca^{++} can influence whether a neuron undergoes LTP or LTD (Lisman, 1994); high internal concentrations will potentiate AMPA receptors and lead to LTP while lower concentrations will activate phosphatases thereby "shutting down" the system resulting in LTD. In this regard, kinases, phosphatases and proteases would be regulated by Ca^{++} levels, which would be regulated by binding proteins.

5.4. R-4: dendritic spines and receptor dynamics

On the postsynaptic side, dendritic spine morphology has been argued to be a critical factor in the bi-directional regulation of plasticity (Chen et al., 2004) as has been the postsynaptic receptor density and composition (Richards et al., 2004). Nicholoson, et al., (2004) recently reported that age-related deficits in spatial learning were associated with a reduction in the number of perforated postsynaptic synapses in the CA1 stratum radiatum. In addition, induction of long-term depression has been found to decrease the number of AMPA receptors located at the synapse but have no effect on NMDA receptor clustering (Carroll et al., 1999). Recently (Massey, et al., 2004) it was reported that activation of extrasynaptic NR2B-containing NMDA receptors resulted in LTD while activation of NR2A-containing receptors resulted in LTP suggesting that the location and composition of NMDA receptors may be a key regulatory mechanism in the dialogue process (R4). These studies suggest the possibility that the composition and/or size of the postsynaptic density serves a regulatory mechanism in the dialogue process.

6. THE RECURRENT, ONGOING DIALOGUE: STEPS 11-18 AND BEYOND

Once the first cycle of steps has been completed (Steps 1 – 10), the dialogue process repeats itself from enhanced neurotransmitter release (Step 11) to interactions between phosphorylated GAP-43 and the EPM (Step 18). This cycle continues under the regulated control of the 4 regulatory steps as described above.

7. CONCLUSIONS

An emergent property of the 'synaptic dialogue' model, constructed to account for molecular events on both sides of the synapse, is a recurrent mechanism that serves to prolong and persist synaptic alterations. Because this does not rely on an instructive protein synthesis mechanism, a strong inference has been drawn that protein synthesis may not be a key event in the storage of long-lasting memories (Routtenberg and Rekart, 2005). This emergent property of the dialogue model contains a regulated, self-sustaining mechanism in that posttranslational modifications of synaptic proteins can underlie long-term memories while retaining the logical co-activity elements of the Hebb synapse through the release of retrograde factors.

One feature of the model is that glutamate receptor activation postsynaptically (NMDA) regulates protein kinases and hence, the phosphorylation state of proteins, presynaptically. A key proposal of the present dialogue model, is that a line of communication is established between extracellular soluble retrograde factors (fatty acids) that target PKC in the presynaptic terminal, which then sum with cytoplasmic factors within the terminal (Ca^{++}) to activate the enzyme synergistically. Thus, the presynaptic element will have pivotal nodal points such as PKC and GAP-43 that, by virtue of protein-protein interactions, will alter the structure of the terminal. The postsynaptic element plasticity is initiated by the NMDA receptor and Ca^{++} sensitive enzymes and their substrates (PKC and neurogranin for example) leading to retrograde signaling, morphological changes and altered receptor density and sensitivity. These alterations in pre- and postsynaptic elements occur in parallel and are choke points through which synaptic change is effected, maintained and regulated.

8. ACKNOWLEDGEMENTS

Supported by T32 AG20506-01 to MRH and research grants NSF IBN-0090723 and NIH RO1 MH65436 to AR.

9. REFERENCES

Aigner L, Arber S, Kapfhammer JP, Laux T, Schneider C, Botteri F, Brenner HR, Caroni P (1995) Overexpression of the neural growth-associated protein GAP-43 induces nerve sprouting in the adult nervous system of transgenic mice, Cell 83:269-278.

Alder J, Kanki H, Valtora F, Greengard P, Poo MM (1995) Overexpression of synaptophysin enhances neurotransmitter secretion at Xenopus neuromuscular synapses, J Neurosci. 15:511-519.

Alder J, Lu B, Valtora F, Greengard P, Poo MM (1992) Calcium-dependent transmitter secretion reconstituted in Xenopus oocytes: requirement for synaptophysin, Science 257:657-661.

Alexander KA, Wakim BT, Doyle GS, Walsh KA, Storm DR (1988) Identification and characterization of the calmodulin-binding domain of neuromodulin, a neurospecific calmodulin-binding protein, J Biol Chem 263:7544-7549.

Basi GS, Jacobson RD, Virag I, Schilling J, Skene JH (1987) Primary structure and transcriptional regulation of GAP-43, a protein associated with nerve growth, Cell 49:785-791.

Bekkers JM, Stevens CF (1990) Presynaptic mechanism for long-term potentiation in the hippocampus, Nature 346:724-729.

Benowitz, LI, Routtenberg A (1987) A membrane phosphoprotein associated with neural development, axonal regeneration, phospholipid metabolism, and synaptic plasticity, Trends Neurosci 10:527-532.

Benowitz LI, Routtenberg A (1997) GAP-43: an intrinsic determinant of neuronal development and plasticity, Trends Neurosci 20:84-91.

Bliss TVP (1990) Maintenance is presynaptic, Nature 346:698-699.

Bliss TVP, Collingridge GL (1993) A synaptic model of memory: long-term potentiation in the hippocampus, Nature 361:31-39.

Bliss TVP, Douglas RM, Errington ML, Lynch MA (1986) Correlation between long-term potentiation and release of endogenous amino acids from dentate gyrus of anaesthetized rats, J Physiol (London) 377:391-408.

Boothe RG, Greenough WT, Lund JS, Wrege K (1979) A quantitative investigation of spine and dendritic development of neurons in visual cortex (area 17) of Macaca nemestrina monkeys, J Comp Neurol 186:473-489.

Breukel AIM, Besseisen E, daSilva FHL, Ghijsen WEJM (1997) Arachidonic acid inhibits uptake of amino acids and potentiates PKC effects on glutamate, but not GABA, exocytosis in isolated hippocampal nerve terminals, Brain Res 773:90-97.

Byrne JH, Kandel ER (1996) Presynaptic facilitation revisited: state and time dependence, J Neurosci 16:425-435.

Cammarota M, Paratcha G, Levi de Stein M, Bernabeu R, Izquierdo I, Medina JH (1997) B-50/ GAP-43 phosphorylation and PKC activity are increased in rat hippocampal synaptosomal membranes after an inhibitory avoidance training, Neurochem Res 22:499-505.

Cantallops I, Haas K, Cline HT (2000) Postsynaptic CPG15 promotes synaptic maturation and presynaptic axon arbor elaboration in vivo, Nat Neurosci 3:1004-1011.

Carlin RK, Siekevitz P (1983) Plasticity in the central nervous system: do synapses divide?, Proc Natl Acad Sci USA 80:3517-3521.

Caroni P (1997) Intrinsic neuronal determinants that promote axonal sprouting and elongation, Bioessays 19:767-775.

Caroni P (2001b) Actin cytoskeleton regulation through modulation of PI(4,5)P_2 rafts, EMBO J 20:4332-4336.

Caroni P (2001a) New EMBO members' review: actin cytoskeleton regulation through modulation of PI(4,5)P(2) rafts, EMBO J 20:4332-4336.

Carroll RC, Lissin DV, von Zastrow M, Nicoll RA, Malenka RC (1999) Rapid redistribution of glutamate receptors contributes to long-term depression in hippocampal cultures, Nat Neurosci. 2:454-460.

Castro-Alamancos MA, Calcagnotto ME (1999) Pre-synaptic long-term potentiation in corticothalamic synapses, J Neurosci 19:9090-9097.

Chalimoniuk M, King-Pospisil K, Pedersen WA, Malecki A, Wylegala E, Mattson MP, Hennig B, Toborek M (2004) Arachidonic acid increases choline acetyltransferase activity in spinal cord neurons thorugh protein kinae C-mediated mechanism, J Neurochem 90:629-636.

Chan SY, Murakami K, Routtenberg A (1986) Phosphoprotein F1: purification and characterization of a brain kinase C substrate related to plasticity, J Neurosci 6:3618-3627.

Chen Y, Bourne J, Pieribone VA, Fitzsimonds RM (2004) The role of actin in the regulation of dendritic spine morphology and bidirectional synaptic plasticity, Neuroreport 15:829-832.

Christie BR, Abraham WC (1994) Differential regulation of paired-pulse plasticity following LTP in the dentate gyrus, Neuroreport 5:385-388.

Clements MP, Rose SPR (1996) Time-dependent increae in release fo arachidonic acid following passive avoidance training in the day-old chick, J.Neurochem. 67:1317-1323.

Cline HT (2001) Dendritic arbor development and synaptogenesis, Curr Opin Neurobiol 11:118-126.

Colley PA, Routtenberg A (1993) Long-term potentiation as synaptic dialogue, Brain Res Rev 18:115-122.

Colley PA, Sheu FS, Routtenberg A (1990) Inhibition of protein kinase C blocks two components of LTP persistence, leaving initial potentiation intact, J Neurosci 10:3353-3360.

Collins DR, Smith RC, Davies SN (1995) Interactions between arachidonic acid and metabotropic glutamate receptors in the induction of synaptic potentiation in the rat hippocampal slice, Eur J Pharmacol 294:147-154.

Commins S, Gigg J, Anderson M, O'Mara SM (1998) The projection from hippocamal area CA1 to the subiculum sustains long-term potentiation, Neuroreport 9:847-850.

Cox PR, Fowler V, Xu B, Sweatt JD, Paylor R, Zoghbi HY (2003) Mice lacking Tropomodulin-2 show enhanced long-term potentiation, hyperactivity, and deficits in learning and memory, Mol Cell Neurosci 23:1-12.

Cunha RA, Ribeiro JA, Malva JO (2004) Presynaptic kainate receptors modulating glutamatergic transmission in the rat hippocampus are inhibited by arachidonic acid, Neurochem Int 44:371-379.

de Graan PNE, Oestreicher AB, de Wit M, Kroef M, Schrama LH, Gispen WH (1990) Evidence for the binding of calmodulin to endogenous B-50 (GAP-43) in native synaptosomal plasma membranes, J Neurochem 55:2139-2141.

DiStefano PS, Friedman B, Radziejewski C, Alexander C, Boland P, Schick CM, Lindsay RM, Wiegnad SJ (1992) The neurotrophins BDNF, NT-3 and NGF display distinct patterns of retrograde transport in peripheral and central neurons, Neuron 8:983-993.

Dolphin AC, Errington ML, Bliss TVP (1982) Long-term potentiation of the perforant path in vivo is associated with increased glutamate release, Nature 297:496-498.

Dumuis A, Sebben M, Haynes L, Pin JP, Bockaert J (1988) NMDA receptors activate the arachidonic acid cascade system in striatal neurons, Nature 336:68-70.

Dunaevsky A, Tashiro A, Majewska A, Mason C, Yuste R (1999) Developmental regulation of spine motility in the mammalian central nervous system, Proc Natl Acad Sci USA 96:13438-13443.

Edwards FA (1995) LTP - a structural model to explain the inconsistencies Trends Neurosci 18:250-255.

Ehrlich YH, Rabjohns RR, Routtenberg A (1977) Experiential input alters the phosphorylation of specific proteins in brain membranes, Pharmacol Biochem Behav 6:169-174.

Ehrlich YH, Routtenberg A (1974) Cyclic AMP regulates phosphorylation of three protein components of rat cerebral cortex membranes for thirty minutes, FEBS Letters 45:237-243.

Engert F, Bonhoeffer T (1999) Dendritic spine changes associated with hippocampal long-term synaptic plasticity, Nature 1:1-5.

Faber ESL, Sah P (2003) Calcium-activated potassium cannels: multiple contributions to neuronal function, The Neuroscientist 9:181-194.

Feinmark SJ, Begum R, Tsvetko E, Goussakov I, Funk CD, Siegelbaum SA, Bolshakov VY (2003) 12-lipoxygenase metabolites of arachidonic acid mediate metabotropic glutamate receptor-dependent long-term depression at hippocampal CA3-CA1 synapses, J Neurosci 23:11427-11435.

Fernández-Chacón R, Königstorfer A, Gerber SH, García J, Matos MF, Stevens CF, Brose N, Rizo J, Rosenmund C, Südhof TC (2001) Synaptotagmin I functions as a calcium regulator of release probability, Nature 410:41-49.

Fischer A, Sananbenesi F, Schrick C, Spiess J, Radulovic J (2004) Distinct roles of hippocampal de novo protein synthesis and actin rearrangements in extinction of contextual fear, J Neurosci 24:1962-1966.

Fitzsimonds RM, Poo MM (1998) Retrograde signaling in the development and modification of synapses, Physiol Rev 78:143-170.

Fong DK, Rao A, Crump FT, Craig AM (2002) Rapid synaptic remodeling by protein kinase C: reciprocal translocation of NMDA receptors and calcium/calmodulin-dependent kinase II, J Neurosci 22:2153-2165.

Foster TC, McNaughton BL (1994) Long-term enhancement of CA1 synaptic transmission is due to increased quantal size, not quantal content, Hippocampus 1:79-91.

Frey D, Laux T, Xu L, Schneider C, Caroni P (2000) Shared and unique roles of CAP23 and GAP43 in actin regulation, neurite outgrowth, and anatomical plasticity, J Cell Biology 149:1443-1454.
Frey U, Krug M, Reymann KG, Matthies H (1988) Anisomycin, an inhibitor of protein synthesis, block late phases of homo- and heterosynaptic long-term potentiation in the hippocampal CA1-region in vitro, Brain Res 452:57-65.
Frey U, Morris RGM (1997) Synaptic tagging and long-term potentiation, Nature 385:533-536.
Frey U, Morris RGM (1998) Synaptic tagging: implications for late maintenance of hippocampal long-term potentiation, Trends Neurosci 21:181-188.
Fukazawa Y, Saitoh Y, Ozawa F, Ohta Y, Mizuno K, Inokuchi K (2003) Hippocampal LTP is accompanied by enhanced F-actin content within the dendritic spine that is essential for late LTP maintenance in vivo, Neuron 38:447-460.
Geinisman Y (1993) Perforated axospinous synapses with multiple, completely partitioned transmission zones: probable structural intermediates in synaptic plasticity, Hippocampus 3:417-433.
Gianotti C, Nunzi MG, Gispen WH, Corradetti R (1992) Phosphorylation of the presynaptic protein B-50 (GAP-43) is increased during electrically induced long-term potentiation, Neuron 8:843-848.
Globus A, Rosenzweig MR, Bennett EL, Diamond MC (1973) Effects of differential experience on dendritic spine counts in rat cerebral cortex, J Comp Physiol Psychol 82:175-181.
Greenough WT, Juraska JM, Volkmar FR (1979) Maze training effects on dendritic branching in occipital cortex of adult rats, Behav Neural Bio 26:287-297.
Greenough WT, Volkmar FR (1973) Pattern of dendritic branching in occipital cortex of rats reared in complex environments, Exp Neurol 40:491-504.
Greenough WT, Volkmar FR, Juraska JM (1973) Effects of rearing complexity on dendritic branching in frontolateral and temporal cortex of the rat, Exp Neurol 41:371-378.
Grosshans DR, Clayton DF, Coultrap SJ, Browning MD (2002) LTP leads to rapid surface expression of NMDA but not AMPA receptors in adult rat CA1, Nat Neurosci 5:27-33.
Hama H, Hara C, Yamaguchi K, Miyawaki A (2004) PKC signaling mediates global enhancement of excitatory synaptogenesis in neurons triggered by local contact with astrocytes, Neuron 41:405-415.
Hanson PI, Schulman H (1992) Neuronal Ca^{2+}/calmodulin-dependent protein kinases, Ann Rev Biochem 61:559-601.
Haruta T, Takami N, Ohmura M, Misumi Y, Ikehara Y (1997) Ca^{2+}-dependent interaction of the growth-associated protein GAP-43 with the synaptic core complex, Biochem J 325:455-463.
Hayashi Y, Shi SH, Esteban JA, Piccini A, Poncer JC, Malinow R (2000) Driving AMPA receptors into synapses by LTP and CaMKII: requirement for GluR1 and PDZ domain interactions, Science 287:2262-2267.
He Q, Dent EW, Meiri KF (1997) Modulation of actin filament behavior by GAP-43 (neuromodulin) is dependent on the phosphorylation status of serine 41, the protein kinase C site J Neurosci 17:3515-3524.
Hebb DO 1949 Organization of Behavior, Wiley, New York.
Hendry IA, Stockel K, Thoenen H, Iversen LL (1974) The retrograde axonal transport of nerve growth factor, Brain Res 68:103-121.
Hering H, Sheng M (2003) Activity-dependent redistribution and essential role of cortactin in dendritic spine morphology, J Neurosci 23:11759-11569.
Huang KP (1990) Role of protein kinase C in cellular regulation, Biofactors 2:171-178.
Huang YY, Kandel ER (1998) Postsynaptic induction and PKA-dependent expression of LTP in the lateral amygdala, Neuron 21:169-178.
Isaac JTR (2003) Postsynaptic silent synapses: evidence and mechanisms, Neuropharmacology 45:450-460.
Isaac JTR, Nicoll RA, Malenka RC (1995) Evidence for silent synapses: implications for the expression of LTP, Neuron 15:427-434.
Izquierdo LA, Vianna M, Barros DM, Mello e Souza T, Ardenghi P, Sant'Anna M, Rodrigues C, Medina JH, Izquierdo I (2000) Short- and long-term memory are differentially affected by metabolic inhibitors given into hippocampus and entorhinal cortex, Neurobiol Learn Mem 73:141-149.
Janz R, Sudhof TC, Hammer RE, Unni V, Siegelbaum SA, Bolshakov VY (1999) Essential roles in synaptic plasticity for synaptogyrin I and synaptophysin I, Neuron 24:687-700.
Kabir N, Schaefer AW, Nakhost A, Sossin WS, Forscher P (2001) Protein kinase C activation promotes microtubule advance in neuronal growth cones by increasing average microtubule growth lifetimes, J Cell Biology 152:1033-1044.
Kamisaki Y, Maeda K, Ishimura M, Omura H, Moriwaki Y, Itoh T (1994) No enhancement by nitric oxide of glutamate release from P2 and P3 synaptosomes of rat hippocampus, Brain Res 644:128-134.
Kandel ER (2001) The molecular biology of memory storage: a dialog between genes and synapses, Biosci Rep 21:565-611.
Kauer JA, Malenka RC, Nicoll RA (1988) A persistent postsynaptic modification mediates long-term

potentiation in the hippocampus, Neuron 1:911-917.

Kinney WR, McNamara RK, Valcourt EG, Routtenberg A (1996) Prolonged alteration in E-box binding after a single systemic kainate injection: potential relation to F1/GAP-43 gene expression, Mol Brain Res 38:25-36.

Kleschevnikov AM, Routtenberg A (2001) PKC activation rescues LTP from NMDA receptor blockade, Hippocampus 11:168-175.

Kleschevnikov AM, Sokolov MV, Kuhnt U, Dawe GS, Stephenson JD, Voronin LL (1997) Changes in paired-pulse facilitation correlate with induction of long-term potentiation in area CA1 of rat hippocampal slices, Neuroscience 76:829-843.

Kotani S, Nakazawa H, Tokimasa T, Akimoto K, Kawashima H, Toyoda-Ono Y, Kiso Y, Okaichi H, Sakakibara M (2003) Synaptic plasticity preserved with arachidonic acid diet in aged rats, Neurosci Res 46:453-461.

Kuhnt U, Voronin LL (1994) Interaction between paired-pulse facilitation and long-term potentiation in area CA1 of guinea-pig hippocampal slices: application of quantal analysis, Neuroscience 62:391-397.

Lan JY, Skeberdis VA, Jover T, Grooms SY, Lin Y, Araneda RC, Zheng X, Bennett MV, Zukin RS (2001) Protein kinase C modulates NMDA receptor trafficking and gating, Nat Neurosci 4:382-390.

Laux T, Fukami K, Thelen M, Golub T, Frey D, Caroni P (2000) GAP43, MARCKS, and CAP23 modulate PI(4,5)P(2) at plasmalemmal rafts, and regulate cell cortex actin dynamics through a common mechanism, J Cell Biol 149:1455-1472.

Liao D, Jones A, Malinow R (1992) Direct measurement of quantal changes underlying long-term potentiation in CA1 hippocampus, Neuron 9:1089-1097.

Liao D, Hessler NA, Malinow R (1995) Activation of postsynaptically silent synapses during pairing-induced LTP in CA1 region of hippocampal slice, Nature 375:400-404.

Linden DJ, Routtenberg A (1989b) Cis-fatty acids, which activate protein kinase C, attenuate Na^+ and Ca^{2+} currents in mouse neuroblastoma cells, J Physiol 419:95-119.

Linden DJ, Routtenberg A (1989a) The role of protein kinase C in long-term potentiation: a testable model, Brain Res Rev 14:279-296.

Linden DJ, Sheu FS, Murakami K, Routtenberg A (1987) Enhancement of long-term potentiation by Cis-unsaturated fatty acid: relation to protein kinase C and phospholipase A_2, J Neurosci 7:3783-3792.

Ling M, Trollér U, Zeidman R, Lundberg C, Larsson C (2004) Induction of neurites by the regulatory domains of PKCdelta and epsilon is counteracted by PKC catalytic activity and the RhoA pathway, Exp Cell Res 292:135-150.

Lisman J (1994) The CaMkinase II hypothesis for storage of synaptic memory, Trends Neurosci 17:406-412.

Lisman JE (2003) Long-term potentiation: outstanding questions and attempted synthesis, Phil Trans Royal Society of London B Bio. Sci. 358:829-842.

Lissin DV, Carroll RC, Nicoll RA, Malenka RC, von Zastrow M(1999) Rapid, activation-induced redistribution of ionotropic glutamate receptors in cultured hippocampal neurons, J Neurosci 19:1263-1272.

Liu YC, Storm DR (1989) Dephosphorylation of neuromodulin by calcineurin, J Biol Chem 264:12800-12804.

Lovinger DM, Akers RF, Nelson RB, Barnes CA, McNaughton BL, Routtenberg A (1985) A selective increase in phosphorylation of protein F1, a protein kinase C substrate, directly related to three day growth of long term synaptic enhancement, Brain Res 343:137-143.

Lovinger DM, Routtenberg A (1988a) Synapse-specific protein kinase C activation enhances maintenance of long-term potentiation in rat hippocampus, J Physiol (London) 400:321-333.

Lovinger DM, Routtenberg A (1988b) Synapse-specific protein kinase C activation enhances maintenance of long-term potentiation in rat hippocampus, J Physiol 400:321-333.

Luo Y, Vallano ML (1995) Arachidonic acid, but not sodium nitroprusside, stimulates presynaptic protein kinase C and phosphorylation of GAP-43 in rat hippocampal slices and synaptosomes, J Neurochem 64:1808-1818.

Lynch G, Baudry M (1984) The biochemistry of memory: a new and specific hypothesis, Science 224:1057-1063.

Lynch MA, Bliss TVP (1986) On the mechanism of enhanced release of [^{14}C] glutamate in hippocampal long-term potentiation, Brain Res 369:405-408.

Lynch MA, Clements MP, Voss KL, Bramham CR, Bliss TVP (1991) Is arachidonic acid a retrograde messenger in long-term potentiation?, Biochem Society Trans 19:391-396.

Lynch MA, Errington ML, Bliss TVP (1985) Long-term potentiation of synaptic transmission in the dentate gyrus: increased release of [^{14}C] glutamate without increase in receptor binding, Neurosci Lett 62:123-129.

Lynch MA, Errington ML, Bliss TVP (1989) Nordihydroguaiaretic acid blocks the synaptic component of long-term potentiation and the associated increases in release of glutamate and arachidonate: an in vivo study in the dentate gyrus of the rat, Neuroscience 30:693-701.

Lynch MA, Voss KL (1990) Arachidonic acid increases inositol phospholipid metabolism and glutamate release in synaptosomes prepared from hippocampal tissue, J Neurochem 55:215-221.

Maletic-Savatic M, Koothan T, Malinow R (1998) Calcium-evoked dendritic exocytosis in cultured hippocampal neurons. Part II: mediation by calcium/calmodulin dependent protein kinase II, J Neurosci 18:6814-6821.

Maletic-Savatic M, Malinow R, Svoboda K (1999) Rapid dendritic morphogenesis in CA1 hippocampal dendrites induced by synaptic activity, Science 283:1923-1927.

Malinow R, Mainen ZF, Hayashi Y (2000) LTP mechanisms: from silence to four-lane traffic, Curr Opin Neurobiol 10:352-357.

Malinow R, Malenka RC (2002) AMPA receptor trafficking and synaptic plasticity, Annu Rev Neurosci 25:103-126.

Malinow R, Tsien RW (1990) Presynaptic enhancement shown by whole-cell recordings of long-term potentiation in hippocampal slices, Nature 346:177-180.

Manabe T, Renner P, Nicoll RA (1992) Postsynaptic contribution to long-term potentiation revealed by analysis of miniature synaptic currents, Nature 355:50-55.

Massey PV, Johnson BE, Moult PR, Auberson YP, Brown MW, Molnar E, Collingridge GL, Bashir ZI (2004) Differential roles of NR2A and NR2B-containing NMDA receptors in cortical long-term potentiation and long-term depression, J Neurosci 24:7821-7828.

Matthies H Jr, Behnisch T, Kase H, Matthies H, Reymann KG (1991) Differential effects of protein kinase inhibitors on pre-established long-term potentiation in rat hippocampal neuons in vitro, Neurosci Lett 121:259-262.

McGahon B, Clements MP, Lynch MA (1997) The ability of aged rats to sustain long-term potentiation is restored when the age-related decrease in membrain arachidonic acid concentration is reversed, Neuroscience 81:9-16.

McGahon B, Holscher C, McGlinchey L, Rowan MJ, Lynch MA (1996) Training in the Morris water maze occludes the synergism between ACPD and arachidonic acid on glutamate release in synaptosomes prepared from rat hippocampus, Learn Mem 3:296-304.

McGahon B, Lynch MA (1996) The synergism between metabotropic glutamate receptor activation and arachidonic acid on glutamate release is occluded by induction of long-term potentiation in the dentate gyrus, Neuroscience 72:847-855.

McGahon B, Lynch MA (1998) Analysis of the interaction between arachidonic acid and metabotropic glutamate receptor activation reveals that phospholipase C acts as a coincidence detector in the expression of long-term potentiation in the rat dentate gyrus, Hippocampus 8:48-56.

Medina JH, Izquierdo I (1995) Retrograde messengers, long-term potentiation and memory, Brain Res Rev 21:185-194.

Menard C, Valastro B, Martel MA, Martinoli MG, Massicotte G (2004) Strain-related variations in AMPA receptor modulation by calcium-dependent mechanisms in the hippocampus: contribution of lioxygenase metabolites of arachidonic acid, Brain Res 1010:134-143.

Meyer G, Feldman EL (2002) Signaling mechanisms that regulate actin-based motility processes in the nervous system, J Neurochem 83:490-503.

Murakami K, Chan SY, Routtenberg A (1986) Protein kinase C activation by cis-fatty acid in the absence of Ca^{2+} and phospholipds, J Biol Chem 261:15424-15429.

Murakami K, Routtenberg A (1985) Direct activation of purified protein kinase C by unsaturated fatty acids (oleate and arachidonate) in the absence of phospholipids and Ca^{2+}, FEBS Letters 192:189-193.

Murakami K, Routtenberg A (2004) The role of fatty acids in synaptic growth and plasticity, in: Phospholipid Spectrum Disorders in Psychiatry and Neurology, M Peet, I Glen, DF Horrobin, eds., Marius Press, pp. 77-92.

Naor Z, Shearman MS, Kishimoto A, Nishizuka Y (1988) Calcium-independent activation of hypothalamic type I protein kinase C by unsaturated fatty acids, Mol Endocrinol 2:1043-1048.

Nelson RB, Linden DJ, Hyman C, Pfenninger KH, Routtenberg A (1989) The two major phosphoproteins in growth cones are probably identical to two protein kinase C substrates correlated with persistence of long-term potentiation, J Neurosci 9:381-389.

Newton AC (1995) Protein kinase C: structure, function and regulation, J Bio Chem 270:28495-28498.

Newton AC (1997) Regulation of protein kinase C, Current Opinion Cell Bio 9:161-167.

Nicholson DA, Yoshida R, Berry RW, Gallagher M, Geinisman Y (2004) Reduction in size of perforated postsynaptic densities in hippocampal axospinous synapses and age-related spatial learning impairments, J Neurosci 24:7648-7653.

Nicoll RA (2003) Expression mechanisms underlying long-term potentiation: a postsynaptic view, Phil. Trans. Royal Society of London B Bio Sci 358:721-726.

Nicoll RA, Malenka RC (1999) Expression mechanisms underlying NMDA receptor-dependent long-term potentiation, Ann NY Acad Sci 868:515-525.

Nielander HB, Schrama LH, van Rozen AJ, Kasperaitis M, Oestreicher AB, Gispen WH, Schotman P (1990) Mutation of serine 41 in the neuron-specific protein B-50 (GAP-43) prohibits phosphorylation by protein kinase C, J Neurochem 55:1442-1445.

O'Dell TJ, Hawkins RD, Kandel ER (1991) Tests of the roles of two diffusible substances in long-term potentiation: evidence for nitric oxide as a possible early retrograde messenger, Proc Natl Acad Sci USA 88:11285-11289.

Paratcha G, Furman M, Bevilaqua L, Cammarota M, Vianna M, de Stein ML, Izquierdo I, Medina JH (2000) Involvement of hippocampal PKC□1 isoform in the early phase of memory formation of an inhibitory avoidance learning, Brain Res 855:199-205.

Pavlidis P, Montgomery JM, Madison DV (2000) Presynaptic protein kinase activity supports long-term potentiation at synapses between individual hippocampal neurons, J Neurosci 20:4497-4505.

Pellerin L, Wolfe LS (1991) Release of arachidonic acid by NMDA-receptor activation in the rat hippocampus, Neurochem Res 16:983-989.

Pilpel Y, Segal M (2004) Activation of PKC induces rapid morphological plasticity in dendrites of hippocampal neurons via Rac and Rho-dependent mechanisms, Eur J Neurosci 19:3151-3164.

Rajan I, Cline HT (1998) Glutamate receptor activity is required for normal development of tectal cell dendrites in vivo, J Neurosci 18:7836-7846.

Rajan I, Witte S, and Cline HT (1999) NMDA receptor activity stabilizes presynaptic retinotectal axons and postsynaptic optic tectal cell dendrites in vivo, J Neurobiol 38:357-368.

Ramírez-Amaya V, Balderas I, Sandoval J, Escobar ML, Bermúdez-Rattoni F (2001) Spatial long-term memory is related to mossy fiber synaptogenesis, J Neurosci 21:7340-7348.

Ramírez-Amaya V, Escobar ML, Chao V, Bermúdez-Rattoni F (1999) Synaptogenesis of mossy fibers induced by spatial water maze overtraining, Hippocampus 9:631-636.

Rekart JL, Meiri K, Routtenberg A (2004) Hippocampal-dependent memory is impaired in heterozygous GAP-43 knockout mice, Hippocampus, DOI: 10.1002/hipo.20045.

Represa A, Deloulm JC, Sensenbrenner M, Ben-Ari Y, Jacques B (1990) Neurogranin: immunocytochemical localization of a brain-specific kinase C substrate, J Neurosci 10:3782-3792.

Reymann KG, Frey U, Jork R, Matthies H (1988) Polymixin B, an inhibitor of protein kinase C, prevents the maintenance of synaptic long-term potentiation in hippocampal CA1 neurons, Brain Res 440:305-314.

Richards DA, de Paola V, Caroni P, Gahwiler BH, McKinney RA (2004) AMPA-receptor activation regulates the diffusion of a membrane marker in parallel with dendritic spine motility in the mouse hippocampus, J Physiol 558:503-512.

Roberts AC, Glanzman DL (2003) Learning in Aplysia: looking at synaptic plasticity from both sides, Trends Neurosci 26:662-670.

Rosenthal A, Chan SY, Henzel W, Haskell C, Kuang WJ, Chen E, Wilcox JN, Ullrich A, Goeddel DV, Routtenberg A (1987) Primary structure and mRNA localization of protein F1, a growth-related protein kinase C substrate associated with synaptic plasticity, EMBO J 6:3641-3646.

Roseth S, Fykse EM, Fonnuum F (1998) The effect of arachidonic acid and free fatty acids on vesicular uptake of glutamate and gamma-aminobutyric acid, Eur J Pharmacol 341:281-288.

Routtenberg A (1990) Trans-synaptophobia, in: Excitatory Amino Acids and Neuronal Plasticity, Y Ben-Ari, ed., Plenum Press, New York, pp. 401-403.

Routtenberg A (1993) Resisting memory storage: activating endogenous protein kinease C inhibitors, in: Phospholipids and Signal Transmission, R Massarelli, LA Horrocks, JN Kanfer, K Löffelholz, eds., Springer-Verlag, Berlin, pp. 151-162.

Routtenberg A (1997) Why the metabotropic glutamate receptor is part of a tripartite synergistic switch for memory storage, in: Metabotropic Glutamate Receptors and Brain Function, F Moroni, F Nicoletti, DE Pelligrini-Giampietro, eds., Portland Press, London, pp. 107-116.

Routtenberg A (1999) Tagging the Hebb synapse, Trends Neurosci 22:255-256.

Routtenberg A, Cantallops I, Zaffuto S, Serrano PA, Namgung U (2000) Enhanced learning after genetic overexpression of a brain growth factor, Proc Natl Acad Sci USA 97:7657-7662.

Routtenberg A, Ehrlich YH (1975) Endogenous phosphorylation od four cerebral cortical membrane proteins: role of cyclic nucleotides, ATP and divalent cations, Brain Res 92:415-430.

Routtenberg A, Rekart JL (2004) Post-translational modification as substrate for long-lasting memory, Trends Neurosci, in press.

Sanes JR, Lichtman JW (1999) Can molecules explain long-term potentiation?, Nat Neurosci 2:597-604.

Sasaki Y, Asaoka Y, Nishizuka Y (1993) Potentiation of diacylglycerol-induced activation of protein kinase C by lysophospholipds. Subspecies differences, FEBS Letters 320:47-51.

Schaechter JD, Benowitz LI (1993) Activation of protein kinase C by arachidonic acid selectively enhances the phosphorylation of GAP-43 in nerve terminal membranes, J Neurosci 13:4361-4371.

Schmidt JT (2004) Activity-driven sharpening of the retinotectal projection: the search for retrograde synaptic signaling pathways, J Neurobiol 59:114-133.

Schulz PE, Cook EP, Johnston D (1994) Changes in paired-pulse facilitation suggest presynaptic involvement in long-term potentiation, J Neurosci 14:5325-5337.

Shearman MS, Naor Z, Sekiguchi K, Kishimoto A, Nishizuka Y (1989) Selective activation of the gamma-subspecies of protein kinase C from bovine cerebellum by arachidonic acid and its lipoxygenase metabolites, FEBS Letters 243:177-182.

Shearman MS, Sekiguchi K, Nishizuka Y (1989) Modulation of ion channel activity: a key function of the protein kinase C enzyme family, Pharmacol Rev 41:211-237.

Shearman MS, Shinomura T, Oda T, Nishizuka Y (1991a) Protein kinase C subspecies in adult rat hippocampal synaptosomes. Activation by diacylglycerol and arachidonic acid, FEBS Letters 279:261-264.

Shearman MS, Shinomura T, Oda T, Nishizuka Y (1991b) Synaptosomal protein kinase C subspecies: A. Dynamic changes in hippocampus and cerebellar cortex concomitant with synaptogenesis, J Neurochem 56:1255-1262.

Sheu FS, Azmitia EC, Marshak DR, Parker PJ, Routtenberg A (1994) Glial-derived S100b protein selectively inhibits recombinant beta protein kinase C (PKC) phosphorylation of neuron-specific protein F1/GAP43, Brain Res Mol Brain Res 21:62-66.

Sheu FS, Marais RM, Parker PJ, Bazan NG, Routtenberg A (1990) Neuron-specific protein F1/GAP-43 shows substrate specificity for the beta subtype of protein kinase C, Biochem Biophys Res Comm 171:1236-1243.

Shi S, Hayashi Y, Esteban JA, Malinow R (2001) Subunit-specific rules governing AMPA receptor trafficking to synapses in hippocampal pyramidal neurons, Cell 105:331-343.

Shi SH, Hayashi Y, Petralia RS, Zaman SH, Wenthold RJ, Svoboda K, Malinow R (1999) Rapid spine delivery and redistribution of AMPA receptors after synaptic NMDA receptor activation, Science 284:1811-1816.

Shinomura T, Asaoka Y, Oka M, Yoshida K, Nishizuka Y (1991) Synergistic action of diacylglycerol and unsaturated fatty acid for protein kinase C activation: its possible implications, Proc Natl Acad Sci USA 88:5149-5153.

Skene JH (1989) Axonal growth-associated proteins, Annu Rev Neurosci 12:127-156.

Skene JH, Virag I (1989) Posttranslational membrane attachment and dynamic fatty acylation of a neuronal growth cone protein, GAP-43, J Cell Bio 108:613-624.

Soderling SH, Langeberg LK, Soderling JA, Davee SM, Simerly R, Raber J, Scott JD (2003) Loss of WAVE-1 causes sensoritmotor retardation and reduced learning and memory in mice, Proc Natl Acad Sci USA 100:1723-1728.

Stevens CF, Wang Y (1994) Changes in reliability of synaptic function as a mechanism for plasticity, Nature 371:704-707.

Strittmatter SM, Fankhauser C, Huang PL, Mashimo H, Fishman MC (1995) Neuronal pathfinding is abnormal in mice lacking the neuronal growth cone protein GAP-43, Cell 80:445-452.

Sudhof TC (2004) The synaptic vesicle cycle, Annu Rev Neurosci 27:509-547.

Tao H, Zhang LI, Bi G, Poo M (2000) Selective presynaptic propogation of long-term potentiation in defined neural networks, J Neurosci 20:3233-3243.

Tarrant S, Routtenberg A (1977) The synaptic spinule in the dendritic spine: electron microscopic study of the hippocampal dentate gyrus, Tiss Cell 9:461-473.

Tashiro A, Dunaevsky A, Blazeski R, Mason CA, Yuste R (2003) Bidirectional regulation of hippocampal mossy fiber filopodial motility by kainate receptors: a two-step model of synaptogenesis, Neuron 38:773-784.

Thomson AM (2000) Facilitation, augmentation and potentiation at central synapses, Trends Neurosci 23:305-312.

Turrigiano GG, Nelson SB (2004) Homeostatic plasticity in the developing nervous system, Nat Rev Neurosci 5:97-107.

Valtora F, Pennuto M, Bonanomi D, Benfenati F (2004) Synaptophysin: leading actor or walk-on role in synaptic vesicle exocytosis? Bioessays 26:445-453.

van Aelst L, Cline HT (2004) Rho GTPases and activity-dependent dendrite development, Curr Opin Neurobiol 14:297-304.

Volgushev M, Voronin LL, Chistiakova M, Singer W (1997) Relations between long-term synaptic modifications and paired-pulse interactions in the rat neocortex, Eur J Neurosci 9:1656-1665.

Volkmar FR, Greenough WT (1972) Rearing complexity affects branching of dendrites in the visual cortex of the rat, Science 176:1145-1147.

Voronin LL, Cherubini E (2003) "Presynaptic silence" may be golden, Neuropharmacology 45:439-449.

Voronin LL, Kuhnt U, Ivanov NV, Gusev AG (1992) Reliability of quantal parameter estimates and their changes during long-term potentiation in guinea pig hippocampal slices, Neurosci Lett 146:111-114.

Voronin LL, Volgushev M, Chistiakova M, Kuhnt U, Singer W (1996) Involvement of silent synapses in the induction of long-term potentiation and long-term depression in neocortical and hippocampal neurons, Neuroscience 74:323-330.

Wang JH, Kelly PT (1997) Attenuation of paired-pulse facilitation associated with synaptic potentiation mediated by postsynaptic mechanisms, J Neurophysiol 78:2707-2716.

Weeber EJ, Atkins CM, Selcher JC, Varga AW, Mirnikjoo B, Paylor R, Leitges M, Sweatt JD (2000) A role for the beta isoform of protein kinase C in fear conditioning, J Neurosci 2016:5906-5914.

Williams JH, Bliss TVP (1988) Induction but not maintenance of calcium-induced long-term potentiation in dentate gyrus and area CA1 of the hippocampal slice is blocked by nordihydroguaiaretic acid, Neurosci Lett 88:81-85.

Williams JH, Errington ML, Lynch MA, Bliss TVP (1989) Arachidonic acid induces a long-term activity-dependent enhancement of synaptic transmission in the hippocampus, Nature 341:739-742.

Wong KL, Murakami K, Routtenberg A (1989) Dietary cis-fatty acids that increase protein F1 phosphorylation enhance spatial memory, Brain Res 505:302-305.

Wong ROL, Ghosh A (2002) Activity-dependent regulation of dendritic growth and patterning, Nat Rev Neurosci 3:803-812.

Yamazaki M, Matsuo R, Fukazawa Y, Ozawa F, Inokuchi K (2001) Regulated expression of an actin-associated protein, synaptopodin during long-term potentiation, J Neurochem 79:192-199.

Young EA, Owen EA, Meiri KF, Wehner JM (2000) Alterations in hippocampal GAP-43 phosphorylation and protein level following contextual fear conditioning, Brain Res 860:95-103.

Young E, Cesna T, Meiri KF, Perrone-Bizzozero N (2002) Changes in protein kinase C (PKC) activity, isoenzyme translocation, and GAP-43 phosphorylation in the rat hippocampal formation after a single-trial contextual fear conditioning paradigm, Hippocampus 12:457-464.

Zalutsky RA, Nicoll RA (1990) Comparison of two forms of long-term potentiation in single hippocampal neurons, Science 248:1619-1624.

Zuber MX, Strittmatter SM, Fishman MC (1989) A membrane-targeting signal in the amino acid terminus of the neuronal protein GAP-43, Nature 341:345-348.

COORDINATED PRE- AND POSTSYNAPTIC CHANGES INVOLVED IN DEVELOPMENTAL ACTIVITY-DEPENDENT SYNAPSE ELIMINATION

Phillip G. Nelson[1], Min Jia[1], Min-Xu Li[1], Rahel Gizaw[1], Maria A. Lanuza[2] and Josep Tomas[2]

1. INTRODUCTION

A central problem for current neuroscience is to develop an unifying mechanistic hypothesis for the different forms of nervous system plasticity. In addition to the short-term adaptations of synaptic activity like facilitation, the various models that have been studied include long term potentiation (LTP), long term depression (LTD), various forms of behavioral conditioning such as eye-blink conditioning, sensitization and desensitization and the synaptic and neuronal pruning or removal that is a prominent aspect of development of the nervous system. It seems to be the case that several different mechanisms are involved in these different forms of activity dependent changes in neural function and circuit performance; certainly the relatively simple Hebb model of 1949 (Hebb, 1949) serves as only the broadest guide to understanding such changes. These variations notwithstanding, some general model for understanding neuro-plasticity is emerging and should be a useful conceptual framework for guiding future experiments.

A fundamental distinction has been made on practical, experimental grounds between the enormous changes (reductions) in neural circuit structure and function that occur during development and the changes in circuit connectivity that occur as a result of input in the mature organism and that are considered to underly learning and memory. LTP, LTD and various conditioning paradigms are experimentally approachable models of the latter. The degree of mechanistic overlap between these models and the massive circuit changes that occur during development is not fully known, but some convergences have been noted and will be dealt with briefly in this paper. Our major purpose will be to summarize recent data on some of the molecular and cell biologic mechanisms that play important roles in activity-dependent synapse elimination at the mammalian neuromuscular junction in vitro and in vivo.

[1] Lab Develop Neurobiology NICHD, NIH, Bethesda, MD 20892, USA. E-mail: pgnelson@codon.nih.gov

[2] Unitat d'Histologia i Neurobiologia, Facultat de Medicina i Ciències de la Salut, Universitat Rovira i Virgili, 43201 Reus, Spain.

2. NEURON AND SYNAPSE LOSS DURING DEVELOPMENT

During the first few postnatal years in the human a substantial reduction in the neural elements making up the brain occurs (Bourgeois and Rakic, 1993, 1996; Conel 1936-1967). This elimination affects both neuronal and synapse number and the elimination process is strongly a function of activation of the cells involved. Furthermore, this activity dependence is such that it mediates a competitive interaction between the elements. Coordinated pre- and postsynaptic activity results in increased survival and growth of neural elements while un-coordinated activity (where presynaptic activity results in a weak response and/or does NOT produce postsynaptic spikes) results in decreased connectivity and possible death of the elements. Thus, strong synaptic connections serve to decrease the effectiveness of other inputs and weak inputs can neither support themselves nor eliminate other inputs. The combinations of inputs that are activated by naturally occurring patterns of stimulation would tend to be strengthened under these rules. Connectivity that is efficient at responding to real-life inputs will preferentially survive the pruning process and these rules are, therefore, selective for an efficiently operating system, in terms of responding well to those stimuli presented during the developmental process. A primary question regarding this process is what are the mechanisms that could subserve this selective process emphasizing the survival of effective connections and the elimination of less effective ones. Figure 1 shows the magnitude of the reductive phenomena that occur during development. Rakic (Bourgeois and Rakic, 1993) has made the provocative calculation that under normal conditions during development, synapses are lost at the rate of about 5,000/second for several years! This is net loss, and some synapses are being strengthened, i.e. are growing so actual loss is presumably even greater.

Despite these dramatic observations, most discussions of plasticity emphasize the importance of 'instructive' mechanisms, by means of which activated synapses are strengthened. We suggest that a general, non-specific growth process coupled to a 'selective' mechanism of activity-dependent synapse elimination (affecting all non-activated inputs to a cell) would produce the required development of functionally optimal circuitry.

3. SYNAPSE LOSS AT THE NEUROMUSCULAR JUNCTION

A particularly well characterized loss of synapses during development is represented by the conversion of the polyneuronal-innervated skeletal muscle fibers occurring at birth to the exclusively singly innervated muscle cells that are seen 3-4 weeks later in rodent. Redfern (Redfern, 1970) described this process in the rodent in 1970 and it has been much studied since then (Brown et al., 1976). A notable feature of this synapse reduction is that it is dependent on activity in nerve and muscle (Thompson et al., 1979); blockade of activity either by tetrodotoxin (which blocks both pre- and postsynaptic activity) or by curare (which blocks only postsynaptic activity) essentially prevents the normal loss of polyinnervation. Furthermore, the

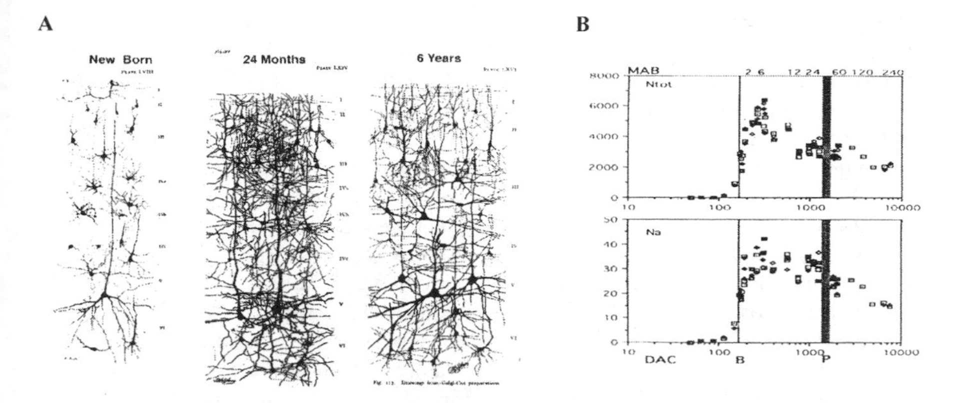

Figure 1. (A) Drawings from Golgi-Cox impregnations of material from the human cerebral cortex at different stages of development (Left: newborn; Center: 24 months postnatal; Right: 6 years post-natal. From J. L. Conel, The Postnatal Development of the Human Cerebral Cortex, Vol 1-8. Harvard University Press, Cambridge MA. 1939-1967. (B) From Bourgeois and Rakic, 1993. (B) Top, Total number (N_{tot}) of synaptic contacts per EM probe. Time is represented on the x-axis on a logarithmic scale. Each point represents the counts in a single vertical probe 13 μm wide and 0.075 μm thick. Bottom, The density (N_a) of synaptic contacts per 100 μm^2 of neuropil as a function of time. At each age, one to six probes were counted and plotted on a semilogarithmic scale. MAC, months after birth; DAC, days after conception; B, birth; P, puberty.

withdrawal of some inputs and the stabilization of others is a competitive process that is depending on activity (Brown et al., 1982; Fladby and Jansen, 1987; Thompson and Jansen, 1977; Van Essen et al., 1990; Balice-Gordon and Lichtman, 1994). A careful examination of the time course of loss of innervating axons and the disappearance of postsynaptic receptors reveals that there is a degree of independence between these two processes. A considerable decrement in polyneuronal innervation occurs at a time when relatively little loss of the postsynaptic acetylcholine receptors can be demonstrated (Figure 2) (Lanuza et al., 2002). On the other hand, there can be a local receptor loss with only a later loss of the corresponding nerve axon (Balice-Gordon and Lichtman, 1993). On the basis of some of these observations and the competitive nature of the nerve and receptor loss, a two-factor model of spatially and temporally discrete synapse stabilizing and synapse reducing effects has been postulated to surround active inputs to a target (Fields and Nelson, 1992; Balice-Gordon and Lichtman, 1993; Nelson et al., 1993). We describe here experiments that have been designed to give some molecular specification to such two-factor theories.

4. AN IN VITRO MODEL SYSTEM OF COMPETITIVE SYNAPSES

A formal requirement for a system capturing the competitive nature of synapse elimination is that two separate populations of presynaptic neurons converge on and innervate a common population of target cells. The Campenot chamber incorporates this property in that the two side chambers of the system contain neurons whose axons project into the common center chamber which can contain the common target populations. In the system as we have constructed it, we have cholinergic motoneurons from the ventral rodent spinal cord in the two side chambers and skeletal muscle cells in the center. Electrodes can be introduced into the system so that selective stimulation of the axons from one side chamber or the other can be accomplished. We can demonstrate that some muscle cells become innervated by axons from both sides either by recording intracellularly from a muscle cell while stimulating axons from one side or the other, or by noting twitch responses from individual muscle cells while stimulating (Jia et al., 1999). In this in vitro preparation it is quite feasible to record responses to selective activation for prolonged periods of time and after treatments of various sorts (electrical stimulation or pharmacological manipulations). Detailed, high resolution morphological analysis of pre- and postsynaptic structures as well as quantal analysis of the synaptic function of the connections under study are possible (Li et al., 2002). Genetically modified animals can be used as the source for the tissue cultured material that goes into the chambers to test some of the molecular hypotheses derived from the pharmacological experiments. Figure 3 shows some of the features of this system along with the equipment for stimulating the preparation chronically.

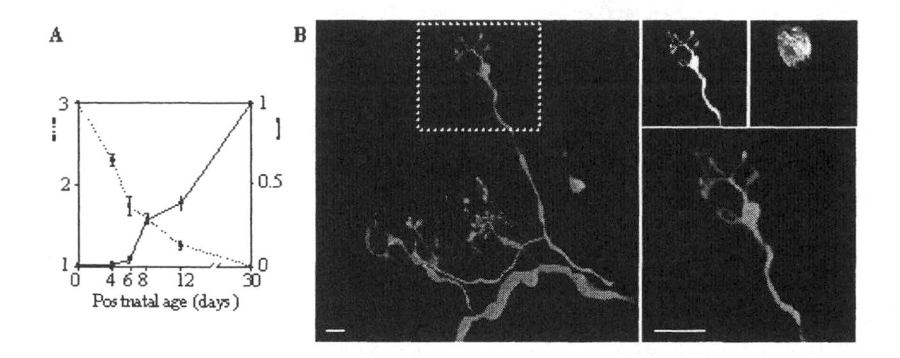

Figure 2. (A) Percentage of synaptic AChR clusters showing AChR loss or redistribution (existence of holes) (continuos line) versus the mean number of axons per synapse (dotted line). From Lanuza et al., 2002. (B) Pool of neuromuscular junctions from a 6-days–old mice in different stages of elimination and a monoinnervated neuromuscular junction with a AChR cluster without holes. Scale bars are 10 μm.

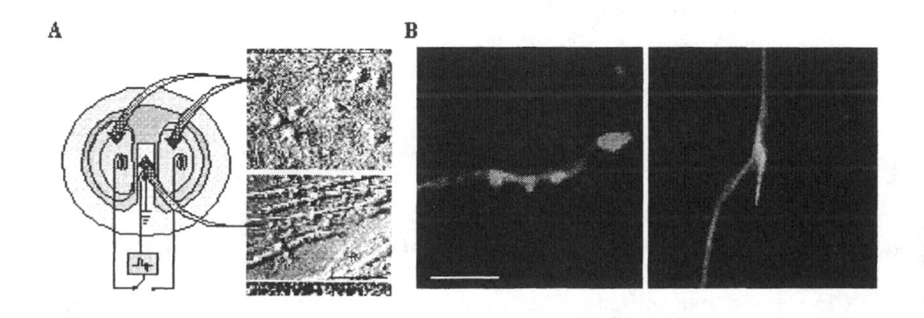

Figure 3. (A) Diagram of Campenot chamber used in our experiments. (B) Two examples of muscle cells innervated by a neurite (green) over a AChR cluster (red). Scale bar in A is 100 μm and in B, 10 μm.

5. HEBBIAN SYNAPSE LOSS CAN BE DEMONSTRATED IN OUR SYSTEM

The basic question concerning the Hebb synapse is this: if a target cell is innervated by two inputs and one of these is stimulated and one is not, what happens to the strength of the connections from the stimulated and from the non-stimulated inputs?. We sought to answer this question regarding our system by establishing that a given myotube was innervated by axons from both side compartments, then stimulating one set of axons and evaluating the changes in synapse strength of axons from the each of the two sides. When we did this by recording evoked end plate potentials (EPPs) as an indicator of synapse strength, we found that unilateral stimulation produced a selective diminution or loss of the NON-STIMULATED input (Jia et al., 1999). There is less or no change in the stimulated input. It may be noted that stimulation of singly innervated synapse produced no detectable loss of synapse efficacy. The loss of the non-stimulated input was primarily due to a decrease in postsynaptic responsiveness. Some of the observations that demonstrate this biology are shown in Figure 4 in which intracellular recordings document the changes in synaptic strength produced by stimulation of one input to a muscle cell. This selectivity in the effect of stimulation on synapse strength was reflected in measurements of changes in the twitch responses and the postsynaptic EPP, as shown in Figure 4. No detectable change in homosynaptic EPP was produced by stimulation, but heterosynaptic down regulation of the EPP did occur (the unstimulated EPP was reduced) (Li et al., 2001). Some diminution of the presynaptic neurite could also be demonstrated, and this seemed to be selective for the unstimulated synapses.

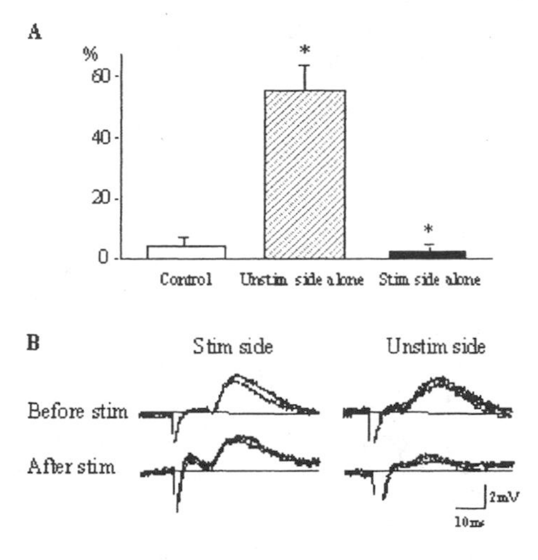

Figure 4. (A) Percentage of functional synapse loss. Those synaptic connections initially effective in producing a twitch in an identified myotube were tested after 20 hours of stimulation. The percentage of those initially effective connections that lost only one input, either the stimulated or the unstimulated input, is plotted. Control: no stim 20 hours: n = 9, N = 475; unstim side and stim side are from same myotube, n = 7, N = 228 pair connections. *p < 0.01, comparing the loss of stimulated and unstimulated inputs. (B) End plate potentials (EPPs) recorded from a doubly innervated muscle. Unilateral electrical stimulation for 1 hour induced EPP reduction primarily of the unstimulated side.

6. THROMBIN IS A SERINE PROTEASE INVOLVED IN SYNAPSE ELIMINATION

Thrombin is the final enzyme in the blood clotting cascade. Its action is to cleave the fibrin precursor fibrinogen to allow polymerization of fibrin to form the blood clot. Thrombin is a highly regulated enzyme, being the product of a multi-enzyme cascade, and in addition to the fibrin production, it has a complex alternative scheme of action that includes a thrombin receptor (a series of integral membrane protease activated receptors or PARs); a peptide fragment produced by thrombin action on these receptors (the fragments are called thrombin receptor activating peptides or TRAPs) and an inhibitor of thrombin, called protease nexin-1 (PN-1). These molecules (Festoff et al., 1991, Liu et al., 1994; Akaaboune et al., 1998, Lanuza et al., 2003) are variously localized to the endplate region and as is the case with thrombin and prothrombin (Kim et al., 1998) are regulated during development and during denervation and re-innervation and are activated by muscle stimulation.

Treatment of muscle cells in vitro with thrombin results in loss of AChR from receptor aggregates (Davenport et al., 2000) and treatment of innervated myotubes with PN-1 or a highly specific inhibitor of thrombin, hirudin, blocks heterosynaptic, activity-dependent synapse loss (Figure 5A) (Jia et al., 1999). Treatment with the TRAP but not an inactive close homologue produces synapse loss. Cholinergic stimulation of muscle induces a secretion of thrombin-like activity into the culture medium (Glazner et al., 1997). Taken together with related findings on the neuromuscular synapse in vivo showing that synapse loss is blocked by hirudin and accelerated by thrombin and the TRAP (Zoubine et al., 1996; Lanuza et al., 2003), these observations suggested to us that the thrombin system might place an important role in mediating activity-dependent synapse reduction. Indeed, it is becoming increasingly evident the system of thrombin and other proteases and a complex series of protease activated receptors are initiating players in a broad range of cell biologic regulatory processes.

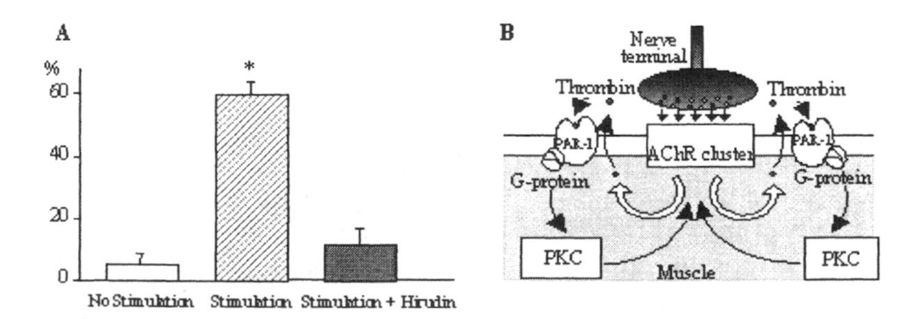

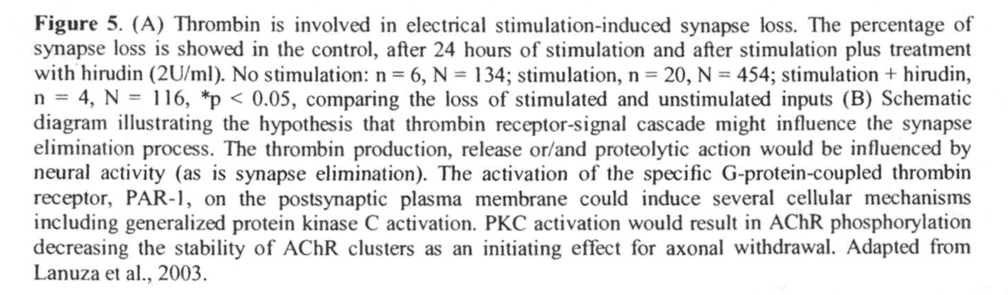
Figure 5. (A) Thrombin is involved in electrical stimulation-induced synapse loss. The percentage of synapse loss is showed in the control, after 24 hours of stimulation and after stimulation plus treatment with hirudin (2U/ml). No stimulation: n = 6, N = 134; stimulation, n = 20, N = 454; stimulation + hirudin, n = 4, N = 116, *$p < 0.05$, comparing the loss of stimulated and unstimulated inputs (B) Schematic diagram illustrating the hypothesis that thrombin receptor-signal cascade might influence the synapse elimination process. The thrombin production, release or/and proteolytic action would be influenced by neural activity (as is synapse elimination). The activation of the specific G-protein-coupled thrombin receptor, PAR-1, on the postsynaptic plasma membrane could induce several cellular mechanisms including generalized protein kinase C activation. PKC activation would result in AChR phosphorylation decreasing the stability of AChR clusters as an initiating effect for axonal withdrawal. Adapted from Lanuza et al., 2003.

Our observations have raised the question of what down-stream agents activated by thrombin might have a more direct effect on the synapse elimination process. One of several cell biologic elements that are influenced by thrombin are some of the G-proteins which are in turn coupled to protein kinase C (PKC) (Figure 5B). We therefore examined the effects of agents regulating PKC as well as another kinase, PKA.

7. PROTEIN KINASE C AND A HAVE ANTAGONISTIC EFFECTS ON SYNAPSE STABILITY

Protein kinase C is activated by the phorbol ester PMA and we found that treatment with PMA produced a profound decrease in synapse strength (Li et al, 2001). This was primarily due to a decrease in postsynaptic responsiveness since the quantal size for evoked EPPs was decreased while the quantal number was unchanged (low Ca^{2+}, by contrast, produces a change solely in quantal number as would be predicted) (Figure 6). It is well known that the action of PMA is transitory, and we found that prolonged, high dose treatment with PMA resulted in a reversal of the down-regulation of synapse strength. In consonance with the decrease in quantal size produced by PMA treatment, such treatment also produced a loss of postsynaptic AChR (Lanuza et al., 2000). We have shown that carbachol stimulation of myotube cultures induces an increase in membrane associated activity of PKC theta (Kim et al., 2002), so that cholinergic stimulation of muscle might be expected to down-regulate postsynaptic responsiveness, and this effect would plausibly be mediated by PKC theta. An elimination scenario can be envisaged in which cholinergic stimulation of muscle releases active thrombin which in turn activates the PAR to initiate a cell biologic cascade involving PKC theta activation (Figure 5B). PKC activity somehow destabilizes the AChR leading to loss of synapse efficacy. The basis for receptor destabilization will be discussed below. This interpretation involving PKC as a crucial mediator of synapse down-regulation is confirmed by the results of experiments with cells from genetically modified mice. We used animals in which the PKC theta gene had been 'knocked out' to obtain muscle cells lacking the kinase. When these myotubes were innervated by normal neurons, the preparations failed to show the heterosynaptic activity dependent synapse reduction seen with wild type muscle (Li et al., 2004).

The scenario presented here can perhaps explain synapse loss but does not provide a rationale for the stabilization of the stimulated input. For this we need to look at other consequences of synapse activation. It is known that a number of molecules beside acetylcholine are released from active nerve terminals at the neuromuscular junction. One of these molecules is the Calcitonin Gene Related Peptide or CGRP. We asked whether this molecule might play a role in selective synapse stabilization. Indeed, treatment with 1 μM CGRP prevents the heterosynaptic loss produced by unilateral stimulation of doubly innervated myotubes (Li et al., 2001). CGRP is known to activate cAMP in myotubes, so we asked whether this molecule could subserve the synapse stabilizing effect and we found that the heterosynaptic decrease in EPP amplitude produced by stimulation was prevented by cAMP treatment (Figure 7). Similarly the loss of synapse efficacy and of AChR density produced by PMA were both prevented by cAMP treatment (Figure 7). cAMP may be presumed to act by activation of PKA and this presumption is confirmed by manipulations of PKA. If PKA is blocked by H-89 or injection of the protein PKA blocker, PKI, the positive effect of synapse activation is lost and

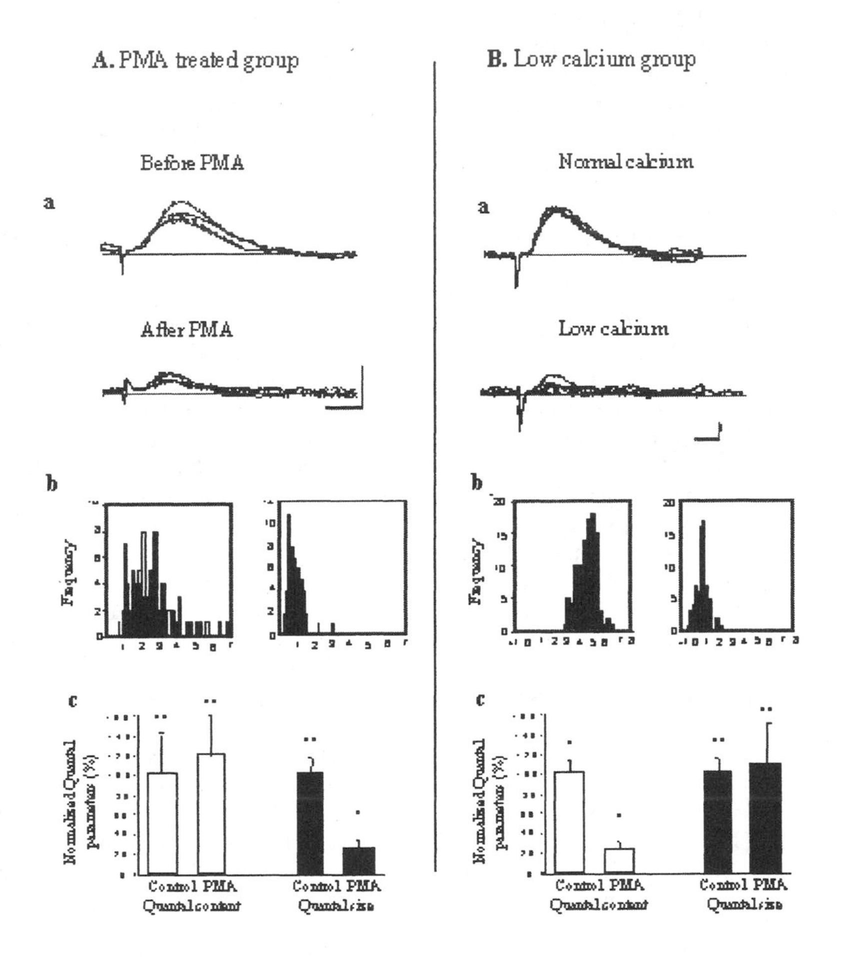

Figure 6. .Effects of PMA and low Ca^{2+} on synaptic transmission. (A, left) PMA effects. (a) Evoked EPPs before and after 2 hours treatment with 100 nM PMA applied to the center chamber. (b) Amplitude histograms of the evoked EPPs. Left: before PMA; right: after PMA. (c) Summary of several experiments similar to that shown in (b) above, confirming the lack of effect of PMA on transmitter output (quantal content) and a major reduction in postsynaptic responsiveness (quantals size). $n = 4$, $*p > 0.8$, $** p < 0.034$. (B, right) Low Ca^{2+} effects. Similar data to those described in (1). $n = 3$, $*p < 0.011$, $**p > 0.0.9$. From Li et al., 2002.

stimulation of an input results in its own down-regulation (Li et al., 2002). Thus, we hypothesize that activation of a synapse has a double effect; a PKC mediated synapse down-regulation and a PKA mediated synapse stabilization. The different spatial distribution of these effects, general and wide-spread, (affecting all synapses) for PKC, and local, (affecting only activated synapses) for PKA explains the stimulus-specific Hebbian quality of the activity-dependent synapse plasticity.

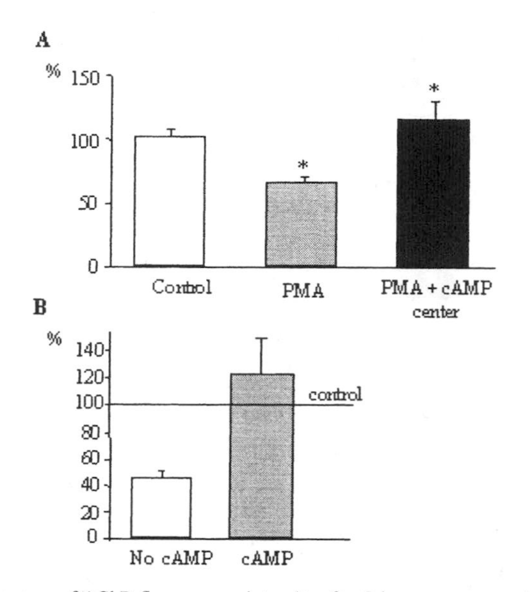

Figure 7. (A) Percentage of AChR fluorescence intensity after 2 hours treatment with PMA (60 nM) and PMA + cAMP (2 mM). cAMP completely prevents the decrement in synaptic AChR produced by PMA ($p < 0.032$). Control: $n = 27$; PMA, $n = 29$; PMA + cAMP, $n = 29$. (B) EPP amplitude of the unstimulated input before and after 2 hours stimulation with and without cAMP (2mM) in the center chamber. For the unstimulated input, cAMP produces a significant block of EPP decrement ($p < 0.004$). Data are expressed in percentage respect to the control.

8. THE KINASES ALSO AFFECT PRESYNAPTIC FUNCTION

The evidence and discussion presented above has involved postsynaptic mechanisms of plasticity, involving the AChR, but by definition, synapse elimination must also involve the presynaptic neurite. We have found evidence for a presynaptic component for the synapse plasticity we are studying (Li et al., 2004). As noted above, PKC theta activity in muscle was essential for expression of activity-dependent synapse down-regulation as judged from experiments with genetically modified animals. When the role of PKC theta in nerve was investigated, we found that preparations in which the muscle was normal, but the nerve was lacking the kinase, also did not show the heterosynaptic down-regulation (Figure 8). We conclude that PKC theta activity in the neuron is essential for the plasticity (Li et al., 2004). However, the activity that is essential is localized in the nerve terminal, or at least kinase activity in the cell body is not involved, since preparations in which the kinase activity in the cell body is increased by incubating the side compartments with PMA show no synapse down regulation. The effect of neuronal PKC theta knockout is dependent on ongoing activity. In preparations lacking neuronal PKC theta, but in which electrical activity is blocked with TTX, PMA treatment does produce synapse down-regulation (Figure 8). This suggests that in the absence of neuronal PKC theta, spontaneous electrical activity produces a synapse sparing effect that blocks the synapse down-regulation resulting from PMA treatment of muscle with normal PKC theta levels.

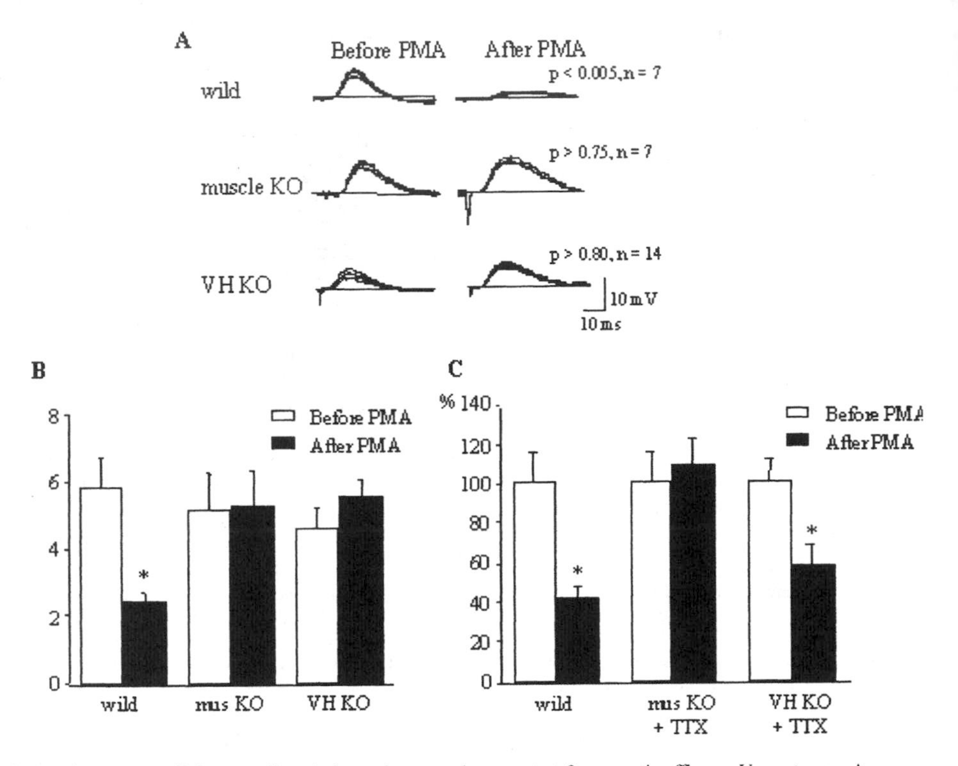

Figure 8. (A) Intracellular recordings in heterologous cultures to test for synaptic efficacy. Upper traces: A preparation made from wild type animals shows marked reduction in EPP amplitude after PMA treatment. Middle traces: A preparation made with normal neurons but with KO muscle cells shows no loss of synapse strength after PMA. Lower traces: A preparation made with normal muscle but KO neurons also fails to show synapse decrement with PMA. (B) Mean EPP amplitude (mV) from several experiments like those shown in (A) confirming the lack of PMA effect in preparations in which either the nerve or the muscle cells are from KO animals. (C) The effect of activity on PMA action in different heterologous culture combinations as revealed by blocking activity with tetrodotoxin. EPPs were measured in myotubes before PMA treatment (open bars) and again, in the same myotubes, after two hours of treatment (filled bars) combining 100 nM PMA and 10^{-7} M TTX. In B and C, the data labeled mus KO were from preparations made with muscle cells from KO animals and nerve from wild animals. VH KO preparations were made with ventral horn neurons form KO animals and muscle from wild animals. *$p < 0.02$. From Li et al., 2004.

While PKC activity in the neuronal cell body is not involved in the expression of our synaptic plasticity, PKA activity in the neuronal cell body is, since application of H-89 (a PKA blocker) in the side compartment produced block of the synaptic stabilization produced by stimulation, with a ca. 15 minutes delay between application of the blocker and loss of stability (Li et al., 2001). That is, stimulation of untreated preparations does not produce down-regulation of the stimulated input, as described above. In preparations incubated with H-89 in the side compartment, stimulation does produce loss of synapse efficacy, but with a slight delay. This suggests that PKA is required in the neuronal soma for production or transport of some material that is necessary to maintain the efficacy of stimulated synapses.

9. ACTIVITY COORDINATES PRE- AND POSTSYNAPTIC CHANGES

As suggested above, we feel that these observations on an in vitro synapse model suggest a partial scenario for the functioning of a Hebbian, activity-dependent synaptic plasticity. The steps in this process can be schematized (Figure 9).

1. Electrical activity in the nerve terminal produces release of acetylcholine and CGRP.
2. This activity also produces an increase in presynaptic PKC which down regulates CGRP release.
3. It also initiates a process in the neuronal soma that serves to maintain transmitter output from the nerve terminal.
4. Released acetylcholine mediates the postsynaptic response, including muscle twitch.
5. The acetylcholine also produces activation generally throughout the muscle cell of PKC theta. This is mediated, at least in part, by activation of thrombin, the thrombin receptor and one or more G-proteins.
6. The CGRP released from the presynaptic neurite activates adenelyl cyclase, cAMP and PKA. This activation is spatially restricted to the immediate vicinity of the active nerve terminal. The increased PKA serves to stabilize the AChR and preserve synaptic efficacy.

In the steady state these several highly regulated processes and molecules are in balance, essentially maintaining a basal synaptic strength. When activity of a given input is altered, these various parameters of the system are all altered since they are a function of activity levels. Depending on the specific kinetics of these processes, any perturbation can result in relatively long lasting changes in synaptic efficacy, and in the limit may result in the elimination of some inputs.

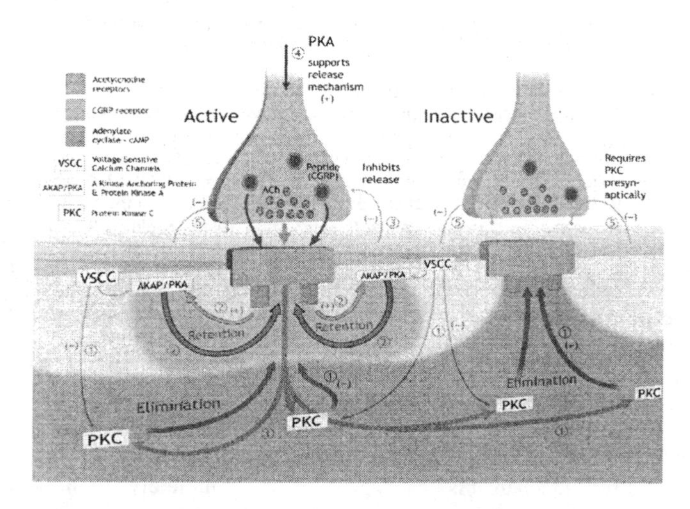

Figure 9. Schematic diagram illustrating the hypothesized opposing kinase actions that generate the selective activity-dependent retention and loss of active and inactive neural inputs, respectively, on a common postsynaptic cell. From Li et al., 2002.

10. MECHANISM OF KINASE ACTION: PHOSPHORYLATION OF THE ACHR

How does activation of the kinases get translated into changes of synaptic efficacy? A good deal of evidence suggests that a partial answer to this question involve the phosphorylation of the AChR itself. It has been well documented that the different subunits of the hetero-pentameric AChR phosphoproteins, are targets of PKA and PKC action and phosphorylation of specific sites on the delta and epsilon subunits affects receptor function or stability. In particular, PKC phosphorylates the delta subunit at a site which decreases stability of the complete receptor when phosphorylated (Nimnual et al., 1998). The physiology of the AChR is affected by the specific phosphorylation mediated by PKA (on the epsilon subunit) (Nishizaki and Sumikawa, 1994), and our observations show that receptor stability is increased by manipulations that increase cAMP or PKA activities. The destabilization produced by PKC activation can be blocked or reversed by PKA activity. It seems rather likely that other mechanisms besides receptor phosphorylation might be involved in activity-dependent changes related with selective synapse elimination. A large number of subsynaptic, cytoplasmic proteins have been identified which serve as a stabilizing structure for the synaptic apparatus (Sanes and Lichtman, 1999). The association of the receptors (direct or indirect) with this synaptic infrastructure may well be critical for maintenance of the synapse and be dependent on a variety of phosphorylation reactions.

This hypothesized PKA/PKC interaction has an inherently competitive aspect. That is, as a given input becomes effective compared to other inputs, its PKC mediated reduction of other ineffective inputs will tend to reduce them even further to the point of complete elimination. Additional nerve-muscle interactions are presumably involved in establishing the final, stable, relatively activity-independent synaptic structure and function.

11. RELATIONSHIP BETWEEN IN VITRO AND IN VIVO RESULTS

Experiments on a neuromuscular preparation in rats and mice give clear evidence for the involvement of thrombin, the thrombin receptor, G-proteins, and PKC in the normal process of synapse elimination in these preparations (Lanuza et al., 2001, 2002, 2003). Treatment in vivo with hirudin delays and treatment with a TRAP accelerates the elimination process. Furthermore, both a thrombin inhibitor and the thrombin receptor are concentrated at the neuromuscular junction. This allows for extensive regulation of thrombin function at the junction. Blockade of the PKC produces an initial blockade of synapse elimination and a later delay (Figure 10A) that is accompanied by a substantial retardation of AChR cluster maturation. This effect of the inhibition of PKC is observed also in PKC theta K.O. animals (the elimination process is slowed by about 3 days) (Figure 10B). In no case, however, does interference or manipulation of the thrombin, thrombin receptor, G-protein, kinase, phosphorylation chain result in permanent alteration of the synapse elimination process. It is clear that there is a substantial redundancy in the system, such that alternative mechanisms can produce a completely mono-neuronal innervation of skeletal muscle when the model we have proposed is not operational.

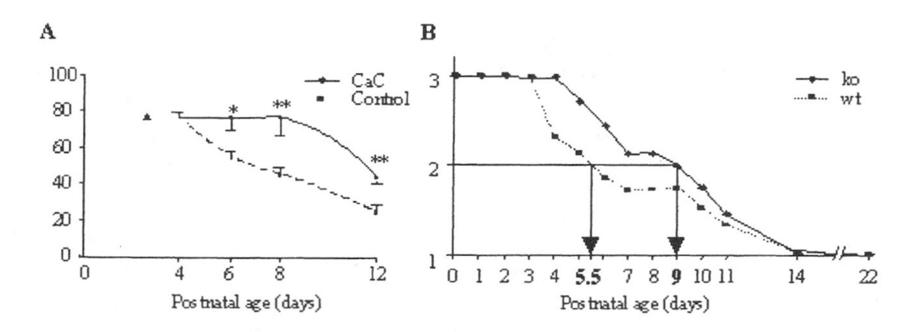

Figure 10. (A) Effect of an extended treatment with CaC on synapse elimination in the period between 4 and 12-days-old. It is represented the percentages of polyneuronal innervation at days 6, 8 and 12 when neonatal Levator auris longus (LAL) muscles were injected daily since day 4 with CaC (200 nM; circles) or PBS (squares). The percentage of polyneuronal innervation on day 4, when the first injection was administered is showed with a triangle. The percentages marked with asterisks are significantly different from the PBS corresponding percentage. In all cases, * $p<0.05$, ** $p<0.005$. From Lanuza et al., 2001. (B) The time course of synapse elimination in normal and PKC theta KO mice in vivo, as measured by the average number of axons contacting the neuromuscular junctions as a function of days postnatal. The arrows show the times at which the half-way point of loss (2 axons per synapse) occurred. From Li et al., 2004.

12. RECEPTOR PHOSPHORYLATION AND CENTRAL SYNAPTIC PLASTICITY

A great deal of evidence has accumulated that central neurotransmitter receptors are phosphoproteins whose functional characteristics are specifically regulated by their state of phosphorylation. Much of this work has focused on the relationship between phosphorylation and changes in receptor function during LTP and LTD, since these are the most widely accepted models for learning and memory. Various members of the glutamate receptor families are thought to be involved in these versions of plasticity and much is now known about the phosphorylation of glutamate receptors on specific amino acid residues by different kinases and under different conditions. For instance, molecular biological methods can produce glutamate receptor 2 molecules that behave as either phosphorylated or non-phosphorylated forms in slice preparations and these different forms of the receptor have an effect on levels of synapse efficacy and on LTP or LTD (Seidenman et al., 2003). PKC phosphorylation of GlutR2 Serine^{880} is a critical event in the induction of cerebellar LTD (Chung et al., 2003) and Serine^{880} of the GlutR2 is critical for a stabilizing interaction between the receptor and the GRIP/ABP proteins in the PDZ (the subsynaptic cytoplasmic proteins that stabilize the postsynaptic structures) (Seidenman et al., 2003). Phosphorylation of Serine^{880} interferes with the receptor/PDZ stabilizing interaction and dephosphorylation facilitates the binding. Producing effectively phosphorylated GlutR2 molecules excludes the receptors from synapses, weakens the synapses functionally and partially occludes with LTD (Seidenman et al., 2003). On the other hand LTD is associated with dephosphorylation of the GluR1 AMPA subunit, while phosphorylation of this subunit accompanies LTP (Lee et al., 2000). Different kinases are involved in these reactions depending on the history of the synapse. Regulation of *N*-methyl-D-aspartate receptors (NMDARs) by protein kinases is critical in synaptic transmission. It has been demonstrated that PKC is involved in NMDAR-dependent LTP and may contribute to enhanced synaptic efficacy at hippocampal CA1 synapses (Hu et al., 1987; Malinow et al., 1989; Hrabetova and Sacktor, 1996; Carroll et al., 1998; Sweatt et al., 1998; Malenka and Nicoll, 1999). In this case, the activation of the PKC occurs through phosphorylation of receptor-associated proteins (Zheng et al., 1999).

An interesting point mentioned above has to do with the relationship between the mechanisms responsible for LTP, LTD or other models of learning and memory and developmental activity-dependent synaptic changes such as the profound alteration of innervation that occurs at the neuromuscular junction. The central nervous system homologue of the NMJ is the change in innervation pattern of layer IV of the visual cortex, in which binocular innervation is replaced by a largely monocular pattern. This is activity dependent and large changes in innervation are produced by manipulating the visual input to the system. One study strongly suggests that some of the same mechanisms are involved in LTD in visual cortex as are responsible for the developmental alterations in connectivity produced by ocular deprivation (Heynen et al., 2003).

Kinase action and phosphorylation reactions they mediate have consequences that may translate into long-term alteration in synaptic structures and functions. As noted above, the stability of neurotransmitter receptors in the postsynaptic membrane is altered by phosphorylation and receptor loss may be irreversible. Clearly additional mechanisms are involved in long-term plasticity. These mechanisms include retrograde signaling from the postsynaptic to presynaptic structures, altering the presynaptic release machinery, as well as postsynaptic cascades resulting in altered gene transcription in response to synaptic activation.

13. REFERENCES

Akaaboune M, Hantai D, Smirnova I, Lachkar S, Kapsimali M, Verdiere-Sahuque M, Festoff BW (1998) Developmental regulation of the serpin, protease nexin I, localization during activity-dependent polyneuronal synapse elimination in mouse skeletal muscle. J. Comp. Neurol. 397:572-579.

Balice-Gordon RJ, Lichtman JW (1993) In vivo observations of presynaptic and postsynaptic changes during the transition from multiple to single innervation at developing neuromuscular junctions. J. Neurosci. 13: 834-855.

Balice-Gordon RJ, Lichtman JW (1994) Long-term synapse loss induced by focal blockade of postsynaptic receptors. Nature 372(6506):519-524.

Bourgeois JP, Rakic P (1993) Changes of synaptic density in the primary visual cortex of the macaque monkey from fetal to adult stage. J. Neurosci. 13(7):2801-2820.

Bourgeois JP, Rakic P, (1996) Synaptogenesis in the occipital cortex of macaque monkey devoid of retinal input from early embryonic stages. Eur. J. Neurosci. 8(5):942-950.

Brown MC, Hopkins WG, Keynes RJ (1982) Short-and long-term effects of paralysis on the motor innervation of two different neonatal mouse muscles. J. Physiol. 329:439-450.

Brown MC, Jansen JK, Van Essen D, (1976) Polyneuronal innervation of skeletal muscle in new-born rats and its elimination during maturation. J. Physiol. 261(2):387-422.

Carroll RC, Nicoll RA, Malenka RC (1998) Effects of PKA and PKC on miniature excitatory postsynaptic currents in CA1 pyramidal cells. J. Neurophysiol. 80(5):2797-800.

Chung HJ, Steinberg JP, Huganir RL, Linden DJ (2003) Requirement of AMPA receptor GluR2 phosphorylation for cerebellar long-term depression. Science 300(5626):1751-1755.

Conel JL (1939-1967) In: The Postnatal Development of the Human Cerebral Cortex. Harvard University Press, Cambridge MA., Vol 1-8.

Davenport RW, Lanuza MA, Kim S, Jia M, Snyder E, Nelson PG (2000) Thrombin action decreases acetylcholine receptor aggregate number stability in cultured mouse myotubes. Brain Res. Dev. Brain Res. 122(2):119-123.

Festoff BW, Rao JS, Hantai D (1991) Plasminogen activators and inhibitors in the neuromuscular system: III. The serpin protease nexin I is synthesized by muscle and localized at neuromuscular synapses. J. Cell Physiol. 147(1):76-86.

Fields RD, Nelson PG (1992) Activity-dependent development of the vertebrate nervous system, Int. Rev. Neurobiol. 34: 133-214.

Fladby T, Jansen JK (1987) Postnatal loss of synaptic terminals in the partially denervated mouse soleus muscle. Acta Physiol. Scand. 129(2): 239-46.
Glazner GW, Yadav K, Fitzgerald S, Coven E, Brenneman DE, Nelson PG (1997) Cholinergic stimulation increases thrombin activity and gene expression in cultured mouse muscle. Brain Res. Dev. Brain Res. 99(2):148-154.
Hebb DO (1949) The organization of behavior. (Wiley, New York).
Heynen AJ, Yoon BJ, Liu CH, Chung HJ, Huganir RL, Bear MF (2003) Molecular mechanism for loss of visual cortical responsiveness following brief monocular deprivation. Nat. Neurosci. 6(8):854-862.
Hu GY, Hvalby O, Walaas SI, Albert KA, Skjeflo P, Andersen P, Greengard P (1987) Protein kinase C injection into hippocampal pyramidal cells elicits features of long term potentiation. Nature 328(6129): 426-429.
Hrabetova S, Sacktor TC (1996) Bidirectional regulation of protein kinase M zeta in the maintenance of long-term potentiation and long-term depression. J. Neurosci. 16(17):5324-5333.
Jia M, Li M, Dunlap V, Nelson PG (1999) The thrombin receptor mediates functional activity-dependent neuromuscular synapse reduction via protein kinase C activation in vitro. J. Neurobiol. 38(3):369-381.
Kim S, Bondeva T, Nelson PG (2002) Activation of protein kinase C isozymes in primary mouse myotubes by carbachol. Brain Res. Dev. Brain Res. 137(1):13-21.
Kim S, Buonanno A, Nelson PG (1998) Regulation of prothrombin, thrombin receptor and protease nexin-1 expression during development and after dennervation in muscle, J. Neurosci. Res. 53:304-311.
Lanuza MA, Li MX, Jia M, Kim S, Davenport R, Dunlap V, Nelson PG (2000) Protein kinase C-mediated changes in synaptic efficacy at the neuromuscular junction in vitro: the role of postsynaptic acetylcholine receptors. J. Neurosci. Res. 61(6):616-625.
Lanuza MA, Garcia N, Santafe M, Nelson PG, Fenoll-Brunet MR, Tomas J (2001) Pertussis toxin-sensitive G-protein and protein kinase C activity are involved in normal synapse elimination in the neonatal rat muscle. J. Neurosci. Res. 63(4):330-340.
Lanuza MA, Garcia N, Santafe M, Gonzalez CM, Alonso I, Nelson PG, Tomas J (2002) Pre- and postsynaptic maturation of the neuromuscular junction during neonatal synapse elimination depends on protein kinase C. J. Neurosci. Res. 67(5):607-617.
Lanuza MA, Garcia N, Gonzalez CM, Santafe M, Nelson PG, Tomas J (2003) Role and expression of thrombin receptor PAR-1 in muscle cells and neuromuscular junctions during the synapse elimination period in the neonatal rat. J. Neurosci. Res. 73(1):10-21.
Lee HK, Barbarosie M, Kameyama K, Bear MF, Huganir RL (2000) Regulation of distinct AMPA receptor phosphorylation sites during bidirectional synaptic plasticity. Nature 405(6789):955-959.
Li MX, Jia M, Yang LX, Dunlap V, Nelson PG (2002) Pre- and postsynaptic mechanisms in Hebbian activity-dependent synapse modification. J. Neurobiol. 52(3):241-250.
Li MX, Jia M, Yang LX, Jiang H, Lanuza MA, Gonzalez CM, Nelson PG (2003) The role of the theta isoform of protein kinase C (PKC) in activity-dependent synapse elimination: evidence from the PKC theta knock-out mouse in vivo and in vitro. J. Neurosci. 24(15):3762-3769.
Li MX, Jia M, Jiang H, Dunlap V, Nelson PG (2001) Opposing actions of protein kinase A and C mediate Hebbian synaptic plasticity. Nat. Neurosci. 4(9):871-872.
Liu Y, Fields RD, Festoff BW, Nelson PG (1994) Proteolytic action of thrombin is required for electrical activity-dependent synapse reduction. Proc. Natl. Acad. Sci. USA 91(22):10300-10304.
Malenka RC, Nicoll RA (1999) Long-term potentiation--a decade of progress? Science 285(5435):1870-1874.
Malinow R, Schulman H, Tsien RW (1989) Inhibition of postsynaptic PKC or CaMKII blocks induction but not expression of LTP. Science 245(4920):862-866.
Nelson PG, Fields RD, Yu C, Liu Y (1993) Synapse elimination from the mouse neuromuscular junction in vitro: A non-Hebbian activity-dependent process. J. Neurobiol. 24:1517-1530.
Nimnual AS, Chang W, Chang NS, Ross AF, Gelman MS, Prives JM (1998) Identification of phosphorylation sites on AchR delta-subunit associated with dispersal of AchR clusters on the surface of muscle cells. Biochemistry 37(42):14823-31482.
Nishizaki T, Sumikawa K (1994) A cAMP-dependent Ca2+ signalling pathway at the endplate provided by the gamma to epsilon subunit switch in ACh receptors. Brain Res. Mol. Brain Res. 24(1-4):341-345.
Redfern PA (1970) Neuromuscular transmission in new-born rats, J. Physiol. Lond. 209:701-709.
Sanes JR, Lichtman JW (1999) Development of the vertebrate neuromuscular junction. Annu. Rev. Neurosci. 22:389-442.
Seidenman KJ, Steinberg JP, Huganir R, Malinow R (2003) Glutamate receptor subunit 2 Serine 880 phosphorylation modulates synaptic transmission and mediates plasticity in CA1 pyramidal cells. J. Neurosci. 23(27):9220-9228.
Sweatt JD, Atkins CM, Johnson J, English JD, Roberson ED, Chen SJ, Newton A, Klann E (1998) Protected-site phosphorylation of protein kinase C in hippocampal long-term potentiation. J. Neurochem. 71(3): 1075-1085.

Thompson W, Jansen JK (1977) The extent of sprouting of remaining motor units in partly denervated immature and adult rat soleus muscle. Neuroscience 2(4):523-535.

Thompson W, Kuffler DP, Jansen JK (1979) The effect of prolonged, reversible block of nerve impulses on the elimination of polyneuronal innervation of new-born rat skeletal muscle fibers. Neuroscience. 4(2):271-281.

Van Essen DC, Gordon H, Soha JM, Fraser SE (1990) Synaptic dynamics at the neuromuscular junction: mechanisms and models. J. Neurobiol. 21(1):223-249.

Zheng X, Zhang L, Wang AP, Bennett MV, Zukin RS (1999) Protein kinase C potentiation of N-methyl-D-aspartate receptor activity is not mediated by phosphorylation of N-methyl-D-aspartate receptor subunits. Proc. Natl. Acad. Sci. USA. 96(26):15262-15267.

Zoubine MN, Ma JY, Smirnova IV, Citron BA, Festoff BW (1996) A molecular mechanism for synapse elimination: novel inhibition of locally generated thrombin delays synapse loss in neonatal mouse muscle. Dev. Biol. 179:447-457.

RECENT ADVANCES IN THE ROLE OF INTEGRINS IN DEVELOPMENTAL AND ADULT SYNAPTIC PLASTICITY

Leslie Sargent Jones*

1. INTRODUCTION

The importance of adhesion molecules like integrins to neural development has been known for a long time (in "neuroscience years"). The role of integrins in adult neural functions such as long term potentiation (LTP), is more recent, only gaining ground in the last decade, not long after the term "integrin" was first coined (Tamkun et al., 1986). The idea that integrins, or something like them, must be participating in the microscopic rearrangements of adult processes like LTP that may recapitulate the larger structural remodeling of embryology seems obvious to many of us now. Interestingly, it was not obvious to everyone initially, and our efforts to publish the first report of an integrin in adult mammalian brain was met with substantial skepticism, requiring further experiments and more evidence before eventual publication (Grooms et al., 1993).

2. BACKGROUND ON INTEGRINS

Integrins, along with selectins, cadherins, and immunoglobulins, are the molecular glue that helps hold tissues together. A moderately complex family of heterodimers, integrins are intramembranous proteins that attach cells to each other, or to the extracellular matrix around the cells. The integrin dimers always consist of an α and β subunit bound together noncovalently. They form a large molecule, like a bobble-head doll, with an oversized, extracellular head, a shorter intramembranous region, and a disproportionately small "tail" protruding into the cytoplasm for linkage with a plethora of possible intracellular molecules that help transduce integrin signaling. The family of possible pairings include at least 18 α subunits and eight β subunits (Giancotti and

* L.S. Jones, Dept. Pharmacology, Physiology & Neuroscience, University of South Carolina School of Medicine, Columbia, SC 29208. E-mail: leslie.jones@schc.sc.edu

Tarone, 2003; Hynes, 2002), though members of each group pair only limitedly with each other (see Figure 1).

Since our own, earlier reviews of the role of integrins in the adult CNS (Jones, 1996; Jones and Grooms, 1997) there has been an exciting explosion of findings on the role of integrins in both the developing and adult CNS, specifically with regard to the structure and function of the synapse. There have also been some other reviews on these and related topics (Hoffman, 1998; Murase and Schuman, 1999; Benson et al., 2000; Milner and Campbell, 2002; Dustin and Colman, 2002), including a very recent and exhaustive review by Clegg et al. (2003), indicating a growing interest in, and understanding of, the importance of integrins to the synaptic dialog.

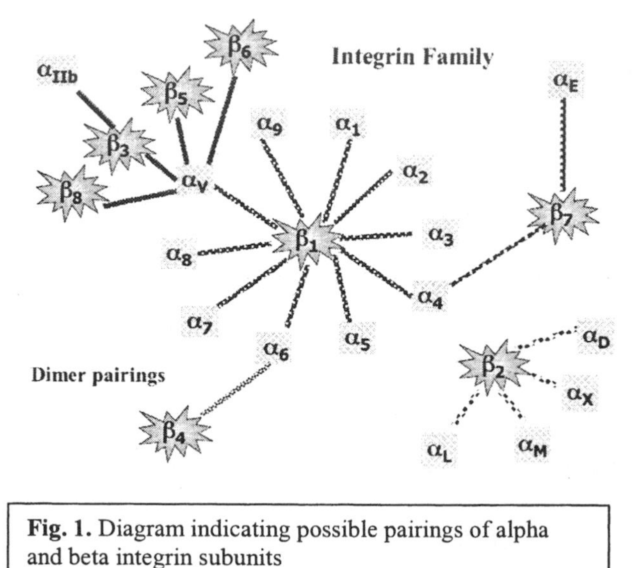

Fig. 1. Diagram indicating possible pairings of alpha and beta integrin subunits

Some of the most tantalizing advances are in the area of integrin signaling. Much more is now known about the structure of integrins and the possible mechanisms for priming and then the activation of integrins for bidirectional signaling between the intra- and extracellular realms. As Humphries et al. (2003) point out in their review, however, the more we learn, the more complex the integrins appear to be, and many new questions have arisen with the new findings. It is now known, for example, that integrins can reside in either an inactive or active state; only when an integrin is in an active state is it available for ligand binding and activation. The switch to the ligand-available state is defined as priming (term used here as in Humphries et al., 2003, to indicate switching from resting, inactive state to the active, available-for-binding state), and details on the priming process and the activation states (denotes that the ligand has bound to the integrin molecule and that the integrin complex is activated to transduce information) indicate that conformational rearrangements occur in each case. Further, cations already known to be critical to synaptic activity, Ca^{++} and Mg^{++}, are also critical to integrin function and these conformational rearrangements between inactive, active, and ligand-bound (Mould et al., 1995). The priming process, whereby the integrin is changed into a ligand-receptive state, is the inside-out signaling; the cell is signaling to the external environment that the cell is contact-ready. Recent reports on this inside-out signaling, however, indicate that switching from the inactive form of an integrin receptor to the active, ligand-available form can depend on the cytoplasmic tails of the α and β components unclasping, in some cases, and can be independent of cationic activation

(Kim et al., 2003). The schematic images of the integrin heads unbending to become available to ligand as the intracellular tails are primed to switch from an attached to a detached and non-interacting conformation (e.g., Kim et al., 2003; Giancotti, 2003; Li et al., 2003) unveil yet another mechanism of cell control of receptor function that is new and intriguing. The implications of this structural mechanism of activation and the potential role of integrin clustering (and perhaps homo-oligomerization) as either a competing or complementary model of integrin activation (Hynes, 2003; Li et al., 2003) is only starting to be explored in the context of synaptic function. This aspect of integrins, the conformational changes and switching from resting, to active, to activated (bound) states is common to integrins generally, of course, and is not specific to the function of integrins at the synapse. It is simply important to remember that integrins have these molecular properties, and to recognize that these signaling mechanisms may have curious ramifications for the synapse.

Since the recent reviews cover the general developments in the field of integrins in nervous system functions, I shall focus here only on the role of integrins in developmental synaptogenesis and adult synaptic plasticity in work that has come out in the new millennium. Hopefully, this narrowly tailored overview of recent advances will illustrate how far we have come in accepting adhesion molecules as critical components of synaptic function.

3. SYNAPTIC INTEGRINS DURING DEVELOPMENT

It has been known for twenty years now that various integrins play essential roles in the normal development of the embryonic nervous system. Neural crest cell migration, in particular, has been well documented as depending on integrin function (Kil et al., 1998; Bronner-Fraser, 1994). Neuroblast migration in the central nervous system also appears to depend on integrins, with examples from the retinal and the olfactory rostral migratory stream serving as exemplary models for this; the recent review of integrin function throughout the nervous system covers this literature in some detail (Clegg et al., 2003). The dependence of embryonic neural development on integrins is evident also from the work of Graus-Porta et al. (2001), where correct development of the laminae and folia of the cerebral and cerebellar cortex was shown to be prevented when β1-integrin expression was disrupted.

Similarly, much is known about the role of integrins in embryonic and post-embryonic neurite outgrowth and axon fasciculation (see Clegg et al., 2003). It is often difficult to be sure of the normal contribution of a particular integrin subunit, however, as cells are adaptable and appear to have "fallback" systems in situations where an integrin, or integrin ligand, is unavailable. For example, in our own study on the role of integrins in regeneration after injury using postnatal, organotypic hippocampal slice cultures, disrupting β1-integrin expression with a function-blocking antibody did not prevent neurite outgrowth in response to a knife cut. Instead, the tissue appeared to switch extracellular matrix molecule expression, and probably the adhesion molecules utilized, and the neurite outgrowth and repair went on unimpaired (Paulman and Jones, unpublished observations). We do not know what the replacement adhesion molecules were, but β3-integrins are a possibility, or completely different adhesion molecules, such as NCAM.

Less is known about integrin function in synaptic development in the postnatal brain. Using the organotypic hippocampal slice culture system as a means of examining neural development in the postnatal timeframe, we have seen that integrins in the β1 family are essential to the normal processes of cell spreading and synaptogenesis, as are any integrins that recognize the Arg-Gly-Asp (RGD) consensus binding sequence (Nikonenko et al., 2003). Blocking β1-integrin function with an antibody, or interfering with RGD-containing ligand binding (with the competitive peptide GRGDSP), resulted in a marked decrease in the number of synapses produced. This decrease was seen despite the fact that the cell density in the tissue was much greater than normal, as the manipulations interfered with the typical spreading and cell migration that is seen in the organotypic slice culture system.

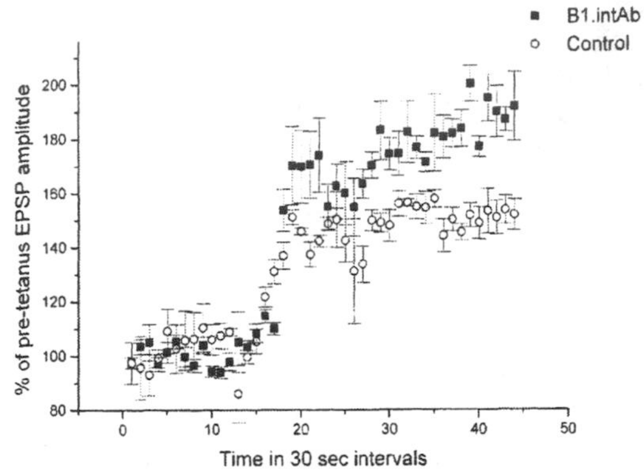

Fig. 2 . Difference in LTP between untreated, control organotypic hippocampal slices and slices exposed to β1-integrin function-blocking antibody during the first "postnatal" week; slices were not in antibody during the experiment. A series of three theta bursts (delivered at 100 Hz) were applied at the 12 minute time-point to induce LTP; potentiation was observed to last for at least 15 minutes in each case (N=3 slices per group).

As an interesting aside, our unpublished studies on the electrophysiology of these cell-dense slices indicated that they were capable of supporting LTP. In fact, the LTP in these slices showed a significantly larger percent change in response amplitude than did LTP in untreated slices (see Figure 2; previously unpublished figure). It is important to note that the integrin expression was not being blocked during the experiment to examine the LTP, only during the development that resulted in the atypical slice anatomy (Nikonenko et al., 1999). Thus, the increase in the amplitude of the LTP is most likely a function of the modified anatomy of the cultured slice. Some of the amplitude difference might be due to the observation that these treated slices exhibit somewhat smaller EPSPs (probably related to lower synapse numbers), and thus the change in EPSP due to LTP, as a percent, appears greater than in normal, larger EPSP-producing slices, but that does not account for all of the increase; it may also be that the tighter packing of the cells makes ephaptic contributions to the LTP of greater impact, and thus offsets the reduction in LTP that was expected given the reduced number of synapses. Another possibility, as yet unexplored, is that blocking access to either β1-containing or RGD-binding integrins during this critical period of "development" (albeit in a culture) caused the neurons and glia to change their adhesion molecule profile, and that the adhesion moieties used instead were actually contributing to a more efficient LTP. Another intriguing possibility is that interfering with RGD-binding integrins prevents normal, activity-dependent synapse maturation, and that this might result in a larger percentage of synapses remaining in the high-probability-of-glutamate-release, immature form, resulting in greater neurotransmitter release upon activation, and therefore greater potentiation. This possibility is based on the findings of Chavis and Westbrook (2001), who reported that the maturation of some glutamate synapses is seen as a switch from a high- to low-probability-of-release state, and that this was dependent on the function of a β3-integrin in the hippocampal neurons they were studying.

Studying the role of integrins in synapse formation during development is somewhat simplified in *C. elegans* and *Drosophila*, where the numbers of subunits and dimers are significantly less. The nematode *C. elegans*, for example, has a single β subunit and only two α subunits, resulting in an integrin dimer that binds RDG containing ligands (α2βpat3) and one that binds laminin specifically (α1βpat3) (Bökel and Brown, 2002). In *Drosophila*, two β subunits and five α subunits provide for more dimer options (Bökel and Brown, 2002), but with so few integrins present, knocking out just one has lethal consequences, so learning the specific functions of integrins in synaptic development is not a simple problem. In *Drosophila*, the so-called "position specific" or βPS integrin (structurally related to β1-integrin) has been shown to be critical to normal neuromuscular junction synapse development postembryonically (Beumer et al., 1999; Rohrbough et al., 2000), and that the integrin effect is mediated through a CaM kinase-II dependent pathway (Beumer et al., 2002). Sone et al. (2000) hypothesized that the βPS-integrin was contributing to synaptic structural development, modification, and control, since the protein is found in the periactive zones surrounding the active zones of neuromuscular junction synapses.

There may be some mechanistic overlap between that hypothesis from *Drosophila* studies and the findings from the more complex organism *Xenopus laevis* that a proteolytic disruption of the integrin-laminin bond, and elimination of the laminin, precedes the formation of neuromuscular synapses in embryonic *Xenopus* tissue in culture, followed by a reformation of β1-integrin bonds, perhaps with other ligands, as new synapses are made (Anderson et al., 1997). This aspect of integrin function in *Xenopus* is further supported by the localization of α3β1 integrin in the active zones of *Xenopus* motor nerve terminals (Cohen et al., 2000). The importance of integrins in *Xenopus* neuromuscular synapse formation was also shown by using an RGD-containing peptide to block the RGD-recognition site of integrins competitively; the results illustrated the role of integrins in modulating acetylcholine release and subsequent synapse formation (Fu et al., 2001). The conserved role of integrins from *C. elegans* to *Xenopus* is further evidence for the integral part these adhesion molecules play in synaptic development.

Further support for the conserved, essential role for integrins comes from studies in mammals where, in mice for example, a specific integrin, α7β1, has been shown to be essential to the formation of the neuromuscular synapse developmentally in an elegant sequence of studies starting before 2000 and continued more recently (Burkin et al, 2000; Burkin et al., 2001). Results from other studies in mouse suggest that β1-integrin may contribute to synapse formation in the cerebellum, as the protein co-localizes with the synaptic protocadherin CNR1 (cadherin-related neuronal receptor 1) in the molecular layer of the cerebellum during the main developmental stage of synaptogenesis (Mutoh et al., 2004). Hama et al. (2004) have also found recently that integrins mediate the astrocyte-neuron contacts that seem to be critical for triggering synaptogenesis (Haydon,

2001) in cultured rat embryonic cortical neurons. At the later time point of the first postnatal week of rats, the presence of RGD-recognizing integrins at the neuromuscular synapse is apparently required for the survival of motor neurons (Wong et al., 1999), pointing to a function for integrins in developing synapses with broader implications than just synaptogenesis. As more information about integrins is gathered from the different stages of development of both vertebrates and invertebrates, and from both neuromuscular and inter-neuronal synapses, additional mechanisms of integrin modulation of synaptogenesis will continue to emerge.

4. INTEGRINS IN ADULT SYNAPTIC PLASTICITY

There is a rapidly growing body of literature that points to an essential part for integrins in the work-horse model for studying changes in synaptic efficacy that may reflect cellular processes occurring in learning and memory: LTP. But first, it has been essential to establish that integrins even exist at the synapse. There has been new evidence every year since 1993 that integrins are indeed present perisynpatically. Recent reviews address the localization of integrins at synapses up through the last few years (Benson et al., 2000; Milner and Campbell, 2002; Dustin and Colman, 2002; Clegg et al., 2003), but a few new findings must be added to the list. In a very interesting study comparing the adhesion molecule profiles of several different synapse types within the hippocampus, Schuster et al. (2001) found a striking difference in the localization of molecules such as NCAM, L1, and β1-integrin. Specifically, the mossy fiber synapses, formed from the dentate granule projections onto CA3 pyramidal neurons, did not contain any β1-integrin that they could detect, whereas β1-integrin was found postsynaptically on many synapses on spines of CA3/CA4 (their terminology) pyramidal neurons, but not on all. The authors could not determine if the integrin was present presynaptically as well. The fascinating question is whether the difference in localization of β1-integrin containing and non-containing synapses has any bearing on the difference in the type of LTP expressed in the two regions, mossy fiber synapse vs. distal CA3 dendritic spine synapse. The authors asked whether the non-NMDA receptor-dependent form of LTP measured at the mossy fiber synapse is also β1-integrin independent, but that the type of LTP that is NMDA-receptor dependent might also be integrin dependent, reflecting the differences in structure, neurotransmitter receptors, and mechanism of these synapses (Schuster et al., 2001).

A valuable line of inquiry about the mechanism of the contribution of integrins to synaptic function focuses on the proteins that help transduce integrin activation within the cell. Integrin-linked kinase (ILK), for example, is essential to the normal function of integrins in neurons, and mutations of ILK result in problems in both *C. elegans* and *Drosophila* like those seen with integrin mutations (for review see Wu and Dedhar, 2001). The ILK protein, through its mediation of the integrin-actin association, may play a key part in the earlier observations of Rosenmund and Westbrook (1993), where blocking actin depolymerization was shown to prevent rundown in NMDA-channel activity. Providing a nexus between these findings and the integrin-dependence of certain types of LTP seems inevitable. The finding from Bahr (2000) that using RGD (to block integrin ligation) resulted in a reduction of the breakdown products usually seen with NMDA-induced cytoskeletal disassembly is a step in that direction. It is also known now that the integrin-laminin interaction essential to neuron survival in hippocampal primary cell culture in turn requires intact ILK function to mediate integrin-induced Akt activation; Akt (named from AKT8 directly-transforming-murine retrovirus) is a kinase linker in this cell survival pathway (Gary et al., 2003). While this function has not been shown to occur through synaptic integrins, the possibility exists that similar mechanisms are in place for mediating some of the mechanisms regulating synaptic survival.

In an ongoing series of studies, Eminy Lee and colleagues have been studying the integrin associated protein (IAP) and its involvement in learning and memory at the behavioral level. Based on their earlier finding that IAP mRNA is increased substantially following a learning task (Lee et al., 2000), they used mice in a follow-up study of inhibitory avoidance to test the effect of blocking IAP function in the dentate gyrus (Chang et al., 2001). They report that this treatment not only significantly impaired memory retention in the mice, but also inhibited both the production or LTP by tetanization, and KCl-stimulated glutamate release (measured by *in vivo* microdialysis); this suggests a potential mechanism of action of IAP in contributing to memory retention. Although IAP is not an integrin *per se*, it is a closely related adhesion protein, and some of the functions of adhesion proteins in synaptic modulation underlying learning and memory may be shared among these proteins. It may be that different adhesion proteins are responsible for modulating synaptic activity at different times in the acquisition/consolidation time frame. The IAP mRNA increase was seen at 3 hours after training, but not later, while the impact of RGD-recognizing integrins on LTP appears to occur at least in the first 30 minutes (Staubli and Scafidi, 1999).

The role of integrins themselves in synaptic plasticity is a developing story with data coming from both invertebrate and vertebrate models. Rohrbough et al. (2000) reported that a *Drosophila* integrin αPS3-integrin (encoded by the gene *volado*) was critical to neurotransmission and plasticity at the neuromuscular junction synapse. They tested this using both genetic mutations of *vol* as well as inhibiting volado protein function acutely with the RGD-containing peptides that competitively block this integrin (as well as many others in mammals). The RGD integrin-receptor antagonist had the effect of interfering with short-term facilitation produced either by Ca^{++} or activity-dependent means, while actually resulting in an increase in synaptic transmission amplitude. Similarly, Wildering et al. (2002) recently examined the role of integrins in synaptic activity in *Lymnaea stagnalis* (pond snail) and reported that, using the fibronectin ligand, or RGD-containing mimics, to activate integrin receptors resulted in significant physiological changes in motor neurons. Agonist binding to integrins produced changes in voltage-gated Ca^{++} currents, intracellular Ca^{++} levels, and pacemaker activity in the neurons. These effects occurred rapidly and resulted in increased neuronal electrical activity. While these effects were measured in whole cells and synapses *per se* were not studied, the increase in Ca^{++} levels in particular fits with the picture of changes that occur with synaptic plasticity and LTP.

In vertebrates the story is largely the same. Chan et al. (2003) used genetically-altered mice to reduce expression of various α integrins, and concluded that α3 and α5 are needed for normal paired-pulse facilitation, but reducing α8 expression in addition to α3 and α5 was also required in order to see maximal impairment of hippocampal LTP and spatial memory. Interestingly, just reducing one or the other integrin had slightly different effects: reduced α3, for example, did not impair the amplitude of NMDA-dependent LTP, but did reduce the duration of the potentiation (making it no longer "long" term). This fits with results from Kramer et al. (2003) who also report that α3 integrin is essential to LTP consolidation in hippocampal slices from rat. Chan et al. (2003) proposed that there is redundancy in the contribution of the various α-integrins known to be present on mammalian neurons in the hippocampus, and that they can limitedly replace each other in producing normal physiological responses, as we concluded in our regeneration studies (*vide supra*). However, the different integrin subunits may normally each be responsible for somewhat different components of the entire LTP response.

Results from studies to determine what stage of LTP integrins might contribute to suggest that the first few minutes following the potentiation-inducing stimulus are when RGD-sensitive, integrin-mediated changes are occurring. Hernandez et al. (2001) and Lebaron et al. (2003) reported that using RGD to block integrin-ligand interactions only worked to impair LTP production if the RGD was present within 5 minutes of the LTP-inducing stimulus. This is somewhat in contradiction to findings also from rat hippocampal slices in which RGD was used to block integrins during potentiation production, but in this case the time period of integrin involvement was reported as more extended (possibly up to 25 minutes) (Staubli et al., 1998). Earlier work from this same, latter group also reported that RGD did not interfere with the early minutes of induction, but only the later stages of LTP consolidation (Xiao et al., 1991). Some of these differences in the reports on the time dependence of the integrin contribution to LTP may be attributable to methodological problems. Given the possibility that multiple integrins contribute to different facets of LTP, it may simply be that different integrins can be perturbed, resulting in the variable observations.

Since Chen and Grinnell (1995) first reported that integrins were responsible for stretch mediated neurotransmitter release at motor neuron terminals of *Rana pipiens* (bull frog) there have been several further reports that help to explain the integrin mediation of both stretch and hypertonicity (Kashani et al., 2001) as synaptic current activators. The findings for hypertonicity have been extended to *Drosophila* (Suzuki et al., 2002) with the elaboration that the RGD-sensitive portion of the hypertonic response (increase in miniature synaptic potential frequency) is dependent on the cAMP/PKA intracellular cascade, and the hypothesis that the hypertonic response actually also is mediated by an integrin-dependent mechanical stress (Kahsani et al., 2001). It is worth exploring the possibility that some of the integrin contribution to changes in synaptic function from other types of stimulation might also be attributable to mechanical stress, and that these observations might obtain in mammals as well.

5. CONCLUSION

The progress in our understanding of the integrin contribution to the synaptic dialog over the last five years has been exciting. From new insights into the part integrins play in synaptogenesis during development to extending the list of functions for integrins in adult synaptic plasticity, the role of the integrin family is expanding rapidly. Particularly intriguing are the findings linking integrins with novel mechanisms of synaptic activation. Also, while the details about the integrins involved in different types of potentiation and depression are not known, nor are the types and locations of the ligands for those integrins, emerging data is filling in those gaps. Some of the sites of adhesion clearly could be across the synaptic cleft, attaching pre- and postsynaptic membranes, but the role of putative neuron-glia attachments has not been examined in this context, and some potentially important mechanisms of synaptic modulation could await discovery in that area. Certainly, the evidence for glial modulation of synaptic function is clear (Newman, 2003; Huang et al., 2004) and becoming increasingly compelling. Probing the role of astrocytes in integrin adhesions and synaptic modulation might uncover new mechanisms for glia-neuron interactions. Questions like these are bound to lead to further findings on the diverse and fascinating functions of integrins at the synapse.

6. ACKNOWLEDGEMENTS

The author gratefully acknowledges Dr. Jeffrey Patton and Dr. Helen Scharfman for their helpful suggestions on this work.

7. REFERENCES

Anderson MJ, Shi ZQ, Zackson SL (1996) Proteolytic disruption of laminin-integrin complexes on muscle cells during synapse formation, Mol Cell Biol 16:4972-84.

Anderson MJ, Shi ZQ, Zackson SL (1997) Nerve-induced disruption and reformation of beta1-integrin aggregates during development of the neuromuscular junction, Mech Dev 67:125-39.

Bahr BA (2000) Integrin-type signaling has a distinct influence on NMDA-induced cytoskeletal disassembly, J Neurosci Res 59:827-32

Benson DL, Schnapp LM, Shapiro L, Huntley GW (2000) Making memories stick: cell-adhesion molecules in synaptic plasticity, Trends Cell Biol 10:473-82.

Beumer K, Matthies HJ, Bradshaw A, Broadie K (2002) Integrins regulate DLG/FAS2 via a CaM kinase II-dependent pathway to mediate synapse elaboration and stabilization during postembryonic development, Development 129:3381-91.

Beumer KJ, Rohrbough J, Prokop A, Broadie K (1999) A role for PS integrins in morphological growth and synaptic function at the postembryonic neuromuscular junction of *Drosophila*,
Development 126:5833-46.

Bökel C, Brown NH (2002) Integrins in development: moving on, responding to, and sticking to the extracellular matrix, Dev Cell 3:311-21.

Bronner-Fraser M (1994) Neural crest cell formation and migration in the developing embryo FASEB J 10:699-706.

Burkin DJ, Kim JE, Gu M, Kaufman SJ (2000) Laminin and alpha7beta1 integrin regulate agrin-induced clustering of acetylcholine receptors, J Cell Sci 113:2877-86.

Burkin DJ, Wallace GQ, Nicol KJ, Kaufman DJ, Kaufman SJ (2001) Enhanced expression of the alpha 7 beta 1 integrin reduces muscular dystrophy and restores viability in dystrophic mice.
J Cell Biol 152:1207-18.

Chan CS, Weeber EJ, Kurup S, Sweatt JD, Davis RL (2003) Integrin requirement for hippocampal synaptic plasticity and spatial memory, J Neurosci 23:7107-16.

Chang HP, Ma YL, Wan FJ, Tsai LY, Lindberg FP, Lee EH (2001) Functional blocking of integrin-associated protein impairs memory retention and decreases glutamate release from the hippocampus, Neuroscience 102:289-96.

Chavis P, Westbrook G (2001) Integrins mediate functional pre- and postsynaptic maturation at a hippocampal synapse, Nature 411(6835):317-21.

Chen BM, Grinnell AD (1995) Integrins and modulation of transmitter release from motor nerve terminals by stretch, Science 269(5230):1578-80.

Clegg DO, Wingerd KL, Hikita ST, Tolhurst EC (2003) Integrins in the development, function and dysfunction of the nervous system, Front Biosci 8:d723-50.

Cohen MW, Hoffstrom BG, DeSimone DW (2000) Active zones on motor nerve terminals contain alpha 3beta 1 integrin, J Neurosci 20:4912-21.

Dustin ML, Colman DR (2002) Neural and immunological synaptic relations, Science 298:785-9.

Fu WM, Shih YC, Chen SY, Tsai PH (2001) Regulation of acetylcholine release by extracellular matrix proteins at developing motoneurons in Xenopus cell cultures, J Neurosci Res 63:320-9.

Gary DS, Milhavet O, Camandola S, Mattson MP (2003) Essential role for integrin linked kinase in Akt-mediated integrin survival signaling in hippocampal neurons, J Neurochem 84:878-90

Giancotti FG, Tarone G (2003) Positional control of cell fate through joint integrin/receptor protein kinase signaling, Annu Rev Cell Dev Biol 19:173-206.

Graus-Porta D, Blaess S, Senften M, Littlewood-Evans A, Damsky C, Huang Z, Orban P, Klein R, Schittny JC, Muller U (2001) Beta1-class integrins regulate the development of laminae and folia in the cerebral and cerebellar cortex, Neuron 31:367-79.

Grooms SY, Terracio L, Jones LS (1993) Anatomical localization of beta 1 integrin-like immunoreactivity in rat brain, Exp Neurol 122:253-9.

Hama H, Hara C, Yamaguchi K, Miyawaki A (2004) PKC signaling mediates global enhancement of excitatory synaptogenesis in neurons triggered by local contact with astrocytes, Neuron 41:405-15.

Haydon PG (2001) GLIA: listening and talking to the synapse, Nat Rev Neurosci 2:185-93.

Hernandez RV, Garza JM, Graves ME, Martinez JL Jr, LeBaron RG (2001) The process of reducing CA1 long-term potentiation by the integrin peptide, GRGDSP, occurs within the first few minutes following theta-burst stimulation, Biol Bull 201:236-7.

Hoffman KB (1998) The relationship between adhesion molecules and neuronal plasticity, Cell Mol Neurobiol 18:461-75.

Huang YH, Sinha SR, Tanaka K, Rothstein JD, Bergles DE (2004) Astrocyte glutamate transporters regulate metabotropic glutamate receptor-mediated excitation of hippocampal interneurons, J Neurosci 24:4551-9.

Humphries MJ, McEwan PA, Barton SJ, Buckley PA, Bella J, Paul Mould A (2003) Integrin structure: heady advances in ligand binding, but activation still makes the knees wobble, Trends in Biochem Sci 28:313-320.

Hynes RO (2002) Integrins: bidirectional, allosteric signaling machines, Cell:110:673-87.

Hynes, RO (2003) Changing Partners Science 300:755-756

Jones LS (1996) Integrins: possible functions in the adult CNS Trends Neurosci 19:68-72.

Jones LS and Grooms SY (1997) Normal and aberrant functions of integrins in the adult central nervous system, Neurochem Int 31:587-95.

Kashani AH, Chen BM, Grinnell AD (2001) Hypertonic enhancement of transmitter release from frog motor nerve terminals: Ca2+ independence and role of integrins, J Physiol 530(Pt 2):243-52.

Kil SH, Krull CE, Cann G, Clegg D, Bronner-Fraser M (1998) The alpha4 subunit of integrin is important for neural crest cell migration, Dev Biol 202:29-42.

Kim M, Carman CV, Springer TA (2003) Bidirectional transmembrane signaling by cytoplasmic domain separation in integrins, Science 301:1720-5.

Kramar EA, Bernard JA, Gall CM, Lynch G (1998) Alpha3 integrin receptors contribute to the consolidation of long-term potentiation, Dev Biol 202:29-42.

LeBaron RG, Hernandez RV, Orfila JE, Martinez JL Jr (2003) An integrin binding peptide reduces rat CA1 hippocampal long-term potentiation during the first few minutes following theta burst stimulation, Neurosci Lett 339:199-202.

Lee EH, Hsieh YP, Yang CL, Tsai KJ, Liu CH (2000) Induction of integrin-associated protein (IAP) mRNA expression during memory consolidation in rat hippocampus, Eur J Neurosci 12:1105-12.

Li R, Mitra N, Gratkowski H, Vilaire G, Litvinov R, Nagasami C, Weisel JW, Lear JD, DeGrado WF, Bennett JS (2003) Activation of integrin alphaIIbbeta3 by modulation of transmembrane helix associations, Science 300:795-8.

Milner R and Campbell IL (2002) The integrin family of cell adhesion molecules has multiple functions within the CNS, Journal of Neuroscience Research 69:286-91.

Mould AP, Akiyama SK, Humphries MJ (1995) Regulation of integrin alpha 5 beta 1-fibronectin interactions by divalent cations. Evidence for distinct classes of binding sites for Mn2+, Mg2+, and Ca2+, J Biol Chem 270:26270-7

Murase S, Schuman EM (1999) The role of cell adhesion molecules in synaptic plasticity and memory, Curr Op Cell Bio 11:549-553.

Mutoh T, Hamada S, Senzaki K, Murata Y, Yagi T(2004) Cadherin-related neuronal receptor 1 (CNR1) has cell adhesion activity with beta1 integrin mediated through the RGD site of CNR1,
Exp Cell Res 294:494-508.

Newman EA (2003) Integrins are involved in synaptogenesis, cell spreading, and adhesion in the postnatal brain, Brain Res Dev Brain Res 140:185-94.

Nikonenko I Toni N Shigeri Y Muller D Jones LS (1999) Effect of an integrin competitive antagonist (GRGDSP) on development in the hippocampal slice culture, Soc Neur Abs 25: 911.2.

Rohrbough J, Grotewiel MS, Davis RL, Broadie K J (2000) Integrin-mediated regulation of synaptic morphology, transmission, and plasticity, Neurosci 20: 6868-6878.
Rosenmund C, Westbrook GL (1993) Calcium-induced actin depolymerization reduces NMDA channel activity, Neuron 10:805-14.
Schuster T, Krug M, Stalder M, Hackel N, Gerardy-Schahn R, Schachner M (2001) Immunoelectron microscopic localization of the neural recognition molecules L1, NCAM, and its isoform NCAM180, the NCAM-associated polysialic acid, beta1 integrin and the extracellular matrix molecule tenascin-R in synapses of the adult rat hippocampus, J Neurobiol 49:142-58.
Sone M, Suzuki E, Hoshino M, Hou D, Kuromi H, Fukata M, Kuroda S, Kaibuchi K, Nabeshima Y, Hama C (2000) Synaptic development is controlled in the periactive zones of *Drosophila* synapses, Development 127:4157-68.
Staubli U, Chun D, Lynch G (1998) Time-dependent reversal of long-term potentiation by an integrin antagonist, J Neurosci 18:3460-9.
Staubli U, Scafidi J (1999) Time-dependent reversal of long-term potentiation in area CA1 of the freely moving rat induced by theta pulse stimulation, J Neurosci 19:8712-9
Suzuki K, Grinnell AD, Kidokoro Y (2002) Hypertonicity-induced transmitter release at *Drosophila* neuromuscular junctions is partly mediated by integrins and cAMP/protein kinase A.
J Physiol 538(Pt 1):103-19
Tamkun JW, DeSimone DW, Fonda D, Patel RS, Buck C, Horwitz AF, Hynes RO (1986)
Structure of integrin, a glycoprotein involved in the transmembrane linkage between fibronectin and actin, Cell 46:271-82.
Wildering WC, Hermann PM, Bulloch AG (2002) Rapid neuromodulatory actions of integrin ligands, J Neurosci 22:2419-26.
Wong KC, Meyer T, Harding DI, Dick JR, Vrbova G, Greensmith L (1999) Integrins at the neuromuscular junction are important for motoneuron survival, Eur J Neurosci 11:3287-92.
Wu C, Dedhar S 2001 (2001) Integrin-linked kinase (ILK) and its interactors: a new paradigm for the coupling of extracellular matrix to actin cytoskeleton and signaling complexes.
J Cell Biol 155:505-10.
Xiao P, Bahr BA, Staubli U, Vanderklish PW, Lynch G (1991) Evidence that matrix recognition contributes to stabilization but not induction of LTP, Neuroreport 2:461-4.

CONSOLIDATION: A VIEW FROM THE SYNAPSE

Christine M. Gall and Gary Lynch*

1. INTRODUCTION

Memories are easily disrupted shortly after their formation but then become progressively more stable over time. Beginning in the late 19th century, researchers (Ribot, 1882; Muller and Pilzecker, 1900) have generally interpreted this to mean that newly acquired material goes through a consolidation period lasting for several minutes or longer. Strong support for this argument, and a general paradigm for testing it, came in the late 1940's with the discovery that electroconvulsive seizures erase recent memory in rodents (Duncan, 1948). When, shortly afterwards, it was found that various drugs reverse memory with an efficiency that is inversely related to time after training, the intuitively attractive idea that memory needs time to stabilize seemed solidly grounded. However, a report in 1968 issued a strong, and what proved to be lasting, challenge to the consolidation hypothesis (Misanin et al., 1968). Specifically, this study found that memory-reversing seizures were fully effective *days* after training if they were preceded by the cue that had been learned, an observation suggesting that reactivation of memory makes it vulnerable to disruption. The reactivation argument, which itself has recently been reactivated by new results (Nader, 2003), led to an extended and continuing debate on whether the disrupting manipulations affect encoding, re-encoding, retrieval, or still other psychological variables that contribute to stable memory. Resolution of these issues, particularly given that animals appear to have multiple memory systems, does not appear to be imminent.

Shortly after its initial description, long-term potentiation (LTP) was discovered to have a consolidation period with temporal properties that align reasonably well with those expected from the initial memory-erasure (seizure) studies in rodents. Specifically, potentiation, generated with high frequency stimulation in anesthetized rats, was quickly

* Christine M. Gall, Dept. Anatomy & Neurobiology, Univ. California, Irvine CA 92697-4292. Gary Lynch, Dept.of Psychiatry & Human Behavior, Univ. California, Irvine CA 92612-1695. E-mail: cmgall@uci.edu

and fully reversed by low frequency stimulation applied within a few minutes of induction (Barrionuevo et al., 1980). This effect is synapse-specific, in that reversing LTP does not disturb neighboring contacts, and the reversing stimulation has only transient effects when applied to control synapses. Normal sized LTP is readily elicited after reversal, as expected if all traces of prior potentiation had been eliminated by the low frequency stimulation (Barrionuevo et al., 1980). Together these results establish that reversal works either by erasing the synaptic state that expresses LTP before that state can be stabilized or by interrupting still active processes that produce stability. Chronic recording confirmed the results from anesthetized animals, established that LTP does not spontaneously recover even after delays of 24 hours (as expected if erased), and showed that the stimulation needed for reversal does not create physiological disturbances in freely moving rats (Staubli and Lynch, 1990). Other studies obtained reversal at various sites in hippocampus and cortex of intact animals (e.g., (Burette et al., 1997; Martin, 1998; Froc et al., 2000).

The first in vitro studies of LTP consolidation mimicked the behavioral paradigms in that global disruptions of physiology were administered at various intervals after potentiation had been induced. Hypoxia of a duration just sufficient to transiently block synaptic responses completely eliminated LTP when applied within the first few minutes after induction but was without effect 30 minutes later (Arai et al., 1990). This effect was blocked by antagonists of the adenosine A1 receptor, indicating that the extracellular build-up of the purine during hypoxia was responsible. As predicted by this hypothesis, infusion of adenosine during the first 5 minutes after high frequency stimulation reversed potentiation (Arai et al., 1990); the link between adenosine and the reversal effect has been obtained in several slice experiments (Abraham, 1997; Huang et al., 2001). Good evidence is also available that reversal depends on NMDA receptors (Huang et al., 2001; Kramár and Lynch, 2003) and group I metabotropic glutamate receptors (Wu et al., 2004), requires a particular isoform of calcineurin (Zhuo et al., 1999), and is prevented by inhibitors of PP1–type protein phosphatases, but not by PP2A or PP2B inhibitors (Huang et al., 2001). Possibly related to this, reversal is accompanied by dephosphorylation of AMPA receptors at Ser 831 (but not Ser 845) (Huang et al., 2001), an effect that could relate to a change in the size of the active AMPA receptor pool (see below).

Reversal is readily obtained with naturalistic theta patterns of afferent stimulation (i.e., 5-7 Hz) (Larson et al., 1993; Barr et al., 1995) but, as expected from the work with less discrete manipulations, such stimulation becomes progressively less effective for reversing LTP over a 30-minute post-induction period (Staubli and Chun, 1996; Huang et al., 1999). With regard to the argument by Sarvey and colleagues that norepinephrine promotes plasticity at various sites in hippocampus (Stanton and Sarvey, 1987), the monoamine was found to significantly enhance reversal by modest trains of low frequency stimulation (Larson et al., 1993). Given that the conditions needed for reversal are not far removed from neuronal activity patterns recorded in freely moving animals, it is possible that reversal occurs during behavior. Two lines of recent evidence accord with this idea. First, exploration of a novel environment reverses LTP in field CA1 if the behavior occurs during the immediate post-induction period (Xu et al., 1998). Second, LTP is deficient in slices expressing spontaneous (autonomous) 1-3 Hz oscillations referred to as 'sharp waves' but is fully restored by adenosine A1 receptor antagonists at concentrations that block LTP reversal (Colgin et al., 2004).

In all, reversal studies support the 'classical' consolidation hypothesis by showing that a form of synaptic plasticity, related by diverse lines of evidence to commonplace memory, does in fact pass through a period, beginning within seconds of induction, in which it becomes progressively more resistant to disruption. Work on LTP is also beginning to explain how consolidation occurs and why it follows the time course it does. This area of research is largely concerned with finding cellular processes set in motion by a few hundred milliseconds of afferent activity and that, within minutes, produce changes that are extraordinarily persistent. It is also the case that the sought after processes must satisfy these requirements without violating LTP's synapse specificity rules. The argument to be made here is that a largely overlooked but vital component of synapses, namely cell-cell and cell matrix adhesion, contains the ingredients needed for an hypothesis that satisfies the above constraints.

2. INTEGRINS: SYNAPTIC ADHESION RECEPTORS THAT CONTRIBUTE TO CONSOLIDATION

Integrins are transmembrane heterodimers (αß) that mediate cell-cell and cell-matrix adhesion throughout the body (Giancotti and Ruoslahti, 1999; Milner and Campbell, 2002). Eighteen α and eight ß subunits have been identified and are known to combine to form over 20 different integrin receptors. Integrin subunit composition determines both the ligand specificity and the signaling properties of the receptors (Miranti and Brugge, 2002). Many integrin ligands are expressed in brain (e.g., tenascin, laminin, fibronectin, reelin, L1, and ADAMs) (Milner and Campbell, 2002) most of which contain the Arginine-Glycine-Aspartate (RGD) consensus sequence (Ruoslahti and Pierschbacher, 1986; Plow et al., 2000) recognized by the integrins so far found in hippocampus. The functional contributions of RGD-binding integrins are commonly studied using synthetic peptide ligands (e.g., GRGDSP, GRGDNP) that bind to the extracellular domain of the receptors and thereby modify their intracellular operations.

Integrins allow cells and their processes to adhere to the extracellular matrix, an operation that is essential to growth, migration, and aggregation. They accomplish this by attaching their extracellular domains to the matrix and their intracellular extensions to the submembrane actin network: Integrins thus crosslink the skeleton of the cell with the stable matrix lying immediately beyond the surface membrane (Geiger et al., 2002). Once attached to the matrix, integrins initiate potent intracellular signaling cascades (Giancotti and Ruoslahti, 1999; Miranti and Brugge, 2002; Schwartz and Ginsberg, 2002). Two related tyrosine kinases -- Proline Rich Tyrosine Kinase 2 (Pyk2) and Focal Adhesion Kinase (FAK) --- physically associate with the c-terminus of the integrin ß subunit and are activated by receptor ligation or clustering (Richardson and Parsons, 1995; Schlaepfer and Hunter, 1998; Vuori, 1998; Wu et al., 2001; Loeser et al., 2003). The two kinases trigger signaling pathways that (a) regulate actin polymerization and cytoskeletal organization, (b) interact with signaling set in motion by trophic factor receptors, and (c) modify gene expression. The first of these functions is intimately related to the formation of adhesion junctions while the second two allow adhesion to interrupt cycling and promote differentiation.

Integrins are dynamic in that they can be converted from a low to a high ligand affinity state in a matter of minutes (Hughes and Pfaff, 1998; Travis et al., 2003). The consequences of this are well illustrated in migration where activated integrins at the front of the cell form adhesions to the matrix, initiate (through FAK signaling) filament formation, and provide the anchors needed for traction. Adhesion disassembly at the trailing edge of the cell, an essential part of migration, is initiated by a rise in intracellular calcium, possibly followed by activation of the calcium-regulated phosphatase calcineurin and the calcium-activated protease calpain (Huttenlocher et al., 1997). Integrin activation and inactivation are thus continuous and routine events in motile cells but ones that can be adjusted to produce extremely persistent structures, as in the case of cells that become stationary (Ridley et al., 2003). The hypothesis described here posits that activation and inactivation are also part of the life of the synapse and provide the means for rapidly producing long lasting changes.

The first evidence for functional integrins at synaptic junctions in adult brain came from studies demonstrating binding of integrin ligands to synaptosomal membrane fractions (Bahr et al., 1991). Specific binding of the matrix protein fibronectin to membranes was more than four-fold greater than binding to homogenate samples and was markedly reduced by co-incubation with GRGDSP peptide. Additional work showed that synaptosomes from adult hippocampus adhere to immobilized vitronectin or fibronectin, much like dissociated cells, but not to fibronectin lacking the RGD sequence (Bahr et al., 1991). Affinity chromatography with fibronectin columns was then used to purify to near homogeneity a polypeptide from synaptosomal membranes that both recognized the RGD sequence and was labeled by antibodies against the $\alpha 5\beta 1$ integrin (the fibronectin receptor) (Bahr and Lynch, 1992). These results provided presumptive evidence that synapses in mature hippocampus are enriched in an integrin-like protein that is competent as an adhesion receptor. Subsequent work confirmed these points and showed that the protein co-purified with synaptic markers including the GluR1 and GluR2/3 subunits of the AMPA receptors (Bahr et al., 1997).

Initial tests for integrin involvement in LTP examined the effects of RGD-containing peptides on potentiation of Schaffer-collateral synapses in the apical dendrites of CA1. Bath application of GRGDSP blocked the stabilization but not the induction of LTP: potentiation was initially comparable to that recorded in control slices but returned to baseline levels by about 60 minutes (Staubli et al., 1990; Xiao et al., 1991; Bahr et al., 1997; Staubli et al., 1998). Similar results were obtained in later studies (Chun et al., 2001) using disintegrins, a collection of very potent RGD-containing integrin ligands purified from snake venom (Huang, 1998). These compounds were effective when applied immediately after induction, confirming that they affected processes other than induction and initial expression (Chun et al., 2001). It has also been reported that RGD peptides block LTP when applied after immediately induction (Bahr et al., 1997; LaBaron et al., 2003) but with decreasing efficiency over the subsequent 30 minutes (Staubli et al., 1998). The similarity in time courses between the disruptive effects of integrin binding and that for LTP consolidation supports the idea that the two effects are related.

3. MULTIPLE INTEGRINS ARE INVOLVED IN LTP CONSOLIDATION

Different types of cells express different subsets of the more than 20 species of integrins and this appears to be an important factor in how they terminally differentiate. Systematic mapping of integrin mRNA expression in brain (Pinkstaff et al., 1999) showed that subunit expression profiles also differ markedly across regions and cell types. Some prominent examples are illustrated in Figure 1 that shows integrin subunit cRNA labeling in sections through hippocampus, cerebellum, and striatum. Clearly, some integrin proteins are narrowly distributed (α4) while others are broadly expressed (αv), and a given neuronal population expresses multiple integrins (see CA3 in panels A and B, and piriform cortex in panels F and G). Even within structures there are notable cell type differences. For example, within hippocampus α1 is expressed by the CA3 pyramidal cells alone, α7 mRNA is expressed by the granule and CA3 pyramidal cells, while α8 mRNA is most prominently expressed in field CA1. In cerebellum, the Purkinje cells express α3 but notα 7 mRNA while the latter subunit is present in high concentrations in the subjacent granule cells. Note also that α7 mRNA is found in both cerebellar and hippocampal granule cells, two neuronal populations that share a number of peculiar features. These observations encourage the idea that differentiation of cell types in brain, as in other parts of the body, is guided by differential expression of integrins. And, given the results described earlier, these distributions raise the possibility that region-specific integrin activities underlie marked regional differences in synaptic plasticity, a point that will be returned to shortly.

Immunocytochemical studies have confirmed that individual neurons contain a suite

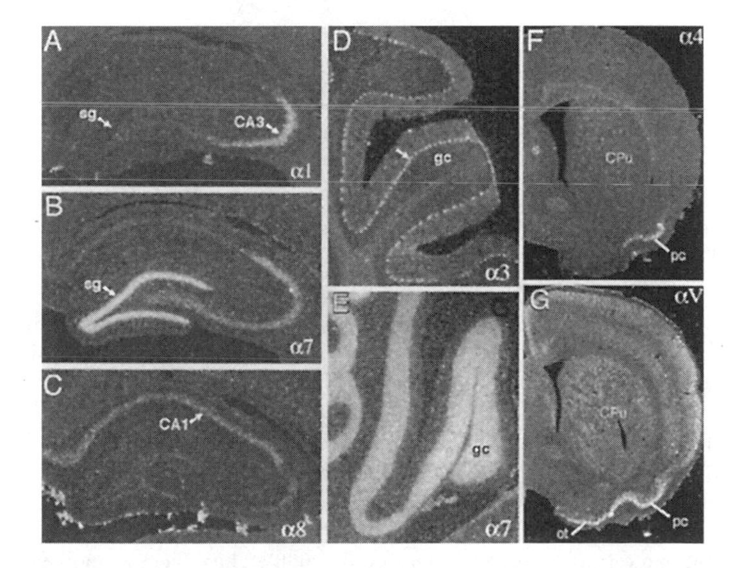

Figure 1. Integrin gene expression is regionally differentiated in adult brain. Photomicrographs show the autoradiographic localization of mRNAs encoding select α integrin subunits in adult rat brain (^{35}S-cRNA in situ hybridization; dark field microscopy, labeling appears as white puncta). A-C) Sections through rostral hippocampus show that α1 (A), α7 (B) and α8 (C) transcripts are differentially distributed across the major subfields with CA1 pyramidal cells prominently expressing α8, CA3 pyramidal cells expressing α7 and α1 and the dentate granule cells (sg) expressing α7. D,E) Sections through cerebellar cortex show that α3 is expressed by the Purkinje cells (arrow in D) but not the granule cells (gc); in contrast α7 (E) is expressed at very high levels in the granule cell layer. F,G) Sections through striatum show α4 (F) and αv (G) cRNA labeling: α4 is expressed by a few discrete neuronal populations in brain including those in layer II piriform cortex (pc) whereas αv mRNA is broadly distributed across much of forebrain including the caudate/putamen (CPu) neocortex, and olfactory tubercle (ot) as well as layer II piriform cortex.

of integrins. Pyramidal cells in field CA1 express ß1 (Grooms et al., 1993), ß8 (Nishimura et al., 1998), α5 (Jones, 1996; Bi et al., 2001; King et al., 2001), α8 (Einheber et al., 1996) and αv (Nishimura et al., 1998) subunits. *In situ* hybridization indicates that α3, α5, α8, αv, ß1 and ß5 mRNAs are present at moderate to high levels while α1, α2, v4, α6, α7, ß2, ß3 and ß4 transcripts are at low to undetectable levels (Pinkstaff et al., 1997; Pinkstaff et al., 1999). More recent studies have identified transcripts for subunits ß3 and ß8 in field CA1 of neonatal and adolescent rats and have shown that α1 and α6 expression can be induced, and mRNA levels for other subunits increased, in that region by seizures (Gall and Lynch, 2004). These findings make the important point that integrin expression in brain is malleable, an effect shown in other tissues to have important consequences to cell functioning. Using known rules for subunit dimerization (Milner and Campbell, 2002), the subunits expressed in field CA1 would be expected to give rise to 8 different receptors of the "ß1" and "αv" integrin families: α3ß1, α5ß1, α8ß1, αvß1, αvß3, αvß5 and αvß8 (Fig. 2 *diagram*). Among these, all but α1ß1 are RGD-binding and all but α1ß1 and αvß5 binds fibronectin (Milner and Campbell, 2002); the 7 RGD-binding integrins all are candidates for mediating the effects on synaptic physiology described above. It bears repeating that the CA1 integrin profile differs from that found in other brain regions or even in other subfields of hippocampus.

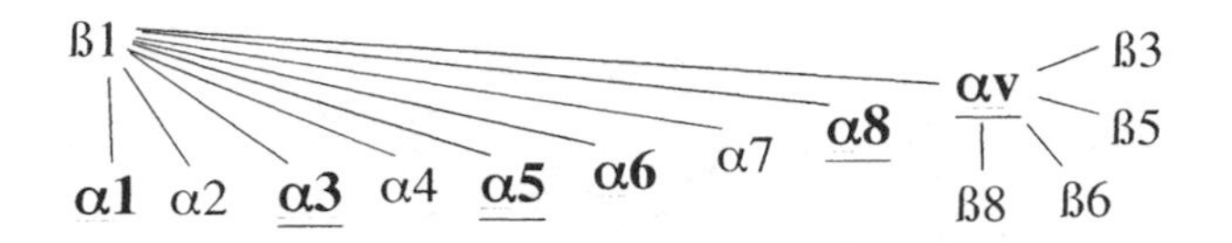

Figure 2. Schematic showing subunit pairings that comprise the ß1 and αv integrin families. Lines indicate integrin αß partners. Pairs with the α subunit with a single underline are expressed in CA1 pyramidal cells: pairs with the α subunit double-underlined are RGD-binding.

It is likely that many, if not most, of the above integrins are present in spine synapses and therefore in the proximity of neurotransmitter and scaffolding proteins involved in LTP. Immunocytochemical studies localized α1, α8, and ß1 immunoreactivity (ir) to all neuronal components including spines (Einheber et al., 1996; Murase and Hayashi, 1996; Murase and Hayashi, 1998) and found punctate αv- and ß8-ir throughout the cortical neuropil (Nishimura et al., 1998) along with dense α5-ir in cortical spines (Bi et al., 2001). Immuno-electron microscopy confirmed that spines contain concentrations of α8 (Einheber et al., 1996), ß8 (Nishimura et al., 1998) and ß1 (Schuster et al., 2001) in hippocampus and α3 in cortex (Rodriquez et al., 2000). Thus, there is good evidence that some number of the CA1 integrins are present in dendritic spines.

The results just described inevitably raise the question of whether multiple integrins contribute to the consolidation effects obtained with RGD peptides and disintegrins. Tests of this used antibodies that block the functioning of specific integrin species. In the first study, local infusion of a neutralizing antibody to α5 integrin, during and after the period of stimulation, had no effect on initial LTP but led to a slow decay (-35% at 45 minutes) in the degree of potentiation thereafter. This decline was much more gradual than that obtained with broad spectrum integrin ligands (e.g., GRGDSP, disintegrins), suggesting that more than one integrin dimer contributes to the stabilization of LTP. This conclusion was supported in a subsequent study (Kramár et al., 2002) that introduced a new paradigm for probing the status of LTP consolidation. Assuming that a given treatment blocks consolidation, then it should leave LTP vulnerable to reversal at those time points (e.g., 30–60 minutes post-induction) at which potentiation is normally stabilized. The degree to which low frequency stimulation reduces LTP at these time points provides a simple measure of how effectively the experimental manipulation stops consolidation. The great advantage of the reversibility-test protocol is that it removes the necessity of recording for protracted periods, thereby avoiding the stability problems inherent in work with acute slices. Accordingly, control sites were locally infused with aCSF and experimental sites with anti-$\alpha3$ integrin, theta burst stimulation was applied to induce LTP, and the rate of potentiation decline was compared at the two sites for 30 minutes. After this interval, one-minute trains of 5Hz (theta pulse) stimulation were applied in an attempt to reverse LTP at control and experimental recording loci. Potentiation at the two recording sites was not detectably different at 5 minutes post-induction but then declined more rapidly in the antibody-treated field over the next 25 minutes. Theta pulse stimulation had no lasting effects on potentiation at the control site but erased more than 50% of LTP at loci treated with anti-$\alpha3$ integrin antibody.

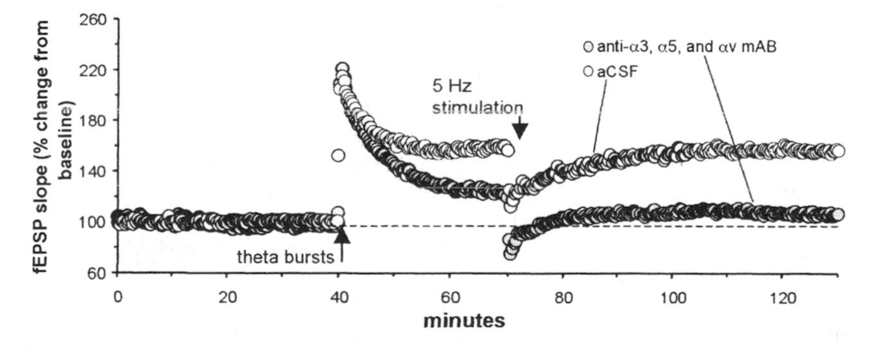

Figure 3. Function blocking antibodies against hippocampal integrins block the consolidation of LTP. Neutralizing antibodies against $\alpha5\beta1$, $\alpha3\beta1$, and $\alpha v\beta3$ integrins were locally applied to a population of Schaffer-commissural synapses in the apical dendrites of hippocampal field CA1. LTP was then induced with a train of theta bursts. LTP decayed steadily throughout the subsequent 30 minutes with its residual component being completely reversed by a 5Hz train of stimulation. (Adapted from Kramár and Lynch, 2002).

These findings suggest that multiple hippocampal integrins, including $\alpha3\beta1$ and $\alpha5\beta1$, work together to consolidate LTP. The reversal paradigm was used to further test this idea (Kramár and Lynch, 2003). Results obtained with a cocktail of neutralizing antibodies against three integrins ($\alpha3$, $\alpha5$, αv) are plotted in Fig. 3; as shown, this treatment caused LTP to decay steadily and left the potentiation remaining at 30 minutes completely vulnerable to reversal by 5 Hz stimulation. These results reinforce the conclusion that multiple integrin species combine their signaling activities to produce the full consolidation effect. Further evidence in support of this was obtained studies using transgenic mice (Chang et al., 1999). Briefly, reduced $\alpha3$ expression resulted in a significant depression in LTP as assessed 60 min after stimulation: potentiation immediately after stimulation was comparable in mutant and wild type mice. Animals with both $\alpha3$+/- and $\alpha5$+/- mutations had much greater impairments in LTP, with no evident loss of initial potentiation, while stable LTP was eliminated in triple transgenics ($\alpha3$, $\alpha5$ and $\alpha8$ heterozygotes). The triple transgenics behaved normally during 7 days of training in the Morris Water Maze but were significantly deficient in spatial memory probe trials. Related to this last point, RGD peptides block consolidation in the marine invertebrate *Hermissenda crassicornis* (Epstein et al., 2004), a result that, with the transgenic mouse findings, significantly strengthens the hypothesis that integrins are critical to the stability of a broad range of memory types.

There are precedents for multiple integrin species working together to produce a biological endpoint, though none are described for brain. It is likely in these cases, both from what is known of integrin biology in general and from specific experimental observations, that the different receptors initiate events in common and separately initiate diverse signaling cascades. How might this work in the case of LTP? Figure 4 describes one possibility. In this model, induction events for LTP cause the synapse to quickly move (rate constant a_1) from its baseline condition into a stage #1 configuration and then, somewhat more slowly (rate constant a_2), into a stage #2 configuration. These events are reversible with the decay rate (b_1) from stage #1 to baseline being substantially faster than the rate constant (b_2) for the return from stage #2 to stage #1. Integrin driven signaling stabilizes the two configurations. If stage #1 is <u>not</u> consolidated, it will decay back to baseline before it can morph into stage #2 (unconsolidated: b_1 is faster than a_2); if stage #1 is consolidated, then it transfers to stage #2 before it can revert to baseline (consolidated: b_1 is slower than a_2). Any of three integrins in the model can activate the homologous kinases Pyk2 and FAK and it is assumed that this common action is sufficient to consolidate stage #1. Note that this means that blocking any one integrin does not prevent consolidation of stage #1 because the remaining two species are still free to activate the necessary kinases. Blocking all three integrins, however, prevents consolidation of stage #1 and thus allows b_1 to bring the altered synapse back to baseline *before it can transition to stage #2*. Under these conditions LTP dissipates at the rate of the unconsolidated b_1 (i.e., steadily over 30-40 minutes).

In addition to their common actions, each of the three integrins in the model activates its own signaling cascade. All three of these must converge in order for stabilization in configuration #2 to occur; blocking any one species will result in partial consolidation and thus cause the synapse to slip slowly back towards configuration #1 (rate constant b_2 is slow) and the baseline.

Ideas of this kind, and several variants can be imagined, bear some resemblance to the multiple stages of consolidation proposed for memory. A later section will consider the types of neurobiological processes that might implement them.

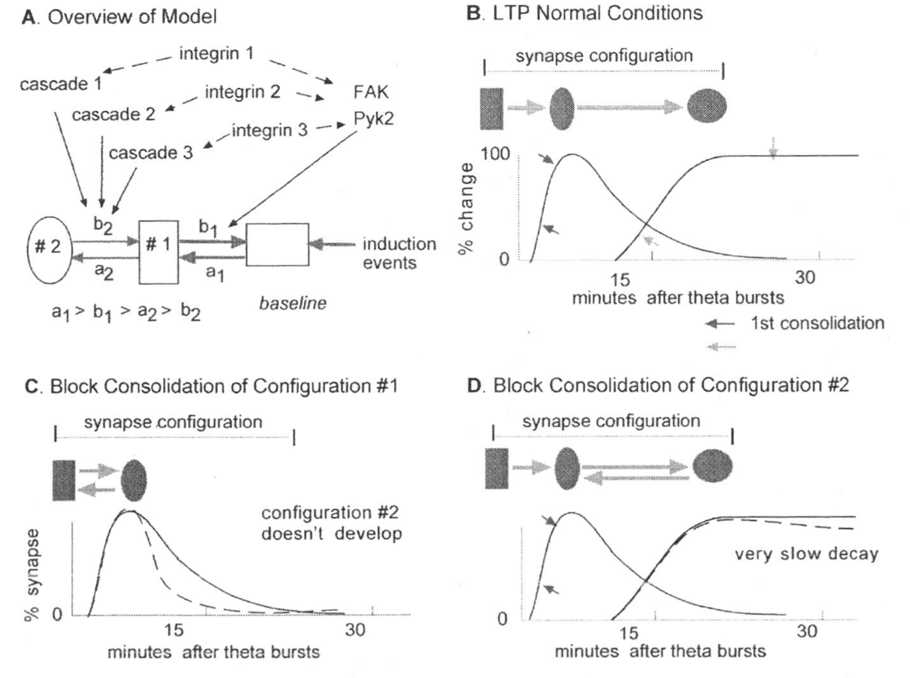

Figure 4. Stages of consolidation in the apical dendrites of pyramidal neurons. (A) Synapses are proposed to have three configurations --- baseline (rectangle), potentiated state #1 (oval), potentiated state #2 (circle) -- interconnected by four rate constants (a_1,a_2,b_1,b_2). The relative speeds of the four transitions are shown at the bottom of the panel; the fastest (a_1) and slowest (b_2) have $t_{1/2}$'s of about 10 minutes and several hours, respectively. Consolidation is proposed to involve three species of integrins. Each of these by itself activates two integrin related tyrosine kinases (Pyk2,FAK), an event that consolidates potentiation state #1. This changes the rate constants such that b_1 is slower than a_2. Activation of *any* of the three integrins will thus allow the system to proceed from potentiation state #1 to potentiation state #2. Each of the three integrins is postulated to set in motion its own signaling cascade; in the model, all three of these are needed to fully consolidate the system in stage #2. (B) This diagram illustrates the suggested time courses for the two stages of consolidation. The left-hand curve describes the degree to which the synapse is modified into the potentiation state #1 configuration; note that the state has a rapid onset and reaches its peak in about 30 seconds (Gustafsson and Wigstrom, 1990) and then *with consolidation* (arrows) decays back to baseline over the next 30-60 minutes. Potentiated state #2 is postulated to develop more slowly from state #1 and has a more protracted consolidation process that is not completed until approximately 30 minutes after induction. State #2 is then essentially permanent. (C) This panel illustrates the effects of blocking consolidation of state #1, something that requires suppression of FAK and Pyk2 activation, and hence blocking the signaling from all three integrins (see panel A). Under these conditions state #1 decays quickly back to baseline and does not persist long enough for state #2 to appear. LTP has its normal initial expression but falls back to baseline within 30-40 minutes. (D) Blocking consolidation of potentiated state #2 is illustrated in this figure, an event that occurs, at least partially, if *any* of the three integrin cascades is interrupted. The rate constant governing the transition from state #2 back to state #1 is hypothesized to be very slow, so that the effects of blocking consolidation take a significant amount of time to become apparent. LTP in this instance will begin to decay within 30 minutes but will take several hours to reach baseline.

4. REGIONAL AND DEVELOPMENTAL VARIATIONS IN LTP CONSOLIDATION

Integrin expression, as noted, differs significantly across brain regions and cell types. Remarkably, this also holds for dendritic domains of individual neurons (Bi et al., 2001; King et al., 2001), as can be seen in sections of hippocampus that have been immunostained for the α5 subunit (Fig. 5). The integrin is present at high levels in the apical dendrites (stratum radiatum) but is virtually absent in the basal dendrites (stratum oriens) of field CA1 in adult rats (Fig. 5A). Polarization is also evident in pyramidal cells of neocortex (Fig. 5C).

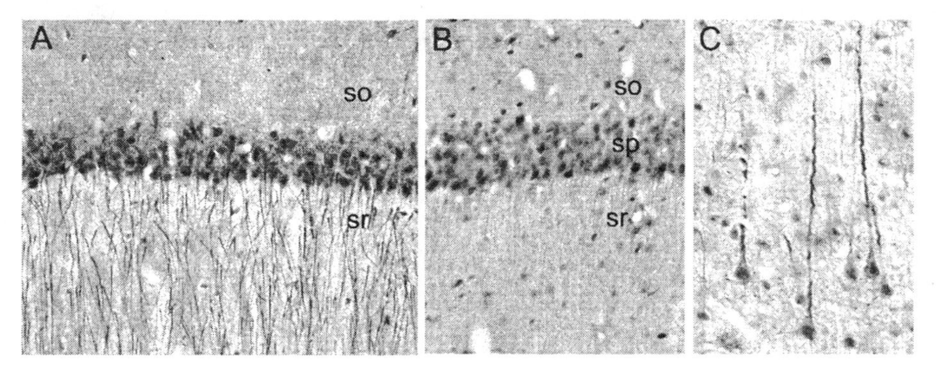

Figure 5. Integrin α5 immunoreactivity (ir) is polarized to apical dendrites in hippocampal and neocortical pyramidal cells. A,B) Photomicrographs showing α5 immunostaining in field CA1 of adult (A) and P14 (B) rats. In the adult (A), the apical dendrites in stratum radiatum (sr) are densely stained whereas basal dendrites in stratum oriens (so) do not contain detectable α5-ir. At postnatal day 14 (B), extremely low levels of α5-ir are evident within s. radiatum (sp, stratum pyramidale). Panel C shows the typical pattern of α5-ir among cortical pyramidal cells (layer V neurons shown): α5-ir is particularly dense in perikarya and apical dendrites whereas only faint immunostaining is localized to proximal basal dendrites of occasional cells.

These observations raise the possibility that differences in integrin concentrations contribute to morphological differentiation within cortical pyramidal cells. And, from the hypothesis described in figure 4, they predict that LTP consolidation will differ substantially between apical and basal dendrites. This has been confirmed (Kramár and Lynch, 2003). Theta bursts were applied to the Schaffer-commissural projections in CA1 stratum oriens (basal dendrites) or stratum radiation (apical dendrites) and field EPSPs were recorded from the same lamina. While LTP in the apical fields was unaffected by 5Hz stimulation delivered at 30 or 60 minutes post-induction, that in the basal dendrites was significantly attenuated. Moreover, spaced (10 min) applications of 5-Hz pulses beginning at 30 min post-TBS completely erased basal dendritic LTP. Thus it appears that LTP in the basal dendrites, though persistent, does not fully consolidate. Beyond this, RGD containing ligands did not increase the decay rate of basal dendritic LTP or enhance the efficacy of low frequency stimulation in reversing it, as expected if there is no integrin driven consolidation in stratum oriens (Kramár and Lynch, 2003).

Figure 7 describes a greatly simplified variant of the earlier model that offers a possible explanation for the properties of LTP in the basal dendrites. In this, synapses do not pass through an intermediate stage but instead are pushed directly into configuration

#2 by the very potent LTP induction conditions found in the basal dendrites. There is no consolidation in the model; instead, the altered connections very slowly revert to baseline following the same b_2 rate constant operating in the apical dendrites. According to this argument, LTP in stratum (s.) oriens resembles that found in s. radiatum after blocking integrins involved in consolidation of potentiated state #2. While LTP in the basal dendrites is persistent, the absence of the consolidation process leaves it vulnerable to disruption.

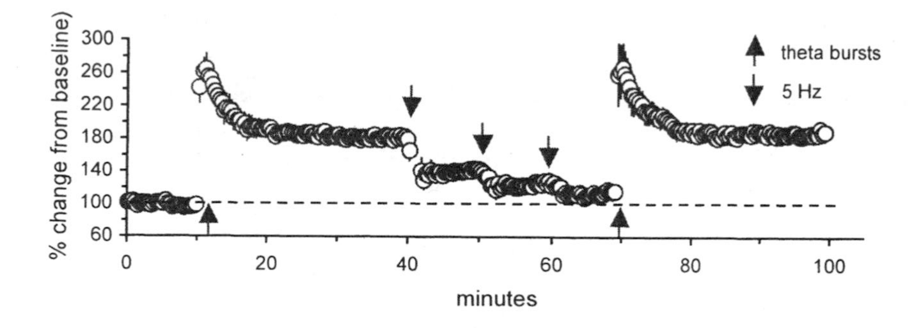

Figure 6. Low frequency stimulation partially reverses LTP when applied 30 or more minutes after induction in the basal dendrites of CA1 pyramidal neurons. LTP was induced in stratum oriens with a single train of theta bursts delivered to the basal branch of the Schaffer-commissural projection system. Three trains of 1Hz stimulation, spaced apart by 10 minutes, were applied starting 30 minutes later. This caused a near complete reversal of potentiation. Theta bursts were then delivered a second time and found to induce the same degree of potentiation as found prior to reversal (adapted from Kramár and Lynch, 2002).

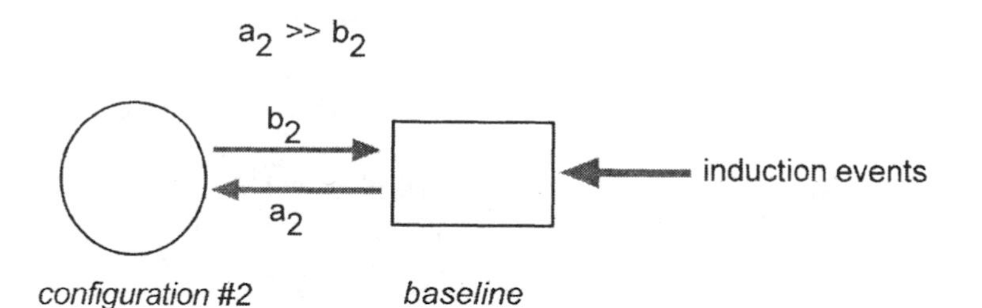

Figure 7. Consolidation in the basal dendrites of field CA1 pyramidal cells. This figure is meant to be compared to Fig. 4A, which describes the hypothesized stages of consolidation in the apical dendrites. As shown, the induction events for LTP in stratum oriens are proposed to move the synapse from baseline to potentiated state #2, without the need for the intermediary state #1. State #1 is essentially the same as that in the apical dendrites and has the same very slow rate constant (b_2) for its return to baseline. It is postulated that there is no consolidation in the basal dendrites, leaving the single decay constant to determine the duration of LTP.

The above results add to the list of features in which potentiation in the basal dendrites differs from that in the apical field. LTP is induced with lower threshold and attains greater magnitude potentiation in s. oriens than in s. radiatum (Capocchi et al., 1992; Arai et al., 1994; Leung and Shen, 1999; Kramár and Lynch, 2003) while induction in the apical dendrites is less sensitive to inhibition by NMDA receptor antagonists (Leung and Shen, 1999). The more potent LTP in the basal dendrites arises from two factors that shunt or counteract the depolarization generated by theta bursts: (a) feedforward IPSPS are smaller in *s.oriens* as are (b) the substantial afterhyperpolarizations (AHPs) that persist throughout the duration of a theta train (Arai et al., 1994). The latter effect appears to be due to reduced concentrations of voltage sensitive potassium channels on the proximal branches of the basal dendrites (Sah and Bekkers, 1996). The non-NMDA receptor dependent form of LTP is both more sensitive to tyrosine kinase inhibition and less sensitive to blockers of voltage sensitive calcium channels in apical as compared to basal dendrites (Cavus and Teyler, 1998). The collection of LTP-related apical vs. basal differences, including what appears to be two levels of consolidation at sites receiving largely the same input, could have important behavioral implications, as will be discussed below.

The broad hypothesis that integrins contribute to the terminal differentiation of central neurons, and then in adulthood provide for the anchoring of synaptic changes, raises questions about when during development these adhesion proteins appear in spines and dendrites. There is very little information on this point but an analysis of $\alpha 5$ yielded surprising results with important implications for the integrin-consolidation hypothesis (Bi et al., 2001). The $\alpha 5\beta 1$ integrin (as mapped with $\alpha 5$ subunit immunostaining) is present in moderate concentrations in pyramidal cell bodies at postnatal day (P) 7 but is missing from their dendrites. Dendritic staining is evident but still low at P14 (Fig. 5B). There is currently no explanation for why the integrin, though clearly expressed in early postnatal life, would remain confined to cell bodies for so much of subsequent development. One consequence of this peculiar phenomenon is that $\alpha 5\beta 1$ arrives in fine dendritic processes and spines only after much of synaptogenesis is completed. This implies that at least some aspects of synaptic adhesion are added after transmission is fully operational; it is tempting to speculate that the integrins contribute to the cessation of growth, much as they serve in other systems to suppress cycling, and thereby drive cells towards their mature state (Streuli and Edwards, 1998). With regard to LTP, potentiation appears in the apical synapses of CA1 at P9 (Baudry et al., 1981) and is well developed by P14-17 (Kramár and Lynch, 2003). However, the integrin hypothesis predicts that potentiation at these ages will not stabilize and should be vulnerable to disruption even after extended post-induction periods. Tests of this prediction were positive: low frequency applied 30-60 min post induction fully reversed LTP in slices from P10 rats and caused substantial reductions in slices from immature (P14-P21) rats (Kramár and Lynch, 2003)

In all, consolidation across regions and stages of development correlates with expression of the $\alpha 5\beta 1$ integrin. Coupled with evidence from studies using integrin ligands, neutralizing antibodies, and transgenic mice, this finding strongly suggests that a causal relationship exists between the adhesion receptor and LTP stabilization. The observed differences in consolidation also imply (a) that brain has multiple memory systems that are distinguishable not so much by the nature of content as by the persistence of the material they encode and (b) that learning, at least in cortical telencephalon, develops well in advance of stable memory.

5. INTEGRINS AS LINKS BETWEEN EXPRESSION AND CONSOLIDATION

Structural changes in spines and the synaptic apposition zone accompany LTP, as described in a collection of studies going back almost to the beginnings of research on the potentiation effect. Early studies showed that prolonged trains of high frequency stimulation delivered to the entorhinal cortex cause a very large degree of spine swelling throughout the dentate gyrus, but it was not clear if this effect was associated with potentiation (Van Harreveld and Fifkova, 1975). The first experiments using LTP induced by bursts of high frequency afferent activity found that potentiation in field CA1 in vivo is accompanied by a decrease in the coefficient of variability and skewedness of the distributions of both spine areas and postsynaptic density (PSD) lengths (Lee et al., 1980). These effects, which were replicated in slice experiments (Lee et al., 1981), were taken as evidence that LTP caused a population of elongated spines to become rounder. Other groups provided evidence that LTP in the apical dendrites of field CA1 is accompanied by rounding of spine heads (Chang and Greenough 1984) and, in dentate gyrus, possibly by transitions between spine configurations (Desmond and Levy, 1986). A very sizeable literature gradually developed on the topic, along with considerable debate concerning the extent to which spines are added or segmented as part of the potentiation process (Geinisman et al., 1991; Toni et al., 1999; Yuste and Bonhoeffer, 2001; Fiala et al., 2002). Much of the confusion likely results from the use of different target dendrites (CA1 vs. dentate gyrus), experimental preparations (slices vs. in vivo), and degrees of maturation (cultured vs. acute slices). In any event, it seems reasonably clear that LTP is associated with changes in spine morphology and that these are likely to increase the total amount of synaptic apposition.

Such changes are attractive candidates for the stable component of LTP because, as structural modifications, they have the potential for extreme longevity in the face of protein turnover. Moreover, rounding (or segmenting) of spines would be expected, and has been observed, to produce similar shifts in PSDs and thus would increase the area available for insertion of transmitter receptors; increasing the size of the receptor pool would be reflected in the increased size of EPSCs. Morphological changes could thus account for both the expression of LTP and its extreme persistence. Recent work provides important support for these arguments. Individual spines of certain types have been shown to persist for months (Grutzendler et al., 2002; Trachtenberg et al., 2002) and the size of the post-synaptic apposition zone correlates well with the number of AMPA-type glutamate receptors it contains (Nusser et al., 1998; Takumi et al., 1999). Moreover, the conversion of individual spines from an elongated to a round configuration in response to high frequency afferent activity, or local release of glutamate, has now been directly observed, albeit in cultured slices (Matsuzaki et al., 2004).

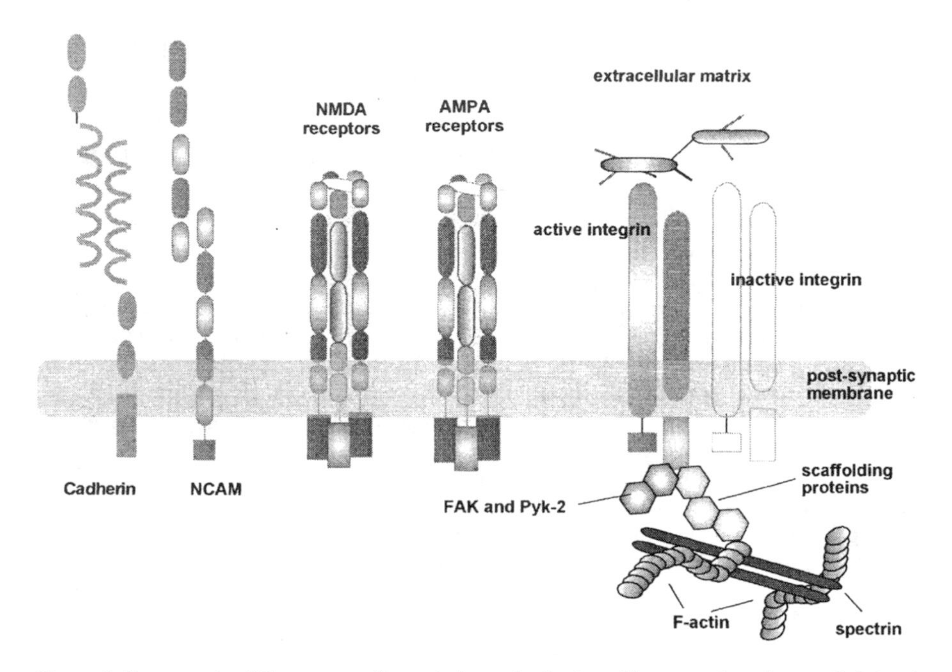

Figure 8. Two aspects of the synapse: Transmission and adhesion. Shown are two classes of glutamate receptors (AMPA and NMDA) and three types of adhesion receptors. NCAM (and other IgG-class adhesion receptors) and cadherins attach to like proteins extending from the pre-synaptic membrane while integrins bind to a consensus sequence in extracellular matrix proteins. A complex collection of intracellular proteins attaches to the cytoplasmic tail of the integrin β unit, prominent among which are two homologous tyrosine kinases (FAK, Pyk2) and a set of scaffolding proteins that form a bridge to the actin cytoskeleton. Note that the latter is also cross-linked by dimeric spectrin. Integrins are present in active and inactive states; in the latter condition, contact with the matrix and cytoskeleton is minimal.

From these results, LTP involves, and perhaps requires, mechanisms that shift the shape of spines and then anchor their new configuration. While these two steps could involve the same process, there are reasons to think that they depend on different chemistries and the following discussion incorporates this assumption. Figure 8 summarizes several of the adhesive features of the synapse so far discussed. Three classes of adhesion receptors -- cadherens, IgG-class cell adhesion receptors (e.g., NCAM), and integrins -- localized to synapses (Fields and Itoh, 1996) are illustrated. The first two of these exhibit homotypic binding of comparable proteins emerging from the pre- and post-synaptic membrane; integrins exhibit heterotypic binding and, for the most part, bind to the extracellular matrix. There is significant evidence implicating all three groups of adhesion receptors in LTP (Luthl et al., 1994; Muller et al., 1996; Tang et al., 1998; Bozdagi et al., 2000) but so far only integrins have been specifically related to consolidation. Three points made earlier about integrins are included in the figure: 1) the receptors exist in active and inactive states; 2) they are attached to the tyrosine kinases FAK and Pyk2, and 3) they are connected via a series of scaffolding proteins to the actin cytoskeleton. As shown, the greatly simplified skeleton involves a connection between polymerized actin to the β integrin subunit and cross-linking of the actin network by dimeric spectrin. The components of the system shown here are known to be present at synapses but the hypothesized organization is largely derived from cells outside brain.

Figure 9 describes the postulated roles for integrins in LTP consolidation. The top left panel is a highly schematic version of the changes in spines and post-synaptic densities (PSD: detached ovals) that accompany LTP. Note that rounding of the spine head generates a more circular PSD with more or less the same circumference. This increases the area available for glutamate receptors; filling of this space is assumed to occur immediately after the shape change. Spine change is shown in the model to occur in two stages. There are no data pointing to this and it is shown here to accommodate the argument that LTP passes through two stages that, absent consolidation, reverse to baseline at very different rates (see Fig. 7). In panel 9A, the triggering events for LTP (initially, an increase in spine calcium) set in motion a set of enzymatic responses including activation of proteases, and especially calpains, that target the cytoskeleton and its linkages to the integrins (Siman et al., 1984; Schoenwaelder et al., 1997). NMDA receptor stimulation and theta burst stimulation both activate calpain in spines (del Cerro et al., 1994; Vanderklish et al., 1995; Vanderklish et al., 2000) and the indicated targets (dashed lines) are all described as substrates for the enzyme in multiple cell types. Calpain activation appears to make a critical contribution to integrin disengagement during cell migration (Huttenlocher et al., 1997) and it is proposed here to play a similar role in relaxing the constraints on spine shape. Integrin activation (panel 9B) can be a very rapid process and this is assumed to be the case in the model. There is a vast literature on how activation (a.k.a., 'inside-out' signaling) occurs (Mould, 1996; Hughes and Pfaff, 1998; Travis et al., 2003) and elements so far identified as involved in the process are present in dendrites and mobilized by synaptic activity (e.g., elevated calcium levels, protein kinase C activity). Still it should be noted that integrin activation has not been described for brain neurons in situ. Activation increases the affinity of the integrins for their matrix ligands and, with ligand binding, causes the integrin related kinases, as well as the enormously complex array of scaffolding proteins, to assemble around the C-terminus of the β subunit. These events are shown as being complete in panel 9B. Ligand binding leads to a wave of outside-in conformational changes in the integrin (Travis et al., 2003) which sets signaling in motion beginning with the activation of tyrosine kinases Pyk2 and FAK. Three pathways, two of which have been described for adult hippocampus, are illustrated in Fig. 9B. *First*, phosphorylation of AMPA (GluR1 serine831) and NMDA (NR2A and NR2B) receptor proteins occurs after integrin ligation. This leads to increases in synaptic currents mediated by the two classes of receptors (Kramár et al., 2003; Lin et al., 2003), possibly a special case of the general phenomenon of integrin modulation of neighboring transmembrane receptors (Porter and Hogg, 1998; Miranti and Brugge, 2002; Schwartz and Ginsberg, 2002).

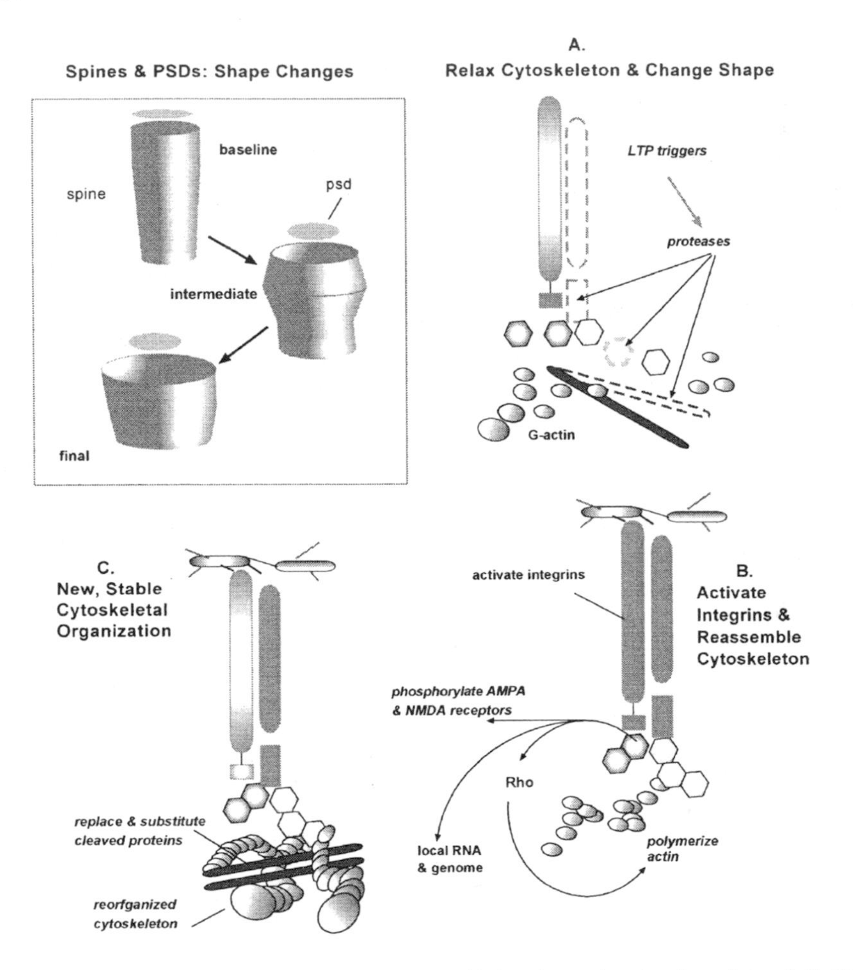

Figure 9. Proposed mechanisms for producing and consolidating changes in spine shape that underlie LTP. The top left panel describes changes in spine shape that accompany LTP. Two effects are illustrated, which revert quickly ('intermediate') or slowly ('final') to baseline unless consolidated. The structural change involves a movement from an elongated configuration to a rounder, shorter shape. The post-synaptic density (psd) mirrors the changes in spine morphology and moves from a thin extended profile to a round shape. This causes an increase in its area and therefore an increase in space available for AMPA receptors. (A) The triggers for LTP activate phospholipases, proteases and kinases; the proteases destabilize the spine by cleaving structural elements (dotted lines) that connect integrins to the cytoskeleton and cross-link the actin network. (B) The above changes engage latent integrins that then bind to the cytoskeleton and activate the integrin related kinases FAK and Pyk2. This last event, along with the assembly of a collection of proteins to the cytoplasmic tail of the β subunit of the integrin dimer, triggers three effects: (i) Phosphorylation of AMPA and NMDA receptors augments LTP expression. (ii) Activation of Rho family GTPases, and related effects, results in actin polymerization. (iii) Signaling cascades are set in motion that affect translation from local mRNA and that influence gene expression. (C) The assembled actin network is stabilized by spectrin and anchoring to the integrins via scaffolding proteins. Replacement copies of cleaved proteins are drawn from extant supplies or from triggered synthesis.

Phosphorylation of glutamate receptor proteins is expected to have two LTP-related effects: (a) an immediate contribution to the potentiation of synaptic currents and (b) facilitation of receptor insertion into the PSD (Tomita et al., 2001). *Second*, integrin activation triggers pathways leading to the neuron's protein synthesis machinery. Recent work using adult hippocampus demonstrated that ligand binding to integrins increases the expression of nerve growth factor, brain-derived neurotrophic factor, Arc, and c-fos without altering the expression of other transcripts, including those for growth-associated protein-43, amyloid precursor protein and fibroblast growth factor-1 (Gall et al., 2003). Changes in mRNA content were not secondary to changes in neuronal activity in that they occurred in the presence of tetrodotoxin and the AMPA antagonist CNQX. These results extend the trophic function of integrins, which are well described for a diverse array of cell types, to neurons in the adult brain. Other work has shown that binding of matrix ligands to integrins prevents the translocation of a critical translation initiation factor to actin-bound mRNA (Smart et al., 2003); given that the factor in question moves into spines in response to appropriate signals, this result could indicate that integrins regulate local protein synthesis in the vicinity of dendritic spines (Smart et al., 2003), as suggested elsewhere (Dong et al., 2003). *Third*, integrins acting on Rho family GTPases trigger actin polymerization and cytoskeletal reorganization in many cell types and experimental circumstances (DeMali et al., 2003). This provides a route whereby integrin signaling could stabilize the spine in a new anatomical configuration and, indeed, it was this capability that first drew attention to integrins as potential mediators of LTP consolidation. Whether integrins do in fact polymerize actin in adult brain is perhaps the single most important assumption made in the model described in Figure 9.

Figure 9C describes a later stage of LTP consolidation. In this, proteins cleaved during the disassembly phase (panel 9A) are replaced by transfers from extant pools, synthesis of new copies, or substitution. Pioneering work by Stanton and co-workers suggested that protein synthesis is needed for the production of stable LTP and presciently raised the possibility that this could be related to reports that synthesis is required for memory formation (Stanton and Sarvey, 1984). Subsequent studies have greatly extended these first results (Deadwyler et al., 1987; Nguyen et al., 1994; Frey et al., 1996; Frey and Morris, 1997). Translation from extant mRNA in dendrites or even the cell body could begin adding new proteins minutes after induction; the observation that integrin ligation affects local translation is germane to this. Also of interest with regard to stabilizing potentiation is the presence in dendrites of mRNA for a cytoskeleton-associated protein Arc (a.k.a., Arg3.1) (Link et al., 1995; Lyford et al., 1995) having domains closely resembling those of spectrin, the actin cross-linking protein found at high levels in spines and cleaved there by modest amounts of theta burst stimulation (Vanderklish et al., 1995; Vanderklish et al., 2000). It is intriguing to consider the possibility that dendrites contain mRNA species that are specialized for rapid translation of proteins that can substitute for other, slowly synthesized proteins degraded during the course of modifying synapses. Finally, gene induction, which is triggered by intense afferent activity as well as by integrins (see above), would generally be expected to be too slow to contribute to the 30-minute LTP consolidation period described above. However, several papers argue for a protein synthesis dependent phase of consolidation beginning hours after induction. The present model does not deal with this but would interpret the delayed, 'late-LTP' phase in terms of slow decay resulting from failure to replace proteins lost in the first few minutes after theta bursts. Beyond this, there are a number of transcriptional and translational events that are activated within minutes of intense neuronal activity and these could conceivably add proteins at the end of the 30-minute consolidation period.

6. CONSOLIDATION BEYOND THE SYNAPSE

Much, perhaps most, of the interest in LTP can be traced to an assumption, held by most who study the phenomenon, that it will in the end provide a satisfactory neurobiological account for the most basic aspect of memory encoding. Testing this proposition is greatly complicated by confusion about the relationships between different forms of memory and what is, and is not, LTP. The first papers asking if conditions that block LTP also selectively impair memory showed that such treatments disrupt spatial and olfactory memory but also leave some forms of learning intact (Staubli et al., 1985; Morris et al., 1986; Morris et al., 1989). While other interpretations are possible, this result certainly suggests the conclusion that the potentiation effect is not the substrate for all types of memory. The point is underlined by the suspicion that LTP may be missing from parts of the brain thought to encode certain kinds of memory. Even within hippocampus, there are forms of long-lasting potentiation (mossy fiber potentiation) that are quite distinct from LTP (Staubli, 1992). This is where confusion about definitions becomes a problem; it has become common practice to label activity-dependent potentiation as LTP, even when it is not at all clear that the observed effect is synapse-specific, persistent, and rapidly induced. If LTP of the type found in hippocampus is not so ubiquitous as suggested by the literature, and if other forms of plasticity have been mistakenly labeled, then it follows almost by necessity that the effect is not associated with all variants of memory.

On the other hand, the work reviewed here adds to the list of ways in which LTP matches the demanding constraints, placed by behavioral observations, on putative substrates for much of memory as it is experienced in everyday life. Commonplace memory forms quickly, can last for indefinite periods, and possesses vast capacity; LTP is induced in seconds, can persist without evident change for weeks (Staubli and Lynch, 1987) with a total duration equal to a large part of a rat's lifespan (Abraham et al., 2002), and has the synapse specificity needed for high capacity. Memory also requires time to become resistant to disruption, with significant gains in stability occurring within the first 30 minutes after its formation (Duncan, 1948); LTP, as discussed, also has these properties. Adding information on how spatio-temporal patterns of afferent activity induce LTP to neural network models results in systems that exhibit more elaborate features of memory such as (i) cells that learn temporal sequences and (ii) storage capacities that scale linearly with network size (Granger et al., 1994). Networks using LTP-based learning rules also place memories into categories and then organize the categories into multi-level hierarchies (Ambros-Ingerson et al., 1990), operations that are embedded in human memory processing. These correspondences and an extensive literature using diverse pharmacological and genetic manipulations make an impressive case that the potentiation effect is the substrate of common forms of memory.

Given the strong probability that LTP is an encoding device for some, though not all, forms of information, can we now use what has been learned of its neurobiology to help explain poorly understood aspects of memory, or even reveal facets of the phenomenon overlooked or poorly appreciated in behavioral analyses? The following sections consider some possibilities relating to consolidation.

6.1. Parallel, Differential Encoding

The differences between apical and basal dendrites strongly suggest that pyramidal neurons carry out two types of memory encoding on the same general information. LTP in the basal dendrites is more pronounced and easily induced than that in the apical tree but now appears to be significantly less stable. These functional differences have been traced to variations in cell processes known to control LTP induction and expression. Given that the overwhelming majority of all synapses in s. radiatum and s. oriens arise from the same hippocampal subfields (ipsi- and contralateral field CA3), it would appear that field CA1 doubly, but differentially, encodes the same information. Why? One possibility is that the dual system allows animals to deal with the problem of knowing what to encode in a stimulus-rich, unfamiliar environment; in other words, knowing how to store enough so that useful information is captured without, at the same time, accumulating meaningless material that may interfere with computation. In the proposed scenario, the basal dendrites record information very broadly and do so with minimal demands on the animal's attention (because of easily induced potentiation), making it possible to carry out rapid exploration while still gaining considerable information about the environment. Over time, objects or events will be recognized as recurrent because they gradually build up potentiation in the basal dendrites, at which point longer sampling sessions will allow for the lengthy theta trains needed to encode them in the apical dendrites. Interestingly, although the point has not been experimentally tested, the presence of substantial potentiation in the basal dendrites will likely lead to action potentials in the target cells which, under appropriate circumstances, invade the apical dendrites and add to the depolarization generated there by afferent bursts: basal LTP will therefore promote the formation of apical LTP (The authors owe this argument to Istvan Mody). However, it is likely that most stimuli in the novel environment will, in the context of other stimuli, lack interesting properties or prove to be transient; according to the present hypothesis, potentiation generated by these cues will gradually decay back to baseline response levels, leaving the pertinent synapses free on other days to again act as a low threshold memory system.

The above argument should not be taken to mean that LTP in the apical dendrites never occurs early in learning; supervised training, in which animals are directed by external agencies to salient cues, would presumably engage the system from the outset. The circumstance here discussed involves unsupervised learning in which animals must find information against a noisy background.

Behavioral researchers have for some time struggled with the question of whether memory is transferred between compartments of different durations as opposed to being multiply encoded in systems of different stability. The above arguments suggest that so far as field CA1 and hours-long vs. very stable memory are concerned, the same storage site makes two copies with the transient variant paving the way for a more stable representation. This would predict that a selective disruption of consolidation would leave animals with memories lasting for several hours after exploring a new and complex environment but with little retention of spatial information on tests conducted days later. Testing this prediction, which could result in a separation of hippocampal memory systems, raises the issue, discussed immediately below, of how to recognize if a given treatment disrupts consolidation as opposed to erasing synaptic modifications.

6.2. Reversal, Consolidation, and Brain Activity Patterns

Treatments usually described as blocking memory consolidation could in fact be eliminating the changes that express the memory; time dependency in this scenario would simply mean that the treatment must be applied before the changes have had time to stabilize. Discussions of electroconvulsive shock (ECS) studies sometimes note that this global manipulation may cause amnesia by disrupting active 'memory traces' rather than interfering with the chemical processes that convert the traces into long-term memory [e.g., (Miller and Matzel, 2000)]. But hippocampal networks are very unlikely to persist in an active or holding state for even a few seconds and hence such states cannot explain why ECS causes substantial losses in memory when applied minutes after learning (Duncan, 1948). It is more likely instead that ECS either reverses the changes that express memory or prevents the progressive increase in stability normally seen during the immediate post-learning period. Since there appear to be no manipulations that distinguish between the two possibilities, it is not possible to explore the hypothesis (above) that hippocampus contains dual memory systems distinguishable by whether they consolidate or not. That is, a treatment that causes reversal will likely affect both LTP variants because both encode using the same induction and expression mechanisms. Initial studies suggest that integrin related manipulations that cause LTP to decay in the apical dendrites have minimal effects in the basal dendrites; additional work may yield treatments of this type that are appropriate for in vivo use, and thus that can be employed to test the prediction that blocking LTP consolidation will leave intact LTP-based forms of memory that last for hours.

The issue of reversal vs. consolidation takes on added significance with the discovery that autonomous oscillations (sharp waves: SPWs) block LTP (Colgin et al., 2004). This provides a first suggestion that hippocampus and perhaps cortex contain specialized processes for suppressing as well as inducing LTP. Whether SPWs prevent induction, reverse newly induced potentiation, or block consolidation will determine their significance to network computations. Adenosine A1 receptor antagonists, agents that are well established to counteract the effects of post-induction low frequency stimulation on LTP (Larson et al., 1993; Huang et al., 1999), prevent the blocking action of sharp waves; possibly, then, the suppression is less of a blockade of LTP than a disruption of consolidation. As discussed, the low frequency stimulation paradigm works as well or better in stratum oriens than in stratum radiatum; within the context of the LTP model described here, this suggests that SPWs operate by rapidly reversing potentiation rather than blocking consolidation. If so, then they would be a device for quickly erasing all traces of a memory as opposed to one that prevents select information from going into long-term storage.

How might such a tool be used? The brain waves in question appear when rats are alert but not moving (Buzsaki, 1986), suggesting that recently encoded LTP could be reversed simply by engaging in a particular behavior. The consolidation period would, in addition to being the time needed to stabilize LTP, constitute an interval during which animals have the option of erasing what has just been learned. Of course, the reversal process could be partial, so that only some of the recent changes are eliminated; learning would then be a process of continuously acquiring and sculpting material, with both activities associated with particular behavioral actions and their concomitant brain activity patterns. The plausibility of these ideas is increased by evidence that behavior can indeed reverse recently induced LTP (Xu et al., 1998), although it is not known if this is linked to particular actions or brain wave patterns. It will be of interest in this context to ask if there are instances of unstructured learning in which rats alternate periods of alert immobility with active sampling. Stated more broadly, are there instances in which active shaping through partial erasure serves to improve the quality of memory?

And it may yet be discovered that there are means for modifying LTP, possibly even with SPWs, long after its consolidation period has passed. The issue arises because surprisingly little is known about the effects of theta bursts on already potentiated synapses. The induction events (NMDA receptors, elevated spine calcium) will presumably still transpire, possibly (though this is less certain) leading to activation of calpain and degradation of elements responsible for organizing the spine cytoskeleton. There is, therefore, no *a priori* reason to assume that potentiated spines cannot be destabilized, and thus rendered vulnerable to LTP reversal, by the very conditions that caused them to enter the potentiated configuration in the first place. There is no lack of plausible reasons why this would not happen but experimental evidence on the point is lacking. The issue is of great importance if for no other reason than that such an effect would allow new learning to place the information encoded by previously potentiated synapses at risk. As noted earlier, there is a long and continuing debate on the degree to which reactivation of memories makes them vulnerable to disruption (Misanin et al., 1968). Studies on whether LTP can be rendered reversible, or dependent on a second round of consolidation, by patterned synaptic activity long after it has been initially stabilized could provide insights into this aspect of memory and the nature of forgetting.

6.3. Developmental Differences in the Stability of Encoding

Consolidation, as discussed, emerges significantly later during maturation than does the induction and expression of LTP. This presumably means that animals can learn spatial cues, and related kinds of information, well in advance of when they can retain them for extended periods. Immature rats exhibit an 'infantile amnesia' syndrome involving seemingly adequate learning but faster forgetting than is the case for adults ((Campbell and Campbell, 1962), and many others). The effect is evident in 17-day old pre-weanling rats, and thus is present in some form during the period in which LTP is fully expressed but does not adequately consolidate. While numerous studies indicate that retrieval failures are part of the phenomenon (Campbell and Jaynes, 1969; Flint and Riccio, 1997), an inability to appropriately stabilize memory remains a potential contributor. Humans also have a childhood amnesia effect such that episodes occurring before an early age, generally estimated as being between years 3-5 (see (Bruce et al., 2000) for a recent evaluation), cannot be recalled as episodic memory. In all, there is ample evidence that memory during an extended period of postnatal development lacks the salience and stability found later in life.

While it is plausible that selective pressures would result in a dual memory system of the type discussed above, and indeed this might even be the reason that dendritic domains evolved on pyramidal neurons, such arguments seem less likely in the case of the slow development of consolidation. There are no evident competitive advantages, and some rather obvious disadvantages, associated with an absence of rapidly induced, long-term memory during the second and third postnatal weeks. However, delaying the onset of consolidation could serve to align the emergence of stable cortical memory with the development of mature cognitive processes that retrieve and utilize memory (Nelson and

Fivush, 2004). As noted earlier, encoding in network models using LTP-based synaptic learning rules results in self-organized hierarchical classification systems, such that an object is first recognized as belonging to a broad category of like objects, then as a member of a considerably smaller group, and finally as either itself or something that is novel. Building such elaborate structures without input from mature cortical processing systems could conceivably create problems for subsequent operations.

An alternative account is that the absence of consolidation during development is a side effect arising from the multiple functions executed by integrins. As discussed, there are several reasons to assume that integrin signaling will prove to be as crucial to the differentiation of neurons as it is to other cell types. Any contributions in this direction by the receptors would have to be timed to allow dendritic growth and synaptogenesis to reach completion, a requirement that is satisfied by the peculiar delay between expression of integrins (or, perhaps, "select" integrins) and their transport into the apical dendrites. According to this argument, the postulated use of integrins to terminate growth and provide for stabilization inevitably results in a postnatal period in which plasticity is present but can't be fully utilized. Infantile and childhood amnesia then become incidental outcomes of a developmental adjustment needed to arrive at a full complement of synapses.

7. ACKNOWLEDGEMENTS

The authors would like to thank the many laboratory members who contributed to studies described here including Drs. Amy Arai, Ben Bahr, Joie Bernard, Xiaoning Bi, Laura Colgin, Eniko Kramár, Kevin Lee, Bin Lin, Ching-Yi Lin, Jason Pinkstaff and Ursula Staubli. Thus work was supported by grants MH61007 and NS37799.

8. REFERENCES

Abraham WC (1997) Induction and reversal of long-term potentiation by repeated high frequency stimulation in rat hippocampal slices. Hippocampus 7:137-145.

Abraham WC, Logan B, Greenwood JM, Dragunow M (2002) Induction and experience-dependent consolidation of stable long-term potentiation lasting months in the hippocampus. J Neurosci 22:9626-9634.

Ambros-Ingerson J, Granger R, Lynch G (1990) Simulation of paleocortex performs hierarchical clustering. Science 247:1344-1348.

Arai A, Black J, Lynch G (1994) Origins of the variations in long-term potentiation between synapses in the basal versus apical dendrites of hippocampal neurons. Hippocampus 4:1-9.

Arai A, Kessler M, Lynch G (1990) The effects of adenosine on the development of long-term potentiation. Neurosci Lett 119:41-4.

Arai A, Larson J, Lynch G (1990) Anoxia reveals a vulnerable period in the development of long-term potentiation. Brain Res. 511:353-357.

Bahr B, Sheppard A, Lynch G (1991) Fibronectin binding by brain synaptosomal membranes may not involve conventional integrins. NeuroReport 2:13-16.

Bahr BA, Lynch G (1992) Purification of an Arg-Gly-Asp selective matrix receptor from brain synaptic plasma membranes. Biochemical Journal 281, 137-142.

Bahr BA, Staubli U, Xiao P, Chun D, Ji ZX, Esteban ET, Lynch G (1997) Arg-Gly-Asp-Ser-selective adhesion and the stabilization of long-term potentiation: Pharmacological studies and the characterization of a candidate matrix receptor. J Neurosci 17:1320-1329.

Barr DS, Lambert NA, Hoyt KL, Moore SD, Wilson WA (1995) Induction and reversal of long-term potentiation by low- and high-intensity theta pattern stimulation. J Neurosci 15:5402-10.

Barrionuevo G, Schottler S, Lynch G (1980) The effects of repetitive low frequency stimulation on control and "potentiated" synaptic responses in the hippocampus. Life Sciences 27:2385-2391.

Baudry M, Arst D, Oliver M, Lynch G (1981) Development of glutamate binding sites and their regulation by calcium in rat hippocampus. Brain Res 227:37-48.

Bi X, Lynch G, Zhou J, Gall CM (2001) Polarized distribution of α5 integrin in dendrites of hippocampal and cortical neurons. J Comp Neurol 435:184-193.

Bozdagi O, Shan W, Tanaka H, Benson DL, Huntley GW (2000) Increasing numbers of synaptic puncta during late-phase LTP: N-cadherin is synthesized, recruited to synaptic sites, and required for potentiation. Neuron 28:245-259.

Bruce D, Dolan A, Phillips-Grant K (2000) On the transition from childhood amnesia to the recall of personal memories. Psychol Sci. 11:360-364.

Burette F, Jay TM, Saroche S (1997) Reversal of LTP in the hippocampal afferent fiber system to the prefrontal cortex in vivo with low-frequency patterns of stimulation that do not produce LTD. J. Neurophysiol. 78:1155-1160.

Buzsaki G (1986) Hippocampal sharp waves: their origin and significance. Brain Res. 398:242-252.

Campbell BA, Campbell EH (1962) Retention and extinction of learned fear in infant and adult rats. J. Comp. Physiol. Psychol. 55:1-8.

Campbell BA, Jaynes J (1969) Effect of duration of reinstatement on retention of a visual discrimination learned in infancy. Dev. Psychol. 1:71-74.

Capocchi G, Zampolini M, Larson J (1992) Theta burst stimulation is optimal for induction of LTP at both apical and basal dendritic synapses on hippocampal CA1 neurons. Brain Res. 591:332-336.

Cavus I, Teyler TJ (1998) NMDA receptor-independent LTP in basal versus apical dendrites of CA1 pyramidal cells in rat hippocampal slice. Hippocampus 8:373-379.

Chang HP, Lindberg FP, Wang HL, Huang AM, Lee EHY (1999) Impaired memory retention and decreased long-term potentiation in integrin-associated protein-deficient mice. Learning and Memory 6:448-457.

Chun D, Gall CM, Bi X, Lynch G (2001) Evidence that integrins contribute to multiple stages in the consolidation of long term potentiation. Neuroscience 105:815-829.

Colgin LL, Kubota D, Jia Y, Rex CS, Lynch G (2004) Long-term potentiation is impaired in rat hippocampal slices that produce spontaneous sharp waves. J. Physiol 558:953-961.

Deadwyler SA, Dunwiddie T, Lynch G (1987) A critical level of protein synthesis is required for long-term potentiation. Synapse 1:90-95.

del Cerro S, Arai A, Kessler M, Bahr BA, Vanderklish P, Rivera S, Lynch G (1994) Stimulation of NMDA receptors activates calpain in cultured hippocampal slices. Neurosci. Lett. 167:149-152.

DeMali KA, Wennerberg K, Burridge K (2003) Integrin signaling to the actin cytoskeleton. Curr Opin Cell Biol. 15:572-582.

Desmond NL, Levy WB (1986) Changes in the postsynaptic density with long-term potentiation in the dentate gyrus. J. Comp. Neurol. 253:476-482.

Dong E, Caruncho H, Liu WS, Smalheiser NR, Grayson DR, Costa E, Guidotti A (2003) A reelin-integrin receptor interactions regulates Arc mRNA translation in synaptoneurosomes. Proc. Natl. Acad. Sci. USA:5479-5484.

Duncan CP (1948) The retroactive effect of electroshock on learning. J. Comp. Physiol. Psychol. 42:32-44.

Einheber S, Schnapp LM, Salzer JL, Cappiello ZB, Milner TA (1996) Regional and ultrastructural distribution of the α8 integrin subunit in developing and adult rat brain suggests a role in synaptic function. J. Comp. Neurol. 370:105-134.

Epstein HT, Kuzirian AM, Child FM, Alkon DL (2004) Two different biological configurations for long-term memory. Neurobiol. Learn Mem 81:12-18.

Fiala JC, Allwardt B, Harris KM (2002) Dendritic spines do not split during hippocampal LTP or maturation. Nat. Neurosci. 5:297-298.

Fields RD, Itoh K (1996) Neural cell adhesion molecules in activity-dependent development and synaptic plasticity. Trends. Neurosci. 19:473-480.

Flint RW, Riccio DC (1997) Pretest administration of glucose attenuates infantile amnesia for passive avoidance conditioning in rats. Dev. Psychobiol. 31:207-216.

Frey U, Frey S, Schollmeier F, Krug M (1996) Influence of actinomycin D, a RNA synthesis inhibitor, on long-term potentiation in rat hippocampal neurons in vivo and in vitro. J. Physiol. 490:703-711.

Frey U, Morris RG (1997) Synaptic tagging and long-term potentiation. Nature 385:533-536.

Froc DJ, Chapman CA, Trepel C, Racine RJ (2000) Long-term depression and depotentiation in the sensorimotor cortex of the freely moving rat. J Neurosci 20(1):438-45.

Gall CM, Lynch G (2004) Integrins, synaptic plasticity, and epileptogenesis, in Recent Advances in Epilepsy Research, Advances. (H. Scharfman, D. Binder, eds). Vol. 548:12-33. Kluwer Adacemic.

Gall CM, Pinkstaff JK, Lauterborn JL, Xie Y, Lynch G (2003) Integrins regulate neuronal neurotrophin gene expression through effects on voltage sensitive calcium channels. Neuroscience 118:925-940.

Geiger B, Bershadsky A, Pankov R, Yamada KM (2002) Transmembrane extracellular matrix - cytoskeleton crosstalk. Nature Cell Biol. 4:793-805.
Geinisman Y, deToledo-Morrel L, Morrel F (1991) Induction of long-term potentiation is associated with an increase in the number of axospinous synapses with segmented postsynaptic densities. Brain Res. 566:77-88.
Giancotti FG, Ruoslahti E (1999) Integrin signaling. Science 285:1028-1032.
Granger R, Whitson J, Larson J, Lynch G (1994) Non-Hebbian properties of long-term potentiation enable high-capacity encoding of temporal sequences. Proc Natl Acad Sci U S A 91:10104-10108.
Grooms S, Terracio L, Jones L (1993) Anatomical localization of ß1 integrin-like immunoreactivity in rat brain. Exp. Neurol. 122:253-259.
Grutzendler J, Kasthuri N, Gan WB (2002) Long-term dendritic spine stability in the adult cortex. Nature 420:751-752.
Gustafsson B, Wigstrom H (1990) Long-term potentiation in the hippocampal CA1 region: its induction and early temporal development. Prog. Brain Res. 83:223-32.
Huang CC, Liang YC, Hsu KS (1999) A role for extracellular adenosine in time-dependent reversal of long-term potentiation by low-frequency stimulation at hippocampal CA1 synapses. J. Neurosci. 19:9728-9738.
Huang CC, Liang YC, Hsu KS (2001) Characterization of the mechanism underlying the reversal of long term potentiation by low frequency stimulation at hippocampal CA1 synapses. J Biol Chem 276(51):48108-17.
Huang TF (1998) What have snakes taught us about integrins? Cell Mol. Life Sci. 54:527-540.
Hughes PE, Pfaff M (1998) Integrin affinity modulation. Trends Cell Biol. 8:359-364.
Huttenlocher A, Palecek SP, Lu Q, Zhang W, Mellgren RL, Lauffenburger DA, Ginsberg MH, Horwitz AF (1997) Regulation of cell migration by the calcium-dependent protease calpain. J. Biol. Chem. 272:32719-32722.
Jones LS (1996) Integrins: possible functions in the adult CNS. Trends Neurosci. 19(2):68-72.
King VR, McBride A, Priestly JV (2001) Immunohistochemical expression of the α5 integrin subunit in the normal adult rat central nervous system. J. Neurocytol. 30:243-252.
Kramár EA, Bernard JA, Gall CM, Lynch G (2002) Alpha3 integrin receptors contribute to the consolidation of long-term potentiation. Neuroscience 110(1):29-39.
Kramár EA, Bernard JA, Gall CM, Lynch G (2003) Integrins modulate fast excitatory transmission at hippocampal synapses. J Biol Chem 278(12):10722-30.
Kramár EA, Lynch G (2003) Developmental and regional differences in the consolidation of long-term potentiation. Neuroscience 118(2):387-98.
LaBaron RG, Hernandez RV, Orfila JE, Martinez JLJ (2003) An integrin binding peptide reduced rat CA1 hippocampal long-term potentiation during the first few minutes following theta burst stimulation. Neurosci. Lett. 339:199-202.
Larson J, Xiao P, Lynch G (1993) Reversal of LTP by theta frequency stimulation. Brain Res. 600:97-102.
Lee K, Schottler F, Oliver M, Lynch G (1980) Brief bursts of high-frequency stimulation produce two types of structural changes in rat hippocampus. J. Neurophysiol. 44:247-258.
Lee KS, Oliver M, Schottler F, Lynch G (1981) Electron microscopic studies of brain slices: The effects of high frequency stimulation on dendritic ultra structure, in Electrical Activity in Isolated Mammalian CNS Preparations. (G. Kerkut, H. V. Wheal, eds). Vol. pp. 189-212. Academic Press, New York.
Leung LS, Shen B (1999) N-methyl-D-aspartate receptor antagonists are less effective in blocking long-term potentiation at apical than basal dendrites in hippocampal CA1 of awake rats. Hippocampus 9:617-630.
Lin B, Arai AC, Lynch G, Gall CM (2003) Integrins regulate NMDA receptor-mediated synaptic currents. J. Neurophysiol. 89:2874-2878.
Link W, Konietzko U, Kauselmann G, Krug M, Schwanke B, Frey U, Kuhl D (1995) Somatodendritic expression of an immediate early gene is regulated by synaptic activity. Proc. Natl. Acad. Sci. USA 92:5734-5738.
Loeser RF, Forsyth CB, Samarel AM, Im HJ (2003) Fibronectin fragment activation of proline-rich tyrosine kinase PYK2 mediates integrin signals regulating collagenase-3 expression by human chondrocytes through a protein kinase C-dependent pathway. J Biol Chem 278(27):24577-85.
Luthl A, Laurent J, Figurov A, Muller D, Schachner M (1994) Hippocampal long-term potentiation and neural cell adhesion molecules L1 and NCAM. Nature 372:777-779.
Lyford GL, Yamagata K, Kaufmann WE, Barnes CA, Sanders LK, Copeland NG, Gilbert DJ, Jenkins NA, Lanahan AA, Worley PP (1995) Arc, a growth factor and activity-regulated gene, encodes a novel cytoskeleton-associated protein that is enriched in neuronal dendrites. Neuron 14:433-445.
Martin SJ (1998) Time-dependent reversal of dentate LTP by 5 Hz stimulation. NeuroReport 9:3775-3781.
Matsuzaki M, Honkura N, Ellis-Davies GC, Kasai H (2004) Structural basis of long-term potentiation in single dendritic spines. Nature 429:761-766.

Miller RR, Matzel LD (2000) Memory involves far more than 'consolidation'. Nat Rev Neurosci. 1:214-216.
Milner R, Campbell IL (2002) The integrin family of cell adhesion molecules has multiple functions within the CNS. J. Neurosci. Res. 69:286-291.
Miranti CK, Brugge JS (2002) Sensing the environment: a historical perspective on integrin signal transduction. Nature Cell Biol. 4:83-90.
Misanin JR, Miller RR, Lewis DJ (1968) Retrograde amnesia produced by electroconvulsive shock after reactivation of a consolidated memory trace. Science 160:554-555.
Morris RG, Anderson E, Lynch GS, Baudry M (1986) Selective impairment of learning and blockade of long-term potentiation by an N-methyl-D-aspartate receptor antagonist, AP5. Nature 319:774-776.
Morris RG, Halliwell RF, Bowery N (1989) Synaptic plasticity and learning. II: Do different kinds of plasticity underlie different kinds of learning? Neuropsychologia 27(27):41-59.
Mould AP (1996) Getting integrins into shape: recent insights into how integrin activity is regulated by conformational changes. J. Cell Sci. 109:2613-2618.
Muller D, Wang C, Skibo G, Toni N, Cremer H, Calaora V, Rougon G, Kiss JZ (1996) PSA-NCAM is required for activity-induced synaptic plasticity. Neuron 17:413-22.
Muller GE, Pilzecker A (1900) Experimentelle beitrage zur lehre vom gedachtnis. Z. Psychol. S1:1-288.
Murase S, Hayashi M (1996) Expression pattern of integrin ß1 subunit in Purkinje cells of rat and cerebellar mutant mice. J. Comp. Neurol. 375:225-237.
Murase SI, Hayashi Y (1998) Integrin α1 localization in murine central and peripheral nervous system. J. Comp. Neurol. 395:161-176.
Nader K (2003) Memory traces unbound. Trends Neurosci 26:65-72.
Nelson K, Fivush R (2004) The emergence of autobiographical memory: a social cultural developmental theory. Psychol. Rev. 111:486-511.
Nguyen PV, Abel T, Kandel ER (1994) Requirement of a critical period of transcription for induction of a late phase of LTP. Science 265:1104-1107.
Nishimura SL, Boylen KP, Einheber S, Milner TA, Ramos DM, Pytela R (1998) Synaptic and glial localization of the integrin αvß8 in mouse and rat brain. Brain Res. 791:271-282.
Nusser Z, Lujan R, Laube G, Roberts JD, Molnar E, Somogyi P (1998) Cell type and pathway dependence of synaptic AMPA receptor number and variability in the hippocampus. Neuron 21:545-559.
Pinkstaff JK, Detterich J, Lynch G, Gall C (1999) Integrin subunit gene expression is regionally differentiated in adult brain. J. Neurosci. 19:1541-1556.
Pinkstaff JK, Lynch G, Gall CM (1997) Localization and seizure regulation of integrin ß1 mRNA in adult rat brain. Mol. Brain Res. 19:1541-1556.
Plow EF, Haas TA, Zhang L, Loftus J, Smith JW (2000) Ligand binding to integrins. J. Biol. Chem. 275:21785-21788.
Porter JC, Hogg N (1998) Integrins take partners: cross-talk between integrins and other membrane receptors. Trends Cell Biol. 8:390-396.
Ribot T (1882) Diseases of Memory. New York, Appleton.
Richardson A, Parsons TJ (1995) Signal transduction through integrins: a central role for focal adhesion kinase? BioEssays 17(3):229-236.
Ridley AJ, Schwartz MA, Burridge K, Firtel RA, Ginsberg MH, Borisy G, Parsons JT, Horwitz AF (2003) Cell migration: integrating signals from front to back. Science 302:1704-1709.
Rodriquez MA, Pesold C, Liu WS, Kriho V, Guidotti A, Pappas GD, Costa E (2000) Colocalization of integrin receptors and reelin in dendritic spine postsynaptic densities of adult nonhuman primate cortex. Proc. Natl. Acad. Sci. USA 97:3550-3555.
Ruoslahti E, Pierschbacher MD (1986) Arg-Gly-Asp: a versatile cell recognition signal. Cell 44:517-518.
Sah P, Bekkers JM (1996) Apical dendritic location of slow afterhyperpolarization current in hippocampal pyramidal neurons: implications for the integration of long-term potentiation. J. Neurosci. 16:4537-4542.
Schlaepfer D, Hunter T (1998) Integrin signalling and tyrosine phosphorylation: just the FAKs? Tnds Cell Biol. 8:151-157.
Schoenwaelder SM, Yuan Y, Cooray P, Salem HH, Jackson SP (1997) Calpain cleavage of focal adhesion proteins regulates the cytoskeletal attachment of integrin alpahIIbbeta3 (platelet glucoprotein IIb/IIa) and the cellular retraction of fibrin clots. J. Biol. Chem. 272:1694-1702.
Schuster T, Krug M, Stalder M, Hackel N, Gerardy-Schahn R, Schachner M (2001) Immunoelectron microscopic localization of the neural recognition molecules L1, NCAM, and its isoform NCAM180, the NCAM-associated polysialic acid, ß1 integrin and the extracellular matrix molecule tenascin-R in synapses of the adult rat hippocampus. J. Neurobiol. 49:142-158.
Schwartz MA, Ginsberg MH (2002) Network and crosstalk: integrin signalling spreads. Nature Cell Biol 4:E65-E68.

Siman R, Baudry M, Lynch G (1984) Brain fodrin: substrate for the endogenous calcium-activated protease, calpain I. Proc. Natl. Acad. Sci., USA 81:3572-3576.

Smart FM, Edelman GM, Vanderklish PW (2003) BDNF induces translocation of initiation factor 4E to mRNA granules: evidence for a role of synaptic microfilaments and integrins. Proc Natl Acad Sci U S A. 100:14403-14408.

Stanton PK, Sarvey JM (1984) Blockade of long-term potentiation in rat hippocampal CA1 region by inhibitors of protein synthesis. J. Neurosci. 4:3080-3088.

Stanton PK, Sarvey JM (1987) Norepinephrine regulates long-term potentiation of both the population spike and dendritic EPSP in hippocampal dentate gyrus. Brain Res. Bull. 18:115-119.

Staubli U (1992) A peculiar form of potentiation in mossy fiber synapses. Epilepsy Res. Suppl. 7:151-157.

Staubli U, Chun D (1996) Factors regulating the reversibility of long term potentiation. J. Neurosci. 16:853-860.

Staubli U, Chun D, Lynch G (1998) Time-dependent reversal of long-term potentiation by an integrin antagonist. J. Neurosci. 18(9):3460-3469.

Staubli U, Faraday R, Lynch G (1985) Pharmacological dissociation of memory: anisomycin, a protein synthesis inhibitor, and leupeptin, a protease inhibitor, block different learning tasks. Behav. Neural. Biol. 43:287-297.

Staubli U, Lynch G (1987) Stable hippocampal long-term potentiation elicited by "theta" pattern stimulation. Brain Res. 435:227-234.

Staubli U, Lynch G (1990) Stable depression of potentiated synaptic responses in the hippocampus with 1-5 Hz stimulation. Brain Res. 513:113-118.

Staubli U, Vanderklish PW, Lynch G (1990) An inhibitor of integrin receptors blocks LTP. Behav. Neural Biol. 53:1-5.

Streuli CH, Edwards GM (1998) Control of normal mammary epithelial phenotype by integrins. J. Mammary Gland Biol. Neoplasia 3:151-163.

Takumi Y, Ramirez-Leon V, Laake P, Rinvik E, Ottersen OP (1999) Different modes of expression of AMPA and NMDA receptors in hippocampal synapses. Nature Neurosci. 2:618-624.

Tang L, Hung CP, Schuman EM (1998) A role for the cadherin family of cell adhesion molecules in hippocampal long term potentiation. Neuron 20:1165-1175.

Tomita S, Nicoll RA, Bredt DS (2001) PDZ protein interactions regulating glutamate receptor function and plasticity. J. Cell Biol. 153:F19-F23.

Toni N, Buchs PA, Nikonenko I, Bron CR, Muller D (1999) LTP promotes formation of multiple spine synapses between a single axon terminal and a dendrite. Nature 402:421-425.

Trachtenberg JT, Chen BE, Knott GW, Feng G, Sanes JR, Welker E, Svoboda K (2002) Long-term in vivo imaging of experience-dependent synaptic plasticity in adult cortex. Nature 420, 788-794.

Travis MA, Humphries JD, Humphries MJ (2003) An unraveling tale of how integrins are activated from within. Trends Pharm. Sci. 24:192-197.

Van Harreveld A, Fifkova E (1975) Swelling of dendritic spines in the fascia dentata after stimulation of the perforant fibers as a mechanism of post-tetanic potentiation. Exp. Neurol. 49:736-749.

Vanderklish P, Saido TC, Gall C, Arai A, Lynch G (1995) Proteolysis of spectrin by calpain accompanies theta-burst stimulation in cultured hippocampal slices. Mol. Brain Res. 32:25-35.

Vanderklish PW, Krushel LA, Holst BH, Gally JA, Crossin KL, Edelman GM (2000) Marking synaptic activity in dendritic spines with a calpain substrate exhibiting fluorescence resonance energy transfer. Proc. Natl. Acad. Sci. USA 97:2253-2258.

Vuori K (1998) Integrin signaling: tyrosine phosphorylation events in focal adhesions. J. Membrane Biol. 165:191-199.

Wu J, Rowan MJ, Anwyl R (2004) Synaptically stimulated induction of group I metabotropic glutamate receptor-dependent long-term depression and depotentiation is inhibited by prior activation of metabotropic glutamate receptors and protein kinase C. Neurosci. 123:507-514.

Wu X, Davis GE, Meininger GA, Wilson E, Davis MJ (2001) Regulation of the L-type calcium channel by α5ß1 integrin requires signaling between focal adhesion proteins. J. Biol. Chem. 276:30285-30292.

Xiao P, Bahr BA, Staubli U, Vanderklish PW, Lynch G (1991) Evidence that matrix recognition contributes to stabilization but not induction of LTP. Neuroreport 2(8):461-464.

Xu L, Anwyl R, Rowan MJ (1998) Spatial exploration induces a persistent reversal of long-term potentiation in rat hippocampus. Nature 394(6696):891-4.

Yuste R, Bonhoeffer T (2001) Morphological changes in dendritic spines associated with long-term synaptic plasticity. Annu Rev Neurosci. 24:1071-1089.

Zhuo M, Zhang W, Son H, Mansuy I, Sobel RA, Seidman J, Kandel ER (1999) A selective role of calcineurin alpha in synaptic depotentiation in hippocampus. Proc Natl Acad Sci U S A 96(8):4650-5.

MORPHOLOGICAL PLASTICITY OF THE SYNAPSE

Interactions of Structure and Function

Diano F. Marrone, Janelle C. LeBoutillier, and Ted L. Petit*

1. INTRODUCTION

The neural mechanisms that underlie learning and memory continue to be one of the most intensive areas of study within behavioral neuroscience. While the study of learning and memory has a long history, many of the most notable advancements have been in the pursuit of its biological basis.

Although alluded to by many previous researchers, the notion that small, coordinated changes in the efficacy of individual synapses may dramatically alter the behavior of neurons, and ultimately organisms themselves, was first stated explicitly by Donald Hebb in 1949. Since that time, a multitude of mechanisms that alter the efficacy of transmission at an individual synapse have been discovered. This included both structural [throughout the following discussion, structural changes will be taken as conformational changes in synapse shape observable by electron microscopy (EM)] and non-structural changes. The challenge for future research is to establish precisely how and when these changes occur, and how they may interact. This interaction between structure and function in mediating learning and memory is of prime importance, since they are inextricably linked – the structure of a given neural region puts constraints on its function, and changes in the functional demands placed on a given region often alter its structure. Unfortunately, despite may significant discoveries about the mechanisms of neural plasticity, the specific nature of the changes that occur across neural regions and during various forms of learning and memory remain unclear. This, however, is not surprising given the complexity of the problem. Learning and memory is a multi-system problem with components that span several levels of complexity in the brain, from molecules to synapses to neural circuits, and information about each of these levels will be crucial to ultimately uncovering of the dynamics of this phenomenon.

There are a number of ways in which a synapse may alter its structure to potentially increase its efficacy (detailed in section 3). In addition to this complexity in

*Diano F. Marrone, Arizona Research Laboratories Div. of Neural Systems, Memory and Aging, University of Arizona, Tucson, AZ, USA, 85724. Janelle C. LeBoutillier & Ted L. Petit, Dept. of Life Sciences, University of Toronto, Toronto, ON, Canada, M1C 1A4. E-mail: petit@scar.utoronto.ca

the form of individual synapses, one must consider the diverse structures in which these synapses are contained may make unique contributions to the processes of learning and memory, mediated by distinct physiological properties. In turn, these physiological changes may drive, or interact with, changes in synapse number and structure in unique ways (section 4). Finally, several changes in synapse function have been uncovered which are not readily observable at the structural level and may interact with these observable changes. Some of these largely unexplored potential interactions in this cascade will be addressed (section 5). Collectively, these data provide a substrate for evaluating what aspects of synaptic structural plasticity are well established, as well as which areas are critical to future research, in order to plot a course towards a full understanding of the relationship of synapse structure and function, and how these dynamics may mediate learning and memory.

2. LONG-TERM POTENTIATION AS A MODEL OF LEARNING

Gathering data regarding the structural changes associated with behavioral plasticity is in several regards problematic. This is because the neural circuits which are presumably activated during such learning are likely to be highly distributed. As such, it becomes difficult to say convincingly that the neuropil reconstructed using any imaging technique (which, by virtue of the labor-intensive nature of most modern forms of imaging, is quite small) reflects those few activated contacts. However, despite these complications, imaging methods such as EM remain crucial to the elucidation of the fine structural changes induced by changes in neural activity. In an effort to balance both the advantages and inherent complications of such imaging techniques, many investigations have turned to one of the most convincing cellular models of learning and memory – long-term potentiation (LTP). The adoption of this model allows very robust changes to be induced in a compact area of neuropil, and thus reconstructing this region maximizes the probability that any conformational changes induced will be captured accurately.

Before proceeding with an in-depth discussion of the data surrounding structural and functional plasticity following LTP, a brief discussion of the model system which has given rise to much of this data is warranted. The process of LTP may be defined as the sustained increase (for hours, days, or weeks) in the amplitude of the response evoked in a cell, or population of cells, by a test pulse delivered to an afferent pathway following stimulation of that pathway (Landfield and Deadwyler, 1988). Although the first example of LTP was induced in the dentate gyrus (Bliss and Lomo, 1973), this phenomenon has been most thoroughly investigated in hippocampal CA1 pyramidal neurons following stimulation of Schaffer collaterals. However, neocortical LTP has also recently become a well-established phenomenon both in slices (e.g., Kirkwood et al., 1993) and *in vivo* (e.g., Trepel and Racine, 1998).

While there is debate about whether LTP truly is a mechanism of learning and memory (e.g., see Holscher, 1999), many researchers consider this form of synaptic plasticity to be a good model of how memory might work at the cellular level (justification of LTP as a memory model is beyond the scope of this discussion, the interested reader is directed to Morris, 2003 for a more detailed discussion). However, because of this debate on how well LTP mimics the plasticity induced by learning and memory, behavioral evidence will also be addressed whenever possible.

3. STRUCTURAL CHANGES IN THE SYNAPSE

Much research has been conducted on the structural correlates of both learning and LTP. As these morphological correlates of synaptic plasticity have been the topic of numerous reviews (e.g., Klintsova and Greenough, 1999; Muller et al., 2000; Marrone and Petit, 2002; Harris et al., 2003), they will not be described in detail. However, the array of potential mechanisms revealed requires brief discussion.

Foremost, changes in the synapse following altered experience may be divided into two broad categories – synaptogenesis and morphogenesis. The former refers to the formation of new synaptic contact, either by *de novo* synapse formation, or through the formation of multi-synaptic boutons (Figure 1a). In contrast, morphogenesis refers to the alteration of the structure of existing synapses to alter their efficacy of transmission (Figure 1b). When considering morphogenesis, several means of structural change, both pre- and postsynaptic, have been uncovered.

Presynaptic structure may change in a number of ways to alter transmission. The vesicular content of a bouton may change in two related ways, based on the identification of two functionally distinct pools of vesicles – the readily releasable pool (RRP) and the reserve pool. The RRP is defined as the population of vesicles that are immediately available for release at the start of prolonged high-frequency stimulation. It is thought that the RRP coincides with the pool of morphologically docked vesicles, i.e., vesicles that appear immediately adjacent to the active zone membrane in electron micrographs. In support of this notion, the average number of morphologically docked vesicles is equal to the average size of the RRP (von Gersdoff et al., 1996; Schikorski and Stevens, 1997), the RRP size and docked vesicle number have the same distribution for a population of synapses (Schikorski and Stevens, 1997; Murthy et al., 1997), and these two pools are the same size at individual synapses (Schikorski and Stevens, 2001). As the RRP helps determine the probability of transmitter release (PTR), changes in the number of morphologically docked vesicles may have a profound impact on synaptic strength.

In addition to the RRP, the reserve pool (the remaining vesicular content of an individual bouton) may change. This may impact synapse strength due to an indirect effect on the RRP. This is because the RRP empties more quickly than it can be refilled during intense presynaptic activity, and short-term depression results. Although recent data (e.g., Stevens and Wessling, 1999) indicates that a secondary process may be involved, several workers have speculated that the longer recovery times from depression caused by extensive activity might reflect the depletion of reserve vesicles (the remainder of non-morphologically docked vesicles) upon which the RRP draws when it refills (e.g., Birks and MacIntosh, 1961; Elmqvist and Quastel, 1965; Liu and Tsien, 1995a, 1995b). Thus, there is evidence for activity-dependent modification of both the RRP and reserve pool, each resulting in changes in the PTR, and consequently synapse strength.

In addition to changes in vesicle content, the presynaptic bouton itself may alter in size. Several studies utilizing combined imaging and electrophysiology (e.g., Mackenzie et al., 1999) suggest that larger boutons have greater quantal size (i.e., release more neurotransmitter and evoke a greater postsynaptic response). However, modeling evidence suggests that while terminal growth increases quantal size, it also lowers PTR. This is because the process of neurotransmitter release is critically dependent upon

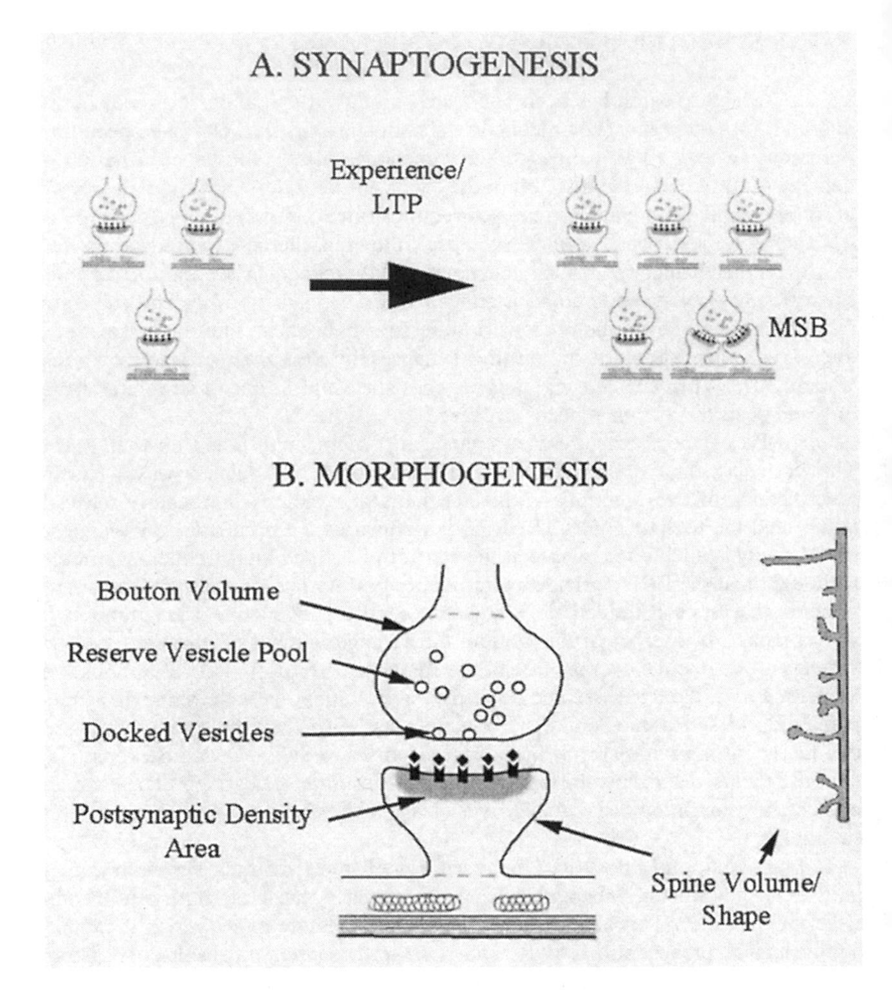

Figure 1. Schematic diagram of the various types of structural plasticity outlined in section 3, categorized as (A) synaptogenesis and (B) morphogenesis. Under the term of synaptogenesis, experience and LTP increase the total number of synaptic contacts by both *de novo* synaptogenesis, as well as by the creation of multi-synaptic boutons (MSB). Under morphogenesis, a schematic synaptic contact (center) illustrates the various means by which synapses may change their structure to alter efficacy. Also shown is a schematic of dendritic spines (right), illustrating the variation in their morphological profiles (adapted from Fiala and Harris, 1999).

residual intracellular calcium concentration ($[Ca^{2+}]_i$) accumulating during repetitive activity in both short- and long-term synaptic enhancement. In fact, all phases of this enhancement are attributable at least in part to a presynaptic increase in the number of transmitter quanta released, and all phases require Ca^{2+} entry for their induction (reviewed in Zucker, 1994; 1999). As diffusion constitutes the primary means of transporting free ions inside living cells, modifying the geometric boundaries of the diffusion medium alters $[Ca^{2+}]_i$ locally, and, thus synaptic function as a whole (Ghaffari-

Farazi et al., 1997; 1999). Several mathematical models (e.g., Bertram, 1997; Bertram et al., 1999) have postulated the effects that changing membrane geometry may have on synaptic function. The most notable of these models, however, is one recently developed by Ghaffari-Farazi et al. (1997, 1999), as this model allows simulation across realistic terminal morphology, revealing several important conclusions This model has shown that, in general, PTR is increased as: (a) the width of the synaptic compartment or partition (in a partitioned/perforated synapse) decreased; (b) the distance from the channel to the membrane wall of the compartment/partition decreased; or (c) the size of the neck connecting a partition to the rest of the bouton decreased. Thus, the more restricted the volume surrounding the release site, the higher the local $[Ca^{2+}]_i$ (since it is less able to diffuse), and the higher the PTR. In a larger active zone, the average distance from the release site to the terminal wall is greater, and thus the average PTR at any individual release site is lower. Thus, it seems that increases in synapse size ***without*** any other change in structure serves to decrease PTR, as an increase in terminal area facilitates Ca^{2+} diffusion.

This issue relates to two other changes in synaptic morphology – synaptic curvature and perforations/partitions. Synapses generally exhibit a great deal of variation in the curvature of the cleft, ranging in the dentate gyrus (for example) from 39° (concave) to -35° (convex) inflection (Marrone et al., 2004). Recent modeling evidence (Wu et al., 2003) suggests that this variation may have a profound impact on PTR with repetitive stimulation. When a vesicle is docked near a point of concave inflection (i.e., the presynaptic membrane protrudes into the postsynaptic cleft), PTR increases, while the opposite is true for a vesicle docked near a convex inflection point. Thus, it seems that the manipulation of synaptic curvature serves both to compensate for the loss of PTR with size, and augment synaptic efficacy in such a way that both quantal size and PTR can increase simultaneously. In addition to curvature, a number of synapses exhibit either perforations (breaks in the postsynaptic PSD), or partitions (membranes invaginating the presynaptic terminal). The addition of a perforation may increase PTR simply by moving release sites away from the center of the terminal (where PTR is lowest) towards terminal walls. Thus, the addition of a perforation serves to increase the total PTR versus a single active zone of the same size. In addition, the creation of a synaptic partition segments the terminal into smaller compartments, which minimizes terminal area in the vicinity of the release site, and functionally creates two smaller, more effective synapses that fire simultaneously. These synapses would have a combined quantal size equal to the original synapse, but with a higher PTR. Thus, there appears to be a delicate balance between bouton shape, size, and synaptic efficacy.

At the postsynaptic level, much structural plasticity has been attributed to the dendritic spine. The addition of dendritic spines increases the potential of synapse packing density, thus playing a critical role in synaptogenesis. This is especially true for the formation of multisynaptic boutons, which are thought to be formed by the outgrowth of filopodia (immature dendritic spines) adjacent to areas that have received large amounts of stimulation. In addition, spine morphology may play a role in the plasticity of synaptic transmission. This is mediated by a similar dynamic between $[Ca^{2+}]_i$, and physiological change at the synapse as described in the presynaptic terminal.

Imaging experiments show that spines compartmentalize calcium such that localized changes in $[Ca^{2+}]_i$ do not spread to neighboring inactive synapses (Yuste and Denk, 1995; Guthrie et al., 1991; Muller and Connor, 1991). Spine shape and size likely alter these dynamics (see Majewska et al., 2000), and translate into different signaling events at the synapse. In addition, many biophysical models suggest that spine necks can slow charge-transfer from the synapse to the parent dendrite (Segev and Rall, 1988; 1998). Hence there is a larger potential in the spine head for a transient period after synaptic activation which facilitates the opening of voltage-dependent channels.

Given this brief description of the various means available for changes in structure to impact synapse function, it is apparent that this is a tremendously complex problem in which many questions still remain to be addressed. For instance, critical understanding is still forthcoming as to which types or patterns of stimulation induce which types of morphological changes. In turn, it is not clear which types of morphological changes (if any) support which types of learning and memory. Moreover, distinguishing those changes which support learning and memory from those that may serve other functions, such as supporting the greater metabolic demands of a synapse with greater levels of activity, remains problematic. Critical understanding in these areas may be achieved by examining which morphological cascades are induced in neural regions that display distinct physiological dynamics.

4. DIVERSE CONTRIBUTIONS TO LEARNING AND MEMORY

Although structural change in an individual synapse is a complex set of phenomena which may work alone or in concert to generate changes in synapse efficacy, to consider these synaptic changes in isolation would be too simplistic. Learning and memory is a multi-faceted process mediated by many distinct neural regions, each acting with unique physiological properties. Uncovering what morphological changes may be preferentially induced in neural regions with distinct physiology may be crucial to linking synapse structure and function. Although a discussion of all of the various regions which contribute to learning and memory is beyond the scope of this discussion, two of the best understood regions – the hippocampus and the neocortex – will be discussed to illustrate these structure/function interactions. Before proceeding with such a discussion, however, the functional roles of these regions must be described.

4.1 Behavioral Contribution of the Hippocampus vs. Neocortex

The neocortex is widely assumed to store traces of experience underlying both explicit (declarative) and implicit (procedural) learning. In the former case, it is generally assumed that structures within the medial temporal lobe, including the hippocampus, are involved in the earliest stages of encoding and storage, somehow guiding the eventual consolidation of information in the cortex. Evidence from both humans and animal models (e.g., Squire et al., 1993) have suggested that the hippocampus and associated structures are involved in the rapid, but nonpermanent, storage of certain types of information, while the neocortex is responsible for the ultimate long-term storage of information in cortico-cortical maps (Buonomano and Merzenich, 1998). The change in dependence from the hippocampus to the neocortex appears to be slow and gradual. Previously learned memories can remain hippocampal-dependent for a prolonged period. Studies in rats (Winocur, 1990; Kim and Fanselow, 1992) have produced retrograde gradients covering a period of days or weeks. Primate experiments (Zola-Morgan & Squire, 1990) show a severe impairment relative to controls for memory acquired 2 or 4 weeks prior to surgery, but not for older memories. Thus, memory in all of these cases depends initially on plastic changes within the hippocampal system, but the knowledge underlying them is eventually stored in the neocortical system via the gradual accumulation of small changes (i.e., the process of consolidation). These dynamics suggest that, while both the neocortex and hippocampus are involved in the acquisition and storage of information, the hippocampus does so utilizing a fast learning rate, while neocortical circuits operate with a slow learning rate (a full description of the evidence for this theory is beyond the scope of this text – see McClelland et al., 1995; Eichenbaum, 2000 for further detail).

Electrophysiological data on hippocampal and neocortical LTP show some interesting parallels with this model. Foremost, acquisition rates of LTP seem to reflect those potentially invoked by learning and memory. For example, Ivanco and Racine (2000) report that repeated daily stimulation of the neocortex in a chronically implanted animal induces an LTP that reaches asymptotic levels by about 10-15 days. This is significantly longer than the 1-3 days typically required to induce lasting LTP in the hippocampus (Ivanco and Racine, 2000). Moreover, neocortical LTP is longer lasting than hippocampal LTP as typically induced (Racine et al., 1983; Trepel and Racine, 1998). These dynamics support neocortical LTP as a suitable model for long-term memory storage.

4.2 Structural Change in the Neocortex vs. the Hippocampus: Relation to Function

As described above, a wide array of evidence suggests that the hippocampus works to rapidly integrate incoming information using a fast learning rate, while neocortical circuits utilize a slow learning rate to consolidate this information into a semi-permanent state (e.g., McClelland et al., 1995). Given these dynamics, consideration may now be given to how unique changes in synapse structure in these two regions may mediate these distinct functional properties.

Quantitative morphometry in the hippocampus suggests that hippocampal synapses tend to follow certain universal trends in the presence of enhanced activation, whether induced by development, experience, LTP, or any other synaptic efficacy-enhancing event. Various morphological, molecular, and behavioral correlates among hippocampal development, lesion recovery, and LTP have appeared in numerous reviews (e.g., Bulinski et al., 1998; Ivanco and Greenough, 2000; Marrone and Petit, 2002). As such, they will only be briefly mentioned here.

Foremost, plastic events in the hippocampus are characterized by a remarkable ***stability*** in synapse number. That is, a wide array of studies, using different induction techniques, imaging methods, and stereology (Desmond and Levy, 1983; 1986a, b; Geinisman et al., 1992; 1993; 1996; Sorra and Harris, 1998; Weeks et al., 1999; 2000; Stewart et al., 2000; Ostroff et al., 2003) have repeatedly reported no overall change in synapse number associated with LTP. This appears to be the most robust morphometric finding regarding LTP in the hippocampus. However, this stability in synapse number does not mean there is a lack of reorganization. In fact, changes in the relative proportions of many types and forms of synapses have been reported, and these will be discussed briefly below.

The most robust structural changes observed in individual synaptic contact seem to involve synapse curvature and perforations. One of the most rapid morphological alterations following appropriate patterns of synaptic activity is a change in synaptic curvature. The alteration of curvature appears to depend on the synapse's proximity to the locus of stimulation: Synapses in the immediate vicinity of heightened activation tend to become concave (i.e., synapses in which the presynaptic bouton invaginates into the postsynaptic spine), while synapses in adjacent regions tend to become convex shaped contacts (see Marrone and Petit, 2002).

In addition, an increase in the relative proportion of perforated synapses has long been associated with the induction of hippocampal LTP. Weeks et al. (1999; 2000; 2001) reported that stimulated synapses appear to become more perforated for only a short duration (1 hr) following LTP induction in the dentate gyrus. Specifically, this increase occurred predominantly in concave perforated synapses and these concave perforated synapses remained prevalent at more distant time points (24 hrs and 5 days). This phenomenon seemed to be an intermediate in a process where stimulated synapses may form new smaller and more numerous concave synapses. Similar to our results, Geinisman et al. (1991; 1993) found an increase in the number of perforated synapses per neuron 1 hr following the induction of LTP in the rat dentate gyrus. This increase in perforated synapses observed at 1 hr post-induction was no longer evident at 13 days post-induction (Geinisman et al., 1996). This observed transience in perforated synapses is consistent with Weeks' finding of no significant changes in the overall density of perforated synapses at 5 days following induction. Recently, these observations were confirmed using calcium accumulation markers to identify activated synapses (Buchs and Muller, 1996; Toni et al., 1999). Following LTP induction, Buchs and Muller (1996) found that the majority (60%) of labeled synapses exhibited perforations, whereas this proportion was significantly lower (only 20%) in synapses not labeled. Also using calcium markers, Toni et al. (1999) reported increased perforations in activated synapses during the first 30 min following in vitro LTP induction in the rat CA1 region. Thus, it seems that the proliferation of perforated synapses is not a generalized phenomenon, but rather restricted to those synapses activated through LTP induction. However, it is not clear whether the observed overall transient nature of perforated synapses mentioned above is due to a global reduction of synaptic perforations, or retention of synaptic perforations in activated synapses accompanied by a reduction of perforations in non-activated synapses.

General increases in synaptic size have been reported following LTP in several studies (e.g., Chang and Greenough, 1984; Desmond and Levy, 1988). In turn, these increases in overall synaptic size have been shown to correlate highly with synaptic perforations, as perforated synapses are generally much larger than macular (i.e., non-perforated) synapses (e.g., Jones and Calverly, 1991; Markus et al., 1994). Desmond and Levy (1983) reported a significant increase in the total postsynaptic surface area per unit volume for only concave shaped synapses. Based on their results, Desmond and Levy (1986a, b) proposed that a rapid structural conversion occurs following the induction of synaptic potentiation, in which synapses change from convex to concave and subsequently grow in size. This type of structural conversion may explain the temporary decrease in synaptic length reported by Weeks et al. (1999; 2000; 2001), who reported an increase in synaptic active zone length at 1 hr following LTP induction. However, Desmond and Levy observed an overall decrease in synaptic length at 24 hr following induction. At 5 days post-induction, synaptic length once again increased. Buchs and Muller (1996) reported that following LTP induction synaptic profiles labeled as activated synapses through calcium accumulation markers were characterized by a larger apposition zone between pre- and postsynaptic structures, and longer postsynaptic densities compared to non-marked synapses. These results add strong support to the idea that ultrastructural modifications may occur only in those synaptic contacts activated through LTP induction.

Thus, based on the accumulated literature regarding the morphological correlates of LTP, it seems that hippocampal plasticity is characterized by rapid yet transient structural changes in individual synaptic contacts without changes in net synapse numbers. The available behavioral evidence largely supports the transience of hippocampal morphological change. For instance, Rusakov et al. (1997) did not report significant long-term changes in density or sizes of synapses, in either dentate gyrus or CA1 areas of hippocampus when morphometric analyses were carried out on spatially trained animals. This transience is best supported, however, by a study conducted recently by Eyre et al. (2003), who assessed retention of the Morris water maze, as well as synapse density in the dentate gyrus, at 3-, 9-, and 24-hr after the beginning of training. The 3-hr group displayed neither retention nor morphological change, while the 9-hr group displayed both. Thus, these early time point data suggest that synapse formation in the hippocampus does coincide with learning in a hippocampal-dependent task. At 24-hr after the beginning of training, however, rats still retained memory for the platform location, but showed control levels of synapse density.

However, as stated above, this lack of net synaptogenesis does not imply a lack of synaptic reorganization. Several experiments have found synaptogenesis in specific synaptic populations. For instance, Ramirez-Amaya et al. (1999) demonstrated a significant increase of mossy fiber terminals in the CA3 stratum oriens region induced by training rats during 3 days in a Morris water maze through histological methods, combined with EM, in connection with spatial learning. In addition, Rusakov et al. (1997), using a multivariate analysis of synapse distribution, found a significant training-associated increase in synapse clustering (shorter distances between active zones) in hippocampal CA1. The key conclusion to be made is that, whatever means of physical reorganization is generally used by synaptic networks to support hippocampal plasticity, long-term changes in net synapse numbers is not among them.

Thus, it seems that morphological plasticity in the hippocampus (as confirmed by both LTP and behavioral evidence) is characterized by a rapid succession of structural changes in individual synapses, including increased synaptic perforations, and synapse growth, without changes in total synapse numbers. This fits well with the theoretical notion that the hippocampus supports the immediate but transient amalgamation of information coming in from all sensory modalities, and participating in the eventual consolidation of this information within the neocortex. Given the distinct functional role of the neocortex, one may expect that the typical morphological cascade observed following plastic events in the neocortex may be equally different. As detailed below, the limited LTP data available, coupled with confirmatory behavioral evidence, suggests that this is the case.

Using the Golgi-Cox preparation to examine dendritic branching and spine number (both of which highly correlate with synapse number) following LTP, increases have been observed in the dendritic length, complexity and spine number of basilar dendrites of cortical Layer V near the site of potentiation (Monfils et al., 2004). Using a similar protocol, increases in both apical and basilar spine number, as well as increased apical dendritic field complexity have been reported (Ivanco et al., 2000). The increase in dendritic material in layer V suggests a greater number of excitatory synapses in the potentiated animals. These conclusions, however, require verification from EM, especially considering that Golgi-based data has suggested an increase in spine density in the hippocampus following LTP induction (Trommald et al., 1996), despite the fact that a wide array of EM evidence has concluded otherwise.

Although, to our knowledge, only one such study utilizing EM-based stereological techniques has been conducted, it does partially support the evidence described above. Using the same chronic implantation protocol as the two EM studies described above, Marrone (2004) has found an increase in the number of synapses per neuron in Layer V, but not layer II/III, in the region of potentiation. It is likely that this increase reflects an LTP-specific mechanism, since current source density analysis suggests that Layer V is the sink for this potentiated response (Chapman et al., 1998). Moreover, this increase was restricted to excitatory axo-spinuous synapses, suggesting that cortical LTP expression is mediated by NMDA-dependent increases in excitatory transmission, rather than a GABA-mediated decrease in inhibition. In addition, this synaptogenesis occurred without significant morphological change in any synaptic component. That is, no consistent changes were observed in the size or curvature of synaptic contacts; the size or vesicular content of boutons; or the size of dendritic spines.

This profile of intense synaptogenesis in the absence of morphogenesis is in essence opposite to the changes typically associated with hippocampal LTP. However, this limited evidence requires confirmation from the wide array of data on the morphological correlates of cortical-dependent behavior.

These studies can be parsed into two broad categories – studies examining exposure to enriched environments, and those which train animals in specific skilled tasks. In enrichment studies, one group of animals is typically placed in impoverished conditions (IC), consisting of standard laboratory cages, while a second group is housed in enriched conditions (EC) – a more stimulating environment containing group-housing conditions, visually stimulating objects and an opportunity to interact freely with the environment. A third group is also usually added receiving social conditions (SC), in which animals are group-housed without the externally stimulating environment in order to control for the effects of social interaction. For brevity, only differences found exclusively in EC (i.e., differences in EC vs. both SC and IC) will be discussed.

The bulk of the evidence indicating that different experiences can lead to differences in synapse numbers involves inference from measures of postsynaptic surface (spines and dendrites) in Golgi impregnated tissue. From this, there is an assumption that dendritic space is correlated tightly with synaptic numbers. In general, these studies have found similar results (e.g., Beaulieu and Colonnier 1987, Floeter and Greenough 1979, Stell and Riesen 1987). Typically, EC-exposed animals showed dramatic increases of approximately 20% in the dendritic fields of neurons in the visual cortex (e.g. Greenough and Volkmar, 1973; Volkmar and Greenough, 1972), along with lesser increases in other neural areas (e.g., Greenough et al., 1973) relative to IC-exposed animals.

Although the synaptic changes observed following exposure to enriched environments support the concept that learning about the environment can be related to synaptic number in the cerebral cortex, there is little direct evidence of this. Training animals in specific tasks and then investigating changes in the morphology of neurons in regions suspected of being involved in their performance provides a more direct approach to this issue. Such studies have been conducted in a variety of modalities.

In the visual domain, Chang and Greenough (1982) took advantage of the fact that the visual pathways of the laboratory rat are about 90% crossed by occluding one eye of rats and then training them in a visual maze. Comparison of the neurons in the two hemispheres revealed that those in the trained hemisphere (contraleral to the open eye) had larger dendritic fields.

In addition, a large number of experiments have taken advantage of the natural talents of rats in completing complex motor tasks. Because the cortical control of the forelimbs is largely crossed, it is possible to train one limb to reach for food and to compare layer V neurons in the forelimb region of motor cortex in the trained and untrained hemispheres. Again, several studies have shown increases in dendritic arborization and spine density in "trained" regions (Greenough et al., 1985; Kolb et al., 1998, Withers and Greenough, 1989). In a recent study, Kleim et al. (2004) investigated how cortical plasticity emerges during motor skill learning by combining intra-cortical micro-stimulation with quantitative EM to measure both the functional and structural plasticity in layer V simultaneously. Trained animals had significantly larger distal forelimb representations only after 10 days of training, and had more synapses per neuron after 7 and 10 d of training only. These results show that the onset of synaptogenesis precedes map reorganization, suggesting a functional role for the newly formed synapses in memory consolidation during late stages of training.

Similar results have also been found with olfactory learning (Knafo et al., 2001). Three days after training rats in an olfactory discrimination task, Knafo et al. reported that spine density along apical dendrites was 15% higher in layer II pyramidal neurons of trained (compared to pseudo-trained or naïve) rats. Knafo et al. suggest that this increase in spine density in trained rats indicate an increased number of excitatory synapses interconnecting pyramidal neurons in the piriform cortex following olfactory learning.

From this accumulated data, the changes in dendritic fields seen in the studies of learning across several modalities are strikingly similar to the changes seen in studies of enriched rearing, suggesting that the observed changes in synaptic connectivity in animals in enriched environments are involved in learning and memory (Greenough and Chang 1989). Although there may ultimately be fine distinctions, it seems that all of these manipulations (i.e., LTP induction, exposure to enriched environments, and skilled tasks training) create the same type of qualitative changes – altered synapse number with limited changes in structure.

Again, this data is at odds with the available data on morphological change associated with hippocampal plasticity. In general, plastic responses in cortical (vs. hippocampal) synapses (a) involve long-term increases in synaptic number; and (b) do not involve morphological change other than in synapse number (at least not on the same scale that they have been observed in the hippocampus). Collectively, this suggests that the distinct electrophysiological properties of cortical synapses versus their hippocampal counterparts are mediated by distinct sets of morphological cascades, and two strong correlations have emerged. A robust correlation has repeatedly been reported between synaptogenesis and late and enduring synaptic plasticity. In addition, many reports have supported a strong correlation between changes in synapse structure (curvature, spine morphology, perforations, and perhaps growth) with quick and transient plastic changes in the synapse. This implies that changes in synapse ultrastructure, while perhaps being crucial to rapid induction of changes in synaptic strength, are not sufficient to create enduring increases in synaptic power. Such enduring changes in synaptic power, it seems, are mediated by (or at least correlated with) changes in synapse number.

5. FUNCTIONAL IMPLICATIONS OF SYNAPTIC STRUCTURAL CHANGE

The current data suggest that changes in synapse ultrastructure are necessary to induce rapid, transient changes in synaptic power, but are not sufficient to maintain these changes. Although it remains to be proven whether such structural changes are necessary for this characteristic rapid induction of hippocampal LTP, the fact that neither a rapid cascade of structural plasticity nor a rapid electrophysiological plasticity are seen in the neocortex suggests that this is so. The critical question this data creates, however, is how these distinct morphological profiles relate to the functional properties of the regions in which they are seen.

The key finding linking these distinct morphological cascades in cortical vs. hippocampal circuits to distinct functional properties stems from earlier (Racine et al., 1983) and more recent (Bolshakov et al., 1997; Winder et al., 1998) research that has indicated that LTP is not a simple unitary temporal process but may involve 2 or 3 overlapping phases that rely on different mechanisms. These families of LTP decay curves have average decay times of 2.1 h, 3.5 days, and 20.3 days, respectively (Abraham and Otani, 1991). The shortest decay curve, LTP1 (2.1 h) is protein-synthesis independent, and thus corresponds to early-phase LTP in the more commonly used terminology, while both LTP2 (3.5 days) and LTP3 (20.3 days) fall in the category of late-phase LTP. Thus, LTP is thought to occur in stages, where the initial induction of the potentiation gives way to two or more maintenance phases which preserve the potentiation for relatively long periods (Winder et al., 1998). Although there is a great deal of overlap between these phases (Otani and Ben-Ari, 1993), there is evidence that the mechanisms involved at each stage are distinct. For instance, gene transcription is linked to the late phase of LTP maintenance (e.g., Nguyen et al., 1994).

This electrophysiological evidence is well complemented by morphological data on the time course of hippocampal LTP. It seems that the results typically reported in connection with hippocampal LTP are tied functionally to this early induction phase, since many of these changes are no longer apparent at later post-induction intervals (e.g., Geinisman et al., 1996; Weeks et al., 2001), despite the fact that these changes **(a)** are robust during early LTP (e.g, Geinisman, 1992; Weeks et al., 1999; 2000), **(b)** can be correlated to the degree of LTP expressed (Weeks et al., 1998), and **(c)** can be blocked by interventions that block LTP (Weeks et al., 2003). Cortical LTP, it seems, lacks this early phase of LTP, and the above data suggest that this lack of early LTP corresponds to a lack of the morphological cascade typically linked to early LTP in the hippocampus.

Thus, an intimate connection can be observed between the dynamics of how synaptic plasticity is expressed and the morphological changes that are induced by such dynamics. This connection, however, is far from concrete. In particular, large gaps exist in our knowledge of precisely which morphological changes are either necessary or sufficient for the induction of plastic changes.

5.1 Morphological Change is not Sufficient for LTP Expression

As stated above, although a strong correlative link can be drawn between changes in synapse function and changes in their structure, little data addresses whether these structural changes are either necessary or sufficient to produce changes in synapse function. In fact, to the authors' knowledge, no data exists which addresses the necessity of such structural changes. The scarce data that is present on the sufficiency of morphological change, however, suggests that synaptogenesis alone is insufficient to produce lasting changes in synaptic transmission. A recent study by Dhanrajan et al. (2004) took advantage of the fact that in aged animals, a large number fail to sustain LTP given stimulation that reliably induces LTP in young animals. When separating "failure" and non-failure" animals, Dhanrajan et al. found that perforant path stimulation produced the same degree of synaptogenesis (observed via an increase in the number of branched spines) in both groups. Although this data has yet to be replicated, it provides important insight that synaptogenesis alone may not be sufficient to mediate the changes in synapse efficacy associated with LTP. Similar results have been obtained in our lab in the neocortex (Marrone, 2004). Using chronic implantation, animals given an amount of low-frequency stimulation insufficient to produce any functional changes in synaptic transmission detectable through electrophysiology resulted in results which mirror the typical anatomical changes observed following LTP induction – namely an increase in excitatory axo-spinuous synapses (without structural changes in individual terminals) in the region of stimulation.

It should be pointed out, however, that although this combined data strongly suggest that synaptogenesis is insufficient to produce increases in synapse power, it may still prove to be necessary for such changes. The addition of new synapses may provide a critical substrate for further changes that ultimately alter synaptic efficacy. Furthermore, the alterations which ultimately differentiate animals which do and do not express LTP following stimulation may not be observable at the EM level. Indeed, candidate expression mechanisms have been identified and supported both pre- and postsynaptically, and their role in the expression of enhanced synaptic efficacy will be discussed briefly below.

5.2 Nonstructural Mediators of LTP Expression

Precisely how LTP and, conversely, long-term depression (LTD) are expressed remains an unresolved issue (see Malenka and Nicoll, 1999). Because the presynaptic terminal does not always release neurotransmitter with an incoming action potential (AP), and PTR is malleable (see Schikorski and Stevens, 1997), LTP or LTD could be expressed by an increase or a decrease in PTR, respectively. Alternatively, changes in synaptic efficacy could be mediated by changes in receptor properties (e.g., conductance) or number. Furthermore, both mechanisms may operate, and may be preferentially invoked by different stimulation protocols. While controversial, it seems that LTP most likely depends on both pre- and postsynaptic events (e.g., Morris, 2003, Antonov et al., 2003). While many candidate expression mechanisms have been uncovered, only a select few of these will be discussed below.

Presynaptically, there is considerable evidence that presynaptic glutamate release is up-regulated during LTP in the dentate gyrus (e.g., Dolphin et al., 1982). Quantal analysis in CA1 has also demonstrated increases in quantal size, potentially reflecting increased probability of quantal release or changes in the number of release sites (Oliet et al., 1996). Although these changes may be observed at the EM-level as a change in synapse curvature or the number of docked vesicles (respectively), a further possibility is that glutamate release is enhanced via an increase in either the time of vesicular fusion or the size of the fusion pore, which are not observable via EM.

Postsynaptically, there is mounting *in vitro* evidence that altering the balance of the activity of protein kinase and phosphatase controls the trafficking of AMPA receptors in and out of synaptic membranes (Malinow and Malenka, 2002). High frequency stimulation induces rapid insertion of GluR1-containing receptors in the postsynaptic membrane (Hayashi et al., 2000; Esteban et al., 2003). Over time (in the absence of further stimulation) GluR1subunits are gradually replaced by GluR2 and GluR3 (Shi et al., 2001). Other evidence suggests that the conductance through the AMPA receptors is increased during LTP (Benke et al., 1998) which, in turn, may be mediated by GluR1 subunits (Malinow and Malenka, 2002). In contrast, LTD induces endocytosis of AMPA receptor subunits from the synaptic membrane (Carroll et al., 1999).

For the sake of brevity, only a small fraction of the ever-growing list of potential LTP mechanisms have been discussed in the paragraphs above. Despite these omissions, the preceding discussion does underscore the fact that many potential mechanisms may be inherently non-structural (at least as the term has been defined here). How these non-structural mediators of LTP may be impacted by (or interact with) changes in synapse structure, however, remains largely unexplored.

5.3. Structure and Function Interactions

Despite the morphological evidence suggesting that structural change is not sufficient for LTP expression (and by extension, experience-dependent plasticity), modeling evidence suggests that these structural changes have important functional consequences. In fact, although recent data suggests that these structural changes alone may have less of an impact on function than non-structural changes, synaptic efficacy may be profoundly altered through interaction between these two types of changes. The dynamics of how these interactions may impact synaptic efficacy are theoretically almost limitless. For the sake of simplicity, however, only two such interactions, for which there is a reasonable degree of evidence, will be described below as examples of how such interactions may affect synapse transmission.

The first of these examples concerns the discovery of silent synapses – synapses lacking AMPA, but not NMDA channels, which would thus be “silent” (i.e., produce no postsynaptic response) unless there was large degree of activity or the postsynaptic membrane was already depolarized. Based on the study of the GluR1 phosphorylation (Lee et al. 2000), it seems that LTP then brings about an increase in the phosphorylation of the calcium/calmodulin-dependent protein kinase II (CaMKII) site on GluR1 (reversed by ‘de-potentiation’), while long-term depression (LTD) induces dephosphorylation of the protein kinase A (PKA) site on GluR1 (reversed by ‘de-depression’). These findings suggest that there must be two separate molecular memories (i.e., for LTP and LTD) at the synapse. In a recent review, Lisman (2003) argued that this suggested two distinct mechanisms, where initially, the unsilencing of synapses produces an ‘active state’; when LTD is then induced, the synapses is silenced. Lisman also suggests that ‘active’ synapses can be further potentiated by an independent process. The combination of morphological data on synaptogenesis without physiological change alongside recent data on LTP saturation may make it possible to synthesize these findings. Recent data (Frey et al., 1995) has revealed that when LTP is saturated (i.e., no further LTP can be induced with further stimulation), it is possible to induce further LTP several hours later. If (as suggested by the lack of physiological change) the synaptogenesis induced by stimulation produces mainly silent synapses, these newly formed, silent synapses may underlie recovery from saturation. Furthermore, it should be possible to test this hypothesis. Since 'unsilencing' occurs by a process that differs pharmacologically from potentiation, it should be possible to make saturation of LTP relatively permanent by blocking this unsilencing by inhibiting PKA. If saturation is truly blocked, this creates a scenario in which the extent, persistence, and localization of synaptogenesis can be inferred by examining the proliferation of "silent" synapses through combined immunocytochemistry and EM. This may provide useful information in describing not only the degree of synaptogenesis (as we have done) but also the extent to which these new synapses are functional in the hippocampus vs. neocortex. This scenario encapsulates a situation in which the ultimate mediation of increases in synapse signaling are induced non-morphologically (by the insertion of AMPA receptors), but morphological change (synaptogenesis) provides the critical substrate for this mediation to occur.

Another interaction stems from Lisman's (2003) conception of the "module" based on several pieces of data. During the rise time of small mEPSCs (50 msec – Magee and Cook, 2000), glutamate remains highly concentrated near the site of vesicle release (100 nm). This activates the AMPA channels within a hotspot approximately 200 nm in diameter (Raghavachari and Lisman 2004) – approximately the same dimension as a module containing a sufficient number of AMPA channels to generate a quantal event (Lisman and Harris, 1993). As glutamate spreads, its concentration falls and AMPA channels are no longer efficiently activated because low concentrations drive desensitization. If some modules are silent (i.e., contain no AMPA receptors), as Lisman (2003) suggests, vesicles released in silent modules would generate only an NMDA component (which does not depend on the proximity of release), while those released over 'active' modules would generate both NMDA and AMPA components. This model explains the dramatic variation of the AMPA/NMDA ratio of evoked responses at single synapses (Renger et al. 2001), as well as the observation that blocking desensitization can dramatically increase the probability of response (Diamond and Jahr 1995; Choi et al. 2000; Gasparini et al. 2000). Lisman (2003) suggests that this is because the glutamate released at silent modules can affect distant AMPA channels when desensitization is blocked.

Although this model elegantly encapsulates a large array of physiological data, it may be further complicated by the contribution of bouton structure. This is because the distribution of vesicles docked and available for release are distributed randomly across the presynaptic active zone (Schikorski and Stevens, 1997; 2001) and the release probability of each of these vesicles is largely dependent on the local terminal morphology (Ghaffari-Farazi et al., 1999), thus the terminal morphology proximal vs. distal to the module may affect the physiological response to an equivalent vesicular release. Although data suggests that the number of vesicles simultaneously released at large synapses are potentially greater than 10 (Oertner et al. 2002; Conti & Lisman, 2003), an illustrative example using only 2 vesicles (with only one released in a single event) will be used to demonstrate these dynamics (Figure 2).

If one vesicle is docked in the immediate vicinity of a module, and one is docked 300 nm away (outside the "effective range" for triggering activation of that module), the response of that module may critically depend on which of these vesicles is released first. In turn, the order of vesicular release depends on several characteristics of the local morphology, such as (for example) presynaptic curvature. At a synaptic contact with little or no inflection in the synaptic cleft, the PTR would be approximately equal, and the order in which these vesicles are released during repetitive activation would be random. If the vesicle proximal to an active module is released first, the AMPA receptors would be activated, and the postsynaptic depolarization would cause the silent module to also be active if the distal vesicle is released soon after. If the vesicle distal to the active module is released first, no postsynaptic response will be induced, since it is over a silent module. Moreover, the low glutamate concentration over the active module will drive sensitization of the AMPA receptors, leading to a decreased response (or perhaps none) to the subsequent release of the proximal vesicle. Such a random process may contribute to the large variation of the AMPA/NMDA ratio of evoked responses cited above.

If, however, the curvature of the synaptic cleft is altered, this may alter the PTR of each release site such that one particular vesicle will usually be released before another. In turn, this order of release may profoundly affect the postsynaptic response. If a concave inflection point is located at the site of the proximal vesicle, and a convex inflection point is located at the distal vesicle [such "irregular" profiles are relatively common, at least in the dentate gyrus (Weeks et al., 1999, 2000, 2001; Marrone et al., 2004)], the proximal release site will accumulate calcium much faster due to enhanced compartmentalization, and release will occur there. At the distal, convex release site, a much larger concentration of local calcium will be necessary to trigger release, and this vesicle will be released several APs later. This, as stated above, may lead to the activation of both the active and silent modules. If the curvature profiles at the proximal and distal release sites are reversed, so is the order of vesicular release. Thus, the release of a vesicle over a silent module will produce no effect postsynaptically, and the low concentration of glutamate flowing to the active module will drive desensitization of the receptors, such that even if the proximal vesicle is released with further APs, the postsynaptic response will be dramatically reduced (or perhaps non-existent). This second scenario again demonstrates that even when plasticity is critically mediated by non-structural events (the insertion of "modules" of postsynaptic AMPA receptors and changes in their conductance), morphological characteristics (presynaptic compartmentalization) may interact to dramatically alter the degree to which these non-structural changes can be expressed.

Collectively, these examples of dynamic interactions between morphological and molecular changes underscore both the importance of continued examination of the anatomy of plasticity in the face of the discovery of new molecular determinants of plasticity, as well as the need to integrate these two areas of research to provide a comprehensive picture of how plastic changes in the synapse are expressed.

6. CONCLUSIONS AND FUTURE DIRECTIONS

The findings described above reaffirm that, although much has been discovered about the ways that synapse structure and function may change as a product of experience, many questions remain unanswered. In particular, the field has yet to move significantly beyond correlating structural changes in the synapse with changes in plasticity. Given the wide array of evidence, the fact that plastic changes in synapse function (i.e., changes induced by learning and/or LTP induction) induce alterations in

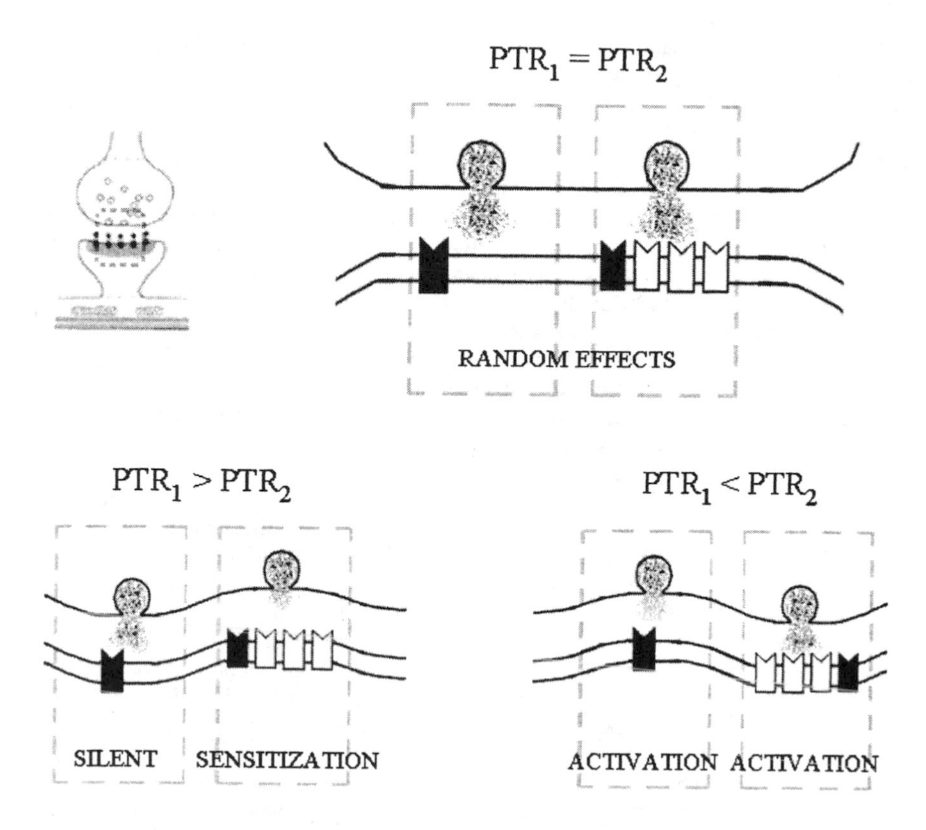

Figure 2. Schematic diagram displaying 2 vesicular release sites, one over a silent [NMDA receptors (black) only] module, and one over an active [both AMPA (white) and NMDA receptors], illustrating how changes in curvature may affect the postsynaptic response to vesicular release. At a flat synapse (top), the PTR at each release site is equal, thus with repetitive APs the order of vesicular release is random, and so are the postsynaptic effects. If the vesicle over a silent module has a greater PTR due to a concave curvature (bottom left), release will occur there first, causing no response at the silent module, and sensitization of the active module. If, however, the vesicle over an active module has a greater PTR due to a concave curvature (bottom right), release will occur over the active module first, causing activation of the active module, and perhaps also of the silent module due to postsynaptic depolarization.

synapse structure is well established. New evidence needs to progress beyond these correlations by linking specific means of inducing and maintaining these physiological changes with their anatomical expression. Many means exist for such a dissection of these mechanisms that have yet to be implemented on a large scale.

For instance, although we know that different types of morphological changes are typically induced in different regions, we do not know how induction protocol affects these changes. There is evidence that hippocampal LTP **can** be made stable. In the right circumstances, LTP can last several weeks in CA1 (Staubli and Lynch, 1987) or up to a year in the dentate gyrus (Abraham et al., 2002). The use of this induction protocol may induce changes in synapse structure that would resemble those typically seen in the cortex. This possibility underscores a more general issue in linking synapse structure and function. As the array of mechanisms that may support the persistence of synaptic plasticity continues to grow, examining which structural changes are induced by each of these mechanisms has the promise of allowing a relationship to be drawn between specific morphological changes (i.e., synaptogenesis, perforation formation, induction of synapse curvature, etc.) to these specific maintenance mechanisms, and the electrophysiological and molecular knowledge already discovered about them. In turn, this would allow for a comprehensive picture of the physiological, molecular, and anatomical expression of that plasticity mechanism. This can be accomplished using two broad perspectives – observing how network dynamics in individual neural regions relates to the morphology of synapses in those networks, and attempting to disrupt these networks in order to address which morphological changes are crucial to network function. As describing the ways in which these issues may be addressed could arguably encompass an additional full discussion, only a select few of the most relevant examples of how experiments may address these concerns will be described.

When addressing the issue of relating network dynamics to morphology, much remains to be discovered. For instance, it is not known whether the net changes in number induced in the neocortex are the result of qualitatively different changes in those synapses which are active, or those which are not. For instance, several studies (described above) have reported evidence of synaptogenesis in specific regions or types of synapse. These changes, however, may be masked by opposing changes in non-activated synaptic contacts. Evidence suggests that synapses adjacent to stimulated contacts which are not themselves stimulated often show opposite changes both structurally (Weeks et al., 1999) and functionally (Coussens and Teyler, 1999).

Thus, while the increases in cortical synapse number remains a valid observation, it is not known whether this synaptogenesis is induced because (a) more synapses are recruited by LTP in cortical vs. hippocampal systems, or (b) because there is less compensatory turnover of non-potentiated contacts. Likewise, the apparent lack of structural change following cortical plasticity could be the result of either (a) fewer synaptic changes in activated cortical synapses, or (b) more compensatory alterations in non-activated contacts. Methods exist for distinguishing between synapse which are active and those which are not by imaging the accumulation of either calcium (Buchs and Muller, 1996) or polyribosomes (Ostroff et al., 2003). Unfortunately, these methods have yet to be used outside of the hippocampus. Such studies may provide critical data on the dynamics of cortical vs. hippocampal plasticity.

In fact, relating the structural changes of synapses within a particular region to the functional dynamics of neural circuits within that region is an issue that needs to be extended beyond the hippocampus and cortex in general. These relationships need to be explored in other regions, such as the cerebellum, which shows drastically different mechanisms for plasticity. In addition, regions of the hippocampus and neocortex show important functional distinctions from other regions within the same structure. These distinctions need to be related to structural change. For example, in the hippocampus, LTP at perforant path-dentate gyrus granule cell synapses and Schaffer collateral-CA1 pyramidal cell synapses is NMDA receptor dependent. LTP at synapses of the perforant path onto CA3, and interconnecting CA3 neurons via the longitudinal-commissural pathway, is also NMDA receptor-dependent, whereas that at mossy fiber synapses onto CA3 neurons is NMDA receptor-independent. *In vivo*, different tetanization frequencies are optimal on these different pathways (Yeckel and Berger, 1998). These functional distinctions may, at least partially, be mediated by distinct morphological changes in individual synapses within those regions. Exploration of these structural changes may yield important information relating synapse structure and function. Such information is crucial considering that modeling evidence suggests that multiple combinations of morphological profiles may yield the same level of function (Wu et al., 2003). For example, a large macular synapse with a high degree of concave inflection may be functionally equivalent to a smaller perforated synapse with little inflection. This 'functional equivalence' further underscores the importance of teasing apart the precise functional role of individual changes in morphology, how they interact, and the circumstances in which they are preferentially induced.

Several methods for uncovering mechanistic relationships by blocking part of the cascade that induces or maintains plastic changes in the synapse exist in the LTP literature, and can be adapted to the study of structural plasticity. Foremost, several methods exist for selectively inhibiting excitatory (e.g., muscimol) or inhibitory (e.g., picrotoxin) transmission, and interventions also exist for blocking LTP *in vitro* (e.g., APV) or long-term memory *in vivo* (e.g., anisomycin). In addition, compounds exist which block many of the molecular signals thought to be crucial to learning-induced plasticity. For example, KN-62 inhibits CaMKII and Rp-cAMPS selectively inhibits PKA. While a wealth of data exists about the effects of these agents on LTP and/or learning, very little data has explored the structural consequences of such disruptions. In fact, to the knowledge of the authors, only two studies (Sorra and Harris, 1998; Weeks et al., 2003) have ever examined synapse structure following any form of LTP blockade. In addition, data has yet to be gathered relating detailed synapse morphology as a consequence of many of the "systemic" factors that affect plasticity, such as neuromodulation. Dissociating the effects of these factors on synapse structure may yield important evidence as to the functional role of specific changes in synapse structure, and reveal the circumstances in which they are triggered.

Such observations may help in moving from correlation to causal relationship. Unfortunately, few studies have attempted to uncover such causal functions. Although the continued observation of correlations between synapse structure and function provide valuable data, they clearly represent only a first step in understanding this complex interaction. This is because the link reflects a statistical correlation rather than a mechanistic connection. The eventual understanding of these many causal relationships may reveal a comprehensive picture of the ways in which the synapse may alter its efficacy, when these changes will be induced, and how they interact – all the knowledge necessary for a comprehensive theory on the plasticity of the synapse.

7. ACKNOWLEDGEMENTS

The research described here has been supported by the Natural Sciences and Engineering Research Council of Canada in the form of an operating grant to TLP and a postgraduate scholarship and postdoctoral fellowship to DFM.

8. REFERENCES

Abraham WC, Greenwood JM, Logan BL, Mason-Parker SE, Dragunow M (2002) Induction and experience-dependent reversal of stable LTP lasting months in the hippocampus. J. Neurosci. 22:9626-9634.

Antonov I, Antonova I, Kandel ER, Hawkins RD (2003) Activity-dependent presynaptic facilitation and Hebbian LTP are both required and interact during claaiscal conditioning in Aplysia. Neuron 37:135-147

Beaulieu C, Colonnier M (1987) Effect of the richness of the environment on the cat visual cortex. J. Comp. Neurol. 266:478-494.

Benke TA, Luthi A, Isaac JTR, Collingridge GL (1998) Modulation of AMPA receptor unitary conductance by synaptic activity. Nature 393:793-797.

Bekkers JM, Stevens CF (1989) NMDA and non-NMDA receptors are colocalized at individual excitatory synapses in cultured rat hippocampus. Nature 341:230-233.

Benshalom G, White EL (1986) Quantification of thalamocortical synapses with spiny stellate neurons in layer IV of mouse somatosensory cortex. J. Comp. Neurol, 253:303-314.

Bertram R (1997) A simple model of neurotransmitter release and facilitation. Neural Comput., 9:515-523.

Bertram R, Smith GD, Sherman A (1999) Modeling study of the effects of overlapping Ca^{2+} microdomains on neurotransmitter release. Biophys. J. 76:735-750.

Birks RI, MacIntosh FC (1961) Acetylcholine metabolism of a sympathetic ganglion. Can. J. Biochem. Physiol. 39:787-827.

Bliss TV, Lømo T (1973) Long-lasting potentiation of synaptic transmission in the dentate area of the anaesthetized rabbit following stimulation of the perforant path. J. Physiol. Lond. 232:331-356.

Bolshakov VY, Siegelbaum SA (1995) Regulation of hippocampal transmitter release during development and long-term potentiation. Science 269:1730-1734.

Buchs PA, Muller D (1996) Induction of long-term potentiation is associated with major ultrastructural changes of activated synapses. Proc. Natl. Acad. Sci. U.S.A. 93:8040-8045.

Bulinski JC, Ohm T, Roder H, Spruston N, Turner DA, Wheal HV (1998) Changes in dendritic structure and function following hippocampal lesions: correlations with developmental events? Prog. Neurobiol., 55:641-650.

Buonomano DV, Merzenich MM (1998) Cortical plasticity: From synapses to maps. Annu. Rev. Neurosci. 21:149-186.

Carroll RC, Lissin DV, von Zastrow M, Nicoll RA, Malenka RC (1999) Rapid redistribution of glutamate receptors contributes to long-term depression in hippocampal cultures. Nature Neurosci. 2:454-460.

Chang FLF, Greenough WT (1982) Lateralized effects of monocular training on dendritic branching in adult split-brain rats. Brain Res. 232:283-292.

Chang FLF, Greenough WT (1984) Transient and enduring morphological correlates of synaptic activity and efficacy change in the rat hippocampal slice. Brain Res. 309:35-46.

Chapman CA, Trepel C, Ivanco TL, Froc DJ, Wilson K, Racine RJ (1998) Changes in field potentials and membrane currents in rat sensorimotor cortex following repeated tetanization of the corpus callosum in vivo. Cereb. Cortex 8:730-742.

Choi S, Klingauf J, Tsien RW (2000) Postfusional regulation of cleft glutamate concentration during LTP at 'silent synapses'. Nature Neurosci. 3:330-336.

Conti R, Lisman J (2003) The high variance of AMPA and NMDA responses at single hippocampal synapses: evidence for multiquantal release. Proc. Natl. Acad. Sci. USA 100:4885-90

Coussens CM, Teyler TJ (1996) Long-term potentiation induces synaptic plasticity at nontetanized adjacent synapses. Learn. Mem. 3:106-114.

Desmond NL, Levy WB (1988) Synaptic interface surface area increases with long-term potentiation in the dentate gyrus. Brain Res., 453:308-314.

Desmond NL, Levy WB (1983) Synaptic correlates of associative potentiation/depression: an ultrastructural study in the hippocampus. Brain Res. 265:21-30.

Desmond NL, Levy WB (1986a) Changes in the numerical density of synaptic contacts with long-term potentiation in the hippocampal dentate gyrus. J. Comp. Neurol. 253:466-475.

Desmond NL, Levy WB (1986b) Changes in the postsynaptic density with long-term potentiation in the dentate gyrus. J. Comp. Neurol. 253:476-482.

Dhanrajan TM, Lynch MA, Kelly A, Popov VI, Rusakov DA, Stewart MG (2004) Expression of long-term potentiation in aged rats involves perforated synapses but dendritic spine branching results from high-frequency stimulation alone. Hippocampus 14:255-264.

Diamond JS, Jahr CE (1995) Asynchronous release of synaptic vesicles determines the time course of the AMPA receptor-mediated EPSC. Neuron 15:1097-1107.

Dolphin AC, Errington ML, Bliss TVP (1982) Long-term potentiation of the perforant path in vivo is associated with increased glutamate release. Nature 297:496-498.

Eichenbaum H (2000) A cortical-hippocampal system for declarative memory. Nature Rev. Neurosci. 1:41-50.

Elmqvist D, Quastel DMJ (1965) A quantitative study of endplate potentials in isolated human muscle. J. Physiol. 178:505-529.

Esteban JA, Shi SH, Wilson C, Nuriya M, Huganir RL, Malinow R (2003) PKA phosphorylation of AMPA receptor subunits controls synaptic trafficking underlying plasticity. Nature Neurosci. 6:136-143.

Eyre MD, Richter-Levin G, Avital A, Stewart MG (2003) Morphological changes in hippocampal dentate gyrus synapses following spatial learning in rats are transient. Eur. J. Neurosci. 17:1973-1980.

Fiala JC, Harris KM (1999) Dendrite structure, in: Dendrites, G. Stuart, N. Spruston, M. Häusser, eds., Oxford University Press, Oxford, pp. 1-34.

Frey U, Schollmeier K, Reymann KG, Seidenbecher T (1995) Asymptotic hippocampal long-term potentiation in rats does not preclude additional potentiation at later phases. Neurosci. 67:799-807.

Gasparini S, Saviane C, Voronin LL, Cherubini E (2000) Silent synapses in the developing hippocampus: lack of functional AMPA receptors or low probability of glutamate release? Proc. Natl. Acad. Sci. USA 97: 9741-9746.

Geinisman Y, Morrell F, de Toledo-Morrell L (1992) Increase in the number of axospinous synapses with segmented postsynaptic densities following hippocampal kindling. Brain Res. 569:341-347.

Geinisman Y, deToledo-Morrell L, Morrell F (1991) Induction of long-term potentiation is associated with an increase in the number of axospinous synapses with segmented postsynaptic densities. Brain Res. 566:77-88.

Geinisman Y, de Toledo-Morrell L, Morrell F, Heller RE, Rossi M, Parshall RF (1993) Structural synaptic correlate of long-term potentiation: formation of axospinous synapses with multiple, completely partitioned transmission zones. Hippocampus 3:435-45.

Geinisman Y, de Toledo-Morrell L, Morrell F, Persina IS, Beatty MA (1996) Synapse restructuring associated with the maintenance phase of hippocampal long-term potentiation. J. Comp. Neurol., 368:413-423.

Ghaffari-Farazi T, Liaw J, Berger TW, (1997) Impact of synaptic morphology on presynaptic calcium dynamics and synaptic transmission. Soc. Neurosci. Abstr. 23:2105.

Ghaffari-Farazi T, Liaw J, Berger TW (1999) Consequence of morphological alterations on synaptic function. Neurocomput. 26:17-27.

Greenough WT, Larson JR, Withers GS (1985) Effects of unilateral and bilateral training in a reaching task on dendritic branching of neurons in the rat motor-sensory forelimb cortex. Behav. Neur. Biol. 44:301-314.

Greenough WT, West RW, deVoogd TJ (1978) Subsynaptic plate perforations: changes with age and experience in the rat. Science, 202:1096-1098.

Greenough WT, Chang FL (1985) Synaptic structural correlates of information storage in mammalian nervous systems. In: Synaptic plasticity, C.W. Cotman, ed., Guilford Press, London, pp.335-372.

Harris KM, Fiala JC, Ostroff L (2003) Structural changes at dendritic spine synapses during long-term potentiation. Philos Trans R Soc Lond B Biol Sci., 358:745-748.

Hayashi Y, Shi SH, Esteban JA, Piccini A, Poncer JC, Malinow R (2000) Driving AMPA receptors into synapses by LTP and CaMKII: Requirement for GluR1 and PDZ domain interaction. Science, 287:2262-2267.

Hebb DO (1949) The Organization of Behavior: A Neuropsychological Theory. New York: Wiley.

Holscher C (1999) Synaptic plasticity and learning and memory: LTP and beyond. J. Neurosci. Res. 58:62-75.

Ivanco TL, Greenough WT (2000) Physiological consequences of morphologically detectable synaptic plasticity: potential uses for examining recovery following damage. Neuropharmacology 39:765-776.

Ivanco TL, Racine RJ (2000) Long-term potentiation in the reciprocal corticohippocampal and corticocortical pathways in the chronically implanted, freely moving rat. Hippocampus 10:143-152.

Ivanco TL, Racine RJ, Kolb B (2000) Morphology of layer III pyramidal neurons is altered following induction of LTP in somatosensory cortex of the freely moving rat. Synapse 37:16-32.

Jones DG, Calverley R (1991) Perforated and non-perforated synapses in rat neocortex: three dimensional reconstructions. Brain Res. 556:247-258.

Kirkwood A, Dudek SM, Gold JT, Aizenman CD, Bear MF (1993) Common forms of synaptic plasticity in the hippocampus and neocortex in vitro. Science 260:1518-1521.

Kleim JA, Barbay S, Cooper NR, Hogg TM, Reidel MS, Remple MS, Nudo RJ (2002) Motor learning-dependent synaptogenesis is localized to functionally reorganized motor cortex. Neurobiol. Learn. Mem. 77:63-77.

Kleim JA, Hogg TM, VandenBerg PM, Cooper NR, Bruneau R, Remple M (2004) Cortical synaptogenesis and motor map reorganization occur during late, but not early, phase of motor skill learning. J. Neurosci. 24:628-633.

Kleim JA, Lussnig E, Schwarz ER, Comery TA, Greenough WT (1996) Synaptogenesis and Fos expression in the motor cortex of the adult rat after motor skill learning. J. Neurosci. 16:4529-4535.
Klintsova AY, Greenough WT (1999) Synaptic plasticity in cortical systems. Curr. Opin. Neurobiol. 9:203-208.
Knafo S, Grossman Y, Barkai E, Benshalom G (2001) Olfactory learning is associated with increased spine density along apical dendrites of pyramidal neurons in the rat piriform cortex. Eur. J. Neurosci. 13:633-638.
Kolb B, Forgie M, Gibb R, Gorny G, Rowntree S (1998) Age, experience and the changing brain. Neurosci. Biobehav. Rev. 22:143-159.
Landfield P, Deadwyler S. (1988) Long-term Potentiation: From Biophysics to Behaviour. New York: Liss.
Lee HK, Barbarosie M, Kameyama K, Bear MF, Huganir RL (2000) Regulation of distinct AMPA receptor phosphorylation sites during bidirectional synaptic plasticity. Nature 405:955-959.
Lisman J (2003) Long-term potentiation: outstanding questions and attempted synthesis. Phil. Trans. R. Soc. Lond. B 358:829-842.
Lisman JE, Harris KM (1993) Quantal analysis and synaptic anatomy – integrating two views of hippocampal plasticity. Trends Neurosci. 16:141-147.
Liu G, Tsien RW (1995a) Properties of synaptic transmission at single hippocampal synaptic boutons. Nature 375:404-408.
Liu G, Tsien RW (1995b) Synaptic transmission at single visualized hippocampal boutons. Neuropharmacology 34:1407-1421.
Magee JC, Cook EP (2000) Somatic EPSP amplitude is independent of synapse location in hippocampal pyramidal neurons. Nature Neurosci. 3:895-903.
Malenka RC, Nicoll RA (1999) Long-term potentiation – a decade of progress? Science 285:1870-1874.
Malinow R, Malenka R (2002) AMPA receptor trafficking and synaptic plasticity. Ann. Rev. Neurosci., 25:103-126.
Markus EJ, Petit TL, LeBoutillier JC, Brooks WJ (1994) Morphological characteristics of the synapse and their relationship to synaptic type: An electron microscopic examination of the neocortex and hippocampus of the rat. Synapse, 17: 65-68.
Marrone DF (2004) Morphological correlates of long-term plasticity in the rat neocortex. PhD dissertation: University of Toronto.
Marrone DF, Petit TL (2002) The role of synaptic morphology in neural plasticity: structural interactions underlying synaptic power. Brain Res. Rev. 38:291-308.
Marrone DF, LeBoutiller JC, Petit TL (2004) Changes in synaptic ultrastructure during reactive synaptogenesis in the rat dentate gyrus. Brain Res., 1005:124-136.
McClelland JL, McNaughton BL, O'Reilly RC (1995) Why there are complimentary learning systems in the hippocampus and neocortex: Insights from the successes and failures of connectionist models of learning and memory. Psychol. Rev. 102:419-457
Monfils MH, VandenBerg PM, Jeffrey A, Kleim JA, Teskey GC (2004) Long-term potentiation induces expanded movement representations and dendritic hypertrophy in layer V of rat sensorimotor neocortex. Cereb Cortex 14:586-593
Morris RG (2003) Long-term potentiation and memory. Philos Trans R Soc Lond B 358:643-647.
Muller D, Toni N, Buchs PA (2000) Spine changes associated with long-term potentiation. Hippocampus, 10:596-604.
Murthy VN, Sejnowski TJ, Stevens CF (1997) Heterogeneous release properties of visualized individual hippocampal synapses. Neuron 18:599-612.
Nguyen PV, Abel T, Kandel ER (1994) Requirement of a critical period of transcription for induction of a late phase of LTP. Science 265:1104-1107.
Oertner TG, Sabatini BL, Nimchinsky EA, Svoboda K (2002) Facilitation at single synapses probed with optical quantal analysis. Nature Neurosci. 5:657-664.
Oliet SH, Malenka RC, Nicoll RA (1996) Bidirectional control of quantal size by synaptic activity in the hippocampus. Science 271:1294-1297.
Ostroff LE, Fiala JC, Allwardt B, Harris KM (2002) Polyribosomes redistribute from dendritic shafts into spines with enlarged synapses during LTP in developing rat hippocampal slices. Neuron 35: 535-545.
Raghavachari S, Lisman JE (2004) Properties of Quantal Transmission at CA1 Synapses. J. Neurophysiol. (in press – e-publication available from the publisher)
Racine RJ, Milgram NW (1983) Long-term potentiation phenomena in the rat limbic forebrain. Brain Res. 260:217-231.
Ramirez-Amaya V, Escobar ML, Chao V, Federico Bermudez-Rattoni F (1999) Synaptogenesis of mossy fibers induced by spatial water maze overtraining. Hippocampus 9:631-636.
Renger JJ, Egles C, Liu G (2001) A developmental switch in neurotransmitter flux enhances synaptic efficacy by affecting AMPA receptor activation. Neuron 29:469-484.

Rusakov DA, Davies HA, Harrison E, Diana G, Richter-Levin G, Bliss TVP, Stewart MG (1997) Ultrastructural synaptic correlates of spatial learning in rat hippocampus. Neurosci. 80:69-77.
Schikorski T, Stevens CF (1997) Quantitative ultrastructural analysis of hippocampal excitatory synapses. J. Neurosci. 17:5858-5867.
Schikorski T, Stevens CF (1999) Quantitative fine-structural analysis of olfactory cortical synapses. Proc. Natl. Acad. Sci. USA 96:4107-4112.
Schikorski T, Stevens CF (2001) Morphological correlates of functionally defined synaptic vesicle populations. Nature Neuroscience 4:391-395
Shi S, Hayashi Y, Esteban JA, Malinow R (2001) Subunit-specific rules governing AMPA receptor trafficking to synapses in hippocampal pyramidal neurons. Cell 105:331-343.
Sorra KE, Harris KM (1998) Stability in synapse number and size at 2 hr after long-term potentiation in hippocampal area CA1. J. Neurosci. 18:658-671.
Staubli U, Lynch G (1987) Stable hippocampal long-term potentiation elicited by 'theta' pattern stimulation. Brain Res. 435:227-234.
Stevens CF, Wesseling JF (1999) Identification of a novel process limiting the rate of synaptic vesicle cycling at hippocampal synapses. Neuron 24:1017-1028.
Stewart MG, Harrison E, Rusakov DA, Richter-Levin G, Maroun M (2000) Re-structuring of synapses 24 hours after induction of long-term potentiation in the dentate gyrus of the rat hippocampus in vivo. Neurosci. 100:221-227.
Toni N, Buchs PA, Nikonenko I, Bron CR, Muller D (1999) LTP promotes formation of multiple spine synapses between a single axon terminal and a dendrite. Nature 402:421-425.
Trommald M, Hulleberg G, Andersen P (1996) Long-term potentiation is associated with new excitatory spine synapses on rat dentate granule cells. Learn. Mem. 3:218-228.
Volkmar FR, Greenough WT (1972) Rearing complexity affects branching of dendrites in the visual cortex of the rat. Science 176:1445-1447
von Gersdorff H, Vardi E, Matthews G, Sterling P (1996) Evidence that vesicles on the synaptic ribbon of retinal bipolar neurons can be rapidly released. Neuron 16:1221-1227.
Weeks ACW, Ivanco TL, LeBoutillier JC, Racine RJ, Petit TL (1998) The degree of potentiation is associated with synaptic number during the maintenance of long-term potentiation in the rat dentate gyrus. Brain Res. 798:211-216.
Weeks ACW, Ivanco TL, LeBoutillier JC, Racine RJ, Petit TL (1999) Sequential changes in the synaptic structural profile following long-term potentiation in the rat dentate gyrus: I. The intermediate maintenance phase. Synapse 31:97-107.
Weeks ACW, Ivanco TL, LeBoutillier JC, Racine RJ, Petit TL (2000) Sequential changes in the synaptic structural profile following long-term potentiation in the rat dentate gyrus: II. Induction/early maintenance phase. Synapse 36:286-296.
Weeks ACW, Ivanco TL, LeBoutillier JC, Racine RJ, Petit TL (2001) Sequential changes in the synaptic structural profile following long-term potentiation in the rat dentate gyrus: III. Long-term maintenance phase. Synapse 40: 74-84.
Weeks ACW, Ivanco TL, Leboutillier JC, Marrone DF, Racine RJ, Petit TL (2003) Unique changes in synaptic morphology following tetanization under pharmacological blockade. Synapse 47:77-86.
Withers GS, Greenough WT (1989) Reaching training selectively alters dendritic branching in subpopulations of layer II-III pyramids in rat motor-somatosensory forelimb cortex. Neuropsychol. 27:61-69.
Wu GY, Kleim J, Marrone DF, Berger TW, Petit TL (2003) Morphological basis of neural plasticity. CNS Drug Rev. 9:217
Yuste R, Denk W (1995) Dendritic spines as basic functional units of neuronal integration. Nature 375:682-684.
Zucker RS (1994) Calcium and short-term synaptic plasticity. Neth. J. Zool. 44:495-512.
Zucker RS (1999) Calcium- and activity-dependent synaptic plasticity. Curr. Opin. Neurobiol. 9:305-313.

ROLE OF THE SPINE APPARATUS IN SYNAPTIC PLASTICITY

Michael Frotscher and Thomas Deller*

1. INTRODUCTION

The dendrites of many neurons are densely covered with spines. These small appendages of dendrites were discovered by neuroanatomists using Golgi impregnation to stain nerve cells. Before the era of electron microscopy, spines were regarded as artifacts or boutons of axon terminals. Ramón y Cajal early on regarded them as sites of synaptic contact (Ramón y Cajal, 1911). Their nature as postsynaptic sites was disclosed in classic electron microscopic studies by George Gray (1959).

The discovery that dendritic spines are major postsynaptic elements raised the possibility that these structures were sites of synaptic plasticity in the brain. In the seventies of last century, a large number of studies were performed that addressed this issue. It could be shown that the number of spines is reduced in pyramidal cells of the visual cortex after dark rearing or enucleation (Valverde, 1967; 1968; 1971; Globus and Scheibel, 1967), suggesting that the number of dendritic spines is regulated by neuronal activity. Other studies showed that these adaptive changes in spine number are not restricted to primary sensory areas but involve the hippocampus, a region known to play an important role in learning and memory processes (Frotscher et al., 1975). Studies on the hippocampus also revealed that the size of spines was altered in the absence of proper afferent input (Frotscher et al., 1977). Together, these studies pointed to a role of afferent fibers and their functional load in the formation and maintenance of dendritic spines (Hámori, 1973). These early studies were confirmed and extended by recent electrophysiological studies which showed that long-term potentiation (LTP), one form of synaptic plasticity, is accompanied by the de novo formation of dendritic spines (Engert and Bonhoeffer, 1999) and by changes in spine size (Matsuzaki et al., 2004).

In addition to neuronal activity, also hormones have been found to control the number of dendritic spines. Woolley and McEwen (1993) showed that estrogen given to ovarectomized rats increased the number of dendritic spines on dendrites of CA1 pyramidal cells in the hippocampus. This increase in spines on CA1 pyramidal cells is likely to be caused by estrogen effects on CA3 pyramidal neurons which give rise to the Schaffer collaterals terminating on CA1 pyramidal cell dendrites (Rune et

* Michael Frotscher, Institute of Anatomy and Cell Biology, University of Freiburg, D-79104 Freiburg, Germany. Thomas Deller, Department of Clinical Neuroanatomy, J.W. Goethe University, D-60590 Frankfurt/Main, Germany. E-mail: michael.frotscher@anat.uni-freiburg.de

al., 2002). Recently, evidence was provided that not exogenous estrogen but estrogen synthesized by hippocampal neurons themselves underlies this hormone-induced synaptic plasticity (Kretz et al., 2004).

What are the mechanisms that control the de novo formation and size changes of spines? How are these structural changes related to the functional plasticity of spine synapses? The discovery of ribosomes near the base of dendritic spines (Steward and Levy, 1982) raised the possibility of local dendritic protein synthesis at specifically activated spine synapses. Local changes in calcium concentration (Miyata et al., 2000; Sabatini et al., 2002), remodeling of the actin cytoskeleton (Matus et al., 1982; Fischer et al., 1998), and changes in AMPA receptor expression (Hayashi et al., 2000; Schnell et al., 2002; Lisman et al., 2003; Rumbaugh et al., 2003) may all contribute to the structural and functional plasticity at spine synapses (for review, see Li and Sheng, 2003).

In this paper we review our recent studies on a role of the *spine apparatus* in synaptic plasticity (Deller et al., 2003). The spine apparatus is a characteristic organelle of many spines on cortical neurons. It consists of sacs of endoplasmic reticulum (ER) interdigitated by electron-dense membranes (Fig. 1), structural characteristics that were already mentioned in the classical paper by Gray (1959) and described in detail later by Spacek (1985) and Spacek and Harris (1997).

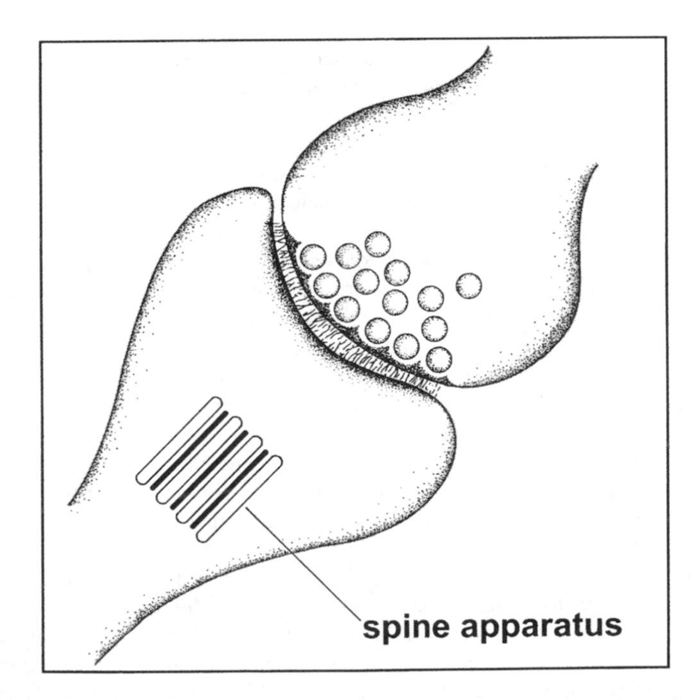

Figure 1. The spine apparatus is a characteristic organelle of many cortical spines. It consists of sacs of smooth endoplasmic reticulum interdigitated by electron-dense plates.

The spine apparatus has been assumed to be a calcium store (Fifkova et al., 1983; Lisman, 1989; Svoboda and Mainen, 1999), likely involved in calcium-dependent changes in synaptic plasticity (Miyata et al., 2000; Sabatini et al., 2002). We report here that mouse mutants deficient in synaptopodin, an actin-associated protein of renal podocytes and telencephalic spines, lack a spine apparatus. Absence of the spine apparatus was found to be accompanied by dramatic changes in LTP and by impairments of the mutant mice in learning and memory tests (Deller et al., 2003).

2. SYNAPTOPODIN IS LOCALIZED TO THE SPINE APPARATUS

Staining for synaptopodin mRNA showed that synaptopodin is expressed in telencephalic structures including the olfactory bulb, neocortex, striatum, and hippocampus (Mundel et al., 1997; Deller et al., 2000a, b). Remarkably, synaptopodin is not expressed in the cerebellum (Mundel et al., 1997). In the hippocampus, the region we have focused on in our studies, synaptopodin is mainly expressed by the principal cells, pyramidal neurons of the hippocampus proper and granule cells of the dentate gyrus.

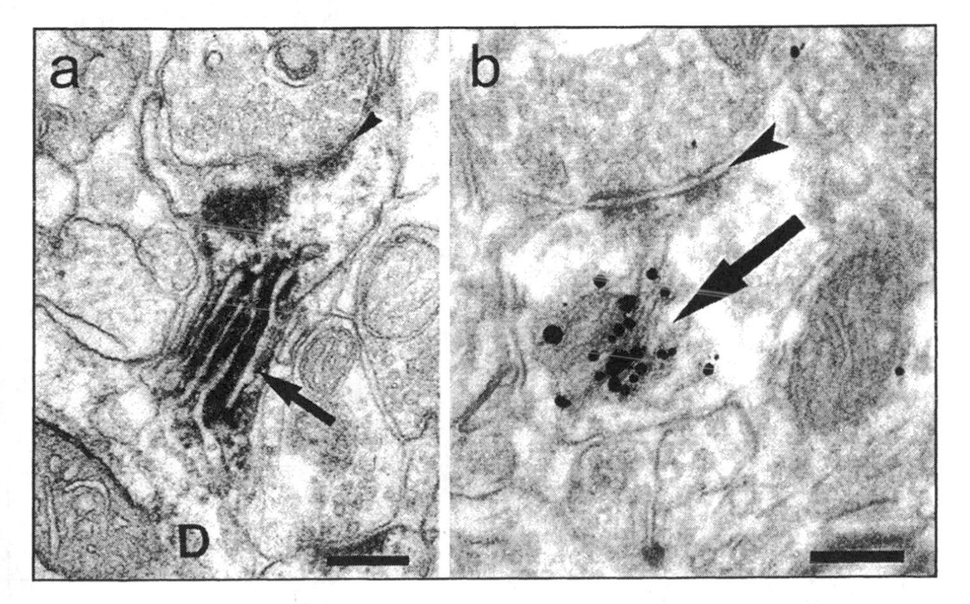

Figure 2. Ultrastructural localization of synaptopodin in the dentate gyrus of rat (a) and mouse (b). In (a) 3,3'-diaminobenzidine was used to visualize the tissue bound antibody. The sacs of smooth endoplasmic reticulum of the spine apparatus (arrow) are unstained and stand out in between the immunoprecipitate (reprinted from Deller et al., 2000a). Copyright (2000) Wiley-Liss, Inc. In (b) a preembedding immunogold procedure (silver-intensified) was employed to visualize the tissue bound synaptopodin antibody (rabbit anti-synaptopodin, see Mundel et al., 1997). Numerous immunogold particles label the spine apparatus organelle (arrow). Note the filaments extending from the spine apparatus to the postsynaptic density. Arrowheads indicate synaptic cleft. Scale bars = 0,2 μm (reprinted from Deller et al., 2002). Copyright (2002) Wiley-Liss, Inc.

While in situ hybridization revealed labeling of cell bodies, immunocytochemistry showed dense staining of dendritic layers, indicating that the protein is sorted to the dendrites of principal cells (Mundel et al., 1997; Deller et al., 2000a, b). At higher magnification we noticed punctate staining indicative of a preferential labeling of synaptic structures (Mundel et al., 1997). This was confirmed by electron microscopy showing diaminobenzidine (DAB) reaction product mainly in dendritic spines. As a rule, the spine apparatus was most heavily labeled. Moreover, serial sections through immunonegative spines showed that these spines did not contain a spine apparatus, suggesting that the immunoreactivity was associated with this characteristic organelle of spines. Finally, in order to study the subcellular distribution of synaptopodin, we employed immunogold labeling. With this procedure, we found an almost exclusive labeling of the spine apparatus (Fig. 2). These morphological findings strongly suggested an association of synaptopodin with the spine apparatus.

3. SYNAPTOPODIN IS AN ACTIN-ASSOCIATED PROTEIN

Initially, synaptopodin was found in renal podocytes (Mundel et al., 1991). In their search for synaptopodin expression in other tissues, Mundel and colleagues found it also in synapses of telencephalic neurons and accordingly termed the protein synaptopodin (from synapse and podocyte). Using immunogold labeling of podocytes, we found synaptopodin very selectively in the podocyte foot processes (Fig. 3, see also Mundel et al., 1991; Ichimura et al., 2003). Both, dendritic spines and the foot processes of podocytes, are motile structures known to contain actin filaments. In fact, in their initial analysis of synaptopodin, Mundel and coworkers (1997) described synaptopodin as actin-associated 100 kD protein without any globular domain structure as indicated by its high amount of prolin. Its presence in dendritic spines suggested an involvement in the remodeling of actin filaments in dendritic spines (Mundel et al., 1997; Fischer et al., 1998; Deller et al., 2000a, b; Capani et al., 2001). In addition, synaptopodin could link the spine apparatus via a cytoskeletal bridge of actin and alpha-actinin2 (Wyszynski et al., 1998) to the postsynaptic density. Such a bridge has been tentatively described by Spacek (1985), who noted cytoskeletal filaments radiating from the outer surface of the spine apparatus to the postsynaptic density of spines (cf. Fig. 2b). Thus, the available data led us to conclude that synaptopodin, by its association with the spine apparatus, links this particular organelle to the actin cytoskeleton and may play a role in the structural plasticity of dendritic spines (Deller et al., 2000a, b).

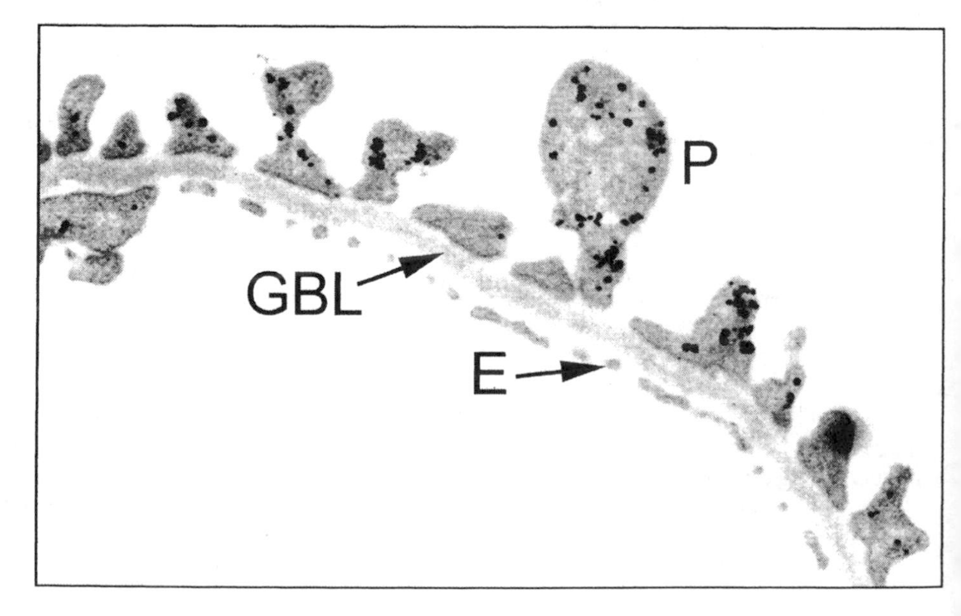

Figure 3. Synaptopodin in podocyte foot processes. Preembedding immunogold-labeling (silver-intensified, rabbit anti-synaptopodin, see Mundel et al., 1997) of kidney glomeruli revealed synaptopodin in podocyte foot processes. E - fenestrated endothelium of glomerular capillaries; GBL - glomerular basal lamina; P - podocyte pedicle. Scale bar: 1 μm.

4. SYNAPTOPODIN-DEFICIENT MICE LACK A SPINE APPARATUS

Next, synaptopodin-deficient mice were generated by gene targeting (Deller et al., 2003). These mice were viable and fertile in a mixed 129/C57BL6 genetic background and could not be distinguished from their wildtype littermates under standard laboratory conditions. Most remarkably, these animals did not show a renal phenotype.

As synaptopodin was found to be closely associated with the spine apparatus, we focused on this particular organelle in our electron microscopic study of brain tissue from synaptopodin-deficient mice. Interestingly enough, we were unable to find spine apparatuses in dendritic spines of synaptopodin mutants. Even the large complex spines or excrescences on CA3 pyramidal neurons which are postsynaptic to the mossy fiber boutons and regularly contain a spine apparatus, did not show this organelle. These initial observations were then substantiated by a comprehensive quantitative analysis of spine apparatuses (>15000 spines analyzed) in wildtype animals and synaptopodin-deficient mice.

While we regularly observed spine apparatuses in thin sections of the neocortex, striatum, and hippocampus (regions CA1 and CA3) from wildtype animals, we were unable to find a single spine apparatus in thin sections of these regions from mutant brains. Besides the loss of spine apparatuses, a comprehensive neuroanatomical study of these animals including immunolabeling for different neuronal and glial subpopulations did not reveal any further abnormalities. In particular, the loss of spine apparatuses was not accompanied by a loss of spines. Spine counts on Golgi-impregnated apical dendrites of layer 5 pyramidal cells in the neocortex and on oblique dendrites of CA1 pyramidal neurons did not reveal differences between genotypes. Likewise, there was no evidence for changes in spine size, and the percentage of different spine length classes was not altered (Deller et al., 2003).

We concluded that deletion of the synaptopodin gene selectively resulted in the loss of spine apparatuses in spines of telencephalic neurons. In turn, the lack of this organelle allowed us to determine the function of the spine apparatus in synaptic plasticity by comparing wildtype mice and synaptopodin-deficient animals. In this context, we found it interesting that a regular spine apparatus is not present in Purkinje cell dendritic spines of wildtype animals (Peters et al., 1991). As synaptopodin is not expressed in the cerebellum (see above), the lack of spine apparatuses in spines of Purkinje cells further indicates that synaptopodin is required for the formation of this organelle. It is the aim of future studies to find out in which way synaptopodin is involved in the formation of spine apparatuses. Recent studies have provided evidence for an increased expression of synaptopodin following the induction of LTP (Yamazaki et al., 2001; Fukazawa et al., 2003). It remains to be shown whether or not LTP also increases the percentage of spines containing a spine apparatus.

5. SYNAPTOPODIN-DEFICIENT MICE SHOW DEFICITS IN SYNAPTIC PLASTICITY

Previous studies provided evidence for the spine apparatus being a calcium store (Fifkova et al., 1983; Svoboda et al., 1999; Emtage et al., 1999). Calcium, entering the cell via NMDA receptors and voltage-gated calcium channels or being released from internal stores, is involved in a variety of processes related to neurotransmission and synaptic plasticity. We thus reasoned that the lack of the spine apparatus in synaptopodin-deficient mice might have changed synaptic plasticity in these animals. Along this line, deletion of the synaptopodin gene might interfere with synaptic plasticity as LTP was found associated with an increased synaptopodin expression (Yamazaki et al., 2001; Fukazawa et al., 2003).

As a first step, we established that basal synaptic transmission was preserved in hippocampal slices of synaptopodin mutants by studying the size of the presynaptic fiber volley and comparing it to the slope of the field excitatory postsynaptic potential (fEPSP). Employing this parameter, basal synaptic transmission was not found to be altered in synaptopodin-deficient mice. Similarly, paired pulse facilitation, a form of synaptic plasticity assumed to be largely presynaptic, was not changed (Fig. 4a).

When we then studied long-term potentiation at the Schaffer collateral-CA1 pyramidal neuron synapse in hippocampal slices, we found striking differences between mutants and wildtype animals (Fig. 4b-d). LTP was significantly reduced following tetanic or theta burst stimulation (Fig. 4b, c). LTP was still clearly impaired in mutant mice 3 hours after stimulation, indicating that both the early (E-LTP) as well as the late phase of LTP (L-LTP) were altered in synaptopodin mutants (Fig. 4d).

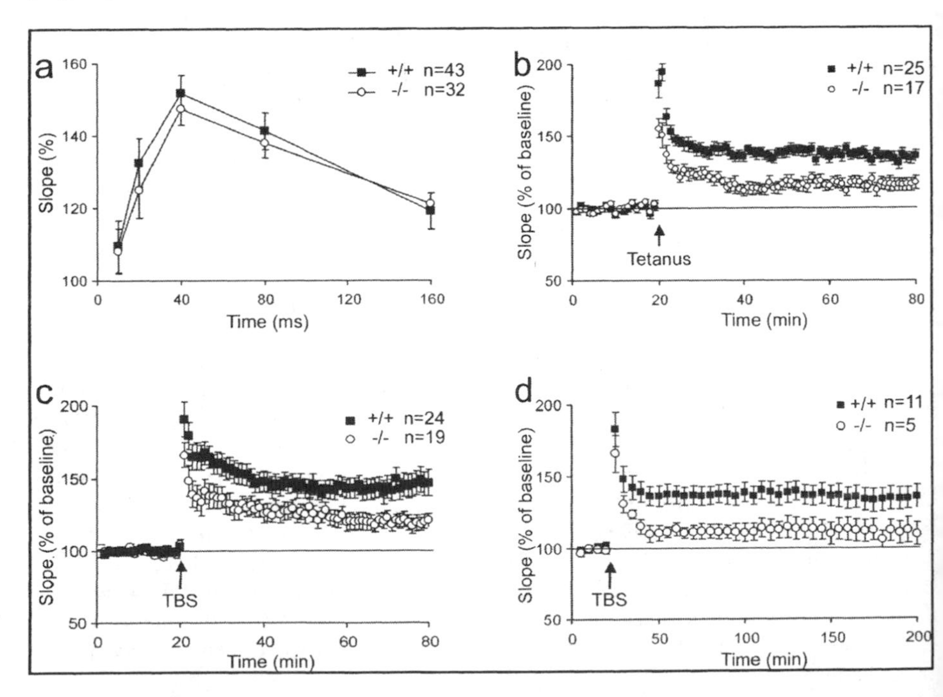

Figure 4. Reduced LTP in the hippocampus of *synaptopodin*-deficient mice. (a) Paired pulse facilitation (PPF) was not significantly different between mutant and wildtype mice. The percentages denote the ratio of the second EPSP slope size to the first EPSP slope. PPF was tested for 10, 20, 40, 80, 160 ms interstimulus interval (ISI). (b) Group data for field EPSP recordings before and after tetanus (100 Hz) application. The difference between *synaptopodin* mutant and wildtype mice is significant ($p<0.01$). Error bars: SEM, n=number of slices. (c) Group data for field EPSP recordings before and after theta burst (TBS; 100 Hz) application. Also for TBS application the difference between mutant and wildtype mice is significant ($p<0.01$; T-test, two-sided). Error bars: SEM, n=number of slices. (d) Group data for field EPSP recordings before and 3 hours after TBS (100 Hz) application. L-LTP is also affected in *synaptopodin*-deficient mice. The difference between mutant and wildtype mice is significant ($p<0.05$; T-test). Error bars: SEM, n=number of slices. Only slices that showed E-LTP were included in the analysis. Reprinted with permission from Deller et al. (2003).

The observed changes in LTP were not caused by a loss of NMDA receptors as established by testing AMPA and NMDA receptor components of the fEPSP. Using the AMPA receptor antagonist DNQX and the NMDA receptor antagonist APV, no significant differences in the NMDA and AMPA receptor components of the fEPSP were observed between wildtype mice and synaptopodin-deficient animals (Deller et al., 2003).

We conclude from these studies that synaptopodin and the spine apparatus are critically involved in synaptic plasticity in the CA1 region of the hippocampus. Further studies need to explore the underlying mechanisms. As the spine apparatus may be linked to the NMDA receptor (Wyszynski et al., 1998; Racca et al., 2000), NMDA receptor-mediated release of calcium from the spine apparatus could be involved in the observed changes in synaptic plasticity. If this scenario holds true and the spine apparatus acts as a calcium store that releases calcium upon NMDA receptor activation, then calcium transients of individual spines should be different between wildtype mice and synaptopodin mutants lacking a spine apparatus. Studies are in progress that address this issue by using 2-photon microscopy and calcium-sensitive dyes (Drakew et al., in preparation). As synaptopodin expression is increased in LTP (and LTP is impaired in synaptopodin-deficient mice lacking a spine apparatus), one is tempted to speculate that LTP is accompanied by an increased formation of spine apparatuses. A spine apparatus is preferentially found in large mushroom spines (Spacek and Harris, 1997), and recent studies have shown that synaptic potentiation is associated with an enlargement of spine heads (Matsuzaki et al., 2004). Accordingly, small spines are preferential sites of LTP induction, whereas large spines, probably the ones containing a spine apparatus, represent physical traces of long-term memory (Matsuzaki et al., 2004).

6. SPATIAL LEARNING IS IMPAIRED IN SYNAPTOPODIN MUTANT MICE

The observed changes in LTP prompted us to study synaptopodin-deficient animals in a variety of behavioral tests. Most relevant to the observed changes in LTP, we noticed significant differences between synaptopodin mutants and controls in the radial arm maze (Deller et al., 2003). Synaptopodin-deficient mice were clearly impaired in their spatial learning ability, i. e., they showed an increased error rate which became statistically significant after the 3rd day of training (Fig. 5). These findings provide an example for parallel changes in hippocampal synaptic plasticity and in hippocampus-associated spatial learning.

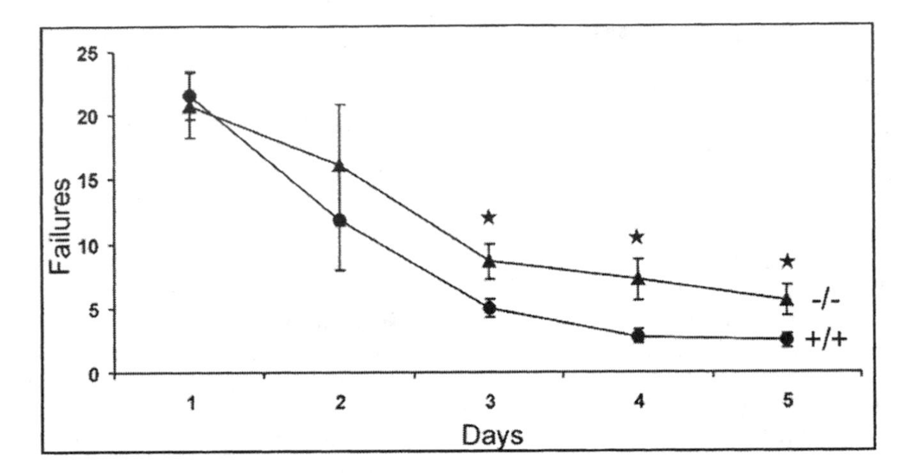

Figure 5. Spatial learning in the radial arm maze is impaired in *synaptopodin*-deficient mice. Mean numbers (+/- SEM) of spatial working memory errors of wildtype (n=16; dots) and mutant mice (n=14; triangles) during the 5 day training period. From day three on, mutant mice show significantly more failures than wildtype mice (*$p<0.05$). Reprinted from Deller et al. (2003). Copyright (2003) National Academy of Sciences, U.S.A.

7. CONCLUSIONS AND OUTLOOK

The studies reviewed here have shown that the lack of a spine apparatus in synaptopodin-deficient mice is accompanied by changes in synaptic plasticity which are relevant for spatial learning. As pointed out at several places, further detailed analysis of synaptopodin-deficient mice is needed to better understand the role of synaptopodin in the formation and function of the spine apparatus and to determine to what extent compensatory changes due to the deletion of synaptopodin are involved. So far, we have established a role for synaptopodin in LTP at Schaffer collateral synapses in CA1. Numerous studies have shown that different mechanisms underlie LTP in CA3 which remains to be studied in synaptopodin-deficient mice. Along this line, we have not yet looked at long-term depression (LTD) in synaptopodin-deficient mice. LTD is a form of synaptic plasticity that is regularly observed in Purkinje cells while LTP can hardly be induced in these neurons. Wildtype Purkinje cells do not express synaptopodin and lack a spine apparatus. Can we change synaptic plasticity in these neurons by transfecting them with synaptopodin cDNA? Can transfection of hippocampal neurons from synaptopodin mutants rescue LTP and the formation of spine apparatuses? We are convinced that further analysis of synaptopodin-deficient mice will allow us to learn more about the function of an interesting protein and a characteristic organelle in dendritic spines of the cerebral cortex.

8. ACKNOWLEDGMENTS

The studies reviewed here have been carried out in close collaboration with Drs Martin Korte, Sophie Chabanis, Alexander Drakew, Herbert Schwegler, Giulia Good Stefani, Aimée Zuniga, Karin Schwarz, Tobias Bonhoeffer, Rolf Zeller, and Peter Mundel and were supported by the Deutsche Forschungsgemeinschaft.

9. REFERENCES

Capani F, Martone ME, Deerinck TJ, Ellisman MH (2001) Selective localization of high concentrations of F-actin in subpopulations of dendritic spines in rat central nervous system: a three-dimensional electron microscopic study, J Comp Neurol 435:156-170.

Deller T, Merten T, Roth SU, Mundel P, Frotscher M (2000a) Actin-associated protein synaptopodin in the rat hippocampal formation: localization in the spine neck and close association with the spine apparatus of principal neurons, J Comp Neurol 418:164-181.

Deller T, Mundel P, Frotscher M (2000b) Potential role of synaptopodin in spine motility by coupling actin to the spine apparatus, Hippocampus 10:569-581.

Deller T, Haas CA, Deissenrieder K, Del Turco D, Coulin C, Gebhardt C, Drakew A, Schwarz K, Mundel P, Frotscher M (2002) Laminar distribution of synaptopodin in normal and reeler mouse brain depends on the position of spine-bearing neurons, J Comp Neurol 453:33-44.

Deller T, Korte M, Chabanis S, Drakew A, Schwegler H, Stefani GG, Zuniga A, Schwarz K, Bonhoeffer T, Zeller R, Frotscher M, Mundel P (2003) Synaptopodin-deficient mice lack a spine apparatus and show deficits in synaptic plasticity, Proc Natl Acad Sci USA 18:10494-10499.

Emptage N, Bliss TVP, Fine A (1999) Single synaptic events evoke NMDA receptor-mediated release of calcium from internal stores in hippocampal dendritic spines, Neuron 22:115-124.

Engert E, Bonhoeffer T (1999) Dendritic spine changes associated with hippocampal long-term synaptic plasticity, Nature 339:66-70.

Fifkova E, Markham JA, Delay RJ (1983) Calcium in the spine apparatus of dendritic spines in the dentate molecular layer, Brain Res 266:163-168.

Fischer M, Kaech S, Knutti D, Matus A (1998) Rapid actin-based plasticity in dendritic spines, Neuron 20:847-854.

Frotscher M, Mannsfeld B, Wenzel J (1975) Umweltabhängige Differenzierung der Dendritenspines an Pyramidenneuronen des Hippocampus (CA1) der Ratte, J Hirnforsch 16:443-450.

Frotscher M, Hámori J, Wenzel J (1977) Transneuronal effects of entorhinal lesions in the early postnatal period on synaptogenesis in the hippocampus of the rat, Exp Brain Res 30:549-560.

Fukazawa Y, Saitoh Y, Ozawa F, Ohta Y, Mizuno K, Inokuchi K (2003) Hippocampal LTP is accompanied by enhanced F-actin content within the dendritic spine that is essential for late LTP maintenance in vivo, Neuron 38:447-460.

Globus A, Scheibel AB (1967) Synaptic loci on visual cortical neurons of the rabbit: the specific afferent radiation, Exp Neurol 18:116-131.

Gray EG (1959) Axo-somatic and axo-dendritic synapses of the cerebral cortex: an electron microscopic study, J Anat 83:420-433.

Hámori J (1973) The inductive role of presynaptic axons in the development of postsynaptic spines, Brain Res 62:337-344.

Hayashi Y, Shi SH, Esteban JA, Piccini A, Poncer JC, Malinow R (2000) Driving AMPA receptors into synapses by LTP and CaMKII : requirement for GluR1 and PDZ domain interaction, Science 287:2262-2267.

Ichimura K, Kurihara H, Sakai T (2003) Actin filament organization of foot processes in rat podocytes, J Histochem Cytochem 51:1589-1600.

Kretz O, Fester L, Wehrenberg U, Zhou L, Brauckmann S, Zhao S, Prange-Kiel J, Naumann T, Jarry H, Frotscher M, Rune GM (2004) Hippocampal synapses depend on hippocampal estrogen synthesis, J Neurosci 24:5913-5921.

Li Z, Sheng M (2003) Some assembly required: the development of neuronal synapses, Nat Rev Mol Cell Biol 4:833-841.

Lisman JA (1989) A mechanism for the Hebb and anti-Hebb processes underlying learning and memory, Proc Natl Acad Sci USA 86:9574-9578.

Lisman J, Schulman H, Cline H (2002) The molecular basis of CaMKII function in synaptic and behavioural memory, Nature Rev Neurosci 3:175-190.

Matsuzaki M, Honkura N, Ellis-Davies GCR, Kasai H (2004) Structural basis of long-term potentiation in single dendritic spines, Nature 429:761-766.

Matus A, Ackermann M, Pehling G, Byers HR, Fujiwara K (1982) High actin concentrations in brain dendritic spines and postsynaptic densities, Proc Natl Acad Sci USA 79:7590-7594.

Miyata M, Finch EA, Khiroug L, Hashimoto K, Hayasaka S, Oda SI, Inouye M, Takagishi Y, Augustine GJ, Kano M (2000) Local calcium release in dendritic spines required for long-term synaptic depression, Neuron 28:233-244.

Mundel P, Gilbert P, Kriz W (1991) Podocytes in glomerulus of rat kidney express a characteristic 44 KD protein, J Histochem Cytochem 39;1047-1056.

Mundel P, Heid HW, Mundel TM, Kruger M, Reiser J, Kriz W (1997) Synaptopodin: an actin-associated protein in telencephalic dendrites and renal podocytes, J Cell Biol 139:193-204.

Peters A, Palay SL, Webster HD (1991) The Fine Structure of the Nervous System. Neurons and their Supporting Cells, Oxford University Press, Oxford.

Racca C, Stephenson FA, Streit P, Robert JD, Somogyi P (2000) NMDA receptor content of synapses in stratum radiatum of the hippocampal CA1 area, J Neurosci 20:2512-2522.

Ramón y Cajal SR (1911) Histologie du Système Nerveux de l'Homme et des Vertébrés, Maloine, Paris.

Rumbaugh G, Sia GM, Garner CC, Huganir RL (2003) Synapse-associated protein-97 isoform-specific regulation of surface AMPA receptors and synaptic function in cultured neurons, J Neurosci 23:4567-4576.

Rune GM, Wehrenberg U, Prange-Kiel J, Zhou L, Adelmann G, Frotscher M (2002) Estrogen up-regulates estrogen receptor alpha and synaptophysin in slice cultures of rat hippocampus, Neuroscience 113:167-175.

Sabatini BL, Oertner TG, Svoboda K (2002) The life cycle of Ca^{2+} ions in dendritic spines, Neuron 33:439-452.

Schnell E, Sizemore M, Karimzadegan S, Chen L, Bredt DS, Nicoll RA (2002) Direct interactions between PSD-95 and stargazin control synaptic AMPA receptor number, Proc Natl Acad Sci USA 99:13902-13907.

Spacek J (1985) Three-dimensional analysis of dendritic spines. II. Spine apparatus and other cytoplasmic components, Anat Embryol 171:235-243.

Spacek J, Harris KM (1997) Three-dimensional organization of smooth endoplasmic reticulum in hippocampal CA1 dendrites and dendritic spines of the immature and mature rat, J Neurosci 17:190-203.

Svoboda K, Mainen ZF (1999) Synaptic [Ca^{2+}]: intracellular stores spill their guts, Neuron 22:427-430.

Valverde F (1967) Apical dendritic spines of the visual cortex and light deprivation in the mouse, Exp Brain Res 3:337-352.

Valverde F (1968) Structural changes in the area striata of the mouse after enucleation, Exp Brain Res 5:274-292.

Valverde F (1971) Rate and extent of recovery from dark rearing in the visual cortex of the mouse, Brain Res 33:1-11.

Woolley CS, McEwen BS (1993) Roles of estradiol and progesterone in regulation of hippocampal spine density during the estrous cycle in the rat, J Comp Neurol 336:293-306.

Wyszynski M, Kharazia V, Shanghvi R, Rao A, Beggs AH, Craig AM, Weinberg R, Sheng M (1998) Differential regional expression and ultrastructural localization of alpha-actinin-2, a putative NMDA receptor-anchoring protein, in rat brain, J Neurosci 18:1383-1392.

Yamazaki M, Matsuo R, Fukazawa Y, Ozawa F, Inokuchi K (2001) Regulated expression of an actin-associated protein, synaptopodin, during long-term potentiation, J Neurochem 79:192-199.

AMYLOID-β AS A BIOLOGICALLY ACTIVE PEPTIDE IN CNS

Georgi Gamkrelidze, Sung H. Yun, and Barbara L. Trommer*

1. INTRODUCTION

Over the last decade, amyloid-β (Aβ), which is detected in both cerebrospinal fluid and plasma in healthy individuals throughout life (Seubert et al., 1992), has been a focus of considerable research. It has attracted attention as a peptide that plays pivotal role in the onset of Alzheimer's disease (AD) (Hardy and Selkoe, 2002) as well as a physiologically relevant messenger in CNS (Kamenetz et al., 2003; Plant et al., 2003). Recent observations indicate that nicotinic acetylcholine receptors (nAChRs) act as receptors for Aβ (Wang et al., 2000b; Wang et al., 2000a). Here we briefly review electrophysiological data concerning Aβ effects that are relevant for AD as well as normal physiology of CNS. We present data obtained in our laboratory comparing the effects of Aβ and the nAChR antagonist, mecamylamine, on long-term synaptic plasticity in hippocampal dentate gyrus (DG), and propose a hypothesis concerning an interaction between nAChRs and Aβ in synaptic transmission and long-term synaptic plasticity as well as AD pathogenesis.

1.1 Structure

The structure and metabolism of Aβ are relatively well characterized (Selkoe, 2001; Turner et al., 2003). Aβ is a peptide that arises from proteolytic processing of the amyloid precursor protein (Selkoe, 2001). Aβ peptides vary in size from 39 to 43 amino acids (Turner et al., 2003). However, the majority of secreted Aβ peptides are 40 or 42 amino acids in length (Selkoe, 2001). Aβ exist as monomers, dimers, and higher order oligomers (Walsh et al., 2000). In general, Aβ peptides are diffusible and soluble (Klein et al., 2001). In the following review we do not distinguish among different species of Aβ peptide, since the existence of multiple forms and configurations of Aβ does not affect the main conclusions presented here.

*Barbara L. Trommer, Georgi Gamkrelidze, Sung H. Yun, Departments of Pediatrics and Neurology, Feinberg School of Medicine, Northwestern University, Chicago, IL 60611; Evanston Northwestern Healthcare Research Institute, Evanston, IL 60201, USA. E-mail: btrommer@northwestern.edu

1.2 Aβ and Alzheimer Disease

The amyloid cascade hypothesis of AD states that the abnormalities of Aβ processing and, consequently, overproduction and aggregation of Aβ are responsible for the disease (Hardy and Selkoe, 2002) and there is considerable evidence that excess Aβ causes the disruption of neuronal networks that leads to cognitive decline (Selkoe, 2002). The presence of amyloid plaques is a pathologic hallmark of AD. However, because amyloid plaques correlate poorly in number, tempo and distribution with the clinical progression of dementia, recent research has focused instead on soluble, easily diffusible Aβ as a root cause of early cognitive deficits in AD patients (Kirkitadze et al., 2002; Klein et al., 2001; Mesulam, 1999; Selkoe, 2002). Indeed, in AD patients the concentration of Aβ peptide in CNS is a predictor of synaptic change associated with the disease (Lue et al., 1999). In addition to human studies, the effects of Aβ have been analyzed extensively in animal models. In particular, the effects of Aβ on synaptic transmission as well as long-term depression (LTD) and long-term potentiation (LTP) have been investigated. LTP and LTD are well established cellular models of learning and memory that respectively increase and decrease synaptic efficacy. In *in vivo* experiments in transgenic mice that exhibited enhanced production of Aβ in the absence of amyloid plaques the suppression of synaptic transmission (Hsia et al., 1999) and LTP (Chapman et al., 1999) as well as the impairment of working memory and learning (Hsia et al., 1999; Kumar-Singh et al., 2000) have all been demonstrated. Similarly, in concentrations that exceeded physiological levels of Aβ in the human cerebrospinal fluid (Walsh et al., 2002), exogenous Aβ has been shown to attenuate synaptic transmission and/or LTP in *in vitro* experiments (Chen et al., 2000; Chen et al., 2002; Lambert et al., 1998; Wang et al., 2002). In contrast to its effect on LTP, Aβ has been shown to facilitate LTD in an in an *in vivo* study (Kim et al., 2001). Thus, animal studies suggest that Aβ at concentrations exceeding the normal physiological levels may disrupt mechanisms of synaptic transmission and LTP while facilitating LTD. Similarly, in the clinical setting, overproduction of Aβ caused by aberrant Aβ processing may lead to decreased synaptic efficacy, diminished synaptic interaction among neurons, and ultimately, the memory impairment characteristic of early AD (Selkoe, 2002).

1.3 A putative physiologic role for Aβ in CNS

In contrast to the accepted role of Aβ in the etiology of AD, relatively little is known about the physiologic function of Aβ in CNS. Recent publications suggest that endogenous Aβ released under physiologic conditions from neurons regulates synaptic plasticity (Kamenetz et al., 2003; Walsh et al., 2002), synaptic transmission (Kamenetz et al., 2003), neuronal excitability (Kamenetz et al., 2003) and neuron viability (Plant et al., 2003). Indeed, it has been demonstrated that the inhibition of Aβ production in cultured cortical neurons severely compromises their survival (Plant et al., 2003). Furthermore, Kamenetz and co-authors (Kamenetz et al., 2003) obtained data indicating that neuronal activity modulates the formation and secretion of Aβ in hippocampal slice neurons. In addition, these authors provide evidence suggesting that Aβ released in response to neuronal activity suppresses synaptic transmission and LTP. In accord with these data, Walsh and co-authors (Walsh et al., 2002) demonstrated that administration of natural cell-derived Aβ in rat cerebral ventricles at concentrations observed in the human cerebrospinal fluid inhibits LTP generation in the hippocampus *in vivo*. Based on their observations, Kamenetz and co-authors (Kamenetz et al., 2003) proposed a model of physiologic activity of Aβ. According to this model, activity-dependent modulation of endogenous Aβ production participates in a negative-feedback that could keep the neuron activity (homeostasis of neuron activity) in check by regulating (suppressing) synaptic transmission and LTP. Thus increases in neuronal firing rate would lead to increased secretion of Aβ which in turn would diminish the synaptic input responsible for neuronal firing, causing the firing rate to be retained in a physiologically "acceptable" range. The model illustrates the possible significance of Aβ peptides in normal physiology of the CNS and provides a conceptual framework for further exploration of the transition of neural network activity from normal states to the onset of AD.

1.4 nAChRs and Aβ

The nAChR is one of few receptors for Aβ that has been identified in the nervous system (Wang et al., 2000b; Wang et al., 2000a). The binding of Aβ to nAChRs has been demonstrated in a number of brain structures including in the hippocampus. In the hippocampus, a structure critical for memory and an early target for AD, the blockade of nAChR currents by Aβ has been demonstrated (Liu et al., 2001; Pettit et al., 2001). As nAChRs play an important role in the modulation of CNS activity (Role and Berg, 1996) interactions between nAChRs and Aβ might mediate some of the reported Aβ effects in the CNS.

nAChRs are heterogeneous family of pentomeric ion channels that modulate several functions in the CNS including synaptic transmission, synaptic plasticity, memory, and learning. Multiple α (α2-α10) and β (β2-β4) subunits of nAChRs have been described in the CNS of vertebrate animals. In the mammalian CNS including the hippocampus, nAChRs containing α4, α2 or α7 subunits are most common (Dani, 2001; Seguela et al., 1993; Wada et al., 1989). Although the current flow through nAChRs is mediated mostly by potassium and sodium ions, calcium makes a significant contribution (Albuquerque et al., 1997). The calcium influx through nAChRs is likely to be important specifically for synaptic transmission and synaptic plasticity (Dani, 2001).

It has been found that nAChRs located on presynaptic terminals increase the release of every tested neurotransmitter including glutamate, serotonin, acetylcholine, dopamine and norepinephrine (Albuquerque et al., 1997; Alkondon et al., 1999; Dani, 2001). The enhancement of neurotransmitter release was observed in a number of brain structures including in the hippocampus (Albuquerque et al., 1997; Gray et al., 1996) and the neocortex (Aramakis and Metherate, 1998). Furthermore, the facilitation of LTP by nAChRs has been demonstrated in the midbrain (Mansvelder and McGehee, 2000) and the hippocampus (Fujii et al., 1999; Ji et al., 2001). In the hippocampal CA1 area, it has been shown that nAChRs located both on presynaptic terminals and postsynaptic neurons contribute to LTP enhancement (Ji et al., 2001). Finally, in behavioral studies it has been documented that nAChR agonists improve working memory and learning, whereas nAChR antagonists impair these cognitive functions (Levin and Simon, 1998; Paterson and Nordberg, 2000). Thus, the activation of nAChRs enhances synaptic transmission and LTP. It is likely that this functional synaptic upregulation is responsible for the cognitive improvements that have been attributed to nAChR agonists.

The studies described above indicate that Aβ and nAChRs affect the same group of neurophysiological processes: synaptic transmission, synaptic plasticity, and working memory. Typically, Aβ suppresses these processes whereas the activation nACRs enhances them. This overlap of targets and the fact that Aβ binds and blocks nAChRs provide the basis for the hypothesis that under physiological conditions, endogenous Aβ regulates synaptic transmission and plasticity by inhibiting nAChRs. The disruption of the balanced interaction between Aβ and nAChRs, i.e. by the overproduction of Aβ, could thus lead to the disruption of synaptic communication among neurons and, consequently, to some of the cognitive symptoms associated with AD.

1.5 Aβ and nAChRs in the dentate gyrus

DG is a particularly appropriate model for investigation of the effects of nAChRs and Aβ on long-term synaptic plasticity. DG acts as a "regulated gate" that controls information inflow into the hippocampus (Heinemann et al., 1992, Lothman et al., 1992). Granule cells (GCs), the principal cells of DG, play a central role in the functioning of the "gate": GCs, along with dentate interneurons, receive a major hippocampal input (perforant path) from the entorhinal cortex and relay information to the remainder of the hippocampus. In addition, DG receives extensive innervation from septo-hippocampal cholinergic afferents. nAChRs play a significant role in the regulation of information processing in DG (Frazier et al., 2003). Synaptic plasticity in DG is vital to normal learning and memory (Naie and Manahan-Vaughan, 2004) and this region is specifically implicated in AD (Palop et al., 2003).

2. EXPERIMENTAL

Here we present data obtained from hippocampal dentate gyrus (DG). To illustrate the similarity between the effects of Aβ and the blockade of nAChRs, we compare the effects of Aβ and mecamylamine, a specific nAChR antagonist, on LTP generation. In addition, we present data showing the absence of an effect of oligomeric Aβ on LTD that contrasts with its inhibitory effect on LTP.

In acute hippocampal slices (350 μm) from rat or mice, field excitatory post synaptic potentials (EPSPs) in extracellular experiments or excitatory postsynaptic currents (EPSCs) in intracellular experiments (whole cell patch) were recorded in response to medial perforant path stimulation as previously described (Trommer et al., 1996; Wang et al., 2002). Artificial cerebrospinal fluid (ACSF) composition was (in mM): NaCl 124, KCl 3, $CaCl_2$ 2.4, $MgSO_4$ 1.3, NaH_2PO_4 1.25, $NaHCO_3$ 26, and glucose 10 (gassed with 95% O_2 / 5% CO_2, pH = 7.4); picrotoxin (100 μM) was included in extracellular experiments. Slices were pre-incubated for 1 hour at room temperature in ACSF alone or ACSF containing oligomeric Aβ1-42 or its corresponding vehicle, and perfused throughout the experiments with the same solution at 32°C. Data were sampled at 10 kHz and low-pass filtered at 3 kHz. LTP was induced by high frequency stimulation (HFS, 3 trains, 100 Hz, 1 sec duration, 20 sec intertrain interval) or by theta burst stimulation (TBS) (10 high frequency bursts (4 pulses at 100 Hz) delivered at 200 ms isi (5 Hz)); 3 trains of TBS applied at intertrain interval 10 seconds. Low frequency stimulation (LFS) to induce LTD consisted of 1 Hz for 15 minutes.

Oligomeric assemblies of Aβ1-42 were prepared as previously described (Stine, Jr. et al., 2003) and resuspended in ACSF at a concentration of 400 or 500 nM. Vehicle for oligomeric Aβ was DMSO in F12 culture media. Final DMSO concentration in ACSF for recording was always $\leq 0.01\%$.

3. RESULTS AND DISCUSSION

We examined the effect of oligomeric Aβ on LTP generation at the perforant path-granule cell synapse (pp-GC synapse) in extracellular and intracellular experiments (Figure 1). In extracellular experiments in response to 500 nM oligomeric Aβ exposure the magnitude of LTP of EPSP was diminished by ~50 % (**Fig. 1A**). Similarly, oligomeric Aβ inhibited LTP of EPSC monitored at the single cell level (**Fig. 1B**).

In a separate group of experiments we tested the effect of the oligomeric Aβ on LTD generation at pp-GC synapse. As illustrated in Figure 2, in contrast to its effect on LTP, Aβ did not affect the generation of LTD.

To test the role of nAChRs in the generation of LTP at pp-GC synapse, we exposed slices to 3 μM mecamylamine (Mec). As illustrated in Figure 3, the application of Mec produced a reduction in LTP magnitude that was similar to the effect of Aβ (compare **Fig. 1 A**).

As shown in Figures 1 and 2, Aβ inhibited LTP while sparing LTD. This pattern of "plasticity imbalance" favoring LTD, in which the demand for restorative plasticity is both enhanced by Aβ and simultaneously rendered more difficult to meet, is consistent with an Aβ-induced accelaration of synaptic loss and consequent disruption of neural networks seen in AD (Selkoe, 2002).

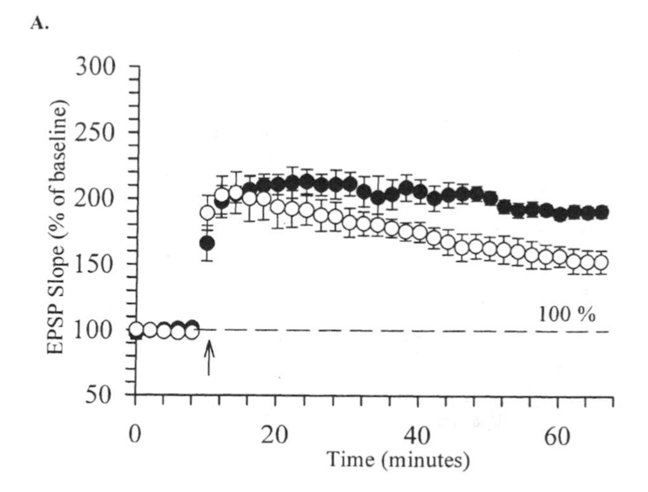

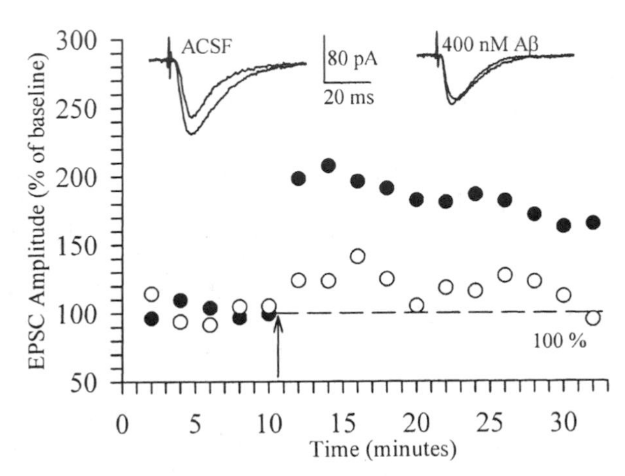

Figure 1. A. LTP in C57/bl6 mice in vehicle (●) and (○) 500 nM oligomeric Aβ. Slices were preincubated in the peptide for one hour and exposed to it during experiments. LTP magnitude in oligomeric Aβ (151.24 ± 13.1 %, N=3) was significantly smaller ($p<0.001$) than in vehicle (190.26 ± 8.03 %, N=4). **B.** LTP in apoe4 transgenic mouse at single cell level (whole cell recording) in ACSF (●) and (○) 400 nM oligomeric Aβ. Slices were preincubated in the peptide for one hour and exposed to it during recording. Representative synaptic current traces prior and 30 minutes after a HFS are shown above the graph.

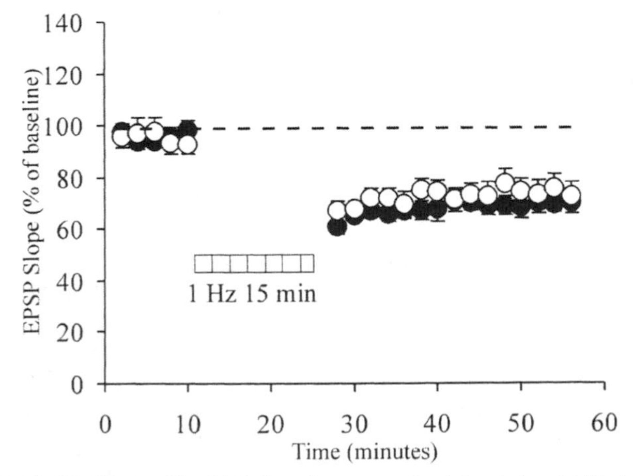

Figure 2. Oligomeric Aβ did not affect LTD. Low frequency stimulation induced LTD in both control (○) (72.3±5.6%, n=5) and 500 nM Aβ (●) (70.0±3.9%, n=5) exposed rat hippocampal slices.

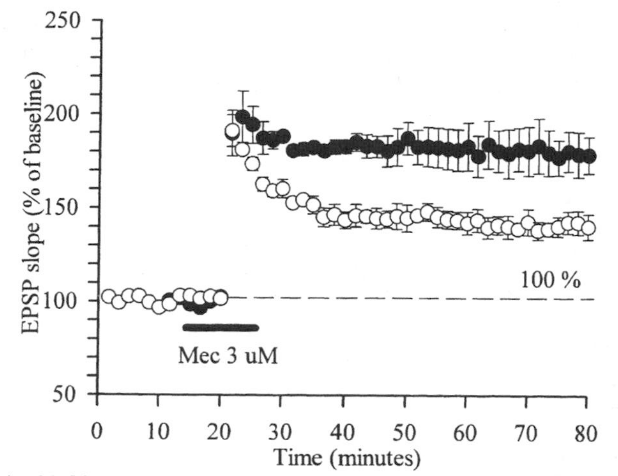

Figure 3. LTP in C57/bl6 mice in ACSF (●) and 3 μM mecamylamine (○). LTP magnitude after mecamylamine exposure (140.0 ± 5.5%,N=3) was significantly smaller ($p < 0.04$) than in ACSF (179.5 ± 11.3% , N=3).

Our experiments with mecamylamine strongly suggest that nAChRs on GCs and/or terminals presynaptic to GCs (possibly perforant path-GC synapses) modulate LTP induction. To date, nAChRs have been identified on DG inhibitory interneurons, but not on GCs. If mecamylamine were acting solely on GABAergic interneurons (decreasing their activity by removing excitatory cholinergic inputs), this would increase GC excitability and enhancement of LTP would be the expected response. Since inhibition of LTP was seen instead, mecamylamine, being a highly specific blocker of nAChRs, is likely to be acting directly at GC and/or terminals presynaptic to GCs. Furthermore, these experiments were performed in presence of the $GABA_A$ receptor antagonist picrotoxin (100 μM), to limit the role of GABAergic inputs, increasing the likelihood that nAChRs are present within the excitatory glutamatergic pathways in this region. The similarity between the effects of mecamylamine and those of Aβ is consistent with the hypothesis that Aβ effects on synaptic plasticity are mediated by Aβ blockade of nAChRs.

4. SUMMARY AND CONCLUSIONS

Aβ plays a critical role in the onset of AD and may also act as a neuromodulator of synaptic function in the normal physiology of the CNS. Aβ and nAChRs have opposite effects on the same group of neurophysiological functions i.e. synaptic transmission, synaptic plasticity, working memory and learning. This overlap of targets and the fact that Aβ binds and blocks nAChRs leads us to hypothesize that under physiologic conditions endogenous Aβ regulates synaptic transmission and plasticity by inhibiting nAChRs to maintain neuron activity homeostasis. The overproduction of Aβ is likely to cause excessive blockade of nAChRs which could lead to the inhibition of synaptic transmission and LTP, and the possibility of unopposed LTD. The resulting alterations in synaptic efficacy would contribute to the cognitive impairment associated with AD. Neurons possessing high affinity nAChRs have been identified as particularly vulnerable to AD, and the α7 nAChR has been proposed as a specific therapeutic target (Kem, 2000). Elucidation of the interaction between nAChRs and Aβ could prove to be fertile ground for the further understanding of Aβ both as a signaling molecule and pathogen in the CNS, and could lead to greater specificity in the development of plasticity rescue strategies in AD therapeutics.

5. REFERENCES

Albuquerque EX, Alkondon M, Pereira EF, Castro NG, Schrattenholz A, Barbosa CT, Bonfante-Cabarcas R, Aracava Y, Eisenberg HM, Maelicke A (1997) Properties of neuronal nicotinic acetylcholine receptors: pharmacological characterization and modulation of synaptic function, J Pharmacol Exp Ther 280:1117-1136.

Alkondon M, Pereira EF, Eisenberg HM, Albuquerque EX (1999) Choline and selective antagonists identify two subtypes of nicotinic acetylcholine receptors that modulate GABA release from CA1 interneurons in rat hippocampal slices, J Neurosci 19:2693-2705.

Aramakis VB, Metherate R (1998) Nicotine selectively enhances NMDA receptor-mediated synaptic transmission during postnatal development in sensory neocortex, J Neurosci 18:8485-8495.

Chapman PF, White GL, Jones MW, Cooper-Blacketer D, Marshall VJ, Irizarry M, Younkin L, Good MA, Bliss TV, Hyman BT, Younkin SG, and Hsiao KK (1999) Impaired synaptic plasticity and learning in aged amyloid precursor protein transgenic mice, Nat Neurosci 2:271-276.

Chen QS, Kagan BL, Hirakura Y, and Xie CW (2000) Impairment of hippocampal long-term potentiation by Alzheimer amyloid beta-peptides, J Neurosci Res 60:65-72.

Chen QS, Wei WZ, Shimahara T, Xie CW (2002) Alzheimer amyloid beta-peptide inhibits the late phase of long-term potentiation through calcineurin-dependent mechanisms in the hippocampal dentate gyrus, Neurobiol Learn Mem 77:354-371.

Dani JA (2001) Overview of nicotinic receptors and their roles in the central nervous system, Biol Psychiatry 49:166-174.

Frazier CJ, Strowbridge BW, and Papke RL (2003) Nicotinic receptors on local circuit neurons in dentate gyrus: a potential role in regulation of granule cell excitability, J Neurophysiol 89:3018-3028.

Fujii S, Ji Z, Morita N, Sumikawa K (1999) Acute and chronic nicotine exposure differentially facilitate the induction of LTP, Brain Res 846:137-143.

Gray R, Rajan AS, Radcliffe KA, Yakehiro M, Dani JA (1996) Hippocampal synaptic transmission enhanced by low concentrations of nicotine, Nature 383:713-716

Hardy J, Selkoe DJ (2002) The amyloid hypothesis of Alzheimer's disease: progress and problems on the road to therapeutics, Science 297:353-356.

Heinemann U, Beck H, Dreier JP, Ficker E, Stabel J, Zhang CL (1992) The dentate gyrus as a regulated gate for the propagation of epileptiform activity, Epilepsy Res Suppl 7:273-280.

Hsia AY, Masliah E, McConlogue L, Yu GQ, Tatsuno G, Hu K, Kholodenko D, Malenka RC, Nicoll RA, Mucke L (1999) Plaque-independent disruption of neural circuits in Alzheimer's disease mouse models, Proc Natl Acad Sci USA 96:3228-3233.

Ji D, Lape R, Dani JA (2001) Timing and location of nicotinic activity enhances or depresses hippocampal synaptic plasticity, Neuron 31:131-141.

Kamenetz F, Tomita T, Hsieh H, Seabrook G, Borchelt D, Iwatsubo T, Sisodia S, Malinow R (2003) APP processing and synaptic function, Neuron 37:925-937.

Kem WR (2000) The brain alpha7 nicotinic receptor may be an important therapeutic target for the treatment of Alzheimer's disease: studies with DMXBA (GTS-21), Behav Brain Res 113:169-181.

Kim JH, Anwyl R, Suh YH, Djamgoz MB, Rowan MJ (2001) Use-dependent effects of amyloidogenic fragments of (beta)-amyloid precursor protein on synaptic plasticity in rat hippocampus in vivo, J Neurosci 21:1327-1333.

Kirkitadze MD, Bitan G, Teplow DB (2002) Paradigm shifts in Alzheimer's disease and other neurodegenerative disorders: The emerging role of oligomeric assemblies, J Neurosci Res 69:567-577.

Klein WL, Krafft GA, Finch CE (2001) Targeting small Abeta oligomers: the solution to an Alzheimer's disease conundrum?, Trends Neurosci 24:219-224.

Kumar-Singh S, Dewachter I, Moechars D, Lubke U, De Jonghe C, Ceuterick C, Checler F, Naidu A, Cordell B, Cras P, Van Broeckhoven C, and Van Leuven F (2000) Behavioral disturbances without amyloid deposits in mice overexpressing human amyloid precursor protein with Flemish (A692G) or Dutch (E693Q) mutation, Neurobiol Dis 7:9-22.

Lambert MP, Barlow AK, Chromy BA, Edwards Freed R, Liosatos M, Morgan TE, Rozovsky I, Trommer B, Viola KL, Wals P, Zhang C, Finch CE, Krafft GA, Klein WL (1998) Diffusible, nonfibrillar ligands derived from $A\beta_{1-42}$ are potent central nervous system neurotoxins, Proc Natl Acad Sci USA 95:6448-6453.

Levin ED, Simon BB (1998) Nicotinic acetylcholine involvement in cognitive function in animals, Psychopharmacology (Berl) 138:217-230.

Liu Q, Kawai H, Berg DK (2001) beta -Amyloid peptide blocks the response of alpha 7-containing nicotinic receptors on hippocampal neurons, Proc Natl Acad Sci USA 98:4734-4739.

Lothman EW, Stinger JL, Bertram EH (1992) The dentate gyurs as a control point for seizures in the hippocampus and beyond. Epilepsy Res Suppl 7:301-13.

Lue LF, Y. Kuo YM, Roher AE, Brachova L, Shen Y, Sue L, Beach T, Kurth JH, Rydel RE, Rogers J (1999) Soluble amyloid beta peptide concentration as a predictor of synaptic change in Alzheimer's disease, Am J Pathol 155:853-862.

Mansvelder HD, McGehee DS (2000) Long-term potentiation of excitatory inputs to brain reward areas by nicotine, Neuron 27:349-357.

Mesulam MM (1999) Neuroplasticity failure in Alzheimer's disease: bridging the gap between plaques and tangles, Neuron 24:521-529.

Naie K, Manahan-Vaughan D (2004) Regulation by metabotropic glutamate receptor 5 of LTP in the dentate gyrus of freely moving rats: relevance for learning and memory formation, Cerebral Cortex 14:189-198 (2004).

Palop JJ, Jones B, Kekonius L, Chin J, Yu GQ, Raber J, Masliah E, Mucke L (2003) Neuronal depletion of calcium-dependent proteins in the dentate gyrus is tightly linked to Alzheimer's disease-related cognitive deficits, Proc Natl Acad Sci USA 100:9572-9577.

Paterson D, Nordberg A (2000) Neuronal nicotinic receptors in the human brain, Prog Neurobiol 61:75-111.

Pettit DL, Shao Z, Yakel JL (2001) beta-Amyloid(1-42) peptide directly modulates nicotinic receptors in the rat hippocampal slice, J Neurosci 21:1-5.

Plant LD, Boyle JP, Smith IF, Peers C, Pearson HA (2003) The production of amyloid beta peptide is a critical requirement for the viability of central neurons, J Neurosci 23:5531-5535.

Role LW, Berg DK (1996) Nicotinic receptors in the development and modulation of CNS synapses, Neuron 16:1077-1085.

Seguela P, Wadiche J, Dineley-Miller K, Dani JA, Patrick JW (1993) Molecular cloning, functional properties, and distribution of rat brain alpha 7: a nicotinic cation channel highly permeable to calcium, J Neurosci 13:596-604.

Selkoe DJ (2001) Alzheimer's disease: genes, proteins, and therapy, Physiol Rev 81:741-766.

Selkoe DJ (2002) Alzheimer's disease is a synaptic failure, Science 298:789-791.

Seubert P, Vigo-Pelfrey C, Esch F, Lee M, Dovey H, Davis D, Sinha S, Schlossmacher M, Whaley J, Swindlehurst C (1992) Isolation and quantification of soluble Alzheimer's beta-peptide from biological fluids, Nature 359:325-327.

Stine WB Jr, Dahlgren KN, Krafft GA, LaDu MJ (2003) In vitro characterization of conditions for amyloid-beta peptide oligomerization and fibrillogenesis, J Biol Chem 278:11612-11622.

Trommer BL, Liu YB, Pasternak JF (1996) Long-term depression at the medial perforant path-granule cell synapse in developing rat dentate gyrus, Dev Brain Res 96:97-108.

Turner PR, O'Connor K, Tate WP, Abraham WC (2003) Roles of amyloid precursor protein and its fragments in regulating neural activity, plasticity and memory, Prog Neurobiol 70:1-32.

Wada E, Wada K, Boulter J, Deneris E, Heinemann S, Patrick J, Swanson LW (1989) Distribution of alpha 2, alpha 3, alpha 4, and beta 2 neuronal nicotinic receptor subunit mRNAs in the central nervous system: a hybridization histochemical study in the rat, J Comp Neurol\ 284:314-335.

Walsh DM, Klyubin I, Fadeeva JV, Cullen WK, Anwyl R, Wolfe MS, Rowan MJ, Selkoe DJ (2002) Naturally secreted oligomers of amyloid beta protein potently inhibit hippocampal long-term potentiation in vivo, Nature 416:535-539.

Walsh DM, Tseng BP, Rydel RE, Podlisny MB, Selkoe DJ (2000) The oligomerization of amyloid beta-protein begins intracellularly in cells derived from human brain, Biochemistry 39:10831-10839.

Wang HW, Pasternak JF, Kuo H, Ristic H, Lambert MP, Chromy B, Viola KL, Klein WL, Stine WB, Krafft GA, Trommer BL (2002) Soluble oligomers of beta amyloid (1-42) inhibit long-term potentiation but not long-term depression in rat dentate gyrus, Brain Res 924:133-140.

Wang HY, Lee DH, D'Andrea MR, Peterson PA, Shank RP, Reitz AB (2000a) beta-Amyloid(1-42) binds to alpha7 nicotinic acetylcholine receptor with high affinity. Implications for Alzheimer's disease pathology, J Biol Chem 275:5626-5632.

Wang HY, Lee DH, Davis CB, Shank RP (2000b) Amyloid peptide Abeta(1-42) binds selectively and with picomolar affinity to alpha7 nicotinic acetylcholine receptors, J Neurochem 75:1155-1161.

INDEX